Heat Pipes: Theory, Design and Applications

Heat Pipes: Theory, Design and Applications

Edited by **Edgar Miller**

Willford Press
New York

Published by Willford Press,
118-35 Queens Blvd., Suite 400,
Forest Hills, NY 11375, USA
www.willfordpress.com

Heat Pipes: Theory, Design and Applications
Edited by Edgar Miller

International Standard Book Number: 978-1-68285-284-2 (Hardback)

Printed in China.

Contents

Preface

This book has been a concerted effort by a group of academicians, researchers and scientists, who have contributed their research works for the realization of the book. This book has materialized in the wake of emerging advancements and innovations in this field. Therefore, the need of the hour was to compile all the required researches and disseminate the knowledge to a broad spectrum of people comprising of students, researchers and specialists of the field.

Heat pipes are used widely in space technology, nuclear power conversions, heat transfer systems, cellular phones, cooling of computers and in several other industrial operations. Heat pipes are based on the models of heat exchangers which are used for transportation of thermal energy over long distances with minimal temperature difference. There are various kinds of heat pipes available with modified designs and techniques like Constant Conductance Heat Pipes (CCHPs), Variable Conductance Heat Pipes (VCHPs) and Diode Heat Pipes, etc. This comprehensive book encompasses sections on all the significant aspects of heat pipes, theoretical principles behind their functioning, modifications & designs and their wide scope of applications. It is a beneficial text for students, researchers, engineers and readers in general who are keen to enhance their knowledge on heat pipes.

At the end of the preface, I would like to thank the authors for their brilliant chapters and the publisher for guiding us all-through the making of the book till its final stage. Also, I would like to thank my family for providing the support and encouragement throughout my academic career and research projects.

Editor

TEMPERATURE UNIFORMITY ANALYSIS OF A MULTI-WELL VAPOR CHAMBER HEAT SPREADER

Shung-Wen Kang[a*], Yu-Tang Chen[b], Chun-Hsiang Hsu[a] and Jun-Ying Lin[a]

[a] *Mechanical and Electro-Mechanical Engineering, Tamkang University, Tamsui Dist., New Taipei City, 25137, Taiwan(R.O.C)*
[b] *Department of Mechanical Engineering, De Lin Institute of Technology, Tucheng Dist., New Taipei City, 236, Taiwan(R.O.C)*

ABSTRACT

In this study, temperature uniformity and heating rate of heat spreader with multi-well are simulated and analyzed by CFD software under natural convection condition. Firstly, multi-well heat spreader made of aluminum, copper, silver and vapor chamber are simulated and compared, when dual and six heat sources are applied at a heating power of 1200W. Secondly, with six heat sources at heating power of 188, 300, 600 and 1200W, the heating rate of heat spreaders are studied. Vapor chamber heat spreader has the better temperature uniformity both in six sources and dual sources mode due to a higher thermal conductivity, followed by silver, copper and aluminum heat spreader. Aluminum and silver heat spreader show the higher heating rate, followed by copper heat spreader, vapor chamber heat spreader due to a larger heat capacity, heat up the slowest.

Keywords: *Multi-well heat spreader, Heating rate, Temperature uniformity, Vapor chamber*

1. INTRODUCTION

The polymerase chain reaction, PCR, was invented by K. Mullis, and includes three basic reactions, denature, annealing, and extension (Adams *et al.*, 2005), all of which have corresponding temperatures, range of temperatures, periods of reaction, and temperature uniformity. As these reactions absolutely influence the final products (Zhang and Xing, 2007), it is important to reduce the period of reaction and raise temperature uniformity. Many domestic and foreign researches are based on micro chips, which use MEMS technology or micro-tunnel polymerase chain reaction biochips (Singh and Ekaputri, 2006; Kreith and Bohn, 2001; Shen and Chen, 2005). Even though, the situation of a reduced area raises temperature uniformity to ± 0.5°C, the cost is high and the situation is not available for mass production. In this research, the size of the multi-well heat spreader is four times that of a micro chip and utilizes the temperature uniformity of a vapor chamber heat spreader, which allows simultaneous operations, in order to meet the needs of PCR.

A Vapor Chamber is a closed hollow object that is full of working fluid, with an inner vacuum pump. The purpose of vacuum pumping is to reduce the boiling point of the working fluid, which makes it easier to achieve the phase transition that allows heat to spread rapidly (Kang *et al.* 2010; Tsai *et al.* 2010; Kang and Huang, 2008). Thermacore, Inc. (2000) simulated copper fins, which they compared with a vapor chamber combined with copper fins. The results show that copper fins have the obvious phenomena of centralizing the heat sources, which can easily generate a situation of uneven temperature; in contrast, a vapor chamber with copper fins has even temperature distribution. Currently, vapor chamber heat spreaders apply electronic cooling, which meet the high requirements of even temperatures in biomedical technology.

This research applies a basic simulation and analysis with copper, aluminum, silver, and vapor chamber heat spreaders. First, we change the number of heat sources, with a fixed heating power of 1200W, and simulate temperature distribution where the heating center of dual and six heating sources on a multi-well heat spreader is from room temperature to 90°C. This study uses six heat sources to operate simulations at different heating speeds of 188, 300, 600, and 1200W, observes the period of reaction and speed of warming, and discusses the a multi-well vapor chamber heat spreader in the conclusion.

2. CFD SIMULATION

2.1 Simulating the model of machine

Figure 1 shows a simplified model of the test equipment, which includes a multi-well vapor chamber heat spreader, heat sources, and cooling fins, which is an exact size according to actual measurements. As this study only simulates the situation of warming in the machine, we simplified the model and ignored fans, nozzle shaped fins, and fan cover.

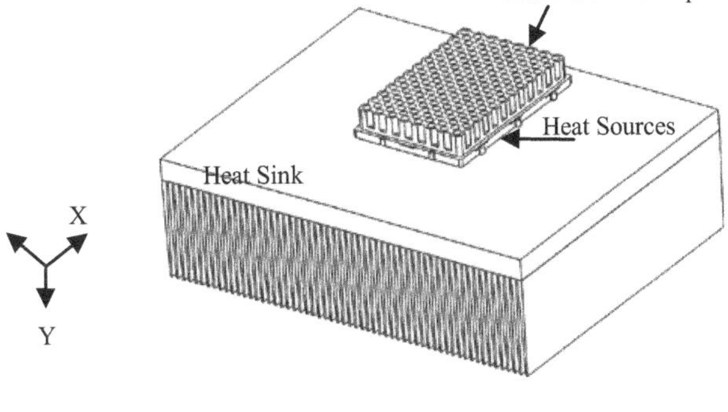

Fig. 1 The model of test equipment

Corresponding author. Email: swkang@mail.tku.edu.tw

1. Multi-well vapor chamber heat spreader (112×75×17.2mm): Figure 2 is composed of a vapor chamber heat spreader (thickness 4mm), with 96 holes (diameter of the holes is 5mm and depth is 10mm).
2. Heat sources: the area of heat sources is 30×30mm. Based on simplified modeling, the heat sources is 2D, and assumes that heat generation is even.
3. Cooling fins (250×200×73mm): the thickness of the substrate is 13mm, the fins are nozzle shaped, the thickness of the substrate is 2mm, and the bottom is 1mm, as seen in Figure 3. In addition, the fins have 17 hollow cylinders, with a diameter of 13mm, and are located in the area with the multi-well heat spreader to allow the multi-well heat spreader, heat sources, and cooling fins to fully cooperate; as the actual nozzle fins are difficult to model, the simulation has simplified fins, with the single size of 1.5mm.

2.2 Exact Formula

Mass-conservation equation, i.e.:

$$\frac{\partial u}{\partial x}+\frac{\partial v}{\partial y}+\frac{\partial w}{\partial z}=0$$

Momentum conservation equation, i.e.:

$$\rho\left(\frac{\partial u}{\partial t}+u\frac{\partial u}{\partial x}+v\frac{\partial u}{\partial y}+w\frac{\partial u}{\partial z}\right)=\rho g_x-\frac{\partial p}{\partial x}+\mu\left(\frac{\partial^2 u}{\partial x^2}+\frac{\partial^2 u}{\partial y^2}+\frac{\partial^2 u}{\partial z^2}\right)$$

$$\rho\left(\frac{\partial v}{\partial t}+u\frac{\partial v}{\partial x}+v\frac{\partial v}{\partial y}+w\frac{\partial v}{\partial z}\right)=\rho g_y-\frac{\partial p}{\partial y}+\mu\left(\frac{\partial^2 v}{\partial x^2}+\frac{\partial^2 v}{\partial y^2}+\frac{\partial^2 v}{\partial z^2}\right)$$

$$\rho\left(\frac{\partial w}{\partial t}+u\frac{\partial w}{\partial x}+v\frac{\partial w}{\partial y}+w\frac{\partial w}{\partial z}\right)=\rho g_z-\frac{\partial p}{\partial z}+\mu\left(\frac{\partial^2 w}{\partial x^2}+\frac{\partial^2 w}{\partial y^2}+\frac{\partial^2 w}{\partial z^2}\right)$$

Energy conservation equation, i.e.:

$$\frac{\partial}{\partial t}(\rho h)=\nabla\cdot(k\nabla T)+S_h$$

where ρ denotes density, u, v, and w are velocity components, p denotes pressure, μ denotes the viscosity coefficient, g denotes gravity, and h denotes enthalpy.

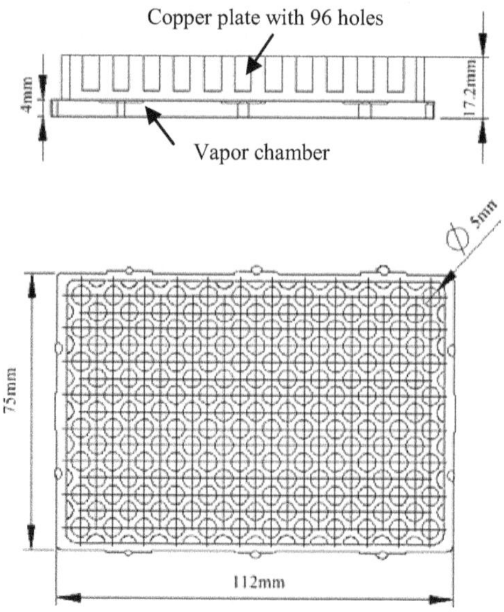

Fig. 2 The size of multi-well vapor chamber heat spreader

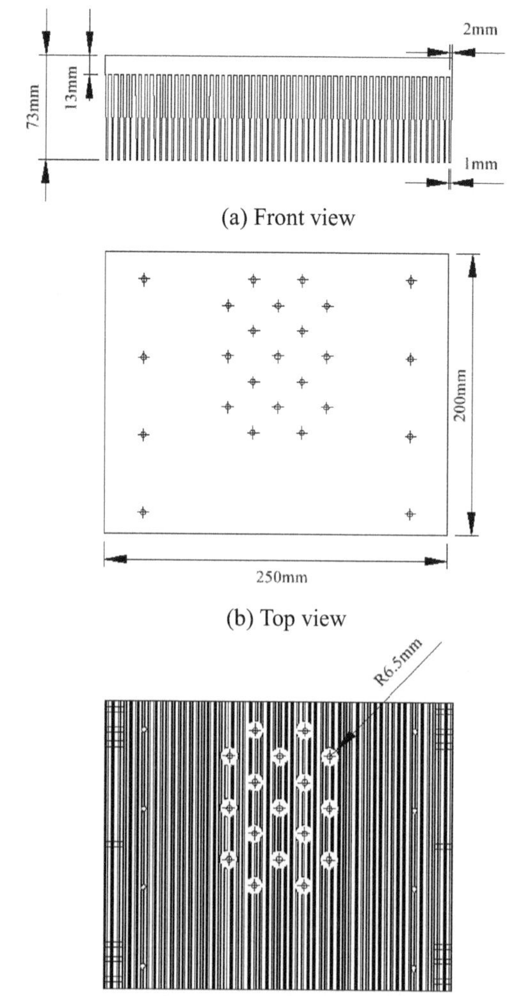

(a) Front view

(b) Top view

(c) Bottom view

Fig. 3 The size of cooling fins

In the natural convection, we used a Rayleigh number to calculate the intensity of buoyant force. i.e.:

$$Ra=\frac{g\beta\beta\Delta T^3\rho}{\mu\alpha}$$

ANSYS Icepak uses the Boussinesq model in the natural convection. All of the equations, with the exception of the momentum equation's buoyant force, assume that density is an identical value. i.e.:

$$(\rho-\rho_0)g\approx-\rho_0\beta(T-T_0)g$$

where ρ_0 is the density of the fluid, which is an identical value; T0 is the temperature of operation.

2.3 Convergence Test

This article sets the convergence of momentum and energy at 10-1, 10-2, and 10-3 in order to maintain the solution in a steady state, and the solution of 10-2 is close to 10-3, and in both of the two situations, energy can weaken to 10-7; therefore, this research set the convergence of momentum at 10-2, and the convergence of energy at 10-7.

2.4 The relaxation factor

As this article intends to resolve the process of transient heating in natural convection, to make the process easier, we set the relaxation factor at 0.7, and the momentum relaxation factor at 0.1, in order to complete the solution. It symbolizes that the difference of pressure from one iteration to the next will be limited under 70%, and the change of momentum will be limited under 10%.

3. RESULTS AND DISCUSSION

3.1 Comparisons of experiments

The simulated test equipment, under heating powers of 188w and 288w, and with dual or six heating sources, are compared for temperature distribution and time, with results as shown in Fig. 4 and 5. The results of the simulations are limited to these experiments, under dual or six heat source situations, with the temperature maintained at 90°C, the deviation becomes very low, approximately 1.72%.

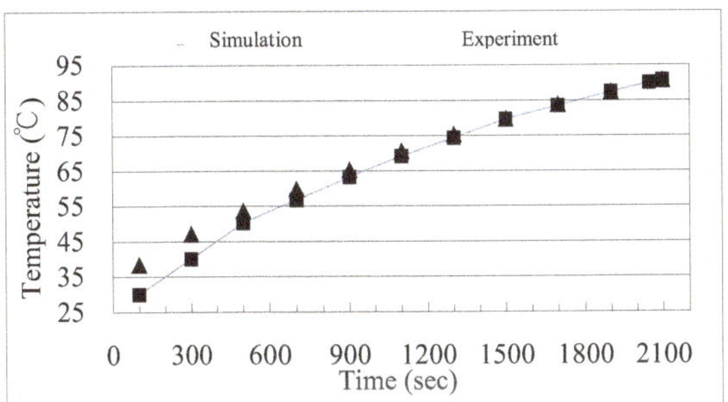

Fig. 4 The comparison of the simulation and experiment with six heat sources in a copper heat spreader

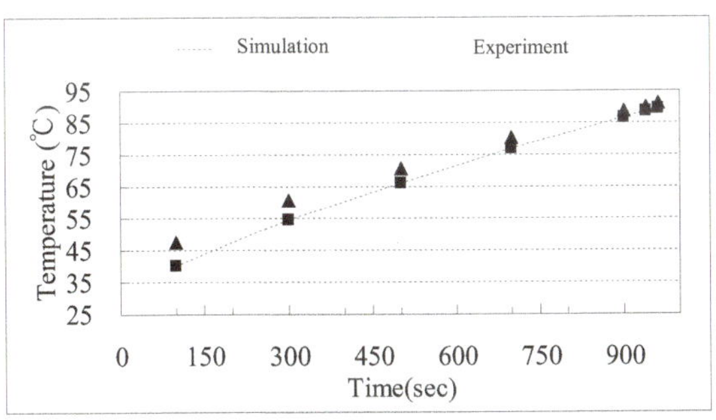

Fig. 5 The comparison of the simulation and experiment of dual heat sources with a copper heat spreader

3.2 Analysis with dual and six heat sources temperature uniformity

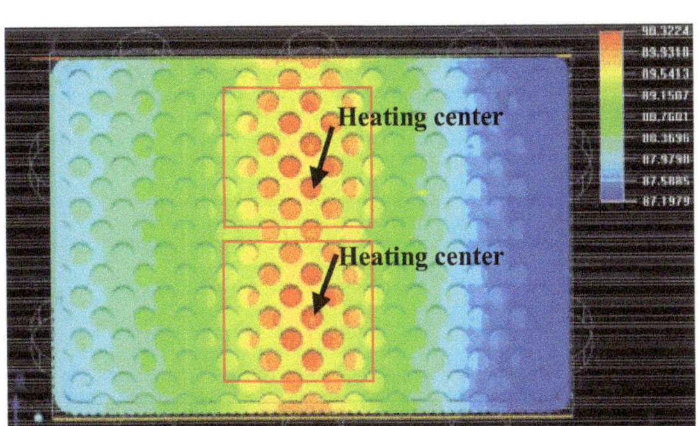

Fig. 6 Temperature distribution of aluminum vapor chamber under dual sources (ΔT=3.12 °C)

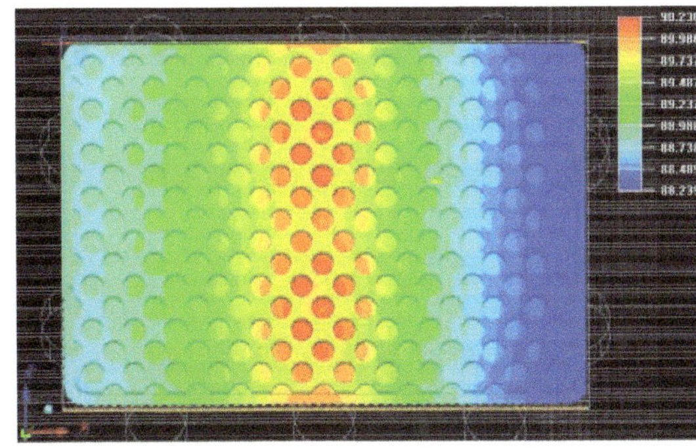

Fig. 7 Temperature distribution of copper vapor chamber under dual sources (ΔT=2.00 °C)

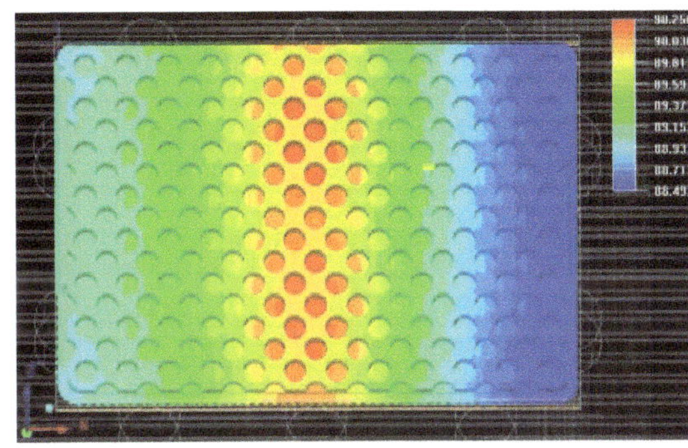

Fig. 8 Temperature distribution of silver vapor chamber under dual sources (ΔT=1.76 °C)

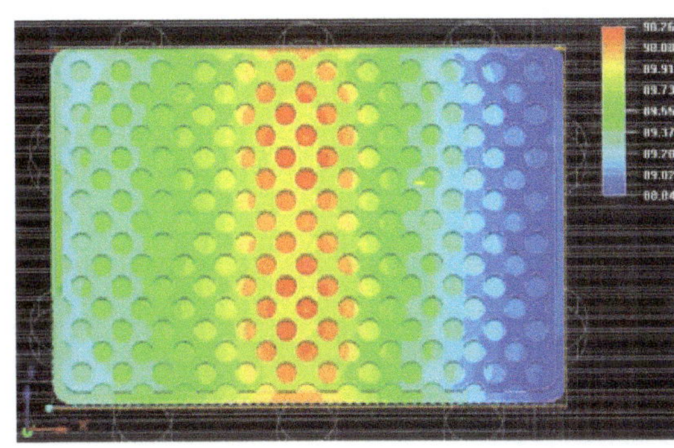

Fig. 9 Temperature distribution of vapor chamber under dual sources (ΔT=1.42 °C)

Fig. 10 Temperature distribution of aluminum vapor chamber under six sources (ΔT=1.58 °C)

Fig. 11 Temperature distribution of copper vapor chamber under six sources (ΔT=1.09 °C)

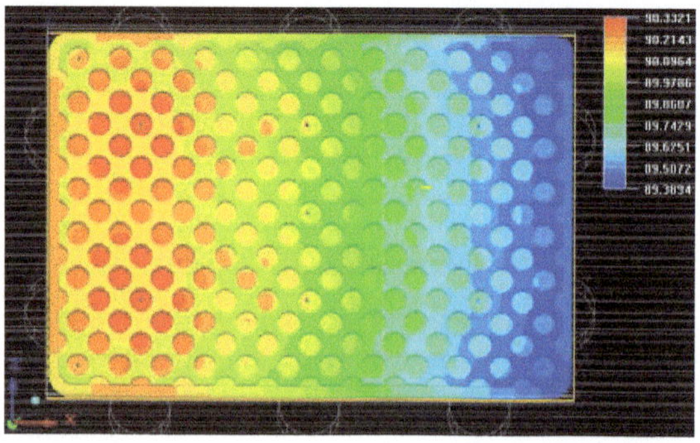

Fig. 12 Temperature distribution of copper vapor chamber under dual sources (ΔT=0.94 °C)

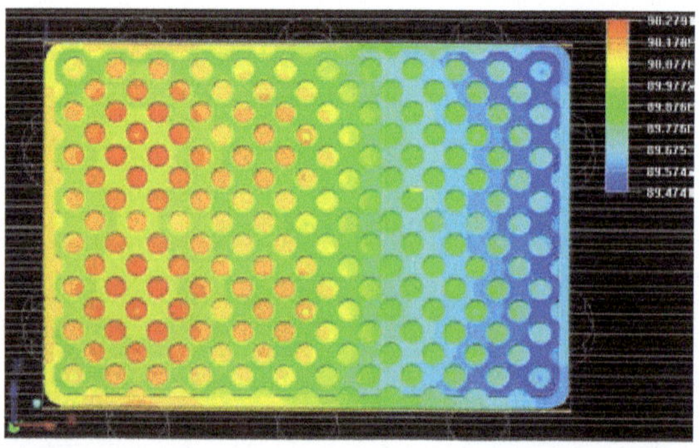

Fig. 13 Temperature distribution of vapor chamber under dual sources (ΔT=0.81 °C)

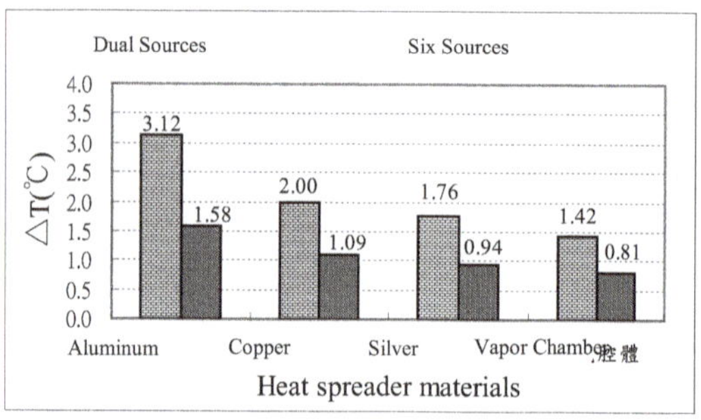

Fig. 14 The comparison of the temperature differential under dual and six heat sources at a heating power of 1200W

The simulation results, as shown in Figs. 6 to 13, compare the different material temperature differences, and find that when under six heat sources, the temperature difference is smaller than in dual heat sources, as shown in Fig. 14. Furthermore, the temperature uniformity is proportional to the material thermal conductivity coefficient. Thus, temperature uniformity could be arranged as vapor chamber, silver, copper, and aluminum.

3.3 The diagram, of different heating power and the period of warming

Figure 15 shows that the simulation of the vapor chamber heat spreader requires a warming period from room temperature to 90°C when testing several heating powers of 188, 300, 600, and 1200W.

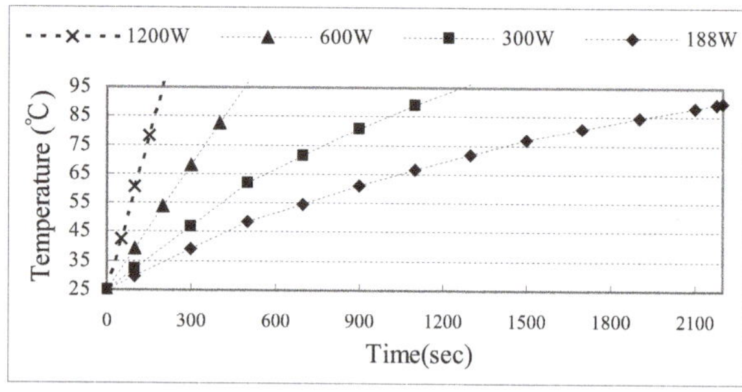

Fig. 15 The temperature distribution of the vapor chamber heat spreader at different watts

With the lower heating power, room temperature is regarded as an important parameter to influence the temperature of the heat source; therefore, the warming line is a curve. If the heating power goes lower, the steady temperature could be lower than 90°C. Prior to increasing the heating power, room temperature has no influence on the temperature of the heat source, and the line of warming tends to be a linear. Table 1 shows the detailed information, including the heating rate of several of kinds of heat spreaders with temperature uniformity.

According to the four heating speeds shown in Table 1, there are different heating rate, with different ingredients heating from room temperature to 90°C. When the temperature in the center of the heat spreader reaches 90°C, it captures the distribution of the temperature in the heat spreader. This study calculates the difference between the high and low temperatures (ΔT), as shown in Fig. 16, which shows different temperature uniformity and heating rate at 188, 300, 600, and 1200W, with different ingredients in the heat spreader. Fig. 16 compares silver, aluminum, copper and vapor chambers, the vapor chamber have the best uniformity; however, the heating speed is also the slowest. In addition, the copper has the second best uniformity, with a faster heating rate.

The heating rate in the reaction of PCR is regarded as the faster, the better, while temperature uniformity is regarded as the lower, the better; therefore, the ratio of the temperature uniformity and the heating rate is the smaller, the better. Fig. 17 presents the relationships between heat spreaders of various materials and temperature uniformity, the ratio of the heating rate, and the heating power, and shows that the ratio of a vapor chamber heat spreader's uniformity and the heating rate is lower than any other material heat spreaders. The results suggest that a vapor chamber heat spreader is the most suitable for application in the reaction of PCR.

4. CONCLUSIONS

In this study, the temperature uniformity and heating rates of heat spreaders with multi-wells were simulated and analyzed by ANSYS-Icepak software under natural convection conditions. We selected four materials for the multi-well heat spreader, including copper, aluminum, silver, and vapor chambers, as well as dual and six heating sources, at heating powers of 188, 300, 600, and 1200W, in order to simulate this research. The simulation found that, when at lower heating powers, the surrounding temperature would have great effect, thus, the temperature rise in a curve line. Otherwise, when at higher heating powers, the effecting by the surrounding temperature is very small, so the temperature was straight up. According to the results, when under six heating sources, temperature uniformity would better than with dual sources. In addition, comparisons of temperature uniformity show that, a vapor chamber is better than silver, silver is better than copper, and copper is better than aluminum. Therefore, a multi-well vapor chamber heat spreader has better temperature uniformity than other materials.

ACKNOWLEDGEMENTS

This research work was done at Department of Mechanical and Electro-Mechanical Engineering, Tamkang University, Taiwan under the auspices of The National Science Council Research Fund (under Contact No. 99-2622-E-032-003-CC3).

Table 1 The material properties of heat spreaders at different watts, with six heat sources

Aluminum

Heat power (W)	Temperature differential (ΔTi)	Time (∂t)	Heating speed (ΔTi)/ ∂t)	Temperature uniformity (ΔT)	The ratio of temperature uniformity and heating speed $\left(\frac{\Delta T}{\Delta Ti/\partial t}\right)$
188	65	2175	0.030	0.504	16.861
300	65	1070	0.061	0.782	12.866
600	65	444	0.146	0.983	6.714
1200	65	185	0.351	1.584	4.509

Copper

Heat power (W)	Temperature differential (ΔTi)	Time (∂t)	Heating speed (ΔTi)/ ∂t)	Temperature uniformity (ΔT)	The ratio of temperature uniformity and heating speed $\left(\frac{\Delta T}{\Delta Ti/\partial t}\right)$
188	65	2184	0.030	0.299	10.056
300	65	1076	0.060	0.404	6.683
600	65	437	0.149	0.725	4.871
1200	65	192	0.339	1.085	3.206

Silver

Heat power (W)	Temperature differential (ΔTi)	Time (∂t)	Heating speed (ΔTi)/ ∂t)	Temperature uniformity (ΔT)	The ratio of temperature uniformity and heating speed $\left(\frac{\Delta T}{\Delta Ti/\partial t}\right)$
188	65	2178	0.030	0.249	8.337
300	65	1080	0.060	0.410	6.819
600	65	421	0.154	0.713	4.617
1200	65	182.5	0.356	0.943	2.647

Vapor chamber

Heat power (W)	Temperature differential (ΔTi)	Time (∂t)	Heating speed (ΔTi)/ ∂t)	Temperature uniformity (ΔT)	The ratio of temperature uniformity and heating speed $\left(\frac{\Delta T}{\Delta Ti/\partial t}\right)$
188	65	2330	0.028	0.208	7.456
300	65	1166	0.056	0.331	5.941
600	65	492	0.132	0.526	3.983
1200	65	212.2	0.306	0.805	2.628

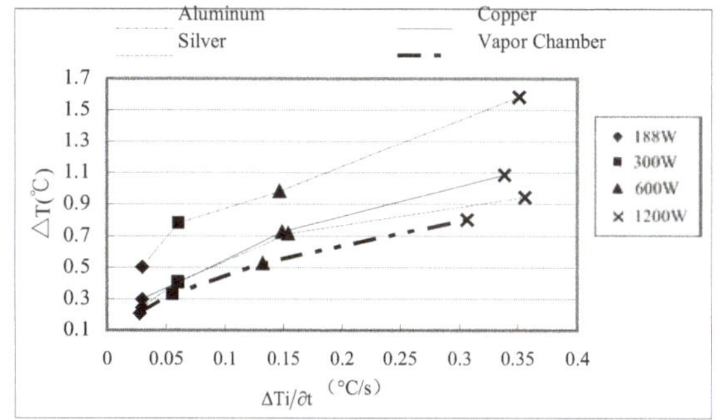

Fig. 16 The temperature uniformity and heating speed of the heat spreader with different watts and different materials

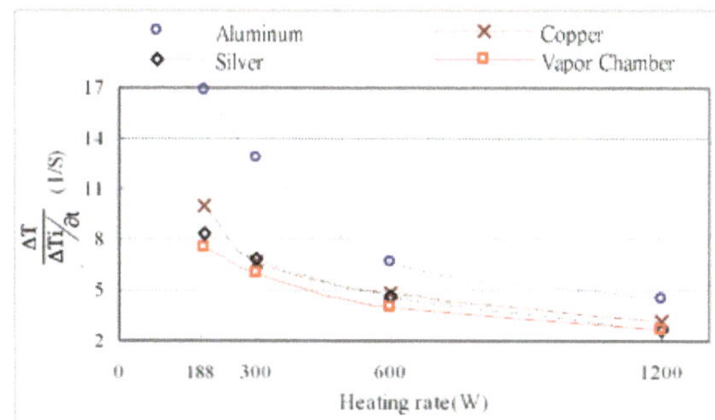

Fig. 17 The ratio of the temperature uniformity and heating rate with different watts and different materials

REFERENCE

Adams, S., and Russ, C. Uribe, M., Marincola, F.M., Stroncek, D., 2005, "PCR Validation Testing Utilizing a Novel Dynamic Measurement System," Cultech, Inc.

Icepak 4.1 User's Guide, August 2003.

Kang, S. W., Hung, Y. H., 2008, "Feasibility study of an aluminum vapor chamber," *Proceeding of 9th International Heat Pipe Symposium*, Malaysia, pp. 103-106.

Kang, S. W., Yu, C. S., Weng, M. T., 2010, "Effective thermal management of multiple electronic components," *IMPACT 2010 & International 3D IC Conference*, Taipei, Oct. 20-22.

Kreith, F., Bohn, M.S., 2001, "Principles of Heat Transfer," 6th Edition, Brooks/Cole, Pacific Grove, California.

Shen, K., Chen, X., 2005, "A microchip-based PCR device using flexible printed circuit technology," *Sensors and Actuators B*, vol. 105, pp. 251-258.
http://dx.doi.org/10.1016/j.snb.2004.05.069

Singh, J., Ekaputri, M., 2006, "PCR thermal management in an integrated Lab on Chip," *J. Phys.-Conf. Series*, **34**, 222–227.

Thermacore, Inc., "Therma-BaseTM Heat Sink", 2000.

Tsai, M. C., Yu, C.S., Kang, S. W., 2010, "A Large Size Vapor Chamber Used on Multiple Heat Sources Blade Server System," *15th International Heat Pipe Conference*, Clemson, SC, April 25-30.

Tsai, M. C., Chien, K. C., Huang, C. Y., Kang, S. W., 2008, "Thermal Performance of a Vapor Chamber," *9th International Heat Pipe Symposium*, Malaysia, pp. 86-89.

Zhang, C., Xing, D., 2007, "SURVEY AND SUMMARY Miniaturized PCR chips for nucleic acid amplification and analysis: latest advances and future trends," *Nucleic Acids Research*, Vol. 35, No. 13, 4223–4237.
http://dx.doi.org/10.1093/nar/gkm389

FLAT PLATE HEAT PIPES: FROM OBSERVATIONS TO THE MODELING OF THE CAPILLARY STRUCTURE

Frédéric Lefèvre[a,b,c,*], Stéphane Lips[a,b,c], Romuald Rullière[a,b,c], Jean-Baptiste Conrardy[a,b,c], Martin Raynaud[d], Jocelyn Bonjour[a,b,c]

[a]*Université de Lyon, CNRS*
[b]*INSA-Lyon, CETHIL, UMR5008, F-69621, Villeurbanne, France*
[c]*Université Lyon 1, F-69622, France*
[d]*Thales Communications, Thales, 92701 Colombes, France*

ABSTRACT

This paper presents a synthesis of observations made inside six different flat plate heat pipes (FPHPs). Silicon and copper systems with different capillary structures and sizes are studied. They are hermetically sealed on their upper face with a transparent plate, which enables liquid-vapor interface observations. A confocal microscope is used to measure the liquid film shapes inside the systems and describe the meniscus curvature radius from the evaporator to the condenser. Physical mechanisms, including the boiling limit and the effect of the interfacial shear stress, involved in different capillary structures such as grooves or meshes are discussed. Experimental results are compared to an analytical model describing both the liquid and vapour flows inside the FPHP and coupled to an analytical solution for the wall temperature. The capillary structure is modeled by considering the permeability and the equivalent thermal conductivity of a porous medium. The comparison between the model and the experimental data makes possible the estimation of the physical properties of the capillary structure. These experimental parameters are introduced in the model to estimate the performance of the heat pipe in a real application with several electronic components.

Keywords: *confocal microscopy, grooves, meshes, electronic cooling, meniscus*

1. INTRODUCTION

The interest in the use of heat pipes for thermal management of electronic components is recognized for both aerospace and terrestrial applications, especially with increasing heat flux requirements. In many applications, heat pipes are circular and are used to transport heat from one single heat source to one single heat sink. Flat plate heat pipes (FPHP) have the same function, but offer a wide cross-section, which enables the reduction of their thickness without reducing their thermal performance. They are particularly interesting when several heat sources have to be cooled, such as several electronic components on an electronic card. The thermal performance of a FPHP depends mainly on the capillary structure, which is usually made of grooves, meshes or sintered powder. FPHP can also be realized using several micro heat pipes placed parallel one to each other at a small distance.

Flat plate heat pipes (FPHP) have been widely studied in the last fifteen years, especially for their applications to electronic cooling (Lallemand and Lefèvre, 2004; Vasiliev *et al.*, 2008). Although the literature on FPHP is abundant, experimental papers are relatively rare and most of the time do not provide measurements inside these systems, but only surface temperature measurements. Besides, theoretical works are generally focused on capillary-driven flows and thermal models in ideal conditions, i.e. the system being optimally filled with liquid (Khrustalev and Faghri, 1995; Kim *et al.*, 2003; Lefèvre *et al.*, 2008).

Since a few years, we developed experimental benches to visualize the flow in the capillary structure of FPHPs closed with transparent plates. Different capillary structures were studied (grooves with constant or variable width, crossed grooves, grooves covered with screen meshes, meshes) for FPHPs made of copper or silicon. Depending on the applications, we tested FPHPs with different sizes and different evaporator, condenser and adiabatic areas. The behavior of those systems filled with various working fluids were studied for different filling ratios and a wide range of applied heat fluxes.

Confocal microscopy has been used with varying degrees of success (depending on various parameters such as the size of the capillary structure, the heat flux applied at the evaporator, …) to characterize the liquid-vapor interface in the capillary structure (Rullière *et al.*, 2007; Lips *et al.*, 2010a,b,c and 2011) and thin films of condensation on the groove fins (Lefèvre *et al.*, 2010). Such measurements had never been provided before although they are necessary to validate most of the theoretical works published on FPHP (Lefèvre *et al.*, 2008; Revellin *et al.*, 2009). Temperature measurements are indeed not sufficient to validate hydrodynamic models that depend mostly on the meniscus curvature radius variation from the condenser to the evaporator. Furthermore, the temperature field in flat heat pipes depends on the thickness of thin films in both the condensing and evaporating areas (Lips *et al.*, 2010a). Thus, the curvature radius of the menisci is a key parameter in both hydrodynamic and thermal models of heat pipes.

While confocal microscopy has been very useful to characterize FPHPs, visual observations have been sufficient to describe different physical phenomena of high importance to understand how FPHPs work. In the first sections of this paper, a synthesis of our observations is presented as well as what they have brought to the knowledge on these systems.

* *Corresponding author. Email: Frederic.lefevre@insa-lyon.fr*

Physical phenomena involved in FPHPs are complex: evaporation and condensation phenomena, in presence of thin liquid films, in capillary geometries, and possibly boiling heat transfer. Models including such a complex physic are useful for the understanding but not convenient for engineering applications. In this paper, we also present results obtained with a very simple model (Lefèvre and Lallemand, 2006), in which the FPHP is characterized by its geometrical parameters, the properties of the capillary structure (permeability, equivalent thermal conductivity and effective radius), and the thermophysical properties of the wall and the working fluid. This model is particularly convenient for engineers since it enables the calculation of both the 3D temperature field and the capillary limit of a FPHP used to cool several electronic components (or any kind of heat sources) and several heat sinks, whatever their locations and their heat transfer rate.

In the last section of the article, the model is compared to experiments realized with a FPHP in order to estimate the permeability and the equivalent thermal conductivity of its capillary structure. The estimated properties are compared to analytical classical laws of the literature. Finally, these experimental parameters are introduced in the model to simulate the cooling of a real electronic card with several electronic components.

2. EXPERIMENTAL BENCHES DEVELOPED TO STUDY FPHPs

This section presents the physical and geometrical characteristics of different FPHPs and the associated experimental benches that were developed to characterize these systems.

2.1 Six different flat plate heat pipes

Six flat plate heat pipes, with different sizes and capillary structures have been tested. They are hermetically sealed on their upper face with a transparent plate, which enables liquid-vapor interface observations. Five of them are made of copper and one of silicon. In all cases, temperature sensors are located in the wall to measure the temperature field in working conditions.

FPHP 1, shown in Fig. 1, is made of a copper plate of thickness 1.5 mm and surface 267×115 mm² (Lefèvre *et al.*, 2011 and 2012). The capillary structure, of dimensions 230 × 90 mm², is made of one or two CuSn 325 square screen mesh layers. The capillary screen meshes are plated against the copper plate by means of a coarse screen mesh, which creates the vapor space (Fig. 2). The heat source is a thick resistor film of dimensions 55.1×96.5 mm² and resistance 90.6 Ω. Two different heat sinks were tested: either two aluminum radiators (140×170 mm²), cooled by natural convection of air at ambient temperature were screwed to each extremity of the heat pipe by thermal wedge locks or water heat exchangers of heat transfer area 30 × 90 mm² were plated against the wall also at each extremity.

FPHP 2 (Fig 3) is made of 109 longitudinal micro-grooves, machined in a copper plate of dimensions 230×90 mm² (Rullière *et al.*, 2007; Lefèvre *et al.*, 2008). Unlike classical FPHP applications, in which the heat source is small compared to the adiabatic and the heat sink regions, this FPHP has been tested with a large heat source (190×90 mm²), a small heat sink (30×90 mm²) and a small adiabatic region (10×90 mm²). The application was the cooling of fuel cells. Each groove has a square cross-section of height and width equal to 400 µm. The distance between two grooves is equal to 400 µm. Good thermal performance was obtained with FPHP 2 in horizontal and in vertical favorable positions, but it was not able to work in unfavorable tilted position because capillary pressure induced by the grooved capillary structure was too low to overcome the hydrostatic pressure drop in vertical position. In order to overcome this limit, screen meshes were plated against the grooves (Lefèvre *et al.*, 2012) to increase the capillary pressure gradient while preserving the high permeability of the grooves. FPHP 3 is similar to FPHP 2, but the grooves are covered by two layers of meshes (Fig. 4).

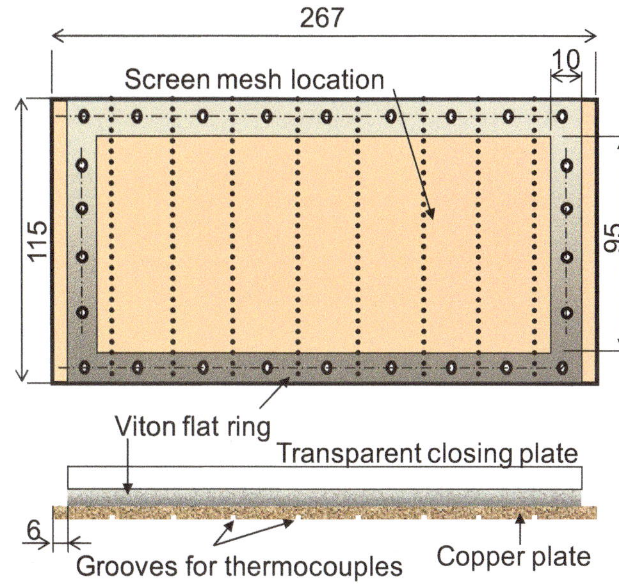

Fig. 1 Schematic of FPHP 1 (Lefèvre *et al.*, 2012); length in mm

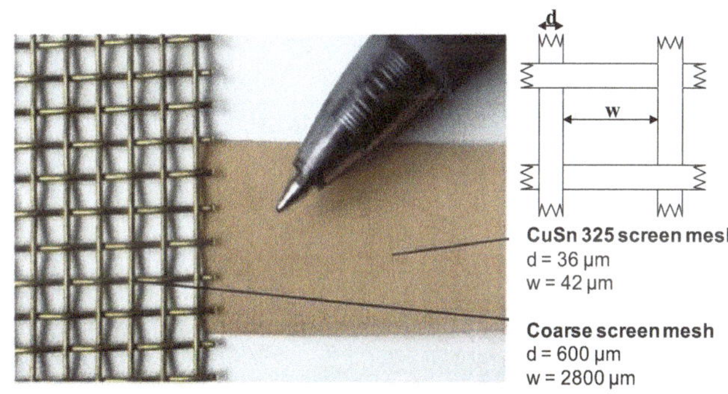

Fig. 2 Coarse and fine screen meshes (Lefèvre *et al.*, 2012)

Fig. 3 Picture of FPHP 2 (Rullière *et al.*, 2007)

FPHP 4 is also made of copper with the same grooves as FPHP 2 (Lips *et al.*, 2009 and 2010). Compared to the previous systems, the dimensions are close to those encountered in electronic cooling applications (overall dimensions of 70×90 mm², evaporator and condenser area of 20×90 mm² and 30×90 mm² respectively).

The fifth heat pipe, FPHP 5, of dimensions 72×68 mm², is shown in Fig. 5. It is made of 85×90 crossed micro-grooves (that can also be viewed as an array of posts) machined in a copper plate. As for FPHP 2, 3 and 4, each groove has a square cross section (400×400 µm²) and the distance between two grooves is equal to 400 µm.

The sixth and last heat pipe, FPHP 6 (Fig. 6) is made of silicon. Its capillary structure, of diameter 50 mm, is etched on a silicon wafer of

100 mm diameter and 525 μm thickness. It is made of 72 radial micro-grooves that are connected by crossed grooves at the centre of the device. The etching depth is equal to 200 μm. The radial grooves have a rectangular cross-section whose width varies from 670 μm at the periphery to 110 μm at the connection with the crossed grooves. These crossed grooves have a width equal to 100 μm and they are separated by fins of same width, which creates a network of 300 small posts of 100×100 μm^2.

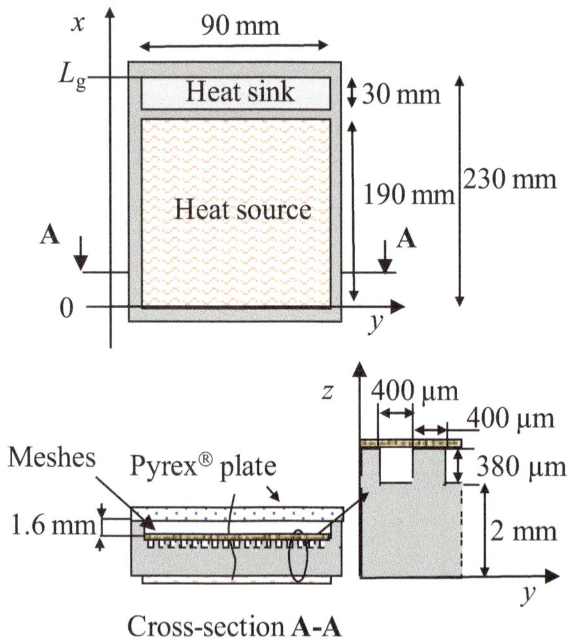

Fig. 4 Schematic of FPHP 2 (grooves) or FPHP 3 (grooves and meshes)

Fig. 5 Picture of FPHP 5 (Lips *et al.*, 2010c)

2.2 Confocal Microscopy

Capillary phenomena, evaporation, and condensation in micro channels have been the focus of many experimental papers these last years. In a heat pipe, these phenomena are linked together and cannot be studied separately. Therefore, it is difficult to perform measurements in such a device. For example, micro-PIV is not helpful in heat pipes because the process of insemination is not convenient. Indeed, any particles introduced in the liquid would be blocked at the evaporator because of their non evaporative nature. Among classical techniques for film thickness measurement applied to micro-scale two-phase flow systems (Tibiriça *et al.*, 2009), confocal microscopy appears to be the most appropriate to characterize FPHPs.

The confocal microscope is a STIL Micromesure 2 system. The optical sensor has a nominal measuring range of 350 μm. The working distance is about 13 mm. The maximum measuring angle for specular reflection is about 27°. The optical sensor velocity is about 1 $mm.s^{-1}$, which is fast enough for not thermally disturbing the measurement. Once the top of the groove is detected by the sensor, the liquid-vapor interface is recorded. The resulting data are used to estimate the curvature of the surface.

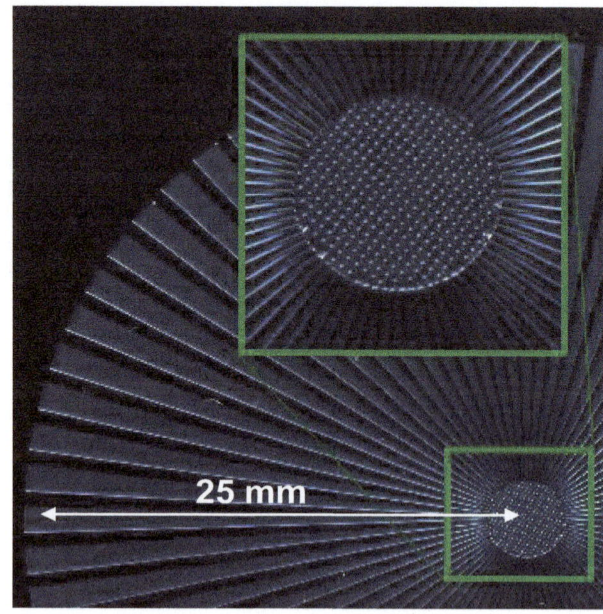

Fig. 6 Picture of FPHP 6 (Lefèvre *et al.*, 2010)

Optical principles of confocal microscopy are described in Fig. 7. A white light point source is imaged by a lens that has not been corrected for chromatic aberration on a series of monochromatic point images in the measurement space. The reflected light passes back through the lens and is directed towards a spectral detector by a partially reflecting mirror. The spectral analysis of the collected lights that are separated by a spatial filter enables the location of the surface to be measured. The monochromatic light having the highest intensity corresponds to the surface to be measured.

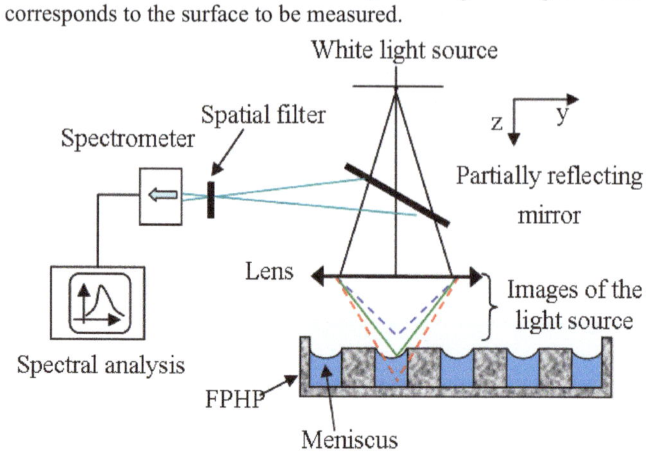

Fig. 7 Schematic diagram of confocal microscopy

3. SYNTHESIS OF THE OBSERVATIONS INSIDE THE CAPILLARY STRUCTURE OF THE FPHPs

3.1 Examples of measurements

Fig. 8 presents an example of measurement obtained with FPHP 6, filled with methanol, in the condenser area. Two grooves are visualized. The width of the grooves and the fins are about 400 μm and 800 μm, respectively (the scale on the y axis is 100 times lower than that on the x axis). This example shows that confocal microscopy permits to measure both the meniscus curvature radius inside the grooves and the condensate film on the fins. This last measurement is not possible with FPHPs made of copper. Indeed, the reflection of white light on a copper surface is larger than on silicon surface. As a result, the intensity of the reflected light is so high compared to that of the light reflected on the liquid-vapor interface that the focalization on the liquid film is not accurate enough for proper measurements. With silicon, the intensity of the reflected light is lower, which enables this measurement.

In example of Fig. 8, the thin condensate film thickness is lower than 5 µm. Circles that fit the experimental data are plotted. The radius of curvature is found to be 1.08 mm in the grooves and 12.4 mm on the fins. The curvatures between these zones are reversed. Therefore, the Young-Laplace law yields to the conclusion that the liquid pressure is higher than the vapor pressure on the fins and lower than the vapor pressure in the grooves. It has to be noted that, from our measurements, the transition between the two films looks abrupt. Thus, a very small region shall be present where the curvature changes from a constant positive value to a constant negative value. Such measurements are useful to understand evaporation and condensation processes that highly depend on the shape of the liquid-vapour interface (Lips *et al.*, 2010a).

For FPHP 6, the microscope was not useful to measure the menisci in the part of the capillary structure where groove widths are to small (< 150 µm). Indeed, for such small grooves, the slope of the meniscus at the junction with the grooves is higher than the maximum measuring angle of the optical sensor. This is a real limitation of this measurement method.

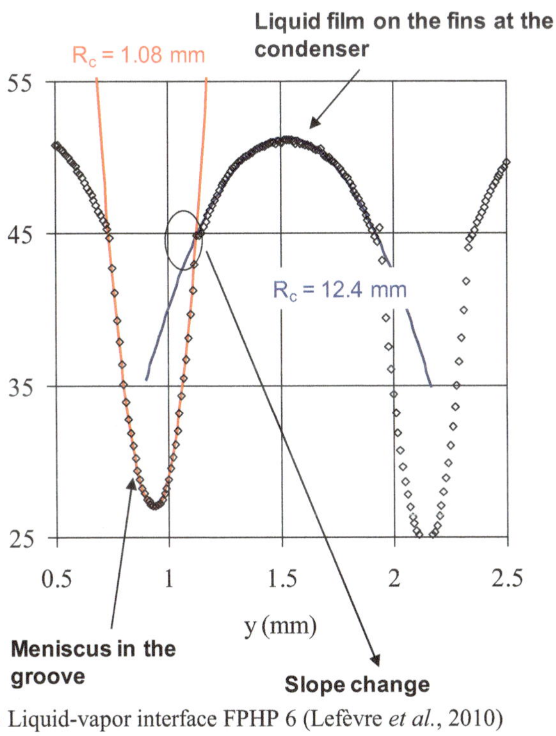

Fig. 8 Liquid-vapor interface FPHP 6 (Lefèvre *et al.*, 2010)

Confocal microscopy has also been used to measure the liquid-vapor interface in crossed-grooves (FPHP 5). Fig. 9 presents a measurement obtained at the junction between two grooves. The top of the fins is not homogeneous because of the copper roughness. The liquid-vapor interface is smoother and has a saddle shape. As the measurement of the confocal microscope is punctual, the surface is obtained by moving the FPHP with a motorized table along x and y directions.

The surface curvature is identified from the experimental data. The capillary pressure P_{cap} is deduced from the Young-Laplace law. The measurement method can be validated in non working conditions, with the FPHP being tilted. In such conditions, the capillary pressure is theoretically equal to the hydrostatic pressure of the liquid. Fig. 10 presents the capillary pressure measured for FPHP 5 filled with methanol and with FC72, with an inclination angle with respect to the horizontal equal to 5° and 2.5° respectively. Solid lines are the hydrostatic pressures. A good agreement is observed between experimental and theoretical results. The accuracy of the method is evaluated to be of the order of 1 Pa.

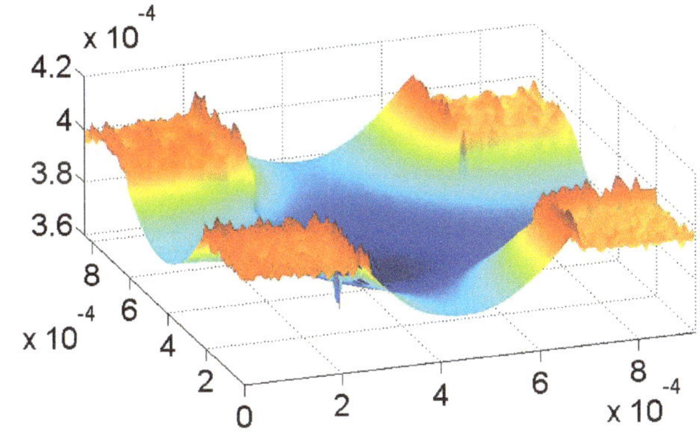

Fig. 9 Liquid-vapor interface in FPHP 5 (dimensions in m) (Lips *et al.*, 2010c)

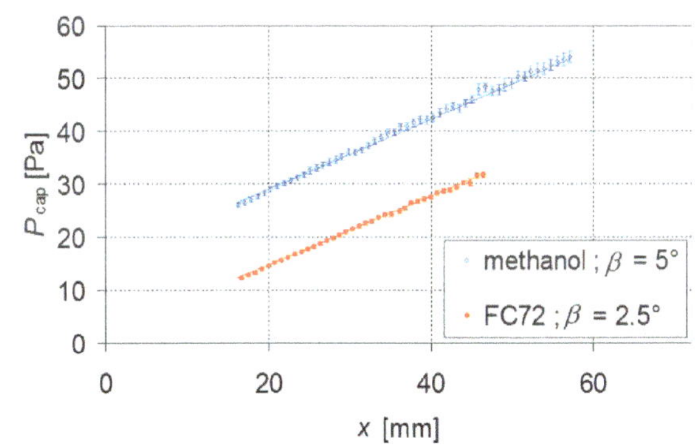

Fig. 10 Validation of the experimental method (Lips *et al.*, 2010c)

3.2 Synthesis of the information brought by the observations

FPHP 2 and FPHP 4 have been studied in different experimental conditions (Rullière *et al.*, 2007; Lips *et al.*, 2010b) in order to obtain a significant experimental database. Experimental parameters are the vapor space thickness (1 mm, 1.6 mm and 2 mm), the inclination, the filling ratio, the heat flux transferred by the system (between 0.5 Wcm^{-2} and 3 Wcm^{-2}) and the working fluid (N-pentane or methanol).

Fig. 11 presents the comparison between this database and a hydrodynamic model based on the balance equations and the Young-Laplace law (Lips *et al.*, 2011). To facilitate and reduce the presentation of the hydrodynamic measurements, the experimental results obtained by confocal microscopy are presented in the form of the mean capillary pressure gradient $\Delta P_{cap}/\Delta x$ in the adiabatic zone. The uncertainty of the capillary pressure gradient is given for each experiment. In all the cases, a good agreement is observed between experimental and numerical results. The numerical model tends to slightly overestimate the capillary pressure gradient in the system. The mean relative error of the model is about 15 %.

In theoretical models developed in the literature, shear stress calculation is often viewed as the most important parameter for pressure drop calculations. In the present model, the liquid-wall and vapor-wall shear stresses are calculated with a classical approach in laminar conditions (Faghri, 1995). In Fig. 11, interfacial shear-stress has been calculated considering the liquid as motionless for the vapor, but in all these experimental conditions, neglecting the interfacial shear stress does not change significantly the results (Lips *et al.*, 2011), even with a small vapor space (1 mm). It has to be noted that these experimental conditions are close to those encountered in the literature (Lallemand and Lefèvre, 2004).

While interfacial shear stress has not a strong influence on FPHP performance, experimental results obtained by confocal microscopy with FPHP 4 show that FPHP performance is strongly influenced by both the filling ratio and the vapor space thickness (Lips *et al.*, 2010). A small vapor space thickness induces liquid retention in the FPHP corners and sides and thus reduces the thermal resistance of the system. Nevertheless, the meniscus curvature radius at the condenser decreases with the decrease of the vapor space thickness, which reduces the maximum capillary pressure of the FPHP. As an example, a vapor space of 1 mm leads to a 35 % decrease of the maximum capillary pressure with respect to a very large vapor space thickness.

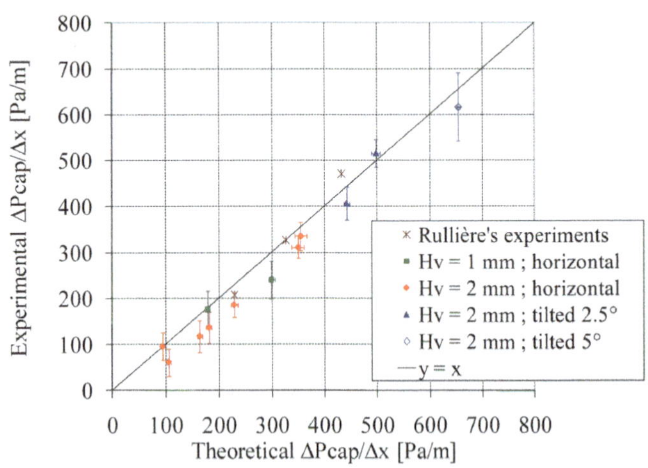

Fig. 11 Comparison between experimental and numerical hydrodynamic results (Lips *et al.*, 2011)

Confocal microscopy is well adapted to characterize FPHP but visual observations are sufficient to describe different physical phenomena of high importance to understand how FPHP works. For example, nucleate boiling (Fig. 12) has been observed in various configurations for relatively small heat fluxes (Lips *et al.*, 2009; Lefèvre *et al.*, 2012).

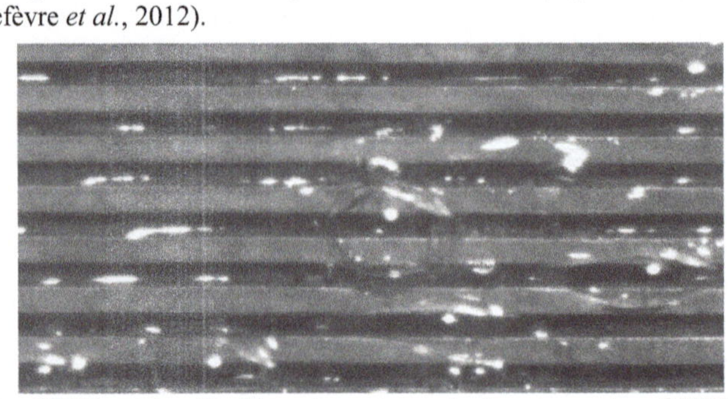

Fig. 12 Example of boiling phenomenon in the evaporator of FPHP 4 (Lips *et al.*, 2009)

For FPHP 4, filled with methanol, nucleate boiling was observed for heat fluxes higher than 3 W/cm² and until 18 W/cm². Unexpectedly, this phenomenon does not prevent the working of the system and conversely increases its thermal performance. An experiment conducted in non horizontal orientation (Lips *et al.*, 2011) showed that the presence of nucleate boiling in the system does not constitute a limit as it is generally stated in the literature. In all cases, dryout of the evaporator is due to the capillary limit. All these investigations have been made under normal gravity conditions. But heat pipes are often used in space applications: it would be interesting to perform identical tests under micro gravity, in order to determine if the conclusions of this study can be extended to this case. In any case, a better understanding of these phenomena is necessary, particularly to theoretically predict the operation limit in case of nucleate boiling. Nucleate boiling was also observed in FPHP 5 and 6 for similar heat fluxes. Conclusions were the same. For FPHP 2, this phenomenon was not observed because the evaporator area is very large and thus capillary limit was observed for heat fluxes lower than 1 Wcm^{-2}, before the onset of nucleate boiling.

On the contrary, the performance of the heat pipe obtained with the association of grooves and meshes (FPHP 3) is not as high as one could expect (Lefèvre *et al.,* 2012) regarding the capillary limit. Indeed, while this capillary structure enables to work in tilted unfavourable position, which was not possible with the grooved heat pipe (FPHP 2), a clear nucleate boiling limitation is observed for rather small heat fluxes (~ 0.4 W/cm²). This dryout is brutal and takes place on all the surface of the FPHP, which was not the case for FPHP2 when the capillary limit was encountered. This dryout is probably due to the rapid expansion of bubbles, which grow between the screen mesh layer and the grooves, rather than outside of it, because of the capillary pressure barrier imposed by the meshes. As a result, bubble growth pushes away the liquid from the grooves. As the screen mesh layers are not able to drain the liquid on the entire surface, the capillary structure suddenly dries out after the boiling incipience.

The observations made inside FPHPs have increased considerably the knowledge on these systems. Confocal microscopy enables to measure the vapor and liquid pressure difference along the capillary structure. These measurements, associated with temperature measurements enable the estimation of the thermophysical properties of the capillary structure by comparison with an appropriate model.

4. AN ANALYTICAL MODEL FOR THE COOLING OF MULTIPLE ELECTRONIC COMPONENTS

This section presents a 2D analytical solution for both the liquid and vapor flows inside a FPHP coupled to a 3D analytical solution for the temperature inside the FPHP wall. The thermal model is able to calculate the part of the heat flux transferred only by heat conduction in the wall from the heat transferred by change of phase.

4.1 Equations of the model

The model was already described in previous articles (Lefèvre and Lallemand, 2006; Revellin *et al.*, 2009) to which the reader should refer for further details.

The geometry of the FPHP is shown in Fig. 13. It consists of a flat plate of $a \times b$ rectangular dimensions. The electronic components and the heat sinks can be located anywhere on the FPHP and have different heat transfer rates. The rest of the surface is adiabatic. The wall of the FPHP is made of a material of thermal conductivity λ_s and its height is equal to c. The capillary structure is modeled by considering a porous medium of permeability K, thermal conductivity λ_p and height H_p. The liquid inside the porous medium has a dynamic viscosity μ_l, a heat conductivity λ_l and a density ρ_l. The equivalent thermal conductivity of both the liquid and the porous medium is equal to λ_{eq}. The vapor space height is equal to H_v, and λ_v and μ_v are the vapor thermal conductivity and dynamic viscosity, respectively. The thermal properties are supposed to be constant. They are calculated at the saturation temperature T_{sat}.

The 3D heat conduction equation is solved in steady state by considering adiabatic conditions in $x = 0$, $x = a$, $y = 0$ and $y = b$. In $z = c$, a heat flux, which depends on the location and the heat transfer rate of the heat sources and the heat sinks is imposed. In $z = 0$, a Fourier boundary conditions is considered with a heat transfer coefficient function of both the thickness and the thermal conductivity of the capillary structure. By considering these boundary conditions, the temperature field $T(x,y,z)$ in the wall of the FPHP can be expressed by Fourier's series:

$$T(x,y,z) = C_1 \sum_{m=0}^{\infty}\sum_{n=0}^{\infty} A_{mn} \cos\left(\frac{m\pi x}{a}\right)\cos\left(\frac{n\pi y}{b}\right) \quad (1)$$

where C_1 is a constant and A_{mn} are functions of the location and of the power of the heat sources and heat sinks. They also depend on the thickness and the equivalent thermal conductivity of the capillary structure.

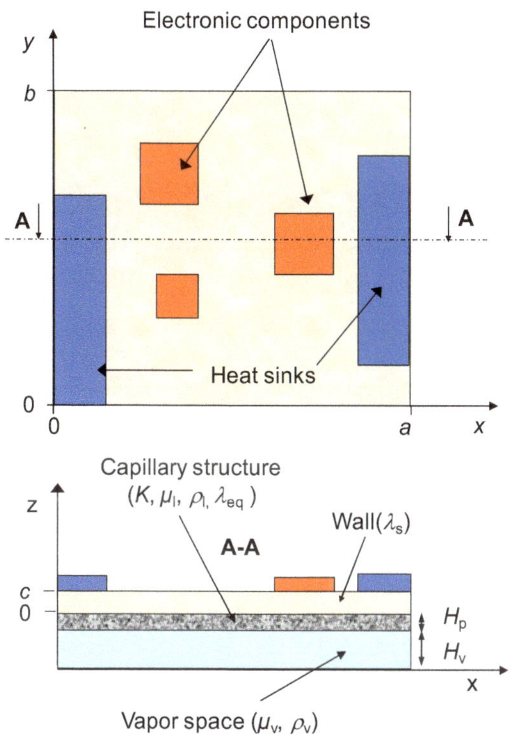

Fig. 13 Schematic of the FPHP for the model (Lefèvre and Lallemand, 2006)

Once the temperature field is calculated, it is possible to calculate the evaporation and condensation mass flow rates at each point of the system by deriving the temperature field in $z = 0$. A 2D hydrodynamic model, based on Darcy's law and on mass balance equations, is used to calculate the liquid and vapor velocities and pressures in the capillary structure. The velocities are equal to zero in $x = 0$, $x = a$, $y = 0$ and $y = b$. By considering these boundary conditions, the pressure field for both the liquid and the vapor, can also be expressed by Fourier's series:

$$P(x,y) = C_2 \sum_{m=0}^{\infty}\sum_{n=0}^{\infty} B_{mn} \cos\left(\frac{m\pi x}{a}\right)\cos\left(\frac{n\pi y}{b}\right) \quad (2)$$

where C_2 is a constant depending on μ_l, ρ_l and K for the liquid and μ_v, ρ_v and H_v for the vapor. B_{mn} are functions derived from the thermal model.

The maximum heat transfer capability of the FPHP is obtained when the maximum capillary pressure is equal to the sum of the pressure drops in the liquid ΔP_l and the vapor ΔP_v and the hydrostatic pressure ΔP_g that can be either positive or negative, depending on the location of the heat sources and the heat sinks:

$$\Delta P_{cap} = \Delta P_l + \Delta P_v + \Delta P_g \quad (3)$$

The capillary pressure can be calculated using the following classical expression (Faghri, 1995), which results from the Young–Laplace law:

$$\Delta P_{cap} = \frac{2\sigma}{r_{eff}} \quad (4)$$

where r_{eff} is the effective pore radius and σ is the surface tension of the fluid.

4.2 Estimation of the permeability and the equivalent thermal conductivity by comparison with experimental data

The comparison between the model and experimental data enables the estimation of both the permeability and the equivalent thermal conductivity of the capillary structure. When temperature measurements are available along the FPHP wall, the thermal conductivity is obtained when the model fits the experimental data. One can note that if the distance between the heat sources and the heat sinks is high enough, it is possible to differentiate the thermal conductivity at the condenser and at the evaporator (Revellin *et al.*, 2009) using the same model. When pressure measurements (deduced from confocal microscopy measurements) are available, the permeability is obtained when the model fits the experimental data. Otherwise, an estimation of the permeability can be obtained by considering the heat transfer rate for which the dryout of the evaporator is reached. If the maximum capillary pressure of the capillary structure is known, an estimation of the permeability is obtained when the sum of the pressure drops in the liquid and the vapour, which is a function of the permeability, is equal to the capillary pressure.

As an example, we present in this section the comparison between the model and experimental results obtained with FPHP 1 (Lefèvre *et al.*, 2012). The capillary structure is made of two CuSn 325 square screen mesh layers. The FPHP is filled with methanol. The heat source of surface 55.1×96.5 mm^2 is located in the middle of the FPHP. Two water heat exchangers of surface 30×90 mm^2 are located on each side of the system (Fig. 14).

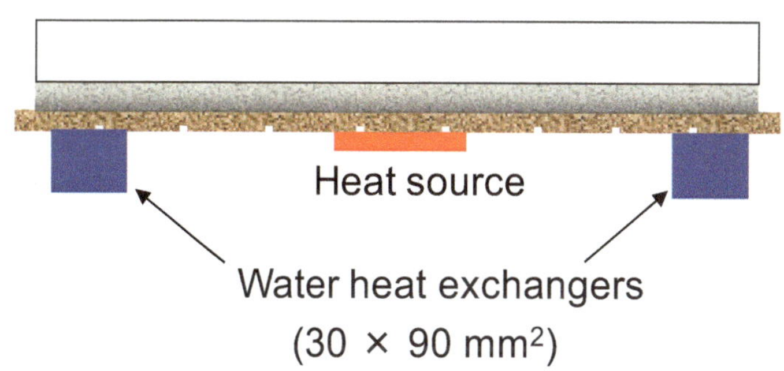

Fig. 14 Locations of the heat sink and the heat source on the FPHP 1

Fig. 15 presents the temperature profile in horizontal orientations for a heat transfer rate equal to 15 W. Fig. 16 presents the maximal temperature difference between the heat source and the heat sinks in horizontal orientation vs. the heat transfer rate of the heat source. Linear characteristics are obtained until a threshold heat transfer rate of 82 W. Below this value, the thermal resistance is equal to 0.14 K/W and can be compared to the thermal resistance of the copper plate alone (0.62 K/W).

In the model, the thickness of the capillary structure is set to 140 μm, which corresponds to twice the thickness of a CuSn mesh screen layers. The maximum capillary pressure is calculated with Eq. (4) by considering that the effective radius is equal to half the distance w between two fibers (Fig. 2). For methanol, ΔP_{cap} is equal to about 2000 Pa. In the model, the permeability is adjusted until the pressure drops in the liquid and the vapor, calculated for the threshold heat transfer rate of 82 W, is equal to the capillary pressure. The permeability of the meshes is estimated to 0.55×10^{-10} m^2.

For a mesh with a fiber diameter equal to d and a distance between two fibers equal to w (Fig. 2), the permeability K can be expressed as (Faghri, 1995):

$$K = \frac{d^2 \varphi^3}{122(1-\varphi)^2} \begin{cases} \varphi = 1 - \dfrac{1.05 \pi N d}{4} \\ N = \dfrac{1}{d+w} \end{cases} \quad (5)$$

where φ is the porosity. For this mesh, d = 35 μm and w = 42 μm. Thus, the permeability calculated with Eq. (5) is equal to 0.18×10^{-10} m², which is 3 times lower than that estimated experimentally.

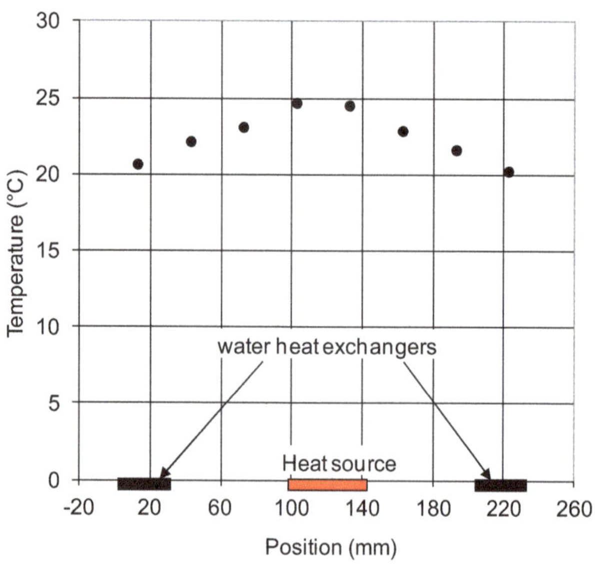

Fig. 15 Temperature field for a heat transfer rate of 15 W in horizontal location

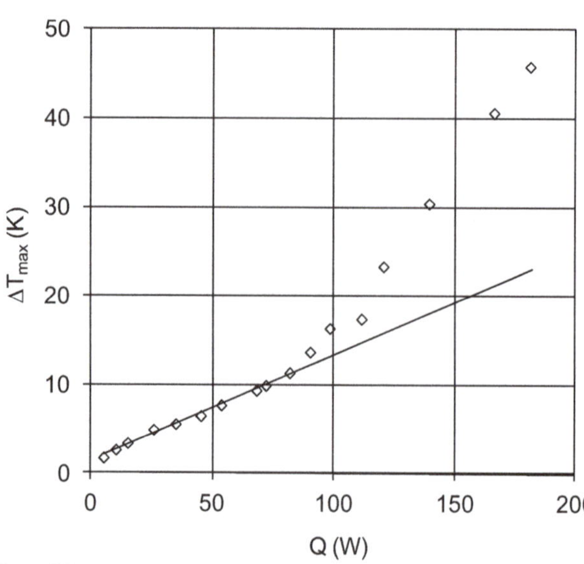

Fig. 16 Maximal temperature difference vs. heat transfer rate; horizontal location (Lefèvre *et al.*, 2012)

Fig.17 presents an example of the comparison between the model and the measured temperatures for a heat flux of 82 W. The equivalent thermal conductivity that fits at best the experimental data is equal to 0.23 W/mK. It has to be noted that this equivalent thermal conductivity is close to that of methanol (λ = 0.2 W/mK).

For a mesh with a fiber diameter equal to d and a distance between two fibers equal to w (Fig. 2), the equivalent thermal conductivity depends on both the liquid and the mesh thermal conductivities λ_l and λ_s (Faghri, 1995):

$$\lambda_{eq} = \frac{\lambda_l \left[(\lambda_l + \lambda_s) - (1-\varphi)(\lambda_l - \lambda_s) \right]}{\left[(\lambda_l + \lambda_s) + (1-\varphi)(\lambda_l - \lambda_s) \right]} \quad (6)$$

The equivalent thermal conductivity calculated with Eq. (6) is equal to 0.44 W/mK, which is twice higher than that estimated experimentally.

The permeability estimated experimentally is 3 times higher than that calculated theoretically while the experimental equivalent thermal conductivity is 2 times lower. Both discrepancies can be partly explained by an underestimation of the real thickness of the screen meshes. Indeed, as the screen meshes are not sintered to the wall, the contact with the wall is most likely imperfect, which leads to an increase of the actual thickness of the capillary structure. However, the underestimation of the real thickness of the capillary structure is not sufficient to explain such important differences. The reliability of the usual laws for both the permeability and the equivalent thermal conductivity can be questioned, as it was already questioned for grooved capillary structures (Revellin *et al.*, 2009). Lips et al (2010a) also showed experimentally that the usual laws strongly overestimate

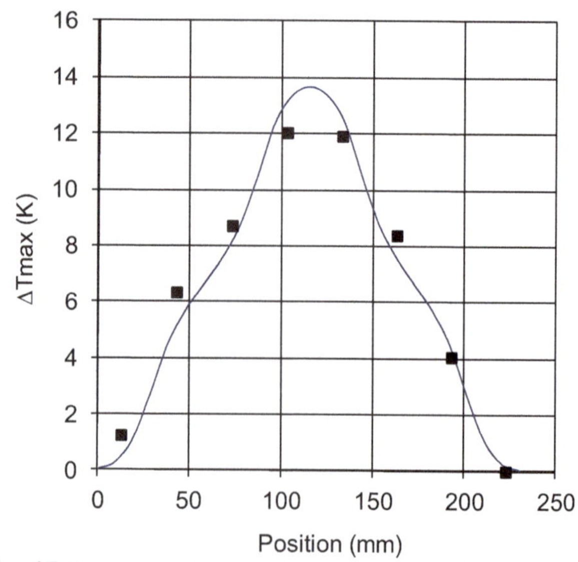

Fig. 17 Comparison between the analytical model and experimental data (Q = 82 W; λ_{eq} = 0.23 W/mK)

Finally, these different comparisons show that there is a real need for accurate equivalent thermal conductivity correlations for the different capillary structures encountered in practice. It has to be noted that for crossed grooves, Lips et al (2010c) could not compare their experimental thermal conductivity to such a theoretical law because no correlation is available in the literature for this kind of geometry. On the other hand, Lips et al (2010c and 2011) showed that satisfactory models are available for the determination of the permeability for grooved capillary structures. However, the present study emphasizes that it is not the case for capillary structure made of metallic meshes.

Fig. 18 presents the equivalent thermal conductivity of the capillary structure as a function of the heat transfer rate Q. These results have been calculated by comparing the analytical model to the experimental data of Fig.16. The equivalent thermal conductivity is rather constant for heat transfer rates between 45 W and 82 W. It decreases beyond 82 W, due to the beginning of the dryout of the capillary structure. For increasing heat fluxes from Q = 0 W to Q = 45 W, the thermal conductivity increases from a very small value (0.07 W/mK) to a constant value, which can be explained by a slight overfilling for small heat fluxes.

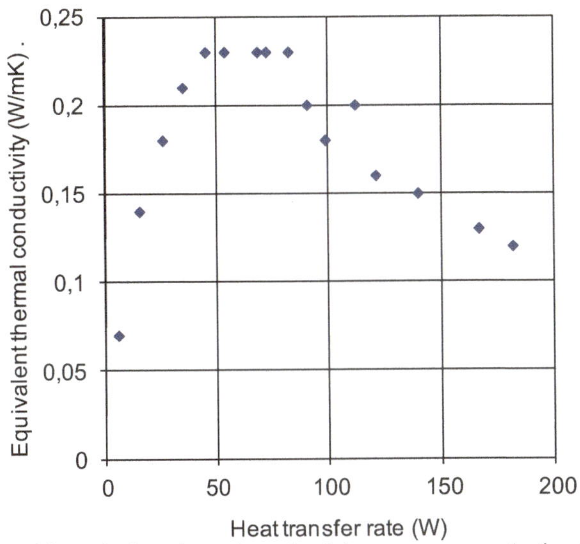

Fig. 18 Equivalent thermal conductivity versus power (estimated from experimental results of Fig. 16)

4.3 Application to the cooling of several electronic components

Fig. 19 presents the temperature field for the FPHP in horizontal orientation with three heat sources and two heat sinks located on its shorter edge. The properties of the capillary structure are those estimated in the previous sections, i.e. an equivalent thermal conductivity equal to 0.23 $Wm^{-1}K^{-1}$ and a permeability of 0.55×10^{-10} m^2. The heat sink size and location are the same as the water heat exchangers on each side of the FPHP. The total heat transfer rate is equal to 75 W and is supposed to be equally distributed into 2 dissipative components of area 10×10 mm^2 each and one dissipative component of area 20×20 mm^2. The maximum temperature difference reaches 20 K. Just as an illustration, one can note that an aluminum plate (λ=165 $Wm^{-1}K^{-1}$) of thickness 13 mm (i.e. more than 4 times thicker than a FPHP of 3 mm) would present the same thermal performance (Fig.20). The maximum temperature is reached under the small heat sources, where the heat flux is the highest.

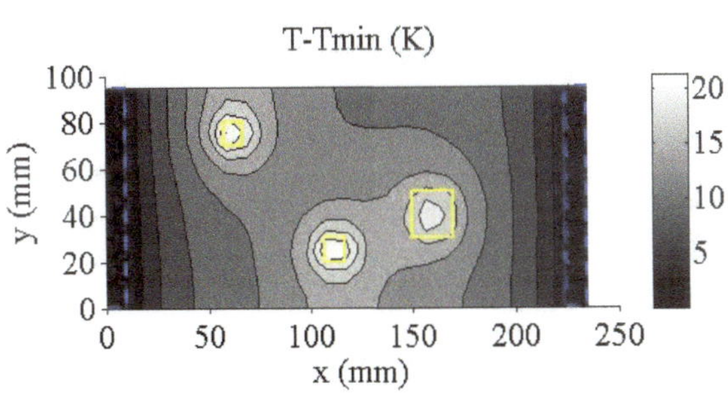

Fig. 19 Temperature field on the FPHP with 3 electronic components

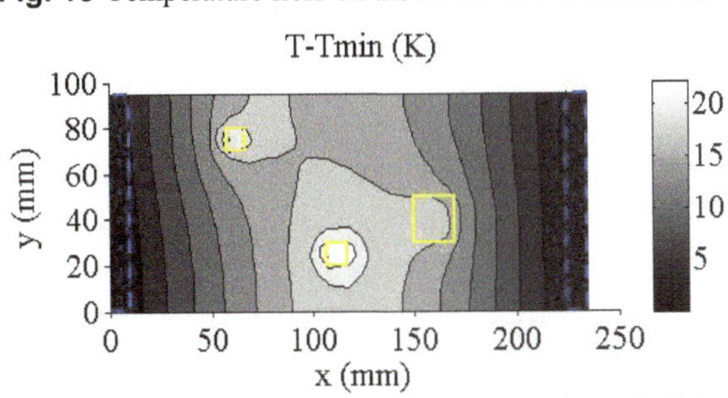

Fig. 20 Temperature field on an aluminum plate of thickness 13 mm

Fig. 21 presents the liquid pressure field inside the screen meshes. The maximum pressure difference inside the capillary structure is lower than 1700 Pa, which is below the capillary limit (2000 Pa).

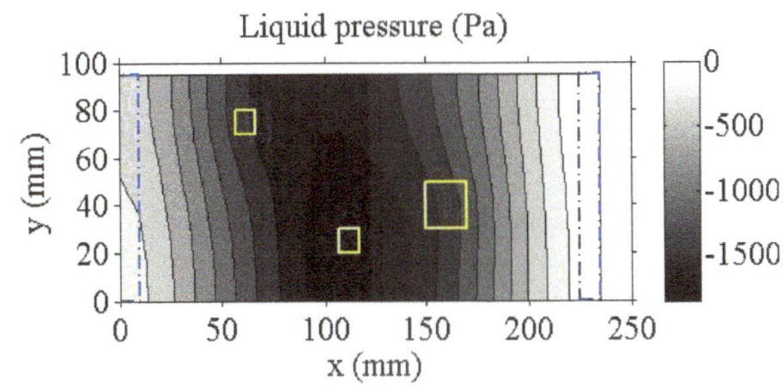

Fig. 21 Liquid pressure in the screen meshes

5. CONCLUSIONS

A synthesis of observations made inside the capillary structure of six different flat plate heat pipes (FPHPs) has been presented. Measurements of the liquid-vapor interface have been provided using confocal microscopy. This non intrusive technique enables the measurements of liquid film of very small thickness (down to 2 µm). These results are very useful to understand the physical mechanisms inside FPHPs that are generally characterized only with temperature measurements, the system being opaque. The results were used to validate FPHP models. The effect of the interfacial shear stress was found to be negligible in all the experimental conditions. On the contrary, we showed that the vapor space thickness has a major influence on the liquid distribution inside the system and thus the thermal performances of FPHPs. It was also shown that nucleate boiling does not constitute an operating limit in grooved heat pipe as it is generally stated in the literature. On the contrary, when meshes are plated against a grooved capillary structure, a strong boiling limit is observed for rather small heat fluxes.

The comparison with an analytical model shows that the equivalent thermal conductivity of the meshes is close to the thermal conductivity of the methanol. This result is not as high as one could expect in comparison with theoretical results. On the contrary, the permeability estimated experimentally is higher than that calculated theoretically. This result confirms that new laws have to be developed to better describe both the permeability and the equivalent conductivity of any kind of capillary structures. Hydrodynamic and thermal simulations show the ability of the flat plate heat pipe to cool several electronic components located on a printed circuit board. The maximum heat transfer rate in horizontal location is around 80 W for a 0.2 m long card.

NOMENCLATURE

a, b, c	dimensions of the FPHP (m)
A_{mn}, B_{mn}	coefficients for the Fourier series
C_1, C_2	constants
d	distance between two fibers (m)
H	height (m)
K	permeability (m^2)
N	number of fibers
P	pressure (Pa)
Q	heat transfer rate (W)
r_{eff}	effective radius (m)
T	temperature (K)
w	distance between two fibers (m)
x, y, z	coordinates (m)

Greek Symbols

λ	thermal conductivity (W/mK)
μ	dynamic viscosity (Pa.s)
φ	porosity
ρ	density (kg/m^3)
σ	surface tension (N/m)

Subscripts

cap	capillary
eq	equivalent
g	gravity
l	liquid
max	maximum
p	porous
v	vapor
s	solid
sat	saturation

REFERENCES

Faghri, A., 1995, Heat Pipe Science and Technology, Taylor and Francis, Washington D.C.

Khrustalev, D., Faghri, A., 1995, "Thermal characteristics of conventional and flat miniature axially grooved heat pipes", Journal of Heat transfer, **117**, 1048-1054. http://dx.doi.org/doi:10.1115/1.2836280

Kim, S.J., Seo, J.K., Do, K.H., 2003, "Analytical and experimental investigation on the operational characteristics and the thermal optimization of a miniature heat pipe with a grooved wick structure", *International Journal of Heat and Mass Transfer,* **46**, 2051-2063. http://dx.doi.org/doi:10.1016/S0017-9310(02)00504-5

Lallemand, M., Lefèvre, F., 2004, "Micro/mini heat pipes for the cooling of electronic devices", *Proceedings of 13th International Heat Pipe Conference*, Shanghai, China.

Lefèvre, F., Lallemand, M., 2006, "Coupled thermal and hydrodynamic models of flat micro heat pipes for the cooling of multiple electronic components", *International Journal of Heat and Mass Transfer*, **49**, 1375-1383. http://dx.doi.org/10.1016/j.ijheatmasstransfer.2005.10.001

Lefèvre, F., Rullière, R., Pandraud, G., Lallemand, M., 2008, "Prediction of the temperature field in flat plate heat pipes with micro-grooves – Experimental validation", *International Journal of Heat and Mass Transfer*, **51**, 4083-4094. http://dx.doi.org/doi:10.1016/j.ijheatmasstransfer.2007.12.007

Lefèvre, F., Rullière, R., Lips, S., Bonjour, J., 2010, "Confocal microscopy applied to capillary film measurements in a radial flat plate heat pipe made of silicon", *Journal of Heat Transfer*, **132**(3), 031502 http://dx.doi.org/doi:10.1115/1.4000057

Lefèvre, F., Conrardy, J.B., Raynaud, M., Bonjour, J., 2011, "A flat plate heat pipe with screen meshes for the cooling of electronic components", *Proceedings of 10th International Heat Pipe Symposium, Taipei, Taiwan.*

Lefèvre, F., Conrardy, J.B., Raynaud, M., Bonjour, J., 2012, "Experimental investigations of flat plate heat pipes with screen meshes or grooves covered with screen meshes as capillary structure", *Applied Thermal Engineering*, **37**, 95-102. http://dx.doi.org/10.1016/j.applthermaleng.2011.11.022

Lips, S., Lefèvre F., Bonjour J., 2009, "Nucleate boiling in a flat grooved heat pipe", *International. Journal of Thermal Science*, **48**, 1272-1278. http://dx.doi.org/doi:10.1016/j.ijthermalsci.2008.11.011

Lips, S., Bonjour, J., Lefèvre, F., 2010a, "Investigation of evaporation and condensation processes specific to grooved flat heat pipes", *Frontiers in Heat Pipes*, **1**, 023001. http://dx.doi.org/10.5098/fhp.v1.2.3001

Lips, S., Lefèvre, F., Bonjour, J., 2010b, "Combined effects of the filling ratio and the vapor space thickness on the performance of a flat plate heat pipe", *International Journal of Heat and Mass Transfer*, **53**, 1113-1121. http://dx.doi.org/doi:10.1016/j.ijheatmasstransfer.2009.10.022

Lips, S., Lefèvre, F., Bonjour, J., 2010c, "Thermo-hydraulic study of a flat plate heat pipe by means of confocal microscopy: application to a 2D capillary structure", *Journal of Heat Transfer*, **132**, 112901. http://dx.doi.org/10.1115/1.4001930

Lips, S., Lefèvre, F., Bonjour, J., 2011, "Physical mechanisms involved in grooved flat heat pipes: experimental and numerical analyses", *International Journal of Thermal Sciences*, **50**, 1243–1252. http://dx.doi.org/10.1016/j.ijthermalsci.2011.02.008

Revellin, R., Rullière, R., Lefèvre, F., Bonjour, J., 2009, "Experimental validation of an analytical model for predicting the thermal and hydrodynamic capabilities of flat micro heat pipes", *Applied Thermal Engineering*, **29**, 1114-1122. http://dx.doi.org/10.1016/j.applthermaleng.2008.06.002

Rullière, R., Lefèvre, F., Lallemand, M., 2007, "Prediction of the maximum heat transfer capability of two-phase heat spreaders - Experimental validation", *International Journal of Heat and Mass Transfer*, **50**, 1255-1262. http://dx.doi.org/doi:10.1016/j.ijheatmasstransfer.2006.09.015

Tibiriça, C. B., Do Nascimento, F. J., Ribatski, G., 2010, "Film thickness measurement techniques applied to micro-scale two-phase flow systems", *Experimental Thermal and Fluid Science*, **34**, 463-473 http://dx.doi.org/10.1016/j.expthermflusci.2009.03.009

Vasiliev, L.L., 2008, "Micro and miniature heat pipes – Electronic component coolers", *Applied Thermal Engineering*, 28, 266–273. http://dx.doi.org/10.1016/j.applthermaleng.2006.02.023

THERMAL PERFORMANCE OF VAPOR CHAMBER WITH NANOFLUIDS

K.N. Shukla*, A. Brusly Solomon and B.C. Pillai

Karunya University, Karunya Nagar, Coimbatore-641 114, India

ABSTRACT

The paper describes experimental studies on the heat transfer performance of a wickless vapor chamber with a heat sink. A vapor chamber of 78 mm length and 64 mm width was fabricated with a thickness of 5 mm and tested for two different working fluids, copper-water and aluminium-water nanofluids with the filling ratio of 30% of the vapor chamber volume. As a baseline comparison de-ionized (DI) water is considered as a working fluid. The vapor chamber was tested for heat inputs ranging from 90 -150 W. The effects of heat input and metal concentration in the working fluids and the effect of particle deposition on the performance of the vapor chamber are presented. The thermal resistance of vapor chamber was decresed with the increase of the heat input for copper-water nanofluids while it remained constant for DI-water. Also the thermal resistance was found to be lower for higher concentration of the nanofluid.

Keywords: Vapor chamber, heat sink, nanofluids, thermal resistance.

1. INTRODUCTION

Future electronic systems especially computers and communication equipment are expected to be of small size, lightweight but with compact components that release very high magnitudes of heat. These miniaturized systems generate large heat fluxes during operation, which warrant the development of efficient thermal management systems. Different modules like heat pipes, heat sinks, heat spreaders and vapor chambers are used to remove inevitable hotspots in the electronic systems as presented by Shukla (1981, 2008, and 2009). Vapor chambers have been mostly used in the base of the heat sinks and they work as heat spreaders to spread the heat from the hotspots and eventually dissipate it to environment. These vapor chambers offer low thermal resistances as compared with solid heat sinks. A vapor chamber is a vacuum vessel with a wick structure lining inside the walls which in turn is saturated with a working fluid. As heat is applied, the fluid at that location immediately vaporizes and the vapor rushes to fill the vacuum. Wherever the vapor comes into contact with a cooler wall surface it condenses, releasing its latent heat of vaporization. The first experimental analysis of a flat plate heat pipe or vapor chamber was reported by Wang and Vafai (2000). Nguyen *et al.* (2000) showed that the performance of the heat sink and vapor chamber is approximately 25 to 45% better than a conventional heat sink with copper and aluminium heat spreaders. Subsequently there have been many investigations on vapor chambers employing different working fluids for various geometries and wick structures. For example, Xuan *et al.* (2004) conducted an experiment with the addition of sintered layer at the evaporator section of the flat heat pipe. It was found that, the sintered layer in the evaporator section of the heat pipe accelerated the boiling process leading to performance enhancement of the flat heat pipe. Go(2005) investigated the thermal performance of an acetone-charged vapor chamber heat sink containing micro wick structures for cooling microprocessors in PC desktop applications. In this experiment, the wick was modified to obtain a high aspect ratio of the micro channel. Three metal-etched wick plates were stacked and inserted into the 2 mm high vapor chamber. Cooling performance was examined by measuring the working temperatures and thermal resistances for the various heat inputs and three different tilt angles. It was concluded that the stacked metal-etched micro wick structure can provide capillary pressure sufficient to exceed the pressure drops. Carbajal *et al.* (2006) analysed the temperature distribution of flat heat pipes with aluminium sandwich wick and compared the performance of the vapor chamber with a copper block. It was found that the temperature at the bottom surface of the vapor chamber is much flatter than that of the solid copper block. Kang *et al.* (2012) carried out CFD simulation to study the temperature uniformity of vapor chamber heat spreader.

In order to improve the axial and radial heat transfer, Ming et al. (2009) constructed a grooved evaporator section in the vapor chamber and studied its performance under various conditions. Wong et al. (2010) tested a novel vapor chamber with modified evaporator and condenser. In this vapor chamber, parallel grooves are made on the inner surface of the top plate (condenser) with inter-groove openings, to replace the conventional porous wick and found that the evaporation and condensation heat transfer coefficients decreased with increasing heat loads. They also found that the evaporation-resistance dominated the vapor chamber resistance. In the tested case the convectional heat transfer coefficients associated with evaporation were a few times larger than those associated with

Corresponding author: Email:kn_shukla@rediffmail.com

condensation. Choi (1995) showed that the thermo physical properties of the fluids were improved by adding ultra-fine metal particles with ordinary fluids. This followed by several studies using nano- particles suspension to improve the performance of heat transfer devices such as heat pipes, pulsating heat pipes, grooved surfaces, flat heat pipes etc. (see, Yang *et al.*, 2008; Liu *et al.*, 2007; Naphon*et al.*, 2008; Qu *et al.*, 2010; Do and Jang 2010). Very recently nanofluids have been used in the micro grooved flat heat pipe (Hyang and Jang, 2010). The thermal performance of a flat micro-heat pipe was investigated for a working fluid consisting of nano particles-alumina suspensions. Shukla *et al.* (2010, 2012) found 14% heat transfer enhancement in cylindrical heat pipes with nanofluids.

Vapor chamber heat sinks can use a variety of working fluids, depending on the operating temperature. The thermodynamic properties of water make it an order of magnitude better than any other fluid for majority of electronics cooling applications (Shukla *et al.* 2009). From these studies, it was understood that the thermal resistance of heat transfer devices can be reduced or heat transfer capacity can be increased by using nanofluids as working fluids.

The paper describes the experimental studies on the heat transfer performance of the vapor chamber with a heat sink using nanofluids as its working fluid. The main purpose of the study is to reduce the total thermal resistance of the heat sink by varying the working fluids of the vapor chamber.

2. ANALYTICAL MODEL OF VAPOR CHAMBER

An analytical model based on the method of equivalent thermal resistances was developed to estimate the overall heat transfer coefficient of the vapor chamber heat sink. The total thermal resistance is sum of the several resistances, expressed by the following equation,

$$R_{tot} = R_{sp} + R_{boil} + R_{cond} + R_{diss} \quad (1)$$

The spreading resistance, R_{sp} is determined by the method given by Simons (2004),

$$R_{sp} = \frac{\psi_{max}}{k_1 r_1 \sqrt{\pi}};\ \psi_{max} = \frac{\varepsilon\tau}{\sqrt{\pi}} + \frac{1}{\sqrt{\pi}}(1-\epsilon)\varphi; \quad (2)$$

$$\varphi = \frac{\tanh(\lambda\tau) + \frac{\lambda}{Bi}}{1 + \frac{\lambda}{Bi}\tanh(\lambda\tau)}; \quad (3)$$

where,

$$r_1 = \sqrt{\frac{A_b}{\pi}}, \quad r_2 = \sqrt{\frac{A_h}{\pi}},$$

$$\lambda = \pi + \frac{1}{\epsilon\sqrt{\pi}}, Bi = \frac{h\,R}{k} \quad (4)$$

The thermal resistance due to boiling heat transfer is determined by the method proposed by Kattan *et al.* (1998). The method is based on two-phase flow pattern map for horizontal evaporating flows:

$$R_{boil} = \frac{1}{A_b h_{boil}} \quad (5)$$

Where

$$h_{boil} = \frac{h_{vap} + h_{wet}}{2} \quad (6)$$

where,

$$h_{wet} = (h_{nb}^3 + h_{cb}^3)^{\frac{1}{3}}; \quad (7)$$

$$h_{nb} = 55pr^{0.12}(-\log_{10}(pr))^{0.55} M^{-0.5} q^{.67}; \quad (8)$$

$$h_{cb} = 0.0133 Re_L^{0.69} Pr_L^{0.4} \frac{k_L}{\delta}; \quad (9)$$

$$h_{vap} = 0.23 Re_G^{0.8} Pr_G^{0.4} \frac{k_G}{\delta}; \quad (10)$$

where Reynolds number, $\mathrm{Re} = \frac{m' d_i}{\mu}$ and *pr* is the reduced pressure. The thermal resistance due to condensation is described as below:

$$R_{cond} = 1/A_c h_{cond} \quad (11)$$

where,

$$h_{cond} = 0.555\left[\frac{g(\rho_l - \rho_v)\rho_l k_l^3}{\mu_l (T_{sat} - T_s)}\left(h_{fg} + \frac{3}{8}c_{pl}(T_{sat} - T_s)\right)\right]^{0.25} \quad (12)$$

for $Re_v < 35000$.

The thermal resistance due to heat dissipation from the extended surfaces is described as below:

$$R_{diss} = \frac{1}{h_0(\eta A_{fin} + A_b)} \quad (13)$$

$$\eta = \frac{\tanh(mL_c)}{mL_c},\ mL_c = \sqrt{\frac{2h_\infty}{kt_f}}L_f \quad (14)$$

3. PREPARATION OF NANOFLUIDS

A two-step method is used to prepare the copper and aluminium nanoparticle/water mixture for the present study. Nano-particles of 99.5% purity were prepared using laser evaporation technique of NaBond technologies Co. and dispersed in the base fluid. De-ionized water with the required wt. per cent of nano particles are kept under ultrasonic cavitation for about 30 minutes using Ultrasonic Processor-UP400S. The nano particles are dispersed completely after a few minutes which can be seen by the change of the colour. It is found that, both copper and aluminium nanofluids remains stable even after 48 hours. The fluid contains particle sizes of 70-130nm.The detail of the preparation of nanofluids are described by Shukla et al. (2010, 2012). The vapor chamber is tested within 8 hours from the time of charging of working fluids.

4. EXPERIMENTAL SET UP

The experimental setup consists of a vapor chamber made up of electrolytic copper of the dimensions 78 x 64 x 5 mm, aluminium heat sink with the dimensions of 78 x 64 x 25 mm with thirteen fins of 23 mm height and 1 mm thickness. The top surface of the vapor chamber and the bottom surface of the heat sink were brought in to contact with the air gap between the two surfaces filled with a thermal interface material (TIM), see Fig 1. A vapor chamber heat sink was fabricated with a filling ratio of 30% of the vapor chamber volume. A copper heater block was fabricated to provide necessary heat to vapor chamber by using a stainless steel cartridge heater. Fourteen numbers K type thermocouples were used to measure the temperature at different locations of the vapor chamber. Five thermocouples were placed near the heater on the bottom surface of the vapour chamber to measure evaporator temperature, out of which one was to measure the heater temperature. Another five thermocouples were placed on the top surface of vapor chamber to measure the condenser temperature. Three thermocouples were placed on the fin surface to measure fin temperature. One thermocouple was used to measure ambient temperature. The adjacent sides of vapor chamber and the heater section were insulated by the Teflon block to ensure adiabatic conditions. All the thermocouples were connected to a PC based data acquisition system. The power supplied to the heater was calculated using the measured voltage and current supplied to the cartridge heater. A commercial 12V DC fan was used for forced convection on the top of the fins.

Experiments were conducted after the installation of the test setup. In the first step the 12V DC fan was switched on to provide necessary cooling to the heat sink. Then the cartridge heater was turned on and the thermocouple readings were recorded by a data acquisition system connected to the computer. The heat input varied in the range of 90-150W.

The accuracy of the temperature measurements (K- type thermocouple) is found to be ±1.5°C .The uncertainty in the heat input is ±5% for the full scale value of 250W.The uncertainty in the measurements of the total resistance of the vapour chamber is determined by Eq. (15) as

$$\frac{\Delta R}{R} = \sqrt{\left(\frac{\Delta Q}{Q}\right)^2 + \left(\frac{\Delta(\Delta T)}{\Delta T}\right)^2} \quad (15)$$

Table 1 presents uncertainties in the thermal resistances at different heat inputs which decrease with the increase of the heat input.

Table 1 Uncertainty in the thermal resistance

S .No	Heat input (W)	Uncertainty (%)
1	90	7.5
2	110	5.4
3	130	4.3
4	150	3.4

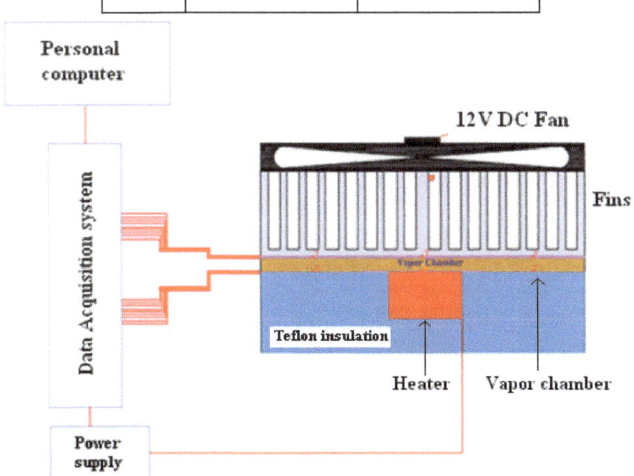

Fig. 1 Experimental setup

5. RESULTS AND DISCUSSIONS

The resistances expressed by the above Eq. (2), Eq. (5), Eq. (11) and Eq. (13) are calculated for Di- water and the nanofluids. These resistances are presented in Figs. (2-4). As a base line comparison, Wei *et al.* (2003) obtained a spreading resistance data of 0.12°C/W for a heat source of dimensions 17x17 mm on a vapor chamber of the dimensions 125x125 x 4.5mm. In the present study, a spreading resistance of 0.038°C/W was obtained for a heat source of dimensions 35x35 mm on a vapour chamber of smaller dimensions. The thermal resistance due to heat sink is much higher than other resistances. Thermal resistance due to boiling is lowest which shows that the liquid attains almost same temperature as wall of the vapour chamber.

The total thermal resistance R_{tot} is plotted in Fig. 4 for DI-water and the copper-water nanofluid. The total thermal resistance is lower by about 5% for the nanofluid. Thus the performance of the vapour chamber is marginally improved by the application of the nanofluid. The experimental value of the thermal resistance of the vapour chamber is calculated as below:

$$R_v = \frac{T_e - T_c}{Q} \quad (16)$$

where T_e and T_c are the average value of the evaporator and the condenser of the vapour chamber.

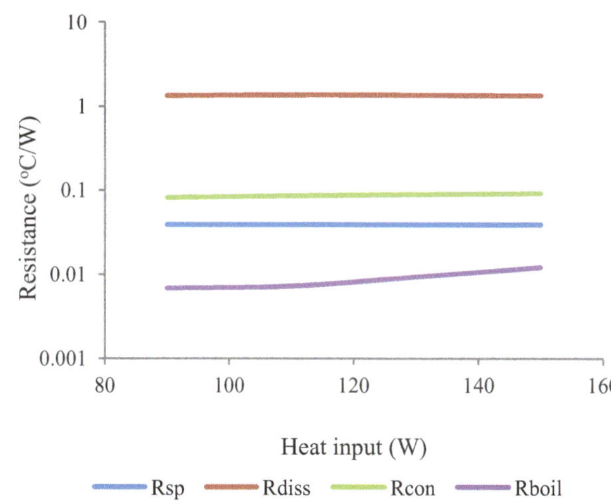

Fig. 2 Thermal resistances in vapour chamber with DI-water

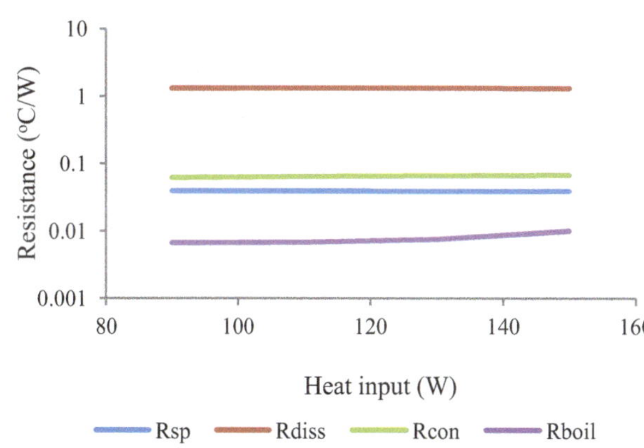

Fig. 3 Thermal resistances in vapour chamber with nanofluid

The thermal performance of the vapor chamber heat sink was investigated for high heat inputs. To ensure steady state operation, the measurements were taken when the temperatures remained steady for about 10 minutes. The total thermal resistance of the vapour chamber is calculated by Eq. (16) under steady state condition. Figure 5 presents a comparison of the experimental and analytical values of the thermal resistances of the vapour chamber without dissipation. The analytical expression under predicts the thermal resistances of the vapour chamber. The experimental values are based on the average value of the four thermocouple readings around the heater on the evaporator and five thermocouple readings on the condenser of the vapour chamber during steady state. The measured temperature around the heater might be higher than the actual evaporator temperature giving higher value of the thermal resistance.

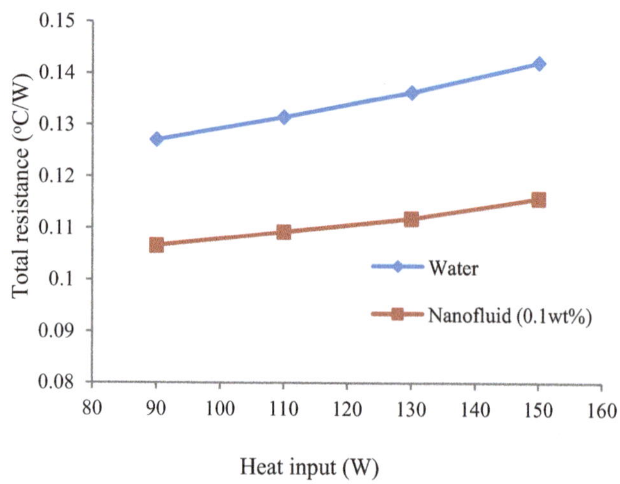

Fig. 4 Total resistance of vapour chamber without dissipation

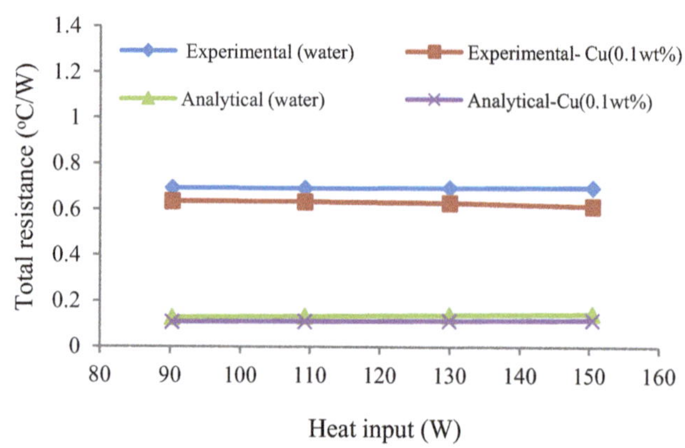

Fig. 5 Comparison of experimental and analytical values of the total thermal resistances of vapour chamber

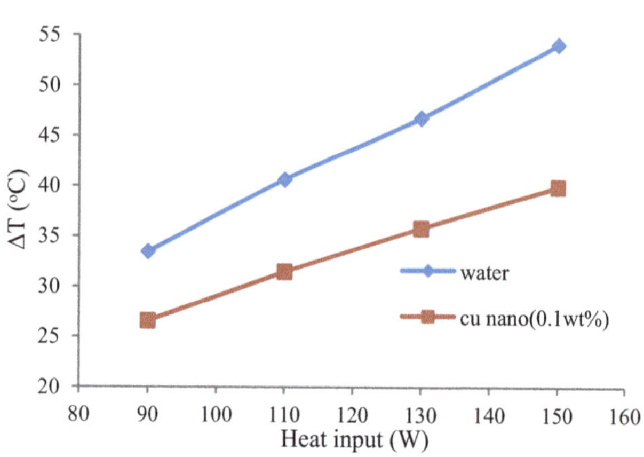

Fig. 6 Temperature drop between liquid and wall of the vapor chamber

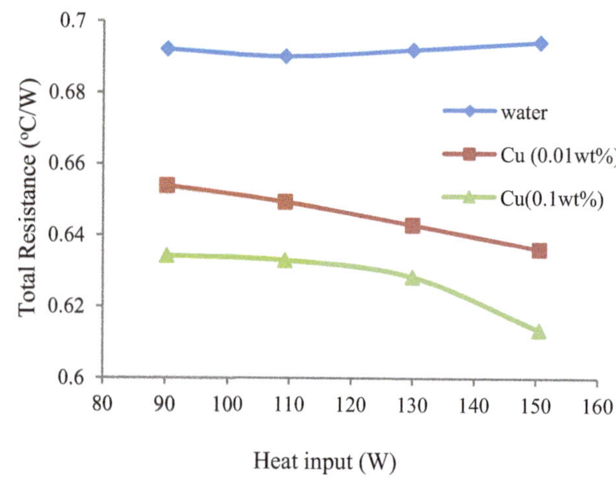

Fig. 7 Thermal resistance of integrated vapor chamber with copper-water nanofluid

The temperature drop between the vapor and the wall is calculated from the sum of the spreading and the boiling resistances and is presented in Fig.6 for various heat input. The temperature drop in DI- water, as expected is higher than that of the nanofluid.

5.1 EFFECT OF NANOPRTICLE CONCENTRATION

The thermal resistances of integrated vapor chamber charged with DI-water and copper-water nanofluid at 0.01 and 0.1 wt.% are presented in Fig.7. The thermal resistance of vapor chamber is almost constant as the heat input increses for DI- water and the same is decresed as the heat input increases for copper-water nanofluids. Also, the thermal resistance is lower for higher concentration of the nanofluid. The maximum reduction in thermal resistance was found to be 8.3 and 11.6 % for 0.01 and 0.1weight % of copper nanofluid respectively. A Similar trend was observed for alminum-water nanofluid as shown in Fig. 8. The maximum thermal resistance reduces to 9.3% and 10.6 % for 0.01 and 0.1 weight percent of alminum- nanofluids respectively.

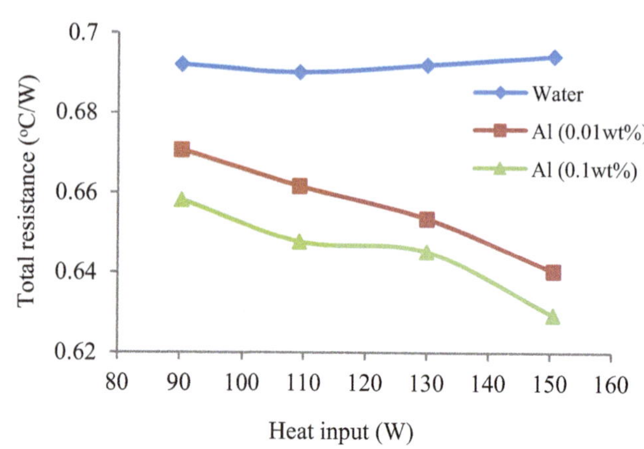

Fig. 8 Thermal resistance of integrated vapor chamber with aluminum- nanofluid.

The improvement in the thermal conductivities leads to improvement in the heat transfer capacities of the nanofluids. Shukla *et al* (2010) observed an enhancement of 3.79 - 35% in thermal conductivity for 0.01 - 0.1 wt. % of nano copper particles in base water. It is quite insightful to consider the naive-idea of modelling nanofluid- thermal conductivity as a series of two layers - one of particle and the other of pure fluid. A very simple calculation shows that the effective thermal conductivity can be expressed as

$$\frac{1}{k_{eff}} = \frac{\beta}{k_p} + \frac{1-\beta}{k_l} \tag{17}$$

where β is the weight fraction of the solid particle. For thermal conductivities shown in Table 2, there is an enhancement of 1 and 11% for 0.01 and 0.1 wt. % respectively for both the aluminium and copper nanofluids. Figures (7) and (8) present almost same improvement in the thermal resistances for 0.1 wt. % of aluminium and copper nanofluids. However for lower concentration 0.01 wt. % of the particle, the improvement in thermal resistance is more than that of thermal conductivity. An addition of non-volatile substance increases the boiling temperature of the liquid. The lower concentration of the particles makes the nanofluid homogeneous with higher boiling point and an enhanced critical heat flux (CHF). You *et al.* (2003) obtained an optimum value of the concentration of the nanoparticles after which no further enhancement of the CHF was possible. The enhanced CHF may be responsible for the improvement of thermal resistance of the 0.01wt% of the nanofluids.

Table 2 Thermal conductivities of water vapor and solid particles

Material	Thermal Conductivity (W/m-K)
Water Vapour	0.016
Aluminium	215
copper	400

5.2 EFFECT OF NANOPRTICLE DEPOSITION

In order to explore the reason for thermal performance enhancement in the vapor chamber, the same was opened and the evaporator surface was viewed under scanning electron microscope (SEM) after the experiments. Figure 9(a) shows the photograph of the evaporator surface of vapor chamber. Nano particle deposition was observed soon after the imitation of the boiling process. Figure 9(b &c) shows the SEM image of porous formation at the bottom surface which is completely covered by deposition of nano particles. An irregular porous layer formed at the evaporator surface enhances wet ability and promotes dry spot- rewet ability, thus delaying the CHF (Kim et al. 2006).

To understand the effect of nanoparticle deposition on the evaporator surface of the vapor chamber, the heat transfer coefficient was evaluated by Eq. 18 at the evaporator. The heat transfer coefficient at the evaporator section of the vapor chamber is calculated as

$$h_e = \frac{Q_{in}}{A_e(T_e - T_b)} \tag{18}$$

where T_e is the average values of the temperatures in the evaporator.

Figure 10 shows the heat transfer coefficient for various heat inputs for DI-water and copper- nano fluid. Comparisons have been made between the evaporation heat transfer coefficient of vapor chamber charged with DI- water and copper- nanofluids. Results show that the heat transfer coefficient increases linearly with the increase of the heat input. There is 15-20% increase in boiling heat transfer coefficient for the vapor chamber charged with 0.01wt% copper- nanofluid as compared with DI- water. Similarly there was 30-40% enhancement in the heat transfer coefficient for 0.1 wt. % of copper nanofluid. As the nano particles with higher concentration form thicker porous layer compared with the nanofluids with less concentration, that causes more dry- spot rewet ability and hence the enhancement in heat transfer coefficient.

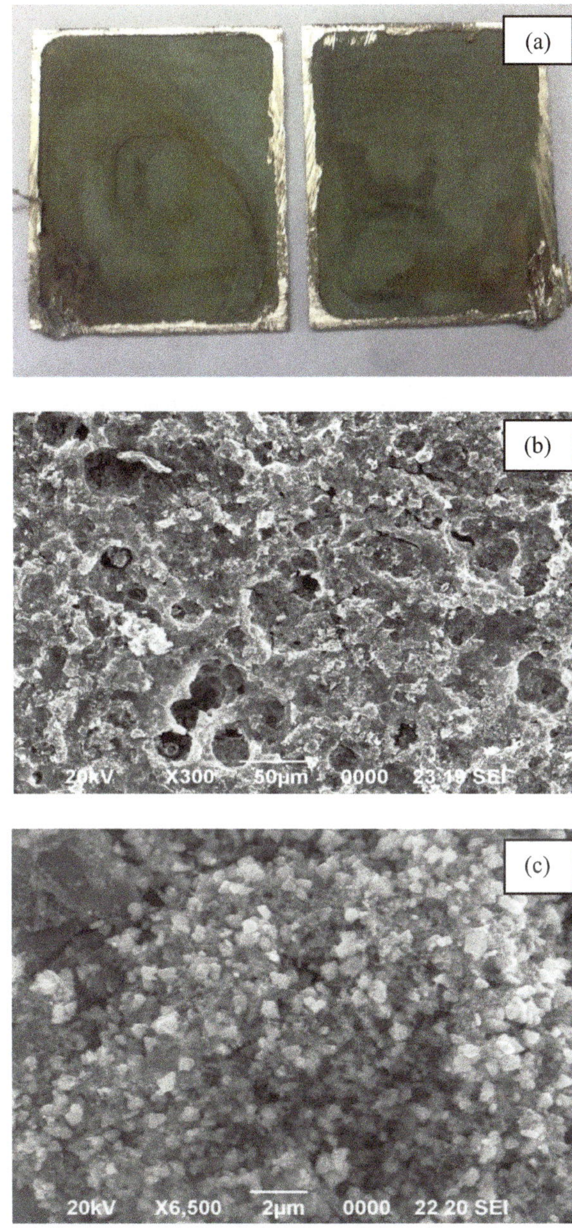

Fig. 9 View of scale formation inside vapor chamber: (a) Photographic view, (b) SEM view at 300 X, (c) SEM views at 6500X

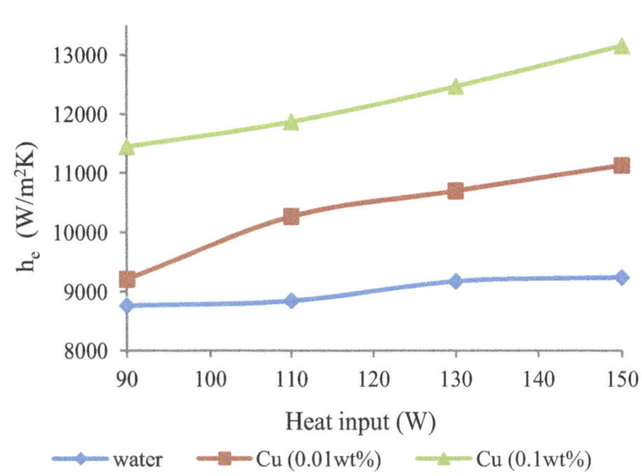

Fig. 10 Heat transfer coefficient at the evaporator

6. CONCLUSION

Thermal performance of a vapor chamber with heat sink was studied by using nanofluids as working fluids. From the studies it is found that the nanofluids charged vapor chamber performs better than that charged with de-ionized water. Also found the decrease of the thermal resistances with the increase of the weight percentage of the nano particles. The copper nanofluids charged vapor chamber performs better than that charged with aluminium nanofluids and de-ionized water. There are many factors responsible for the heat transfer enhancement of the nanofluids. Thermal conductivity, concentration and particle densities play major role in the enhancement of heat transfer. It was also found that, the thin porous layer formed on the evaporator surface causes dry spot rewet ability and is also responsible for the heat transfer enhancement of the vapor chamber using nano fluids. The vapour chamber - heat sink with nanofluids can easily be integrated with heat dissipating components, it will have wide applications in electronics cooling.

NOMENCLATURE

A surface area (m^2)
C_{pl} specific heat of liquid (J/kg-K)
d vapour chamber thickness (m)
g acceleration due to gravity (m/s^2)
h heat transfer coefficient (W/m^2-K)
h_{fg} Latent heat of evaporation (J/kg)
k thermal conductivity (W/m-K)
L length (m)
$\dot{m}$ mass flow rate ($kg.m^2/s$)
pr reduced pressure
Pr Prandtl number
r_1 equivalent radii of the base of the vapour chamber
r_2 equivalent radii of the heat source
Q heat transferred (W)
Re Reynolds number
T Temperature (K)
t thickness of the plate (m)

Greek Symbols
ρ density (kg/m^3)
μ dynamic viscosity (Pa.s)
η fin efficiency
τ dimensionless thickness of the base plate (t/r_2)
ε dimensionless contact radii (r_1/r_2)
δ film thickness (m)

Subscripts
amb ambient
b base
boil boiling condition
cond condensation
c condenser
e evaporator
f fin
h heater block
l liquid
p particle
v vapor
s sink
sat saturation
v vapor
∞ environment
β weight fraction
diss dissipation
sp spreading
wet wetting

REFERENCES

Carbajal, C., Sobhan, C. B., Peterson, G. P., Queheillalt, D. T., Wadley, H. N .G., 2006, "Thermal Response of a Flat Heat Pipe Sandwich Structure to a Localized Heat Flux," *International Journal of Heat and Mass Transfer*, **49**, 4070–4081.

Choi, U. S., 1995, "Enhancing thermal Conductivity of Fluids with Nanoparticles," *ASME FED*, 231, 99-103.

Cooper, M. G., 1984, "Saturation Nucleate Pool Boiling," *Ist UK National Conference on Heat Transfer*, 2, 785-793 (Industrial and Chemical Engineering Symposium Series No. 86)

Do, K. H., Jang, S. P., 2010, "Effect of Nanofluids on the Thermal Performance of a Flat Micro Heat Pipe with a Rectangular Grooved Wick," *International Journal of Heat and Mass Transfer,* **53**, 2183-2192.

Go, J. S., 2005, "Quantitative Thermal Performance Evaluation of a Cost-Effective Vapor Chamber Heat Sink Containing a Metal-etched Micro Wick Structure for Advanced Microprocessor Cooling," *Sensors and Actuators: A Physical*, **121,** 549–556.

Kattan, N., Thome, J.R., Favrat, D., 1998, "Flow Boiling on Horizontal Tubes, Part I: Development of Adiabatic Two-Phase Flow Patterns," *ASME Journal of Heat Transfer*, **120**, 156-165.

Kang, S. W.,Chen, Y. T., Hsu, C.H., Lin, J. Y., 2012,"Temperature Uniformity Analysis of a Multi-well Vapor Chamber Heat Spreader," *Frontiers in Heat Pipe FHP*, **3**, 013004.

Kim, S.J., Bang, I. C., Baongiorno, J., Hu, L. W., 2006, "Effects of Nanoparticle Deposition on Surface Wettability Influencing Boiling Heat Transfer in Nanofluids," *Applied Physics Letters.*, **89**(15), 153107-1-3.

Liu, Z., Xiong, J. G., Bao, R., 2007, "Boiling Heat Transfer Characteristics of Nanofluids in a Flat Heat Pipe Evaporator with Micro-Grooved Heating Surface," *International Journal of Multiphase Flow,* **33**, 1284–129.

Ming, Z., Zhongliang, L.,Guoyuan, Ma , 2009, "The Experimental and Numerical Investigation of a Grooved Vapor Chamber," *Applied Thermal Engineering* , **29**, 422–430.

Naphon, P., Assadamongkol, P., Borirak, T., 2008, "Experimental Investigation of Titanium Nanofluids on the Heat Pipe Thermal Efficiency," *International Communications in Heat and Mass Transfer,* **35**, 1316–1319.

Nguyen, T., Mochizuki, M, Mashiko, K, Saito, Y, 2000, "Use of Heat Pipe/ Heat Sink for Thermal Management of High Performance CPU's," *Proc. 16th IEEE Semi-Therm Symposium*, 76-79.

Qu, J. H., Wu, Y., Cheng, P., 2010, "Thermal Performance of an Oscillating Heat Pipe with Al2O3–Water Nanofluids," *International Communications in Heat and Mass Transfer,* **37**, 111–115.

Shukla, K. N., 1981," Transient Response of a Gas-Controlled Heat Pipe," *AIAA Journal*, **19**(8), 1063-1070.

Shukla, K. N., 2008," Thermo Fluid Dynamics of Loop Heat Pipe Operation," *International Communications in Heat and Mass Transfer*, **35**, 916-920.

Shukla, K. N., 2009, "Heat Transfer Limitation of a Micro Heat Pipe," ASME *Journal of Electronic Packaging*. **13**, 024502.

Shukla, K. N., Solomon, A. B., Pillai, B. C., 2009, "Experimental Studies on Rotating Heat Pipe," *Heat Transfer-Asian Research*, **38**(8), 475-484.

Shukla, K. N., Solomon, A. B., Pillai, B. C., Ibrahim, M., 2010, "Thermal Performance of Cylindrical Heat Pipe Using Nanofluids," *Journal of Thermophysics and heat Transfer*, **24**,796-802.

Shukla, K. N., Solomon, A. B., Pillai, B. C., Jacob Ruba Singh, B., Kumar, S. S., 2012, "Thermal Performance of Heat Pipe with Suspended Nanoparticle," *Heat and Mass Transfer*, **48** (11), 1913-1920.

Simons, R. E., 2004, "Simple Formulas for Estimating Spreading Thermal Resistance," Electronics Cooling, 10(2), 8-10

Wang, Y., Vafai, K., 2000, "An Experimental Investigation of the Thermal Performance of an Asymmetrical Flat Plate Heat Pipe," *International Journal of Heat and Mass Transfer,* **43**, 2657-2668.

Wei, J., Cha, A., Copeland, D., 2003, "Measurement of Vapour Chamber Performance," *IEEE, Semi-Therm Symposium*, 191-194.

Wong, S-C., Hsieh, K- C., Wu, J.-D. , Han, W.-L., 2010, "A Novel Vapor Chamber and Its Performance," *International Journal of Heat and Mass Transfer*, **53**, 2 377–2 384.

Xuan, Y., Hong,Y., Li, Q, 2004, "Investigation on Transient Behaviours of Flat Plate Heat Pipes," *Experimental Thermal and Fluid Science,* **28**, 249–255.

Yang, X. F., Liu, Z. H., Zhao, U. J. F., 2008, "Heat Transfer Performance of a Horizontal Micro Grooved Heat Pipe Using CuO Nanofluid," *Journal Micromechanics and Microengineering*, **18**, 035038.

You, S. M., Kim, J. H., Kim, K. H., 2003, "Effect of Nanoparticles on Critical Heat Flux of Water in Pool Boiling Heat Transfer," *Applied Physics Letters.* **83**, 3374-3376.

PERFORMANCE EVALUATION OF THERMOSYPHON INTEGRATED HEAT SINK FOR CPU COOLING

Shanmuga Sundaram Anandan*, Anirudh Bhaskaran

Department of Mechanical Engineering, Rajalakshmi Institute of Technology, Chennai, Tamil Nadu, India-600124

Abstract

An experimental study on Thermosyphon Integrated Heat Sink (TIHS) for CPU cooling has been carried out for both natural and forced convection to analyze the performance characteristics of the CPU cooling module (TIHS) at steady state condition. A thermal model based on the experimental values has been developed to study the thermal behavior of Thermosyphon Integrated Heat Sink during CPU cooling. The thermal model is employed to determine the actual heat transfer and the effectiveness of the CPU cooling module and compared it with the conventional cooling method and found that there is an appreciable improvement in the proposed CPU cooling module.

Keywords: *Thermosyphon, Heat Sink, Effectiveness, CPU cooling*

1. Introduction

A Thermosyphon is a two phase closed heat transfer system containing small amount of liquid in the evaporator section. The evaporator section is heated causing the liquid to vaporize and the vapor rises to the cold end of the tube to get condensed and thereby releasing its latent heat to the cooler medium. The condensate is returned to the evaporator section by gravity and the cycle repeats. The amount of heat that can be transported by these systems is normally several orders of magnitude greater than pure conduction through a solid metal as cited by Dunn et.al (1994). Since the latent heat of evaporation is large, considerable amount of heat can be transported from end to end with a very small temperature difference, hence such reliable heat transfer devices can be used in thermal management systems with limited space applications and other heat recovery systems.

The Thermosyphon Heat Sink integrated system works on both conduction and convection mechanism. The cooling module consisting of a parallel plate fin heat sink bonded to the array of thermosyphon tubes enhances the heat transfer within the system and with the surroundings through the larger heat transfer area, lower thermal resistance (between the source and the sink) and high effective thermal conductance of thermosyphon). The heat dissipated from the source is conducted through the heat sink base attached with a Thermal Interfacing Material (TIM) in order to reduce the contact thermal resistance between heater and the base. The heat from the base is transferred to heat sink and to the thermosyphons through conduction. The thermosyphon filled with R-134a as a refrigerant convects heat in to it and thereby causes the liquid to vaporize as it reaches its saturation temperature. The vapor carries the latent heat to the condenser kept inside the air duct with a tube axial fan mounted on it for forced convection. The relatively cooler medium at the condenser causes the vapor to condense and form film condensation which flows back to the
evaporator through the wall of the thermosyphon due to gravity. In order to enhance the heat transfer at the condenser, a u-shaped longitudinal fin is bonded with the thermosyphons. This entire cycle repeats again and again to maintain the operating temperature of the CPU within the limit.

Tardy et.al (2009) developed a mathematical model for heat pipes in thermal storage. Finned heat pipes were used to transfer heat from the evaporator placed in an air stream flow to the condenser placed in a storage tank containing ice. The thermal behavior of heat pipes have been studied experimentally and analyzed under different conditions. H.Jouhara et.al (2008) developed a thermal model for a small diameter thermosyphon charged with water, FC-84, FC-77, FC-3283 and an experimental investigation were carried out for water and the dielectric heat transfer liquids and the thermosyphon thermal performance have been presented.

2. THERMAL RESISTANCE NETWORK MODELING

2.1. Calculation of Thermal Resistance

The actual heat transfer for the Thermosyphon Heat Sink integration system can be found out by considering the thermal resistance shown in Figure1.

$$R_{tot} = \frac{1}{R_{HB}+R_{BS}+R_{HS}+R_{TF}+R_{FA}+R^*} \quad (1)$$

$$where, R^* = \left(\frac{1}{\frac{1}{R_{axial}} + \frac{1}{R_{tube,E}+R_{boiling}+R_{PD}+R_{cond}+R_{tube,C}} + \frac{1}{R_{HA}}} \right)$$

The thermal resistance given in Equation (1) represents the total thermal resistance across the heat sink, thermosyphon and the U–shaped longitudinal fin. The explanation of the thermal resistance parameters are given in Table 1.

* *Corresponding author. Email:speakdrss@gmail.com*

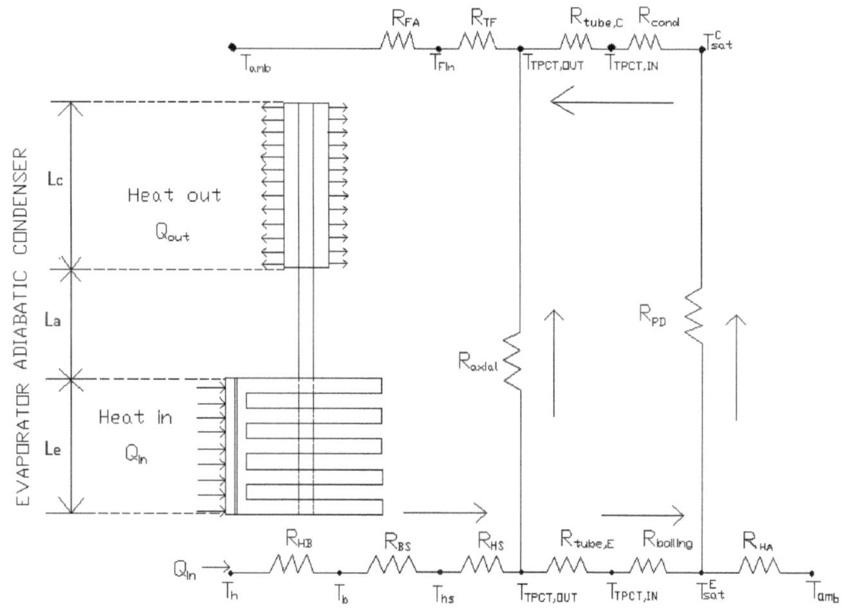

Fig. 1 Thermal resistance network of the Thermosyphon Heat Sink integrated system

Table.1 Thermal resistance equations of the THSIT

S.no	*Thermal Resistance*	*Equations*
1	Heater to base conduction (TIM) resistance	
2	Base to Heat sink conduction resistance	
3	Parallel plate fin Heat Sink conduction resistance	
4	TPCT tube radial resistance at evaporator section	
5	Evaporative boiling convective resistance	
6	Heat sink to ambient convective resistance at evaporator section	
7	TPCT tube axial resistance	
8	Resistance due to pressure drop from evaporator to condenser section	
9	Film condensation resistance at condenser section	
10	TPCT tube radial resistance at condenser section	
11	U-shaped longitudinal fin conduction resistance	
12	Fin to ambient convective resistance at condenser section	

2.1.1 Heat Transfer within the Thermosyphon

The heat transfer taking place within the thermosyphon can be described by calculating the convective heat transfer coefficient for boiling as provided by Imura et al. (1972)

$$\mathbf{h_{evap}} = \mathbf{0.32 q_e^{0.4}} \times \left\{\frac{\rho_l^{0.65} k_l^{0.3} c_{pl}^{0.7} g^{0.2}}{\rho_g^{0.25} h_{fg}^{0.4} \mu_l^{0.1}}\right\} \times \left(\frac{P}{P_{air}}\right)^{0.3} \quad (2)$$

Thermal resistance of the evaporative section is given as

$$R_{boiling} = \frac{1}{h_{evap} \times A_{evap,int}} = \frac{1}{h_{evap} \pi D_{int} L_e N} \quad (3)$$

Heat transfer within thermosyphon at the condenser section can be described by calculating the convective heat transfer coefficient due to film condensation as proposed by Nusselt (2005)

$$h_{cond} = 0.943 \left\{ \frac{\rho_l g k_l^3 (\rho_l - \rho_v)[h_{fg} + 0.68 C_{pl}(T_{sat} - T_w)]}{\mu_l L_c (T_{sat} - T_w)} \right\}^{0.25} \quad (4)$$

Thermal resistance of the condenser section is given as

$$R_{cond} = \frac{1}{h_{cond} \times A_{cond,int}} = \frac{1}{h_{cond} \pi D_{int} L_c N} \quad (5)$$

The resistance due to the pressure drop of the vapor as it flows from evaporator to condenser is given as

$$R_{PD} = \frac{T_v^E - T_v^C}{Q_{latent}} \quad (6)$$

2.1.2 Thermosyphon wall thermal resistances

Thermal resistance across the thickness of the thermosyphon can be calculated as

$$R_{tube,E} = \frac{\ln\left(r_{tube,ext} / r_{tube,int} \right)}{2 \pi k_{tube} L_e} \quad (7)$$

$$R_{tube,C} = \frac{\ln\left(r_{tube,ext} / r_{tube,int} \right)}{2 \pi k_{tube} L_C} \quad (8)$$

Thermal resistance along the axial length of the thermosyphon can be calculated as

$$R_{axial} = \frac{0.5 L_e + L_a + 0.5 L_c}{k_{tube} A_{cs}} \quad (9)$$

2.1.3 Thermal Resistance of the Heat Sink

Thermal resistance of the heat sink proceeds from the contact resistance between the heat source and the base plate at its interface. A thermal pad (T-250) as a Thermal Interfacing Material (TIM) of thickness 0.25mm and thermal conductivity of 1 $Wm^{-1o}C^{-1}$ is attached between heat source and base plate in order to reduce the contact thermal resistance occurring at the interface. TIM has its ability to change its physical characteristics when it reaches the case operating temperature by filling the interfacial air gaps and surface voids in the base plate and heater surface.

$$R_{HB} = \frac{t_{TP}}{k_{TP} A_{TP}} \quad (10)$$

Thermal resistance between the base and the heat sink is given as

$$R_{BS} = \frac{L_b}{k_b A_b} \quad (11)$$

Thermal resistance across the heat sink can be described as

$$R_{HS} = \frac{1}{h(A_b + N_f \eta_f A_f)} \quad (12)$$

2.1.4 Fin-air Thermal Resistance

The parallel plate fins and the u-shaped longitudinal fins exists at the evaporator and condenser sections respectively and the heat transfer coefficient as proposed by Brigg's and young (1963)

$$h_{air} = \frac{k_{air}}{D_{air}} \times 0.1378 \times Re_{air}^{0.718} Pr_{air}^{0.333} \left(\frac{S_f}{L_f} \right)^{0.296} \quad (13)$$

Thermal resistance between the air and the fins at the evaporator and condenser section is given as

$$R_{FA} = \frac{1}{h_{air} N_f \eta_f A_f} = R_{HA} \quad (14)$$

2.1.5 Calculation of Heat Transfer rate and Effectiveness of the system

Actual heat transfer rate includes the latent heat received from the condensing vapor and the sensible heat from the TPCT.

$$Q_{latent} = \frac{T_v^E - T_v^C}{R_{PD}} \quad (15)$$

Sensible heat transfer due to the change in surface temperature of TPCT

$$Q_{sensible} = \frac{T_b - T_\infty}{R_{tot} - R_{HB} - R_{boiling} - R_{PD} - R_{cond}} \quad (16)$$

Actual heat transfer rate is given by,

$$Q_{actual} = Q_{latent} + Q_{sensible}$$

$$= \frac{T_v^E - T_v^C}{R_{PD}} + \frac{T_b - T_\infty}{R_{tot} - R_{HB} - R_{boiling} - R_{PD} - R_{cond}} \quad (17)$$

Effectiveness of the system is given as,

$$\varepsilon = \frac{\text{Actual heat transfer}}{\text{Maximum heat transfer}} \quad (18)$$

Maximum heat transfer is attained when the condenser temperature is equal to the outside evaporator temperature.

3. EXPERIMENTAL SETUP AND PROCEDURE

The experimental setup of the cooling module consists of three sections namely: 1) Evaporator 2) Adiabatic and 3) Condenser as shown in Figure2. The evaporator section consists of the heat sink with 24 parallel plate fins each of 100×100 mm2 area and 0.8mm thickness bonded to a base plate of 100×100 mm2 area and 6mm thickness made of copper. A plate heater acts as a heat source of 400W capacity is attached to the base plate by means of a Thermal Interfacing Material T-250 of thickness 0.25mm with a thermal conductivity of 1W/m°C and a thermal resistance of 0.49 °C/W. Six TPCT of 9.52mm outer diameter and 7.92mm inner diameter and having a thickness of 0.8mm is bonded along with the parallel plate fins in a staggered arrangement. This evaporator section has a chosen length of 100mm. The adiabatic section has a chosen length of 50mm and the condenser section having a length of 150mm is fitted with the longitudinal u-shaped fins made of copper and bonded along with the TPCT.

Each TPCT is bonded with 12 fins of 0.8mm thickness and height of 8.3mm for the entire length of the condenser section. Thereby, the condenser section consists of 72 fins in total for 6 TPCT's. The condenser section is equipped with a cylindrical duct and a tube axial fan mounted on it to provide the forced convection condition. The tube axial fan has a volumetric flow rate of 127 CFM at 2000rpm.

The experiments were conducted for both natural convection and forced convection in which the later was carried out for the volumetric flow rates of 102 CFM, 115 CFM and 127 CFM for various heat inputs (40-70W). The temperatures at the surface of the thermosyphon and the vapor temperatures at the corresponding sections have been measured by 26 K-type Chromel-Alumel Teflon coated thermocouples as shown in fig 3 and the readings were recorded by a data acquisition system. An air flow sensor was also incorporated to measure the flow rate of the air from the tube axial fan. The accuracy of the thermocouples was of ±1.5°C up to 200°C. The accuracy of air flow arrangement was ±3% of full scale. The maximum possible error in measured quantities of temperature is calculated from the minimum values of the temperature measured and accuracy of the instrument. Therefore for an accuracy of 0.1°C of the instrument and ±1.5°C for thermocouples, the calculated error is 0.414%.

The temperatures at the heat sink were measured in such a way that the arithmetic averages of these values are taken as the mean base plate temperature T_{base}. Ambient temperature is measured at one location in the test environment. During testing, the uniform spreading of heat along the surface of the base plate is validated (i.e) the maximum temperature difference between the three measurements is less than 10% of the average of the base plate temperature.

The average of each TPCT surface and vapour temperatures at evaporator and condenser section is taken for evaluating the performance of the system. The working fluid in the thermosyphon was R134a. All the tests were performed under steady state conditions.

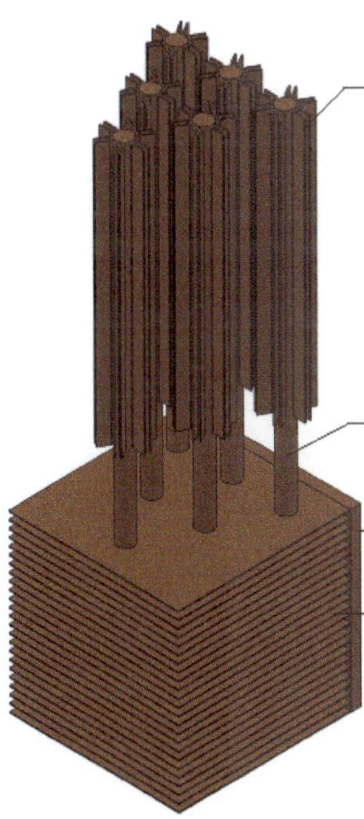

Fig. 2 Experimental Model of Thermosyphon Heat Sink Integrated System

4. RESULTS AND DISCUSSION

The aforementioned equations in the thermal resistance network modeling section were used to calculate the actual heat transfer rate and effectiveness of the system with the average experimental values. The Figure3 shows the graphical variation of total thermal resistance versus the heat input for natural convection and forced convection for air flow rate of 102, 115, 127 CFM respectively. The total thermal resistance of the system for natural convection varies from 0.32K/W to 0.33K/W for the corresponding heat inputs. The total thermal resistance of the system for forced convection for the air flow rates of 102,115,127 CFM varies from 0.227K/W to 0.233K/W. The lowest thermal resistance was found to be at 127CFM air flow rate at 70W heat input. This shows that when the heat input is increased, the total thermal resistance of the system decreases by an insignificant amount.

Similarly, the Figure4 shows the graphical variation of effectiveness of the system versus the heat input for natural and forced convection for an air flow rate of 102,115,127 CFM respectively. The effectiveness of the system for natural convection varies from 0.69 to 0.636 for the corresponding values of heat inputs. The effectiveness of the system for forced convection for air flow rates of 102,115 and 127 CFM varies from 0.803 to 0.878 for corresponding heat inputs. The maximum effectiveness of the system was to be 0.878 at 127CFm and 40W of heat input. This shows that the system's performance will be maximum for forced convection at 127CFM air flow rate and the corresponding thermal resistance of the system is 0.23K/W. The cooling module will dissipate maximum amount of heat from the condenser to the surroundings when the air flow rate along the condenser is maximum.

Therefore, the performance evaluation shows a satisfactory result of appreciable increase in the system performance in the form of effectiveness and found to be 7% increase for natural convection when compared with the conventional cooling method and 25% increase for forced convection. A quantitative comparison on the conventional cooling method and the present cooling method is given in table 2. The design specifications of the conventional cooling module are of a Pentium IV desktop CPU whose tube axial fan has a volumetric flow rate of 8CFM [8]. Moreover, the experimental cooling module presented is a scaled up model for experimental purpose and hence it required higher CFM tube axial fan.

5. CONCLUSION

The thermal modeling of Thermosyphon Heat Sink Integrated system for CPU cooling have been modeled, presented and discussed. An experimental setup has been constructed and tested at steady state condition from which the experimental values have been used to determine the effectiveness of the system. The cooling module proved to be fairly better than the conventional cooling module as discussed previously. The results obtained from experimental studies indicated that as the heat input value increases, both the thermal resistance and effectiveness decreases in a very small amount. Thus in order to enhance the performance of the cooling module, attention has to be taken on dominating individual resistances that constitute the R_{tot} , such as the resistance of power input to the evaporator and the resistance on the condenser side.

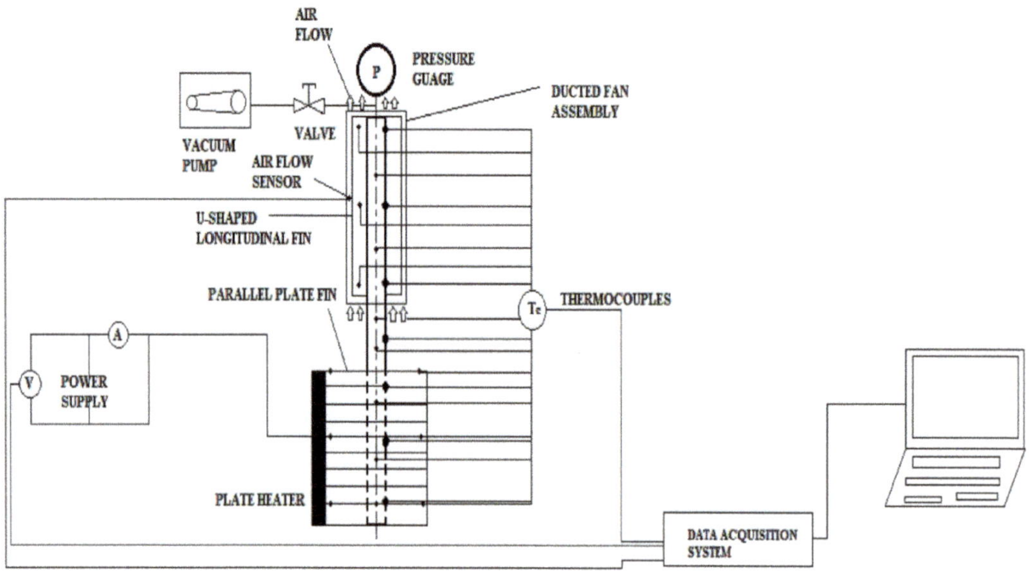

Fig. 3 Experimental setup of Thermosyphon Heat Sink Integrated System with 1 TPCT

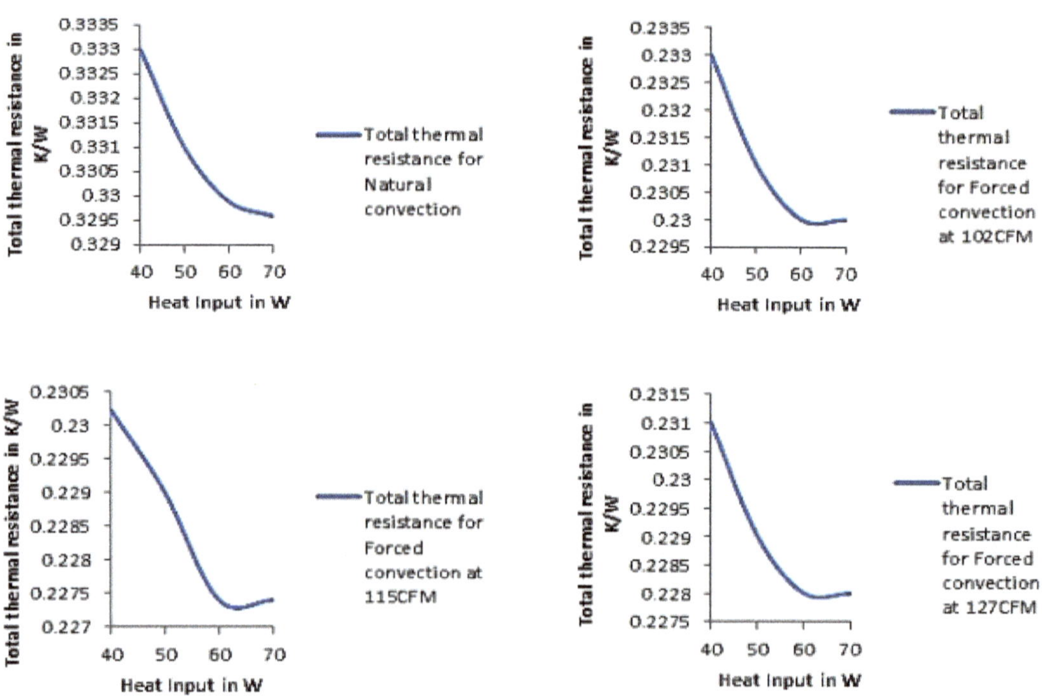

Fig. 4 Total thermal resistance of the system for various heat inputs

Table. 2 Comparison of performance evaluation results between the conventional and the present model

System / Power,W	Conventional Heat Sink Effectiveness	Thermosyphon Heat Sink Integrated System Effectiveness			
	Forced Convection, 8CFM	Natural Convection	Forced Convection, CFM		
			127	115	102
40	0.65	0.69	0.878	0.874	0.87
50	0.62	0.652	0.869	0.875	0.864
60	0.59	0.641	0.832	0.826	0.825
70	0.56	0.636	0.812	0.81	0.803

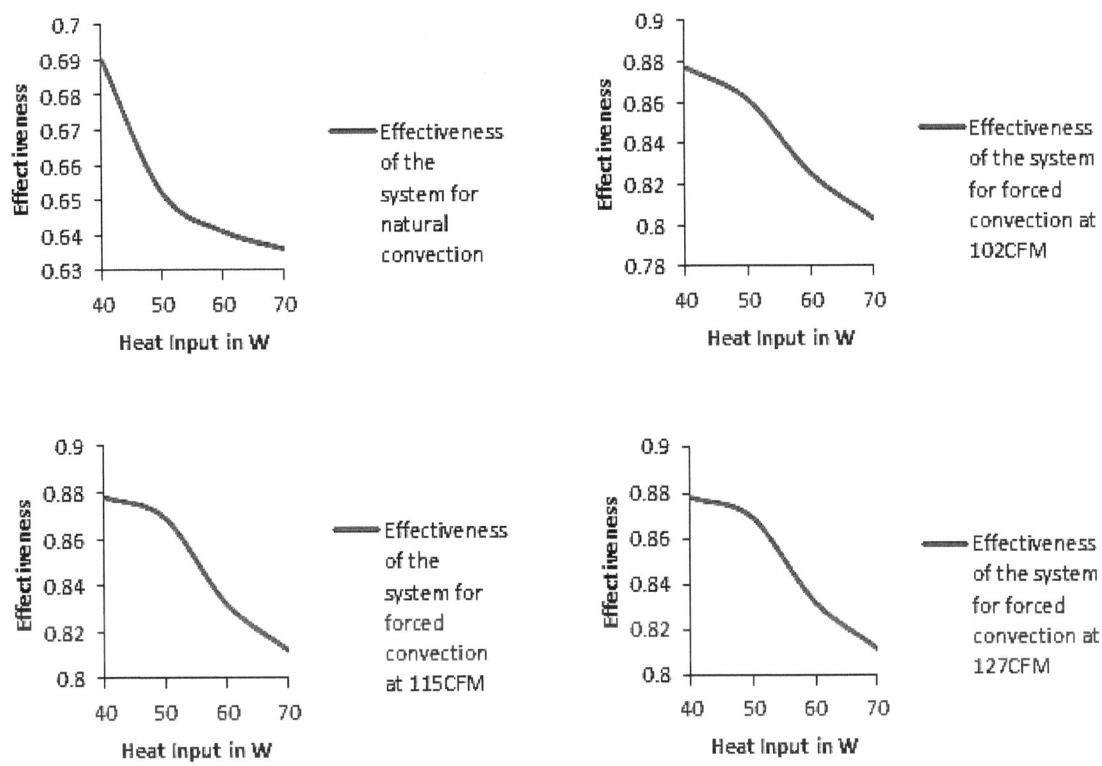

Fig. 5 Effectiveness of the system for various heat inputs

NOMENCLATURE

A, A_{cs}	Cross Sectional area of TPCT, m^2
A_b	Area of base plate, m^2
A_{TP}	Area of Thermal Pad, m^2
C_{pl}	Specific Heat, $Jkg^{-1}K^{-1}$
h	Heat transfer Coefficient, $Wm^{-2}K^{-1}$
h_{fg}	Latent heat of vapourization, J/kg
k	Thermal Conductivity, $Wm^{-1}K^{-1}$
L_a	Adiabatic length, m
L_b	Length of base plate, m
L_c	Condenser length, m
L_e	Evaporator length, m
N	Number of TPCT
N_f	Number of fins
P	Saturation Pressure, bar
q_e	Heat Flux, W/m^2
Q_{in}	Heat Input, W
Q_{out}	Heat Output, W
R_{axial}	Axial resistance of TPCT, K/W
$R_{boiling}$	Evaporator Boiling Resistance, K/W
R_{BS}	Base to Sink Resistance, K/W
R_{cond}	Film Condensation Resistance, K/W
R_{FA}	Fin to air resistance, K/W
R_{HB}	Heater to Base Resistance, K/W
R_{HS}	Heat Sink Thermal Resistance, K/W
R_{TF}	TPCT to fin resistance, K/W
R_{tot}	Total thermal resistance, K/W
$R_{tube,C}$	Condenser side tube Resistance, K/W
$R_{tube,E}$	Evaporator side tube Resistance, K/W
S_f	Fin pitch, m
T_b	Base Temperature, °C
T^C_v	Condenser vapour temperature, °C
T^E_v	Evaporator vapour Temperature, °C
T_h	Heater Temperature, °C
T_{hs}	Heat Sink Temperature, °C
$T_{TPCT,in}$	TPCT inner wall Temperature, °C
$T_{TPCT,out}$	TPCT outer wall Temperature, °C
T_{amb}, T_∞	Ambient Temperature, °C
T_w	Tube wall temperature, °C
T_f	Fin temperature, °C

Greek Symbols

ε	Effectiveness
μ	Dynamic Viscosity, Nsm^{-2}
η	Efficiency of the fin, %
ρ	Density of the fluid, kg/m^3

Subscripts

a	Adiabatic
b	Base plate
C,c,cond	Condenser
E, e,evap	Evaporator
cs	Cross sectional
TP	Thermal Pad

f	fin
in	Input
out	Output
BS	Base to Sink
HB	Heater to Base
HS,hs	Heat Sink
FA	Fin to air
TF	TPCT to fin
tot	total
tube,E	Evaporator tube
tube,C	Condenser tube
sat	Saturation
TPCT,in	TPCT inner wall
TPCT,out	TPCT outer wall
w	wall
v	Vapour

Superscripts

E	Evaporator
C	Condenser

Non Dimensional Numbers

Re	Reynolds Number
Pr	Prandtl Number

Abbreviations

CFM	Cubic Feet per Minute
TIM	Thermal Interfacing Material
TPCT	Two Phase Closed Thermosyphon

ACKNOWLEDGEMENT

We take immense pleasure to express our gratitude to the management of Rajalakshmi Institute of Technology for their cooperation and encouragement for the completion of this project. We would like to express our special gratitude and thanks to Dr.R.V.Seeniraj for extending his technical expertise in completing this project.

REFERENCES

Dunn, P & D.A.Reay 1994, *Heat pipes.4th ed.,* Pergamon.

F.Tardy, Samuel M.Sami, 2009, "Thermal analysis of heat pipes during thermal storage", *Applied Thermal Engineering* 29,329-333. http://dx.doi.org/10.1016/j.applthermaleng.2008.02.037.

H.Jouhara, O.Martinet, A.J. Robinson,2010, "Experimental study of small diameter Thermosyphons charged with with water, FC-84, FC-77 & FC-3283", *Applied Thermal Engineering*,30, issue2-3, 201-211. http://dx.doi.org/10.1016/j.applthermaleng.2009.08.007

Imura, H., Kusada, H., Oyata, J., Miyazaki, T.,and Sakamoto, N., 1972, "Heat Transfer in Two-PhaseClosed Thermosyphon", International Journal of Heat and Mass Transfer, 15(9), 1395-1707. http://dx.doi.org/10.1016/0017-9310(72)90098-1.

S. H. Noie, M. H. Kalaei, M. Khoshnoodi, 2005, " Experimental investigation of boiling and condensation heat transfer of a two phase closed thermosyphon", *International Journal of Engineering*, 18(4),100-112. http://www.sid.ir/En/VEWSSID/J_pdf/856200501A01.pdf.

Briggs, D.E and Young, E.H., 1963, "Convection heat transfer and pressure drop of air flowing across triangular pitch banks of finned tubes", *Chem Eng. Prog. Symp.* Ser., Vol59, No.41, pp1-10

Aniruddha Pal and Yogendra Joshi ,2002, "Design and performance evaluation of a compact thermosyphon", Components and Packaging Technologies, IEEE Transactions 25, PP 601-604. DOI: 10.1109/TCAPT.2002.807997

R.A.Wirtz, Weiming Chen, Ronghua Zhou, "Effect of flow by – pass on the performance of longitudinal fin heat sinks", Journal of electronics packaging,Transactions of the ASME, vol.116, Sep 1994. http://dx.doi.org/10.1115/1.2905687

Ralph L. Webb, "Next Generation Devices for Electronic Cooling with Heat Rejection to Air". ASME J. Heat Transfer, Vol.127, 2005, 2-10 http://dx.doi.org/10.1115/1.1800512

Florio, Harnoi. A, "Combination technique for improving natural convection cooling in electronics", International journal of thermal sciences, vol 46,issue1,2007,pp:76-92 http://dx.doi.org/10.1016/j.ijthermalsci.2006.03.007

PERFORMANCE ENHANCEMENT OF TWO PHASE THERMOSYPHON FLAT-PLATE SOLAR COLLECTORS BY USING SURFACTANT AND NANOFLUID

Sandesh S. Chougule[a,*], Santosh Kumar Sahu[a], Ashok T. Pise[b]

[a] *Indian Institute of Technology, Indore, M.P., 453446, India*
[b] *Government College of Engineering, Karad, M.S., 415124, India*

ABSTRACT

In this study the thermal performance of three different wickless heat pipe solar collectors were investigated by using pure water, water-surfactant and CNT-water nanofluid for different coolant mass flow rates and tilt angles (20^0, 31.5^0, 40^0, 50^0 and 60^0). First collector uses only pure water, the second one utilizes CNT nanoparticle of diameter 10-12 nm and 0.1-10 μm length with water as a base fluid. While the third collector uses water with 150 ppm 2-ethyl-hexanol as coolant for the investigation. Experiments were carried out for the three collectors under the same experimental conditions. It is observed that the performance of the solar collector that uses 2-ethyl-hexanol surfactant with water as coolant is better compared to solar collectors that utilize pure water and CNT/water nanofluid.

Keywords: *Wickless Heat Pipe, Nanofluid, Flat Plate collector, Heat Transfer, Surfactant*

1. INTRODUCTION

In a conventional solar collector, tubes containing water as working fluid are attached to the absorber plate. The solar energy is absorbed by the working fluid and transferred to the storage tank either by natural or forced circulation. The drawbacks associated with conventional solar collectors include the pump and its power requirement, more space to obtain the natural circulation of working fluid, night cooling due to reverse flow of cooled water, pipe corrosion and limited heat carrying capacity of working fluid. Heat pipes are found to provide an alternative solution to the above problems.

Heat pipes, contain small amount of vaporizable fluid, transfers heat by evaporating the working fluid in a heating zone and condensing vapor in the cooling zone. Subsequently, the return flow of condensate to the heating zone occurs through the capillary structure which lies in the inner wall of the heat pipe. However, in certain applications, the return flow of the condensate to the heating zone occurs due to the gravity force and termed as gravity assisted heat pipes, wickless heat pipes or thermosyphon. Because of its ability to transfer large quantity of heat energy, the heat pipes has been widely used in several industrial applications, namely boilers, air-conditioning system, heat exchangers and solar water heating system. The wickless heat pipes are usually integrated with flat absorber plate of the solar heating system, to enhance the thermal performance. Numerous studies have been carried out to understand the thermal performance of two phase thermosyphon used in solar water heaters. Mathioulakis and Belessiotis (2002) reported the results obtained through both theoretical simulation and experimental investigation of a solar collector with heat pipes. They have used ethanol as working fluid and obtained the maximum instantaneous efficiency up to 60%.

The thermal performance of a two-phase thermosyphon solar collector with various refrigerants was reported by Esen and Esen (2005). The thermal performance of the heat pipe with R410A as working fluid was found to be better compared to other refrigerants. The maximum collection efficiencies of solar collector throughout the day were found to be 50.84%, 49.43% and 48.72% for R410A, R407C and R-134a, respectively. The thermal performance of solar collector by using various refrigerants such as: R-134a, R12, and ethanol was reported by Enaburekhan and Yakasai (2009). The heat pipe that utilizes R-134a as working fluid exhibits maximum collection efficiency compared to other working fluids. Since thermosyphon utilizes phase change of the working fluid to transfer heat, selection of a working fluid is essential to achieve the maximum heat transfer capacity. Initially, Choi (1995) had made an attempt to use nanofluid (nanofluid-suspension) in thermal engineering due to its anomalous heat transfer characteristics. The colloidal suspension of nanoparticles in the base fluid with uniform dispersion is termed as nanofluid. Because of its enhanced heat transfer characteristics, nanofluids have been used in thermosyphon to improve the thermal performance. In an experimental study, Liu et al. (2007) used CuO nanofluids in a closed two-phase thermosyphon to study the thermal performance. The maximum enhancement in heat transfer coefficient and critical heat flux was found to be 160% and 120%, respectively at the optimal nanoparticle concentration of 1%. Liu et al. (2010) carried out experiments to study the thermal performance of closed thermosyphon by using carbon nanotubes (CNT) suspensions as working fluid. The maximum enhancement in heat transfer rate was obtained at optimal nanoparticle mass concentration of 2%. The thermal performance of the closed two-phase thermosyphon by using various nanofluids (titanium-ethanol and titanium-water suspension) was reported by Naphon et al. (2008). The heat transfer enhancement of the thermosyphon with titanium-ethanol suspension was found to be 10.6% higher compared to

* *Corresponding author. Email: sandesh_chougule@yahoo.com*

the heat pipe with ethanol. The thermal performance of the heat pipe with titanium-R11 suspension as working fluid was reported by Naphon et al. (2009). The efficiency of the heat pipe, with optimal nanoparticle mass concentration (0.1%), found to be 1.4 times higher compared to that with the pure refrigerant. Gabriela et al. (2011) reported the comparison of thermal performance by using different working fluids, namely iron oxide-water suspension and deionized water. For 5.3% (by volume) of iron oxide nanoparticle, the heat transfer performance was found to be better compared to the DI water. Liu et al. (2011) carried out experiments to study the thermal performance of open thermosyphon with CuO nanofluids in indoor environment. The CuO nanofluid at 1.2 % mass concentration exhibits maximum enhancement in heat transfer rate compared to the deionized water. Chougule and Pise (2012) reported that CNT/water nanofluid posses excellent heat transfer characteristics and can be used in wickless solar heat pipe collector. The effect of solar tracking system on the performance of solar heat pipe collector was studied through experimental investigation (Chougule and Pise, 2011; Chougule et al., 2012). It is evident from the literature that only a few studies (Yuan et al., 2001 and Cheng et al., 2013) has been reported that utilize the water with surfactant as working fluid in the heat pipe solar collector. In view of this, an attempt has been made to study the thermal performance of heat pipe solar collector using 2-ehyl-hexanol surfactant with water as working fluid. The performance of heat pipe with CNT/water nanofluid as working fluid has also been studied in the present investigation. The comparison of thermal performance for nanofluid (CNT/water) and surfactant with water has been made for the given operating condition.

2. EXPERIMENTAL METHOD

2.1 Experimental Set-up

Figure 1 shows the schematic of the solar collector with wickless heat pipes. In this study, attempt has been made to evaluate the performance of two phase thermosyphon solar collector using water, CNT/water nanofluid and water/2-ethyl-hexanol surfactant. In view of this, three same size identical collectors have been fabricated for the investigation. Each collector consists of three absorber plate and each absorber plate contains one heat pipe filled with working fluid. The design parameters of various components heat pipe flat plate collectors are summarized in Table 1. Wickless heat pipes were made of copper tubes with an outer diameter of 12 mm and length of 625 mm. The thickness, width and length of each copper absorber plate are 1 mm, 100mm, and 510 mm, respectively. In order to enhance the ability to absorb the solar radiation, the absorber plates were painted dull black in the present investigation. Each individual wickless heat pipe was brazed to the absorber plate at 120 mm pitch. The heat exchanger, exchanges heat energy from heat pipe to coolant, consists of three interconnected steel collars with an OD of 40 mm, ID of 36 mm and 90 mm long attached around the condensing section of the heat pipes. The three absorber plates with heat pipes and heat exchanger were housed in a collector case of 630 mm length and 360 mm breadth and 100 mm thickness. Thick insulation glass wool behind the absorber plate was used in order to avoid any loss of energy to the surroundings. Transparent glass was used as the upper glazing for the collector.

2.2 Experimental Procedure

In the present study PT100 thermo sensors, pyranometer, ambient air sensor and flow meter was used to measure inlet and outlet coolant temperature, solar irradiation ambient temperature and coolant flow rate, respectively. The mass flow rate of cooling water was measured by using a rotameter with an accuracy of ±0.1 l/min. The quantity of cooling water through the heat exchanger of each collector was adjusted by using flow control valve and measured by using a stop watch with ±0.01 s accuracy and by using a graduated glass bottle of tolerance ±1 cm^3. The temperatures of cooling water, plate and ambient air were measured by using PT100 thermo sensors with an accuracy of ±0.10 ^{0}C. Data logger was used to measure the temperatures. The instantaneous value of global solar radiation intensity was measured by the pyranometer (Model-DWR 8101, Make-DynaLab).

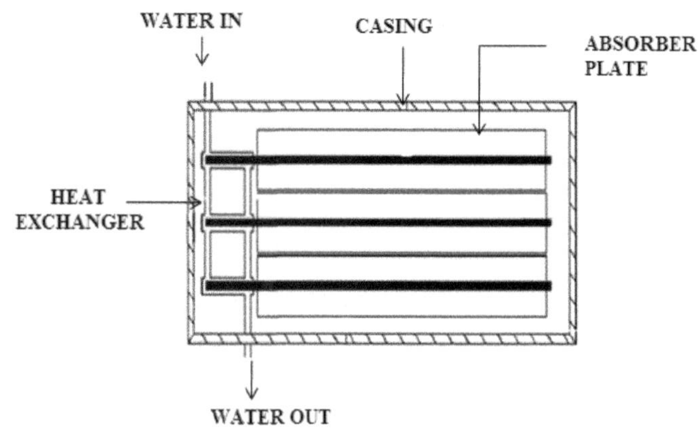

Fig. 1 Schematic of thermosyphon solar collector

The solar collector was installed on a tiltable stand facing south and tested under outdoor conditions of Kolhapur, India (latitude 16.69^0 N; longitude 74.22^0 E). Experiments were carried out throughout the day from 10:00 a.m. to 4:00 p.m. and values were recorded at each half hour interval. During each test run the solar radiation intensity (I_t), ambient air temperature (T_a), inlet cooling water temperature (T_i) and outlet cooling water temperature (T_o) of each collector were recorded for a fixed mass flow rate.

The experiments were carried into the two groups:

- Tests were carried out throughout the day for various tilt angles, namely, 20^0, 31.5^0, 40^0,50^0 and 60^0 for a given mass flow rate (0.00125 kg/sec).
- Tests were conducted for varied range of mass flow rates (0.00125 kg/sec and 0.0022 kg/sec) for a given tilt angle

Table 1 Specifications of the different components of the collectors.

Gross dimensions		
Length 0.63 m	Width 0.40 m	Depth 0.1 m
Transparent cover		
Material: white glass		
Absorber plate		
Material: copper	Length 0.51m	Pitch distance 0.12m
Thickness 0.001m	Coating: black	
Wickless heat pipes		
Material: copper	Outer diameter 0.012 m	
Inner diameter 0.010 m	Evaporator length 0.51 m	
Condenser length 0.085 m	Adiabatic length 0.030m	
Total length 0.625 m		
Number of Heat pipes in each collector 3		
Working fluid:	Collector No.1:- D.I. water	
	Collector No.2:-CNT-water Nanofluid (φ=0.15%)	
	Collector No.3:- water-surfactant	
Insulation		
Material: glass wool	Thickness 0.05 m	Position: Back and sides
Box material		
Frame: aluminum	Base: galvanized iron	

The rate of thermal energy input (Q_{in}), the rate of thermal energy gain (Q_g) and the instantaneous efficiency (η) of each collector were calculated as below,

$$Q_{in} = I_t A_{coll} \quad (1)$$

$$Q_g = \dot{m} C_w (T_o - T_i) \quad (2)$$

$$\eta = \frac{Q_g}{Q_{in}} \quad (3)$$

The energy gain (Q_g) and the efficiency (η) of the collector were calculated by using various the basic quantities such as: mass flow rate of coolant, temperature of inlet and outlet coolant and solar radiation intensity. Considering the errors of various basic quantities, the maximum uncertainty of the energy gain (Q_g) and the efficiency (η) of the collector were found to be less than 2.5% and 6%, respectively.

3. RESULT AND DISCUSSION

Experiments were carried out to investigate the thermal performance of heat pipe solar collector by using various working fluids for heat pipes such as: pure water, water/2-ethyl-hexanol, CNT/water nanofluid. The results are depicted in Figs. 2-6. The variation of various parameters , namely, the rate of solar energy input, ambient air temperature, inlet cooling water temperature, outlet water temperature and plate temperature for three different wickless heat pipe flat plate solar collectors is shown in Figs. 2-3. It is observed that the solar intensity increases, reached maximum and then falls with time (Fig.2). Peak value of solar radiation intensity is found to be different for various tilt angles. The maximum value average solar intensity was observed at 31.5^0.

3.1 Effect of tilt angle

Tests were performed for several tilt angles in three different solar heat pipe collectors under same climatic conditions. First collector uses pure water, second one utilizes CNT/water nanofluid and third collector uses 2-ethyl-hexanol surfactant with water as working fluid. The variation on instantaneous efficiency of three collectors for various tilt angles were shown in Figs. 4(a)-(e).

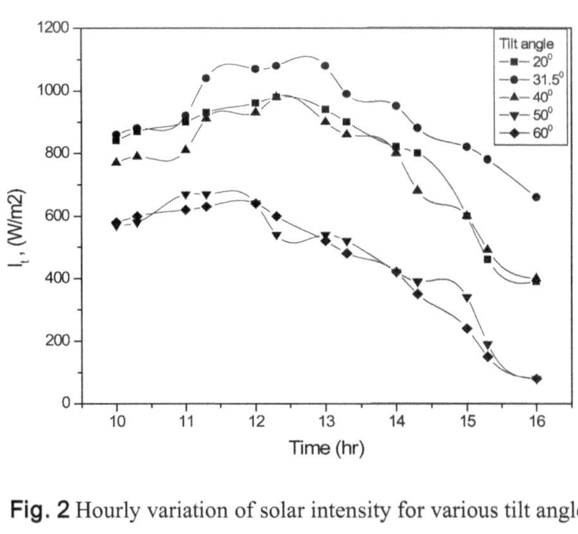

Fig. 2 Hourly variation of solar intensity for various tilt angles.

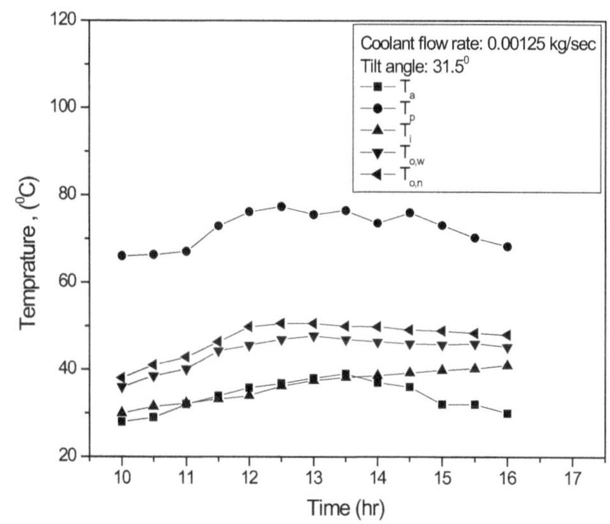

Fig. 3 Instantaneous variation of T_a, T_p, T_i, $T_{o,w}$,and $T_{o,n}$

The water and water/surfactant used collectors show enhancement in thermal performance with increase in tilt angle of collector. The average collector efficiency of water and water/surfactant used collectors are low at lower tilt angles and it increases with increase in the tilt angle up to 50^0 angles then it decreases. The average collector efficiency that uses water working fluid is found to be 24%, 27%, 31%, 43% and 35% at 20^0, 31.5^0, 40^0, 50^0 and 60^0 tilt angle, respectively. While, average collector efficiency that uses surfactant/water working fluid is found to be 35%, 38%, 43%, 54% and 49% at 20^0, 31.5^0, 40^0, 50^0 and 60^0 tilt angles respectively. The average collector efficiencies recorded for CNT/water working fluid collector at 20^0, 31.5^0, 40^0, 50^0 and 60^0 tilt angles are 41%, 36%, 41%, 51% and 45%. The average collector efficiency of the solar collector that uses CNT/water nanofluid decreases with the tilt angle ranging from 20^0-31.5^0, increases with tilt angle varying from 31.5^0- 50^0 and found to decrease beyond 50^0. Table 2 shows variation of average efficiency at various tilt angles with different working fluid for solar heat pipe collector. The thermal performance of solar heat collector with water/surfactant as working fluid for heat pipe is found better as compared to water and CNT/water nanofluid used solar heat pipe collector.

The percentage enhancement in thermal performance of CNT/water nanofluid working fluid collector as compared to water working fluid collector at 20^0, 31.5^0, 40^0, 50^0 and 60^0 tilt angle are 66%, 34%, 32%, 18% and 31%. The percentage enhancement in thermal performance for water/surfactant working fluid collector as compared to water collector at 20^0, 31.5^0, 40^0, 50^0 and 60^0 tilt angle are 45%, 41%, 38%, 25%, and 37%. It shows that, percentage enhancement of water/surfactant based heat pipe collector is observed higher than CNT/water nanofluid used solar collector at angle 31.5^0, 40^0, 50^0, 60^0 when compared with water working fluid solar heat pipe collector. But at 20^0 tilt angle CNT/water nanofluid used solar collector show higher thermal performance than water/surfactant based heat pipe collector at 20^0, because of low entrainment limit in CNT-water heat pipes at low tilt angle. The percentage enhancement in CNT/water nanofluid collector at 20^0 is found more than water/surfactant working fluid solar heat pipe collector when compared to water working fluid collector, because of at these tilt angle after vapor condensation in condenser section when condensed vapors are travelling back to evaporator section of heat pipe CNT working fluid offer lower resistance to condensed vapour in case of CNT nanofluid heat pipe collector.

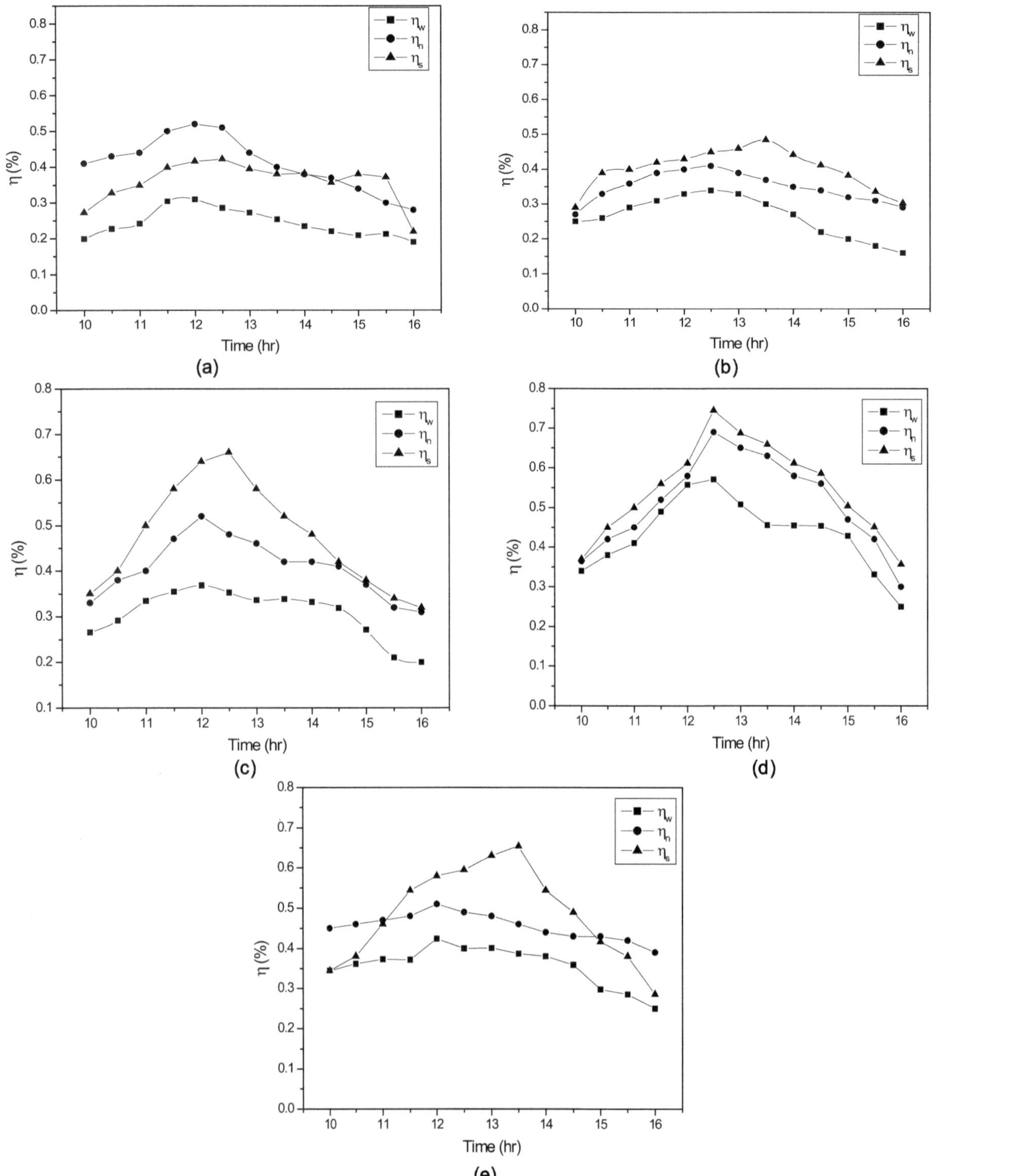

Fig. 4 Daily variation of instantaneous efficiency of different working fluids at (a) $\theta = 20^0$, (b) $\theta = 31.5^0$, (c) $\theta = 40^0$, (d) $\theta = 50^0$, (e) $\theta = 60^0$.

For 50^0 tilt angle, the water, CNT/water nanofluid and surfactant/water exhibit the maximum instantaneous efficiency. The maximum instantaneous efficiency of the collector for CNT/water nanofluid and surfactant/water was found to be 69 %(at 50^0) and 75 %(at 50^0), respectively.

3.2 Effect of coolant rate

The test data have been collected from several test runs under quasi-steady state condition. The maximum variations in ambient, inlet and outlet temperature for this test period are found to be 0.5^0 C, 0.2^0 C and 0.4^0 C, respectively.

Figures 5(a)-5(c) depict the variation of instantaneous efficiency of various solar heat pipe collectors throughout the day. These collectors use different working fluids such as: water, CNT/water nanofluid and surfactant/water. Tests were conducted with three different collectors for different flow rate, namely, 0.00125 kg/sec and 0.0022 kg.sec for a given tilt angle of 31.5^0. It is observed that all collector exhibit better performance at higher coolant flow rate. This may be due to fact that, higher coolant flow rate of coolant enhances the condensation amount of vapors inside the heat pipe. This enhances the heat transfer rate for all collectors. With the use of water as working fluid, better performance was obtained because of condensation of more amount of vapour in the condensing section. For CNT/water nanofluid, the performance increases because of higher effective thermal conductivity of nanoparticles in the nanofluid. This enhances the vapour formation and condensation rate in the evaporator and condensing section, respectively.

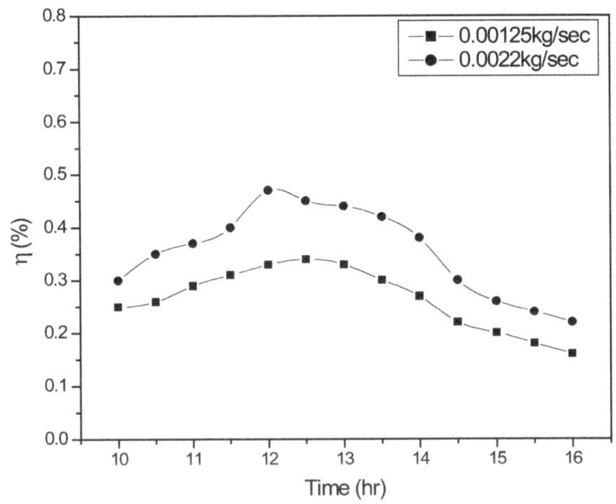

(a)

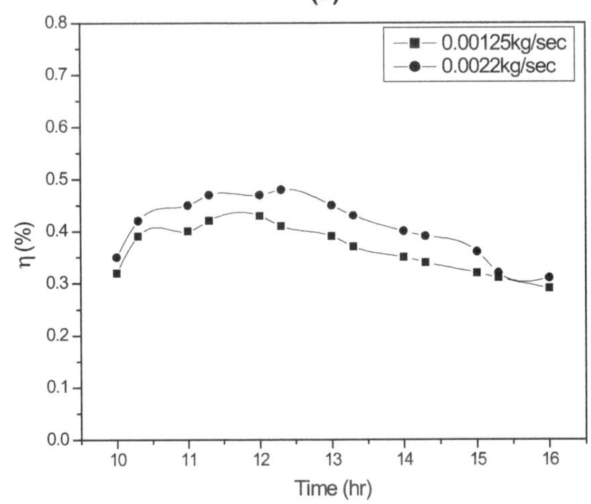

(b)

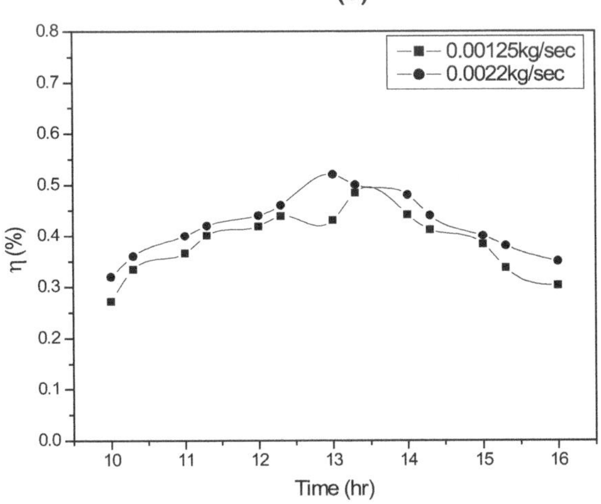

(c)

Fig. 5 Daily variation of instantaneous efficiency for different working fluid (a) water (b) CNT/water (c) water/2-etehyl-hexanol.

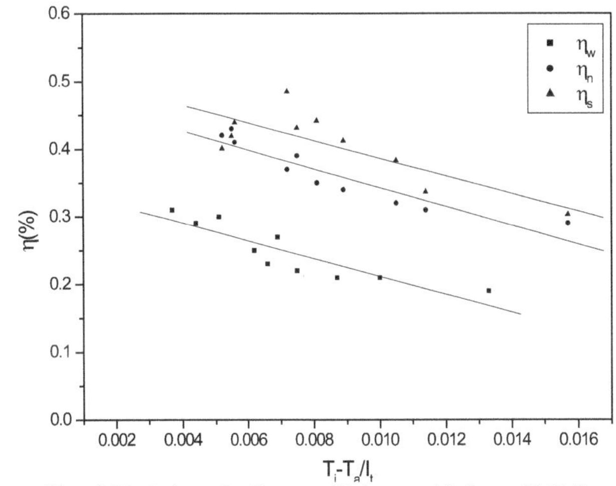

Fig. 6 Variation of collector efficiency with factor T_i-T_a/I

Table 2 Average efficiencies (%) at various tilt angle for various working fluid solar thermosyphon collector

Tilt angle	Working fluid		
	water	CNT-water nanofluid	Water-surfactant
20^0	24	41	35
31.5^0	27	36	38
40^0	31	41	43
50^0	43	51	54
60^0	35	45	49

3.3 Effect of Heat Loss Factor

Figure 6 shows the variation of collector efficiency for various working fluids such as: pure water, CNT/water, water/surfactant in the heat pipe of the collector. The experiments were carried out for the given mass flow rate of 0.00125 kg/sec. It may be noted that by curve fitting technique, the collector efficiency with various working fluids (pure water, CNT/water, water/surfactant) can be expressed as a linear function of (T_i - T_a)/I_t. The intersection of the trend line with the vertical efficiency axis gives as absorbed energy parameter ($F_R(\tau\alpha)$). At this point, the temperature of the inlet fluid to the collector equals to the ambient temperature and the collector efficiency is maximum. The slope of the trend line indicates the amount of energy removed from the solar collector and equals to $F_R U_L$. The intersection point of the trend line with the horizontal axis indicates the zero efficiency point and termed as stagnation point. This situation arises when no fluid flow occurs in the collector. The magnitude of heat removal factor ($F_R U_L$) and heat absorption factor ($F_R U_L$) for various working fluids are obtained in the present investigation and summarized in Table 3. It is observed that the collector that uses water/surfactant and working fluid exhibits highest $F_R(\tau\alpha)$ and lowest $F_R U_L$ among various fluids. This indicates that the thermal performance of solar collector with water/2-ethyl-hexanol is higher followed by CNT/water nanofluid and water as a working fluid in the heat pipe.

Table 3 $F_R(\tau\alpha)$ and $F_R U_L$ values of wickless solar heat pipe collector at different working fluid

Working fluid	$F_R U_L$	$F_R(\tau\alpha)$
Pure water	13.93	0.34
CNT/water nanofluid	13.19	0.48
Water/surfactant	13.03	0.50

4. CONCLUSIONS

An experimental study has been carried out to investigate the thermal performance of heat pipe solar collector by using different working fluids for heat pipes such as: pure water, water/2-ethyl-hexanol, CNT/water nanofluid. Following conclusions are made from the experimental study and is detailed below:

- The thermal performance of solar heat pipe collector is higher by using water/2-ethyl-hexanol followed by CNT/water nanofluid, water as a working fluid for the heat pipe
- The heat transfer rate is found to increase by increasing the tilt angle from 20^0 to 50^0 for both cases, namely, pure water and water/2-ethyl-hexanol. While, the heat transfer rate decreases by increasing the tilt angle beyond 50^0.
- The heat transfer performance of the solar collector that uses CNT/water nanofluid decreases with the tilt angle ranging from 20^0-31.5^0, increases with tilt angle varying from 31.5^0- 50^0 and found to decrease beyond 50^0.
- Solar heat pipe collector that uses nanofluid as working fluid provides better performance at higher tilt angles (at 50^0). The collector will provide improvement in the thermal performance by locating solar heat pipe collector at the place where the angle of getting maximum total solar radiation matches with higher performance tilt angle of solar heat pipe collector.

ACKNOWLEDGEMENTS

The authors would like to thank Dr. S.H. Pawar Vice-Chancellor, D.Y.Patil University, Kolhapur for providing the facility to use Nanotechnology lab. Also thankful to V.S.Paralikar, Golden Star Technical Services Pvt. Ltd., Pune for providing help in manufacturing the heat pipes.

NOMENCLATURE

A	collector effective area, m^2
C	specific heat , J/kgK
I	total solar radiation intensity, W/m^2
F_R	heat removal factor
$\dot{m}$	mass flow rate through heat exchanger, kg/sec.
T	temperature, ^{0}C
Q_g	rate of energy gain, W
Q_{in}	rate of energy input, W
U_L	overall loss coefficient of solar collector (W/m^2 K)
θ	tilt angle
η	collector instantaneous efficiency ,dimensionless
$\tau\alpha$	absorptance-transmittance product

Subscripts

a	ambient
coll	collector aperture
i	inside
n	nanofluid
o	outside
t	total
p	plate
w	water
s	surfactant

REFERENCES

Choi, S.U.S., 1995, "Enhancing Thermal Conductivity of Fluids with Nanoparticles, Developments and Applications of Non-Newtonian Flows," FED-231/MD-66, 99–105.

Cheng, W., Xie, B., Han, F., Chen, H., 2013, "An Experimental Investigation of Heat Transfer Enhancement by Addition of High-Alcohol Surfactant (HAS) and Dissolving Salt Additive (DSA) In Spray Cooling," *Experimental Thermal and Fluid Science*, **45**, 198–202. http://dx.doi.org/10.1016/j.expthermflusci.2012.11.005

Chougule, S. S., and Pise, A.T., 2012, "Studies of CNT Nanofluid in Two Phase system," *International Journal of Global Technology Initiatives*, **1**, F14-F20.

Chougule, S. S., and Pise, A.T., 2011, "Experimental Investigation Heat Transfer Augmentation of Solar Heat Pipe Collector by Using Nanofluid," *21st National and 10TH ISHMT-ASME Heat and Mass Transfer Conference, Madras, India.*

Chougule, S.S., Pise, A.T. and Madane P.A., 2012, "Performance of Nanofluid-Charged Solar Water Heater by Solar Tracking System," *Proc. of the IEEE-ICAESM 2012 Nagapattiam, India*, **VI**, 247-254.

Esen, M. and Esen, H., 2005, "Experimental Investigation of A Two-Phase Closed Thermosyphon Solar Water Heater," *International Journal of Solar Energy*, **79**, 459–468. http://dx.doi.org/10.1016/j.solener.2005.01.001

Enaburekhan, J., Yakasai, U., 2009, "Performance Evaluation of A Refrigerant-Charged Integrated Solar Water Heater in Northern Nigeria," Desalination, **243**, 208–217. http://dx.doi.org/10.1016/j.desal.2008.05.014

Gabriela, H., Angel, H., Ion, M., Florian, D., 2011, "Experimental Study of The Thermal Performance of Thermosyphon Heat Pipe Using Iron Oxide Nanoparticles", *International Journal of Heat and Mass Transfer*, **54**, 656–661. http://dx.doi.org/10.1016/j.ijheatmasstransfer.2010.09.005

Lin, Lu, Zhen-Hua, Liu, Hong-Sheng, Xiao, 2011, "Thermal Performance of an Open Thermosyphon Using Nanofluids for High-Temprature Evacuated Tubular Solar Collectors Part 1: Indoor Experiments", Solar energy, 85, pp. 379-387. http://dx.doi.org/10.1016/j.solener.2010.11.008

Liu, Z.H., Yang, X.F., Wang, G.S., Guo, G.L., 2007, "Effect of Nanoparticles in Nanofluid on Thermal Performance in A Miniature Thermosyphon," *J. Appl. Phys.*, **102**, 013526. http://link.aip.org/link/doi/10.1063/1.2748348

Liu, Z.H., Yang, X.F., Wang, G.S., Guo, G.L., 2010, "Influence of Carbon Nanotube Suspension on the Thermal Performance of a Miniature Thermosyphon," *International Journal of Heat Mass Transfer*, **53**, 1914–1920. http://dx.doi.org/10.1016/j.ijheatmasstransfer.2009.12.065

Mathioulakis, E., Belessiotis, V., 2002, "A New Heat-Pipe Type Solar Domestic Hot Water System," *International Journal of Solar Energy*, **72**, 13-20. http://dx.doi.org/10.1016/S0038-092X(01)00088-3

Naphon, P., Assadamongkol, P., Borirak, T., 2008, "Experimental Investigation of Titanium Nanofluids on the Heat Pipe Thermal Efficiency," *Int. Commun. Heat Mass Transfer*, **35**, 1316–1319. http://dx.doi.org/10.1016/j.icheatmasstransfer.2008.07.010

Naphon, P., Thongkum, D., Assadamongkol, P., 2009, "Heat Pipe Efficiency Enhancement with Refrigerant-Nanoparticles Mixtures," *International Journal of Energy Conversion Management*, **50**, 772–776. http://dx.doi.org/10.1016/j.enconman.2008.09.045

Z., Yuan, K.E., Herold, 2001, "Surface Tension of Pure Water and Aqueous Lithium Bromide With 2-Ethyl-Hexanol," *International Journal of Applied thermal Engineering*, **21**, 881-897. http://dx.doi.org/10.1016/S1359-4311(00)00088-0

6

NON-CONDENSABLE GASES AND OSCILLATING HEAT PIPE OPERATION

Brent S. Taft*

Air Force Research Laboratory Space Vehicles Directorate, Kirtland AFB, NM, 87117, USA

ABSTRACT

The Oscillating Heat Pipe (OHP) is a novel, wickless heat pipe that relies on the phase change induced motion of a contained working fluid to transport heat between the evaporator and condenser. This paper investigates the effects of non-condensable gas (NCG) on OHP operation. A brief synopsis of the existing literature is presented before discussing the experimental setup, method, results, and conclusion. This paper clearly shows that: 1) NCG injection into an OHP produces an overall rise in the steady-state operating temperature, pressure, and thermal resistance; 2) Like LHPs OHPs are more tolerant of NCG than conventional heat pipes.

Keywords: *OHP, Pulsating Heat Pipe, PHP, NCG, Reliability.*

1. INTRODUCTION

With the technological advances in semiconductor devices and the ever-increasing power levels and power densities associated with modern satellites, moving and rejecting waste heat in order to control component temperatures has become a progressively challenging problem. Future satellite missions are expected to need heat-rejection capability measured in the 100's of W/cm^2, consistent with the general electronics trends of miniaturization and increasing power densities. Passive thermal control solutions are generally preferable for transporting waste heat since they are more reliable than active alternatives and do not lead to additional parasitic heat loads. Current passive technologies cannot deal with these extreme heat levels and heat fluxes; however, there are several novel passive technologies that have the potential to meet these future thermal management needs.

Oscillating or Pulsating Heat Pipes (PHPs) are one of the emerging passive technologies that may be capable of meeting these extreme heat rejection needs. The OHP was developed in 1990 by Hisaeru Akachi; however, the OHP received very little attention until the turn of the century. During the last ten years, a great deal of experimental and numerical work has been carried out in order to advance our understanding of the unique fluid flow and heat transfer characteristics of OHPs.

There are many variants on the basic OHP, shown in Fig. 1, but in every case the fluid flow and heat transfer are governed by the same fundamental physics. An OHP is a simply formed, wickless heat pipe that relies on the phase change induced motion of a contained working fluid to transport heat between the evaporator and condenser. The primary OHP driving force is the result of the difference in working fluid saturation pressure that exists in the evaporator and in the condenser. So long as the evaporator is kept sufficiently hot and the condenser sufficiently cold, the oscillatory motion is self-sustaining.

OHPs are different from traditional heat pipes in that they do not require a wick structure to move the condensed working fluid back to the evaporator. Unlike thermosyphons, however, OHPs are gravity-independent above a critical heat flux (Taft et al., 2013).

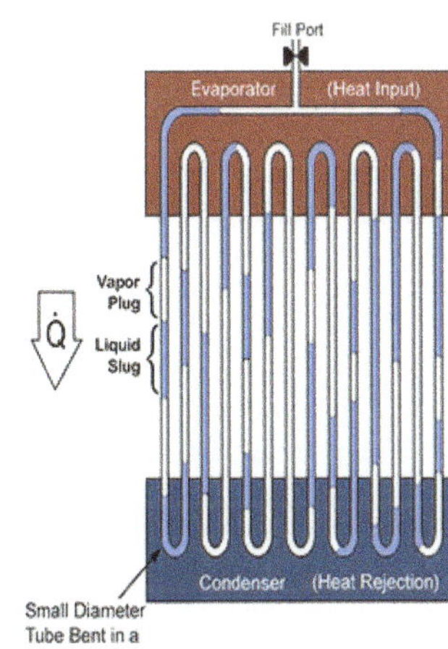

Fig. 1 Schematic of a typical OHP.

In addition to being simple and gravity-independent, OHPs have the potential for better heat transfer capability than standard heat pipes because the majority of the heat transfer associated with OHPs is due to forced convection and the transfer of sensible heat, whereas the heat transfer in a standard heat pipe is dominated by latent heat transfer (Shafii et al., 2001). In an OHP, the effect of latent heat is primarily to drive the oscillatory motions. The simplicity of OHPs may result in a considerable cost and weight savings. The improved heat transfer capability and reduced mass of OHPs have led many to realize that they are ideal for some spacecraft thermal control applications (Shafii et al., 2001; Gu et al., 2004; Gu et al., 2005; Zhang and Faghri, 2003; Taft et al., 2013).

Researchers around the globe are working to transition OHPs to spacecraft thermal control systems. As additional researchers independently verify OHP operation in microgravity, research must turn to answering the reliability and repeatability concerns of space systems engineers. Some of the potential OHP reliability concerns include start-up, non-condensable gas (NCG) effects, freeze-tolerance, material

* *Corresponding author, Email: afrl.rvsv@kirtland.af.mil*

compatibility, and hermeticity. Material compatibility and hermeticity are problems faced with traditional heat pipes, and are relatively well understood for common heat pipe working fluids and envelope materials. Start-up, non-condensable gas effects, and freeze-tolerance, however, have some subtleties, potentially both pro and con, in the OHP as a result of the lack of a wick. JAXA has, at least partially, addressed the start-up issue by using a condenser mounted heater to ensure that fluid can be driven to the evaporator prior to start-up if needed (Maeda et al., 2011). This paper investigates the effects of NCG on OHP operation.

The paper starts by presenting a brief synopsis of the existing literature, before discussing the experimental setup, method, results, and conclusion. This paper clearly shows that NCG does not prevent OHP operation; however, it does cause some performance degradation.

2. BACKGROUND

2.1 NCG Formation

According to Nikitkin et al. the three most important factors in determining heat pipe reliability are micrometeoroid protection, NCG formation, and leakage of the working fluid to space (1998). Singh et al. note that NCG is "one of the most common causes for the heat pipe failure which can occur during the fabrication as well during the operational lifespan of the device" (2010).

There are several causes of NCG formation, but the most prevalent is chemical compatibility issues between the working fluid and the heat pipe envelope and/or wick. Improperly chosen working fluid, wick, envelope combinations can lead to chemical reaction and the formation of NCG, typically hydrogen in the case of ammonia as a working fluid. Fortunately, chemical compatibility issues have mostly been solved for the common working fluid, wick, and envelope combinations. The major heat pipe manufacturers have documented lifetime compatibility of common heat pipe and Loop Heat Pipe (LHP) working fluids, wick and envelope combinations for more than 4 decades.

Unfortunately, minor changes in manufacturing and cleaning processes can lead to unforeseen chemical reactions resulting in the formation of NCG. Additionally, while OHPs typically implement familiar envelope materials (e.g. copper, aluminum, stainless steel), they are introducing new working fluids, whose chemical compatibility is not as well understood.

Although it is rare, and often ignored by domestic spacecraft manufacturers, Nikitkin et al. also cite ionizing radiation as a potential gas generation mechanism (1998). They state that within the former Soviet Union, the breakdown of the heat pipe's working fluid due to ionizing radiation is considered to be significant. Singh et al. add that although very rare, working fluid breakdown can also occur as a result of the dissimilar electrolytic potentials of the container and wick material (2010).

In addition to NCG formation, NCG can be inadvertently introduced to the system during charging. An improperly degassed working fluid is likely to have NCGs in solution.

2.2 NCG in Heat Pipes and Loop Heat Pipes (LHPs)

The sensitivity and effect of NCG on two-phase thermal control systems varies widely as a result of the system design. Heat pipes and Loop Heat Pipes (LHPs) are of particular interest for spacecraft thermal control, and will be briefly discussed here as a comparison to the OHP. The sensitivity of traditional heat pipes to NCG, which can cause performance degradation and complete failure, is well known. On the other hand, there is relatively little literature documenting the effect of NCG on LHPs.

Bienert et al. and Nikitkin et al. showed that an aluminum-ammonia LHP is insensitive to the presence of NCG at up to six times the projected end-of-life NCG inventory (1997; 1998). However, at 10 times the projected end-of-life NCG inventory, the NCG was shown to increase startup time and operating temperature in LHPs. For this case, the evaporator temperature was ~20 °C above tests in which there was no NCG.

Singh et al. investigated the effect of NCG on a miniature copper-water LHP (2010). Singh et al.'s findings are generally in agreement with Bienert et al. and Nikitkin et al. Singh et al. found that "the overall effect of the NCG is to elevate the steady-state temperature of the loop and increase the start-up time required by the evaporator to achieve stable conditions for a given heat load" (2010). With an estimated 0.0436 μg of NCG, start-up time increased by 1 min 10 sec and the steady-state evaporator temperature increased by ~10 °C.

2.3 NCG in OHPs

Saturation pressure plays an important role in virtually all OHP models as the primary driving force. However, there appears to be very little research in which OHP pressure measurements were collected, and even less attention has been paid to NCG effects.

Miyazaki and Akachi, and separately Kim et al., varied the heat flux and charging ratio to study the pressure oscillation characteristics of OHPs; however, the base OHP pressure was not altered (1996; 2003).

In his paper "Characteristics of an Open Loop Pulsating Heat Pipe," Riehl points out that plug/slug flow dynamics are directly dependent on the amount of non-condensable gas in the system (2004). He discusses the vacuum level, outgassing process, and final working fluid purity of his experimental setup. Taft et al. build on Riehl's comments, noting that few researchers document the extent to which they pull a vacuum on their OHP before filling, and even fewer go through a rigorous degassing of the working fluid before fill (2012). It was hypothesized, that inconsistency in the way in which researchers fill their OHP experiment setups may account for some of the variability in predicting startup temperatures and heat fluxes, as it is likely that many, if not most, researchers are inadvertently introducing NCG to their test setups.

O'Conner et al. showed that the effect of adding air to a system is to increase the degree of subcooling of the working fluid (1996). It was shown that the total system pressure can be found using Eq. (1):

$$P_{total} = P_{partial\ vapor} + P_{partial\ gas} \quad (1)$$

where

$$P_{partial\ vapor} = P_{sat}[T_{bulk\ liquid}] \quad (2)$$

and P_{sat} is the saturation pressure taken at the bulk liquid temperature. $P_{partial\ gas}$ is the partial pressure of any NCG in the system. The effective saturation temperature, $T_{sat\ eff}$, of the system can then be found using Eq. (3):

$$T_{sat\ eff} = T_{sat}[P_{total}] \quad (3)$$

Finally, the degree of subcooling is

$$\Delta T_{sub-cooling} = T_{sat\ eff} - T_{bulk\ liquid} \quad (4)$$

3. EXPERIMENTAL APPARATUS

A Flat Plate OHP (FP-OHP) design, as shown in Fig. 2, 3, and 4, was constructed as an embedded OHP in an aluminum structural panel. The OHP has a total of 20 turns and covers a footprint of ~25x25cm^2. The entire FP-OHP structural panel is 30x30x0.6cm^3. Traditionally, an OHP may be manufactured in two separate pieces, later brazed together. However, ultrasonic consolidation (UC) was used to manufacture the OHP. The OHP core, seen in Fig. 3, was machined from Al-6061; the ~1.3mm facesheet was added, sealing the OHP to the bottom surface, and consists of Al-1100 and Al-3003. UC is an additive/subtractive manufacturing technique that is ideally suited for the production of structurally integrated OHPs. UC allows for the creation of high strength materials with internal features that significantly exceed the strength to weight ratio of many current aerospace materials.

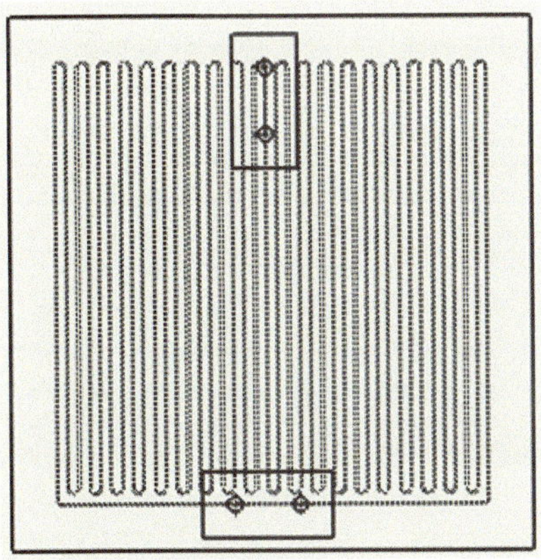

Fig. 2 Schematic of the FP-OHP design.

Fig. 3 FP-OHP Top.

Fig. 4 FP-OHP Bottom.

In order to read pressure measurements inside the OHP, Swagelok® VCR fittings and Omega® pressure transducers were integrated into the panel in both the condenser (0-206 kPa ± 0.689 kPa) and evaporator section (0-3.45 MPa ± 0.002 MPa). After using a helium leak detector to ensure a leak rate of <10E-7atm-cc/sec, vacuum (3E-3 Torr) was pulled on the sealed OHP and acetone (high performance liquid chromatography grade) was added with a 0.8 ±0.01 fill ratio, by mass.

The square channel ($1.3x1.3mm^2$) adheres to the maximum allowable diameter, D_{max}, for liquid to form vapor bubbles in a capillary tube when using acetone as a working fluid in a standard gravitational environment:

$$D_{max} = 1.84\sqrt{\frac{\sigma}{g(\rho_l-\rho_v)}} \tag{5}$$

where σ is the working fluid surface tension, ρ_l is the working fluid liquid density, ρ_v is the working fluid vapor density, and g is gravity.

The OHP performance was measured using the experimental setup shown in Fig. 5. The system consisted of two power supplies (one powering the data acquisition system and a second powering a heater), an ice bath, heater, aluminum water block, and data acquisition system (DAQ). Water was circulated from the ice bath through the aluminum water block at 1.0 GPM using a submersible pump. Heating and cooling were conducted on an area of $7.6x30.5cm^2$ and $12.7x30.5cm^2$, respectively, as shown in Fig. 6. Thermal paste (Artic Silver 5) was used as the thermal interface material between all contact surfaces. Four C-clamps were used with insulating pads to apply clamping pressure between the OHP and both the water block and heater.

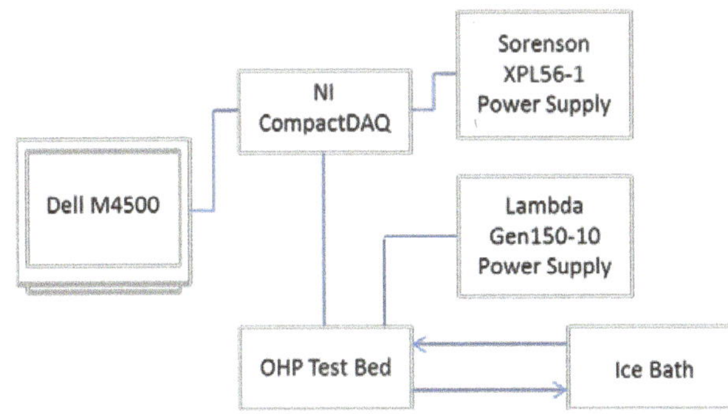

Fig. 5 Schematic of the experimental setup.

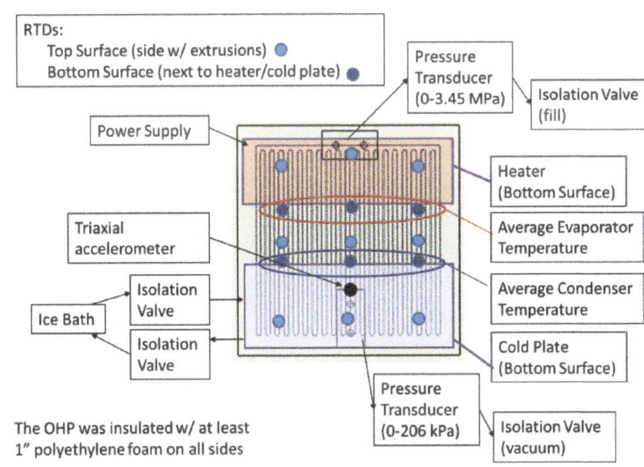

Fig. 6 Location of heater, cold plate, pressure transducers, and RTDs.

A total of 16 RTDs (100 Ohm), with uncertainty of 0.75°C, were placed at the surface of the OHP as shown in Fig. 6. Temperature measurements were recorded at 10 Hz using a National Instruments CompactDAQ (cDAQ-9178) with a customized virtual instrument (VI). The heat input was controlled with a Lambda Gen150-10 power supply, and measured by the DAQ with an uncertainly of less than ±0.1W. The temperature and power measurements were used to calculate the OHP conductance, and its inverse, the thermal resistance. The OHP conductance, UA, is

$$UA = \frac{Q}{T_{evap}-T_{cond}} \tag{6}$$

Where Q is the heat input and T_{evap} and T_{cond} are the evaporator and condenser temperatures, respectively.

The averages consisted of the 3 RTD's most closely located to the heater and chilled water block, respectively. The overall calculated conductance uncertainty is less than ±0.24 W/K, or approximately ±2.2%. Likewise, the total thermal resistance uncertainty is less than ±2.2%.

4. METHOD

This FP-OHP was originally built for microgravity performance testing. Between two parabolic aircraft flights, the system was put through an extensive ground test campaign. Between July 2nd, 2012 and July 11th, 2012 the OHP was tested at multiple orientations and power input levels. However, since this OHP was previously shown to be gravity independent above ~300W, this paper focuses on test results from the horizontal orientation (Taft et al., 2013). Heat input ranged from 100W to a maximum power input of 500W. Testing was conducted both low-power to high-power and high-power to low-power, but in either case the power input was adjusted in 50W increments. Each power level was held until a steady state condition was achieved and was then recorded for approximately five minutes. This procedure was performed three separate times to effectively produce a range of thermal resistances at each power level and orientation.

After the flight campaign, the OHP and test fixture were left to sit in an ambient environment for ~10 months. The clamping pressure between the OHP and both the water-block and heater was not adjusted, and the high temperature polyethylene foam insulation surrounding the OHP, water block, and heater was left in place. Between April 3rd, 2013 and April 10th, 2013 the OHP was again tested. The only difference between the tests was the OHP non-operating, ambient-temperature pressure, which had risen from 49.6 kPa to 77.2 kPa.

It's worth noting that even at 49.6 kPa the OHP pressure is higher than would be predicted for a pure working fluid. At room temperature, or 24 °C, the saturation pressure of acetone is 29.6 kPa, indicating that even during the first test series NCG was accounting for 20.0 kPa of the measured pressure. This initial NCG loading is likely the result of an incomplete degassing process. Due to the operational constraints of microgravity performance testing the OHP was charged on the tarmac, and the working fluid was not fully degassed prior to fill. It has been noted that acetone can hold a significant volume of air in solution, according to Kretschmer et al. as much as 20% at 25 °C (1946).

Using the ideal gas equation, Eq. (7), the mass, m, of NCG can be estimated.

$$m = \frac{PV}{RT} \tag{7}$$

where V is volume and R is the gas constant of air. Using the measured pressure and temperature, and assuming that the working fluid is incompressible (i.e. the volume available to the NCG is the loop volume not occupied by liquid), it was estimated that ~2.71 mg of NCG was in the system during the initial test series. It is worth noting that this is significantly more NCG than would be expected at the end-of-life in a commercially manufactured aerospace-grade heat pipe or LHP. As OHPs continue to gain popularity, commercial manufacturing processes must be transferred and validated for OHP manufacture. It is not unreasonable to expect aerospace-grade OHPs to be manufactured to the same stringent tolerances currently being used for heat pipes and LHPs.

As the OHP sat idle for ~10 months additional NCG bled into the system. As opposed to brazing off the OHP fill ports, Swagelok® valves were used to temporarily seal the OHP while Swagelok® VCR plugs were used to plug the OHP. This sealing method provided more flexibility in filling/re-filling the OHP for various tests; however, it did not provide the long-term hermeticity that a crimped and brazed connection would provide. It is believed that the additional pressure rise can be attributed to leakage of some of the air contained between the valves and VCR plugs around the valve stems into the OHP. During the second test series, NCG accounted for 47.6 kPa of the measured pressure at room temperature. Again using Eq. (7), it was estimated that ~4.23 mg of NCG was trapped in the system during the second test series.

5. EXPERIMENTAL RESULTS AND DISCUSSION

It has previously been reported that variation in OHP performance between tests is larger than typical experimental errors (Taft et al., 2013). It is hypothesized that this variation in performance could be the result of an OHPs chaotic operating nature, which has been well documented in the literature (Qu et al., 2009; Maezawa et al., 1996; Maezawa et al., 2000). In this test sequence, it is also possible that some variation is the result of slight differences in the ice-bath temperature. In either case, it was observed that performance variation at ~6% exceeded the measurement uncertainty during this test sequence. As such, error bars have been omitted from the charts as the measurement uncertainty is small compared to the natural performance variation. Instead, the average values for each of three separate tests in the two test series were plotted.

Figure 7 shows the OHP thermal resistance as a function of power input and OHP non-operating, ambient-temperature pressure. As the OHP internal pressure is increased, the OHP performance decreases. On average, this performance degradation is 0.03 K/W or 16.5%. It's worth noting that at 100 W power input, the OHP is not oscillating, and the measured resistance is almost entirely the result of conduction through the panel and working fluid. The absence of oscillations results in there being little difference in the measured resistance, regardless of the increased system pressure. The empty panel resistance is 0.44 K/W.

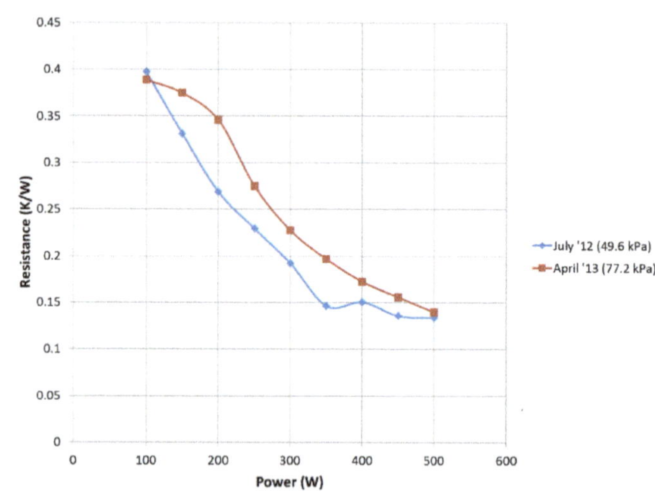

Fig. 7 OHP resistance as a function of power input and OHP non-operating, ambient-temperature pressure.

Figure 8 shows the OHP evaporator pressure as a function of power input and OHP non-operating, ambient-temperature pressure. The OHP evaporator pressure rises with increased ambient-temperature pressure. On average, the pressure rises 33.4 kPa or 27%. The average pressure difference is slightly more than the NCG attributed 27.6 kPa difference at non-operating, ambient-temperature conditions. The additional 5.8 kPa difference is likely the result of the system's increased operating temperature.

Figure 9 shows the OHP evaporator temperature as a function of power input and non-operating, ambient-temperature pressure. Above 100 W power input, the OHP evaporator temperature rises with increased ambient-temperature pressure. On average the temperature rises 7.3 °C or 11%. This temperature rise is on-par or better than the LHP performance demonstrated by Singh et al., Bienert et al, and Nikitkin et al.

Using Eq. (3) and Eq. (4), the effective saturation temperature and subcooling are calculated for the two different NCG pressure states. Figure 10 shows the effect that NCG has on the saturation temperature.

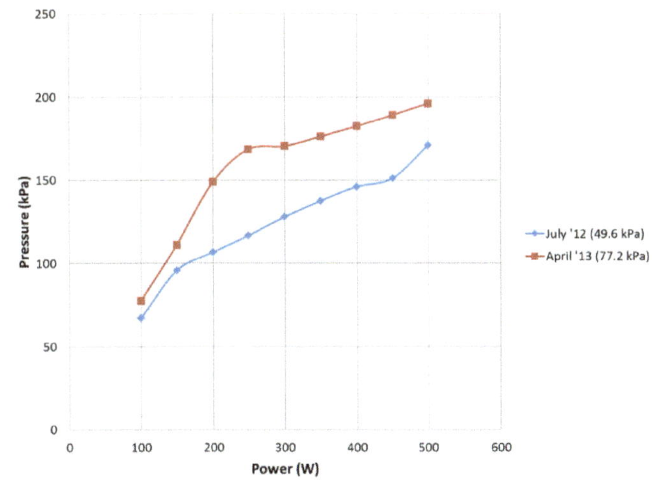

Fig. 8 OHP evaporator pressure as a function of power input and OHP non-operating ambient-temperature pressure.

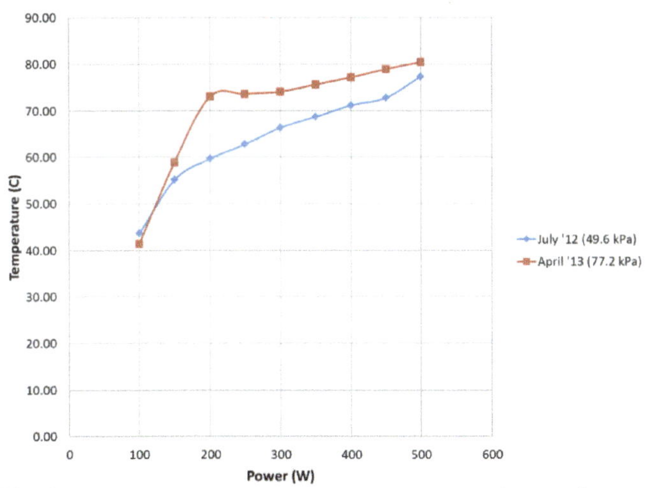

Fig. 9 OHP evaporator temperature as a function of power input and OHP non-operating, ambient-temperature pressure.

The degradation of OHP performance as the base system pressure is increased is not surprising, as the addition of NCG to the system serves to sub-cool the system.

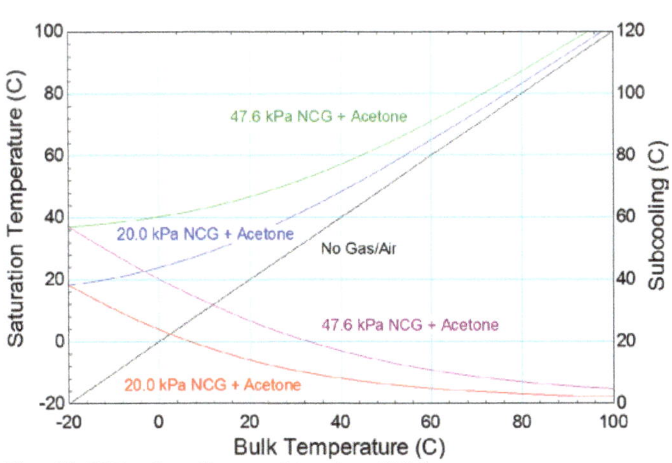

Fig. 10 OHP subcooling resulting from NCG.

6. CONCLUSIONS

The results of this investigation clearly show that the addition of NCG to an OHP will not prevent operation. In fact, since there is so little documentation and almost no direct pressure measurement, it seems likely that many of the OHPs discussed in the literature contain some level of NCG. Although it does not prevent operation, NCG, and the corresponding increase in system base pressure, do degrade performance.

As cleaning/charging processes are further developed and validated for commercial manufacturing of aerospace-grade OHPs, and chemical compatibility of typical OHP working fluids is better understood, the quantity of NCG injected into and or formed in a typical OHP will be vastly reduced. This improvement should provide increased consistency in the thermal performance results of the many OHP studies being conducted worldwide.

Regardless, the ability of an OHP to continue operation, even with the formation of non-condensable gases, provides for a resilient thermal control system. Resiliency may be another perk in using the OHP for spacecraft thermal control; however, NCG formation is only one piece of the OHP reliability puzzle. Future work must address the other factors affecting reliability.

In summary, it was shown that:

1. NCG injection into an OHP produces an overall rise in the steady-state operating temperature, pressure, and thermal resistance, but does not prevent operation.
2. Like LHPs, OHPs are more tolerant of NCG than conventional heat pipes.

NOMENCLATURE

D	Diameter (m)
g	Gravity (m/s^2)
P	Pressure (Pa)
Q	Heat Input (W)
R	Gas Constant of Air (kPa-m^3/kg-K)
T	Temperature (K)
UA	Conductance (W/K)
V	Volume (m^3)

Greek Symbols

ρ	Density (kg/m^3)
σ	Surface Tension (N/m)

Subscripts

cond	Condenser
evap	Evaporator
max	Maximum
l	Liquid
sat	Saturation
sat eff	Effective Saturation
v	Vapor

REFERENCES

Akachi, H., 1990, US Patent Number 4921041, *US Patent*

Bienert, W. B., Wolf, D. A., Nikitkin, M. N., Goncharov, K. A., 1997, "Experimental Study of Loop Heat Pipe Behavior in the Presence of Noncondensible Gases," 10th International Heat Pipe Conference, Stuttgart, Germany. PMCid:1184409

Gu, J., Kawaji, M., and Futamata, R., 2004, "Effects of Gravity on the Performance of Pulsating Heat Pipes," Journal of Thermophysics and Heat Transfer, 18(3), pp. 370-378 http://dx.doi.org/10.2514/1.3067

Gu, J., Kawaji, M., and Futamata, R., 2005, "Microgravity Performance of Micro Pulsating Heat Pipes," Microgravity Science and Technology, 16(1) pp. 181-185 http://dx.doi.org/10.1007/BF02945972

Kim, J.S., Bui, N., Jung, H.S., and Lee, W.H., 2003, "The Study on Pressure Oscillation and Heat Transfer Characteristics of Oscillating

Capillary Tube Heat Pipe," *Journal of Mechanical Science and Technology*, Vol. 17, No. 10, October 1, pp. 1533-1542.

Kretschmer, C. B., Nowakowska, J., and Wiebe, R., "Solubility of Oxygen and Nitrogen in Organic Solvents from -25 to 50 C," Industrial and Engineering Chemistry, Vol. 38, No. 5, 1946, pp. 506-509. http://dx.doi.org/10.1021/ie50437a018

Maeda, M., Okamoto, A., Kawasaki, H., Sugita, H., 2011, "Development of Flat Plate Heat Pipe and the Project of On-orbit Experiment," 41st International Conference on Environmental Systems, 17-21 July 2011, Portland, OR. http://dx.doi.org/10.2514/6.2011-5142

Maezawa, S., Nakajima, R., Gi, K., and Akachi, H., 1996, "Experimental Study on Chaotic Behavior of Thermohydraulic Oscillation in Oscillating Thermosyphon," *Proceedings of the 5th International Heat Pipe Symposium*, Melbourne, pp. 131-137.

Maezawa, S., Sato, F., Gi, K., and Akachi, H., 2000, "*Chaotic Dynamics of Looped Oscillating Heat Pipes (Analysis on Single Loop),"* pp. 25-26.

Miyazaki, Y. and Akachi, H., 1996, "Heat Transfer Characteristics of Looped Capillary Heat Pipe," *Proceeding of the 5th International Heat Pipe Symposium*, Melbourne, pp. 378--383.

Nikitkin, M. N., Bienert, W. B., Goncharov, K. A., 1998, "Non Condensable Gases and Loop Heat Pipe Operation," *SAE Paper No. 981584.*

O'Conner, J.P., You, S. M., and Chang, J. Y., 1996, "Gas-Saturated Pool Boiling Heat Transfer from Smooth and Microporous Surfaces in FC-72," Journal of Heat Transfer, Vol. 188, No. 3, pp. 662-667. http://dx.doi.org/10.1115/1.2822683

Qu, J., Wu, H., Cheng, P., and Wang, X., 2009, "Non-linear analyses of temperature oscillations in a closed-loop pulsating heat pipe," International Journal of Heat and Mass Transfer, 52, pp. 3481-3489. http://dx.doi.org/10.1016/j.ijheatmasstransfer.2009.03.012

Riehl, R., 2004, "Characteristics of an Open Loop Pulsating Heat Pipe," 34th *International Conference on Environmental Systems (ICES)*, SAE paper 2004-01-2509, Colorado Springs, CO.

Shafii, M., Faghri, A., and Zhang, Y., 2001, "Thermal Modeling of Unlooped and Looped Pulsating Heat Pipes," Journal of Heat Transfer, 12(6), pp. 1159-1172. http://dx.doi.org/10.1115/1.1409266

Singh, R., Akbarzadeh, A., Mochizuki, M., 2010, "Operational Characteristics of the Miniature Loop Heat Pipe with Non-Condensable Gases," International Journal of Heat and Mass Transfer, 23(2010), pp. 3471-3482. http://dx.doi.org/10.1016/j.ijheatmasstransfer.2010.04.008

Taft, B.S., Williams, A.D., Drolen, B.L, 2012, "Review of Pulsating Heat Pipe Working Fluid Selection," *Journal of Thermophysics and Heat Transfer,* **26**(4), pp. 651-656. http://dx.doi.org/ 10.2514/1.T3768

Zhang, Y., and Faghri, A., 2003, "Oscillatory Flow in Pulsating Heat Pipes with Arbitrary Number of Turns," Journal of Thermophysics and Heat Transfer, 17(3), pp. 340-347. http://dx.doi.org/10.2514/2.6791

PERFORMANCE EVALUATION OF SODIUM HEAT PIPES THROUGH PARAMETRIC STUDIES

R. Manoj[a], M. C. Kumar[a], R. NarasimhaRao[a], K. Rama Narasimha[b*], P. V. S. Suresh[a]

[a]*Avasarala Technologies Ltd, Bangalore, Karnataka, 560099, India*
[b]*Centre for Emerging Technologies, Jain University, Bangalore, Karnataka, 562112, India*

ABSTRACT

The performance of alkali heat pipe greatly depends on the properties of the wick material. In the present work, an experimental investigation is conducted to study the performance of high temperature sodium heat pipe with annular and composite wick structures. The operating limits of the heat pipe for these wick structures are evaluated and the effective working region of the heat pipe is identified. The experiments are conducted for annular and composite wick, different heat inputs, fill ratio and orientation. The results showed that the composite wick heat pipe is performing better in terms of its lower thermal resistance. The sodium heat pipe showed better heat transfer characteristics at a fill ratio of 40% and at an orientation of 30°.

Keywords: wick, operational limits, thermal resistance

1. INTRODUCTION

Heat pipes are the passive cooling devices used for both low and high temperature applications. Liquid-metal heat pipes are considered for thermal energy transport in many high temperature, high power density space and terrestrial power systems. Alkali metals are commonly used as the working fluid with refractory metals. This is partly because of the low solubility and dissolution rate of the metal. High temperature heat pipes are being used in energy conversion applications such as fuel cells, gas turbine re-combustors and Stirling cycle heat sources. With the technology development in space nuclear power, additional applications include nuclear reactor heat removal elements and radiator elements. Sodium being one of the important alkali metals, it can be used as the working fluid in a nuclear heat pipe.

Several experimental and numerical works have been reported in the literature related to sodium heat pipe. Deverall et al. (1975) studied sodium heat pipes operating in the gravity-assist mode. They considered the sodium heat pipe incorporated into irradiation capsules to provide a means for establishing and controlling a desired specimen temperature. Ivanovskii et al. (1976) investigated the evaporation process and sonic limit in sodium heat pipes. Ivanovskii et al. (1980) examined problems in building vertical high-temperature heat pipe with simple and composite wicks. Faghri (1990) studied the transient and start-up behavior of sodium heat pipe and energy storage module. Amir Faghri and Cao (1991) tested sodium heat pipe with multiple heat source and sinks for a heat input of 850 W and concluded that the startup behavior of the heat pipe greatly depends on the heat removal at the condenser section and hence condenser is cooled using water calorimeter. Large heat losses were observed between the source and the heat pipe which led to miscalculation of the heat transfer. Adkins et al. (1999) developed sodium heat pipe for solar energy collector. The purpose of the heat pipes is to transport high thermal energy from the solar collectors to energy conversion system to generate power. Their experiments showed the degradation of the heat pipe performance due to corrosion of wick material but it was also stated that with metal sintered wick, this problem can be overcome and the heat pipe can work for long duration. The work was focused on the cleaning methods of the wick and the pipe to avoid corrosion during usage. The system for handling alkali metals for the production of flight qualified heat pipe module was developed by Patrick et al. (1999) which emphasized on maintaining the purity of the liquid metals during the development of the heat pipe. The processing approach used to introduce pure alkali metal into the heat pipe modules was presented. Reid et al. (2000) developed sodium heat pipe with high purity sodium and better cleaning methods for long life space nuclear application for NASA. Multiple mesh wick material was used for better transport of liquid sodium from the condenser to the evaporator. This helped in avoiding evaporator burn out due to insufficient supply of sodium from the condenser region. Near isothermal conditions were achieved for 90% length of the heat pipe at 850 K for a heat input of 660 W. Lee and Lee (2001) tested sodium heat pipe for higher heat inputs up to 5 kW. In this investigation, a sodium heat pipe which is operating between 500°C and 1000°C has been fabricated and its performance tests were carried out in a room under free convection conditions around the condenser zone of the heat pipe. It was observed that transport capacity depends strongly on the working fluid and its temperature. They concluded that the heat transport capability of this heat pipe largely depends on vapour core region and the wick composition. They also found that for effective startup, the surface temperature of the heat pipe must be higher than the transition temperature of the working fluid. Near isothermal conditions were achieved but after long hours of start up as the condenser was exposed to air.

Rosenfeld et al. (2004) gave an overview of the sodium heat pipe performance for long durations. Four heat pipes were tested with various wick materials for high heat input of more than 4kW for long durations. These life test results collectively have demonstrated the

* Corresponding author. Email: k.ramanarasimha@gmail.com

potential for high temperature heat pipes to serve as reliable energy conversion system components for power applications that require long operating lifetime with high reliability. Degradation of performance in the heat pipe was observed which was due to welding crack after 1150 hours of operation. With the ability of the sodium heat pipes to perform for long hours, Mullet (2008) designed sodium heat pipe for fast-spectrum heat pipe reactor, called the Heat Pipe-Encapsulated Nuclear Heat Source (HP-ENHS), which utilizes sodium heat pipes to remove heat from a solid molybdenum reactor core fueled by Uranium and Plutonium Nitride. The focus was on the ability of the heat pipe to remove heat effectively using a model that predicts the maximum throughput based on capillary limits. Sabharwall et al. (2008) studied thermosyphon heat pipe and gave the theoretical design of thermosyphon.

Yan and Duan (2010) gave a detailed procedure for fabricating sodium heat pipe. Mochzuki et al. (2011) gave a proposal on the need of the heat pipe for cooling a nuclear reactor. This proposal came after the natural disaster of earth quake and Tsunami which have caused serious potential nuclear reactor melt down in Fukushima, Japan. The heat pipe cooling system proposed was passive and was applicable to Boiling Water Reactor (BWR), Pressurized Water Reactor (PWR) and Fast Breeder Reactor (FBR).

El-Genk and Tournier (2011) confirmed that the startup of alkali metal heat pipes from a frozen state is practically possible and safe, though the input power should be kept sufficiently low until the working fluid is fully thawed and the vapour flow along the heat pipe transitions to the continuum regime. The very low vapour pressures of alkali metals prolong the startup time of the heat pipes to many hours. However, this time could be shortened by adding a small amount of noble gases and increasing the length of the condenser section to accommodate these gases during steady-state operation. Mireles and Houts (2011) studied the influence of extreme condenser cooling conditions on the thermal stability of an operating heat pipe by conducting series of experimental demonstrations where a heat pipe operating at neat prototypic conditions experienced rapid cooling of the condenser. The condenser section of a stainless steel sodium heat pipe was enclosed within a heat exchanger. The heat pipe – heat exchanger assembly was housed within a vacuum chamber held at a pressure of 50 torr of Helium. After the heat pipe achieved steady state condition, high flow of gaseous Nitrogen was passed through the heat exchanger for rapid cooling.

The facts and figures reported in the literature related to sodium heat pipes are taken into consideration while defining the objectives of the present work. It is seen from the literature that the performance of sodium heat pipe greatly depends on the properties of the wick material. Hence in the present work, an experimental investigation is conducted to study the performance of high temperature sodium heat pipe with different wick structures. The operating limits of the heat pipe for annular wick and composite wick are evaluated and the effective working region of the heat pipe is identified. The experiments are conducted for annular and composite wick structure, different heat inputs, fill ratio and orientation. The performance of the heat pipe is determined and analyzed in terms of its frozen startup time and thermal resistance.

2. FROZEN START-UP PROCESS IN SODIUM HEAT PIPE

The start-up process for sodium heat pipes from the frozen state has been experimentally investigated by Deverall et al. (1975), and Ivanovskii et al. (1982). Vapour temperatures measured by Ivanovskii et al. (1982) and Faghri et al. (1991) indicated that the different flow regimes exist in the vapour space.

The start-up process is a very important phenomenon also complicated, as it involves phase change of the working fluid. Initially the working fluid is in the solid state and the vapour density is very low. So, free molecular flow condition prevails in vapour core area. As the input heat flux to evaporator section is increased, evaporator temperature starts to increase. Thus the working fluid in this region starts to melt but the heat transported to adjacent section of evaporator by axial conduction process through heat pipe wall is very slow as the thermal conductivity of the steel is low. Also the heat transported by the vapour is almost negligible. Because of this, there exists a large temperature gradient between the evaporator and the condenser section of the heat pipe.

On continues addition of heat flux to the evaporator, the working fluid in this region will melt from frozen state and starts to evaporate in liquid - vapour interface. By this, the vapour density increases and the molecular free mean flow become small and continuum flow regime are established. But in cold vapour region of the condenser section, free molecular flow regime exists.

As the pressure gradient between the evaporator and condenser section is increased, the vapour flows into the continuum flow region of the condenser section. The heat transport is by latent heat of the working fluid. As the working fluid moves towards the condenser region, the temperature gradient decreases and the heat is transported from the evaporator section to the condenser section.

3. EVALUATION OF EFFECTIVE PORE RADIUS, PERMEABILITY AND EFFECTIVE THERMAL CONDUCTIVITY OF THE WICKS

The performance of the heat pipe depends on the properties of wick like porosity, permeability and effective thermal conductivity. In the present work, these properties of the composite and annular wick are evaluated and tabulated in table 1.

3.1 For composite wick

Porosity is given by

$$\varphi = 1 - \frac{\pi N d}{4} \quad (1)$$

Wick permeability is evaluated using the equation

$$K = \frac{d^2\varphi^3}{122(1-\varphi)^2} \quad (2)$$

Effective radius or capillary radius is given by

$$r_c = r_{eff} = \frac{1}{2N} \quad (3)$$

Effective thermal conductivity is

$$k_{eff} = \frac{k_l[(k_l+k_s)-(1-\varphi)(k_l-k_s)]}{(k_l+k_s)+(1-\varphi)(k_l-k_s)} \quad (4)$$

3.2 For annular wick

Porosity of the annular wick is evaluated using equation (1). Wick permeability for annular wick is given by

$$K = \frac{2t^2}{fR_e} \quad (5)$$

fR_e is read from the graph (Faghri, 1995). Effective radius or capillary radius of annular wick is the annular gap between the wall and the wick. The effective thermal conductivity of the annular wick is the liquid thermal conductivity at the specified temperature. The wick properties thus evaluated are tabulated in table 1.

Table 1: Properties of wicks

Wick Structure	φ	K (m²)	r_{eff} (m)	k_{eff}(W/mK)
Composite	0.61	1.53 x 10^{-9}	0.085×10^{-3}	38
Annular	1	7.5 x10^{-9}	0.3×10^{-3}	57

The performance limits of the heat pipe are estimated with the help of these tabulated values and are presented in the next section.

4. PERFORMANCE LIMITS

Heat pipes undergo various heat transfer limitations depending on the working fluid, the wick structure, the dimensions of the heat pipe and the heat pipe operational temperature. Some of the important heat transfer limits of the sodium heat pipe are evaluated in the present work and listed at the end of this section.

4.1 Capillary Limit

The capillary limit relates to the fundamental phenomenon governing heat pipe operation which is the development of capillary pressure differences across the liquid-vapor interfaces in the evaporator and condenser. When the driving capillary pressure is insufficient to provide adequate liquid flow from the condenser to the evaporator, dry out of the evaporator wick will occur. Generally, the capillary limit is the primary maximum heat transport limitation of a heat pipe and is given by

$$Q_c = \frac{\sigma \rho_l h_{fg} K A_w}{\mu L_{eff}} \left[\frac{2}{r_{eff}} + \frac{g L \rho_l}{\sigma} \right] \tag{6}$$

4.2 Sonic Limit

The sonic limit is due to the fact that at low vapour densities, the corresponding mass flow rate in the heat pipe may result in very high vapour velocities and the occurrence of chocked flow in the vapour passage may be possible. The sonic limit is evaluated by using the relation

$$Q_s = A_v \rho_v h_{fg} \left[\frac{\gamma_v T_v R_v}{2(\gamma_v + 1)} \right] \tag{7}$$

4.3 Viscous Limit

The viscous limit occurs at low operating temperatures, where the saturation vapour pressure may be of the same order of magnitude as the pressure drop required to drive the vapour flow in the heat pipe. This results in an insufficient pressure available to drive the vapour. The viscous limit is sometimes called the vapour pressure limit and is given by

$$Q_v = \frac{\pi r_v h_{fg} \rho_v P_v}{12 \mu_v L_{eff}} \tag{8}$$

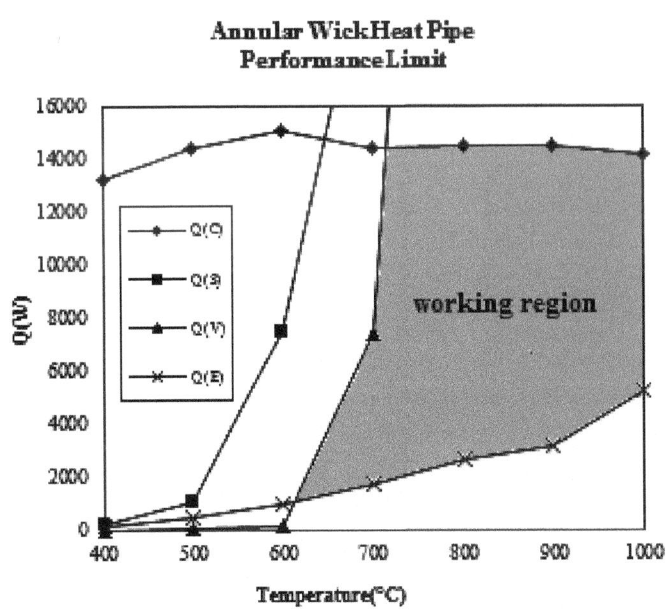

Fig. 1 Performance limits of annular wick heat pipe

4.4 Entrainment Limit

The entrainment limit refers to the case of high shear forces developed as the vapour passes in the counter flow direction over the liquid saturated wick, where the liquid may be entrained by the vapour and returned to the condenser. This results in insufficient liquid flow to the wick structure. The Entrainment limit is given by

$$Q_e = \pi \left[\frac{r_v^2 h_{fg} 2\pi \rho_v \sigma}{r_{eff}} \right]^{0.5} \tag{9}$$

In the present work, annular and composite wicks are considered for the sodium heat pipe and hence the above limits are evaluated to find the operating region of the sodium heat pipe for annular and composite wick structures.

The performance limits for annular and composite wick heat pipe are plotted as shown in Figs. 1 and 2, respectively. The effective working region of the sodium heat pipe is thus identified from Figs. 1 and 2 for annular and composite wick heat pipes respectively.

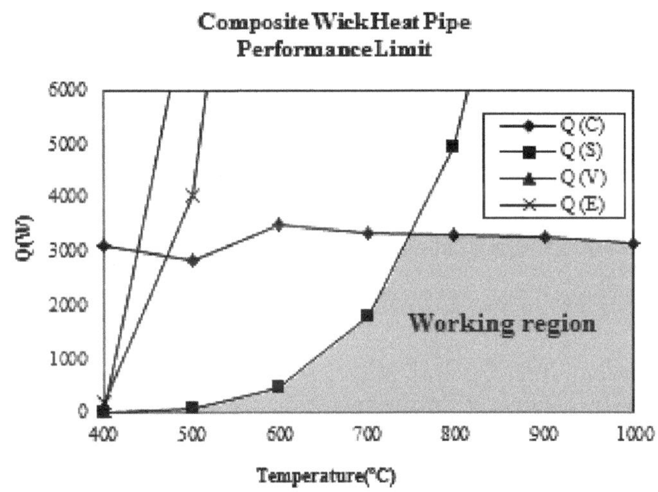

Fig. 2 Performance limits of composite wick heat pipe

5. FABRICATION OF STAINLESS STEEL SODIUM HEAT PIPE

The stainless steel material of SS-316 grade of 10 mm OD and 1 m length is used as the container of the present sodium heat pipe. Two wraps of 60 mesh and one wrap of 150 mesh are used to make composite wick and one wrap of 150 mesh is used to make annular wick. Also four 0.3mm OD stainless steel rods are inserted to get a uniform annular space as shown in Fig. 3.

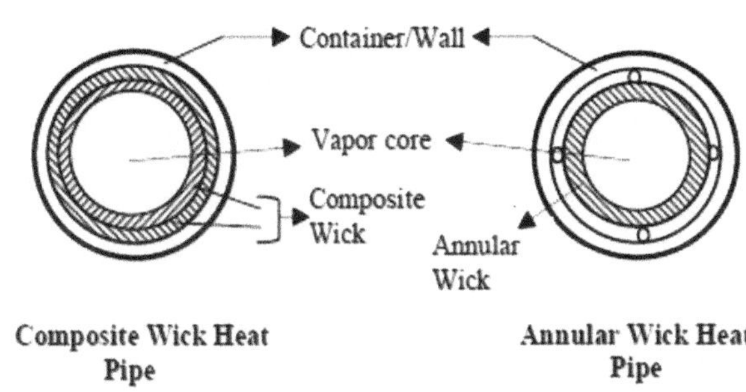

Fig. 3 Cross Sectional view of annular and composite wick sodium heat pipe

The steps followed in the fabrication of the sodium heat pipe are given in Fig. 4.The stainless steel mesh with selected mesh size is rolled to the required dimensions. To get good strength and retain its circular shape, the leading edge of the rolled mesh is spot welded. This forms the wick. The chemical cleaning process both prickling and passivation are carried out in order to ensure cleanliness of the material. Then the wick is inserted into the container. The valve and end cap is welded on each of the container. The heat pipe is evacuated to 10^{-6} mbar. After maintaining the vacuum level of 10^{-6} mbar in the pipe, the empty weight of the pipe with valve is measured. Pipe is fitted to canister with required amount of sodium in it. The canister, heat pipe and the pipe line is heated to about 250ºC and is maintained at that temperature for 15 minutes.

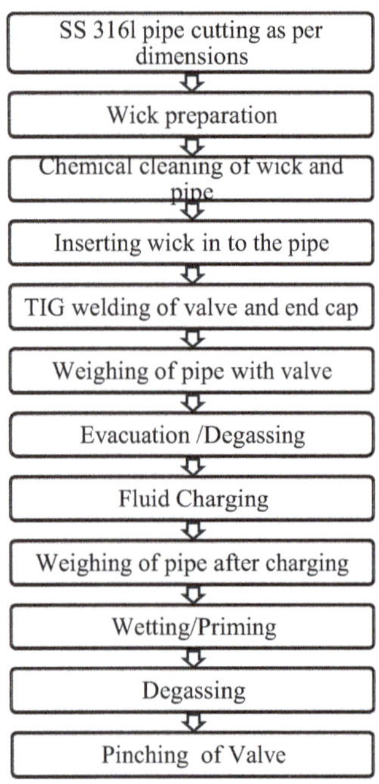

Fig. 4 Fabrication process Flow chart

The sodium is now in liquid form and hence can easily be charged. The canister valve and the heat pipe valve are opened and the valve connected to vacuum line is closed. The sodium is pressurized slightly by argon from the top of the canister for easy flow of the liquid sodium in to the heat pipe. Heaters are switched off after charging and the heat pipe is allowed to cool down gradually. The heat pipe valve and the canister valve are closed and heat pipe is removed from the canister. Heat pipe is weighed to know the exact amount of sodium charged.

Then the entire heat pipe is heated up to 600ºC using ceramic heater. The heat pipe is kept vertical and this process is continued for 48 hours. Then the heater is switched off and the heat pipe is allowed to cool down gradually. Again the heat pipe is evacuated to 10^{-6} mbar to remove any non-condensable gases like argon which was present during charging.

6. EXPERIMENTAL PROCEDURE

In the experimentation, stainless steel (SS-316) tube of 1 m length with 8 mm inside diameter and 1mm thickness were employed, with sodium as the working fluid. The working part of the heat pipe consists of evaporator section of 300 mm length, adiabatic section of 400 mm length and condenser section of 300 mm length. Eight K type thermocouples are spot welded to the required positions of the heat pipe as shown in Fig. 5 to measure the temperature distribution.

The adiabatic section of the heat pipe is surrounded by 50 mm thick glass wool as thermal insulation. A personal computer and data logger were used to record the measured temperature readings. As a source of heat, a 1.5 kW ceramic heater of 300 mm length is used to give heat input to the evaporator section of the heat pipe. The accuracy of controlling the heat input was ±5%. Heat input was measured using digital multi-meter. The condenser section of the heat pipe was left open to the atmosphere. The Experimental setup of the heat pipe is shown in Figs. 6 and 7.

The steps followed in the experimentation are as follows:

1. The heat pipe along with thermocouple setup is positioned inside the heater.
2. The thermocouples are connected to the data logger.
3. The adiabatic region is insulated using glass wool and the condenser region is left free to the atmosphere.
4. The heat input to the evaporator is controlled using dimmer stat.
5. The heat input is supplied to the evaporator section of the heat pipe.
6. The different temperatures are measured using data logger with a time interval of ten minutes.
7. The transient experiments are repeated with these steps for different fill ratio, orientation, heat input and the wick structure.

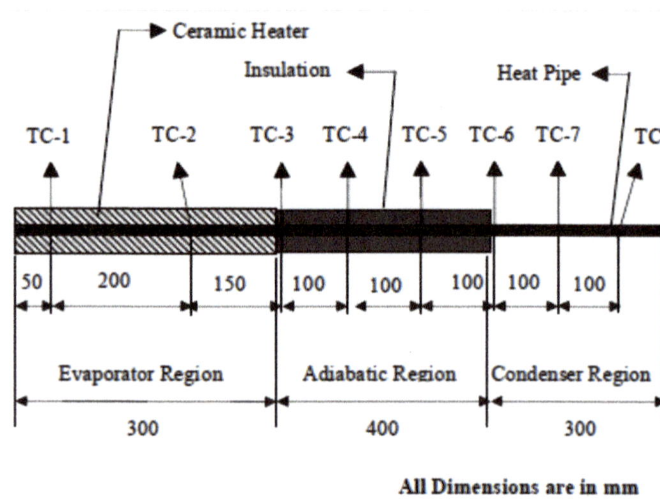

Fig. 5 TC positioning on the heat pipe

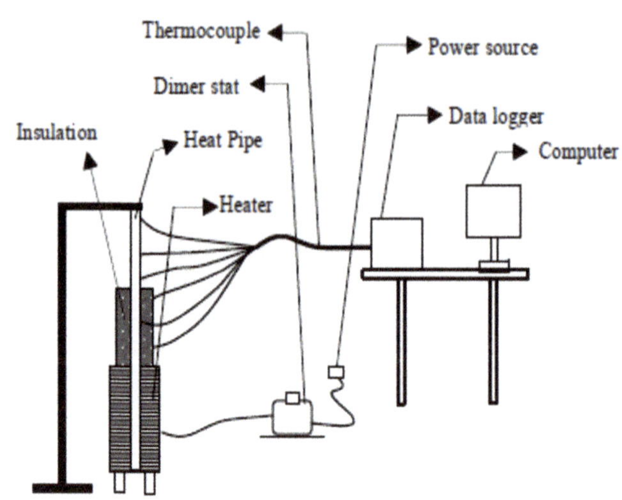

Fig. 6 Schematic of experimental set up

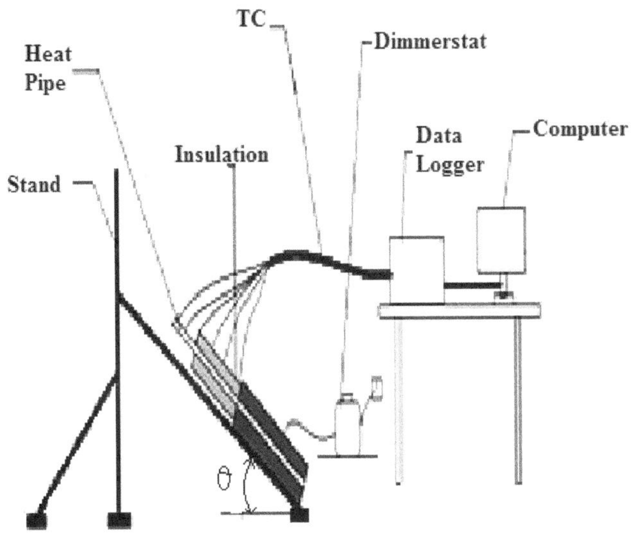

Fig. 7 Schematic of experimental set up for different orientations of heat pipe

7. RESULTS AND DISCUSSION

7.1 Effect of Heat Input

7.1.1 Annular Wick Heat Pipe

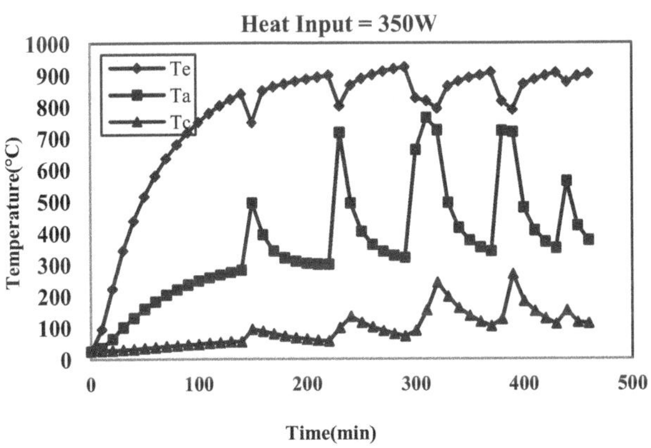

Fig. 8 Temperature Variation in annular Wick Heat Pipe at various positions at 350 W

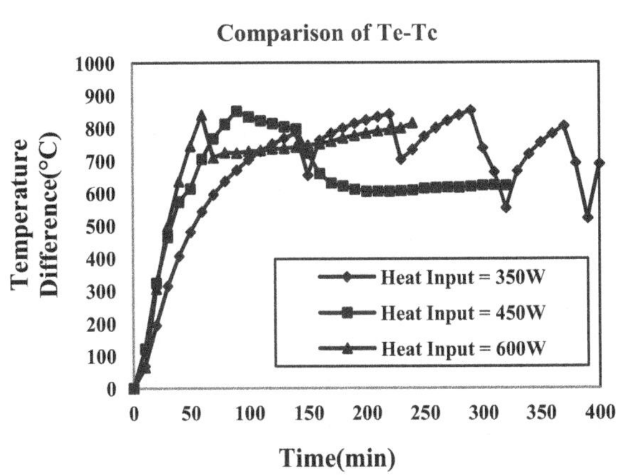

Fig. 9 Temperature difference between Evaporator and Condenser in annular Wick Heat Pipe at different heat inputs

Fig. 8 shows the variation of temperature at different zones of annular wick heat pipe with respect to time at 350 W heat input. One can observe a temperature difference of 500°C between the condenser and the evaporator region at 350 W heat input. This large difference in temperature between the three zones is due to the fact that the heat input given to the evaporator region is insufficient. So the time taken by the condensed fluid to evaporate in the evaporator region and to create required pressure difference between evaporator region and condenser region is more. So one can observe continues unsustained shoot-up in the heat pipe temperature. Also there is large temperature difference of 500°C and 700°C between the condenser and the evaporator section of heat pipe at a heat input of 450W and 600W as shown in Fig. 9. This large difference in temperature between the three zones is due to the fact that the fluid is not flowing back to the evaporator region which indicates that there is an insufficient capillary pumping.

7.1.2 Composite Wick Heat Pipe

Figure 10 shows the temperature variations with respect to time for composite wick heat pipe at 350 W heat input. From Fig. 10, it can be seen that there is a large temperature difference of 750°C between the condenser and the evaporator region. This is because the heat input given to the evaporator region is insufficient and this takes more time to evaporate the condensed fluid in the evaporator region. Figure 11 shows the difference between evaporator and condenser temperature variations with respect to time for composite wick heat pipe at different heat input. The temperature difference at 600 W heat input is low when compared to other heat inputs considered which indicates that there is a smooth flow of the working fluid and proper heat transfer in the heat pipe.

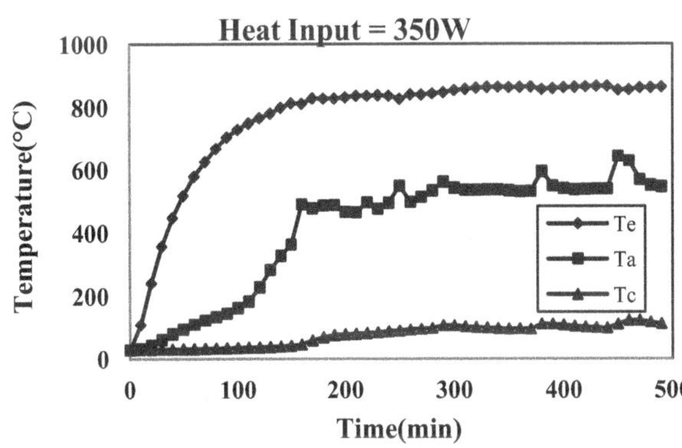

Fig. 10 Temperature Variation of composite Wick Heat Pipe at various positions at 350W Heat Input

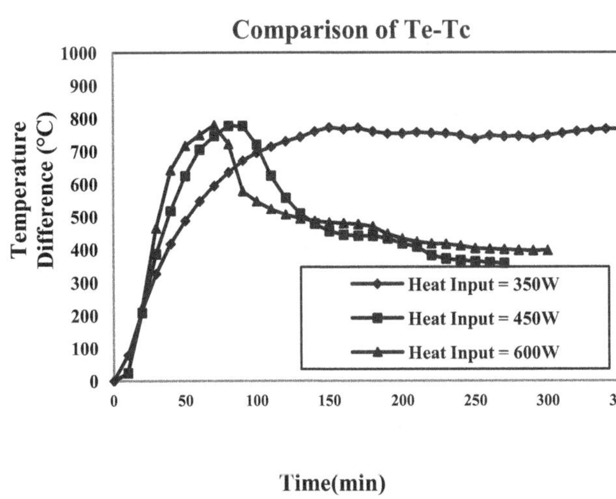

Fig. 11 Temperature difference between Evaporator and Condenser of composite wick heat pipe at different heat inputs

7.1.3 Effect of Wick Structure

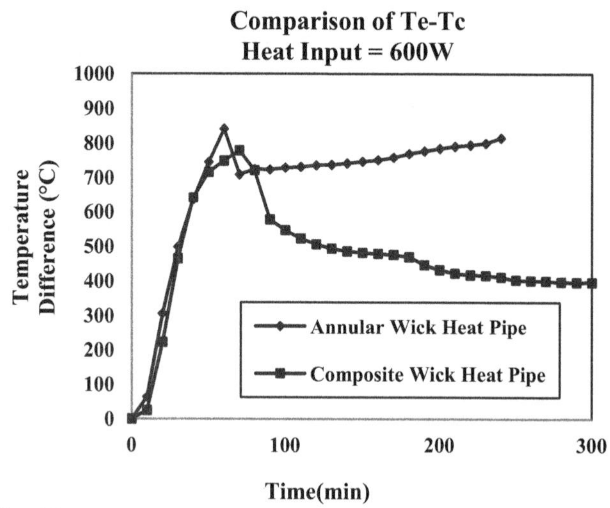

Fig. 12 Temperature difference between Evaporator and Condenser at 600W Heat Input

Figure 12 illustrates the temperature difference between the evaporator and the condenser region with respect to time at a heat input of 600 W for annular and composite wick heat pipes. There is a steady raise in the temperature difference curve profile up to a time period of 75 minutes. This indicates the change in flow regime from molecular free flow regime to continuum flow regime. Also this indicates that the working fluid has started to melt from its frozen state and starts to evaporate in liquid-vapour interface. By this, the vapour density increases. After a time period of 75minutes, it is seen that the temperature difference curve starts to come down gradually. At the same time, the flow regime in the vapour core space gradually changes from molecular free flow regime to continuum flow regime. In this region, it is observed that the temperature difference between the evaporator and the condenser is less in case of composite wick heat pipe compared to the annular wick heat pipe.

7.1.4 Effect of Wick Structure on Thermal Resistance

The thermal resistance of the heat pipe is given by

$$R_{th} = \frac{T_e - T_c}{Q} \quad \text{K/W} \qquad (10)$$

If the temperature difference between the condenser and the evaporator is less, then the thermal resistance is also less and hence better will be the heat transfer performance of the heat pipe

.

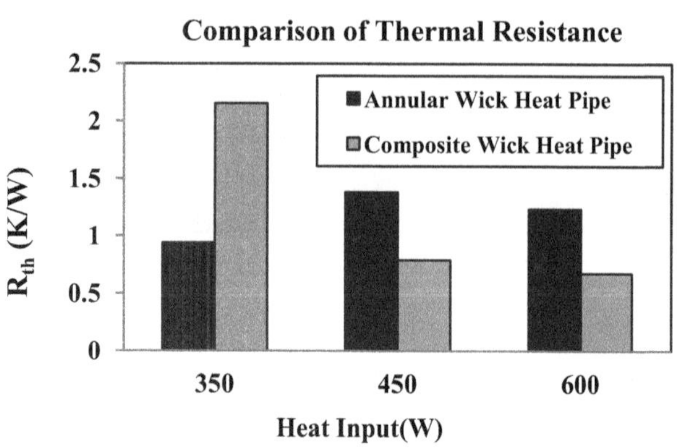

Fig. 13 Thermal Resistance of annular and composite wick Heat Pipe at different heat input

Fig. 13 shows the variation of thermal resistance at different heat inputs for annular and composite wick heat pipe at steady state. From Fig. 13 it is clear that the thermal resistance of the composite wick heat pipe is least at all heat inputs considered when compared to the annular wick heat pipe except for 350 W.

7.1.6 Frozen Startup Time for different Wick Structures

The frozen startup time indicates the time in which the required pressure difference inside the heat pipe is achieved because of which the sodium vapour starts to move towards the condenser region and carrying latent heat of evaporation along with it. Fig. 14 shows the time taken for frozen startup for annular and composite wick heat pipes at different heat inputs. One can see that the startup time required by composite wick heat pipe is less when compared to annular wick heat pipe. This indicates that composite wick sodium heat pipe could attain mere isothermal condition in shorter time period.

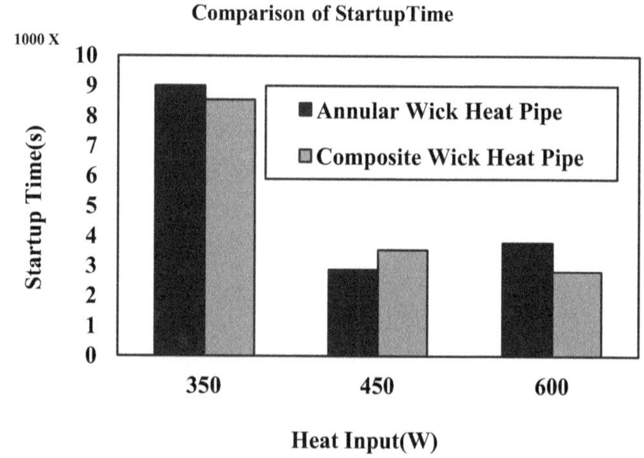

Fig. 14 Frozen Startup Time of different Heat Pipe

7.2Effect of Fill Ratio

7.2.1 Temperature Distribution

Fig. 15 shows the temperature variations of composite wick heat pipe at various locations with respect to time for 40% fill ratio. The heat input of 600 W is provided at the evaporator end. From Fig. 15 it is observed that the wall temperature at the evaporator section of the heat pipe rises at much faster rate when compared to the adiabatic section. This indicates that the sodium is melting from frozen state and starts to evaporate from liquid- vapour interface. This leads to increase of vapour density in the evaporator section. There is a sudden shoot up in the adiabatic temperature and this shoot up is due to increase in the vapour pressure of sodium from 0.5 mbar at 400°C to 100 mbar at 700°C. Also, the second shoot up in temperature can be seen in the condenser section. The sudden raise in temperature indicates the initiation of frozen startup process in the heat pipe.

In the present work, experiments are conducted on sodium heat pipe at fill ratios of 30%, 36% and 40%. At the steady state condition, it is observed that the temperature difference between the evaporator region and condenser region of the composite wick heat pipe is around 130°C for 40% fill ratio as shown in Fig. 16. For 36% and 30% fill ratio, the temperature difference of 200°C and 500°C between the evaporator and condenser was observed. Hence it is found that the thermal effectiveness of the heat pipe is high at 40% fill ratio. The presence of sufficient amount of Sodium will transfer the heat from the evaporator to the condenser section at faster rate. Also the frozen startup of the heat pipe at 40% fill ratio is better when compared with other fill ratios.

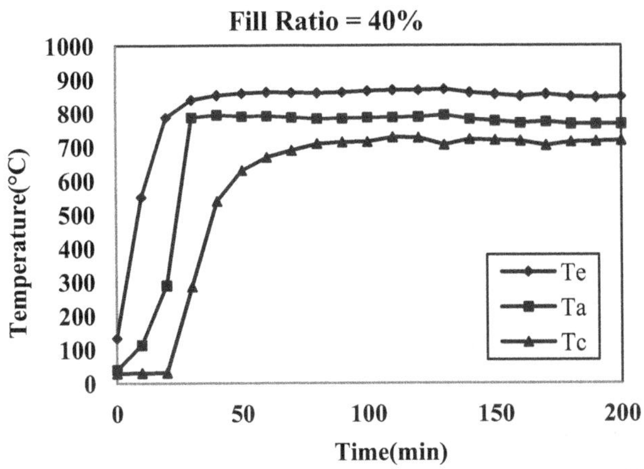

Fig. 15 Temperature variation of Heat Pipe at various positions for 40% Fill Ratio

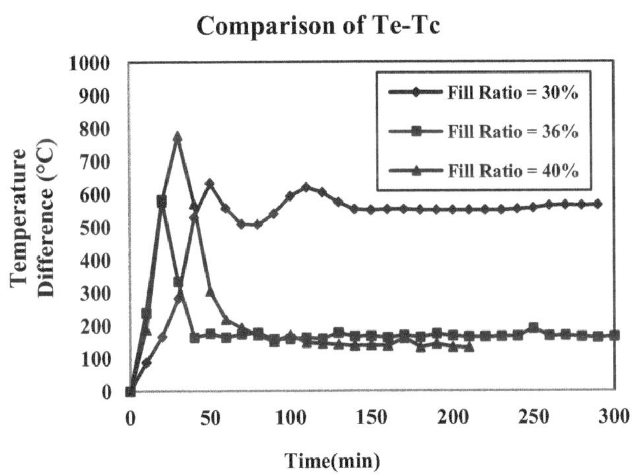

Fig. 16 Temperature difference between Evaporator and Condenser at different Fill Ratio

7.2.2 Effect of Fill Ratio on Thermal Resistance

Fig. 17 shows the thermal resistance of the composite wick sodium heat pipe at different fill ratios. From the figure it is clear that at 40% fill ratio, heat pipe has the least thermal resistance of 0.1303 K/W when compared to the other fill ratios. As the temperature difference between the evaporator and condenser section is less at 40% fill ratio due to the smooth flow of the working fluid, the thermal resistance at 40% fill ratio is lower compared to the other fill ratios considered.

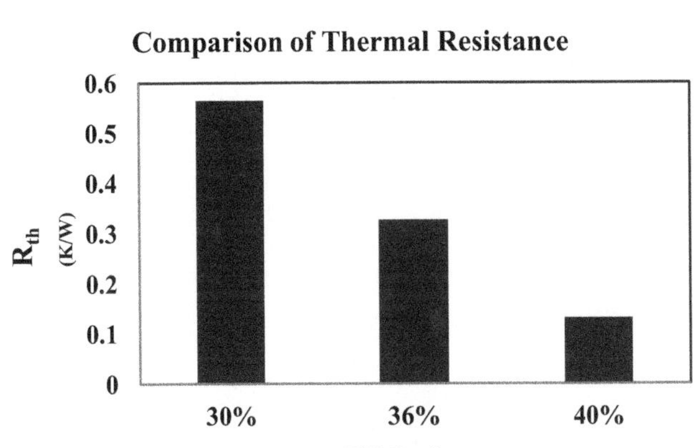

Fig. 17 Thermal Resistance of Heat Pipe at different Fill Ratio

7.2.4 Frozen Startup Time at Different Fill Ratio

Fig. 18 shows the time taken for frozen start up for composite wick heat pipe at different fill ratios. One can see that the startup time required by composite wick heat pipe at 40% fill ratio is less when compared to other fill ratios considered in the study. This is because of the reason that the required pressure difference between evaporator section and condenser section is created at much faster rate by which the vapour shoot-up is seen much early at 40% fill ratio compared to other fill ratios.

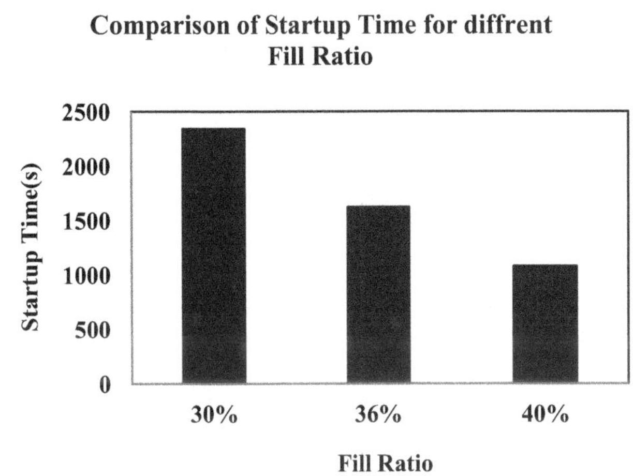

Fig. 18 Frozen Startup Time at different fill ratio

7.3 Effect of Orientation

7.3.1 Temperature Distribution

Fig. 19 illustrates the temperature variations of evaporator, adiabatic and condenser region with respect to time at an orientation of the composite wick heat pipe equal to 30^0. Heat input of 600W is provided at the evaporator end. At 40 minutes from the starting of experimentation, there is a sudden raise in the adiabatic temperature. The temperature shoots up from 143°C to 711°C in the next 20 minutes. This shoot up is due to increase in vapour pressure of sodium from 0.5 mbar at 400°C to 100 mbar at 700°C. Also the second shoot up in temperature can be seen at a time period of 60 minutes from the starting of experimentation in the condenser section from 102°C to 606°C in 30 minutes duration.

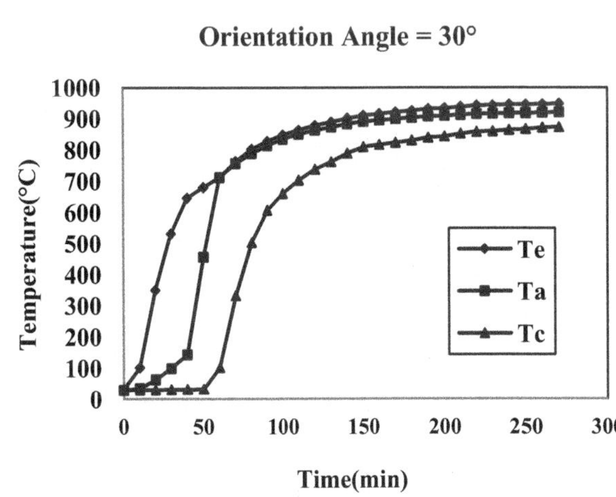

Fig. 19 Temperature Variation of Heat Pipe at various positions at 30° Orientation

The sudden raise in temperature indicates the initiation of frozen startup process of the heat pipe. It also indicates the end of free molecular flow regime and the continuum flow regime being established in the vapour space. During the shoot up, the evaporator temperature steadily drops while the adiabatic and condenser temperature steadily rises. After 265 minutes time period, a steady state condition is achieved and the evaporator temperature reaches 946ºC while the adiabatic and the condenser temperature reaches 919ºC and 872ºC resulting in 74ºC temperature gradient between the evaporator and condenser section.

6.3.2 Temperature Difference between Evaporator and Condenser at different Orientations

Figure 20 illustrates the temperature difference between evaporator and condenser region of composite wick sodium heat pipe for different orientations. From Fig. 20 it is clear that there is a steady raise in the temperature difference curve up to a time period of 50 minutes. This indicates that the working fluid has started to melt from its frozen state and starts to evaporate in liquid-vapour interface. By this the vapour density increases. The peak point in the curve indicates the startup of the heat pipe and this marks the change in flow regime from molecular free flow regime to continuum flow regime.

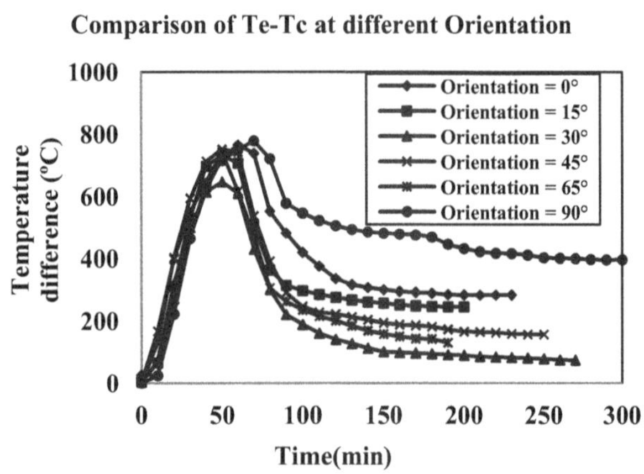

Fig. 20 Temperature difference between Evaporator and Condenser at different Orientation

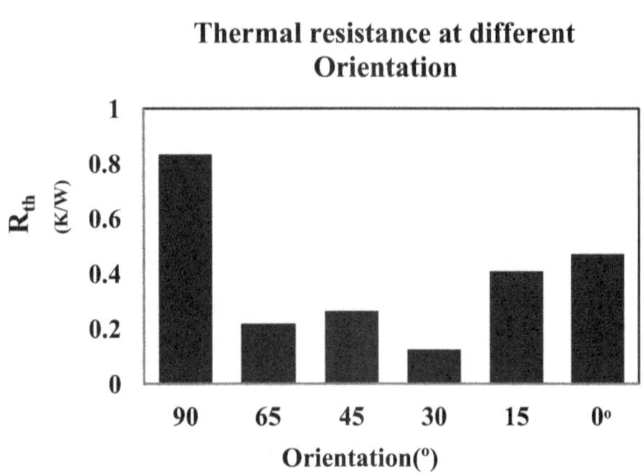

Fig. 21. Thermal Resistance of the Heat Pipe at different Orientation

After this, the temperature difference curve starts to come down gradually. At the same time, the flow regime in the vapour core space gradually changes from molecular free flow regime to continuum flow regime. After the steady state is reached, the temperature difference been the evaporator and the condenser section at 30º orientation was found to be 70ºC which is the least when compared to other orientations considered in the study. The heat transfer performance of the present composite wick sodium heat pipe is thus found to be better at 30^{0}orientation.

7.3.3 Effect of Orientation on Thermal Resistance

Figure 21 shows the comparison of the thermal resistance of composite wick sodium heat pipe at different orientations. If the temperature difference between the condenser and the evaporator is less, then the thermal resistance is also less and hence better will be the heat transfer rate. When compared to other orientations of sodium heat pipe, 30^{0}orientation has shown least thermal resistance of 0.1233 K/W for a heat input of 600 W.

7.3.5 Frozen Startup Time at Different Orientations

The frozen startup time signifies that the required pressure difference excites inside the heat pipe because of which the sodium vapour starts to move up towards the condenser region and carrying latent heat of evaporation. This will be dissipated at the condenser region. Fig. 22 shows the startup time of the composite wick sodium heat pipe at different orientations. When compared to other orientations the sodium heat pipe has shown better results at 30° orientation with 36 minutes of startup time.

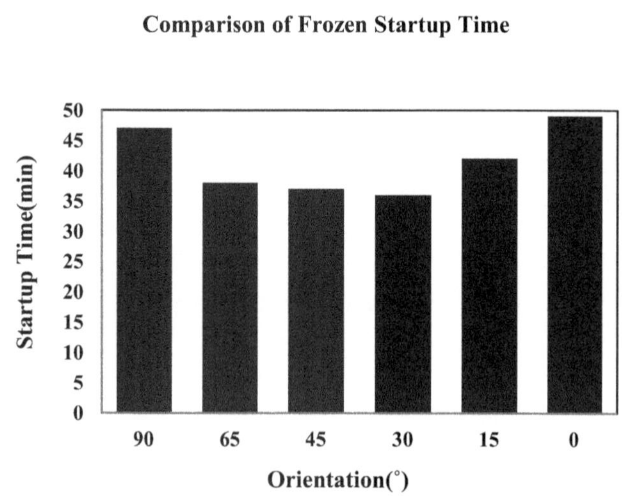

Fig. 22 Frozen Startup Time at different orientations

8. CONCLUSIONS

The following conclusions can be drawn from the present studies on sodium heat pipe:

1. In annular wick heat pipe and composite wick heat pipe at 350W heat input, the presence of un sustained shoot-up confirms that the heat input is insufficient.
2. The composite wick sodium heat pipe is found to be better than the annular wick heat pipes in terms of its lower thermal resistance and higher heat transfer co-efficient.
3. The composite wick heat pipe has good capillary action and permeability resulting in better fluid transfer in the heat pipe which has resulted in achieving near isothermal conditions.
4. Composite wick Heat pipe with 40% fill ratio shows lower temperature difference between Evaporator and Condenser and 66.86% of heat input is being transferred.
5. From the studies, it can be concluded that the composite wick Heat pipe with 40% fill ratio has performed better compared to 30% and 36% fill ratios.
6. It is found that the composite wick sodium heat pipe has performed better at 30º orientation with 70ºC temperature difference between the Evaporator and the Condenser sections.
7. The Frozen startup time of the heat pipe is found to be less in case of composite wick, at a heat input of 600 W, 40% fill ratio and at an orientation of 30°.

ACKNOWLEDGEMENT

This work is carried out at M/s Avasarala Technologies Ltd, Bangalore. The authors would like to acknowledge Mr. T. T Mani. Managing Director &CEO,Avasarala Technologies Ltd., Mr. H C Jagadeesha, Unit Head, Strategic product division & Mr. K S Shashidhar, Senior Engineer for their constant support. The authors also would like to acknowledge the encouragement and support of Dr. Chenraj Jain, Chairman, Jain University and Dr. K. Venkatesh, Director, Centre for Emerging Technologies, Jain University.

NOMENCLATURE

A	Area (m^2)
K	Wick permeability (m^2)
L	Length (m)
P	Pressure (Pa)
Q_c	Capillary limit (W)
Q_s	Sonic limit (W)
Q_v	Viscous limit (W)
Q_E	Entrainment limit (W)
T	Temperature (˚C)
r	radius (m)
h_{lv}	Latent heat (J/Kg)

Greek Symbols

φ	Porosity
σ	Surface tension (N/m)
ρ_l	Liquid Density (Kg/m^3)
μ_l	Liquid Viscosity ($N\text{-}s/m^2$)
ρ	Density (kg/m^3)
Υ	specific heat ratio for sodium =1.67
M	viscosity ($N\text{-}s/m^2$)
ε	Emissivity

Subscripts

a	adiabatic
c	condenser
e	evaporator
l	liquid
i	inner
v	vapour
w	wick
pc	Pipe cross section
sc	condenser surface area
eff	effective

REFERENCES

Adkins, D. R., Artdraka, C.E., Moreno, J. B., Rawlinson, K. S., Showalter, S. K., and Moss, T. A., 1999, "Heat Pipe Solar Receiver Development Activities at Sandia National Laboratories," *Renewable and Advanced Energy Conference for the 21 Century Conference April 1999, Maui, HA*

Deverall, J. E, Keddy E. S. Kemme J. E, and Phillips J. R, 1975, "Gravity –assisted heat pipes for Thermal control systems"*LA-5989-MS Informal Report,* Los Alamos scientific laboratory of University of California Los Alamos, New Mexico

El-Genk, M.S., Tournier, J.M.P., 2011, "Uses of Liquid-Metal and Water Heat Pipes in Space Reactor Power Systems," *Frontiers in Heat Pipes*, **2**(1), 013001

Faghri, A., 1990, "Analysis of Transient and Start-Up Behavior of Heat Pipes and an Energy Storage Module," *Interim Report for Period January 1989 - December 1989,* Wright State University, Dayton, OH.

Faghri, A., 1995, *Heat Pipe Science and Technology*, Taylor & Francis, Washington.

Faghri. A., and Cao, Y., 1991, "A Study of High Temperature Heat Pipe with Multiple Heat Sources and Sinks," *ASME Journal of Heat Transfer,* **113**(4), 1003-1009

Ivanovskii, M. N., Prosvetov, V.V., Sorokin, V.P., Chulkov, B.A., and Yagodkin, I. V, 1978 "Investigated the Evaporation Process and Sonic Limits in Sodium Heat pipes,"*Translated from Inzhenerno-Fizicheskii Zhurnal,* **33**(5), 832-837

Ivanovskii M.N, Sorokin V.P, and Chulkov B.A, 1980, "Problems in Building Vertical High-Temperature Heat Pipes," *Translated from Inzhenerno-Fizieheskii Zhurnal,* (38) 3,389-395

Lee, B.I., Lee, S.H., 2001, "Manufacturing and temperature measurements of sodium heat pipe," *KSME International Journal,* **115** (11), 1533- 1540

Martin, J.J., and Reid, R.S., 2004, "Sodium Based Heat Pipe Modules for Space Reactor Concepts: Stainless Steel SAFE100 Core" *Proceeding of ICAPP,*04 June 13-17, paper 4275, Pittsburgh, PA.

Mochzuki, M., Nguyen, T., Sugihara, S., and Wuttijiumnong, V., 2011, "Nuclear reactor must need heat pipe for cooling" *10th IHPS, Taipei,* Taiwan, Nov 6-9, 2011

Mireles, O.R., and Houts , M.G., 2011,"Transient Response to Rapid Cooling of a Stainless Steel Sodium Heat Pipe," *Nuclear systems,* NASA Marshall Space Flight Center, Huntsville, AL.

Mullet, S., 2008, "A steady state model for the heat pipe-encapsulated nuclear heat source" *Department of Nuclear Engineering University of California, Berkeley, Joint International* January 6-8, 2008, Berkeley Faculty Club, UC Berkeley Campus

Reid, R.S., Sena, J,T., Martinez, A.L., 2000, "Sodium Heat Pipe Module Test for the SAFE-30 Reactor Prototype," submitted to *Institute for Space and Nuclear Power Studies STAIF-2001 Forum* prepared by Los Alamos National Laboratory Los Alamos, New Mexico, USA

Rosenfeld, J.H., Ernst, D.M., Lindemuth, J.E., and Sanzi, J.L., 2004, "An Overview of Long Duration sodium heat pipe Tests," *Prepared for the Space Technology and Applications International Forum (STAIF–2004)* sponsored by the American Institute of Physics Albuquerque, New Mexico, February 8–12

Sabharwall, P., Patterson, M., and Gunnerson, F., 2008, "Theoretical Design of Thermosyphon for Process Heat Transfer from NGNP to Hydrogen Plant," *Proceedings of HTR-2008ASME 4th International Topical Meeting on High Temperature Reactor Technology*, Washington, D.C., USA

Salvail, P.G., and Carter, R.R., 1999, "Alkali Metal Handling Practices," NASA MSFC *NASA Marshall Space Flight Center, Huntsville*

Yan X.K., Duan Y.N., 2011, "Construction of Sodium Heat-Pipe Furnaces and the Isothermal Characteristics of the Furnaces," *International Journal of Thermophysics,* 32, 494-504

8

THERMAL PERFORMANCE OF CLOSED LOOP PULSATING HEAT PIPE USING PURE AND BINARY WORKING FLUIDS

Pramod R. Pachghare[a,*] Ashish M. Mahalle[b]

[a] *Assistant Professor, Department of Mechanical Engineering, Government College of Engineering, Amravati-444 604, India*
[b] *Associate Professor, Department of General Engineering, Laxinarayan Institute of Technology, Nagpur-444 001, India*

ABSTRACT

This paper presents preliminary experimental results on thermal performance of closed loop pulsating heat pipe (CLPHP). The copper capillary tube was used having internal and external diameter 2.0 mm and 3.6 mm respectively. For all experimentation, filling ratio (FR) was 50 %, number of turns was 10 and different heat inputs of 10 to 100W were supplied to PHP. For all PHPs, Vertical bottom "heat mode" (+90^0) position is maintained. The equal lengths of evaporator, adiabatic and condenser sections were 50 mm each. Working fluids are selected as Methanol, ethanol, acetone, water and different binary mixtures. The graphs are plotted, in order to study, characteristics of the thermal resistance and average evaporator temperatures at different heat input for various working fluids. Experimental study on PHP indicated that working fluid is an important factor for the performance of PHPs. The result shows that, the thermal resistance decreases more rapidly with the increase of the heating power from 20 to 60 W, whereas slowly decreases at input power above 60 W. Pure acetone gives best thermal performance in comparison with the other working fluids. No measurable difference has been recorded between the PHPs running with pure and binary mixture working fluids.

Keywords: *Pulsating heat pipe, binary mixture, heat flux, natural convection*

1. INTRODUCTION

Pulsating or Looped type Heat Pipes proposed and patented by Akachi (1990). This is the new member of wickless heat pipes. Their operation is based on the principle of oscillation for the working fluid and a phase change phenomena in a capillary tube. The diameter of the tube must be small enough such that liquid and vapor plugs exist. Due to its excellent features, such as high thermal performance, rapid response to high heat load, simple design and low cost, PHP has been considered as one of the promising technologies for electronic cooling, heat exchanger, cell cryopreservation, the spacecraft thermal control system, etc, Dadong and Cui (2010).

Various mathematical models have also been developed in recent years to predict the oscillating motion and heat transfer performance of the PHPs. Qu and Ma (2007) presented a mathematical model to describe the startup of a PHP. They found that the inner wall surface condition, evaporation in the hot section, superheat, bubble growth, and the amount of vapor bubble trapped in cavities affected the startup of a PHP. Shafii *et al.* (2001; 2002) concluded that the majority of the heat transfer (95%) is due to sensible heat, not due to the latent heat of vaporization. Latent heat serves only to drive the oscillating flow. They also demonstrated that the gravity force has an insignificant effect on the PHP performance.

Although extensive studies have been carried out, certain key aspects of the PHP remain poorly understood, and some analytical results obtained by different investigators are even contradictory to the experimental data. A number of researchers have conducted experimental investigations on PHPs, and the results indicated that the heat transfer capability of PHPs mainly depends on the working fluids, evaporation/condensation lengths, inner diameters, number of turns, etc. Khandekar *et al.* (2003). Charoensawan *et al.* (2003) indicated that in vertical orientation for the 2.0 mm devices, water filled devices showed higher performance as compared to R-123 and ethanol. Whereas in 1.0 mm devices, R-123 and ethanol filled devices showed comparable performance but water showing very poor results. Khandekar *et al.* (2003) demonstrated the effect of input heat flux of the working fluid on the thermal performance of the device. Although the Eötvös number of water and ethanol was much below the prescribed maximum limit of Eö = 4, gravity forces were definitely seen to affect the performance. There is a smooth decrease of the thermal resistance with increasing heat power input.

Kammuang-lue *et al.* (2008) studied that, the higher latent heat of the working fluid, the higher critical heat flux. Yang *et al.* (2009) pointed out that increasing heat load clearly improves the thermal performance. Meena *et al.* (2009) concluded that as working fluids change from R123 to Ethanol and water the critical heat flux decreased. The latent heat of vaporization affects the critical heat flux. The working fluid with the lower latent heat of vaporization exhibits a higher critical heat flux. Dadong and Cui (2010) indicated that the thermal resistance decreases with the increase of the heating power at the same filling ratio. The thermal resistance decreases more slowly for the power inputs larger than 60W. For the pure working fluid PHPs, the thermal resistance is decreases in the sequence of water, ethanol, methanol and acetone. Mameli *et al.* (2011) conducted experiments on the thermal performance of a PHP working with an azeotropic mixture of water (4.5% wt.) and ethanol (95.5% wt.), in comparison to pure ethanol. No measurable difference has been recorded between the PHP

* Corresponding Author
Email address: pramod.pachghare@gmail.com

running with the azeotropic mixture and the PHP running with pure ethanol, in terms of overall thermal resistance.

At different situations, different pure working fluids have their advantages Pachghare *et al.* (2012). But till now, mixtures used as working fluids in PHP have not been thoroughly investigated. The non-azeotropic mixtures, which have the characteristics of phase transition with temperature floating, can make heat source and working fluids match well in temperature CUI Xiaoyu *et al.* (2006). Binary fluid is quite a new developing topic in recent years. Therefore, the working fluids are used as pure and binary mixture (1:1 by volume) for the PHPs. Experimental investigation has been done to find the thermal performance of CLPHP by using different pure and binary mixture of working fluid. The various temperatures are recorded for analyzing the thermal resistance of the PHP.

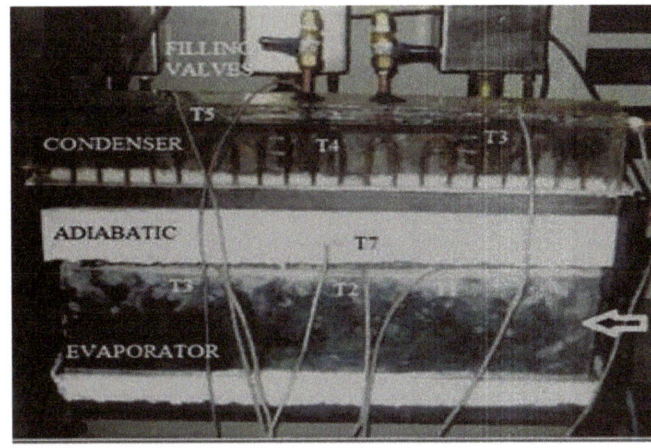

Fig. 1. Experimental setup photograph

2. EXPERIMENTATION

2.1. Experimental set-up

The experimental setup of CLPHP was developed and tested in the laboratory. The photograph of PHP set-up is shown in Fig. 1. All the experimentations are carried out at controlled conditions. The details of the experimental setup are shown in Fig. 2. The setup consists of a closed loop PHP, temperature recorder, power supply unit, and water tank cooling system for condenser. Both the evaporation and adiabatic sections were well thermally insulated by the proper insulation materials. The PHP, consisting of 10 turns, is made of copper capillary tube having inner diameter is 2.0 mm; the outer diameter is 3.6 mm. The pitch distance between tubes was maintained 15 mm. The PHP consists of evaporation, adiabatic and condensation sections with the height of 50 mm for each section. The heating power is provided by a carefully designed power supply unit. Heating was done by oil bath and cooling by water tank. The power meter measures the AC voltage, the current and the corresponding power simultaneously. The Filling Ratio was maintained at 50%. The heating configuration was bottom heat orientation (+90^0). Ten K-type thermocouples were attached to the wall of the PHP, positions are shown in Fig. (2). Flow meter was also recorded the mass flow rate of the cooling water.

About the tube design, the most important condition of PHP is the creation of the liquid slug. According to the literature review, Akachi [15] proposed that the appearance and movement of bubbles are affected by surface tension and buoyancy in the channel. Relation of surface tension and buoyancy could be explained by the dimensionless formula shown in Eq. (1):

$$E\ddot{o} = \frac{g.D^2.\ \rho_{liq} - \rho_{vap}}{\sigma} \tag{1}$$

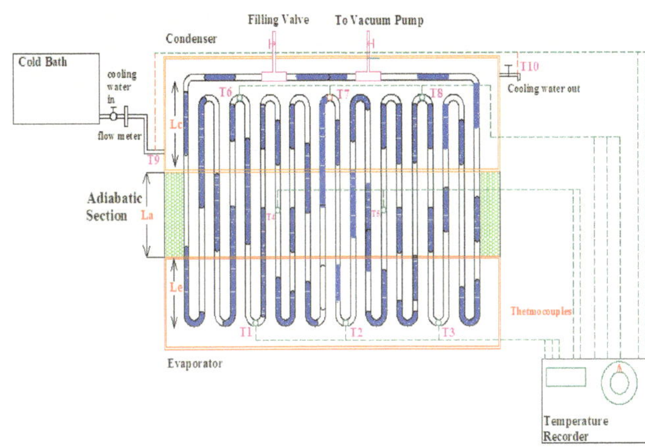

Fig. 2. Experimental setup Details.

When Eö ≈ 4, the bubble will get seized on both side of the wall. At this condition, the terminal velocity becomes zero and the liquid slug flow is formed. The dimension formula for the PHP diameter 'D', is given in Eq. (2).

$$D \leq 2.\sqrt{\frac{\sigma}{g.\ \rho_{liq} - \rho_{vap}}} \tag{2}$$

PHP loop is divided into evaporator, adiabatic and condensation composing of three sections. The intact structure of PHP loop and the experimental apparatus structure are in Fig. 2.

Table 1: The thermo-physical properties of pure and binary# working fluid at 1 atm.

Working Fluid	T_b (^0C)	h_{fg} (kJ/kg)	ρ (kg/m^3)		μ (10^{-7} Ns/m^2)		k (W/m-K)		σ (10^{-3} N/m)	C_p (kJ/kg-K)	
			liq	*vap*	*liq*	*vap*	*liq*	*vap*		*liq*	*vap*
Methanol	64.7	1119.59	750.8	0.566	3291.4	109.64	0.201	0.0018	18.87	2.52	1.601
Ethanol	78.3	962.45	758.1	1.372	4452.6	102.39	0.169	0.0197	17.46	0.73	1.604
Acetone	56.2	520.56	748.5	2.123	2340.6	89.25	0.169	0.0140	19.09	2.28	1.385
Water	100.0	2251.20	958.7	0.597	2790.0	121.00	0.680	0.0248	58.91	4.22	2.034
Water-Acetone	78.1	1385.88	853.6	1.360	2565.3	105.13	0.425	0.0194	39.00	3.25	1.709
Water-Ethanol	89.2	1606.83	858.4	0.985	3621.3	111.70	0.425	0.0223	38.19	2.47	1.819
Water-Methanol	82.4	1685.40	854.8	0.582	3040.7	115.32	0.441	0.0133	38.89	3.37	1.817

Calculated by: Algebraic mean value method.

2.2. Experimental procedure

Put K-type thermocouples fixed in suitable position, i.e. three thermocouples in the evaporator and three thermocouples in the condenser, two thermocouple in the adiabatic section respectively, which are connected with the temperature recorder from T_1 to T_8 as shown in Fig. 2. The heat is supplied by means of oil bath by using dielectric oil for uniform heating. Two heater of 500W each was connected in series to the power supply. It was cover with foam insulation spray type in the adiabatic section, which can prevent heat from being lost and influence in the experimental result. The vacuum is created by means of vacuum pump up to 70cm of Hg. Then, PHP is filled with the necessary working fluid. Turn on the temperature recorder and power supply, adjust the heating power initially to 10 W, and then begin to observe and note down the temperature at each point through the recorder. After the temperature curve is balanced, adjust the heating power to 20 W (plus 10 W). Again, after the temperature curve is balanced and the heating power is adjusted to 30 W. Repeat this step until 100 W. Fill in with the different working fluids at 50% filled ratio. Steps 3–6 are repeated.

3. RESULT AND DISCUSSION

The PHP performance data (for pure and binary mixture of working fluid) were obtained according to the following procedure: Heat input was stepwise increased until a quasi thermal equilibrium was established. Then, the spatial temperatures and heat input were recorded, so the thermal resistances could be determined. The thermal resistance is defined by Eq. (3).

$$R_{th} = \frac{\overline{T}_e - \overline{T}_c}{\dot{Q}} \qquad (3)$$

where, $\overline{T}_e$ and $\overline{T}_c$ are the average evaporator and condenser surface temperatures, whch is the average of three temperatures, calculated by using Eq. (4).

$$\overline{T}_i = \frac{\overline{T}_{i,1} + \overline{T}_{i,2} + \overline{T}_{i,3}}{3} \qquad (4)$$

where, $i = e$ or c

$\dot{Q}$ is heat input power to the PHP. Considering thermal losses $\dot{Q}$ can be determined by Eq. (5) as:

$$\dot{Q} = P - \dot{Q}_{loss} \qquad (5)$$

where, P is the input electrical power (measured with accuracy of ±1.5%). $\dot{Q}_{loss}$ is the heat loss, which was verified to be about 4% to 9%, depending on the heat load. In order to keep a good operation of working fluid in PHP, vertical bottom heat mode condition, 50% filling ratio was chosen for comparison with different working fluid as in the following test.

3.1 Effect of pure and binary mixture working fluid PHPs on the thermal resistance

From Fig. 3 and 4, it is clear that, for pure and binary working fluids of PHPs, the thermal resistances are smoothly decreases with the increasing heat input power. With increasing heat input to the device, the evaporator temperature rises resulting in a greater density gradient in the tubes. Simultaneously the liquid viscosity also drops diminishing the wall friction. Initially at low heat input upto 20W, PHP is not sufficient to sustain a stable behavior. During the starting period fluid motion is chaotic, the circulation is not initiated. The respective adjacent tubes of the PHP become and remain alternatively hot and cold thereafter. The direction of circulation for the liquid is arbitrary; remaining fixed for a given experiment but may be changes with different experimental runs. The PHP is best suitable for all working fluid at the range of 30 to 80 W. The dry-out conditions of the water, ethanol, methanol and acetone PHPs are at 100W, 80W, 60W and 70W heat input respectively. The thermal resistances are according to the boiling temperature of the respective working fluids, shown in Table 1. The boiling point and the latent heat of vaporization of water are more as compared with other working fluid used for PHPs. Thus, water PHP can boil hardly in low power inputs.

The trends of thermal resistance over heat input for all binary mixture of PHPs are shown in Fig. 4. The PHP get dry-out for the water-methanol, water-acetone and water-ethanol is around at 85W, 80W and 90W heat input respectively, which is approximate the algebraic mean values of the boiling point of binary mixture. The temperatures are recorded for all binary mixture of working fluid PHPs. There is no quantifiable difference. It is also clear that, behavior of the thermal resistance is strongly depending on thermo-physical properties of the working fluid of PHPs. Although there are many properties of working fluid, the latent heat of vaporization is the main property that strongly affects the motion of liquid slug and vapor bubble in the tube and the heat transfer of the PHP. Therefore, the latent heat of vaporization of the working fluid will be concentrated.

In Fig. 4, water-methanol binary fluid PHP shows the better thermal performance as compared to other binary fluids up to 80W heat input. The variations of thermal resistance is according to latent heat of vaporization, see Table 1. Latent heat of vaporization for water-methanol binary fluid is high, but dynamic viscosity for water-ethanol binary fluid is more. Therefore the heat carries by water-ethanol binary fluid is low, so the thermal performance of water-ethanol is less as compared to water-acetone. Thereafter 80 W heat inputs, the behavior of all PHPs are according to boiling temperature of the binary fluids.

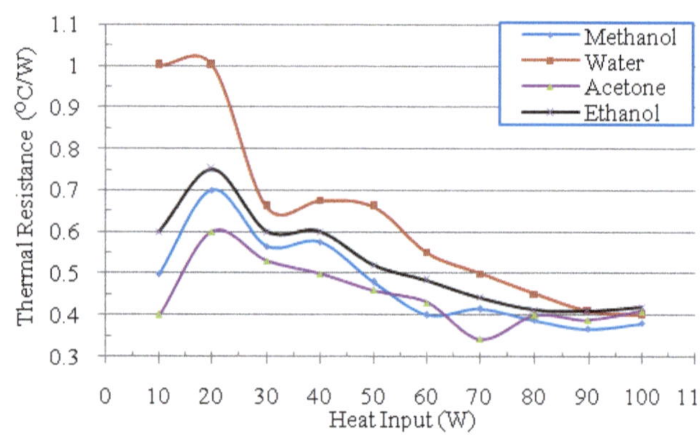

Fig. 3. Thermal resistance of pure working fluid PHP

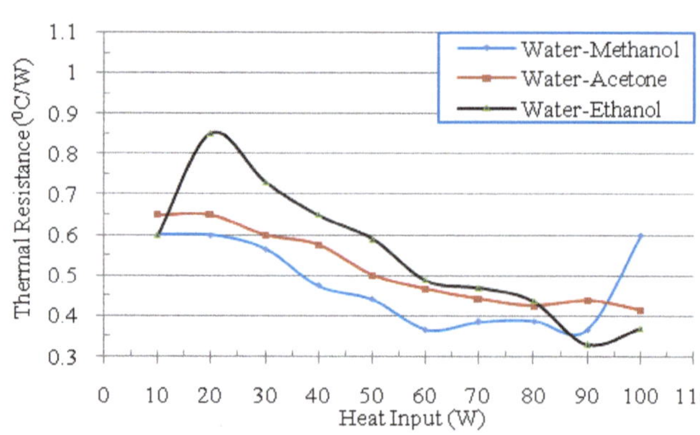

Fig. 4. Thermal resistance of binary working fluid PHP.

During starting up to 30W heat input, the thermal resistance of water PHP is more because of the more surface tension and dynamic viscosity. Therefore, the evaporation section temperature of water PHP is high and the condensation section temperature of water PHP is low, and accordingly, the thermal resistances are large. The behavior of the thermal resistance is according to the algebraic mean values of the latent heat of vaporization.

3.2 Effect of pure and binary mixture working fluids PHP on an average evaporator temperature

The behavior of the average evaporator temperature along with the heat power input in the PHP for all pure working fluids is shown in Fig. 5. In high power inputs, the temperature of the evaporation section is high enough to keep the working fluids of high boiling points can boil vehemently and smoothly flow in one direction. Therefore, all the thermal resistances reduce smoothly and the differences between the working fluids are smaller with the increasing of heating power inputs. The evaporation section temperature of methanol is low for the high $(dp/dT)_{sat}$ and the specific heat. Thus, the evaporation section temperature of methanol is low in high power inputs.

Also Fig. 6, represents the behavior of the average evaporator temperature along with the heat power input of the binary mixture working fluid in the PHP. The trends of all the binary mixture are almost same. During starting up to 20W, all the binary mixture PHPs are having same average evaporator temperature because no circulation of the fluids at low heat input. In 20 to 80 W heat input, the average evaporator temperature of all binary mixture PHPs are according to the average boiling temperature of its constituents. Thereafter 80 to 100W, mixed trend is found because of the dry-out condition at the various heat input. The trends of average evaporator temperature for Water-ethanol and water-acetone PHPs are certainly mixed. Thus, the evaporation section temperature of water-methanol is low in high power inputs.

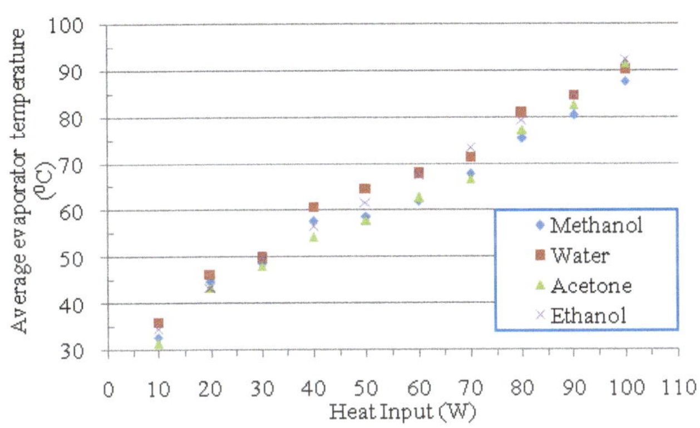

Fig. 5. Average evaporator temperature of pure working fluid PHP

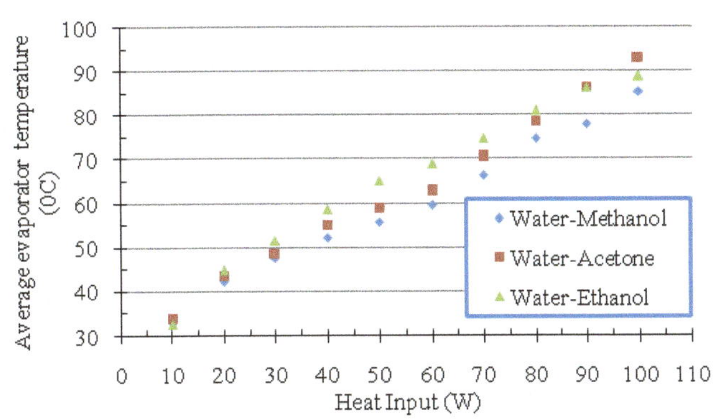

Fig. 6. Average evaporator temperature of binary working fluid PHP.

3.3 Effect of pure and its binary mixture working fluids PHP

In Fig. 7, water-methanol PHP gives the better thermal performance than pure working fluids of same constituent. It is also clear that during starting at 10 to 30W, the thermal resistance of the pure working fluids is varying drastically. But in 30 to 80W heat input, all working fluids behaves same, in 80 to 100W heating zone, water-methanol binary mixture PHP is get dry-out. Whereas in Fig. 8, water- acetone binary mixture working fluid pure acetone gives the better thermal performance than pure water and binary mixture working fluids of same constituent. Initially, thermal resistances of pure working fluid PHPs are fluctuating due to no circulation of liquid. In the high heating zone 80 to 100W, all pure and binary working fluids are stable. But in Fig. 9, water-ethanol binary mixture PHP, pure ethanol gives the better thermal performance than pure water and binary mixture of same constituent in low and moderate heat zone from 10 to 80W. As the binary fluid is in the high heat zone the PHPs gets dry-out and situation is reverse.

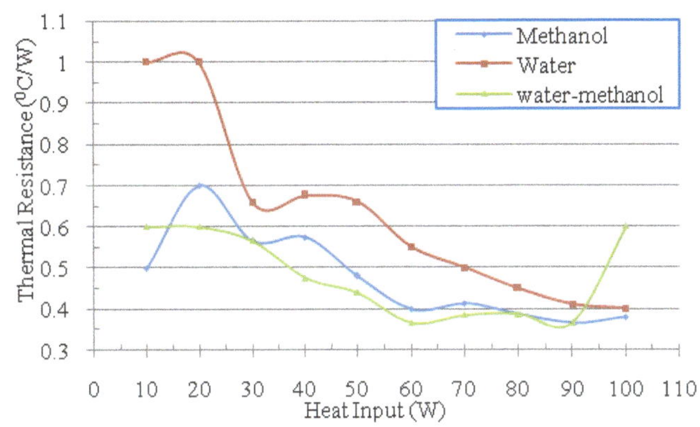

Fig. 7. Thermal resistance of water-methanol PHP

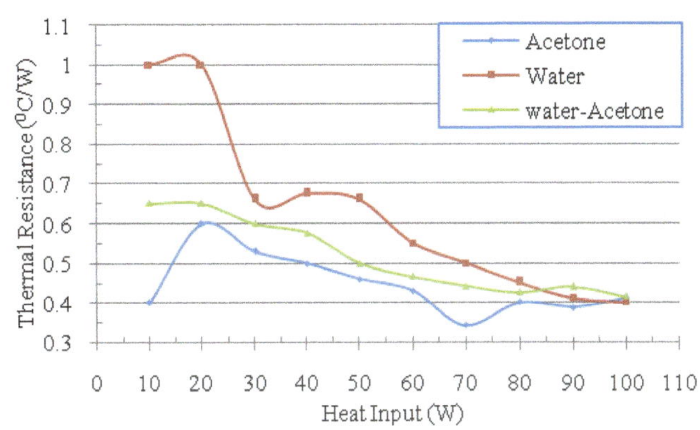

Fig. 8. Thermal resistance of water-acetone PHP

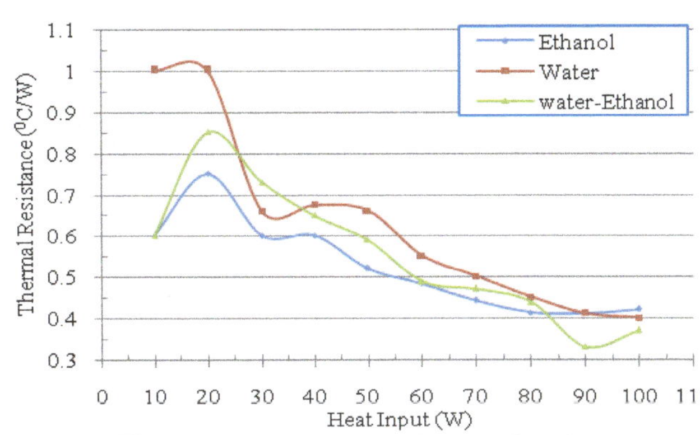

Fig. 9. Thermal resistance of water-ethanol PHP

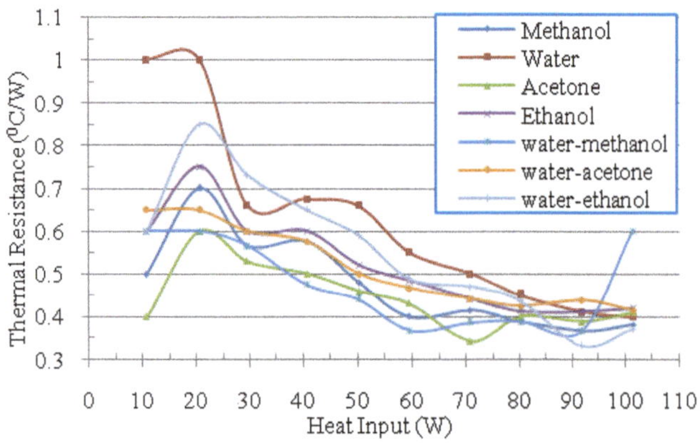

Fig. 10: Thermal resistance of all working fluid PHPs.

From Fig. 10, it is clear that trends of thermal resistance for all working fluids are approximately same in nature, as the difference in thermophysical values is minimal. In all the working fluid PHPs, pure water has more thermal resistance whereas pure acetone has lesser thermal resistance. In all PHPs, pure acetone gives best thermal performance in comparisons with the other pure and binary mixtures working fluid.

4. CONCLUSIONS

A closed loop pulsating heat pipes has been experimentally investigated to study the effects of pure and binary mixture working fluid on the thermal performance. The various working fluids (Methanol, ethanol, acetone, water and different binary mixtures) have been scrutinized. Different fluids are beneficial under different operating conditions. The binary fluid PHP shows more thermal performance as compared to the pure working fluids of same constituents. So, binary fluids are more beneficial to use in PHP for low heat input applications. The following main conclusions can be drawn from the study:

1. For pure and binary working fluids of PHP, thermal resistance is decreases with the increasing heat input. The dry-out for the water-methanol, water-acetone and water-ethanol is at 85W, 80W and 90W heat input respectively, which is approximate the algebraic mean values of the boiling point of binary mixture.
2. The evaporator temperature of methanol is low in high power inputs. A PHP of water-methanol binary mixture gives good thermal performance over other working fluids.
3. In all working fluids, pure water is having more thermal resistance whereas pure acetone is having lesser thermal resistance. So in this set-up, pure acetone gives best thermal performance in comparisons with the other pure and binary mixtures working fluid.
4. No measurable difference has been recorded between the PHP running with pure and binary mixture working fluids, in terms of overall thermal resistance. Working fluid Behavior is strongly depends on the thermo-physical properties, but latent heat of vaporization is the main property that strongly affects the thermal performance.

ACKNOWLEDGEMENTS

The research has been supported generously by Government College of Engineering, Amravati, (M.S.) India. The authors express their sincere appreciation for all of the support provided.

NOMENCLATURE

C_p	: Specific heat	(kJ/kg·K)
D	: Diameter	(m)
Eö	: Eötvös number = $(Bo)^2$	(----)
g	: Gravitational acceleration	(N/m^2)
P	: Electrical input power	(W)
$\dot{Q}$	: Heat input	(W)
R	: Resistance	(^{0}C/W)
h	: Latent heat of phase change	(kJ/kg)
k	: Thermal conductivity	(W/m·K)
q''	: Heat flux	(W/m^2)
T	: Temperature	(^{0}C)
$\bar{T}$	: Average temperature	(^{0}C)
u	: Interfacial velocity	(m/s)

Greek Symbols

σ	: Surface tension	(N/m)
ρ	: Density	(kg/m^3)
μ	: Dynamic viscosity	(Ns/m^2)
υ	: Kinematic viscosity	(Pa.s)

Subscripts

a	: adiabatic section
b	: boiling
c	: condenser section
e	: evaporator section
fg	: fluid to gas
liq	: liquid
sat	: saturation
th	: thermal
vap	: vapor

REFERENCES

Akachi, H., 1990, "Structure of a heat pipe," U.S. Patent 4,921,041.

Akachi, H., Polášek, F., Štulc, P., 1996. "Pulsating Heat Pipes," *Proceedings of the 5th International Heat Pipe Symposium*, 208–217.

Carey, V. P., 2007, "Liquid-Vapor Phase-Change Phenomena," *Taylor and Francis (2nd Edition)* ISBN: 1-59169-035-8, 572-63.

Charoensawan, P., Khandekar, S., Groll, M., Terdtoon, P., 2003, "Closed loop pulsating heat pipes Part A: parametric experimental investigations," *Applied Thermal Engineering*, 23, 2009–2020. http://dx.doi.org/10.1016/S1359-4311(03)00159-5

CUI Xiaoyu, WU Xianzhong, LI Meiling, 2006, "Varying composition capacity control air conditioning system with ternary refrigerant mixture," *Journal of Chemical Industry and Engineering* 57(3), 515-520. http://www.hgxb.com.cn/EN/Y2006/V57/I3/515

Dadong WANG, Xiaoyu CUI, 2010, "Experiment research on pulsating heat pipe with different mixtures working fluids," *The 21st International Symposium on Transport Phenomena*, Kaohsiung City, Taiwan. http://www.see.eng.osaka-u.ac.jp/seeea/seeea/sstudy/2010_nakagawa_taiwan.pdf

Faghri, Amir, 1995, "Heat Pipe Science and Technology," *Taylor and Francis*, ISBN: 1-56032-383-3, 803-855.

Kammuang-lue, N., Charoensawan, P., Ritthidech, S., Booddachan, K. and Terdtoon, P., 2008, "Effects of Working Fluids on Heat Transfer Characteristics of a Closed-Loop Pulsating Heat Pipe at Critical State," *International Heat Pipe Conference.*

Khandekar, S., Charoensawan, P., Groll, M., Terdtoon, P., 2003, "Closed loop pulsating heat pipes Part B: visualization and semi-empirical modeling," *Applied Thermal Engineering*, 23, 2021–2033. http://dx.doi.org/10.1016/S1359-4311(03)00168-6

Khandekar, S., Dollinger N., Groll M., 2003, "Understanding operational regimes of closed loop pulsating heat pipes: an experimental study," *Applied Thermal Engineering,* 23, p. 707–719. http://dx.doi.org/10.1016/S1359-4311(02)00237-5

Khandekar, S., Panigrahi, P., Lefèvre, F., Bonjour, J., 2010, "Local hydrodynamics of flow in a pulsating heat pipe: A review," *Frontiers in Heat Pipes* (FHP), 1, 023003
http://dx.doi.org/10.5098/fhp.v1.2.3003

Mameli, M., Khandekar, S., Marango, M., 2011. "An exploratory study of a pulsating heat pipe operated with a two component fluid mixture," *Proceedings of the 21st National and 10th ISHMT-ASME Heat and Mass Transfer conference*, IIT Madras, India.

Meena, P., Rittidech, S., Tammasaeng, P., 2009, "Effect of Evaporator Section Lengths and Working Fluids on Operational Limit of Closed Loop Oscillating Heat Pipes with Check Valves (CLOHP/CV)," *American Journal of Applied Sciences*, 6 (1), 133-136.
http://dx.doi.org/10.3844/ajassp.2009.133.136

Pachghare, P., Mahalle M., Khedkar, S., 2012, "Effect of Working Fluid on Thermal Performance of Closed Loop Pulsating Heat Pipe: A Review," *International Journal of Computer Applications,* 9, 27-31.
http://www.ijcaonline.org/proceedings/ncipet/number9/5260-1072

Qu, W., Ma, H. B., 2007, "Theoretical analysis of startup of a pulsating heat pipe," *International Journal of Heat and Mass Transfer* 50, 2309-2316.
http://dx.doi.org/10.1016/j.ijheatmasstransfer.2006.10.043

Shafii, M. B., Faghri, A., Yuwen, Z., 2002, "Analysis of heat transfer in unlooped and looped pulsating heat pipes," *International Journal of Numerical Methods Heat Fluid Flow* 12(5), 585–609.
http://dx.doi.org/10.1108/09615530210434304

Shafii, M. B., Faghri, A., Yuwen, Z., 2001, "Thermal modeling of unlooped and looped pulsating heat pipes," *ASME Journal of Heat Transfer,* 123, 1159– 1172.
http://dx.doi.org/10.1115/1.1409266

Shukla, K.N., 2006, "Thermofluid dynamics of loop heat pipe operation," *International Communications in Heat and Mass Transfer*, 35(8), 916-920.
http://dx.doi.org/10.1016/j.icheatmasstransfer.2008.04.020

Yang, H., Khandekar, S., Groll, M., 2009, "Performance characteristics of pulsating heat pipes as integral thermal spreaders," *International Journal of Thermal Sciences*, 48, 815–824.
http://dx.doi.org/10.1016/j.ijthermalsci.2008.05.017

9

VISUALIZATION EXPERIMENT ON THE EVAPORATION CHARACTERISTICS IN NANOPARTICLE-LADEN MESH-WICKED HEAT PIPES

Shwin-Chung Wong*, Ya-Chi Ho

Department of Power Mechanical Engineering, National Tsing Hua University, Taiwan, ROC

ABSTRACT

Visualization experiments are conducted on horizontal heat pipes to investigate the effects of Al_2O_3 or Cu nanoparticles laden in mesh-wick evaporator. The nanoparticle-laden evaporator appears quiescent with no sign of boiling. With 0.01g Cu nanoparticles sintered in the evaporator, the maximum heat load Q_{max} and the minimum evaporator resistance $R_{e,min}$ are only slightly improved. With 0.01g or 0.04g sedimented Al_2O_3 nanoparticles, $R_{e,min}$ reduces by 20%. Q_{max} increases by 20% with a larger amount of Al_2O_3 nanoparticles. $R_{e,min}$ improvement may be attributed to additional heat paths with numerous micro menisci for thin-film evaporation. Q_{max} increases with enhanced capillarity by nanoparticle agglomerates.

Keywords: *Heat pipe, mesh wick, nanofluids, nanoparticles, evaporator resistance*

1. INTRODUCTION

Nanofluids have received tremendous attention because they have considerably enlarged thermal conductivities (Keblinski, *et al.*, 2002). They have been adopted as the working fluid of mesh-wicked heat pipes (Tsai *et al.*, 2004; Do *et al.*, 2010; Liu and Zhu, 2011; Putra *et al.*, 2012), grooved heat pipes (Kang *et al.*, 2006; Liu and Lu, 2009; Yang *et al.*, 2008), sintered-powder heat pipes (Kang *et al.*, 2009). A number of recent review papers have been available (Liu and Li, 2012; Kumaresan and Venkatachalapathy, 2012; Sureshkumar et al., 2013). All these studies reported significant performance enhancement. However, the mechanism for such improvements has not been well understood. While it has been attributed to the enlarged conductivity in some studies (Kang et al., 2006; Liu and Lu, 2009; Yang *et al.*, 2008; Kang *et al.*, 2009), a different view suggested that the thin porous layer formed by sedimented nanoparticles, which increases the surface wettability and capillarity in the evaporator wick, may play a primary role (Do *et al.*, 2010). Unlike pulsating heat pipes in which strong oscillating motions keep the nanoparticles suspended during operation (Ma *et al.*, 2006; Lin *et al.*, 2008), the nanoparticles in regular heat pipes tend to sediment into a thin porous layer. Especially for mesh-wicked and sintered-powder heat pipes, it is expected that the sedimented nanoparticles would agglomerate and become trapped and immobile in the wicks. On the other hand, nanoparticles have been found to improve the pool boiling performance over a grooved surface (Liu *et al.*, 2007) and a flat surface (Kwark *et al.*, 2010) under certain conditions. Direct visualization of the evaporation characteristics in the nanoparticle-laden evaporator of operating heat pipes can be helpful in identifying the mechanism for performance enhancement. In the present study, we will apply our technique of visualization experiment on operating heat pipes (Liou *et al.*, 2010; Wong *et al.*, 2010; Wong *et al.*, 2012) to nanoparticles-laden evaporators.

2. EXPERIMENTAL METHODS

The experimental methods follow our recent works (Liou *et al.*, 2010; Wong *et al.*, 2010; Wong *et al.*, 2012). A detailed description of the heat pipe structure and experimental procedure is available in Ref. (Liou *et al.*, 2010). Here is given a brief description. Fig. 1 shows the overall test setup. The 22 mm × 11 mm heated surface of the heat pipe is connected to the uniform-temperature top surface (20 mm × 10 mm) of a heating post via a layer of thermal grease. The 22 mm width of the heated area is near the 25 mm width of the wick, making the thermal distribution of the heat pipe approximately one-dimensional. A cut-off trench, leaving a plate thickness of 1 mm, is made around the heated area and across the plate width (cf. Fig. 2) to reduce lateral conduction through the base plate. The heating post is carefully insulated except at its top end. The bottom of the other end of the heat pipe is cooled by a cold plate with running water at 20 ℃. Except for these heated and cooled regions, the heat pipe is well insulated during operation. To prevent condensation on the visualization window, the insulation material over the window is removed shortly for recording only. A CCD camera equipped with a microscopic lens (Optem, Zoom 125) shoots vertically downward through the observation window. High magnification of the evaporation zone can be obtained. Illumination is provided by high-intensity LED lights.

Figure 2 shows the flat-plate heat pipe designed for visualization and thermal resistance measurement, along with the positions of seven implanted thermocouples. These thermocouples, K-type 1 mm-diameter stainless-steel sheathed probe (Omega, Inc.), are denoted by T1 to T7, respectively. Thermocouple signals are recorded by a data logger (Fluke, Hydra Series II) with a resolution of 0.1 □. T1 and T2 are used to calculate the total heat load Q_t through the copper heating post. T3 measures in the middle of the copper plate at the center of the heated area. Suspended about 1 mm above the center of the evaporator, T4

* *Corresponding author. Email: scwong@pme.nthu.edu.tw*

measures the vapor temperature leaving the evaporator. T5–T6 are managed to evaluate the lateral conduction through the copper plate (Liou *et al.*, 2010), Q_{cond}, as will be discussed later. T7 measures the copper plate temperature in the cooled area.

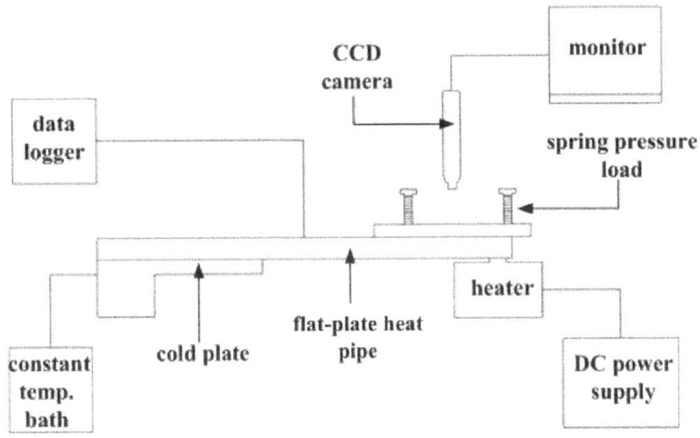

Fig. 1 Overall experimental setup.

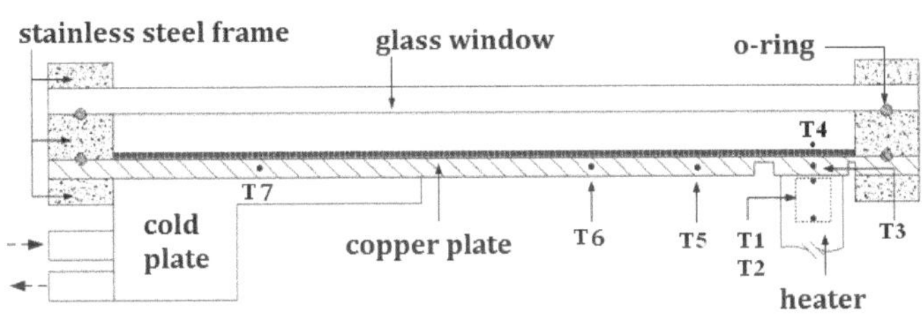

Fig. 2 The flat-plate heat pipe along with thermocouple positions.

The 100 + 200 mesh wick, with an upper 100 mesh and a lower 200 mesh woven screen, is sintered on a 4 mm thick C1020 oxygenless copper base plate under a fixed pressure in a 800 ℃ hydrogen/nitrogen atmosphere for 2 h. To accommodate the wick, a region of 117 × 25 mm^2 on the upper surface of the base plate is cut lower by a depth of 1 mm. This is done to prevent the possibility that a small amount of the working liquid may cycle back via the narrow path between the o-ring and the copper plate, rather than through the wick. Before sintering, the screens and the copper plate are carefully cleaned. The heat pipe has a top glass window for observation. The internal space of the heat pipe is 120 × 30 × 7 mm^3. With the contact surfaces between different pieces sealed with o-rings, the whole structure, including the wicked copper base plate, the top glass window, and the stainless-steel frames, is tightened with bolts. It is noted that excess charge is needed in the present experiments to compensate for the liquid accumulation at the corners, gaps and walls in the heat pipe. A fixed amount of 0.8 cm^3 (the saturate charge is 0.5 cm^3) of degassed de-ionized water or nanofluid is filled into the heat pipe right after it is evacuated down to a pressure of 1×10^{-2} Torr. The heat load is increased stepwise from a low heat load, and data are taken under a thermally stable condition for each heat load.

Usually, the capillarity of the sintered copper wick degrades in air with time due to surface oxidation. In this work, the copper surface wettability is quantified by the contact angle of a sessile *water* drop on a smooth surface. The contact angle is measured on another plate, which has been taken out of the sintering oven simultaneously with the wicked copper plate, when the heat pipe is evacuated to 1 Torr. For all the tests, the static contact angles of water on a smooth copper surface are controlled at 10°–12°.

The evaporator resistance R_e is determined as

$$R_e = (T3 - T4)/Q, \qquad (1)$$

where $Q = Q_t - Q_{cond}$. The net heat flux $q = Q/A_e$, with the heated area A_e = 2.42 cm^2. In this study, the ratios of Q_{cond}/Q_t prior to the occurrence of local dryout are about 1.1–3%, smaller for a smaller evaporator resistance. The uncertainty in the measurements of Q is 7%. This is determined by comparing the output of the DC power supply and the estimated value based on the temperature difference (T1 - T2). The uncertainty in R_e values is also 7%, since the uncertainties primarily arose from those of Q, according to the definition of R_e in Eq. (1).

Two kinds of nanoparticles, Al_2O_3 and Cu, are laden in the evaporator wick. The Al_2O_3 nanoparticles are a commercial product with a monodisperse size of about 50 nm. The Cu nanoparticles, smaller than 100 nm, have been chemically made from $Cu(NO_3)_2$ in our lab. The two kinds of nanoparticles are laden into the evaporator in different manners. The Al_2O_3 nanoparticles are well blended in de-ionized water using an ultrasonic blender to form a nanofluid without surfactant. The nanofluid is then filled into the evacuated heat pipe. The Cu nanoparticles are weakly sintered within the wick in a 575 ℃ hydrogen/nitrogen atmosphere for 0.5 h before water filling. The mass of Cu nanoparticles sintered in the wick is 0.01 g. Fig. 3 shows an SEM image of Cu nanoparticles on a flat copper plate under the same sintering condition. The Cu nanoparticles can be seen to partially bonded into numerous large porous agglomerates. Fig. 4 shows a magnified image of a dried 100+200 mesh wick with Al_2O_3 nanoparticles dried after the filling. The Al_2O_3 nanoparticles distribute non-uniformly with large agglomerates due to inter-particle adhesion. With a fluid charge of 0.8 cm^3, the fraction of the Cu nanoparticles is 1.25wt% (or 0.14vol%). However, because all Cu nanoparticles are located in the evaporator, such concentration values are not to be directly compared with the concentrations of nanofluids for which the nanoparticles may be distributed over the heat pipe. The amount of Al_2O_3 nanoparticles charged into the heat pipe is 0.01 g or 0.04 g. The fraction of the Al_2O_3 nanoparticles are 1.25wt% (or 0.31vol%) and 5wt% (or 1.25vol%), respectively. During liquid filling, the heat pipe is vertically placed with the evaporator at the bottom. Thus, the Al_2O_3 nanoparticles may be distributed over the heat pipe but somewhat denser at the evaporator.

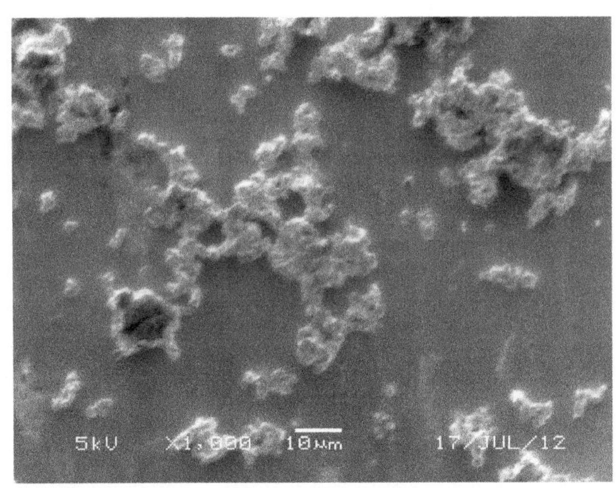

Fig. 3 Cu nanoparticles sintered on a copper plate.

3. RESULTS AND DISCUSSION

Experiments were conducted for four different test cases with the 100+200 mesh wick and a fixed liquid charge of v = 0.8 cm^3: (1) pure water without nanoparticles, (2) water with 0.01 g Cu nanoparticles sintered in the evaporator, (2) water with 0.01 g Al_2O_3 nanoparticles, and (4) water with 0.04 g Al_2O_3 nanoparticles. Visualization was made for the evaporation process and the evaporator resistances were measured versus heat load until local dryout occurred in the evaporator.

Fig. 4 Al_2O_3 nanoparticles in a dry 100+200 mesh wick.

3.1 Visualization Results

For all these cases, the evaporator appears quiescent during operation with no sign of nucleate boiling. The water layer in the evaporator becomes thinner with the increase of heat load. In response to the thinning water layer, the evaporator resistance gradually decreases to a minimum value, $R_{e,min}$, before it rises as a result of local dryout in the evaporator. Representatively, Fig. 5 compares the images near the center of the evaporator at $Q = 0$ W and 79.3 W for a test charged with pure water. At $Q = 0$ W, the wick is mostly immersed in the liquid, with exposed peaks of the upper screen. At $Q = 79.3$ W, when $R_{e,min}$ occurs, most of the upper 100 mesh screen has been exposed but the lower 200 mesh screen is still immersed. Under heat load, the thickness of the liquid layer in the evaporator decreases toward the evaporator end of the heat pipe because the supply of the condensed liquid is gradually consumed in the evaporator. At the condition with $R_{e,min}$, the liquid layer in the wick near the end-edge of the evaporator usually is hardly visible, as illustrated in Fig. 6 for the case with 0.01 g Al_2O_3 nanoparticles at $Q = 84.3$ W. The above characteristics are similar to our previous results for sintered multi-layer copper mesh wicks (Liou *et al.*, 2010) and sintered irregular-powder wicks (Wong *et al.*, 2010) using water as the working fluid. It is noted that nucleate boiling was not observed with a superheat as large as 13.7 K in our previous work (Wong *et al.*, 2010). The present tests indicate that the nanoparticles laden in the evaporator fail to help activate nucleate boiling when the working fluid is water.

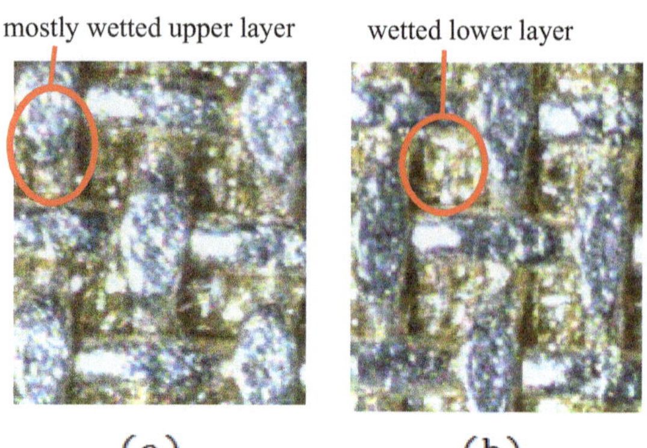

Fig. 5 Wetted 100+200 mesh evaporator with pure water, (a) at $Q =$ 0W; (b) at $Q = 79.3$W with $R_{e,min}$.

Figure 7 presents the images at Q_{max}, when $R_{e,min}$ occurs, near the evaporator center for three different cases. In Fig. 7a and 7b, the lower screen is immersed in water. In Fig. 7c for the case with 0.04 g Al_2O_3 nanoparticles, the mesh wires of the lower 200 mesh screen are buried in the nanoparticle layer. The liquid seems absorbed in the thick porous nanoparticle layer, which will be shown to result in a larger Q_{max}.

Fig. 6 Nearly dried wick at the evaporator end-edge when $R_{e,min}$ occurs for the case with 0.01g Al_2O_3 nanoparticles, $Q = 84.3$ W.

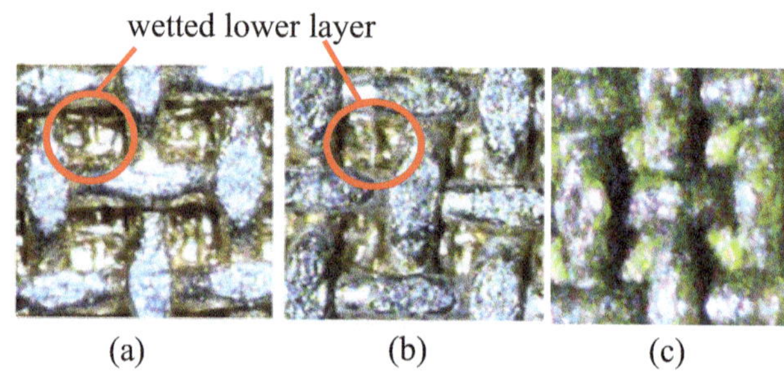

Fig. 7 Evaporator images at $R_{e,min}$, (a) water with 0.01g sintered Cu nanoparticles, $Q_{max} = 81.0$ W; (b) water with 0.01g Al_2O_3 nanoparticles, $Q_{max} = 84.3$ W; (c) water with 0.04g Al_2O_3 nanoparticles, $Q_{max} = 93.9$W.

3.2 Evaporator Performance

Figure 8 compares the evaporator resistances R_e versus heat load Q for the four cases. As mentioned earlier, the present experimental technique has a drawback that a small amount of working liquid would be trapped on the inner corners, gaps, and wall of the chamber. According to our experience through tests without nanoparticles, this drawback, along with finite heat load increment, introduces disperse in Q_{max} of roughly ±8%, whereas $R_{e,min}$ is hardly influenced (Liou, *et al.*, 2010; Wong *et al.*, 2010; Wong *et al.*, 2012). Similar to the characteristics observed in these tests, the evaporator resistances in the present cases reduce with increasing heat load until local dryout occurs. For the case with 0.01 g sintered Cu nanoparticles, although there is a slight increase in Q_{max}, it is limited when the uncertainty associated with the present experimental technique is considered. Slight reduction in $R_{e,min}$ is shown. Since $R_{e,min}$ occurs when the liquid layer in the evaporator wick is thinner, the nanoparticles near the evaporator end where the liquid layer is thinnest can be exposed. In this situation, smaller $R_{e,min}$ may be attributed to the increased thermal paths provided by the nanoparticles. On the numerous additional micro- and nano-scale menisci, thin-film evaporation occurs. Since the copper nanoparticles are sintered on the base plate, the thermal resistances through these new paths should be pretty low. As a result, a lower overall evaporator resistance $R_{e,min}$ is reached. For the case with 0.01 g Al_2O_3 nanoparticles, the $R_{e,min}$ is reduced by 20%, even more than the case with 0.01 g sintered Cu nanoparticles. The reason may be explained as follows. Both sintered Cu nanoparticles and sendimented Al_2O_3 nanoparticles can provide additional thermal paths for thin-film evaporation. Between the copper surfaces and the Al_2O_3 nanoparticles there exists contact resistance which introduces higher thermal resistance. However, the volume fraction of Al_2O_3 nanoparticles (0.31 vol%) is larger than that of Cu nanoparticles (0.14 vol%) due to the lower density of Al_2O_3. Besides, the sintered Cu nanoparticles are largely sintered together, as shown in Fig. 3. Therefore, a much larger number of micro- and nano-scale menisci can be offered by the Al_2O_3 nanoparticles. With much more area available for thin-film evaporation, the $R_{e,min}$ for the case with 0.01 g Al_2O_3 nanoparticles appears lower than the sintered Cu-nanoparticle case. With 0.04 g Al_2O_3 nanoparticles, the $R_{e,min}$ is about the same as that with 0.01 g Al_2O_3 nanoparticles. Comparing Figs. 7b and 7c, we can see that the upper mesh layer is covered with nanoparticles for the 0.04 g case. It is likely that the particle densities near the bottom are similar in the two cases. The $R_{e,min}$s are therefore similar. As to the Q_{max} of these two cases, the increase is 6% and 20%, respectively, in comparison with the case without nanoparticles. In spite of the ±8% uncertainty in the present experiments, the increasing trend in Q_{max} with increased amount of Al_2O_3 nanoparticles reliably indicates the stronger effect at a larger nanoparticle loading. The reasons for improved Q_{max} may be related to the enhanced capillarity offered by the nanoparticle agglomerates (Do *et al.*, 2010).

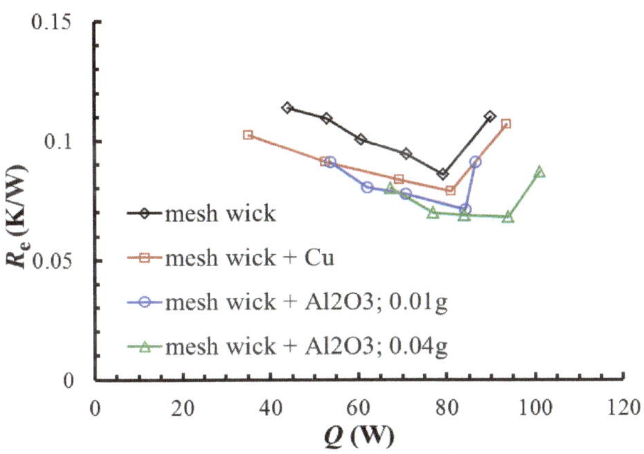

Fig. 8 Evaporator resistances R_e versus heat load Q for various working fluids, $v = 0.8$ cm^3.

4. CONCLUSIONS

Visualization experiments have been conducted to investigate the evaporation characteristics in the nanoparticle-laden mesh-wick evaporator of heat pipe with water as the working fluid. In one case, an amount of 0.01 g Cu nanoparticles is sintered in the evaporator; in two other cases, 0.8 g nanofluid respectively containing 0.01 g (0.31 vol%) or 0.04 g (1.25 vol%) Al_2O_3 nanoparticles is filled into the heat pipe. The tests are conducted as the Al_2O_3 nanoparticles have sedimented. The following conclusions are reached:

1. The evaporation process of nanoparticle-laden mesh wicks is similar to that of mesh wicks without nanoparticles. With increasing heat load, the liquid layer gradually recedes and the evaporator resistance decreases to a minimum value before a rise in response to the occurrence of local dryout in the evaporator. In all the test cases, the evaporator appears quiescent with no nucleate boiling observed.
2. The minimum evaporator resistances are slightly improved in the case with 0.01 g sintered Cu nanoparticles. Stronger improvements are obtained in the minimum evaporator resistance with 0.01 g or 0.04 g sedimented Al_2O_3 nanoparticles. In the latter case, the minimum evaporator resistance is reduced by about 20%. This may be because the nanoparticles provide additional heat paths with a large number of micro- and nano-scale menisci for thin-film evaporation.
3. The maximum heat load is slightly increased in the case with 0.01 g sintered Cu nanoparticles. The increases are more significant in the cases with 0.01 g or 0.04 g Al_2O_3 nanoparticles. In the latter case the increase reaches about 20%. This may result from the enhanced capillarity offered by the nanoparticle agglomerates in the evaporator.

ACKNOWLEDGEMENTS

This work was financially supported by National Science Council, ROC under Contract NSC101-2221-E-007-058.

NOMENCLATURE

A_e	heated area (cm^2)
q	heat flux (W/cm^2)
Q	heat flow (W)
R_e	evaporator resistance (K/W)
v	liquid charge (cm^3)
T	thermocouple

Subscripts

cond	lateral conduction
e	evaporator
max	maximum
min	minimum
t	total

REFERENCES

Asirvatham, L.G., Nimmagadda, R., and Wongwises, S., 2013, "Heat Transfer Performance of Screen Mesh Wick Heat Pipes Using Silver-Water Nanofluid," *International Journal of Heat and Mass Transfe*, **60**, 201–209.

Do, K.H., Ha, H.J., and Jang, S.P., 2010, "Thermal Resistance of Screen Mesh Wick Heat Pipes Using the Water-Based Al_2O_3 Nanofluids," *International Journal of Heat and Mass Transfer*, **53**, 5888-5894.

Kang, S.W., Wei, W.C., Tsai, S.H., and Yang, S.Y., 2006, "Experimental Investigation of Silver Nanofluid on Heat Pipe Thermal Performance," *Applied Thermal Engineering*, **26**, 2377-2382.

Kang, S.-W., Wei, W.-C., Tsai, S.-H., and Huang, C.-C., 2009, "Experimental Investigation of Nanofluids on Sintered Heat Pipe Thermal Performance," *Applied Thermal Engineering*, **29**, 973-979.

Keblinski, P., Phillpot, S.R., Choi, S.U.S., and Eastman, J.A., 2002, "Mechanisms of Heat Flow in Suspensions of Nano-Sized Particles (Nanofluids)," *International Journal of Heat and Mass Transfer*, Vol. 45, pp. 855-863.

Kole, M. and Dey, T.K., 2013, "Thermal Performance of Screen Mesh Wick Heat Pipes Using Water-Based Copper Nanofluids," *Applied Thermal Engineering*, **50**, 763-770.

Kumaresan G., Venkatachalapathy, S., 2012, "A Review on Heat Transfer Enhancement Studies of Heat Pipes Using Nanofluids," *Frontiers in Heat Pipes*, **3**, 043001.

Kwark, S.M., Kumar, R., Moreno, G., Yoo, J., and You, S.M., 2010, "Pool Boiling Characteristics of Low Concentration Nanofluids," *International Journal of Heat and Mass Transfer*, **53**, 972-981.

Lin, Y.H., Kang, S.W., and Chen, H.L., 2008, "Effect of Silver Nanofluids on Pulsating Heat Pipe Thermal Performance," *Applied Thermal Engineering*, **28**, 1312-1317.

Liou, J.-H., Chang, C.-W., Chao, C., and Wong, S.-C., 2010, "Visualization and Thermal Resistance Measurement for the Sintered Mesh-Wick Evaporator in Operating Flat-Plate Heat Pipes," *International Journal of Heat and Mass Transfer*, **53**, 1498-1506.

Liu, Z.H., Xiong, J.G., and Bao, R., 2007, "Boiling Heat Transfer Characteristics of Nanofluids in a Flat Heat Pipe Evaporator with Micro-Grooved Heating Surface," *International Journal of Multiphase Flow*, **33**, 1284-1295.

Liu, Z.H., and Lu, L., 2009, "Thermal Performance of an Axially Microgrooved Heat Pipe Using Carbon Nanotube Suspensions," *Journal of Thermophysics and Heat Transfer*, **23**, 170-175.

Liu, Z.H., and Zhu, Q.Z., 2011, "Application of Aqueous Nanofluids in a Horizontal Mesh Heat Pipe," *Energy Conversion and Management*, **52**, 292-300.

Liu, Z.-H., Li, Y.-Y., 2012, "A New Frontier of Nanofluid Research – Application of Nanofluids in Heat Pipes," *International Journal of Heat and Mass Transfer*, **55**, 6786–6797.

Ma, H.B., Wilson, C., Park, K., Choi, S.U.S., and Tirumala, T., 2006, "An Experimental Investigation of Heat Transport Capability in a Nanofluids Oscillating Heat Pipe," *ASME Journal of Heat Transfer*, **128**, 1213-1216.

Putra, N., Septiadi, W.N., Rahman, H., and Irwansyah, R., 2012, "Thermal Performance of Screen Mesh Wick Heat Pipes with Nanofluids," *Experimental Thermal and Fluid Science*, **40**, 10–17.

Sureshkumar, R., Mohideen, S.T., Nethaji, N., 2013, "Heat Transfer Characteristics of Nanofluids in Heat Pipes: A Review," Renewable Sustainable Energy Review, **20**, 397–410.

Tsai, C.Y., Chien, H.T., Ding, P.P., Chan, B., Luh, T.Y., and Chen, P.H., 2004, Effect of Structural Character of Gold Nanoparticles in Nanofluid on Heat Pipe Thermal Performance, Materials Letters, **58**, 1461-1465.

Wang, G.-S., Song, B., Liu, Z.-H., 2010, "Operation Characteristics of Cylindrical Miniature Grooved Heat Pipe Using Aqueous CuO Nanofluids," *Experimental Thermal Fluid Science*, **34**, 1415–1421.

Wong, S.-C., Liou, J.-H., and Chang, C.-W., 2010, "Evaporation Resistance Measurement and Visualization for Sintered Copper-Powder Evaporator in Operating Flat-Plate Heat Pipes," *International Journal of Heat and Mass Transfer*, **53**, 3792-3798.

Wong, S.-C., Lin, Y.-C., and Liou, J.-H., 2012, "Visualization and Evaporation Resistance Measurement in Heat Pipes Charged with Water, Methanol or Acetone," *International Journal of Thermal Sciences*, **52**, 154-160.

Yang, X.F., Liu, Z.H., and Zhao, J., 2008, "Heat Transfer Performance of a Horizontal Microgrooved Heat Pipe Using CuO Nanofluids," *Journal of Micromechanics and Microengineering*, **18**, 035038.

EFFECT OF COPPER NANOFLUID CONCENTRATION ON THERMAL PERFORMANCE OF HEAT PIPES

R. Senthilkumar*, S. Vaidyanathan, B. Sivaraman

Department of Mechanical Engineering, Annamalai University, Annamalai Nagar, India, 608 002

ABSTRACT

The effect of filling ratio in heat pipe on the thermal performance like thermal efficiency was experimentally studied using copper nanofluid. Heat pipes are widely used for the thermal control of electronic device due to this high performance. The nanofluid used in this study was copper nano particles of size 40 nm with a various concentrations of 25 mg/lit, 50 mg/lit, 100 mg/lit and 125 mg/lit. The base working fluid used in this experimental analysis was de-ionized water. An experimental system is set up to measure the temperature distribution of heat pipes along the surface and calculate the thermal efficiency of copper nanofluid under different concentration. Concerning heat transport limitations, the copper nanofluid show the advantages over the conventional working fluids. The experimental results show that the higher efficiency of the heat pipe obtained with a concentration of 100 mg/lit than the other concentration for all orientations.

Keywords: *Heat pipe, copper nanofluid, efficiency, resistance.*

1. INTRODUCTION

Heat pipes are tremendously efficient heat transfer devices that utilize a phase change of the working fluid inside of the container and it quickly transport large amount of heat from evaporator section to the condenser section. In evaporator section, the heat is absorbed by the working liquid via evaporation and this vapour condenses at a condenser section to release the latent heat. The condensed liquid is then drawn back to the evaporator by the capillary force by means of wick structure which is kept at the inside of the heat pipe to complete a thermal cycle. A good heat pipe is characterized by a low thermal resistance and a high dry-out tolerance. The heat pipe has been widely applied for electronics cooling, air conditioning, power generation, chemical engineering and spacecraft cooling.

The heat transfer characteristics of the heat pipe have been studied by a number of researchers. Lin et al. (2001) experimentally analyzed the two-phase flow and heat transfer of R141b in a small tube. Song et al. (2004) discussed about the heat transfer performance of axial rotating heat pipes at steady state conditions. Xuan et al. (2004) performed experiment on a flat heat pipe under different heat fluxes, inclinations and the amount of the working fluid.

The requirements of high efficiency and compact size in such electronic devices strongly affect their thermal management that is critical and directly influences cost, reliability, and performances. A good heat pipe is characterized by a low thermal resistance and a higher dry-out limit in evaporator section. The type of heat pipe, thickness of wall, wick porosity and type of structure, and even the contact condition between the wick and the container wall will affect the performance of the evaporator (Mughal. 1996 ; Brautsch, 2002). The most important factor to be considered for the design of heat pipe is related to the characteristics of the evaporator.

All of the heat pipes, including conventional heat pipes and micro heat pipes have common problem of heat transfer limitation. The heat transfer limitations determine the maximum heat transfer rate for a particular heat pipe under the normal working conditions. The various limitations of the heat pipe are continuum flow limit, frozen startup limit, viscous limit, sonic limit, entrainment limit, capillary limit, condenser limit and the boiling limit. Among them the capillary limit and boiling limits are important for heat pipe under the normal working conditions. The heat transfer rate of heat transfer devices like heat pipe can be enhanced by adding additives to the working fluids to change the fluid transport properties and flow features.

One of the methods is to improve the heat transfer limitations use of the nanofluids in the heat transfer equipments to enhance the thermal performance of heat transfer devices. An innovative way to enhance liquid thermal conductivity is the dispersion of highly conductive solid nanoparticles within the base fluid. These new generations of conductive fluids with nanoparticles are referred to as nanofluids (Choi, 2005). The nanoparticles within the fluid change its thermal conductivity, viscosity and density. It has been shown experimentally that, for a given concentration level, the thermal conductivity of the nanofluids increases with a decrease in particle diameter (Anoop et al, 2009 ; Jung et al, 2009). Riehl (2006) has observed that a higher heat transfer coefficient can be seen when using nanoparticles in water under low heat input conditions. Tsaia (2004) investigated the influence of particle size on the heat pipe thermal performance using gold nano particles.

A theoretical model and an experimental setup are proposed to describe the heat transfer performance of nano-fluids flowing inside a tube. Xuan and Li (2000) studied experimentally that the thermal conductivity of nano-fluids remarkably increases as the volume fraction of ultra-fine particles increases. Dos et al. (2003) investigated the increase in thermal conductivity with temperature for nano-fluids with water as the base fluid and Al_2O_3 or CuO particles as suspension material. The results indicated an increase in enhancement characteristics with temperature, which makes the nano-fluids even more attractive for applications with high energy density. Wei et al. (2005) used a cylindrical micro-grooved heat pipe with the inner diameter and the length of 6 and 200 mm, respectively. The width and

* *Corresponding author: Email: rsklme@yahoo.com*

the depth of the rectangular groove were 221 and 217 lm, respectively. The working fluid consisted of silver nanoparticles with an average particle size of 10 nm and DI water. The total thermal resistance of the heat pipe using nanofluids could decrease by 28–44% compared with that of the heat pipe using DI water. Kang et al. (2006) employed the dilute dispersion of silver nanoparticles with 10 and 35 nm diameters into DI water as a working fluid. They showed that the total thermal resistance decreased 10–80% compared to DI water at a heat input of 30–60W under the fixed charge volume. Furthermore, they suggested that the heat pipes were properly operated at the higher power compared with the operating power of the heat pipes with DI water.

Xuan, Qiang Li (2003) studied the convective heat transfer feature and flow performance of Cu-water nanofluids in a tube has experimentally been investigated. The results shows the suspended nanoparticles remarkably enhance heat transfer process and the nanofluid has larger heat transfer coefficient than that of the original base liquid under the same Reynolds number.

From the analysis, the nano fluids have a better thermal performance than the conventional working fluids like water, ethylene glycol etc. The heat pipes are made up of copper container with a two-layered stainless steel wire is used as a wick material. The concentrations of copper nanofluid in the DI water are 25 mg/lit, 50 mg/lit, 100 mg/lit and 125 mg/lit. The nanofluids are prepared using the ultrasonic homogenizer for copper nanofluid and size of the copper particle are 40 nm. The experiments are conducted for various inclinations of heat pipe to the horizontal with different heat inputs. The objective of this work is to study about the effect of nanofluid concentration in thermal efficiency improvement of heat pipe using copper nanofluid.

Table 1. Specification of the heat pipe

Heat pipe material	Copper
Wick material	Stainless steel
Total length of pipe	0.6 m
Evaporator length	0.15 m
Adiabatic length	0.3 m
Condenser length	0.15 m
Outer diameter of the pipe	0.020 m
Inner diameter of the pipe	0.0176 m
Wire mesh Diameter	0.183 mm
Heat Input	30, 40, 50, 60 and 70 W
Heat pipe inclination	0°, 15°, 30°, 45°, 60°, 75° and 90°
Wick mesh size, m	2365
Type of thermocouple	copper constantan (T type)
Condenser outer diameter, mm	36
Condenser inner diameter, mm	30
Flow rate of water in the condenser jacket, kg/min	0.08

Table 2. Limitations of heat pipe

Limitation	Values in W
Capillary Limit	77
Sonic Limit	65400
Entrainment Limit	4638
Boiling Limit	642

2. EXPERIMENTAL SETUP

The schematic diagram of experiment is shown in fig.1 and the thermocouple locations are shown in fig. 2. The amount of working fluid needed to fill the heat pipe is nearly 40 ml, which is nearly equal to the amount required to fill the evaporator. The surface temperature of

Table 3. Operating parameters and its uncertainties

Variable	Unit	Operating range	uncertainty
Water flow meter	kg/sec	0.01 - 1.0	± 1%
Temperature indicator	°C	0 - 200	± 1°C
Power transducer	W	0 - 300	± 1 W

temperature distribution along the evaporator surface (three locations), the condenser section (three locations), inlet and outlet of the condenser jacket also measured using thermocouples. The electrical power input is applied at the evaporator section using cylindrical electric heater attached to it with proper electrical insulation and the heater is energized with 230V AC supply using a variac and measured using a power transducer.

Concentric type of water jacket has been used as a condenser at the end to remove the heat from the pipe. The condenser section of the heat pipe is cooled using water flow through a jacket from the overhead. The water flow rate is measured using a rotameter on the inlet line to the jacket. The inlet and outlet temperatures of the cooling water are measured using thermocouples. The surface of the heat pipe is fully insulated with the glass wool in order to reduce temperature impact from the environment. The heat pipes have their vacuum pumped out, and are charged with the copper nanofluids. The amount of heat loss from the evaporator and condenser surface is negligible. The initial vacuum pressure inside the heat pipe is 70 to 75 cm of Hg.

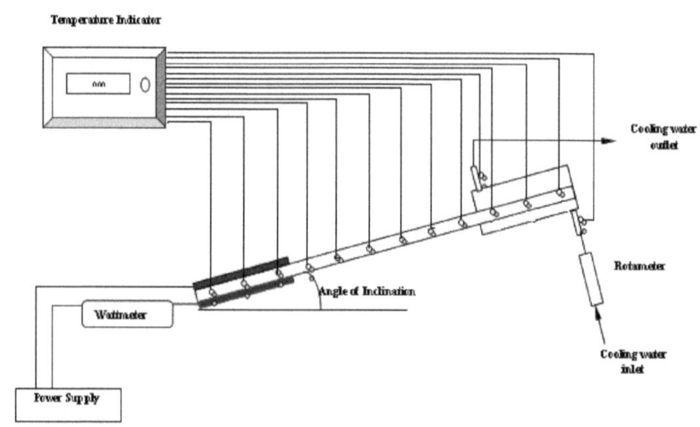

Fig. 1 Experimental setup

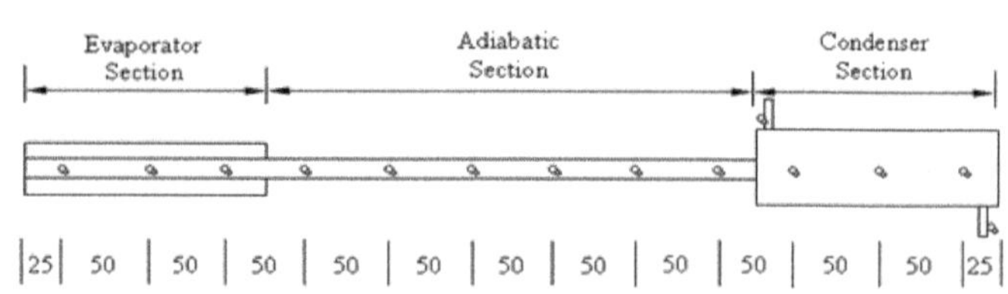

Fig. 2 Thermocouple locations of heat pipe

The experiments are conducted using five identical heat pipes filled with different working fluids which are manufactured as per mentioned dimensions. The power input to the heat pipe is gradually raised to the desired power level. The surface temperatures at six different locations along the adiabatic section of heat pipe, the evaporator wall temperatures, condenser wall temperatures, water inlet and outlet temperatures in the condenser zone are measured at regular time intervals of five minutes until the heat pipe reaches the steady state condition. Once the steady state is reached, the input power is turned off and cooling water is allowed to flow through the condenser to cool the heat pipe and to make it ready for further experimental purpose. Then the power is increased to the next level and the heat pipe is tested for its performance. This procedure is repeated for different heat inputs and different inclinations of pipe to the horizontal and observations are recorded. The output heat transfer rate from the condenser is computed by applying an energy balance to the condenser flow. The vacuum pressure in the inner side of the heat pipe is monitored by vacuum gauge, which is attached in the condenser end of the heat pipe.

3. RESULTS AND DISCUSSION

The heat pipe thermal efficiency is calculated from the ratio to the cooling capacity rate of water at the condenser section and supplied power at the evaporator section (Senthil et al., 2011). Fig. 3-7 shows that the variations of thermal efficiency of the heat pipe with inclination angle. The thermal efficiency increases with increasing the angle of inclination up to 45° afterwards it starts decreases. The efficiency is higher for 100 mg/lit concentration than the other three concentrations. The reason behind that is the formation of the liquid film in the inner side of condenser section which is at higher rate results in the increased values of the thermal resistance between the vapour of the working fluid and the cooling medium (water) in the condenser. Even though adding the more nanoparticles in the working fluid, the thermal efficiency is decreases due to the set down of the nano particles inside of the container of the heat pipe and also it gives the higher resistance to the fluid flow. Therefore the required pumping power increased and its performance deteriorates.

Figures 8 to 14 show that the variations of thermal efficiency of the heat pipe with heat input given to the evaporator section. The thermal efficiency of the heat pipe increases with increasing the heat input in the evaporator section. It is due to the fact that the temperature difference between the evaporator and condenser sections increases which results in higher evaporation heat transfer rate of working fluid. At higher heat input in the evaporator section, the heat transfer from its surface to the working medium is higher and it causes the working medium which is in the form of vapour to move strongly into the condenser section. The cooling water in the condenser absorbs this excessive heat from the working fluid and as a result the efficiency of the heat pipe increases. Generally, the nanoparticles suspension in the fluid has significant effect on the enhancement of heat transfer due to its higher heat capacity and higher thermal conductivity of working fluid. Therefore, the heat pipe thermal efficiency heat pipe increases with nanofluids as compared to that of the base working fluids like water, ethylene glycol etc..

The trial results reveal that the heat pipe efficiency gets reduced when the heat pipe is kept in vertical direction. The gravitational force which assists the flow of working fluid to flow back to the evaporator may accelerate the process which may hinder the heat transfer process at the condenser end and the fluid might have returned to the evaporator section with higher temperature end. This may be the reason why the performance of heat pipe deteriorates when the inclination was increased. This problem may not arise if the condenser section is at the base and evaporator at the top and only the capillary action, forces the liquid to move upward towards the evaporator.

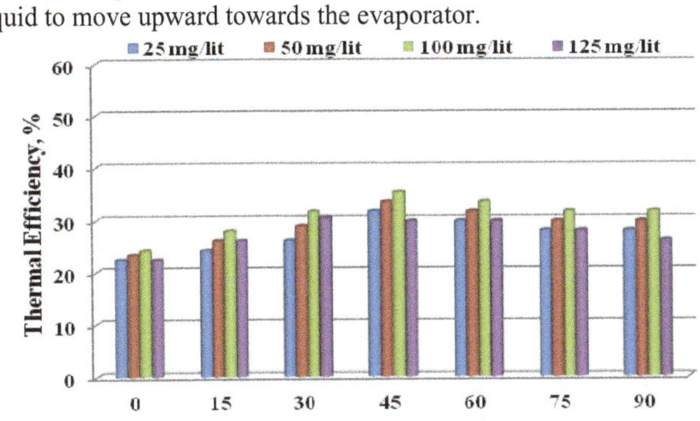

Fig. 3 Variations of heat pipe efficiency for various inclinations at 30 W heat input

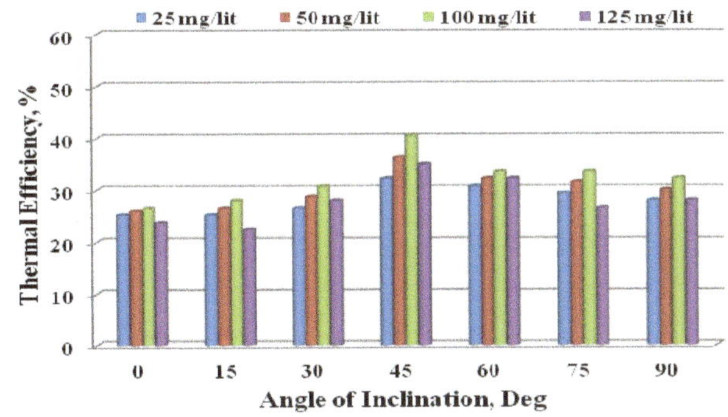

Fig. 4 Variations of heat pipe efficiency for various inclinations at 40 W heat input

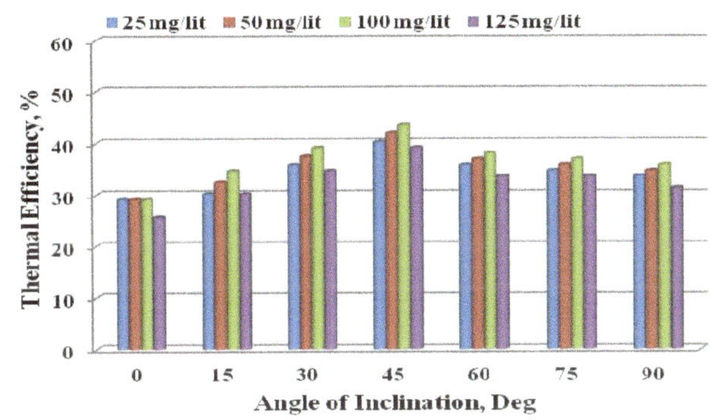

Fig. 5 Variations of heat pipe efficiency for various inclinations at 50 W heat input

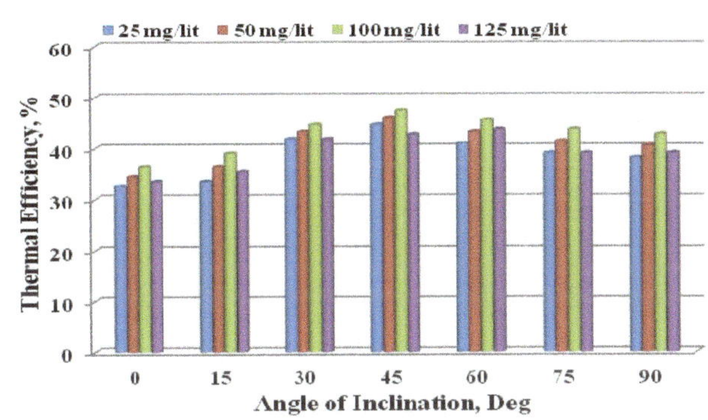

Fig. 6 Variations of heat pipe efficiency for various inclinations at 60 W heat input

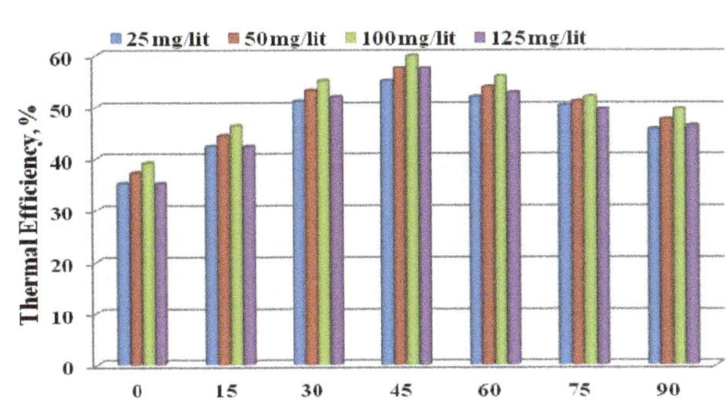

Fig. 7 Variations of heat pipe efficiency for various inclinations at 70 W heat input

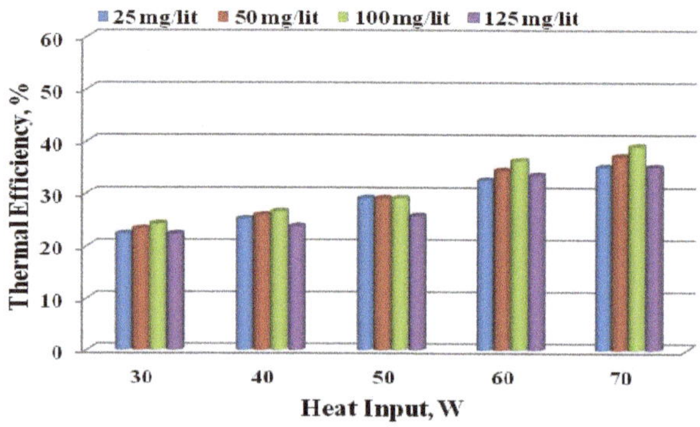

Fig. 8 Variations of heat pipe efficiency with different heat inputs at 0° inclination

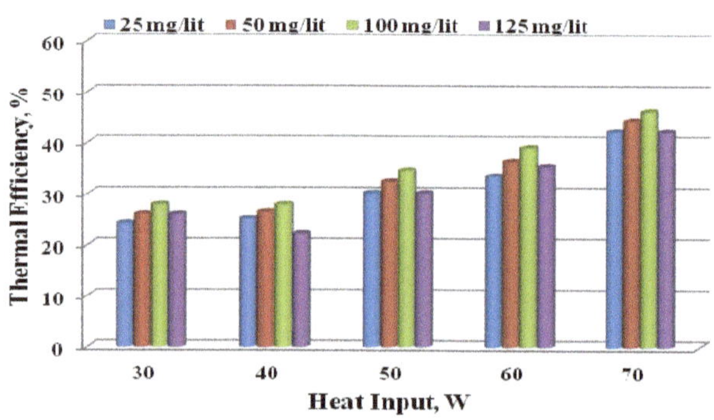

Fig. 9 Variations of heat pipe efficiency with different heat inputs at 15° inclination

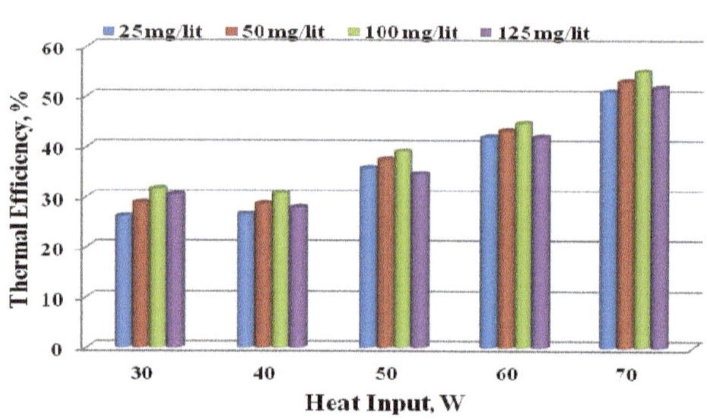

Fig. 10 Variations of heat pipe efficiency with different heat inputs at 30° inclination

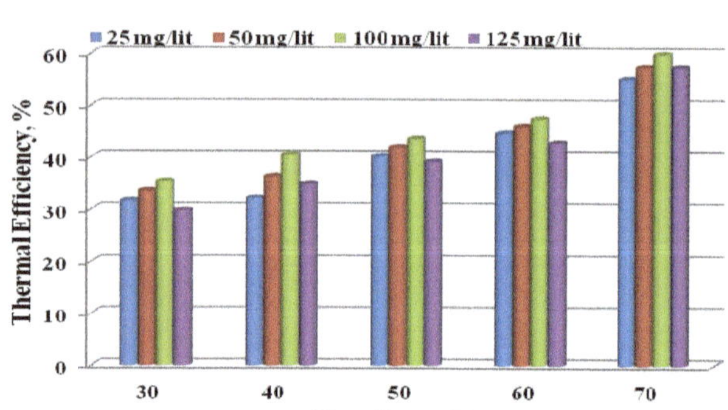

Fig. 11 Variations of heat pipe efficiency with different heat inputs at 45° inclination

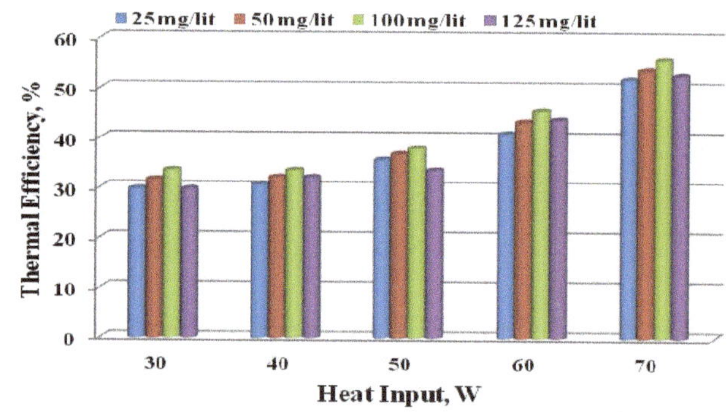

Fig. 12 Variations of heat pipe efficiency with different heat inputs at 60° inclination

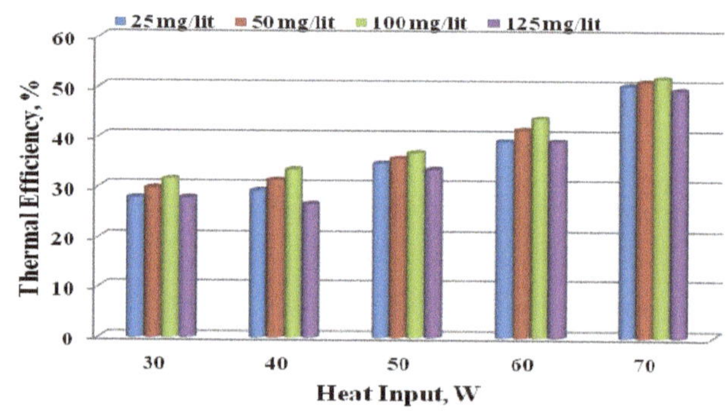

Fig. 13 Variations of heat pipe efficiency with different heat inputs at 75° inclination

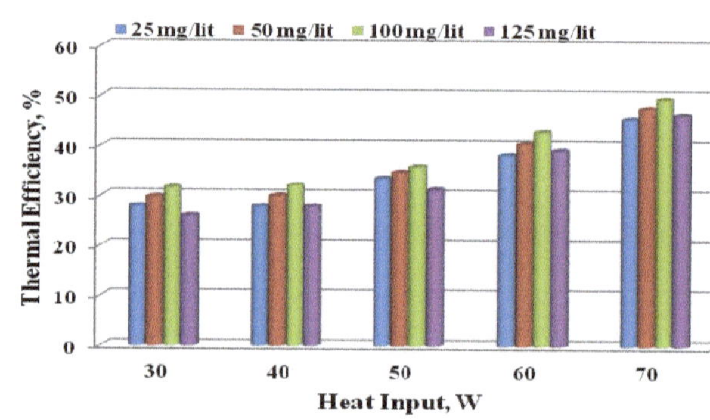

Fig. 14 Variations of heat pipe efficiency with different heat inputs at 90° inclination

4. CONCLUSION

The main aim of this work is to highlight the importance of the copper nanofluid concentrations in the heat pipe under various loads and angle of inclinations. The experiments are conducted with different heat loads and angle of inclinations. The experimental results show that the optimum level of copper nanofluids in the base fluid (DI water). At higher concentrations the copper nanofluids are agglomerates with water; hence its sizes are increased**.** The agglomerated nano particles get deposited gradually and reduce the stability and heat transfer capabilities of the nanofluid. This increased size particles gives the higher resistance to the fluid flow. As a result, the higher thermal performances of nanofluids indicate nano-fluid potential as a substitute for conventional pure water in heat pipes. This finding makes nano-fluids more attractive as a cooling fluid for devices with high energy density. To reveal this phenomenon, further studies on nano-fluid behavior in heat pipes and the properties of nano-fluids may be performed.

ACKNOWLEDGMENT

The authors thank the authorities of Annamalai University for providing the necessary facilities in order to accomplish this piece of work.

REFERENCES

Lin, S., Kew, P.A., Cornwell, K., 2001, "Two-Phase Heat Transfer to a Refrigerant in a 1 mm Diameter Tube", *International Journal of Refrigeration,* **24**(1), 51–56.
http://dx.doi.org/10.1016/S0140-7007(00)00057-8

Song, F., Ewing, D., Ching, C.Y., 2004, "Experimental Investigation on the Heat Transfer Characteristics of Axial Rotating Heat Pipes", *International Journal of Heat and Mass Transfer*, **47**(22), 4721–4731.
http://dx.doi.org/10.1016/j.ijheatmasstransfer.2004.06.001

Xuan, Y., Hong, V., Li, Q., 2004 "Investigation on Transient Behaviors of Flat Plate Heat Pipes", *Experimental Thermal and Fluid Science,* **28**(2-3), 249–255.
http://dx.doi.org/10.1016/S0894-1777(03)00047-5

Mughal, M.P., Plumb, O.A., 1996 "An Experimental Study of Boiling on a Wicked Surface", *International Journal of Heat Mass Transfer,* **39**(4), 771–777.
http://dx.doi.org/10.1016/0017-9310(95)00159-X

Brautsch, A., Kew, P.A., 2002 "Examination and Visualization of Heat Transfer Processes During Evaporation in Capillary Porous Structures", *Applied Thermal Engineering,* **22**(7), 815–824.
http://dx.doi.org/10.1016/S1359-4311(02)00027-3

Choi, S.U.S., 1995, "Enhancing Thermal Conductivity of Fluids with Nanoparticles", *ASME, FED*, **231**, 99–105.

Anoop. K.B., Sundararajan, T., Das, S.K., 2009, "Effect of Particle Size on the Convective Heat Transfer in Nanofluid in the Developing Region", *International Journal of Heat Mass Transfer,* **52**(9-10), 2189–2195.
http://dx.doi.org/10.1016/j.ijheatmasstransfer.2007.11.063

Jung, J., Yoo, J.Y., 2009, "Thermal Conductivity Enhancement of Nanofluids in Conjunction with Electrical Double Layer (EDL)", *International Journal of Heat Mass transfer,* **52**(1-2), 525–528.
http://dx.doi.org/10.1016/j.ijheatmasstransfer.2008.07.016

Riehl, R.R., 2006, "Analysis of Loop Heat Pipe Behavior Using Nanofluid, in Heat Powered Cycles" *International Conference (HPC)*, New Castle, UK, Paper 06102.

Tsaia, C.Y., Chiena, H.T., Ding, P.P., Chanc, B., Luhd, T.Y., Chen, P.H., 2004, "Effect of Structural Character of Gold Nanoparticles in Nanofluid on Heat Pipe Thermal Performance", *Material Lett*ers **58**(9), 1461–1465.
http://dx.doi.org/10.1016/j.matlet.2003.10.009,

Xuan, Y., Li, Q., 2000, "Heat Transfer Enhancement of Nano-fluids", *International Journal of Heat and Fluid Flow,* **21**(1), 58–64.
http://dx.doi.org/10.1016/S0142-727X(99)00067-3

Das, S.K., Putra, N., Thiesen, P., Roetzel, W., 2003, "Temperature Dependence of Thermal Conductivity Enhancement for Nano-fluids", *Journal of Heat Transfer*, **125**(8), 567–574.
http://dx.doi.org/10.1115/1.1571080.

Wei, W,C., Tsai, S.H., Yang, S.Y., Kang, S.W., 2005, " Effect of Nanofluid Concentration on Heat Pipe Thermal Performance", *International Associations of Mechanical Engineers Transactions* **2**, 1432– 1439.

Kang, S.W., Wei, W.C., Tsai, S.H., Yang, S.Y., 2006 "Experimental Investigation of Silver Nanofluid on Heat Pipe Thermal Performance", *Applied Thermal Engineering,* **26**(17-18), 2377–2382.
http://dx.doi.org/10.1016/j.applthermaleng.2006.02.020

Xuan, Y., Li, Q., 2003, "Investigation on Convective Heat Transfer and Flow Features of Nanofluids", *Journal of Heat Transfer*, **125**(1), 151 – 155.
http://dx.doi.org/10.1115/1.1532008.

Senthilkumar, R., Vaidyanathan, S., Sivaraman, B., 2011, "Performance Investigation of Heat Pipe Using Aqueous Solution of n-pentanol with Different Inclinations", *Journal of Mechanical Science and Technology*, **25** (4), 1253-1260
http://dx.doi.org/10.1007/s12206-011-0207-4.

HEAT PIPE SIMULATION UNDER CRITICAL CONDITIONS

Taoufik Brahim,* Abdelmajid Jemni

Laboratory of Thermal and Energy Systems Study (LESTE), National School of Engineers of Monastir (ENIM), University of Monastir (UM), Road Ibn Jazzar 5019 Monastir, Tunisia

ABSTRACT

Flow, heat and mass transfer in heat pipes has been studied extensively. This is due to the increasing need in understanding the complicated transport process for application of diverse fields. Recently one of the heat pipes which aroused the interest of research are the three dimensional heat pipes which incorporates the phase change mechanisms at the liquid and vapour interface. In this paper, an originally three-dimensional steady state model is developed coupling the hydrodynamic and thermal models in sintered heat pipe in cylindrical polar coordinate. The model predicts the heat pipes behaviors under critical conditions were a high heat input is exposed to its evaporator region. Such study has not been well investigated before and can be useful for academic, engineering and industrial purposes. Compared with its normal working conditions were heat pipes limits are not reached, the results illustrate well the function of the heat pipe and the circulation of the working fluid inside the heat pipe regions. Results show also that, for high heat input the circumferential velocity components increase specially in the liquid-vapor interface and the tangential velocity component has a significant effect on the phase change phenomena.

Keywords: *Three-dimensional, Conventional Heat Pipe, Numerical Simulation, Visualization.*

1. INTRODUCTION

Heat pipes behavior under critical conditions were it is exposed to higher heat fluxes or from the frozen state has been not well investigated. Deverall et al (1970) studied the transient behavior of water and metal liquid heat pipes from the frozen state. They concluded that successful star-up for higher heat input was possible if the heat rejection in the condenser region was low enough to enable the heat to melt the working fluids in the condenser and allow liquid to return to the evaporator. Issacci et al (1988) studied the transient behavior of the vapor flow in a rectangular heat pipe. Flow reversal was detected in the condenser and adiabatic regions with water as the working fluid. However, only some axial and radial velocity profiles were presented in the paper. They detected that for high input heat flux, a reverse flow in both adiabatic and condenser region may occurs and suggested that a parametric study in needed to explore the reverse flow versus design parameters like the evaporator, condenser, adiabatic and total length of the heat pipe for different input heat fluxes and condenser temperatures. Reverse flow in the condenser section was also observed by (Ooijen and Hoogindoorn 1979) who performed an experimental and numerical studies on heat pipe vapor flow, their analysis was confined to a steady state, incompressible, two dimensional flows. They also found that the total pressure drop over the heat pipe, for high radial Reynolds numbers, was higher than that approximated by a Poiseuille flow model. The numerical model developed for incompressible, steady state, two dimensional vapor flows in a heat pipe by (Banskton and Smith 1972 and 1973) show that flow reversal in the condenser can occurs for radial Reynolds numbers larger than two. Indication of a reverse flow is important in design of a heat pipe in calculating the shear forces acting on the wick structure and affects the entrainment limit. Indeed, shear force existing at the liquid-vapor interface may tear the liquid from the wick surface and entrain it into the vapor flow steam. This phenomenon limits the heat transport capability of the device. Considering a wick structure of low pore dimensions, the shear forces could very low.

A three dimensional numerical modeling of circumferentially heated and block heated heat pipes was presented by Faghri (1991) to study the startup behavior of high-temperature heat pipes from the frozen state for various heat loads and input locations. Zhu and Vafai (1998) analyzed a three dimensional vapor and liquid flow in an asymmetric flat plate heat pipe. They studied the vapor flow by finite element method using FIDAP code. Vadakkan and al (2003, 2004) developed a three-dimensional numerical model to study the performance of flat heat pipes with multiple discrete heat sources. The effects of heat source strength and separation on steady as well as transient performance were studied. A transient, three-dimensional model for thermal transport in flat heat pipes considering wick microstructure effects is developed recently (Carbajal and al 2007). The Marangoni convection effects and the accommodation coefficient during phase change at the liquid–vapor interface are investigated for a heat pipe with a screen-mesh wick. Xiao and Faghri (2008) developed also a three-dimensional model for flat heat pipe which accounted for heat conduction in the wall and fluid flow in the vapor core and porous wicks.

Most three-dimensional previous studies are performed for flat heat pipe due their applications in electronic cooling or/and spacecraft radiator segments. However, convectional heat pipes have many industrial applications such as heat exchanger systems and a three dimensional analysis with own tool simulation is necessary to predict the heat pipe behavior for the complexity of the system.

* Corresponding author. Email: taoufik.brahim@yahoo.fr

In this paper, a three dimensional model has been performed in cylindrical polar coordinate for the conventional heat pipe coupling the heat pipe three regions (wall, wick, vapor).
A code written with FORTRAN was developed using the finite volume method to solve the governing equations and the wick-level micro-model using the kinetic theory of gases is utilized only to correct the phase change mass flow rates at the liquid-vapor interface (Eq. (14)) in the macro-model.
The Navier–Stokes equations along with the energy equation are solved numerically. Darcy-Brinkman-Forchheimer model is used in the wick structure and evaporation-condensation at the liquid–vapor interface is modeled using the gas kinetic theory model. The heat pipe is exposed to normal and critical heat input to visualize the different velocities within and the results illustrate well the function of the heat pipe and the circulation of the working fluid inside the heat pipe regions.

2. NUMERICAL MODEL

In cylindrical polar coordinate system, the axis " z " is the axis of symmetry lies along the centre of the heat pipe and taken parallel to the base flow velocity. " r " is the radial distance and " φ " is the azimuthal angle. A schematic description of the cylindrical heat pipe is shown in the Fig.1:

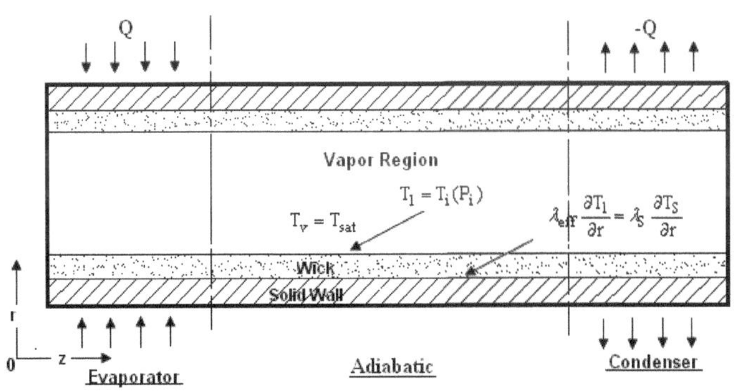

Fig. 1 Heat pipe description

The temperature distribution is a function of (r, z). The Navier-Stokes equations, the energy equation and the continuity equation for the case of the steady axi-symetric motion of an incompressible fluid in a porous media which accounts for the boundary and inertial effects are (Ranjan and al 2011):

- *Continuity equation:*

$$\frac{\partial U_r}{\partial r}+\frac{U_r}{r}+\frac{\partial U_z}{\partial z}=0 \tag{1}$$

- *Navier-Stokes equations:*

$$\rho_l\left(U_r\frac{\partial U_z}{\partial r}+U_z\frac{\partial U_z}{\partial z}\right)=-\varepsilon\frac{\partial P}{\partial z}+\mu\left(\frac{\partial^2 U_z}{\partial r^2}+\frac{1}{r}\frac{\partial U_z}{\partial r}+\frac{\partial^2 U_z}{\partial z^2}\right) -\frac{\mu\varepsilon}{K}U_z-\frac{C_F}{\sqrt{K}}|V|U_z \tag{2}$$

$$\rho_l\left(U_r\frac{\partial U_r}{\partial r}-\frac{U_\phi^2}{r}+U_z\frac{\partial U_r}{\partial z}\right)=-\varepsilon\frac{\partial P}{\partial r} +\mu\left(\frac{\partial^2 U_r}{\partial r^2}+\frac{1}{r}\frac{\partial U_r}{\partial r}-\frac{U_r}{r^2}+\frac{\partial^2 U_r}{\partial z^2}\right)-\frac{\mu\varepsilon}{K}U_r-\frac{C_F}{\sqrt{K}}|V|U_r \tag{3}$$

$$\rho_l\left(U_r\frac{\partial U_\phi}{\partial r}+\frac{U_r U_\phi^2}{r}+U_z\frac{\partial U_\phi}{\partial z}\right)=\mu\left(\frac{\partial^2 U_\phi}{\partial r^2}+\frac{1}{r}\frac{\partial U_\phi}{\partial r}-\frac{U_\phi}{r^2}+\frac{\partial^2 U_\phi}{\partial z^2}\right) -\frac{\mu\varepsilon}{K}U_\phi-\frac{C_F}{\sqrt{K}}|V|U_\phi \tag{4}$$

- *Energy equation:*

$$U_r\frac{\partial T}{\partial r}+U_z\frac{\partial T}{\partial z}=\alpha_{eff}\left(\frac{\partial^2 T}{\partial r^2}+\frac{1}{r}\frac{\partial T}{\partial r}+\frac{\partial^2 T}{\partial z^2}\right) \tag{5}$$

where U_z, U_r, U_φ is the axial, radial and azimuthal velocity components respectively. $|V|=\sqrt{U_r^2+U_z^2+U_\varphi^2}$, and $\sqrt{K}$ is the effective pore diameter in the porous media. C_F is the Egun's constant which value is approximately 0.55.

$\alpha_{eff}=\frac{\lambda_{eff}}{\varepsilon\rho_l C_{pl}}$ is the effective thermal diffusivity.

K is the permeability of the sintered powder wick given by (Chi 1976):

$$K=\frac{r_s^2\,\varepsilon^3}{37.5(1-\varepsilon)^2} \tag{6}$$

where, $r_c = 0.41r_s$ represent the effective capillary radius.
In the vapor core, the equations governing the heat and mass transfers are given by considering the above equation in which the permeability and the porosity are respectively taken equal to infinity and one. Also the other thermo-physical properties like density, thermal conductivity and viscosity are changed to that of the vapor.
The heat pipe wick (sintered powder wick) was assumed to be a homogeneous and isotropic porous medium (constant porosity and permeability). The dissipation effects were neglected (Faghri 1995). The advantage of the sintered powder wick is its ability to handle higher heat flux. Indeed, typical sintered wicks handle 50 W/cm2, and have been tested to up 250 W/cm^2.

$$\lambda_{eff}=\frac{\lambda_l[(2\lambda_l+\lambda_w)-2(1-\varepsilon)(\lambda_l-\lambda_w]}{[(2\lambda_l+\lambda_s)+(1-\varepsilon)(\lambda_l-\lambda_w]} \tag{7}$$

λ_{eff} is the effective thermal conductivity in the wick structure (Chi 1976). In the solid region, λ_{eff} become λ_s.

λ_w is the is the thermal conductivity of the wick material (copper).

2.1 Boundary Conditions

- At liquid-solid wall interface:

$$\begin{cases} U_z=U_r=U_\varphi=0 \\ \lambda_{eff}\frac{\partial T_l}{\partial r}=\lambda_s\frac{\partial T_s}{\partial r} \end{cases} \tag{8}$$

- At heat pipe ends $z=0$ and $z=L$:

$$\begin{cases} U_z = U_r = U_\varphi = 0 \\ \dfrac{\partial T_l}{\partial z} = 0 \end{cases} \tag{9}$$

- At liquid-vapor interface:

$$\begin{cases} U_r = V_i ; U_z = U_\varphi = 0 \\ T_l = T_i(P_i) \end{cases} \tag{10}$$

A constant heat flux " Q" is exposed at the outer surface of the heat pipe wall; positive in the evaporator region and negative in the condenser region:

$$-\lambda_s \frac{\partial T_s}{\partial r} = Q \tag{11}$$

- At heat pipe ends $z=0$ and $z=L$:

$$\frac{\partial T_s}{\partial z} = 0 \tag{12}$$

The energy balance at the liquid /vapor interface yields:

$$\rho_l V_i C_{pl} T_i - \lambda_l \frac{\partial T_l}{\partial r} = \rho_v V_i C_{pv} T_i - \lambda_v \frac{\partial T_v}{\partial r} + \rho_v V_i h_{fg} \tag{13}$$

V_i is the injection velocity corrected until the mass flux residual reduces to an acceptable value. To account for corrected injection velocity at the liquid vapor interface, the evaporation and condensation mass fluxes are calculated from the kinetic theory of gases (Eames and al 1997).

$$V_i = \frac{a_{cc}}{\rho_v} \left(\frac{M}{2\pi T_i} \right)^{1/2} \left[P_i(T_i) - P_v \right] \tag{14}$$

where " a_{cc} "is the accommodation coefficient assumed to unity, " M " is mass molar and " T_i " is the liquid interface temperature calculated by the Clausius-Clapeyron equation as follow:

$$\frac{R}{h_{fg}} \ln\left(\frac{P_i}{P_0} \right) = \frac{1}{T_0} - \frac{1}{T_i} \tag{15}$$

Since the density of the vapor is linked to saturation temperature and vapor pressure to surface tension, an iterative schedule is then added to calculate the vapor saturation temperature.
The simulations were done for water as working fluid. The reference pressure and temperature are respectively P_0 = 1.019710^{5}Pa and T_0 = 273.15K.The effect of wall temperature gradient is neglected and the heat pipe characterizations were summered in the Table 1.
The above system of equations (Eq. (1-5)), are transformed to dimensional form to simplify the resolution. Therefore, all the calculated velocity components are transformed in dimensional form: U/V_i (Carbajal and al 2007).

Table 1: Heat pipe specific characterizations

HP container material Copper: $\lambda_s = 387.6$ W/mK	porous media: Sintered powder wick
Wick thickness: 1 mm	Vapor diameter: 0.0127 m
HP length: 1 m	Porosity: 0.5
Condenser length: 0.4 m	Evaporator length: 0.4 m
Permeability: 0.5810^{-11}m^2	Working fluid: Distilled water
Vapor conductivity: $\lambda_v = 0.0251$W/mK	Liquid density: $\rho_l = 960.63$ kg/m^3
Vapor density: $\rho_v = 0.59$kg/m^3	Surface tension: $\sigma = 5.8910^{-2}$Pam
Heat latent: $h_{fg} = 2251.2$ kJ/kg	Liquid heat capacity: $C_{Pl} = 2034$ J/kg K

3. RESULTS AND DISCUSSION

The code developed to solve the above system of coupled equations used the finite volume method with the power law scheme and SIMPLE algorithm. Many numerical tests were made in order to establish the best grid size for the numerical calculations by varying the number of control volume in both directions. The results indicated that a grid size of 12x360x262 and 62x360x262 (radial, azimuthal, axial) was good enough for the present calculations in the porous media and the vapor regions respectively.

To make sure of the validity of our computational code, a comparison was made in two dimensional analyses with the analytical earlier work of Zhu and Vafai (1999) and the numerical results of Tournier and El-Genk (1994) since luck of three-dimensional studies on conventional heat pipe founded in the literature. In Fig.2 and Fig.3 the outer wall surface temperature and the pressure profiles within the wick structure and the vapour phase are compared with both numerical and analytical previous works. A good agreement is found.

Fig.4 compares the velocity profiles in the liquid and the vapor regions. The positive velocity value denotes that the flow is from the evaporator to the condenser region and the negative velocity value denotes that the flow is from the condenser to the evaporator region. A good agreement is found too when considering the two dimensional model with low Reynolds number.

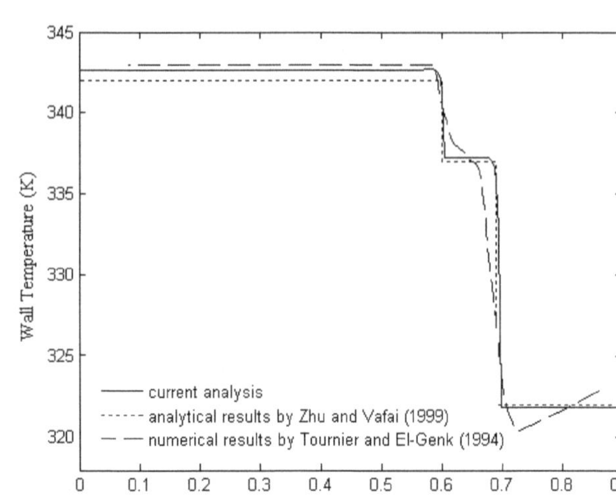

Fig. 2 Wall surface temperature distribution compared with previous works

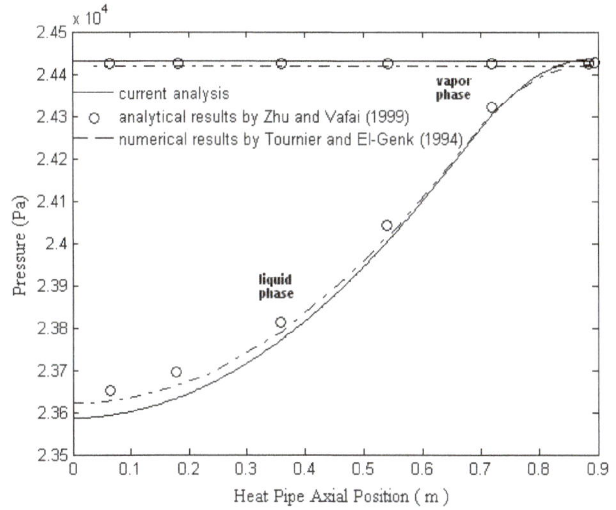

Fig. 3 Pressure profiles in the liquid and the vapor regions of heat pipe compared with previous works

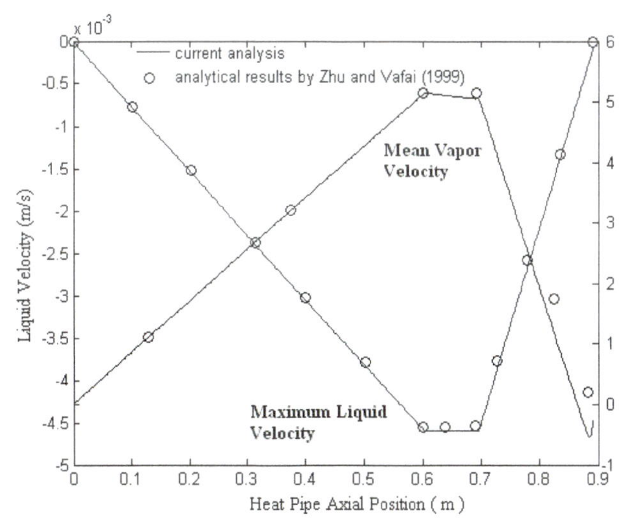

Fig. 4 Validation results of maximum liquid velocity and mean vapor velocity in bidirectional analysis.

Fig.5 and Fig.6 shows the dimensional velocity profile in the vapor region as seen looking down the pipe axis. NZ means the number of grid points along the heat pipe axial direction (same method used by Faghri 1991). The Fig.5 shows that for a normal heat input/output of 5000 W/m2, a symmetry profiles can be concluded in the evaporator and the condenser regions and fully developed profile in the adiabatic region. The maximum radial vapor velocity in the evaporator region is about 8.16910^{-4} m/s and about 0.001 m/s in the condenser region and slows in the adiabatic region (1.8010^{-4} m/s). The azimuthal velocity is very slow in both heat pipe regions (4.1810^{-5} m/s in the evaporator, 3.6610^{-6} m/s in the adiabatic and 4.06410^{-5} m/s in the condenser region); it has no significant effect in the phase change phenomena. In the condenser region the axial vapor velocity decreases slowly in the centerline region and reach a maximum of about 0.0911 m/s.

In Fig.6 and with critical heat flux density (50000W/m^2) a reverse flow was clearly appeared in the condenser end cap. The heat pipe length was insufficient for the flow field to develop. In condenser region the vapor decreases much slowly in the centerline region and first recirculation flow appears for NZ = 227 to the end of the condenser region; witch correspond to a condenser length of nearly 0.13 m in where reverse flow take place. The maximum axial vapor velocity is about 0.9544 m/s and the maximum reverse axial vapor velocity is -0.0368m/s. the maximum radial vapor velocity are 0.0087m/s, 0.004m/s and 0.0495m/s in the evaporator, adiabatic and condenser regions respectively. The maximum azimuthal velocities are 0.0054m/s, 0.0052 and 0.0028 in the evaporator, adiabatic and condenser regions respectively. One can see that this velocity is greater than the radial velocity in adiabatic region. It can contribute to the reverse flow in this region. The maximum reverse flow velocity in the condenser for the azimuthal velocity is -0.0046m/s. Note that such visualizations which in three dimensional analyses (see Faghri 1991) can be very useful when modeling heat pipe to see the effect of the heat pipe parameters (wick and vapor properties, heat pipe dimensions, etc…).

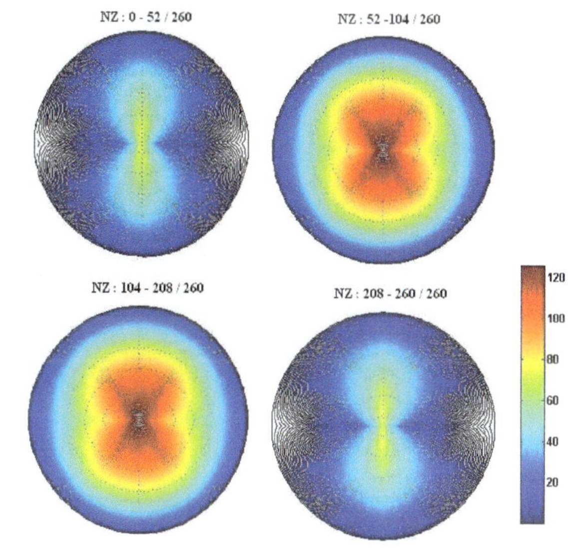

Fig. 5 Dimensional vapor velocity for heat input Q = 5000 W/m^2

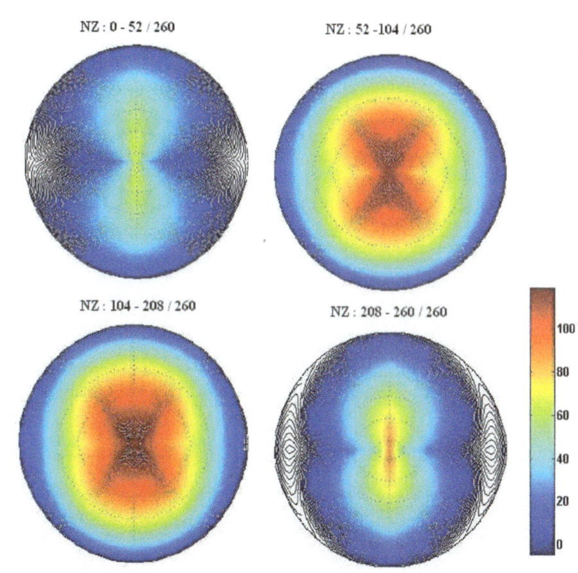

Fig. 6 Dimensional vapor velocity for heat input Q = 50000 W/m^2

Fig.7 shows the liquid circumferential components velocity as seen looking down the pipe axis for heat input/output density of 5000W/m2. The liquid profile is fully developed in the porous structure. Near the wall/liquid interface the velocity is zero along the heat pipe length because the no-slip condition. The maximum axial liquid velocity is reached in the adiabatic section (-1.234210^{-4} m/s) and it is the dominant velocity. The uniform blowing/suction about the circumference of the liquid-wick/vapor interface can be seen in this Fig.7. For high heat input, the axial liquid velocity increases since it depends on of the heat flux and the injection velocity. Its maximum axial liquid velocity is about -1.210^{-3} m/s (Zhu and Vafai 1999).

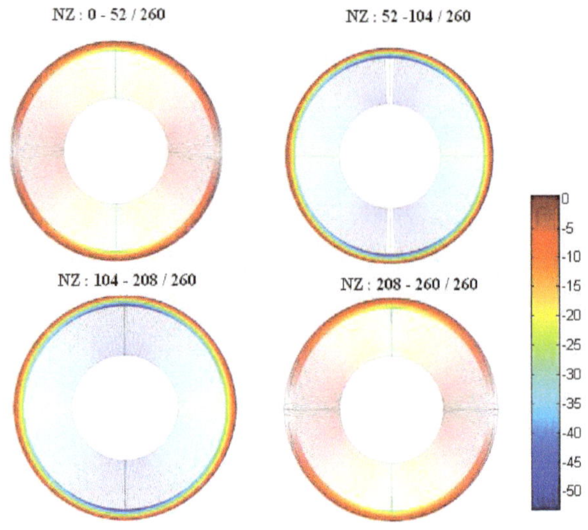

Fig. 7 Dimensional liquid velocity for heat input Q = 5000 W/m^2

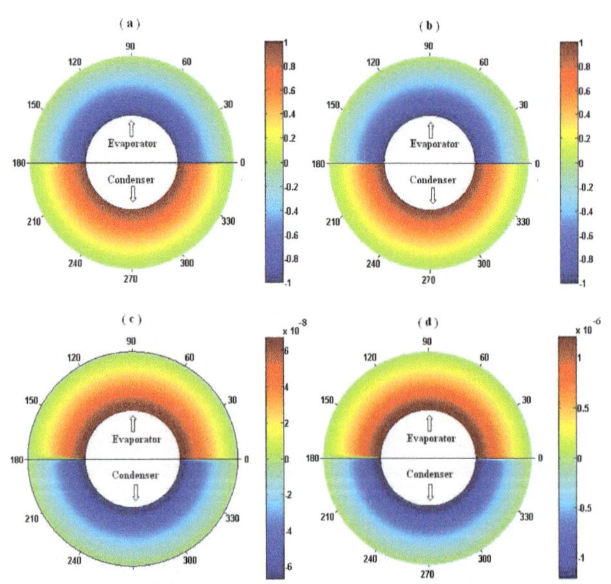

Fig. 8 Dimensional radial velocity components (a, 5000W/ m^2), (b, 50000W/m^2) and azimuthally velocity components (c, 5000W/ m^2), (d, 50000W/ m^2) in the liquid region for two different heat loads.

Fig.8 shows that the dimensional radial velocity for two heat loads (a-b) in the liquid structure has the same profile but increase nearly linearly when the heat loads increase. For normal heat flux (a) the maximum radial velocities are 1.069210^{-6} m/s and 1.05340^{-6} m/s in the evaporator and the condenser regions respectively. For critical heat flux (b), they equal to 1.07010^{-5} m/s 1.04910^{-5} m/s in the evaporator and the condenser regions respectively. In the adiabatic region, the radial velocity in low (magnitude order of 10^{-8}) since no phase change can occur in this region. One can see that the maximum radial velocity occurs near the interface liquid-vapor and become zero in the interface solid wall / wick structure. As seen looking down the pipe axis, the adiabatic region is removed since no significant effect has been noted and this is because no phase change can occurs. The azimuthal velocity (c-d) increase also when the heat load input increase but remain very low (magnitude order between 10^{-14} and 10^{-11}). We can conclude that a two dimensional model in the porous media can describe well the heat pipe wick region since the tangential velocity has no significant effect in this heat pipe situation.

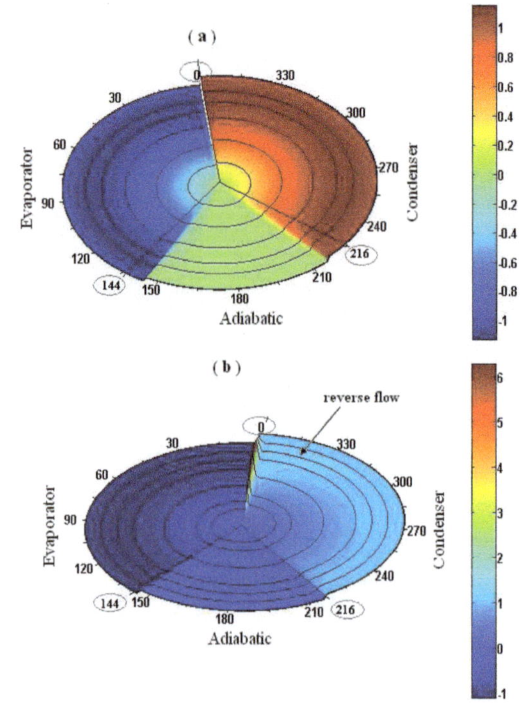

Fig. 9 Dimensional radial velocity components in the vapor region for two different heat loads (a, 5000W/m^2; b, 50000W/m^2)

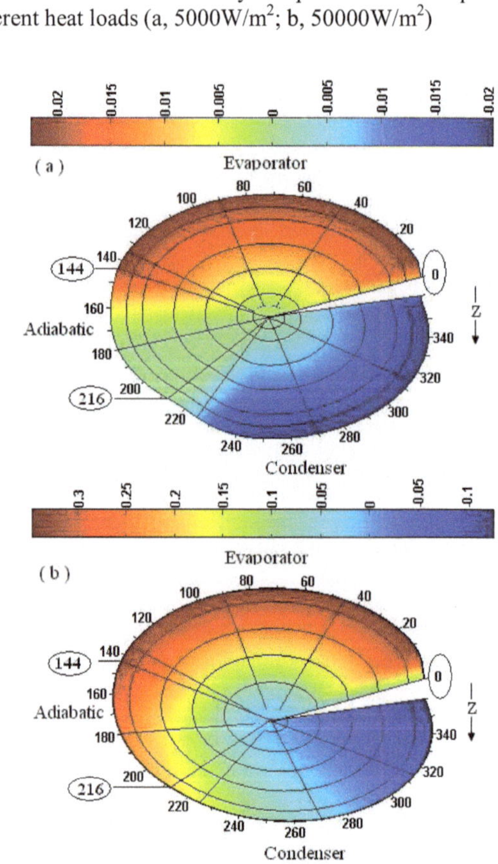

Fig. 9 Dimensional azimuthal velocity components in the vapor region for two different heat loads (a, 5000W/m^2; b, 50000W/m^2)

Note that, the radial velocity depends also on the dimensions and the specific characterizations of the porous media (Eames and al 1997).

The effect of the heat flux on the radial vapor velocity (a-b) is shown in Fig.9 in both heat pipe regions. The radial vapor velocity in accelerated in the condenser end corner (where a reverse flow occurs) and in the adiabatic region ends (Fig.9 (b)); it can contribute to phase change when the heat pipe is exposed to high heat flux density.
The effect of the azimuthal vapor component velocity U_φ is shown in the Fig.10 (a) for normal heat flux and Fig.10 (b) for high heat flux. This velocity increases when the heat increase. One can see that the maximum value of U_φ is reached near the liquid-vapor interface in the vapor region. Its effect on the liquid-vapor interface shape has been not yet well investigated as our knowledge. It becomes greater than the radial velocity in the adiabatic region (a maximum of 0.0052 m/s for the azimuthal velocity and a maximum of 0.004 m/s for the radial velocity). The azimuthal vapor velocity component discovered to be not zero in the adiabatic region when the heat fluxes increase (Fig. 10 (b)). A phase change can occur in this region because the condenser area becomes incapable to reject the great amount of heat to the surrounding ambient. Also, when the tangential velocity increases, the entrainment phenomena increase at the liquid-vapor interface. This result was detected by Bankston and al (1972, 1973) which observed that flow reversal may affect the entrainment limit. We can note also, that when the reverse flow occurred in the end condenser corner of the heat pipe the azimuthal velocity also increase as shown in the Fig.10 (b) when the heat input increase. Note that, in Fig.9 and Fig.10, we are trying to separate the heat pipe regions, so a special visualizing in made taking into account the axial direction on the heat pipe to better visualization. The evaporator region is between 0 and 143 °, the adiabatic region is between 144° and 216° and the condenser region is between 217° and 360°.

4. CONCLUSION

An originally three-dimensional steady state model for thermal and hydrodynamic transport in heat pipe is developed in cylindrical polar coordinate coupling the three regions in the heat pipe. Kinetic theory of gases is used to account the evaporation and the condensation process at the liquid-vapor interface. A code using the finite volume method was developed to solve the governing equations. The results are compared to two dimensional analytical results for liquid and vapor axial velocities. The good agreement confirms the numerical methods. Following these results, we can confirm that results illustrate well the function of the heat pipe and the circulation of the working fluid inside the heat pipe regions. The maximum circumferential velocity occurred near the liquid vapor interface and increase when the heat loads increase. Results show also that the circumferential velocity is not zero in the adiabatic region when the heat pipe is exposed to higher heat input. This result explain why the at higher heat flux the entrainment limit is affected. The three dimensional analysis yields to more accurate results and better visualizations especially under different external thermal solicitations and boundary conditions, or for different heat pipe characterizations and dimensions. This may be useful in heat pipe modeling and optimization for industrial or academic purpose.

NOMENCLATURE

a_{cc}	accommodation factor
C_p	specific heat (J/kg.K)
C_F	Ergun's constant=0.55
h_{fg}	latent heat (KJ/kg)
K	permeability (m^2)
L	heat pipe length (m)
P	pressure (Pa)
Q	heat flux density (W/m^2)
r	radial coordinate (m)
r_s	average sphere radius = 2.110^{-5}m
r_c	contact radius (m)
T	temperature (K)
U	velocity (m/s)
V_i	injection velocity (m/s)
z	axial coordinate (m)

Greek Symbols

α	thermal diffusivity (m^2/s)
ε	porosity
φ	azimuthal angle (rad)
λ	thermal conductivity (W/m.K)
μ	fluid absolute viscosity (kg/m.s)
ρ	fluid density (kg/m^3)
σ	surface tension (Pa.m)

Subscripts

eff	effective
i	injection, interface
l, v, s	liquid, vapor, solid wall
0	reference
sat	saturation
w	wick

REFERENCES

Deverall, J.E., Kemme, J.E., and Florschuetz, L.W., 1970, "Sonic Limitations and Startup Problems of Heat Pipes," *LA-4518. Doit W-7405-ENG-36.*

Issacci, F., Catton, I., Ieiss, A., and Ghoniem, N.M., 1988, "Analysis of Heat Pipe Vapor Dynamics," *Proc. 25th ASME National Heat Transfer Conf.*, Vol. 1, pp. 361-365.

Ooigen, H. V. and Hoogendoorm, C. J., 1979, Vapor Flow Calculations in a Flat-Plate Heat Pipe", *AIAA J.,* 17, No. 11, pp 1251-1259.

Bankston, C. A. and Smith, H. J., 1972,"Incompressible Laminar Vapor Flow in Cyliderical Heat Pipes", *ASME Paper No. 71-WA,,HT-15.*

Bankston, C. A. and Smith, H. J., 1973 "Vapor Flow in Cylindrical Heat Pipes':, *J. Heat Transfer,* 95, no. 8, pp 371-376.

A.Faghri., 1991, "Analysis of Frozen Startup of High – Temperature Heat Pipes and Three Dimensional Modeling of Block –Heated Heat Pipes, "*Defense Technical Information Center.*

N. Zhu and K. Vafai., 1998, "Vapour and Liquid Flow in an Asymmetric Flat Plate Heat Pipe: A Three Dimensional Analytical and Numerical Investigation," *Int. J. Heat Mass Transfer*, Vol. 41, No. 1, pp. 159-174.

U. Vadakkan, J.Y. Murthy, S.V. Garimella., 2003, "Transient analysis of flat heat pipes," *in: Proceedings of the ASME Summer Heat Transfer Conference*, July 21–23, LasVegas, Nevada.

U. Vadakkan, S.V. Garimella, J.Y. Murthy, 2004, "Transport in flat heat pipes at high fluxes from multiple discrete sources," *ASME J. Heat Transfer*26, 347–354

G. Carbajal, C.B. Sobhan, G.P. 2007 ''Bud" Peterson, D.T. Queheillalt , Haydn N.G. Wadley," A quasi-3D analysis of the thermal performance of a flat heat pipe," *International Journal of Heat and Mass Transfer* 50, 4286–4296.

B. Xiao, A. Faghri, 2008, "A three-dimensional thermal-fluid analysis of flat heat pipes," *Int. J. Heat Mass Transfer* 51, 3113–3126.

R. Ranjan , J. Y. Murthy , S. V. Garimella, U. Vadakkan, 2001, "A numerical model for transport in flat heat pipes considering wick microstructure effects", *Int. J. Heat and Mass Transfer* 54,153–168.

Chi,S. W. 1976, Heat Pipe Theory and Practice. *Hemisphere Publishing Company*, Washington, D.C., New York, McGraw-Hill Book Co., 1976. 256 p.

A.Faghri, 1995,"Heat Pipe Science and Technology", *Taylor and Francis*, PP. 136-138, pp. 268-281.

Eames, I. W., Marr, N. J., and Sabir, H., 1997,"The evaporation coefficient of water",: a review, *int. journal heat and mass transfer*, 40, 2963 2973.

N. Zhu., K. Vafai, 1999,"Analysis of cylindrical heat pipes incorporating the effects of liquid-vapor coupling and non-Darcian transport a closed form solution," *Int. J. Heat Mass Transfer* 37 3405-3418.

J.M.Tournier, M.S. El-Genk, 1994, "A heat pipe transient analysis model," *Int. J. Heat Mass Transfer* 37 (1994) 753-762.

12

HYDRODYNAMICS OF A CONFINED MENISCUS IN A SQUARE CAPILLARY TUBE AT LOW CAPILLARY NUMBERS

Gyana Ranjan Rana, Basant Singh Sikarwar, Sameer Khandekar*, P. K. Panigrahi

Department of Mechanical Engineering, Indian Institute of Technology Kanpur, Kanpur 208016 India

ABSTRACT

In microchannel fluid flow, understanding of the liquid-gas interface behavior is vital for developing a wide range of microfluidic devices. The dynamic contact angle of the liquid-air meniscus varies with its velocity and the ensuing meniscus shape has profound effect on the local transport characteristics in its vicinity. Depending on the application, dynamic menisci shapes eventually control the momentum, heat and mass transport coefficients in two-phase microchannel flow geometries, where such conditions are often encountered. To better understand the effect of dynamic contact angle on meniscus shape, high speed visualization of menisci of four different liquids (water, ethanol, glycerin and silicon oil) has been undertaken at different Capillary numbers. Quantitative information of the velocity field and its distribution near the moving liquid-air interface has been done using micro-PIV measurements in a 1 mm × 1 mm dry square capillary having deionized water as the working fluid. This provides vital information on the local flow transport characteristics (two-dimensional velocity fields on a longitudinal plane) in the wake of the meniscus. To augment and complement the study, three-dimensional simulation of the flow field near the liquid-air meniscus has also been performed on Comsol®, applying the two-phase flow level-set method. The results clearly demonstrate that, while the u-velocity profile in the liquid domain is parabolic (Poiseuille-type flow) in nature away from the interface, it drastically changes (become flatter) as we approach the meniscus. Close to the meniscus the flow becomes three-dimensional with both v and w velocities showing a double-vortex, the strength of the latter being lower than the former. This observation is clearly noted both by PIV data as well as the simulations. The characteristic of the flow field in the meniscus wake is the most important parameter which affects the viscous stress generated due to the meniscus motion. The study reveals that controlling the wettability of the liquid can be an effective tool to control the overall transport behavior of the moving confined meniscus.

Keywords: *Micro-PIV experiments, Meniscus shape, Capillary number, Interface dynamics, Square capillary, Flow simulation*

1. INTRODUCTION

Transport phenomenon at liquid-air interface inside microscale devices has emerged as an important research field due to its relevance in many upcoming systems and devices. For example, interfacial flows occur in micro-total analysis system (Juncker et al., 2002), in lab-on-a-chip devices (Walker and Beebe, 2002), micro fuel cells (Pattekar and Kothare, 2004) and micro-/pulsating heat pipes (Khandekar et al., 2010). From a transport perspective, creation of two-phase fluid flows (gas-liquid, vapor-liquid or liquid-liquid systems) inside micro-channel geometries is considered to be one of the most suitable techniques available for enhancement of heat/mass transfer. Transportation of small fluid volume inventory with the advantage of large available interfacial area and small diffusion paths, all lead to attractive advantages from such systems. Understanding the fundamental hydrodynamics of such flows inside confined microchannel geometries is helpful for enhancing the performance parameters in applications like high speed coating, dying, adhesive printing, flow through porous media (petroleum industry), passive heat pipes, laser devices, bio-fluid engineering and bio-medical diagnostics and in chemical and material synthesis etc. Systematic physical insight into the local fluid flow and transport characteristics of dynamic liquid-air interfaces is thus vital for developing a wide range of such emerging micro-devices and systems.

While many of the reported experimental studies have been carried out using circular capillary tubes, interfacial flows inside a square capillary is more complex due of the difficulty in general treatment of motion (Thulasidas et al., 1997; Ajaev and Homsy, 2006). For many applications involving array of microchannels on flat plate substrates, a square or trapezoidal cross sectional geometry is preferred due to manufacturing routes employed. Furthermore, from a fundamental internal convective flow consideration, all other conditions remaining identical, a square channel provides lesser pressure drop than an equivalent circular channel (Shah and London, 1978).

Till now, a variety of experimental and numerical techniques have been adopted to analyze the flow field inside microchannels (Torralba et al., 2005). One popular contemporary technique is the micro-Particle Image Velocimetry (μ-PIV), which offers a powerful, nearly non-intrusive tool to study the liquid flow field. Santiago et al. (1998) developed a micron-resolution μ-PIV to measure ensemble-average and instantaneous liquid flow field in micro-scale devices. They measured velocities in a Hele-Shaw flow around a 30 μm elliptical cylinder, with a bulk velocity of 50μm/s using an epi-fluorescent microscope and 100-300 nm diameter seeding particles. They also specified that the spatial resolution and accuracy of fluid velocity measurements is limited by

* Corresponding author: samkhan@iitk.ac.in

the diffraction limit of the recording optics, noise in the particle image field, the interaction of fluid with finite size seed particles and Brownian motion. Meinhart et al. (1999) have presented the ensemble averaged velocity measurement in a 30 µm x 300 µm rectangular microchannel. Measurements were obtained using low spatial resolution away from the wall and a higher spatial resolution close to the wall. Reported PIV measurements were within 2% of the analytical solution for the flow inside a rectangular channel. Lima et al. (2007) also obtained nearly parabolic velocity profiles for pure water and in-vitro blood, flowing through a 100 µm square-microchannel at low Reynolds number. Gunther et al. (2004) used µPIV to characterize the segmented liquid-gas flow at low Capillary number. The recirculation patterns inside the liquid slug and the symmetric nature of this recirculation was reported. It was pointed that the surface roughness and the compressibility of gas phase are the main reasons behind the loss of symmetry for the velocity inside liquid slugs. This enhances mixing across the centerline of straight micro-channels. Li and Olsen (2006) used micro-PIV to measure instantaneous velocity profiles in a square Poly-di-methyl-siloxane (PDMS) micro-channel with hydraulic diameter ranging from 200 µm to 640 µm and for Reynolds number ranging from 200 to 3971, having water as the working fluid. They calculated mean velocity profiles, velocity fluctuations, Reynolds stresses and spatial correlations of velocity fluctuations. Silva et al. (2008), using micro-PIV technique, observed the effect of the wall surface roughness on velocity profiles inside micro-capillary tubes and compared their results with numerical CFD, without taking surface roughness into account. They found that not considering the surface roughness leads to underestimation of friction factor by about 11%. Kinoshita et al. (2007) used confocal micro-PIV system to measure and investigate the internal flow of a droplet, transported in a square PDMS microchannel having a size of 100 µm x 58 µm. They showed that the droplets have three-dimensional flow structures. This originates due to the drag force on the contact surfaces with the surrounding walls of the channel when the droplet moves inside a square microchannel.

Most of the experiments have been carried out to study the velocity profiles, either far away from the moving liquid-air interface or, alternately, bulk velocity profiles inside liquid plugs using micro-PIV (Meinhart et al., 1999; King et al., 2007, Malsch et al. 2008; Zimmerman, 2010) have been reported. These studies generally indicate that the liquid velocity profile is parabolic, satisfying the no-slip boundary condition at the wall. However, the flow phenomenon very close to the liquid-air interface has not been clear due to its complexity and three-dimensional nature. While numerical and analytical work has been reported (for example, by Chebbi, 2003; Egorov et al., 2003; He et al., 2010; Mehdizadeh et al., 2011), extensive experimental data on near meniscus flow fields is not easily available.

For micro-fluidics, the effect of surface forces becomes prominent during two-phase flow which may also result in the formation of three-phase contact lines with the tube/duct wall. Thus, the surface interaction refers not only to the interfacial interactions between two fluid phases, but also to the interaction of each fluid with channel walls (Huh et al. 2009). Wang et al. (2013) found that the surface wettability of micro-channel walls profoundly influenced both, the nature of two-phase flow and the flow regime transition. Lee and Lee (2008a) studied the pressure drop of two-phase plug flows in round small channels with diameters ranging from 1.62 to 2.16 mm and for tube materials of different wetting properties. They concluded that the pressure drop in small channels was significantly affected by the wetting condition of channel wall, but the effect of wettability was not introduced in their prediction model for pressure drop profile. Yamada et al. (2011) found that the product of friction factor and Reynolds number for hydrophobic channels increased as Reynolds number increased in the laminar flow regime. Bajpai and Khandekar (2013) reported with the help of three-dimensional simulations that the variation of dynamic contact angle of a confined meniscus moving inside a capillary tube affects the local heat transfer coefficient close to three-phase triple line.

When liquid plug flows in a small diameter tube (Bo < 2), the capillary forces play a major role and gravity becomes relatively insignificant. In such a case, the interfacial tension between solid-liquid (σ_{sl}) and solid-gas (σ_{sg}) should also be taken into account during flow analysis. Considering an isolated moving liquid plug inside a capillary tube which is enclosed by two menisci, the dynamic contact angle formed between the flowing liquid meniscus and the solid surface is achieved by a balance between the capillary and the viscous forces (Berthier, 2008). Thus, the moving menisci of the liquid plug are characterized by dynamic advancing and receding contact angles, which are different from the static equilibrium contact angle, because of the interplay between surface tension and viscous forces. The relative scaling of these two forces is represented by the Capillary number ($Ca = \mu U/\sigma$). Rose and Heins (1962) were among the first who highlighted the relation between the advancing contact angle and average velocity U_{avg} of three-phase contact line. Later, various researchers like Hoffman (1975), Tanner (1979) and Blake (2006) studied the wetting behavior and dependency of the advancing contact angle on Ca and the static contact angle.

Advent of microfluidic devices has directed considerable research attention to interfacial dynamics and transport phenomena at small scale devices. The complex flow phenomenon close to the liquid-air interface in a moving confined meniscus needs further attention and clarity. Heat and mass transfer rates are greatly affected by the meniscus shape, which, in turn is a function of local flow dynamics. The three-dimensionality of the flow in close proximity of the meniscus needs to be understood for proper design equations for the visco-capillary dissipation and resulting local transport. In this background, µ-PIV study has been carried in the liquid domain (D.I. water) of a moving liquid-air meniscus inside a dry square glass capillary tube of size 1 mm x 1 mm. As the liquid enters the dry capillary from one end with a constant velocity (i.e., fixed Ca), it creates a confined liquid-air meniscus inside it. It must be noted that the dynamic contact angle is a function of the Capillary number of the meniscus and thus, depending on the flow condition, which includes the effect of wettability, the shape of the meniscus changes, affecting the flow field close to it. This has been scrutinized by high speed visualization in conjunction with the PIV flow field measurements, to understand the mechanism of enhancement of local and average transport coefficient. The numerical difference between the incompressible three-dimensional continuity equation and the summation of the gradients of u and v velocity obtained by PIV (i.e., $(\partial u/\partial x + \partial v/\partial y)$, indicates the three-dimensional nature of the near-meniscus flow field.

To augment and complement the present study, three-dimensional numerical flow field simulations near the liquid-air meniscus have also been carried out on commercial finite element solver, i.e., Comsol®. Wettability of the liquid and the flow Capillary number are the required parameters of interest. The experimental flow fields close to the moving liquid-air interface obtained from PIV measurements are compared with the simulations. After benchmarking of the data, the wettability parameter is varied in the simulations to highlight its isolated and explicit effect on local flow transport characteristics. The results clearly demonstrate that, very close to the interface, the u-velocity profile is not parabolic in nature (becomes flatter), while away from the interface (going in the liquid domain) it satisfies the Poiseuille flow condition. Close to the meniscus, the flow clearly becomes three-dimensional with both v and w velocities showing a double-vortex, the strength of the latter being an order-of-magnitude lower than the former. This three-dimensionality leads to enhanced local transport of momentum. The simulations also reveal that controlling the wettability of the liquid can be an effective tool to control the viscous shear and therefore the pressure drop required to move the meniscus; the meniscus region of liquids with comparatively higher wettability manifests a stronger shear stress as compared to a non-wetting liquid meniscus moving at the same Capillary number.

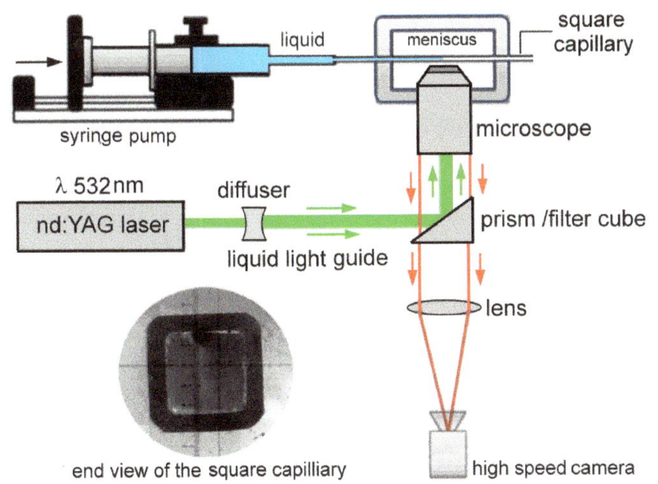

Fig. 1 Schematic diagram of the videography and μ-PIV setup for fluid flow velocity measurement inside glass capillary tubes of size 1 mm × 1 mm.

2. EXPERIMENTAL DESCRIPTION

The schematic of experimental setup for high speed video imaging (by Photron-Fastcam-SA3®) and flow field measurement by micro-PIV (Dantec Dynamics®) close to the moving liquid-air meniscus is shown in Figure 1. It consists of a syringe infusion pump (Cole Parmer-#74900-00) for supply of the working fluid, the output of which is connected to the horizontally placed square capillary tube of cross sectional size 1 mm x 1 mm. Before starting each new run of the experiment, the glass capillary is thoroughly cleaned in an ultrasonic bath by a standard protocol and completely dried in an oven maintained at 120°C. The working fluid is supplied at a constant feed rate to the capillary tube; the liquid-air meniscus thus formed travels at a fixed Capillary number inside the (already dry) capillary tube. The liquid-air meniscus visualization has been performed for four different liquids: de-ionized water, pure ethanol, silicon oil and glycerin. While qualitative visual data is reported here, the detailed numerical data of advancing apparent contact angle with the Capillary number is being reported elsewhere. The arrangement of Photron-Fastcam-SA3 camera for visualization of variation of the advancing contact angle with respect to the given Capillary number is shown in Figure 1.

Micro-PIV experiments were performed with deionized water, both under bright field mode and fluorescent mode (both results compared well with each other, however, for lack of space, only bright field data will be reported here). The dry glass square capillary was placed horizontally over the stage of the inverted microscope (5*X* objectives) of the micro-PIV unit (Leica® Micro-system; DMI 5000). The maximum operating frequency of this micro-PIV was 6 Hz. The working fluid (de-ionized water), having sufficient seeding particles (glass particles of size 10 μm in concentration 12.5 mg/ml of liquid for bright field imaging and fluorescent carboxylate particles of size 1 μm in concentration 1 μl/ml of liquid for florescent mode imaging), was supplied to the square capillary at specified calibrated flow rate. Successive images of the region of interest, just behind the dynamic liquid-air meniscus interface, were taken, using a 12 bit, 1344 × 1024 pixels digital CCD Camera (geometric resolution 1 pixel = 0.75 microns), for PIV flow analysis. All the measurements were taken at the mid plane of the meniscus, a plane which was 0.5 mm up from the bottom wall (relate this to Figure 2). Proper masking of the acquired PIV image was carried out by carefully tracking the seeding particles movement near the wall and at the interface to remove the non-flow outside region from the captured image field. The micro-PIV system produced two successive images and an adaptive cross correlation algorithm was applied to estimate the instantaneous flow field. An interrogation area of 64 × 64 pixels was chosen with a 50% overlap both horizontally and vertically. For removing unreasonable velocity vectors, appropriate filtration was applied after masking the spurious velocity vectors. The acquired PIV images were processed using Dantec® Dynamic software to obtain the velocity at each point in the flow field in the liquid domain.

The micro-PIV analysis was done for two different average water flow rates, 0.6 ml/hr and 1.0 ml/hr (corresponding to $U_{avg} = 0.166$ mm/s and 0.277 mm/s), the higher flow rate being limited by the upper frequency response of the micro-PIV system. Velocity profiles across the channel width, at a fixed *z*-plane (as shown in Figure 2), were plotted for different stream wise locations just near and behind the dynamic liquid-air meniscus. For validation and benchmarking of the data away from the interface, the following analytical solution for fully developed velocity profile inside a square channel (Eq. 1), as given by Bruus (2008), has been used,

$$u(y,z)=\frac{48Q}{\pi^3 ab}\frac{\sum_{n,odd}^{\infty}\frac{1}{n^3}\left[1-\frac{\cosh(n\pi y/a)}{\cosh(n\pi b/2a)}\right]\sin(n\pi z/a)}{1-\sum_{n,odd}^{\infty}\frac{192a}{n^5\pi^5 b}\tanh(n\pi b/2a)} \quad (1)$$

where, *a* and *b* are the height and width of the rectangular channel, respectively, which in the present case of a square channel are identical.

The micro-PIV measurement data provides two components of velocity, i.e. *u* and *v* in the directions *x* and *y*, respectively, on a definite plane of observation. In all regions where the flow remains essentially two-dimensional in nature, the 2-D continuity equation for the flow field, i.e., $(\partial U/\partial x)+(\partial V/\partial y)=0$, must be satisfied. If there remains a non-zero residual of the two-dimensional continuity equation, it is a definitive indication that the flow field is no more two-dimensional but is indeed three-dimensional, wherein the residual points towards the gradient of the *w* velocity in the z-direction. Thus, after obtaining the 2D velocity field (*u* and *v*) from the micro-PIV experiments, a four-point central difference method having a fourth order truncation error was applied at each grid point inside the domain to estimate the residual in the 2D continuity equation. The corner points of the domain were handled by either a forward difference or a backward difference scheme, as applicable. This methodology provides a simple plan for checking the fluid flow behavior and deviations, if any, from two-dimensional nature.

3. NUMERCAL FORMULATION AND SIMULATION METHODOLOGY

Simulation of the isolated liquid-air meniscus traveling with constant velocity inside a square capillary tube has been done using the commercial platform COMSOL® (Comsol 4.3a documentation, 2012). The evolution of the gas–liquid interface, and associated flow in the liquid domain and the adjoining air domain, as the interface moves, has been simulated using the pseudo 'single-fluid' approach. While the flow field is shared by both the fluids, the interface is tracked by an additional advection equation. The hydrodynamics of such flows can then be described by the usual equations for the conservation of mass and momentum, coupled with the additional interface advection which assigns level '1' to the pure liquid domain and level '0' to the pure gas phase, the interface being described by range of values in between 0 and 1 [Tornberg et al., 2000; Olsson and Kreiss, 2005]. The thermo-physical properties of the interface are also scaled using the level-set. For an incompressible, Newtonian fluid, these equations can be written in tensor notation as follows:

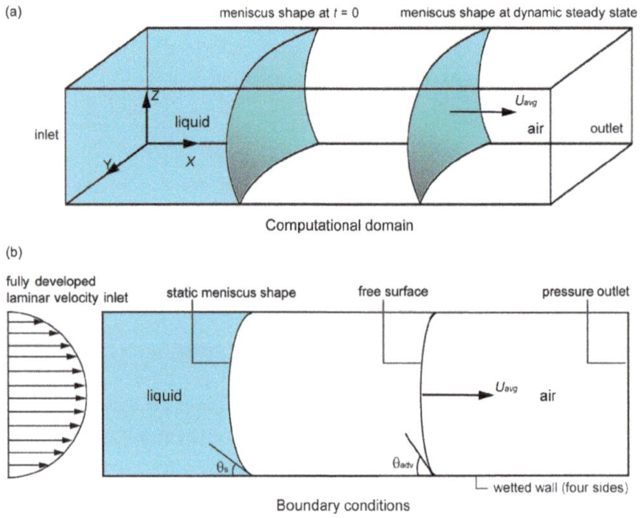

Fig. 2 (a): 3-D view of the computational domain of square capillary containing water and air (b) 2-D view with boundary conditions (at inlet: laminar velocity profile, at meniscus: interface boundary condition, at outlet: zero pressure condition and at wall: wetted no slip). The contact angle varies from static to dynamic at a given Capillary number as the meniscus moves from rest and attains stead-state motion.

Continuity equation:

$$\frac{\partial \rho}{\partial t}+\frac{\partial \rho u_j}{\partial x_j}=0 \tag{2}$$

Momentum equation:

$$\frac{\partial \rho u_i}{\partial t}+\frac{\partial \rho u_j u_i}{\partial x_j}=-\frac{\partial p}{\partial x_i}+\mu\frac{\partial u_i}{\partial^2 x_j}+(F_s)_i+(F_g)_i \tag{3}$$

The advection of the level-set function is described by:

$$\frac{\partial \phi}{\partial t}+\mathrm{u}_i\frac{\partial \phi}{\partial x_j}=\gamma\frac{\partial}{\partial x_j}\left(\varepsilon\frac{\partial \phi}{\partial x_j}-\frac{\phi(1-\phi)\frac{\partial \phi}{\partial x_j}}{|\frac{\partial \phi}{\partial x_j}|}\right) \tag{4}$$

where, i is free indices and j is the repeating indices. u_i denotes the velocity vector in i^{th} direction , p the pressure, ρ the density and μ the dynamic viscosity of the fluid. $(F_s)_i$ is the surface tension force approximated as a body force in the vicinity of the interface, and ϕ is the level-set function of one of the phases. The bulk properties, such as density and viscosity, are determined from the volume-fraction weighted average of the properties of the two fluids. Hence, the density and viscosity are calculated from:

$$\rho=\rho_l+(\rho_l-\rho_{air})\phi \tag{5}$$

$$\mu=\mu_l+(\mu_l-\mu_{air})\phi \tag{6}$$

Figure 2 shows a schematic of the square micro-capillary and the appropriate boundary conditions applied. At the channel inlet, fully-developed laminar inflow conditions, as per Eq. (1), with prescribed volume flows are used. At the outflow boundary, the pressure is specified with no viscous stress; the latter is a reasonable assumption at low Capillary numbers. Initially, the liquid and air domain have to be specified along with any arbitrary shape of the interface. Once the static contact angle between the liquid and the channel wall is specified during initialization, the static shape of the interface is determined. Subsequently, the simulation for the dynamic interface commences. The dynamic contact angle also needs to be specified as input condition, the estimation of which for a given Ca will be explained later. The singularity at the three phase contact line is avoided by taking a wetted wall boundary condition, wherein a slip length equal to the minimum grid size has been employed. The effect of gravity has been neglected as it plays a minor role as compared to the large surface tension forces due to microchannel geometry. An explicit geometric reconstruction scheme is used to represent the interface by using a piecewise-linear approach. The time-step used for the volume fraction equation (Eq. 4) in Comsol® is not necessarily the same as that used for the other equations (Eqs. 2 and 3) but is calculated based upon the characteristic transit time of a fluid element across a control volume and is limited by a specified maximum value for the Courant number (Co). The time taken to empty a cell is calculated by dividing the volume of each cell by the sum of the outgoing fluxes in the region near the fluid interface. The smallest such time is used as the characteristic time of transit for a fluid element across a control volume. The Courant number (Co) is defined by Eq. (7) as

$$\frac{\Delta t}{\Delta x/(U_{avg})_l} \tag{7}$$

where, Δx is the grid size and $(U_{avg})_{\text{fluid}}$ is the fluid velocity. A maximum Courant number of 0.001 is set in the present calculations and a variable time-step based on a fixed Courant number of 0.001or less was used to force the time-step to be the same for all the equations.

Therefore, the terms on the left hand side (Eq. 4) give the correct motion of the interface, while those on the right hand side are necessary for numerical stability. The parameter ε, determines the thickness of the region where, ϕ goes smoothly from zero to one and is typically of the same order as the size of the element of the mesh. By default, ε is constant within each domain and typically equals half the mesh size within the domain passed by the interface. The parameter γ determines the amount of re-initialization and stabilization of level-set function. It needs to be tuned for each specific problem. If γ is too small, the thickness of the interface might not remain constant, and oscillations in ϕ may appear because of numerical instabilities. On the other hand, if γ is too large, the interface moves incorrectly. Recommended value of γ scales with the magnitude of velocity of the specified problem.

The continuum surface force (CSF) model proposed by Brackbill et al. (1992) is used to model the surface tension effects. The CSF model approximates the surface tension induced-stress by a body force, which acts throughout a small but finite fluid region surrounding the interface. The surface tension force per unit area is represented as (Olsson and Kreiss, 2005):

$$F_s=\sigma\,\kappa\,\delta(r-r_{\text{int}})n \tag{8}$$

where, σ is the coefficient of surface tension, κ is the radius of interface curvature, n denotes the unit normal vector on the interface and $\delta(r)$ is the Dirac delta function, nonzero only at the fluid-air interface. The normal n and curvature κ in Eq. (8) are defined in terms of the level-set function (ϕ) via:

Table1: Properties of the gas and the liquid used in the simulations

Fluid	Density (kg/m^3)	Viscosity (kg/m s)	Surface tension (N/m)	Static contact angle θ_s (deg)
Water	998	8.899×10^{-4}	0.0732	101.5
Air	1.175	1.751×10^{-4}	-	

$$n = \frac{\nabla \phi}{|\nabla \phi|} \tag{9}$$

and

$$\kappa = \nabla n \tag{10}$$

The level-set parameter (ϕ) is also used to approximate the delta function by a smooth function defined by (Sussman et al., 1994; Sussman et al., 1998):

$$\delta = 6(|\phi(1-\phi)|)(|\nabla \phi|) \tag{11}$$

This implementation of the surface tension force can induce unphysical velocity near the interface known as spurious currents. This occurs because the pressure and viscous force terms do not exactly balance the surface tension force. The smaller time step and smaller size grid can help in minimizing these spurious currents. A second order, non-iterative fractional step scheme is used for the time marching of the momentum and continuity equations. The residual tolerance for both the equations was set to 10^{-6}.

The simulation is carried out in several steps. At first, the liquid-air interface shape for the specified static contact angle of θ_s at the wetted wall of capillary tube is determined. The second step involves the determination of the dynamic contact angle as a function of the applied Ca. This is achieved by linearization of the classical Hoffman-Tanner law (Bonn et al., 2009). For dynamic systems with non-zero static contact angle, the Hoffman-Tanner law can be slightly modified and extended as (Bertheir, 2008; Khandekar et al., 2010):

$$\theta_d^3 - \theta_s^3 = A \cdot \mathrm{Ca} \tag{12}$$

where, A is a constant. If Ca is small, linearization of Eq. (12) leads to (Berthier, 2008):

$$\theta_d = \left(\theta_s^3 + A \cdot \mathrm{Ca}\right)^{1/3} \approx \theta_s\left(1 + \frac{1}{3}\frac{A \cdot \mathrm{Ca}}{\theta_s^2}\right)$$

or

$$\theta_d - \theta_s \approx \frac{1}{3}\frac{A \cdot \mathrm{Ca}}{\theta_s^2} \tag{13}$$

It should be noted that Ca takes the sign of the velocity in the reference frame of the problem. Thus, ($\theta_d - \theta_s$) takes the sign of Ca and the values of dynamic advancing contact angle is as:

$$\theta_{adv} = \theta_s + \frac{1}{3}\frac{A\,|Ca|}{\theta_s^2} \tag{14}$$

This equation informs us that the advancing contact angle is larger than the static contact angle. As a last step, the dynamic steady state of meniscus motion, at a constant flow velocity, is simulated. Under such conditions, the dynamic contact angle remains constant, as given by Eq. 14.

4. RESULTS AND DISCUSSION

Results obtained both experimentally as well as by simulation, are reported hereunder in the following sequence:

(i) Shape of liquid-air meniscus formed inside the tube at different flow Capillary numbers when dynamic steady-state motion of the meniscus is achieved inside the capillary tube,

(ii) Fluid flow field obtained by micro-PIV close to meniscus at different flow Capillary numbers, and,

(iii) Parametric study highlighting the effect of wettability of the liquid-tube material combination on the near-meniscus hydrodynamics.

4.1 Static and dynamic meniscus shape

The static to dynamic steady-state meniscus shapes are captured by using high speed camera at different Capillary numbers in the square glass capillary. A range of Capillary numbers is obtained by using four different fluids, i.e., water, ethanol, glycerin and silicon-oil ($\rho = 760$ kg/m^3, $\mu = 0.0046$ Pa.s, $\sigma = 0.0188$ N/m). Inside the dry capillary tube, the static contact angle of water and glycerin was found to be higher than 90° (101.5° for water and 107.2° for glycerin,) and it is less than 90° for ethanol and silicon oil. Figure 3 clearly depicts that the contact angle of the meniscus drastically changes (static to dynamic) when the applied flow Capillary number is altered. At low Capillary number, the dynamic water-air meniscus remains nearly flat, close to being perpendicular to the channel wall. However, increase in Capillary number results in convexity of the meniscus shape, due partly to the variation of dynamic pressure difference between the two sides of the water-air meniscus. The behavior of glycerin-air meniscus is qualitatively similar to water-air system for the chosen range of Capillary number. For ethanol, the meniscus remains concave in shape for the entire range of applied Capillary numbers. In contrast, for the case of silicon oil, a drastic change of contact angle occurs with respect to the applied Capillary number; the shape of the interface changes from concave (at low Ca) to convex (at high Ca). These results clearly indicate the strong and unequivocal connection between the wetting characteristics, local interface velocity, and the resulting dynamic pressure, which eventually is intrinsically connected to the local meniscus curvature and thus the resulting shape. It is clear that only Capillary number as a parameter fails to predict the steady-state dynamics shape of the meniscus. It logically follows, as will be demonstrated in the following results that near meniscus transport of momentum (so also, thermal energy in the case of heat transfer systems and mass transport for chemically reacting fluids with the tube wall) will be quite different for these cases.

The contact angle at dynamic steady state is approximately equal to the dynamic contact angle obtained from Eq. 14. In addition, no change in the dynamic contact angle has been observed during the steady motion of meniscus, for glycerin, ethanol and silicon oil. Only for high surface tension system, like water, the steady meniscus motion can show some tendency of intermittent motion (stick-slip tendency) due to local pinning of the three-phase contact line. For a given static contact angle corresponding to the experimental condition, the shape of meniscus at various Capillary numbers has been captured by simulation. A sample result is as shown in Figure 4, which compares the experimental and simulation menisci shapes inside dry-smooth mini-capillary of size 1 mm × 1 mm at four different flow Capillary numbers. The average difference in the curvature obtained by experiments and simulation is within 11%; similar results were obtained with the other fluids also.

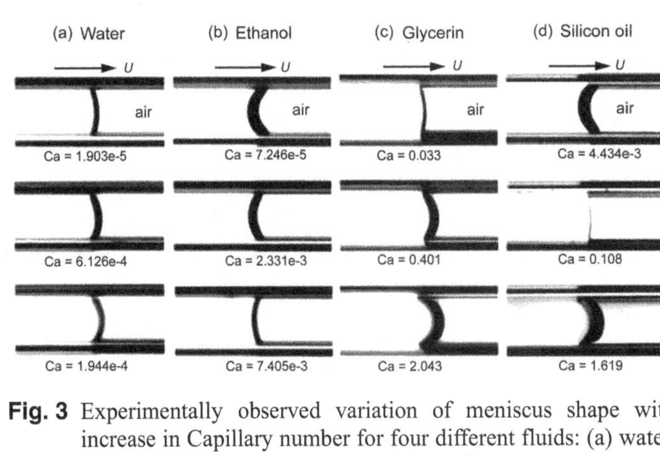

Fig. 3 Experimentally observed variation of meniscus shape with increase in Capillary number for four different fluids: (a) water, (b) ethanol, (c) glycerin, and (d) silicon oil.

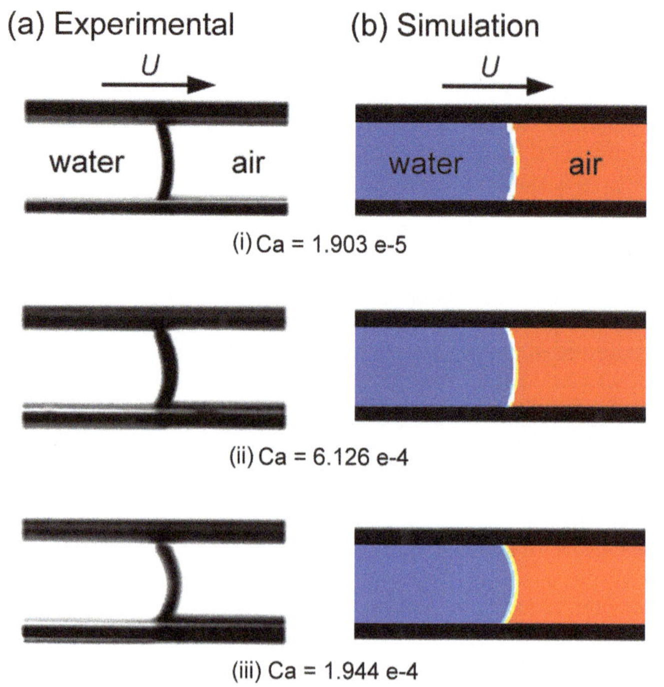

Fig. 4 Comparison of meniscus shape of water-air interface using (a) Experimental images of the moving meniscus and, (b) Simulations at corresponding Capillary numbers.

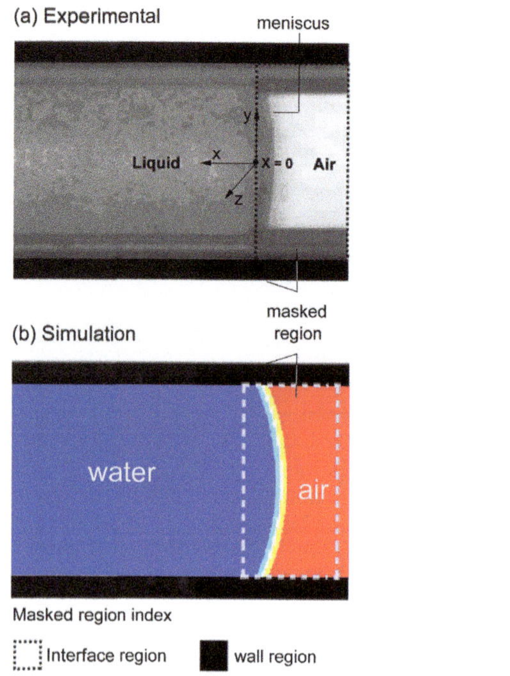

Fig. 5 (a) View of liquid-air meniscus inside square capillary in the bright field mode. This image was taken at the mid-plane of the channel. The masked region at the interface and wall is defined to remove spurious velocity vectors. The gravity vector points towards the z-direction. (b) Image obtained from simulation, at the same conditions; masked region is marked with dotted line.

4.2 Near-meniscus hydrodynamics

Figure 5 (a) shows a typical image obtained by the PIV objective microscope with the relevant coordinate system. This image has been obtained at the mid-plane (*x-y* plane) from the top view of the horizontally placed capillary tube (gravity acts in the *z*-direction). Here, the meniscus motion is towards the *x*-direction at Ca = 2.27×10^{-6}. Due to the curvature of the meniscus, capturing the seeded particles extremely close to the meniscus is difficult and leads to weak signal for post-processing. Therefore, some finite area just close to the meniscus needs to be masked for the cross-correlation analysis to obtain the flow velocities so that spurious signals are filtered. These have been highlighted in Figure 5(a) for the PIV image. The corresponding image obtained by simulation is shown in Figure 5(b). While there are no such physical restrictions during three-dimensional simulations, as encountered in the experiments, for the sake of one-to-one direct comparison of the two strategies, the corresponding masked area is also shown in Figure 5(b). All experimentally obtained values reported here are typically in the un-masked portions of the flow.

Figure 6 (a) shows the contour of velocity components (u, v and $\partial w/\partial z$) of liquid moving in the square capillary obtained by micro-PIV measurements at Ca = 2.27×10^{-6} (working fluid: water) moving inside the square capillary tube. The velocity analysis has been done for the mid-plane (x-y plane) of the capillary tube, as detailed in Figure 4. Figure 6(a) –(i) and (ii), give the experimentally obtained values of velocities u and v, respectively, in the directions x and y. As will be appreciated, micro-PIV analysis only provides two-dimensional (u, v) data at any given plane. Invoking the 2D continuity equation, the three-dimensionality of the flow, if present, can be ascertained – a finite residual of the 2D continuity indicates the presence of $\partial w/\partial z$. This is shown in Figure 6(a) –(iii). The corresponding simulation results for u and v velocities are shown in Figure 6(b) –(i) and (ii), respectively. Figure 6 (b) – (iii) shows the simulation results for the velocity w in the third-direction, i.e. the z-direction. Exactly corresponding set of results, but at a higher U_{avg} = 0.277 mm/s (Ca = 3.7×10^{-6}), are shown in Figure 7 (a) and (b), respectively.

From the u-contours it is clearly observed that the fluid velocity is higher at the centre of the channel away from the interface and it gradually decreases towards the liquid-air interface. The v-contour near to the interface has two dense regions. These regions are having a higher magnitude of v-velocity, albeit in opposite directions, as expected due to fluid circulation near the interface. A vortex pair is clearly visible in the v-velocity distribution. As this is a planer 2D data, away from the interface, the mass continuity is nearly satisfied; however, just close the meniscus region the residual error in the 2D continuity $\left(error = \left|(\partial U/\partial x) + (\partial V/\partial y)\right|\right)$ is more, clearly indicating the existence of gradient of the third component of velocity, w. This three-dimensionality in the flow arises due to the drag force on the contact surface with the sidewalls of the micro-channel and the fact that the meniscus must simultaneously obey the free surface boundary condition. These results clearly support the existence of gradient of the third component of velocity, w, near the meniscus. The corresponding simulation result shows the actual value of w components near the interface. Experimental data and simulation data for u and v closely match each other. The results also simultaneously establish the three-dimensionality of the flow.

To further explore the distribution of the velocity components of the liquid flow near the moving meniscus, Figure 8 and Figure 9 show the variation of velocity components u and v along the channel height at various locations from the liquid-air interface, corresponding to average velocity of 0.166 mm/s (Ca = 2.27×10^{-6}) and 0.277 mm/s (Ca = 3.7×10^{-6}), respectively. Non-dimensionalization of both u and v velocities has been done using the respective U_{avg}, which is the average liquid supply velocity based on the total cross-section of the channel. Close to the interface a significant drop in the magnitude of the u-velocity is seen at the centre region of the channel and this behavior is more prominent at higher flow rates, as can be seen from Figure 8 (a) and 9 (a).

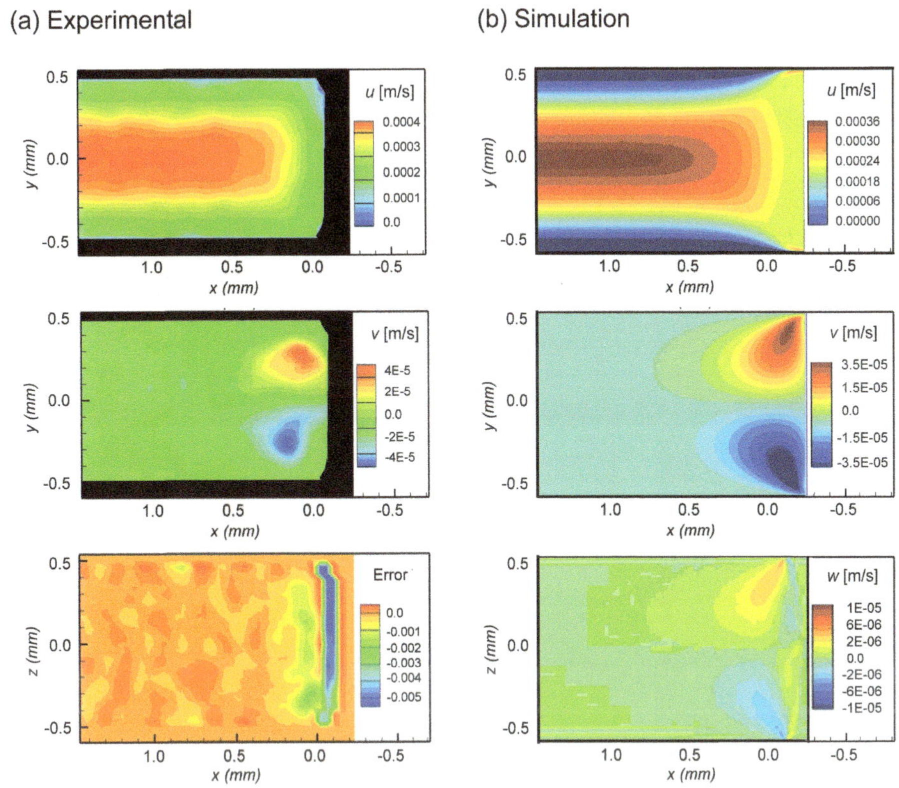

Fig. 6 Flow field at the moving liquid-air interface inside the square-microchannel for $U_{avg} = 0.166$ mm/s. (a) u, v and $\partial w / \partial z$ contour obtained by micro-PIV and (b) u, v and w contour obtained by simulation.

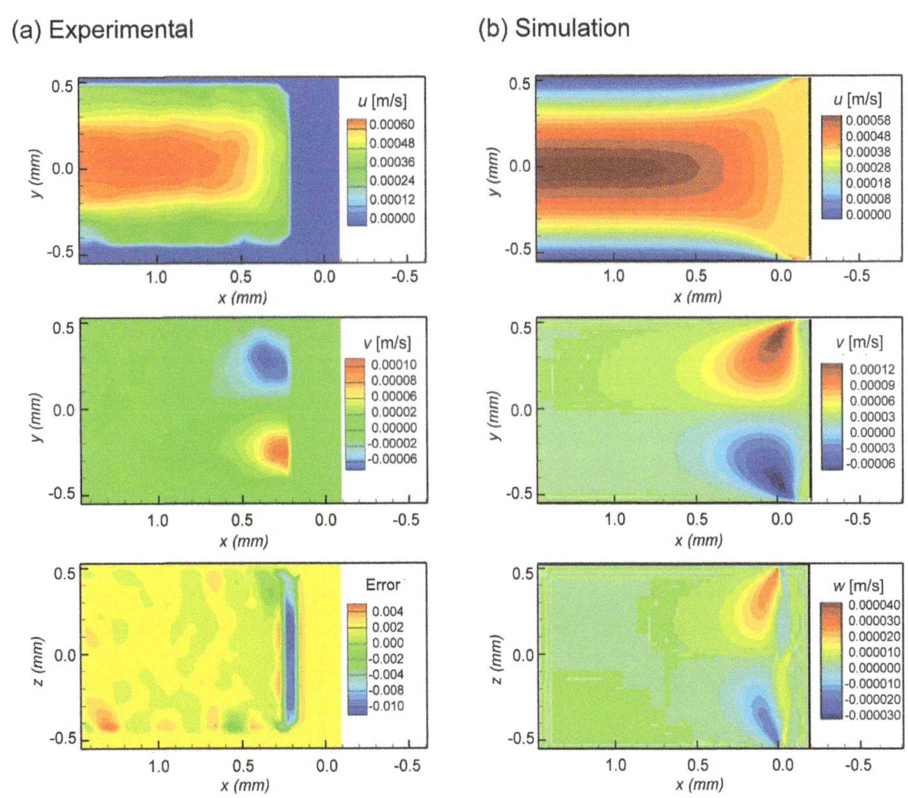

Fig. 7 Flow field at the moving liquid-air interface inside the square-microchannel for $U_{avg} = 0.277$ mm/s. (a) u, v and $\partial w / \partial z$ contour obtained by micro-PIV and (b) u, v and w contour obtained by simulation.

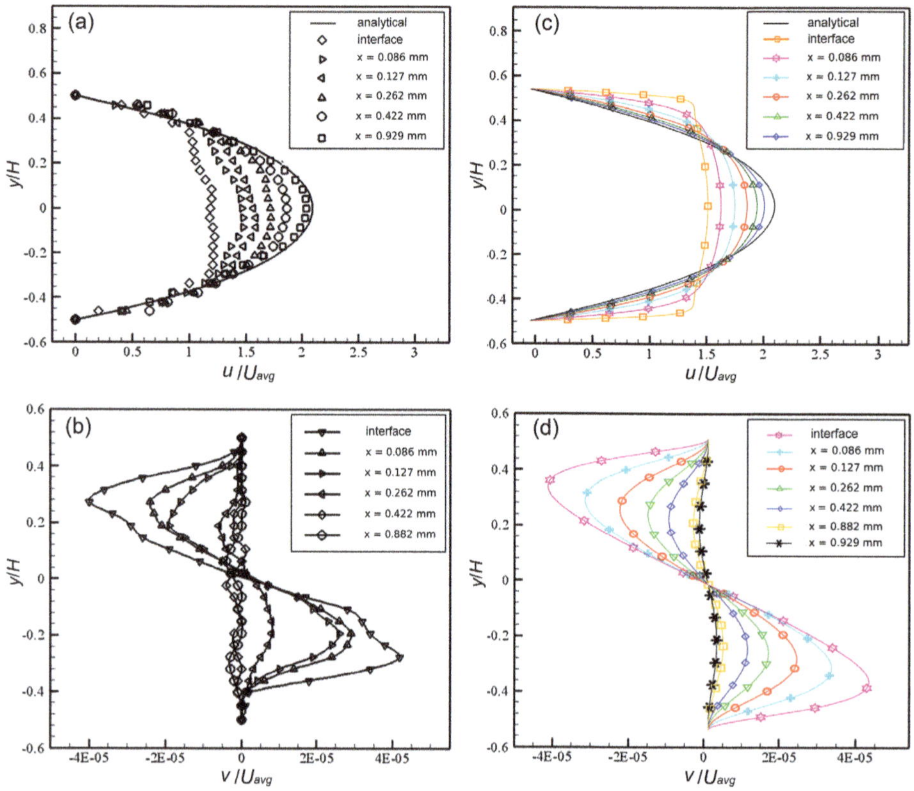

Fig. 8 Variation of non-dimensional u/U_{avg} and v/U_{avg} components of velocities with respect to distance from the liquid-vapor interface for average velocity of flow (U_{avg}) is = 0. 166 mm/s (Ca = 2.27 × 10^{-6}). (a) and (b) are observed experimentally and (c) and (d) are obtained from simulation, respectively. All the measurements have been performed at the mid-plane of the channel cross-section. x is the streamwise axial distance from the interface (x = 0 corresponds to the interface).

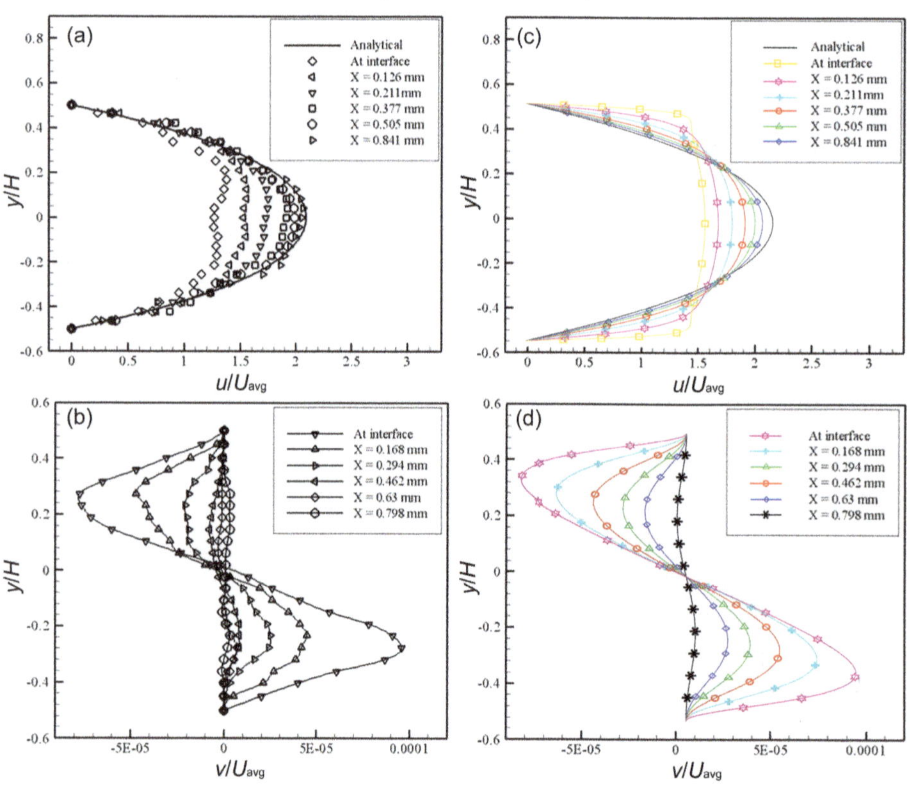

Fig. 9 Variation of non-dimensional u/U_{avg} and v/U_{avg} components of velocities with respect to distance from the liquid-vapor interface for average velocity of flow (U_{avg}) is = 0.277 mm/s (Ca = 3.7 × 10^{-6}). (a) and (b) are observed experimentally and (c) and (d) are obtained from simulation, respectively. All the measurements have been performed at the mid-plane of the channel cross-section. x is the streamwise axial distance from the interface (x = 0 corresponds to the interface).

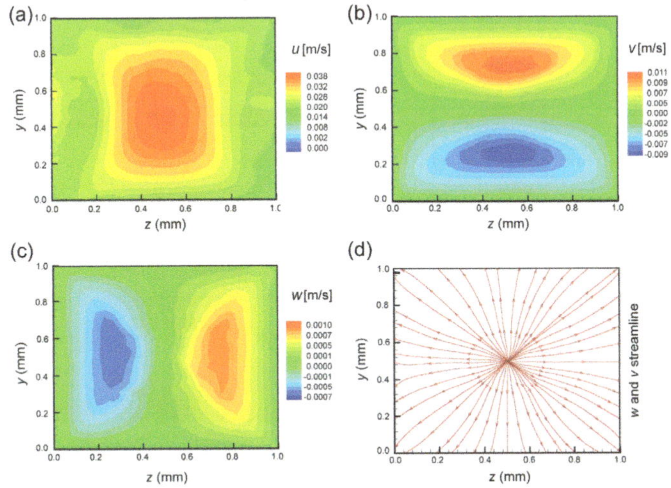

Fig. 10 Liquid flow pattern inside a square capillary tube of size 1.0 mm × 1.0 mm, in the cross-sectional plane (*z*-*y*) close to the meniscus, at $x = 0.02$ mm. The dynamic contact angle is 105° and applicable flow Ca = 5 × 10^{-4} (a) contour of *u*-velocity component, (b) contour of *v*-velocity component, (c) contour of w-velocity component, and (d) streamline of *w* and *v* velocity components.

This behavior is a direct outcome of the obstruction of the fluid flow at the liquid-air interface, the meniscus giving a free-surface boundary condition with only tangential velocities possible. As we move away from the interface, the u-velocity distribution asymptotically adheres to the parabolic Poiseuille distribution. For benchmarking, the analytical profile, given by Eq. 1 has also been plotted. Figures 8 (b) and 9 (b) show the v/U_{avg} with respect to height of channel at various points away from the liquid-air interface. Close to interface the *v*-velocity magnitude considerably increases (obviously at the expense of the decrease in the *u*-velocity) and qualitatively varies like a sinusoid. The magnitude of this wave like variation of the *v* component along the height of channel decreases as the point of observation moves away from the interface. The *v*-velocity eventually becomes negligible away from the interface. As has been pointed out earlier, close to the interface, the flow actually becomes three-dimensional in nature, the evidence of which can only be captured indirectly via the residual error in the 2D continuity, as was shown earlier in Figure 6 (a) –(iii) and Figure 7 (a) –(iii).

Figure 8 (b) and Figure 9(b) show the corresponding simulation results for *u* and *v* velocity profiles. The comparison with the experiments suggest that while majority of the flow physics is nicely captured, both qualitatively and quantitatively; however, some minor discrepancies between the simulation and experiments are due primarily because of: (a) the glass capillary tube can have local non-homogeneity in the wall topography, which leads to local pinning of the three phase contact line, enhancing friction intermittently (b) errors also creep in during the image processing of the PIV data, wherein the data quality close to the wall and just behind the meniscus is limited.

Figure 10 shows the contours of the velocity components (*u*, *v* and *w*) and the streamline of *w* and *v* components in cross section of flow (*z*-*y*) at $x = 0.02$ mm from the interface (near the water-air interface), for dynamic contact angle 105° at Ca = 5 × 10^{-4}. The contour of *u*, *v* and *w* in cross-section plane, as shown in figure 10(a-c), clearly depicts circulation of liquid near the interface. Note the order of magnitudes of the *u*, *v*, and *w* components, respectively; u-component being an order of magnitude higher. The flatness of the *u*-contour, the near symmetric nature of the vortex-pair circulatory patterns, and the three-dimensionality of the flow close to the meniscus is clearly seen from these simulation results.

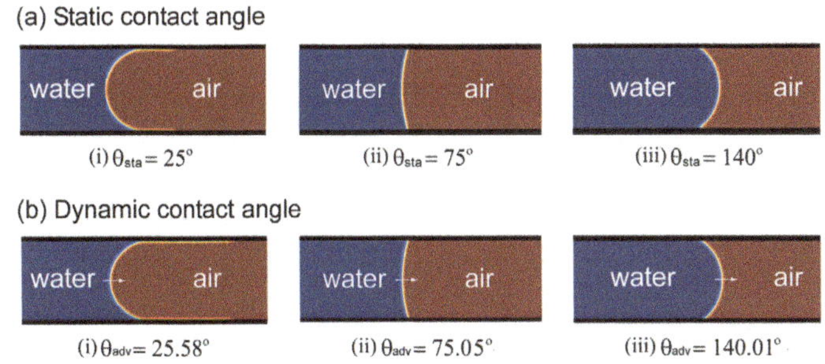

Fig. 11 Time evolution of shape of drop change from static contact angle to contact angle at criticality (when the meniscus just commences to move) (a) contact angle 25°, (b) contact angle 70° and (c) contact angle 140° for average velocity of flow (U_{avg}) is = 0.277 mm/s (Ca = 3.7 × 10^{-6}).

4.3 Effect of substrate wettability on flow hydrodynamics

Having established the simulation procedure and benchmarked against experimental data, one can proceed to study the effect of a range of parameters on the flow. The effect of wettability of the capillary tube (with water) on the hydrodynamics close to the meniscus is presented in this sub-section. Figure 11 shows the time evolution of the shape of meniscus from static contact angle to critical contact angle (the angle at which the meniscus commences its movement) for various wettability scenarios of the capillary tube; the imposed Capillary number is 5×10^{-4}. For capillary material having high wettability (i.e. static contact angle of 25°, hydrophilic case), the critical contact observed is 25.6°. As the wettability of the capillary tube decreases, the difference between the static contact angle and the critical contact angle diminishes, as can be seen in Figure 11 (a) to (c). Intrinsic wettability also changes the resulting meniscus curvature as can be clearly seen, and this shape directly affects the hydrodynamics of the flow in the wake of the meniscus. Figure 12 (a) shows meniscus shape under dynamic steady state and the resulting flow pattern of fluid close to meniscus inside tubes having a range of wettability (static contact angle for case (i) 25°; case (ii) 70° and case (iii) 140°), at flow Ca = 5 × 10^{-4}. Inside capillary tubes having higher wettability, very thin precursor liquid film layer has been observed close to the three-phase contact line, as shown in Figure 12(a)-i. It is observed that the thickness of this fluid layer decreases as contact angle increases and it approaches zero at a contact angle of 90°. Hydrodynamics of the fluid close to interface has been depicted in Figure 12(b)-(d) for the three cases of different wettability, respectively. Figure 12 (b) represents the streamline of flows while Figures 12 (c) to (e) depicts the velocity contours for *u*, *v* and *w* velocity components respectively, in the *xy* plane. While all the three cases manifest three-dimensional flow pattern and present a vortex pair, their respective strengths are different. The zone behind the meniscus which gets affected by the vortex flow is a clear function of the wettability of the fluid-tube pair. The fluid perturbation depth is larger for the meniscus formed with a higher wetting fluid. It is clear from these results that the transport of momentum (and so will be the case for heat and mass transport) is a strong function of wetting characteristics. It will directly affect the momentum transport coefficient (skin friction) in the meniscus region, as is explored next.

The variation of Poiseuille number ($C_f Re$) experienced along the wall due to the steady motion of the meniscus, for the three cases of wettability respectively is shown in Figure 13. Many interesting features can be concluded from this result, in conjunction with Figure 12. Firstly, momentum flux coefficient rapidly increases as we move closer to the three-phase contact line. Higher the wettability of the fluid, higher is the value of C_f, close to the meniscus. As we move away from the meniscus viscous stress reduces and it becomes asymptotically closer to the Poiseuille flow situation.

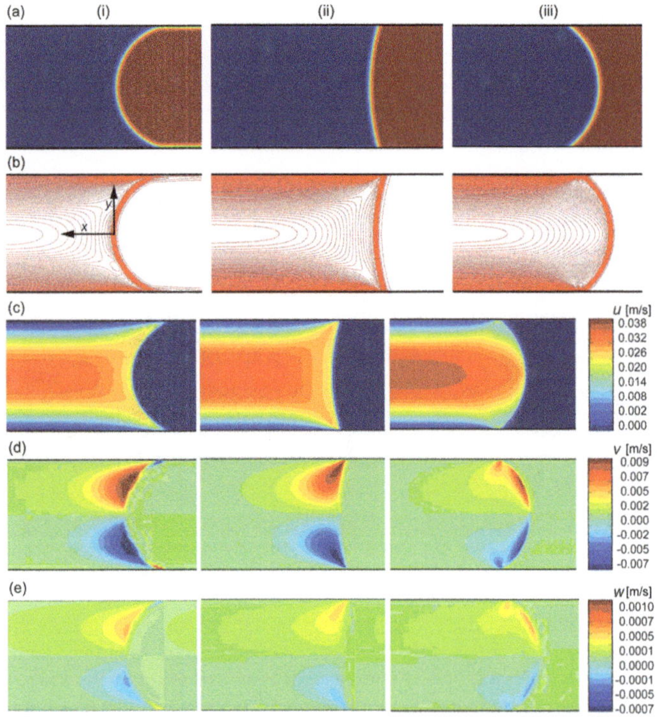

Fig. 12 (a) Meniscus shape for various wettability capillary tube at commencement of motion, (b) *u-v* streamline in *x-y* plane, (c) contour of *u* components of velocity in *x-y* plane, (d) contour of *v* components of velocity in the *x-y* plane and (e) contour of *w* components of velocity in *x-z* plane for various wettability of capillary tube, (i) contact angle 25°, (ii) contact angle 70° and (iii) contact angle 140° for average velocity of flow (U_{avg}) is = 0.038 mm/s (Ca = 5×10^{-4}). (Note: All data representation is in the moving frame of reference)

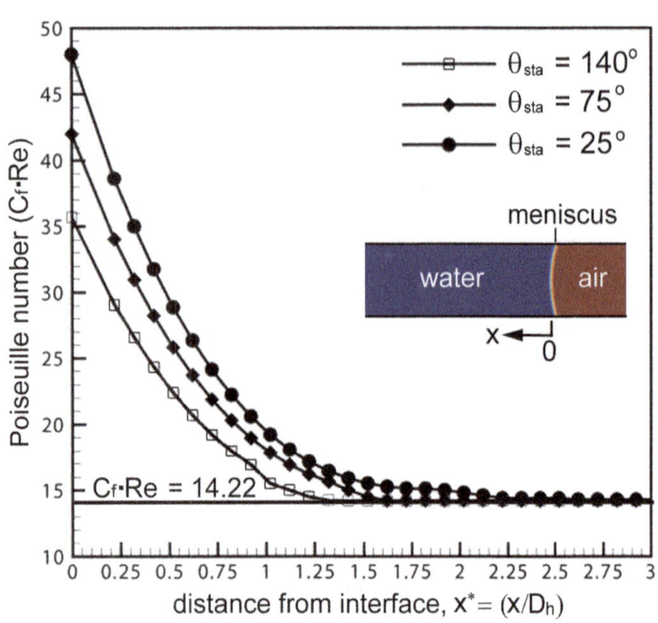

Fig. 13 Variation of Poiseuille number (C_f·Re) experienced along the wall due to the steady motion of the meniscus, for the three cases of wettability respectively.

For reference, the skin friction coefficient for fully developed single-phase liquid flow in a rectangular channel, i.e., C_f = 14.22, is also shown on the figure. The effect of the meniscus is felt at a distance of about $2D_h$ to $3D_h$, this distance being a function of the wettability too. Since a small contact angle is related to a large dispersion force, a stronger dispersion force enhances the interaction between liquid phase and solid substrate and improves the wetting (Churaev 1995). With similar surface roughness of the micro-channels, the surface energy dissipation of the hydrophobic micro-channel is lower, while the interaction at the liquid-solid interface is weaker. Significant drag reduction has also been reported using hydrophobic surfaces in many studies (Tretheway and Meinhart 2002; Choi et al. 2003).

5. SUMMARY AND CONCLUSIONS

In the present work, dynamic shape of an isolated moving meniscus formed between a liquid-air system inside a square mini-channel has been studied for water, ethanol, glycerin and silicon-oil, with varying Ca. It is shown that a drastic change in the dynamic shape of the meniscus can occur due to the imposed bulk velocity. Subsequently, using micro-PIV, liquid velocity distributions near the moving water-air meniscus were measured at the mid-plane of the channel, at low Ca. It was shown that near the liquid-air interface the non-dimensional *u*-velocity profiles are flatter in nature, while moving away from the meniscus the profiles follow the fully developed flow profile. The *v*-velocity is moderately dominant only close to the interface, although its absolute value is rather small as compared to the *u*-velocity. Formation of opposite vortices is clearly indicated by the *v*-profiles. Checking the continuity close to the interface indicated that the flow is indeed three-dimensional in nature in the proximity of the interface. It is highlighted that near meniscus local transport phenomena are critical in understanding the transport behavior of micro-scale multi-phase systems. The experimental data for the flow field close to moving liquid-air interface are compared with numerical simulations. After benchmarking, the wettability parameter is varied in the simulations to highlight its isolated effect on local flow transport characteristics. The results clearly demonstrate that, close to the meniscus, the flow clearly becomes three-dimensional with both *v* and *w* velocities showing a double-vortex, the strength of the latter being an order-of-magnitude lower than the former. This three-dimensionality leads to enhanced local transport. The simulations also reveal that controlling the wettability of the liquid can be an effective tool to control the viscous shear; the meniscus region of wetting liquids manifest stronger shear stress as compared to a non-wetting liquid moving at the same Capillary number. It is expected that the heat and mass transfer near the meniscus region will also be qualitatively affected in a similar fashion as the momentum flux transfer, taking the cue from Prandtl and Schmidt number analogy. Thus, it is concluded from this study that wettability of the fluid is an important parameter for species transfer under dynamic conditions near the meniscus regions in microscale geometries.

ACKNOWLEDGEMENTS

Financial support from the Indo-French Centre for the Promotion of Advanced Research (IFCPAR/CEFIPRA), New Delhi (Project #: 4408-1/2010), under the aegis of Indian Department of Science and Technology and the French Ministry of Foreign Affairs is gratefully acknowledged.

NOMENCLATURE

A	constant (-)
Ca	Capillary number, $(\mu U_{avg}/\sigma)$
C_f	skin friction coefficient (-)
Co	Courant number $\left(\Delta t/(\Delta x/(U_{avg})_l)\right)$
F	force density (N/m^3)
h	height (m)
Q	flow rate (m^3/s)
n	unit normal vector (-)
Po	Poiseuille number (C_f·Re).
Re	Reynolds number $(\rho U_{avg}D/\mu)$
U	velocity (m/s)
u	x-component of velocity (m/s)
v	y-component of velocity (m/s)
w	z-component of velocity (m/s)
x,y,z	directions in Cartesian coordinates (-)

Greek symbols

ϕ	level-set parameter (-)
δ	dirac-delta function (-)
μ	dynamic viscosity (Pa-s)
κ	radius of interface curvature (m)
ρ	density (kg/m^3)
σ	surface tension (N/m)
θ	contact angle (deg or rad)
ξ	non-dimensional distance (-)

Subscripts

adv	advancing
air	air/gas-phase
avg	average
h	hydraulic
l	liquid
d	dynamic
s	static, surface
g	gravitational, volumetric

REFERENCES

Ajaev, V.S., and Homsy, G.M., 2006 "Modelling Shapes and Dynamics of Confined Bubbles," *Annual Review of Fluid Mechanics*, **38**, 277-307.

Bajpai, A.K., and Khandekar, S., 2012, "Thermal Transport Behavior of a Liquid Plug Moving Inside a Dry Capillary Tube," *Heat Pipe Science and Technology*, **3**, 97-124.

Berthier, J., 2008, Microdrops and Digital Microfluidics. William Andrew Inc., New York, USA.

Blake, T.D., 2006, "The Physics of Moving Wetting Lines," *Journal of Colloids and Interface Science*, **299**, 1-13.

Bonn, D., Eggers, J., Indekeu, J., Meunier, J., and Rolley, E., 2009, "Wetting and Spreading," *Review of Modern Physics*, **81**,739-805.

Brackbill, J.U., Kothe, D.B., and Zemach, C., 1992, "A Continuum Method for Modeling Surface Tension," *Journal of Computational Physics*, **100(2)**, 335-354.

Bruus, H., 2008, Theoretical Microfluidics. Oxford Master Series in Physics, ISBN-13: 978-0199235087, Oxford University Press, New York, 48-52.

Chebbi, R., 2003, "Deformation of Gas-Liquid Interfaces in Capillary Tubes," *Journal of Colloid and Interface Science*, **265**, 166-173.

Choi, C.H, Westin, K.J.A, and Breuer, K.S., 2003, "Apparent Slip Flows in Hydrophilic and Hydrophobic Micro-Channels," *Physics of Fluids*, **15(10)**, 2897-2902.

Churaev, N.V., 1995, "Contact Angles and Surface Forces," *Advances in Colloid and Interface Science*, **58(2-3)**, 87-118.

Egorov, A.G., Kornev, K.G., and Neimark, A.V., 2003, "Meniscus Motion in a Prewetted Capillary," *Physics of Fluids*, **15**, 3134-3143.

Gunther, A., Khan, S.A., Thalmann, M., Trachsel, F., and Jensen, K.F., 2004, "Transport and Reaction in Microscale Segmented Gas-Liquid Flow," *Lab on Chip*, **4**, 278-286.

He, Q., Hasegawa, Y., and Kasagi, N., 2010, "Heat Transfer Modeling of Gas-Liquid Slug Flow without Phase Change in a Micro-Tube," *International Journal of Heat and Fluid Flow*, **31**, 126-136.

Hoffman, R.L., 1975, "A Study of Advancing Interface," Journal of Colloid and Interface Science, **50**, 228-241.

Huh, D., Kuo, C-H., Grotberg, J.B., and Takayama1, S., 2009, "Gas–Liquid Two-Phase Flow Patterns in Rectangular Polymeric Microchannels, Effect of Surface Wetting Properties," *New Journal of Physics*, **11**, 075034(14).

Juncker, D., Schmid, H., Drechsler, U., Wolf, H., Michel, B., Rooij, N.D., and Delamarche, E., 2002, "Autonomous Microfluidic Capillary System," *Analytical Chemistry*, **74(24)**, 6139-6144.

Khandekar, S., Panigrahi, P.K., Lefevre, F., and Bonjour, J., 2010, "Local Hydrodynamics of Flow in a Pulsating Heat Pipe: A Review," *Frontiers in Heat Pipes*, **1**, 1-20.

King, C., Walsh, E., and Grimes, R., 2007, "PIV Measurement of Flow within Plugs in a Microchannel," *Microfluidics Nanofluids*, **3**, 463-472.

Kinoshita, H, Kaneda, S, Fujii, T, and Oshima, M, 2007, "Three-Dimensional Measurement and Visualization of Internal Flow of a Moving Droplet Using Confocal Micro-PIV," *Lab on Chip*, **7**, 338-346.

Lee, C.Y., and Lee, S.Y., 2008, "Pressure Drop of Two-Phase Plug Flow in Round Mini-Channels: Influence of Surface Wettability," *Experimental Thermal and Fluid Science*," **32(8)**, 1716-1722.

Li, H., and Olsen, M., 2006, "Micro-PIV Measurement of Turbulent Flow in Square Microchannels with Hydraulic Diameters from 200 μm to 640 μm," *International Journal of Heat and Fluid Flow*, **27**, 123-134.

Lima, R., Wada, S., Takeda, M., Tsubota, K., and Yamaguchi T., 2007, "In vitro Confocal Micro-PIV Measurements of Blood Flow in a Square Microchannel: the Effect of the Haematocrit on Instantaneous Velocity Profiles," *Journal of Biomechanics*, **40**, 2752-2757.

Malsch, D., Kielpinski, M., Merthan, R., Albert, J., Mayer, G., Kohler, M., Stahl, M., and Henkel, T., 2008, "Micro-PIV Analysis of Taylor Flow in Microchannels," *Chemical Engineering Journal*, **135S**, 166-172.

Mehdizadeh, A., Sherif, S.A., and Lear, W.E., 2011, "Numerical Simulation of Thermofluid Characteristics of Two-Phase Slug Flow in Microchannels," *International Journal of Heat and Mass Transfer*, **54**, 3457-3465.

Meinhart, C.D., Wereley, S.T., and Santiago, J.G., 1999, "PIV Measurement of a Microchannel Flow," *Experiments in Fluids*, **27**, 414-419.

Olsson, E., and Kreiss, G., 2005, "A Conservative Level-Set Method for Two-Phase Flow," *Journal of Computational Physics*, **210**, 225-246.

Pattekar, A.V., and Kothare, M.V., 2004, "A Microreactor for Hydrogen Production in Micro Fuel Cell Applications," *Journal of Micro-electro-Mechanical Systems*, **13**, 7-18.

Rose, W., and Heins, R.W., 1962, "Moving Interfaces and Contact Angle Rate-Dependence," *Journal of Colloid and Interface Science*, **17**, 39-48.

Santiago, J.G., Wereley, S.T., Meinhart, C.D., Beebe, D.J., and Adrian, R.J., 1998, "A Particle Image Velocimetry System for Microfluidics," *Experiments in Fluids*, **25**, 316-319.

Silva, G., Leal, N., and Semiao, V., 2008, "Micro-PIV and CFD Characterization of Flows in a Microchannel: Velocity Profiles, Surface Roughness and Poiseuille Numbers," *International Journal of Heat and Fluid Flow*, **29**, 1211-1220.

Sussman, M., Fatemi, E., Smereka, P., and Osher, S., 1998, "An Improved Level-Set Method for Incompressible Two-Phase Flows," *Computers & Fluids*, **27**, 663-680.

Sussman, M., Smereka, P., and Osher, S., 1994, "A Level-Set Approach for Computing Solutions to Incompressible Two-Phase Flow," *Journal of Computational Physics*, **114**, 146-159.

Tanner, L.H., 1979, "The Spreading of Silicone Oil Drops on Horizontal Surfaces," *Journal of Physics–D Applied Physics*, **12**, 1473-1484.

Shah, R.K., and London, A.L., 1978, Laminar Flow Forced Convection in ducts. Academic Press, New York, 1978.

Thulasidas, T.C., Abraham, M.A., and Cerro, R.L., 1997, "Flow Patterns in Liquid Slugs During Bubble Train Flow Inside Capillaries," *Chemical Engineering Science*, **52**, 2947-2962.

Tornberg, A.K., and Engquist, B., 2000, "A Finite Element Based Level-Set Method for Multiphase Flow Applications," *Computing and Visualization in Science*, **3**, 93-101.

Torralba, M., Pita, J.R., Pita, A.A., Huelsz, G., Rio, A.D., and Ortin, J., 2005, "Measurement of the Bulk and Interfacial Velocity Profiles in Oscillating Newtonian and Maxwellian Fluids," *Physical Review-E*, **72**, 0163081-0163089.

Tretheway, D.C., and Meinhart, C.D., 2002, "Apparent Fluid Slip at Hydrophobic Microchannel Walls," *Physics of Fluids*, **14**, L9-L12.

Walker, G.M., and Beebe, D.J., 2002, "A Passive Pumping Method for Microfluidic Devices," *Lab on Chip*, **2(3)**, 131-134.

Wang, X., Yong, Y., Yang, C., Mao, Z.S., and Li, D., 2013, "Investigation on Pressure Drop Characteristic and Mass Transfer Performance of Gas-Liquid Flow in Micro-Channels" *Microfluidics and Nanofluidics*, **16(1-2)**, 413-423.

Yamada, T., Hong, C., Gregory, O.J., and Faghri, M., 2011, "Experimental Investigations of Liquid Flow in Rib-Patterned Microchannels with Different Surface Wettability, *Microfluidics and Nanofluidics*, **11**, 45-55.

Zimmerman, J.D., Weislogel, M.M., and Tretheway, D.C., 2010, "Micro-PIV Measurements Near a Moving Contact Line," *in Proc. of ASME International Mechanical Engineering Congress and Exposition (IMECE)*, Vancouver, Canada, 1295-1297.

ANALYSIS OF HEAT PIPE OPERATING PARAMETERS BY RESPONSE SURFACE METHODOLOGY USING COPPER NANOFLUID AS WORKING FLUID

R Senthilkumar*, J Prabakaran, M Karthikeyan

[a] *Department of Mechanical Engineering, Annamalai University, Annamalai Nagar, India 608 002*

ABSTRACT

The performance of a heat pipe is analysed using Response Surface Methodology (RSM) with various input parameters namely heat input, angle of inclination, filling ratio, concentration of copper nanofluid and flow rate of coolant in the condenser over the output response i.e. the thermal efficiency of the heat pipe. The experimental design is developed based on Box - Behnken design method. The influence of vital input parameters and interaction among them are investigated using analysis of variance (ANOVA). The proposed mathematical model in this study has proven to fit and in line with experimental values with a 95% confidence interval.

Keywords: *Heat pipe, Copper nanofluid, Thermal efficiency, RSM.*

1. INTRODUCTION

Heat pipes are considered as passive heat transfer devices that transfer the heat from heat source to sink. The heat pipes can be designed in a myriad of shape, sealed, evacuated vessels with a little amount of working fluid. The vessels are lined with a capillary structure (wick) which allows the condensate to travel through it (Reay and Kew, 2006). The heat pipes are able to operate against gravity and tolerate very low temperature drops between the evaporator and the condenser. The performance of the heat pipe depends on numerous factors like heat input, angle of inclination, filling ratio of the working fluid in the heat pipe, etc. All the operating parameters play a key role and selection of these parameters ultimately depends on the application.

The conventional working fluids used in the heat pipe have poor heat transfer properties compared to solid metals. A new class of heat transfer fluids that are called as ''Nanofluids'', which are engineered by suspending ultra-fine metallic or non metallic nano particles in conventional fluids such as water, engine oil, ethylene glycol (Choi, 1995). The experimental results show that the nanofluids have remarkably higher thermal conductivity and greater heat transfer characteristics than conventional pure fluids (Xuan and Li, 2000; Daungthongsuk and Wongwises, 2007; Xuan et.al, 2003). The thermo physical and transport properties of the conventional fluids are improved by adding the nanoparticles in base fluid. The effective thermal conductivity of nanofluids increases with increase in temperature (Das et.al, 2003). This finding makes nanofluids even more attractive as a cooling fluid for heat transfer devices with high energy density.

The thermal enhancement using silver nano-fluid as the working fluid in the grooved heat pipe was studied with 10 and 35 nm size and reported that the increase in heat pipe wall temperature was lesser than that for a pure water filled heat pipe under various heat loads (Kang et.al, 2006). The enhancement of heat pipe efficiency with R-11refrigerant–nanoparticles (titanium) mixtures was studied. The effects of the charge amount of the working fluid, heat pipe tilt angle on the efficiency of heat pipe has been measured. It has been found that the heat pipe with 0.1% nanoparticles concentration gives efficiency 1.40 times higher than that with pure refrigerant (Naphon et.al, 2009). The impacts of aqueous copper oxide nanofluids on the thermal performance of a horizontal heat pipe working at steady sub-atmospheric pressures are investigated and it was reported that the copper nanofluid can significantly enhance heat transfer coefficients of both evaporator and condenser sections and also the maximum heat flux of the heat pipe (Liu and Zhu, 2011).

Response surface methodology (RSM) is a collection of mathematical and statistical techniques for developing, improving and optimizing the process parameters by careful design of experiments. The objective is to optimize a response (output variable) which is influenced by several independent variables (input variables). An experiment is a series of tests, called runs, in which changes are made in the input variables in order to identify the reasons for changes in the output response (Montgomery, 2005; Mohammed, 2010; Gunaraj and Murugan, 1999; Ramkumar and Ragupathy, 2012; Prabakaran et al, 2012). In general the relationship is

$$Y = f(X_1, X_2, X_3, \ldots) \pm \varepsilon. \quad (1)$$

Literature review on heat pipe performance reveals that the optimization of heat pipe using RSM does not have reported so far. Hence the present study employs RSM technique to optimize the operating parameters of heat pipe. The objective of the present investigation is to optimize the heat pipe operating parameters like heat input, angle of inclination, filling ratio, concentration of copper nanofluid and flow rate of the coolant on the condenser by response surface methodology of a copper–water heat pipe with screen mesh wick.

* *Corresponding author: Email: rsklme@yahoo.com*

2. EXPERIMENTAL ARRANGEMENT AND TEST PROCEDURE

The experimental setup of heat pipe and thermocouple locations are shown in Fig. 1 and 2 respectively. The evaporator of the heat pipe is heated by a cylindrical electric heater which is attached in the evaporator section of the container. The temperature distributions along the length of the heat pipe have been measured using copper-constantan (T-type) thermocouples. The surface temperature of the heat pipe in adiabatic region is measured using six evenly spaced thermocouples with an uncertainty of ±0.1°C. In addition, three thermocouples are provided each on the evaporator section, on the condenser section and one each in inlet and outlet of the condenser jacket. The heat pipe is charged with working fluid, which approximately corresponds to the amount required to fill the evaporator which is nearly to 40 ml. At the beginning of the experiment, the evaporator temperature is 30°C for all experiments for comparison of results.

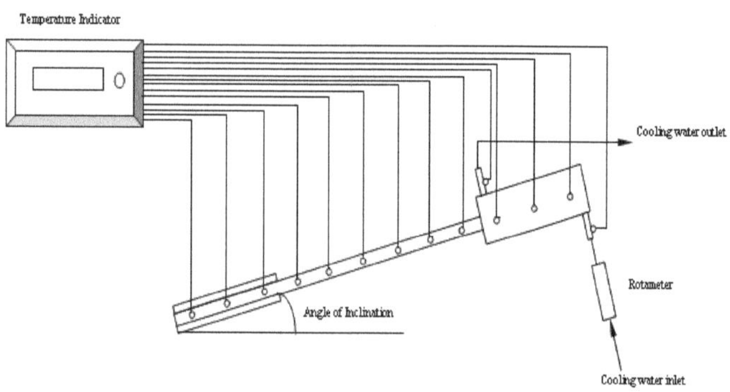

Fig. 1 Experimental arrangement

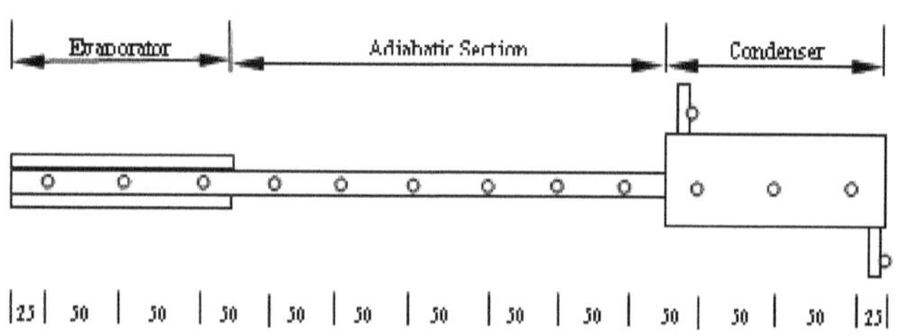

Fig. 2 Thermocouple locations of heat pipe

The cooling water is circulated in the cooling jacket to remove the heat of the working fluid. This cooling jacket is attached to the condenser section which is located at the end of the heat pipe. The heat pipe has the capability to transmit the heat through the internal structure. Therefore, an abrupt increase in the wall temperature could harm the heat pipe if the heat is not rejected at the condenser properly. Hence, the cooling water is first dispersed through the condenser jacket, before the heat is supplied to the evaporator. The condenser section of the heat pipe is cooled using water flowing through a circumferential water jacket. The cooling water flow rate is measured by a rotameter with an uncertainty of ±1% and the flow rate is kept constant at 0.08 kg/min. The initial temperature of cooling water is maintained at 28°C at the time of starting the experiments. The inlet and outlet temperatures of the cooling water are computed using two copper constantan thermocouples.

The experimental specifications are given in Table 1. The adiabatic section of the heat pipe is entirely covered with glass wool insulation to avoid the heat loss. The amount of heat loss from the evaporator and condenser surface is negligible due to the proper insulation in the heat pipe surface. The power input given to the evaporator section of the heat pipe is gradually increased to the preferred power level. The surface temperatures at six different locations along the adiabatic section of heat pipe are measured at regular time intervals until the heat pipe attains the steady state condition. Concurrently, the evaporator wall temperatures, condenser wall temperatures, water inlet and outlet temperatures in the condenser zone are also measured.

Once the steady state is reached, the input power is turned off and cooling water is permitted to flow through the condenser to cool the heat pipe and to make it ready for further experimental purpose. The steady state is defined as the variation in temperature which is less than ±0.1°C for subsequent time intervals. Then the power is increased to the next level and the heat pipe is tested for its performance. The experimental procedure is repeated for different heat inputs and different inclinations of pipe to the horizontal and observations are recorded. The output heat transfer rate from the condenser is calculated by applying an energy balance to the condenser flow. The vacuum pressure in the inner side of the heat pipe is monitored by a vacuum gauge, which is attached in the condenser end of the heat pipe. The initial vacuum pressure inside the heat pipe is 70 to 75 cm of Hg, because at higher vacuum condition the heat pipe performs better. When the heat pipe is placed in the reverse direction, i.e. the evaporator is kept above the condenser, the condensate has to return back to the evaporator by the capillary action of wick only and against the gravity. This result in the reduction in the performance of the heat pipe compared to the performance of gravity assisted heat pipe.

Table 1 Specifications of heat pipe

Properties	*Specifications*
Heat pipe material	Copper
Wick material	Stainless steel (two layers)
Total length of pipe, m	0.6
Evaporator length, m	0.15
Adiabatic length, m	0.3
Condenser length, m	0.15
Condenser outer dia, m	0.036
Condenser inner dia, m	0.030
Outer diameter of the pipe, m	0.020
Inner diameter of the pipe, m	0.0176
Wick permeability, m^2	2.16×10^{-6}
Wire mesh diameter, m	0.183
Heat Input, W	30, 40, 50, 60 and 70.
Heat pipe inclination	0°, 15°, 30°, 45°, 60°, 75° and 90°
Working fluid	Copper nanofluid, 100 mg/lit
Wick mesh size	60/sq. inch
Wick Porosity (ε)	0.6431

3. RESULTS AND DISCUSSIONS

In this experimental study the performance of the heat pipe is analyzed by response surface methodology (RSM). Box Behnken design method is employed with five input parameters namely heat input (A), angle of inclination (B), filling ratio (C) concentration of copper nano particle in the base fluid (D) and the flow rate of the coolant in the condenser (E) over the output response as thermal efficiency. DESIGN-EXPERT software is used for this analysis. Table 2 shows the range of input parameters. In RSM -1 means the minimum and +1 denotes the maximum value of each input parameters.

The computed values of the thermal efficiency are entered in the software design matrix. The Box–Behnken response surface methodology is used to develop the relationship between the experimental variables and the response that is thermal efficiency. A regression analysis is carried out to develop a best fit of the model to the experimental data, which are used to generate response surface plots. Table 3 shows the analysis of variance (ANOVA). The values of "Prob>F" less than 0.05 indicates model terms are significant. The value 0.05 shows that the regression is statistically significant at a 95% confidence level ($P < 0.05$). For the present case heat input (A), angle of inclination (B) and filling ratio (C) are playing significant effect than the concentration (D) and flow rate (E). The square values of angle of inclination (B), filling ratio (C), concentration (D) and flow rate (E) are also having significant effect on the performance of the heat pipe. The interaction effect between heat input and angle of inclination (AB) also has significant effect. The interaction effect of heat input with concentration (AD) and heat input with filling ratio (AC) also made some significant effect on the efficiency. The other effects like AE, BC, BD, BE, CD, CE and DE are insignificant.

Table 2 Process Parameters

Parameter	*Level*		
	-1	*0*	*+1*
Heat input, W	30	50	70
Angle of inclination, Deg	0	30	60
Filling ratio, %	25	75	125
Concentration, mg/lit	25	75	125
Flow rate, kg/s	0.06	0.1	0.14

The "Pred R-Squared" value of 0.9843 (coefficient of determination) is in responsible agreement with "Adj R-Squared" of 0.9717. Adequate precision measures the signal to noise ratio. A ratio greater than 4 is desirable. Here the ratio of 30.5142 indicates an adequate signal. This model can be used to navigate the design space. The model F value is 78.34 and a probability value is less than 0.001 which indicates that the model is significant for to finding the thermal efficiency of the heat pipe. The sum of squares due to lack of fit, or more tersely a lack-of-fit sum of squares, is one of the components of a partition of the sum of squares in an analysis of variance, used in the numerator in an F-test. Based on ANNOVA, different coefficients were estimated to develop the following empirical relation to predict the thermal efficiency of the heat pipe. The values are Prob > F less than the 0.05 indicates that the terms are significant and values are greater than the 0.1 indicates that the insignificants. The lack of fit F value is 3.558 means the lack of fit not significant and it relative to the pure error. Non significant of lack of fit is good for model to fit. Based on the ANNOVA, the following empirical relation was developed to predict the thermal efficiency of the heat pipe. The df denotes the degree of freedom of the model and that value is 20. It is the number of values in the final calculation of a statistic that are free to vary. Mean squares are used in analysis of variance and are calculated as a sum of squares divided by its appropriate degrees of freedom.

$$\text{Thermal Efficiency} = -33.985 + 0.40925\,A + 0.3164\,B + 0.3729\,C + 0.4217\,D + 264.6823\,E + 0.00263\,AB + 0.00119\,AC - 0.00124\,AD - 0.025\,AE + 0.000345\,BC + 0.00062\,BD + 0.15417\,BE - 0.00002\,CD + 0.47875\,CE - 0.2475\,DE + 0.0001\,A^2 - 0.007442\,B^2 - 0.002803\,C^2 - 0.002348\,D^2 - 1484.896\,E^2. \quad (2)$$

The thermal efficiency of the heat pipe is calculated as the ratio of heat rejection in the condenser section to the heat input at the evaporator section (Naphon, 2009). The working fluid in this analysis is copper nanofluid. The base fluid used in the copper nanofluid is DI water. The equation 2 valid for the experiments within the range of heat pipe operation mentioned in the table 2.It is observed from Fig. 3-6 that the thermal efficiency of heat pipe increases linearly with an increase the heat input in the evaporator section. The thermal efficiency of the heat pipe increases with increase in heat flux, due to the fact that the temperature gradient between the evaporator section and condenser sections increases. For higher values of heat input in the evaporator section, the heat generated in the surface is more and the working medium which is in the form of vapour moves vigorously into the condenser section. The cooling water in the condenser absorbs this excessive heat and as a result, the efficiency of the heat pipe increases. The maximum heat input given to the evaporator based on the heat pipe heat transfer limitations (Table 4).

From the figures (3, 7, 8 &9) it is evident that the heat pipe efficiency increases with the increase in the tilt angle up to certain angle and then decreases. This is due to the fact that, the gravitational force has a significant effect on the flow of working fluid between the evaporator section and the condenser section in addition to capillary action of wick. The thermal efficiency of the heat pipe reaches a maximum at 45° tilt angle afterwards the heat pipe thermal efficiency tends to decrease. This may be due to the strong formation of the liquid film at the inner side of condenser section resulting in the high thermal resistance between the vapour of the working fluid and the cooling medium in the condenser. From the figures 4, 7, 10 &11, the thermal efficiency is high at a 100% filling of the evaporator. If the filling ratio more than the volume of the evaporator, the space available for the flow of the vapour is reduced. At lower filling ratio, the amount of working fluid used to carry away the heat from the evaporator is reduced, so that the efficiency heat pipe is lowered. Figure 5, 8, 10 & 12 show the variations of thermal efficiency of the heat pipe with concentration of the copper nanofluid. Thermal efficiency increases with the concentration of copper oxide nano fluid in the base fluid. The value of efficiency is maximum at 100 mg/lit of concentration of copper oxide nano particle in the base fluid. More than that value, the performance of the heat pipe gets reduced due to the agglomeration of nano particles which block of mesh holes in the wick. The variation effect of flow rate of coolant in the condenser is very less than the other parameters.

Figure 13 shows the optimization plot with a desirability of 0.993 generated by RSM. It shows that the optimum value of the thermal efficiency is 50% when the heat input is 69.94W, at 43.34° inclination, filling ratio of 100.15%, at a concentration of CUO 82.93 mg/lit and flow rate 0.10 kg/min.

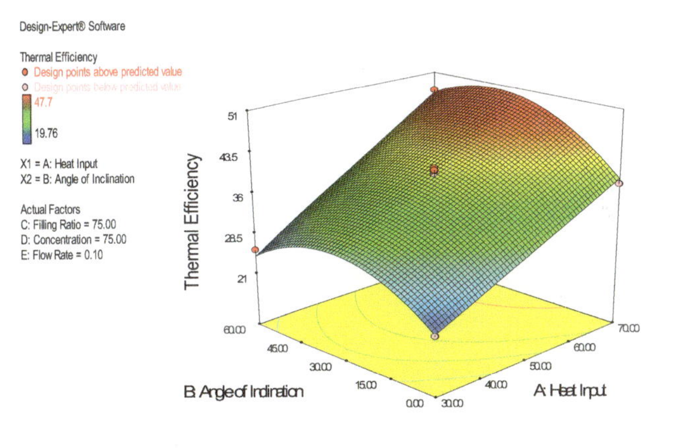

Fig. 3 Effect of heat input and angle of inclination

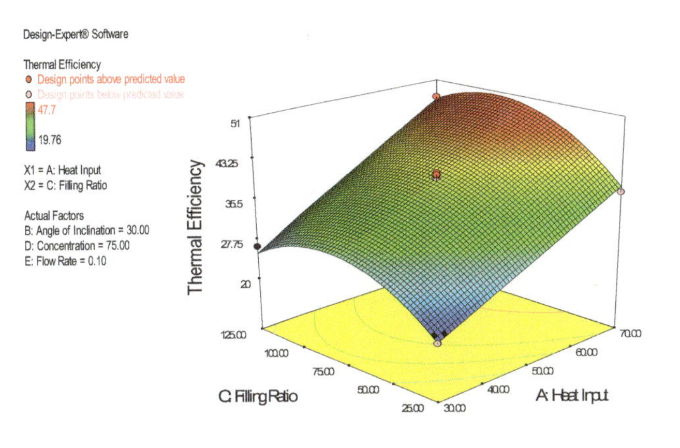

Fig. 4 Effect of heat input and filling ratio

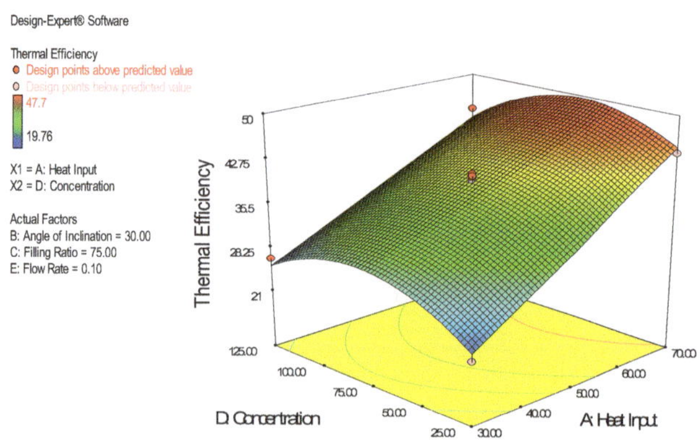

Fig. 5 Effect of heat input and concentration

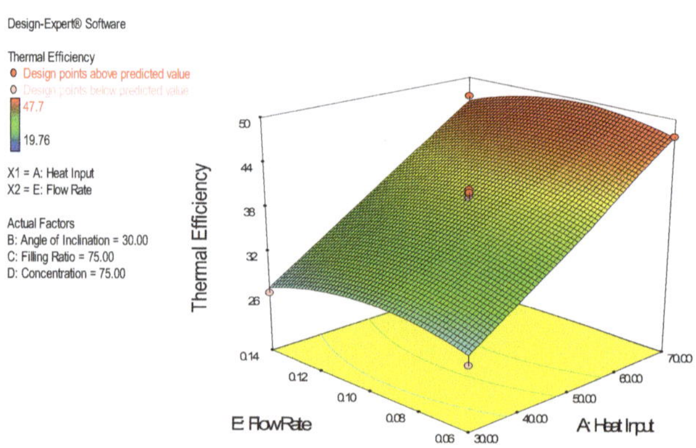

Fig. 6 Effect of heat input and flow rate

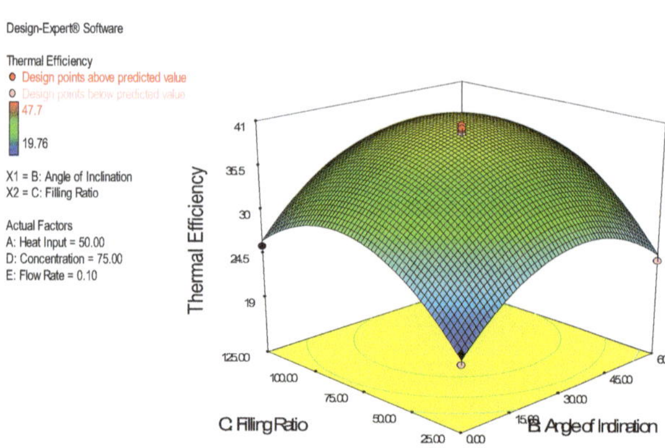

Fig. 7 Effect of angle of inclination and filling ratio

Table 4 Heat Transfer Limitations

Limitation	*Values in W*
Capillary Limit	77
Sonic Limit	65400
Entrainment Limit	4638
Boiling Limit	642

Table 3 ANOVA for Response Surface Quadratic Model - Thermal Efficiency

Source	*Sum of Squares*	*df*	*Mean Square*	*F Value*	*p-value Prob > F*
Model	2744.54	20	137.23	78.34	< 0.0001
A-Heat Input	1549.21	1	1549.2	884.42	< 0.0001
B-Angle of Inclination	114.87	1	114.87	65.58	< 0.0001
C-Filling Ratio	188.65	1	188.65	107.7	< 0.0001
D-Concentration	0.001	1	0.0012	0.001	0.9791
E-Flow Rate	3.432	1	3.4318	1.959	0.1739
AB	9.986	1	9.9856	5.701	0.0248
AC	5.664	1	5.6644	3.234	0.0842
AD	6.150	1	6.1504	3.511	0.0727
AE	0.002	1	0.0016	0.001	0.9761
BC	1.071	1	1.0712	0.612	0.4415
BD	3.460	1	3.4596	1.975	0.1722
BE	0.137	1	0.1369	0.078	0.7821
CD	0.010	1	0.0100	0.006	0.9404
CE	3.667	1	3.6672	2.094	0.1603
DE	0.980	1	0.9801	0.560	0.4614
A^2	0.014	1	0.0140	0.008	0.9296
B^2	391.47	1	391.48	223.49	< 0.0001
C^2	428.451	1	428.45	244.6	< 0.0001
D^2	300.715	1	300.72	171.67	< 0.0001
E^2	49.262	1	49.26	28.123	< 0.0001
Residual	43.792	25	1.7517		
Lack of Fit	40.917	20	2.0458	3.558	0.0820
R-Squared	0.9843				
Adj R-Squared	0.9717				
Pred R-Squared	0.9398				
Adeq Precision	30.5142				

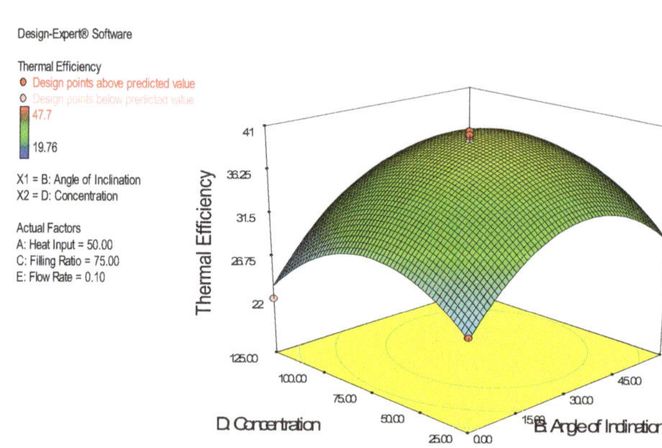

Fig. 8 Effect of angle of inclination and concentration

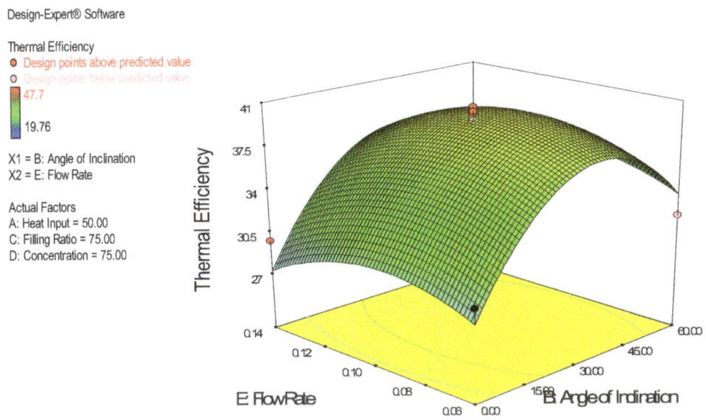

Fig. 9 Effect of angle of inclination and flow rate

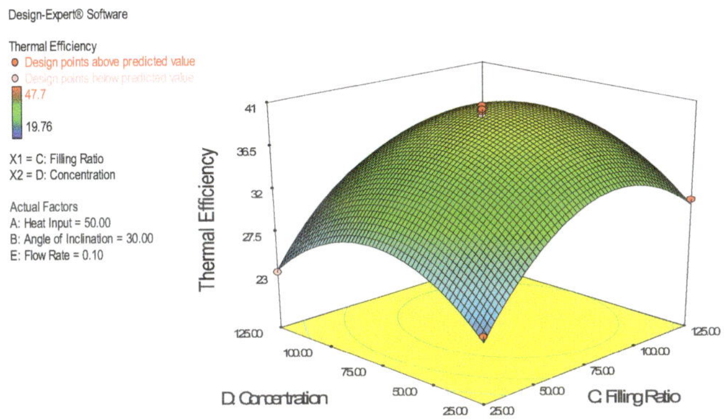

Fig. 10 Effect of filling ratio and concentration

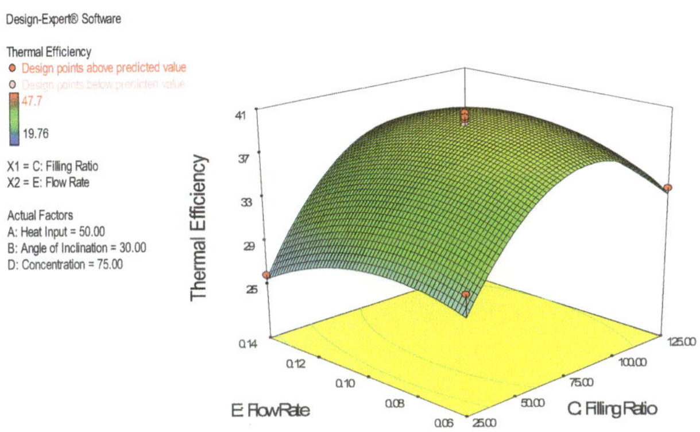

Fig. 11 Effect of filling ratio and flow rate

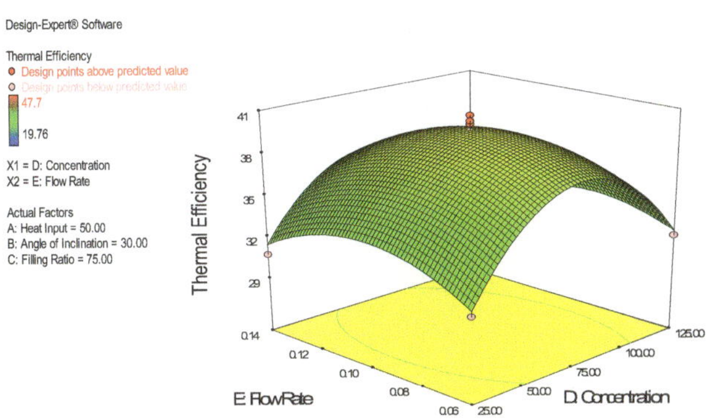

Fig. 12 Effect of concentration and flow rate

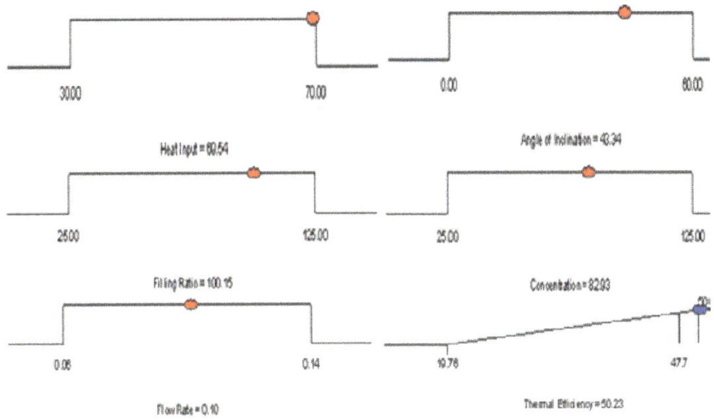

Fig. 13 Optimization plot

4. VALIDATION OF MODEL

In order to confirm the optimization result, the experiment was conducted with 70 W heat input at 45° inclination of the heat pipe, filling ratio of 100%, concentration 85 mg/lit with a flow rate of 0.1 kg/min. The thermal efficiency of the heat pipe have been found as 50.42% which is closer to the optimum value. The values of the efficiency obtained from the experimental and predicted values are compared to examine the validity of the model and compare the relative factors. The confirmation tests are conducted and the errors are tabulated in the Table 5. The findings show that the errors between the experimental and predicted values are less than 2%. It denotes that the predicted model is efficient.

Table 5 Comparison of predicted and experimental values

Sl.No	*Factors*					*Efficiency*		*Error (%)*
	A	B	C	D	E	Predicted	Experimental	
1	30	0	75	75	0.1	22.06	21.80	1.19
2	50	60	75	75	0.06	33.54	32.80	2.26
3	70	30	125	75	0.1	47.18	47.52	0.72
4	50	60	25	75	0.1	24.67	23.85	3.44
5	50	30	125	75	0.14	34.23	33.20	3.01
Confirmation test	70	45	100	85	0.1	50.29	50.42	0.26

5. CONCLUSION

In this paper, the experimental results showed the thermal efficiency of the heat pipe and the statistical analysis results demonstrated that the effects of various operating parameters on the thermal efficiency. Empirical relation for thermal efficiency was developed to correlate dominant input parameters like heat input, angle of inclination, filling ratio, concentration of copper nanofluid and the flow rate of coolant using Response Surface Methodology. The influence of input parameters on the performance of heat pipe was analyzed based on the developed relation by RSM. The predicted value of RSM are very closer to the experimental results and it has reduced the number of experiments, because RSM provides useful interaction between different variables of the system. In the desirability function approach, the value of desirability was 0.993 for the RSM model very close to 1. The predicted RSM model is found to be capable of predictive thermal efficiency of heat pipe.RSM showed a better accuracy and capability of generalization with the design of experiments.

ACKNOWLEDGMENT

The authors thank the authorities of Annamalai University for providing the necessary facilities in order to accomplish this piece of work.

NOMENCLATURE

A	*heat input, W*
B	*angle of inclination. Deg*
C	*filling ratio, %*
D	*concentration, mg/lit*
E	*flow rate. kg/s*
Y	response of RSM
X	input variables
f	mathematical function)

Greek Symbols

ε measurement error

Subscripts

1, 2 ... variable at points

REFERENCES

Choi, S.U.S., 1995, "Enhancing Thermal Conductivity of Fluids with Nanoparticles, Developments and Applications of Non-Newtonian Flows", *FED*, **231**, 99–105.

Das,S.K., Putra, N., Thiesen, P., and Roetzel, W., 2003, "Temperature Dependence of Thermal Conductivity Enhancement for Nanofluids", *Journal of Heat Transfer,* **125**, 567–574. http://dx.doi:10.1115/1.1571080

Daungthongsuk, W., and Wongwises, S., 2007, "A Critical Review of Convective Heat Transfer of Nanofluid", *Renewable and Sustainable Energy Reviews,* **11,** 797–817. http://dx.doi.org/10.1016/j.rser.2005.06.005

Gunaraj, V., and Murugan, N., (1999) Application of Response Surface Methodology for Predicting Weld Bead Quality in Submerged Arc Welding of Pipes. *Journal of Materials Processing Technology*, 88(1-3),pp. 266-275. http://dx.doi.org/10.1016/S0924-0136(98)00405-1

Kang ,S.W., Wei, W.C., Tsai, S.H., and Yang S.U., 2006, "Experimental Investigation of Silver Nano-Fluid on Heat Pipe Thermal Performance", *Applied Thermal Engineering*, **26**(17-18), 2377–2382. http://dx.doi.org/10.1016/j.applthermaleng.2006.02.020

Liu, Z.H., and Zhu, Q.G., 2011, "Application of Aqueous Nanofluids in a Horizontal Mesh Heat Pipe", *Energy Conversion and Management*, **52**(1), 292–300. http://dx.doi.org/10.1016/j.enconman.2010.07.001

Mohammed, J.K., Aziz, H.M., Yosoff, M.S., and Adlan, N., 2010, "Application of Response Surface Methodology (RSM) for Optimization of Ammonical Nitrogen Removal From Semi-Aerobic Landfill Leachate Using Ion Exchange Resin", *Desalination*, **254**, 154-161. http://dx.doi.org/10.1016/j.desal.2009.12.002

Montgomery, C.C., 2005, *Design and Analysis of Experiments*, John Wiley & sons, USA.

Naphon, P., Thongkum, D., and Assadamongkol, P., 2009, "Heat Pipe Efficiency Enhancement with Refrigerant–Nanoparticles Mixtures", *Energy Conversion and Management*, **50**(3), 772–776. http://dx.doi.org/10.1016/j.enconman.2008.09.045

Prabakaran, J., Vaidyanathan, S., and Kanagarajan. D., 2012, "Establishing Empirical Relation to Predict Temperature Difference of Vortex Tube Using Response Surface Methodology", *Journal of Engineering Science and Technology*, **7**(6), 722 – 731.

Ramkumar, R., and Ragupathy, A., 2012, "Application of Response Surface Methodology (RSM) for Optimization of Operating Parameters and Performance Evaluation of Cooling Tower Cold Water Temperature", *An International Journal of Optimization and Control: Theories & Applications*, **2**(1), 39-50. http://dx.doi.org/10.11121/ijocta.01.2012.0093

Reay, D.A., and Kew P.A., 2006, Heat Pipes, Butterworth–Heinemann, Oxford press.

Xuan, Y., and Li, Q., 2000, "Heat Transfer Enhancement of Nano-fluids", *International Journal of Heat and Fluid Flow*, **21**, 58–64. http://dx.doi.org/10.1016/S0142-727X(99)00067-3

Xuan, Y. and Li, Q., 2003, "Investigation on Convective Heat Transfer and Flow Features of Nano-Fluids", *Journal of Heat Transfer,* **125**, 151–155. http://dx. doi:10.1115/1.1532008

HEAT PIPES: REVIEW, OPPORTUNITIES AND CHALLENGES

Amir Faghri*

Department of Mechanical Engineering, University of Connecticut, Storrs, CT 06269, U.S.A.

ABSTRACT

A detailed overview of heat pipes is presented in this paper, including a historical perspective, principles of operations, types of heat pipes, heat pipe performance characteristics, heat pipe limitations, heat pipe frozen startup and shutdown, heat pipe analysis and simulations, and various applications of heat pipes. Over the last several decades, several factors have contributed to a major transformation in heat pipe science and technology . The first major contribution was the development and advances of new heat pipes, such as loop heat pipes, micro and miniature heat pipes, and pulsating heat pipes. In addition, there are now many new commercial applications that have helped contribute to the recent interest in heat pipes, especially related to the fields of electronic cooling and energy. For example, several million heat pipes are now manufactured each month since all modern laptops use heat pipes for CPU cooling. Numerical modeling, analysis, and experimental simulation of heat pipes have also significantly progressed due to a much greater understanding of various physical phenomena in heat pipes, as well as advances in computational and experimental methodologies.

Keywords: *heat pipes, performance characteristics, types of heat pipes, numerical modeling, heat pipe applications*

1. INTRODUCTION

The heat pipe (Faghri, 1995) is a highly effective passive device for transmitting heat at high rates over considerable distances with extremely small temperature drops, exceptional flexibility, simple construction, and easy control with no external pumping power. Engineers, scientists and graduate students interested in heat pipe science often times struggle and spend considerable time poring through archival publications or the contents of heat pipe books in order to understand and predict a heat pipe system. Being aware of this situation, this review is a self-contained document of the state-of-the-art heat pipe science and technology. The physical significance and performance characteristics, as well as the technological and experimental issues related to this technology are covered in this review. The purpose of this review article is also to make a thorough presentation of the basic and advanced techniques for analyzing various heat pipe systems under a variety of operating conditions and limitations.

The subject of heat pipe science has immense importance in a large variety of traditional engineering disciplines. The sub-discipline of heat pipe science has its foundation in several classical fields, such as fluid mechanics, heat transfer, thermodynamics and solid mechanics. Heat pipe science also provides an opportunity for scientists and engineers to apply a variety of complex physical phenomena and fundamental laws in the thermal-fluids area to a relatively simple system, such as the heat pipe. This includes the steady and unsteady forced laminar and turbulent convective heat and mass transfer, compressible vapor effects, phase-change phenomena, boiling, condensation/evaporation, two-phase flow, rotating flows, thin film flows, liquid flow in porous media, rarefied gases, interfacial heat and mass transfer, magneto-hydrodynamic flows, and conjugate heat transfer effects.

This review provides detailed information about the following topics related to heat pipe science and technology:

Section 2: Historical Development of Heat Pipes
Section 3: Principles of Operations
Section 4: Types of Heat Pipes
Section 5: Working Fluids and Temperature Ranges
Section 6: Capillary Wick Designs and Structures
Section 7: Heat Transfer Limitations
Section 8: Heat pipe Start Up
Section 9: Heat Pipe Characteristics
Section 10: Heat Pipe Analysis and Simulation
Section 11: Heat Pipe Applications
Section 12: Conclusions

2. HISTORICAL DEVELOPMENT OF HEAT PIPES

Of the many different types of systems which transport heat, the heat pipe (Faghri, 1995) is one of the most efficient systems known today. The advantage of using a heat pipe over other conventional methods is that large quantities of heat can be transported through a small cross-sectional area over a *considerable* distance with no additional power input to the system. Furthermore, design and manufacturing simplicity, small end-to-end temperature drops, and the ability to control and transport high heat rates at various temperature levels are all unique features of heat pipes.

The predecessor of the heat pipe, the Perkins tube, was introduced by the Perkins family from the mid-nineteenth to the twentieth century through a series of patents in the United Kingdom. Most of the Perkins tubes were wickless gravity-assisted heat pipes (thermosyphons), in which heat transfer was achieved by a change of phase (latent heat of evaporation). The Perkins tube design closest to the present heat pipe was patented by Jacob Perkins (1836). This design was a closed tube containing a small quantity of water operating as a two-phase cycle.

The introduction of the heat pipe was first conceived by Gaugler (1944) of the General Motors Corporation in the U.S. Patent No. 2350348. Gaugler, who was working on refrigeration problems at that time, envisioned a device which would evaporate a liquid at a point above the place where condensation would occur without requiring any

* Email: faghri@engr.uconn.edu

additional work to move the liquid to the higher elevation. His device consisted of a closed tube in which the liquid would absorb heat at one location causing the liquid to evaporate. The vapor would then travel down the length of the tube, where it would recondense and release its latent heat. It would then travel back up the tube via capillary pressure to start the process over. In order to move the liquid back up to a higher point, Gaugler suggested the use of a capillary structure consisting of a sintered iron wick. A refrigeration unit proposed by Gaugler used a heat pipe to transfer the heat from the interior of a compartment to a pan of crushed ice below. His idea, however, was not used by General Motors for the refrigeration problem.

In 1962, Trefethen (1962) resurrected the idea of a heat pipe in connection with the space program. Serious development started in 1964 when the heat pipe was independently reinvented and a patent application was filed by Grover at Los Alamos National Laboratory in New Mexico. Grover et al. (1964) and Grover (1966) built several prototype heat pipes, the first of which used water as a working fluid, and was soon followed by a sodium heat pipe which operated at 1100 K. Grover and his co-workers also demonstrated the effectiveness of heat pipes as a high performance heat transmission device and proposed several applications for their use. In a U.S. patent application filed by Grover on behalf of the U.S. Atomic Energy Commission, Grover (1966) coined the phrase "heat pipe" and described a device almost exactly the same as Gaugler's, stating, "with certain limitations on the manner of use, a heat pipe may be regarded as a synergistic engineering structure which is equivalent to a material having a thermal conductivity greatly exceeding that of any known metal." In the patent application, Grover (1966) gave a very limited theoretical analysis of heat pipes, but presented experimental results obtained from stainless steel heat pipes that incorporated a screen wick with sodium, silver and lithium as working fluids.

The recognition of the heat pipe as a reliable thermal device was initially due to the preliminary theoretical results and design tools that were reported in the first publication on heat pipe analysis by Cotter (1965). Following this publication, research began worldwide. The United Kingdom Atomic Energy Laboratory at Harwell started experimenting with sodium heat pipes to use as thermionic diode converters. In addition, scientists started conducting similar work at the Joint Nuclear Research Center in Ispra, Italy, which soon became the most active research center outside the U.S. Shortly thereafter, other countries such as Germany, France, and the former USSR initiated efforts in this regard.

The early development of terrestrial applications of heat pipes proceeded at a slow pace. Due to capillary action, heat pipes can operate in microgravitational fields without any external force field or pump. Because of this, most early efforts were directed toward space applications. However, due to the high cost of energy, especially in Japan and Europe, the industrial community began to appreciate the significance of heat pipes and thermosyphons in energy savings applications. Today, all developed countries have been actively involved in research, development, and commercialization of heat pipes.

Within the last decade, a major transformation regarding heat pipe technology and application has occurred due to the critical need of electronic cooling and energy systems, as well as the invention of new heat pipes. Several million heat pipes per month are now being manufactured since all modern laptop computers use heat pipe technology to transfer heat away from the processor. Furthermore, research and development for new heat pipes such as loop heat pipes, micro and miniature heat pipes, and pulsating heat pipes, has matured enough for use in various applications.

There is a wealth of published literature: several heat pipe books and monographs, as well as the proceedings of eleven international heat pipe symposiums and seventeen international heat pipe conferences. In addition, there are a large number of archival and non-archival publications and reports related to heat pipes, dating back over the last four decades. A significant amount of basic and applied research & development has been performed since 1985 in the area of heat pipes due to the great potential use of this technology for various applications.

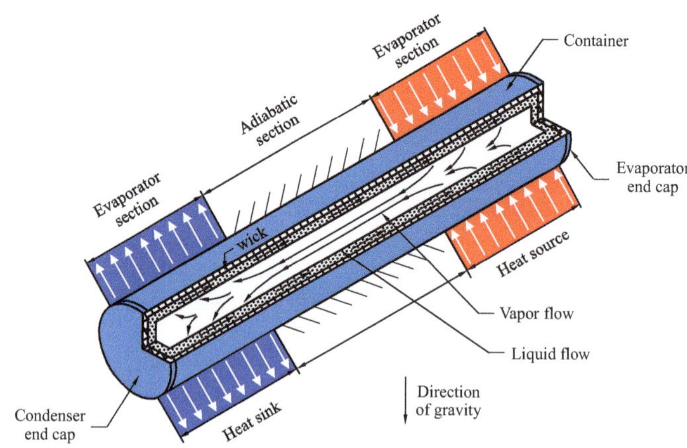

Fig. 1 Schematic of a conventional heat pipe showing the principle of operation and circulation of the working fluid.

3. PRINCIPLES OF OPERATION

The operation of a heat pipe (Faghri, 1995) is easily understood by using a cylindrical geometry, as shown in Fig. 1. However, heat pipes can be of any size or shape. The components of a heat pipe are a sealed container (pipe wall and end caps), a wick structure, and a small amount of working fluid which is in equilibrium with its own vapor. Different types of working fluids such as water, acetone, methanol, ammonia or sodium can be used in heat pipes based on the required operating temperature. The length of a heat pipe is divided into three parts: the evaporator section, adiabatic (transport) section and condenser section. A heat pipe may have multiple heat sources or sinks with or without adiabatic sections depending on specific applications and design. Heat applied externally to the evaporator section is conducted through the pipe wall and wick structure, where it vaporizes the working fluid. The resulting vapor pressure drives the vapor through the adiabatic section to the condenser, where the vapor condenses, releasing its latent heat of vaporization to the provided heat sink. The capillary pressure created by the menisci in the wick pumps the condensed fluid back to the evaporator section. Therefore, the heat pipe can continuously transport the latent heat of vaporization from the evaporator to the condenser section. This process will continue as long as there is a sufficient capillary pressure to drive the condensate back to the evaporator.

The menisci at the liquid-vapor interface are highly curved in the evaporator section due to the fact that the liquid recedes into the pores of the wick. On the other hand, during the condensation process, the menisci in the condenser section are nearly flat. A capillary pressure exists at the liquid-vapor interface due to the surface tension of the working fluid and the curved structure of the interface. The difference in the curvature of the menisci along the liquid-vapor interface causes the capillary pressure to change along the pipe. This capillary pressure gradient circulates the fluid against the liquid and vapor pressure losses, and adverse body forces such as gravity or acceleration.

The vapor pressure changes along the heat pipe are due (Faghri, 1995) to friction, inertia and blowing (evaporation) and suction (condensation) effects, while the liquid pressure changes mainly as a result of friction. The liquid-vapor interface is flat near the condenser end cap corresponding to a zero local pressure gradient at very low vapor flow rates. A typical axial variation of the shape of the liquid-vapor interface and the liquid and vapor pressures for low vapor flow rates are shown in Figs. 2(a) and 2(b), respectively.

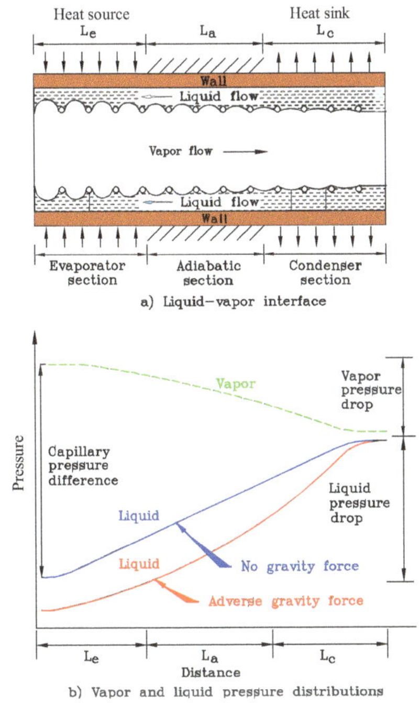

Fig. 2 Axial variation of the liquid-vapor interface, and the vapor and liquid pressures along the heat pipe at low vapor flow rates.

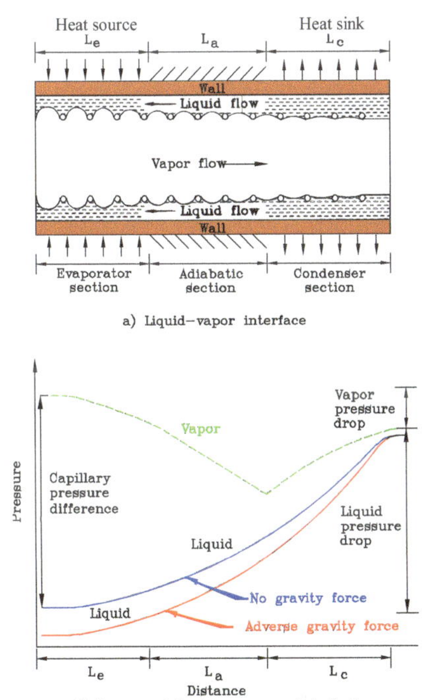

Fig. 3 Axial variation of the liquid-vapor interface, and the vapor and liquid pressures along the heat pipe at moderate vapor flow rates.

The maximum local pressure difference occurs near the evaporator end cap. This maximum local capillary pressure should be equal to the sum of the pressure drops in the vapor and the liquid across the heat pipe in the absence of body forces. When body forces are present, such as an adverse gravitational force, the liquid pressure drop is greater, indicating that the capillary pressure must be higher in order to return the liquid to the evaporator for a given heat input. At moderate vapor flow rates, dynamic effects cause the vapor pressure drop and recovery along the condenser section, as shown in Fig. 3(b).

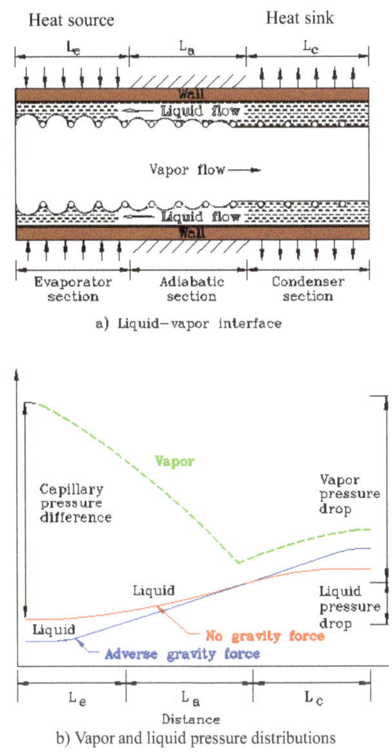

Fig. 4 Axial variation of the liquid-vapor interface, and the vapor and liquid pressures along the heat pipe at high vapor flow rates.

The local liquid-vapor pressure difference is small, but this pressure gradient approaches zero at the condenser end cap similar to the low vapor flow rate case. Again, the capillary pressure difference at the evaporator end cap should be balanced by the sum of the total pressure drop in the vapor and liquid across the heat pipe.

The general trend at high vapor flow rates with low liquid pressure drops is different from the other two cases. The vapor pressure drop can exceed the liquid pressure drop in the condenser section. In such a case, the liquid pressure would be higher than the vapor pressure in the condenser section if the pressure in the liquid and vapor are equal at the condenser end cap. In reality, the wet point is not situated at the condenser end cap as shown in 4(b), and the menisci at the liquid-vapor interface in the condenser are curved.

Traditionally, heat pipe theory consists of fundamental analyses related to hydrodynamic and heat transfer processes. Fluid mechanics theory is generally used to describe the axial liquid pressure drop in the wick structure, the maximum capillary pumping head and the vapor flow in the vapor channel. Heat transfer theory is used to model the transfer of heat into and out of the heat pipe. Phenomena such as conjugate heat conduction in the wall and wick, evaporation and condensation at the liquid-vapor interface, and forced convection in the vapor channel and wick structure are described. The various thermal resistances or elements in a conventional heat pipe are shown in Fig. 5. The thermal processes such as solidification and liquefaction, and those related to rarefied gases can play important roles in modeling transient heat pipe operation during startup from the frozen state.

Fundamentally, one expects to analyze the internal thermal processes of a heat pipe as a thermodynamic cycle subject to the first and second laws of thermodynamics (Zuo and Faghri, 1998; Khalkhali *et al.*, 1999). Zuo and Faghri (1998) predicted the transient heat pipe performance by using a simple thermal resistance model similar to that of Fig 5. However, classical heat pipe theory does not consider a thermodynamic approach. The idealized cycle is shown in Fig. 6(a). A quantity of heat, Q_{in}, is applied to the heat pipe system at an average evaporator temperature, T_e. Under steady operation, the same quantity of heat is rejected at a lower average condenser temperature T_c. Work is produced inside the heat pipe, but is then completely used in overcoming the hydrodynamic losses of the system. The thermal energy is converted to mechanical energy due to phase change at the liquid-vapor interface by producing a pressure head.

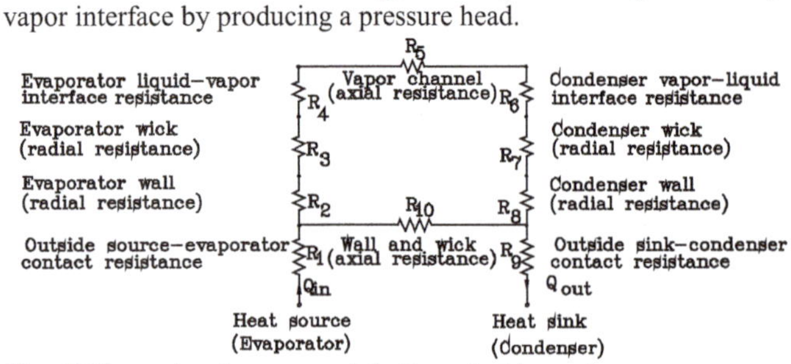

Fig. 5 Thermal resistance model of a typical heat pipe.

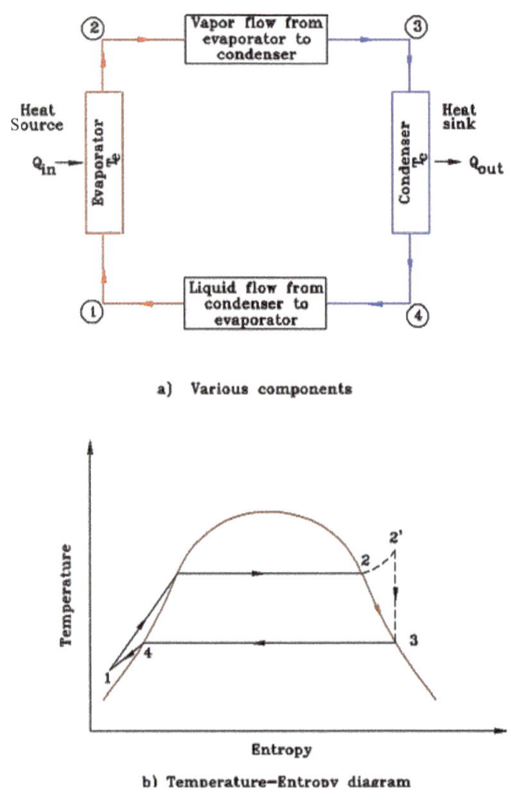

Fig. 6 Thermodynamic cycle of a typical heat pipe.

The thermodynamic cycle of the heat pipe is shown in Fig. 6(b) (Khalkhali *et al.*, 1999). The fluid enters the evaporator as a compressed liquid at temperature T1 and leaves at temperature T2 or T2′ as saturated or superheated vapor, respectively. The vapor flows through the vapor channel from the evaporator to condenser due to the vapor pressure differential in the evaporator and condenser sections (2-3, or 2-2′-3). The vapor enters the condenser section as a saturated vapor or mixture. The condensate enters the adiabatic section as a saturated liquid (4). Finally, the liquid leaves the adiabatic section to enter the evaporator as a compressed liquid to complete the cycle. The work done on the working fluid during its circulation through the heat pipe is the area enclosed by the temperature versus entropy diagram shown in Fig. 6(b). As expected from the second law of thermodynamics, the conversion of thermal energy to kinetic energy is associated with heat rejection at a temperature below the high temperature reservoir in the system with efficiency less than 100%. It should be noted that in most heat pipes, the end-to-end temperature difference is small compared to other conductive systems. Nevertheless, the most ideal heat pipe can never be completely isothermal because this would violate the second law of thermodynamics. The capability of the simple thermodynamic analysis is very limited. In most cases, the heat transfer and fluid mechanics methodologies are needed to solve a heat pipe problem, especially when a quantitative solution is required.

4. TYPES OF HEAT PIPES

Heat pipes have been designed and built with various cross-sectional areas as small as 30 µm width × 80 µm depth and 19.75 mm in length (micro heat pipes), and heat pipes as large as 100 m in length. All heat pipes have an evaporator and condenser section where the working fluid evaporates and condenses, respectively. Many heat pipes also have a transport or adiabatic section which separates the evaporator and condenser sections by an appropriate distance, intended to satisfy the heat pipe limitations and/or the design constraints of the application. A given heat pipe may have multiple evaporators, condensers and adiabatic sections. The working fluid is usually circulated by capillary forces in a wick. However, gravitational, centrifugal, electrostatic, and osmotic forces can also be used to return the liquid from the condenser to the evaporator.

For simplicity of design and manufacturing, heat pipe containers are generally circular cylinders. Other shapes, however, such as rectangular (flat heat pipes), conical (rotating heat pipes), corrugated flexible heat pipes, and nosecap (leading edge heat pipes) geometries have been studied.

Two-Phase Closed Thermosyphon

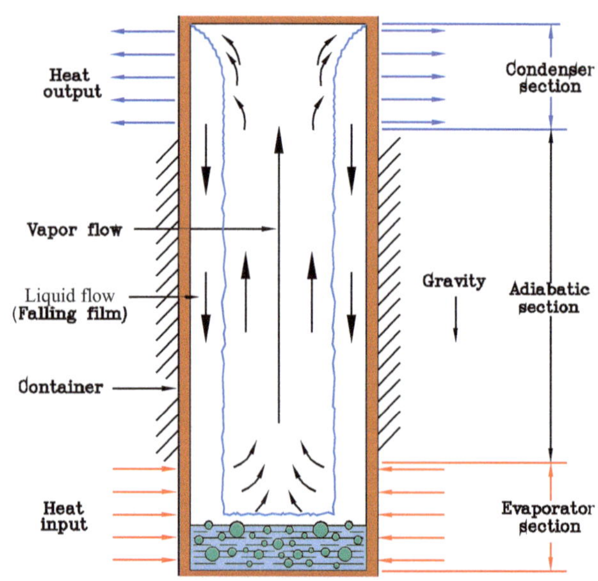

Fig. 7 Gravity-assisted wickless heat pipe (two-phase closed thermosyphon).

A two-phase closed thermosyphon (Faghri, 1995) is a gravity-assisted wickless heat pipe (Fig. 7) (Faghri *et al.*, 1989a). The condenser section is located above the evaporator so that the condensate is returned by gravity. The sonic and vapor pressure limits are constraints to the operation of the thermosyphon as with capillary-driven heat pipes. The entrainment limit is more profound in the thermosyphon than in capillary-driven heat pipes due to the free liquid surface. The counterpart of the entrainment limit in thermosyphons is called flooding (Faghri *et al.*, 1989a), which is the most severe limitation in the operation of these systems. A sudden oscillating wall temperature and vapor pressure rise will occur at the flooding limit. The boiling limit in thermosyphons is due to film boiling, rather than nucleate boiling as in capillary-driven heat pipes. The boiling limit in thermosyphons occurs when a vapor film forms between the pipe wall and the liquid in the evaporator section of the heat pipe. For small liquid fill volumes, the dryout limit may be reached, where all of the working fluid is held in the liquid film, and no liquid pool exists. In this case, any further increase in the input heat will cause a severe temperature increase at the bottom of the evaporator section.

The operation of the thermosyphon is sensitive to the working fluid fill volume. For thermosyphons without wicks, it has been shown experimentally that the maximum rate of heat transfer increases with the amount of the working fluid up to a certain value. A wick structure is sometimes included in the design of thermosyphons to postpone flooding and improve the contact between the wall and the liquid. The capillary limit is generally of no concern in the operation of the thermosyphon due to the fact that gravity is the major driving force for condensate return.

Capillary-Driven Heat Pipe

The capillary-driven heat pipe (conventional heat pipes) consists of a sealed container, in which a wick is placed on the inner radius of the pipe wall (Fig. 8). The purpose of the wick is to provide a capillary-driven pump for returning the condensate to the evaporator section. Enough working fluid is placed inside the sealed pipe to saturate the wick with liquid. The operation of the capillary-driven heat pipe is as follows. Heat input to the evaporator section evaporates the liquid in the wick. The vapor then enters the vapor space and travels to the condenser section due to the higher vapor pressure in the evaporator. Heat removal from the condenser causes the vapor to condense, releasing its latent heat of vaporization. The condensate is then pumped back to the evaporator section by the capillary force generated at the liquid-vapor interfaces of the pores in the wick. Due to the two-phase nature of the capillary heat pipe, it is ideal for transferring heat over long distances with a very small temperature drop, and for creating a nearly isothermal surface for temperature stabilization. Generally, the most commonly encountered limitation to the performance of a capillary-driven heat pipe is the capillary limit. This occurs when the wick cannot return enough liquid to the evaporator section to keep it saturated. At this point, the evaporator wall experiences a sudden, continuous increase in temperature. Conventional capillary heat pipes are used in almost all laptop/notebook computers nowadays to channel heat away from the processors. The capillary heat pipe has also been widely applied to various commercial and aerospace applications.

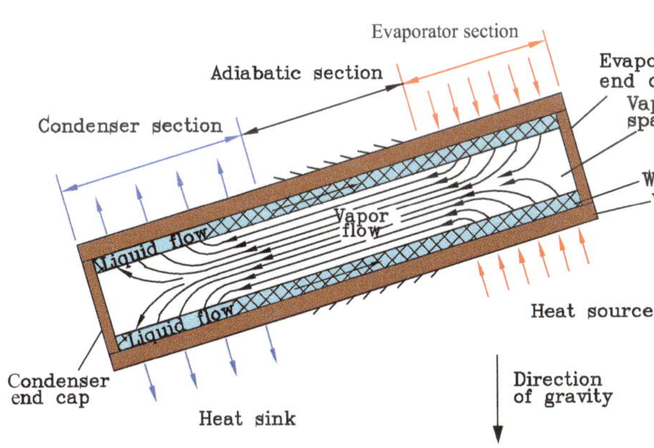

Fig. 8 Conventional capillary-driven heat pipe.

Annular Heat Pipe

The annular heat pipe is similar to the conventional capillary-driven heat pipe in many ways. The main difference between the two is that the cross-section of the vapor space in an annular heat pipe is annular instead of circular (Fig. 9) (Faghri and Thomas, 1989; Faghri, 1989). This enables the designer to place wick material both on the inside of the outer pipe and on the outside of the inner pipe. In this manner, the surface area for heat input and output can be significantly increased without increasing the outer diameter of the pipe. Therefore, the capillary limit of the annular heat pipe is greater than that of a conventional heat pipe having the same outer dimensions. The annular heat pipe has been used as an isothermal furnace, with excellent results due to its temperature flattening capabilities and fast response time to a cold charge. This type of heat pipe can also be used to connect two circular cross-section heat pipes end-to-end by inserting the circular heat pipe into the inner pipe.

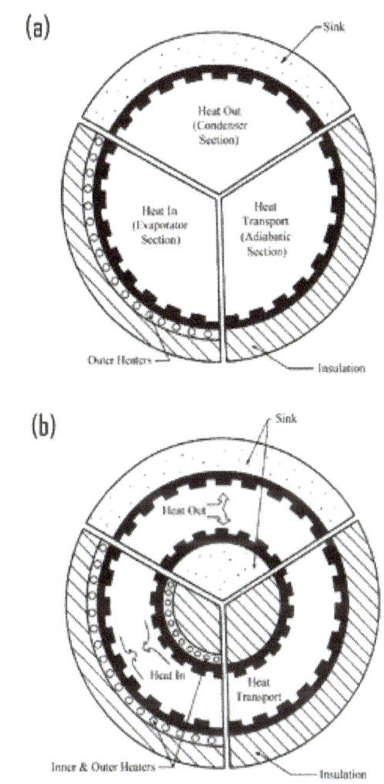

Fig. 9 Conventional and concentric annular heat pipe design: (a) conventional; (b) concentric annular.

Vapor Chamber

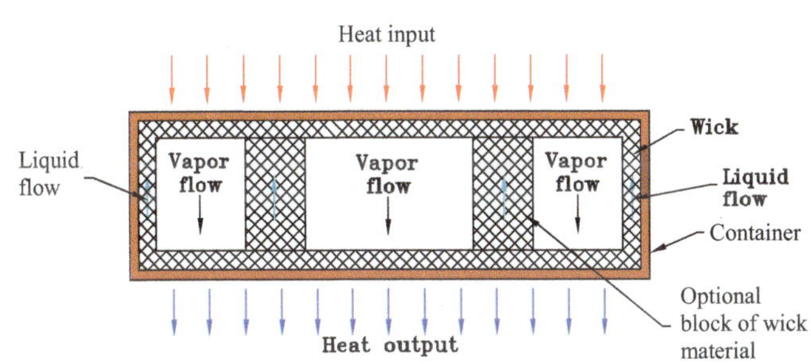

Fig. 10 Vapor chamber.

The vapor chamber is a capillary-driven planar (flat-plate heat pipe in a rectangular or disk-shaped) with a small aspect ratio (Faghri, 1995; Mochizuki *et al.*, 2006; Mochizuki *et al.*, 2007; Xiao and Faghri, 2008), as shown in Fig. 10. Additional wick blocks between the evaporator and condenser aid in condensate return, especially when the condenser is below the evaporator in a gravity field. If the condenser is above the evaporator, there is no need to have a wick in the condenser section, since the condensate on the upper plate will drip back to the evaporator. A wick is needed over the evaporator section, however, in order to uniformly distribute the liquid over the entire surface so to prevent dryout. The vapor chamber is an excellent candidate for use in electronic cooling applications especially with high heat flux applications, including desktops and servers. Vapor chambers are preferred over conventional heat pipes (CHP) for electronic cooling with heat fluxes higher than 50 W/cm^2, since heat flow is two or three dimensional, compared to one dimensional in CHPs. Furthermore, vapor chambers can be placed in direct contact with CPUs using thermal interface materials, thereby reducing the overall thermal resistance of heat sinks.

Rotating Heat Pipe

Rotating heat pipes can be designed in two configurations (Gray, 1969; Faghri, 1995). First, the heat pipe can be in the shape of a circular cylinder with or without an axial taper, which rotates either about its own axis of symmetry or revolves off-axis. Secondly, the heat pipe can be manufactured in the shape of a disk, where two parallel disks are joined together at the outer and inner radii to form the vapor space (Fig. 11). Cylindrical rotating heat pipes operate in the same manner as conventional cylindrical heat pipes except that internal tapers are commonly used to aid in returning the condensate to the evaporator by centrifugal force. With disk-shaped heat pipes, heat is input at the outer radii and extracted at the inner radii, which allows the condensate to be driven back to the evaporator by centrifugal force with the aid of an internal taper. Therefore, capillary wicks are not used in rotating heat pipes. Cylindrical rotating heat pipes have been used to cool the rotating parts of electric motors and metal-cutting tools such as drill bits and end mills. Disk-shaped heat pipes have been proposed to effectively cool turbine components and automobile brakes.

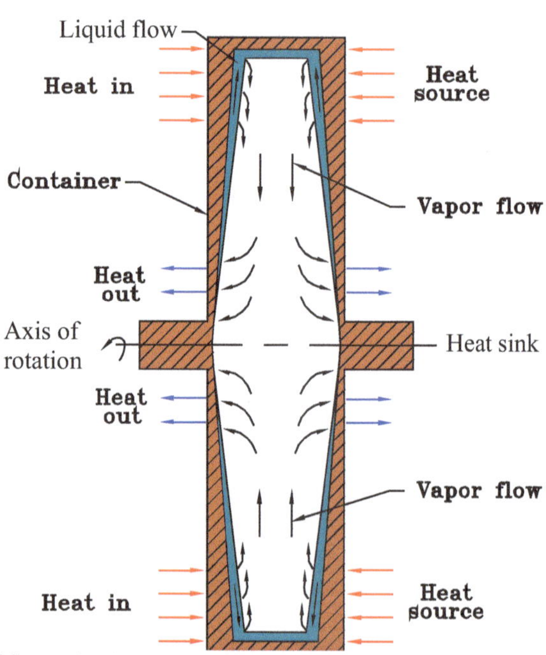

Fig. 11 Rotating heat pipe (radial flow).

Gas-Loaded Heat Pipe

Gas-loaded heat pipes are one type of variable conductance heat pipes. The gas-loaded heat pipe, shown in Fig. 12, is the same as the capillary-driven heat pipe or the two-phase thermosyphon except that a noncondensing gas is introduced into the vapor space. During operation, this gas is swept down the length of the heat pipe by the working fluid vapor to the condenser section. Since condensation of the working fluid does not take place where the noncondensing gas is present, a portion of the condenser is blocked from transferring heat to the heat sink (Fig. 12(a)). If the heat input to the evaporator section is increased, the vapor and gas temperatures both increase, which leads to the compression of the inert gas. This in turn increases the amount of condenser surface area available to transfer heat (Fig. 12(b)). This phenomenon results in the gas-loaded heat pipe being able to maintain a nearly constant evaporator temperature regardless of the heat input. Gas-loaded heat pipes have been used in the annular arrangement previously mentioned for an isothermal furnace and for electronic cooling.

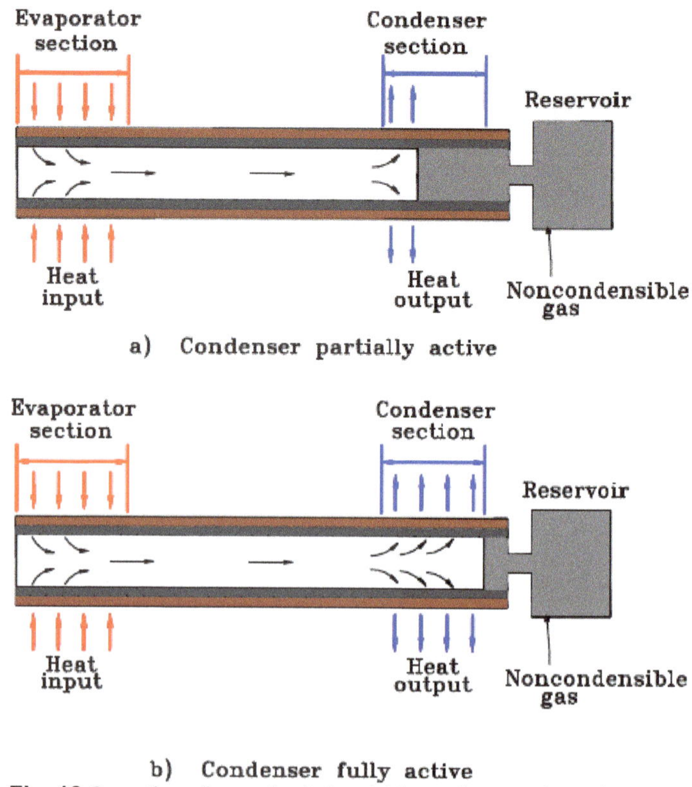

Fig. 12 Operation of a gas-loaded variable conductance heat pipe.

Loop Heat Pipe

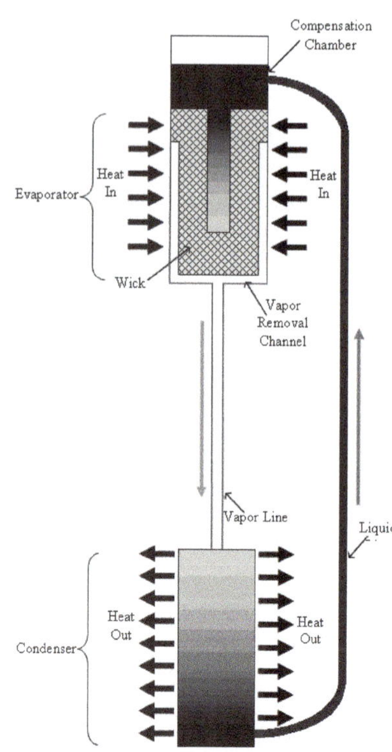

Fig. 13 Loop heat pipe (LHP).

The loop heat pipe (LHP) (Faghri, 1995; Maydanik, 2005) was invented in Russia in 1971 by Gerasimov and Maydanik (Maydanik *et al.*, 1991). The LHP is based on the same physical processes (evaporation and condensation) as those of conventional heat pipes.

The LHP in its simplest form (Fig. 13) consists of a capillary pump (or evaporator), a compensation chamber (or reservoir), a condenser, and liquid and vapor lines. The wicks are only present in the evaporator and compensation chamber. The high capillary force is created in the evaporator due to fine-pored wicks (primary wicks) such as sintered nickel, titanium and copper powder with an effective pore radius of 0.7-15μm and a porosity of 55-75%. The compensation chamber is an important component in the LHP and is often an integral part of the evaporator. The purpose of the compensation chamber is to accommodate excess liquid in an LHP during normal operation. A secondary wick (usually made of larger pores) physically connects the evaporator to the compensation chamber in order to supply the primary wick with liquid, particularly when the compensation chamber is below the evaporator, or when the LHP is operating in microgravity conditions. The motion of vapor and liquid flow in the primary wick proceeds mainly in a radial direction. The evaporator meniscus is inverted down toward the wall being heated. Both the liquid and vapor lines are made of small diameter tubing with no wicks. LHPs can be made flexible and bendable. LHPs provide heat removal over long distances without sensitivity to gravity. Several factors make the LHP an attractive option for spacecraft cooling over conventional heat pipes. Since the wick structure is only in the evaporator, the rest of the container walls can be smooth, which reduces pressure drops in the vapor and liquid flows. The pressure drops throughout the system are also reduced because the vapor and liquid flow are co-current, as opposed to the counter-current flow in conventional heat pipes. For these reasons, the LHP is a more effective thermal bus, and the heat source and sink can be separated by a longer distance than with conventional heat pipes.

Capillary Pumped Loop Heat Pipe

Spacecraft and electronic systems require heat rejection systems which can remove quantities of heat on the order of 10 to 100 times that which present single-phase systems can reject. Another special type of heat pipe has been proposed and extensively tested for this purpose, the capillary pumped loop (CPL) (Faghri, 1995). The CPL was first proposed by Stenger (1966) at the NASA Lewis Research Center. A basic schematic of the CPL is given in Fig. 14. Heat is applied to the evaporator component, which consists of a hollow rod of wick material capped at one end and force fitted into an internally axially-grooved pipe. The heat applied to the outer surface of the evaporator vaporizes the liquid working fluid, which then travels down the length of the axially-grooved vapor channels and into the vapor header. The vapor travels to the condenser, where it is condensed first as a film on the inner wall of the pipe, and then as a liquid slug flow. Before reaching the evaporator, the liquid passes through a subcooler to collapse any remaining vapor bubbles as well as to provide additional subcooling if necessary. The capillary pressure generated in the wick structure continuously pumps the working fluid through the cycle. A two-phase reservoir is provided to control both the working fluid through the cycle and the working temperature of the system, which is similar in function to the variable conductance heat pipe. For multiple evaporator systems, isolators are provided to prevent the depriming of one evaporator from depriming the other evaporators. Despite the wide use of the CPL and LHP in space technology, fundamental confusion still exists regarding their similarities and distinctions in operation, as well as the limitations of these two devices.

Pulsating Heat Pipe

The pulsating heat pipe (PHP) (Faghri, 1995) is made from a long capillary tube bent into many turns, with the evaporator and condenser sections located at these turns. There are two main types of PHPs--looped and unlooped (Fig. 15)—which are classified according to whether or not the two ends of the PHP connect.

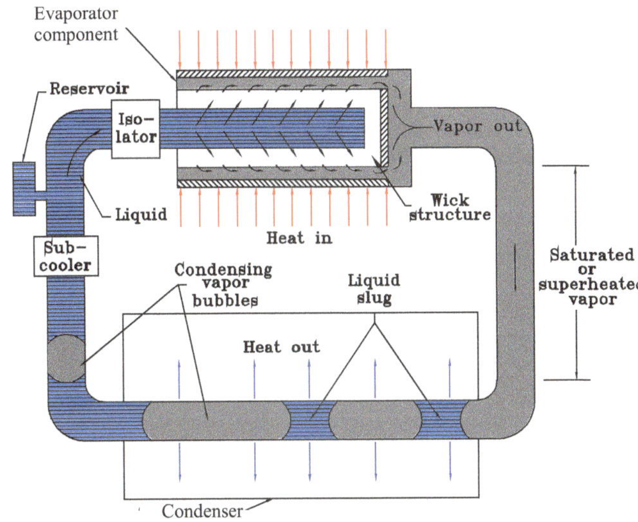

Fig. 14 Capillary pumped looped heat pipe (CPL).

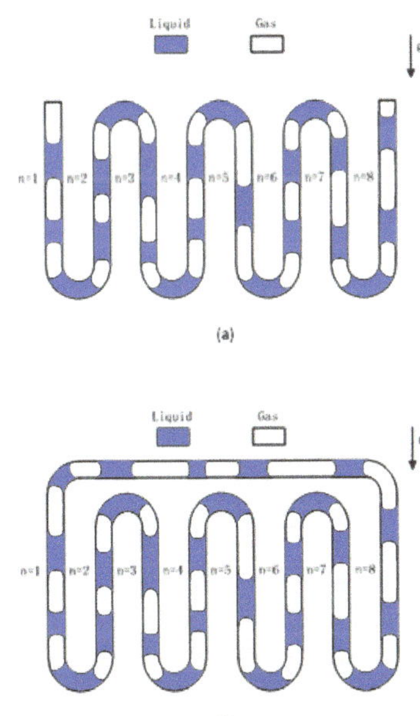

Fig. 15 Pulsating heat pipe (a) unlooped, (b) looped.

A PHP is usually partially charged with a working fluid, with a charge ratio between 40% and 60%. Since the diameter of a PHP is very small (less than 5mm), vapor plugs and liquid slugs are formed as a result of capillary action. Heat input either causes evaporation or boiling, which increases the pressure of the vapor plug in the heating section. Simultaneously, the pressure in the cooling section decreases due to condensation. This pressure difference pushes the liquid slug and vapor plug into the cooling section. The liquid slug and vapor plug in the cooling section are then pushed into the next heating section, which will push the liquid slug and vapor plug back to the cooling section. This process enables the self-excited oscillatory motion of liquid slugs and vapor plugs. Heat is transported from the heating section to the cooling section via the pulsation of the working fluid in the axial direction of the tube.

The unique feature of PHPs, compared to conventional heat pipes, is that there is no wick structure returning the condensate to the heating section. There is therefore no countercurrent flow between the liquid and vapor. The entrainment limit in the conventional heat pipe does not have any effect on the capacity of heat transport by a PHP. With this simple structure, the PHP weighs less than a conventional heat pipe, making it an ideal candidate for space applications. Since the diameter of the PHP is very small, surface tension plays a greater role in the dynamics of the PHP than gravitational force does, enabling successful operation in a microgravity environment. Other applications of PHPs include thermal control of electrical and electronic devices and components, as well as thyristors, diodes and ceramic resistors. Zhang and Faghri (2008) made a comprehensive review related to existing experimental and theoretical research in pulsating heat pipes.

Monogroove Heat Pipe

The monogroove heat pipe developed by Grumman utilizes an extruded cross section which separates the vapor space from the liquid return path (Fig. 16) (Alario *et al.*, 1981; Alario, 1984). Heat input and output are through the flat face on top of the vapor space. Since heat is not conducted through the wall of the liquid return path, the chance of arterial blockage due to vapor bubbles is significantly reduced. Wick structures are placed along the interior walls to deliver condensate from the artery to the evaporator. Monogroove heat pipes are proposed to be used for multi-kilowatt space radiator heat rejections systems.

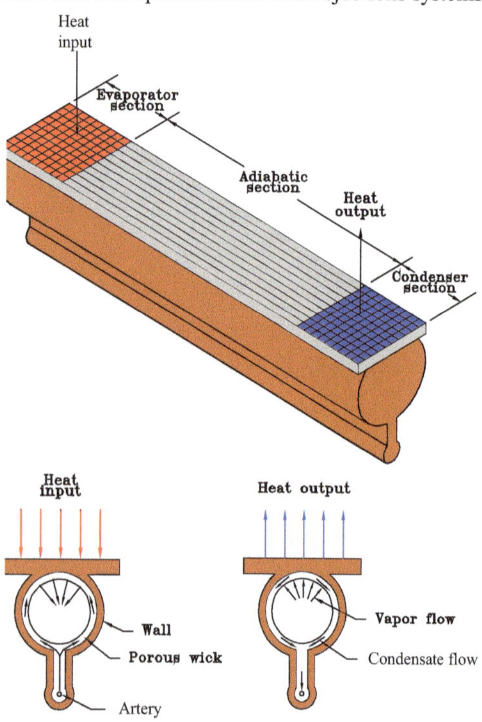

Fig. 16 Monogroove heat pipe.

Micro and Miniature Heat Pipes

Cotter (1984) was the first to propose the micro heat pipe concept for cooling of electronic devices. The micro heat pipe was defined as a heat pipe in which the mean curvature of the liquid vapor interface is comparable in magnitude to the reciprocal of the hydraulic radius of the total flow channel. Typically, micro heat pipes (Fig. 17) have convex but cusped cross sections (Peterson, 1992) (for example, a polygon), with hydraulic diameter in range of 10 to 500 μm (Cao and Faghri, 1994a). Miniature heat pipe is defined as a heat pipe with a hydraulic diameter in the range of 0.5 to 5 mm (Hopkins *et al.*, 1999).

References to micro and miniature heat pipes in open literature are sometimes not properly differentiated as defined above. For example, miniature heat pipes with micro grooves are sometimes improperly referred to as micro heat pipes. It is also worth noting the other structural differences between micro and miniature heat pipes in addition to the size ranges noted above. A true micro heat pipe, for both metallic and silicon applications, has a single non-circular channel for both liquid and vapor phases where the liquid return from condenser to evaporator by capillary force occurs in the sharp corner of the channel. These types of micro heat pipes are generally placed in an array of parallel micro heat pipes on the substrate surface in order to increase the overall heat transfer area. Miniature heat pipes can be designed based on micro axially grooved structure (1D capillary structure), meshes or cross grooves (2D capillary structure).

Attempts have been made to etch micro heat pipes directly into the silicon and use them as thermal spreaders (Peterson and Ma, 1999; Le Berre *et al.*, 2003; Launay *et al.*, 2004a). The performance characteristics of the micro heat pipes are different from those conventional heat pipes (Khrustalev and Faghri, 1994; Launay *et al.*, 2004a).

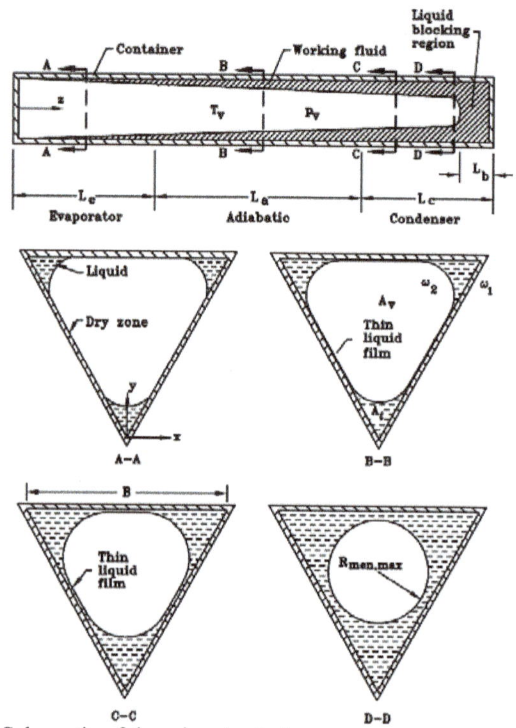

Fig. 17 Schematic of the micro heat pipe

In order for the micro heat pipe to find more commercial applications in microelectric cooling, it must compete with other cooling methods, such as forced convection, impingement and two phase direct cooling in areas such as manufacturing cost and reliability. Because of the extremely small dimensions of micro heat pipes, difficulties in manufacturing and the subsequent degassing and charging are expected. Furthermore, the "true" micro heat pipe built directly into the silicon must compete with the miniature heat pipe attached on the chip surface in this area, which is much less costly.

Miniature heat pipes with micro axial grooves can be successfully used in electronic components cooling systems. Fabrication and experimental data on the performance characteristics of the flat water-copper heat pipe with external dimensions 2x7x120 mm have been reported by Hopkins et al. (1999) with radial heat fluxes of 90 W/cm^2 and 150 W/cm^2 for horizontal and vertical applications, respectively.

Inverted Meniscus Heat Pipe

Advanced miniature heat pipes are needed to meet the demand for dissipating high heat fluxes (over 100 W/cm^2) for electronic components. Several techniques were proposed in this regard, including ivadizing (Khrustalev and Faghri, 1996a), which is the deposit of a porous structure on the top of the lands between the grooves or "inverted meniscus scheme" (Khrustalev and Faghri, 1995c; Khrustalev and Faghri, 1996b).

With extremely high heat fluxes, a vapor blanket appears inside the uniform structure in the evaporator along the heated solid surface (Khrustalev and Faghri, 1996b), as shown in Figs. 18(a) and 18(b). In this case, evaporation takes place in the dry region of the porous structure at the liquid-vapor interface, the location of which shifts depending on the operational conditions. The heat is conducted to this interface from the heated surface through the dry region of the porous element, and the vapor flows mainly along the solid surface through this porous region towards the triangular vapor channel. The vapor flow is provided by the capillary pressure gradient due to the difference in the curvature of the menisci along the liquid-vapor interface inside the porous structure. Therefore, with high heat fluxes, part of the capillary pressure is spent on the compensation of the pressure drop in the vapor flow through the dry porous region. Hence, the maximum heat flux for this configuration should be calculated with respect to the formation of this vapor blanket and can be less than that estimated from the traditional capillary limit. Numerical results show that miniature copper water heat pipes with an inverted meniscus type evaporator of estimated dimensions of 2x7x120 mm is capable of withstanding high heat fluxes on the evaporator wall, which can be about 200 W/cm^2 for horizontal orientation at 120°C (Khrustalev and Faghri, 1996b).

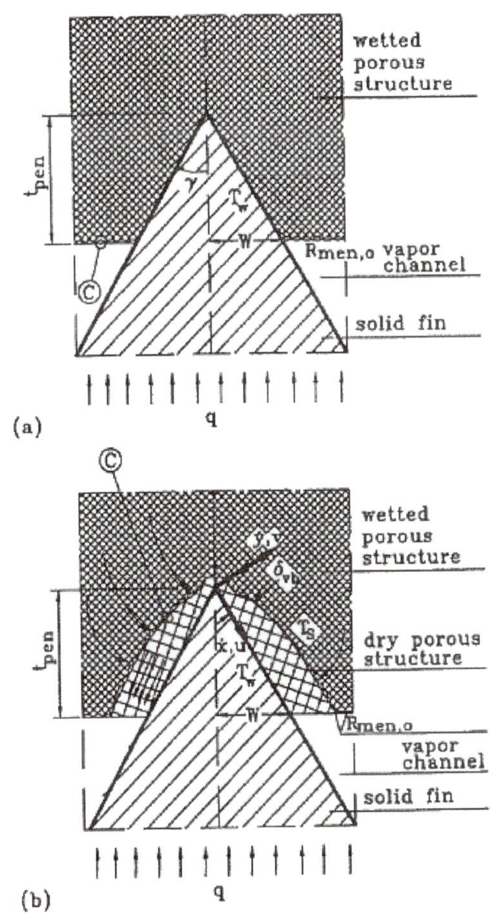

Fig. 18 Schematic of an inverted meniscus type evaporator with the triangular fin: (a) with low heat fluxes; (b) with high heat fluxes.

Nonconventional Heat Pipes

While the majority of heat pipes have a cylindrical or rectangular cross section, specific applications require the use of heat pipes with other geometries. Examples include heat pipe cross sections with polygonal shapes, such as micro/miniature heat pipes, or cross sections that vary along the length of the heat pipe. An example of this latter type of cross section is shown in Fig. 19, where nosecap and leading edge heat pipes are shown schematically. Another example that standard coordinate systems, such as Cartesian or cylindrical coordinates, are not applicable is heat pipe turbine vane cooling (Zuo *et al.*, 1998) as shown in Fig. 20.

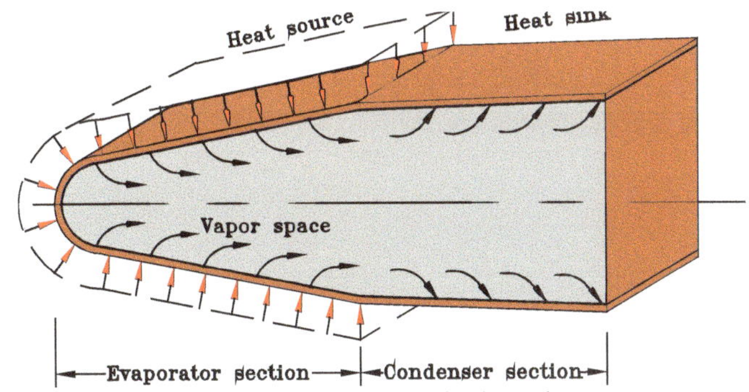

Fig. 19 Conceptual design of a leading edge heat pipe.

The leading edge heat pipe has been proposed to cool the leading edges of future hypersonic aircraft and reentry vehicles (Fig. 19) (Cao and Faghri, 1991). The heat pipe will cover the leading edges of the wings and engine nacelles, where the aerodynamic heating is the most intense. The heat will be absorbed and transported through the heat pipe to the condenser, where it will be rejected by radiation and/or convection. The extremely high heat fluxes and temperatures which will be experienced by the proposed leading edge heat pipe will make its design and manufacturing one of the most challenging design problems faced by heat pipe engineers to date.

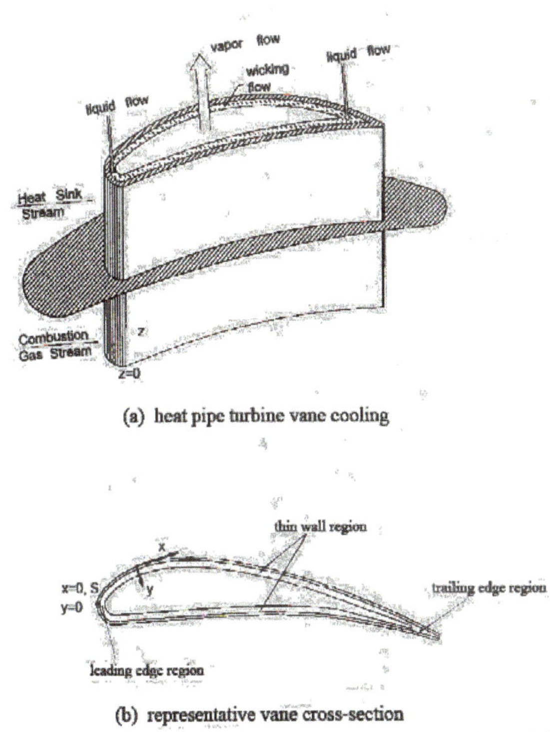

Fig. 20 A sketch of the heat pipe turbine vane: (a) heat pipe turbine vane cooling; (b) representative vane cross section.

5. WORKING FLUIDS AND TEMPERATURE RANGES

Each heat pipe application has a particular temperature range in which the heat pipe needs to operate. Therefore, the design of the heat pipe must account for the intended temperature range by specifying the proper working fluid. Table 1 (Faghri, 1995) lists some of the commonly used and proposed working fluids, their melting and boiling points at atmospheric pressure, and their useful ranges. As a rule of thumb, the useful range extends from the point where the saturation pressure is greater than 0.1 atm and less than 20 atm. Below 0.1 atm, the vapor pressure limit may be approached. Above 20 atm, the

Table 1 Working fluids and temperature ranges of heat pipes

Working Fluid	Melting Point, K at 1 atm	Boiling Point, K at 1 atm	Useful Range, K
Helium	1.0	4.21	2-4
Hydrogen	13.8	20.38	14-31
Neon	24.4	27.09	27-37
Nitrogen	63.1	77.35	70-103
Argon	83.9	87.29	84-116
Oxygen	54.7	90.18	73-119
Methane	90.6	111.4	91-150
Krypton	115.8	119.7	116-160
Ethane	89.9	184.6	150-240
Freon 22	113.1	232.2	193-297
Ammonia	195.5	239.9	213-373
Freon 21	138.1	282.0	233-360
Freon 11	162.1	296.8	233-393
Pentane	143.1	309.2	253-393
Freon 113	236.5	320.8	263-373
Acetone	180.0	329.4	273-393
Methanol	175.1	337.8	283-403
Flutec PP2	223.1	349.1	283-433
Ethanol	158.7	351.5	273-403
Heptane	182.5	371.5	273-423
Water	273.1	373.1	303-550
Toluene	178.1	383.7	323-473
Flutec PP9	203.1	433.1	273-498
Naphthalene	353.4	490	408-623
Dowtherm	285.1	527.0	423-668
Mercury	234.2	630.1	523-923
Sulphur	385.9	717.8	530-947
Cesium	301.6	943.0	723-1173
Rubidium	312.7	959.2	800-1275
Potassium	336.4	1032	773-1273
Sodium	371.0	1151	873-1473
Lithium	453.7	1615	1273-2073
Calcium	1112	1762	1400-2100
Lead	600.6	2013	1670-2200
Indium	429.7	2353	2000-3000
Silver	1234	2485	2073-2573

container thickness must increase to the point where the heat pipe becomes limited by the thermal resistance through the container.

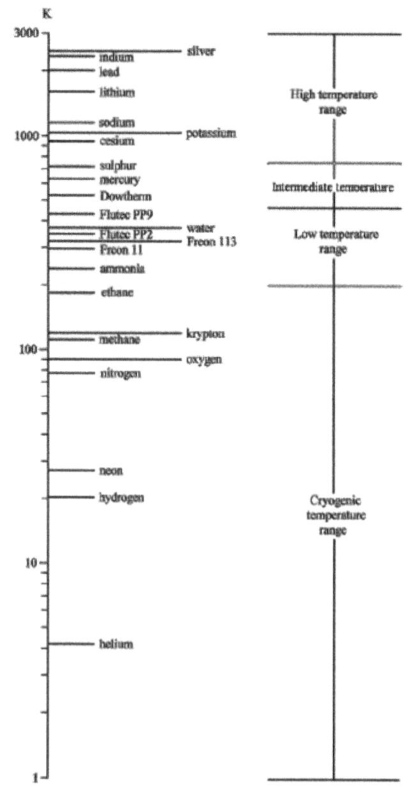

Fig. 21 Operating temperature ranges of various working fluids on a logarithmic temperature scale.

Longevity of a heat pipe can be assured by selecting a container, a wick and welding materials that are compatible with one another and with the working fluid of interest. Performance can be degraded and failures can occur in the container wall if any of the parts (including the working fluid) are not compatible. For instance, the parts can react chemically or set up a galvanic cell within the heat pipe. Additionally, the container material may be soluble in the working fluid or may catalyze the decomposition of the working fluid at the expected operating temperature. A compilation of the most up-to-date information concerning the compatibility of metals with working fluids for heat pipes is given in Table 2 (Faghri, 1995). Figure 21 presents various working fluid boiling points and classifies them into four categories: cryogenic, low, intermediate and high temperature ranges. The working-fluid inventory of a heat pipe is the sum of the masses of the vapor and liquid phases, assuming the wick is full of liquid. This criterion is slightly over the optimum requirement because the meniscus recedes into the evaporator wick during normal operation. However, this situation is more advantageous than underfilling the heat pipe,

which may significantly reduce the maximum heat transfer. With extreme overfill, however, any excess fluid might collect as liquid in the condenser section and increase the thermal resistance, thereby decreasing the heat transport capability of the heat pipe.

Heat pipe fabrication, processing, and testing involve several detailed procedures which are recommended to be strictly followed in order to achieve the highest quality possible. Faghri (1995) provided a detailed procedure for the fabrication, processing, and testing of low, moderate, and high temperatures.

Cryogenic Temperature Range

Cryogenic heat pipes operate between 4 to 200 K. Typical working fluids include helium, argon, oxygen, and krypton. The amount of heat that can be transferred for cryogenic heat pipes is quite low due to the small heats of vaporization, high viscosities, and small surface tensions of the working fluids.

Table 2 Generalized results of experimental compatibility tests

Working Fluid	Compatible Material	Incompatible Material
Water	Stainless Steel[a], Copper, Silica, Nickel, Titanium	Aluminum, Inconel
Ammonia	Aluminum, Stainless Steel, Cold Rolled Steel, Iron, Nickel	
Methanol	Stainless Steel, Iron, Copper, Brass, Silica, Nickel	Aluminum
Acetone	Aluminum, Stainless Steel, Copper, Brass, Silica	
Freon-11	Aluminum	
Freon-21	Aluminum, Iron	
Freon-113	Aluminum	
Heptane	Aluminum	
Dowtherm	Stainless Steel, Copper, Silica	
Lithium	Tungsten, Tantalum, Molybdenum, Niobium	Stainless Steel, Nickel, Inconel, Titanium
Sodium	Stainless Steel, Nickel, Inconel, Niobium	Titanium
Cesium	Titanium, Niobium, Stainless Steel, Nickel-based superalloys	
Mercury	Stainless Steel[b]	Molybdenum, Nickel, Tantalum, Inconel, Titanium, Niobium
Lead	Tungsten, Tantalum	Stainless Steel, Nickel, Inconel, Titanium, Niobium
Silver	Tungsten, Tantalum	Rhenium

[a]Sensitive to cleaning; [b]with Austenitic SS

Low Temperature Range

The low temperature range is from 200 to 550 K. Most heat pipe applications fall within this range. Commonly used fluids are ammonia, acetone, the Freon compounds, and water. Water, which is perhaps the most widely used working fluid, has good thermophysical properties such as large heat of vaporization and surface tension, and has the added benefit of being safe to use during handling.

Intermediate Temperature Range

The working fluids in the medium temperature range, 450 to 750 K, are mercury and sulphur. Compounds such as Thermex or Dowtherm-A (diphenyl/diphenyl oxide eutectics) are also employed in this range. Mercury has extremely attractive properties inherent in a liquid metal such as its high thermal conductivity. However, problems with wetting the wick and wall present difficulties in using mercury in capillary heat pipes. The toxicity of mercury is also a significant problem. A review of intermediate temperature fluid life tests experiments was reported by Anderson et al. (2007).

High Temperature Range

Sodium, lithium, cesium, silver and a sodium-potassium compound (NaK) are often used in the high temperature range (750 K and above). The heat transport rates for liquid-metal heat pipes are generally much higher than those in the other temperature ranges because the surface tension coefficients, latent heats of vaporization, and thermal conductivities of liquid metals are very high.

Faghri (1995) presents thermophysical property data for most heat pipe working fluids and container materials along with polynomial temperature-property relations for the working fluids.

6. CAPILLARY WICK DESIGNS AND STRUCTURES

The wick structure within the heat pipe is present to return condensate to the evaporator section. While small pores are needed at the liquid-vapor interface to develop high capillary pressures, large pores are preferred within the wick so that the movement of the liquid is not restricted too greatly. For this reason, many different types of wick structures have been developed in order to optimize the performance of the capillary heat pipe. The types of wick structures can be divided into two categories: homogeneous and composite wicks. Homogeneous wicks (Table 3) (Faghri, 1995) have the benefit of being relatively simple to design, manufacture and install. Composite wicks (Table 4) (Faghri, 1995), however, can significantly increase the capillary limit of the heat pipe, but have the drawback of high manufacturing costs. When selecting a wick structure for a particular application, one must keep in mind the benefits and drawbacks of each type of wick.

There are three properties of wicks that are important in heat pipe design:

Minimum capillary radius: This parameter should be small if a large capillary pressure difference is required, such as in terrestrial operation for a long heat pipe with the evaporator above the condenser, or in cases where a high heat transport capability is needed.

Permeability: Permeability is a measure of the wick resistance to axial liquid flow. This parameter should be large in order to have a small liquid pressure drop, and therefore, higher heat transport capability.

Effective thermal conductivity: A large value for this parameter gives a small temperature drop across the wick, which is a favorable condition in heat pipe design.

A high thermal conductivity and permeability, and a low minimum capillary radius are somewhat contradictory properties in most wick designs. For example, a homogeneous wick may have a small minimum capillary radius and a large effective thermal conductivity, but have a small permeability. Therefore, the designer must always make trade-offs between these competing factors to obtain an optimal wick design.

Homogeneous Wicks

Homogeneous wicks (Table 3) (Faghri, 1995) are constructed with one type of material or machining technique. The screen wick is seemingly the simplest and most common type of wick structure. It consists of a metal or cloth fabric which is wrapped around a mandrel and inserted into the heat pipe. After placement, the mandrel is removed, leaving the wick held by the tension of the wrapped screen, in the case of a metal fabric. For a cloth fabric, a spring may be inserted into the heat pipe to hold the wick against the inside of the pipe wall. The capillary pressure generated by a screen wick is determined by the size of the rectangular pores between the individual threads. The permeability is determined by the number of wraps and the looseness of the wraps, which create annular gaps through which the condensate can flow.

Table 3 Typical homogeneous wick designs

Wick type	Capillary pumping	Thermal conductivity	Permeability	Comments
 A. Wrapped Screen	High	High	Low–average	Single or multiple wraps of wire screen mesh
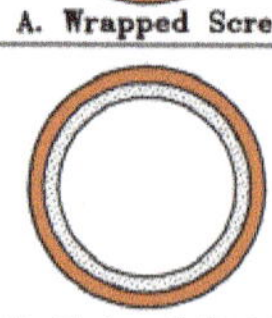 B. Sintered Metal	High	Average	Low–average	Packed spherical particles, felt metal fibers or powder
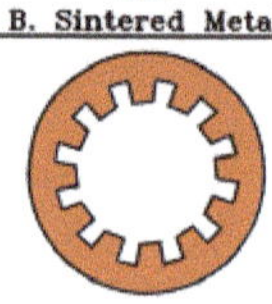 C. Axial Grooves	Low	High	Average–high	Rectangular, circular, triangular, or trapezoidal grooves
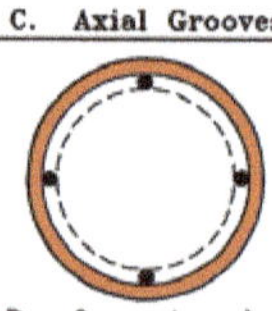 D. Open Annulus	Low	Low	High	Wire screen mesh spaced from wall
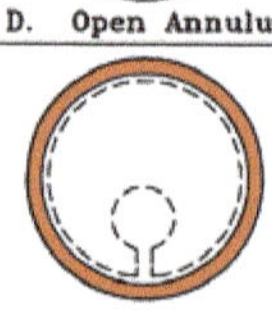 E. Open Artery	Low	High	High	Wire screen mesh formed into artery and wall lining
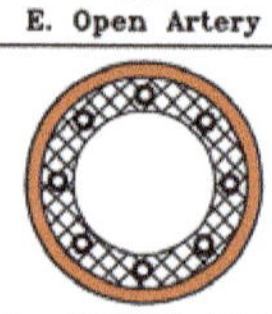 F. Integral Artery	High	High	Average–high	Homogeneous material with built–in arteries

Sintered metal wicks are manufactured by packing tiny metal particles between the inner heat pipe wall and a mandrel in powder form. This assembly is then heated until the metal spheres are sintered to each other and to the inner wall of the heat pipe. Special materials are used for the mandrel so that it can be removed, leaving an open vapor space. This type of wick is obviously more difficult to manufacture compared to the simple screen wick. However, the capillary pressure developed by the sintered metal wick (and therefore capillary limit) is more easily predicted, as the annular gaps present in a screen wick present uncertainties in the permeability. Also, since a metal powder is sintered, the effective thermal conductivity is much higher than a comparable screen wick due to the poor thermal contact between the screen wraps.

Axial groove wicks are formed by the extrusion or broaching of grooves into the inner radius of the pipe. Several different types of grooves have been used and proposed, which have rectangular, triangular, trapezoidal, or nearly circular cross-sections. Trapezoidal grooves are currently the most common type. The performance of axial groove wicks is excellent, provided that the application does not call for a significant adverse elevation against gravity. Since the size of the grooves are large compared to those of a screen or sintered metal wick, the capillary pumping pressure is quite small. However, the permeability and the effective thermal conductivity are very high. Drawbacks to this type of wick are the difficulty in machining the grooves for long heat pipes, which might prove to be excessively expensive, and the liquid-vapor interaction at the opening of the groove at high vapor velocities. Other attractive features of axial groove wicks include the field-tested performance, reliability, and simplicity of design.

The open annulus wick is simply a single-wrap screen wick, held away from the inner pipe wall by stand-offs. This provides an unimpeded return flow path between the screen and the pipe, which greatly increases the permeability over the simple screen wick, while maintaining the high capillary pressure. However, the effective thermal conductivity of this wick is very low for most working fluids due to the low liquid thermal conductivity. Difficulties in priming during startup, and depriming during near-dryout operation are common in this type of wick. The idea of an open condensate return path is also present in the artery wick, but the problem of a low effective thermal conductivity, as is present in the open annulus wick, is significantly reduced.

The artery wick combines the two necessities of having a small pore radius for capillary pressure generation and having a large pore radius for high permeability. In the artery wick structure, the interior of the heat pipe is either covered by a screen wick or a sintered metal powder in the usual manner, but a hollow passage(s) running the length of the pipe is fashioned and is in communication with the rest of the wick structure. Condensate is collected within this passage(s) or artery and is pumped back to the evaporator section by the capillary forces generated at the liquid-vapor interface. Since the inner diameter of the artery is much larger than the effective pore radius of the wick, the liquid is able to easily traverse the length of the heat pipe with a minimum pressure drop. As with any design, there are difficulties associated with the artery wick. First, problems associated with the manufacturing of arteries must be addressed. The startup of artery wick heat pipes is also critical due to the inevitable presence of vapor bubbles within the artery, which effectively block liquid return. Methods of collapsing these bubbles are complicated and not always effective.

Composite Wicks

Composite wick structures (Table 4) (Faghri, 1995) employ the benefits of having small pores for generating high capillary pumping pressures and having large pores for increasing the permeability of the liquid return path. Again, the simplest type of composite wick is the screen wick, except that two screens with different pore sizes are used. Several wraps of a screen with a large pore size is used against the inner pipe wall for the liquid return path, and a single wrap of a screen with a much smaller pore size is placed adjacent to the vapor space to develop high capillary pressures. Similarly, axial grooves covered by a single wrap of a small-pore screen can solve many of the problems associated with the homogeneous axial groove wick. Since the screen effectively separates the liquid and vapor flows, the entrainment of the liquid into the vapor flow by the interfacial shear is nearly eliminated. Also, this composite wick can be used in adverse gravity fields because the screen generates the needed capillary pressures.

A slab wick in conjunction with circumferential grooves is often used for applications where vapor velocities are generally low. The slab wick, which is usually a felt or several layers of screen, is made in the shape of a bar with a rectangular cross section. The longest side of the cross section is the same as the inner diameter of the heat pipe. Circumferential grooves distribute the condensate around the entire circumference. Obviously, this type of wick is not appropriate for very long heat pipes due to the difficulty in machining circumferential grooves on the inside of the pipe. The tunnel or spiral artery also uses a felt or several layers of screen, but is fashioned in the shape of a tube, which is smaller than the inside diameter of the pipe. The artery is then located on the pipe axis by slab wick stand-offs, which provide condensate communication between the hollow artery and the circumferential grooves on the inside diameter of the pipe. In the monogroove heat pipe design, the vapor and liquid condensate flow in different channels separated by a narrow groove. As with other arterial type wick designs, circumferential grooves in the vapor channel help distribute the condensate. The double-walled artery consists of an inner tube with external grooves, which is perforated at the evaporator and condenser sections to allow vapor to escape into the interior. This inner tube is concentrically placed in an outer tube with a screen mesh wick at the inner wall. The double-walled artery wick design produces high capillary pressures, and since condensate returns to the evaporator by both a screen wick and arteries on the exterior of the inner pipe, total blockages of the liquid return are unlikely.

Table 4 Typical composite wick designs

Wick type	Capillary pumping	Thermal conductivity	Permeability	Comments
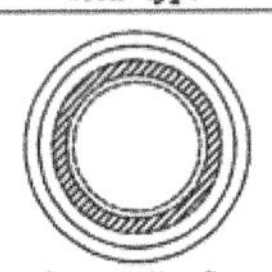 A. Composite Screen	High	Low-average	Average	Two or more layers of homogeneous material. The material next to wall has largest pore size
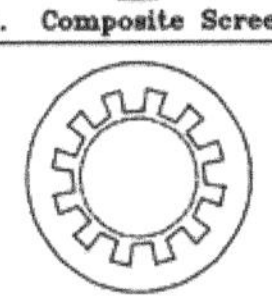 B. Screen Covered Grooves	High	High	Average-high	Axial grooves covered with one screen.
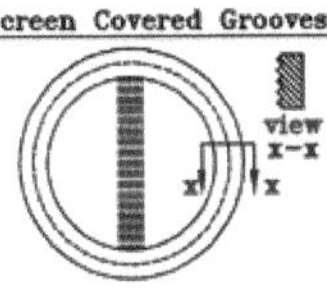 C. Composite Slab	High	High	High	Slab of homogeneous or non-homogeneous material with circumferential grooves
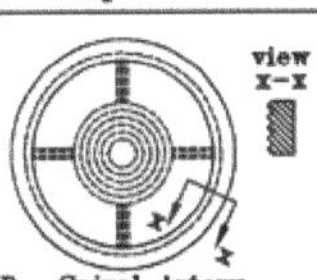 D. Spiral Artery	High	High	Average-high	Spiral wire screen mesh and circumferential grooves
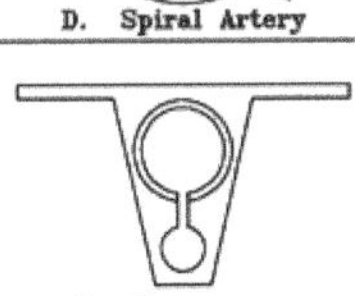 E. Monogroove	Average	High	High	Slab of non-homogeneous material with circumferential grooves
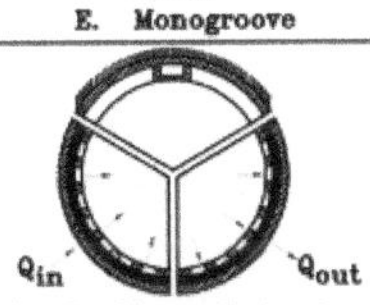 F. Double-walled artery	High	High	Average-high	Concentric perforated externally grooved inner tube wall. Inner-annuli screen mesh wick.

7. HEAT TRANSFER LIMITATIONS

There are various parameters that put limitations and constraints on the steady and transient operations of heat pipes (Faghri, 1995). In other words, the rate of heat transport through a heat pipe is subject to a number of operating limits. Physical phenomena that might limit heat transport in heat pipes are due to capillary, sonic, entrainment, boiling, frozen startup, continuum vapor, vapor pressure and condenser effects. The heat transfer limitation can be any of the above limitations depending on the size and shape of the pipe, working fluid, wick structure, and operating temperature. The lowest limit among the eight constraints defines the maximum heat transport limitation of a heat pipe at a given temperature. The physical phenomena for each limitation are briefly presented below. A detailed presentation of the criteria for the heat transfer limitations for heat pipes are provided in Faghri (1995).

Capillary Limit

The ability of a particular capillary structure to provide the circulation for a given working fluid is limited (Faghri and Thomas, 1989; Faghri, 1995). This limit is commonly called the capillary limitation or hydrodynamic limitation. The capillary limit is the most commonly encountered limitation in the operation of low-temperature heat pipes. It occurs when the pumping rate is not sufficient to provide enough liquid to the evaporator section. This is due to the fact that the sum of the liquid and vapor pressure drops exceed the maximum capillary pressure that the wick can sustain. The maximum capillary pressure for a given wick structure depends on the physical properties of the wick and working fluid. Any attempt to increase the heat transfer above the capillary limit will cause dryout in the evaporator section, where a sudden increase in wall temperature along the evaporator section takes place.

Sonic Limit

The evaporator and condenser sections of a heat pipe represent a vapor flow channel with mass addition and extraction due to the evaporation and condensation, respectively. The vapor velocity increases along the evaporator and reaches a maximum at the end of the evaporator section. The limitation of such a flow system is similar to that of a converging-diverging nozzle with a constant mass flow rate, where the evaporator exit corresponds to the throat of the nozzle. Therefore, one expects that the vapor velocity at that point cannot exceed the local speed of sound. This choked flow condition is called the sonic limitation. The sonic limit usually occurs either during heat pipe startup or during steady state operation when the heat transfer coefficient at the condenser is high. The sonic limit is usually associated with liquid-metal heat pipes due to high vapor velocities and low densities. Unlike the capillary limit, when the sonic limit is exceeded, it does not represent a serious failure. The sonic limitation corresponds to a given evaporator end cap temperature. Increasing the evaporator end cap temperature will increase this limit to a new higher sonic limit. The rate of heat transfer will not increase by decreasing the condenser temperature under the choked condition. Therefore, when the sonic limit is reached, further increases in the heat transfer rate can be realized only when the evaporator temperature increases. Operation of heat pipes with a heat rate close to or at the sonic limit results in a significant axial temperature drop along the heat pipe.

Boiling Limit

If the radial heat flux in the evaporator section becomes too high, the liquid in the evaporator wick boils and the wall temperature becomes excessively high. The vapor bubbles that form in the wick prevent the liquid from wetting the pipe wall, which causes hot spots. If this boiling is severe, it dries out the wick in the evaporator, which is defined as the boiling limit. However, under a low or moderate radial heat flux, low intensity stable boiling is possible without causing dryout. It should be noted that the boiling limitation is a radial heat flux limitation as compared to an axial heat flux limitation for the other heat pipe limits. However, since they are related through the evaporator surface area, the maximum radial heat flux limitation also specifies the maximum axial heat transport. The boiling limit is often associated with heat pipes of non-metallic working fluids. For liquid-metal heat pipes, the boiling limit is rarely seen.

Entrainment Limit

A shear force exists at the liquid-vapor interface since the vapor and liquid move in opposite directions. At high relative velocities, droplets of liquid can be torn from the wick surface and entrained into the vapor flowing toward the condenser section. If the entrainment becomes too great, the evaporator will dry out. The heat transfer rate at which this occurs is called the entrainment limit. Entrainment can be detected by the sounds made by droplets striking the condenser end of the heat pipe. The entrainment limit is often associated with low or moderate temperature heat pipes with small diameters, or high temperature heat pipes when the heat input at the evaporator is high.

Vapor Pressure Limit

At low operating temperatures, viscous forces may be dominant for the vapor moving flow down the heat pipe. For a long liquid-metal heat pipe, the vapor pressure at the condenser end may reduce to zero. The heat transport of the heat pipe may be limited under this condition. The vapor pressure limit (viscous limit) is encountered when a heat pipe operates at temperatures below its normal operating range, such as during startup from the frozen state. In this case, the vapor pressure is very small, with the condenser end cap pressure nearly zero.

Frozen Startup Limit

During the startup process from the frozen state, the active length of the heat pipe is less than the total length, and the distance the liquid has to travel in the wick is less than that required for steady state operations. Therefore, the capillary limit will usually not occur during the startup process if the heat input is not very high and is not applied too abruptly. However, for heat pipes with an initially frozen working fluid, if the melting temperature of the working fluid and the heat capacities of the heat pipe container and wick are high, and the latent heat of evaporation and cross-sectional area of the wick are small, a frozen startup limit may occur due to the freezing out of vapor from the evaporation zone to the adiabatic or condensation zone.

Condenser Heat Transfer Limit

In general, heat pipe condensers and the method of cooling the condenser should be designed such that the maximum heat rate capable of being transported by the heat pipe can be removed. However, in exceptional cases with high temperature heat pipes, appropriate condensers cannot be developed to remove the maximum heat capability of the heat pipe. Due to the presence of noncondensible gases, the effective length of the heat pipe is reduced during continuous operation. Therefore, the condenser is not used to its full capacity. In both cases, the heat transfer limitation can be due to the condenser limit.

Vapor Continuum Limit

For heat pipes with very low operating temperatures, especially when the dimension of the heat pipe is very small such as micro heat pipes, the vapor flow in the heat pipe may be in the free molecular or rarefied condition. The heat transport capability under this condition is limited, and is called the vapor continuum limit.

Flooding Limit

The flooding limit is the most common concern for long thermosyphons with large liquid fill ratios, large axial heat fluxes, and small radial heat fluxes (Faghri *et al.*, 1989a). This limit occurs due to the instability of the liquid film generated by a high value of interfacial shear, which is a result of the large vapor velocities induced by high axial heat fluxes. The vapor shear hold-up prevents the condensate from returning to the evaporator and leads to a flooding condition in the condenser section. This causes a partial dryout of the evaporator, which results in wall temperature excursions or in limiting the operation of the system.

8. HEAT PIPE START UP

The procedures typically used in designing heat pipes are based on the fact that the heat pipe is in its normal operating state, i.e., nominal temperature at the steady state. However, the heat pipe must be started from the ambient temperature, which is either lower or higher than the operating temperature (Faghri, 1995). If the startup is too fast, the possibility of overheating the evaporator section can result in damage to the heat pipe.

Ideally, the heat pipe should first be started by increasing (or decreasing) the temperature of the entire pipe to its operating temperature. The heat input to the pipe is then slowly increased from zero to the operating input, maintaining a uniform temperature across the length of the pipe. This procedure is not usually feasible in practice. Normally, the heat input increases suddenly from zero to the operating input, without the benefit of increasing the temperature of the entire pipe. This may result in problems occurring in the startup period, which will be discussed here for three different common situations.

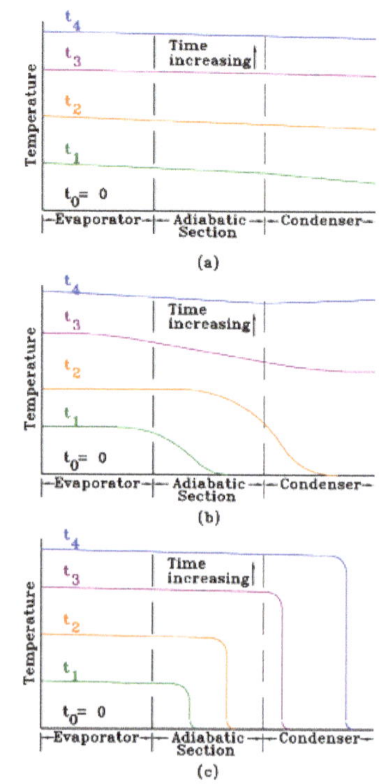

Fig. 22 Transient axial temperature profiles during heat pipe startup: (a) Uniform startup; (b) Frozen startup; (c) Gas-loaded heat pipe startup.

Uniform Startup

Figure 22(a) shows the startup of a heat pipe in which the axial temperature is nearly uniform throughout the startup period. This type of startup profile is normally seen in heat pipes in which the working fluid in the wick is in the liquid state, and the vapor is in the continuum state. An example is a copper-water heat pipe starting from ambient temperature to some higher operating temperature. As can be seen, no problems are encountered during the uniform startup.

Frozen Startup

Figure 22(b) presents the case of a heat pipe in which the working fluid within the wick structure is initially frozen, and the vapor space is essentially evacuated. This type of frontal startup is usually found in liquid-metal heat pipes, since liquid metals are in the solid state at room temperature. Frozen startup proceeds as follows: Heat is first conducted through the pipe wall and into the wick structure, increasing the temperature in the evaporator section only. After the working fluid in the evaporator wick is liquefied, evaporation begins to fill the vapor space with vapor. The vapor travels to the adiabatic section and condenses, releasing its latent heat and increasing the adiabatic section temperature. This front continues down the length of the pipe until it reaches the condenser end cap. At that time, the axial temperature distribution starts to become uniform, and the startup process is completed. The presence of large axial temperature gradients indicates that the sonic limit is occurring during frozen startup. In actuality, this is not a limit per se, since the heat pipe will not be damaged during this process. Another more important limit is the frozen startup limit, as previously discussed. The frozen startup limit occurs when more working fluid is evaporated than can be resupplied by the wick structure, due to the fact that the working fluid in the adiabatic and condenser sections is completely frozen. When this occurs, the evaporator section is depleted of liquid and overheats due to dryout.

Gas-Loaded Heat Pipe Startup

The noncondensible gas present in gas-loaded heat pipes gives the startup a frontal character. Figure 22(c) shows the case in which the working fluid in the wick is initially liquid in a gas-loaded heat pipe. Before the heat input is applied in the evaporator, the noncondensible gas is evenly distributed throughout the vapor space. When evaporation occurs, the working fluid vapor drives the gas to the condenser end of the heat pipe. Since the gas and the vapor are essentially separated, the effect of the gas is to block the condenser section from transferring heat to the heat sink. The presence of the gas can be seen in the axial temperature profile, which has an abrupt drop at the vapor-gas interface. More heat is then fed into the pipe, increasing the wall and vapor temperatures within the evaporator and adiabatic sections. This increases the vapor pressure, which compresses the gas in the condenser section. The vapor-gas front moves downstream into the condenser section, eventually unblocking part of the condenser, so that normal operation can take place.

9. HEAT PIPE CHARACTERISTICS

High Effective Thermal Conductivity

The heat pipe is a device of very high thermal conductance (Faghri, 1995). For example, a temperature difference of 900°C is needed to transfer 1 KW heat across a 30-mm-diameter 1-m-long copper rod. A heat pipe of the same size can transfer the same amount of heat with a temperature difference of less than 10°C. This indicates that the heat pipe can have a thermal conductivity 90 times higher than that of a copper bar of the same size. Numerous heat pipe designs have been developed for various applications with varied heat transport capabilities.

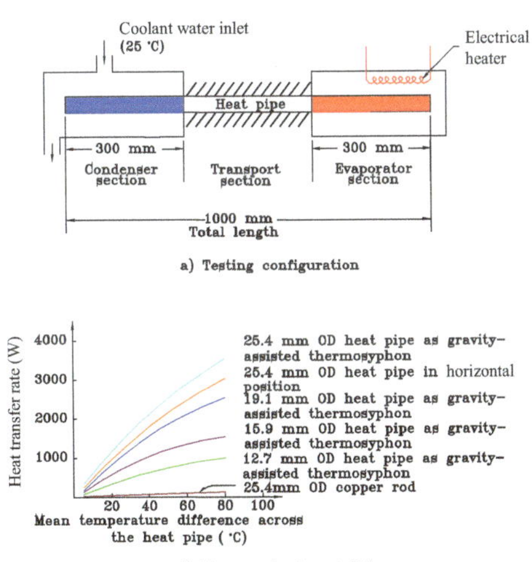

Fig. 23 Heat transfer characteristics of some typical commercial copper-water screen mesh heat pipes.

The heat transfer characteristics of a heat pipe can be measured using a conventional configuration as shown in Fig. 23(a). Figure 23(b) shows the heat transfer rate versus the mean temperature difference between the evaporator and condenser sections of commercially available copper-water heat pipes with a screen mesh wick and different diameters. Most of the tests were made in a vertical position (gravity-assisted wicked heat pipe), where the driving force is due to both capillary pumping and gravity. A typical test for horizontal operation, as well as a comparison with a copper rod of 24.5 mm OD, are also shown in Fig. 23(b).

The thermal response of a conventional commercial screen mesh copper-water heat pipe with 6.35 mm OD and an inner diameter of 5.85 mm in comparison with a copper pipe of the same dimension is shown in Fig. 24. There are commercial companies all over the world (with more concentration in the Far East) who manufacture heat pipes of different shapes and sizes for various operating temperature ranges. A typical commercial off-the-shelf performance characteristic for several copper water heat pipes (Fujikura, Ltd) is presented in Fig. 25.

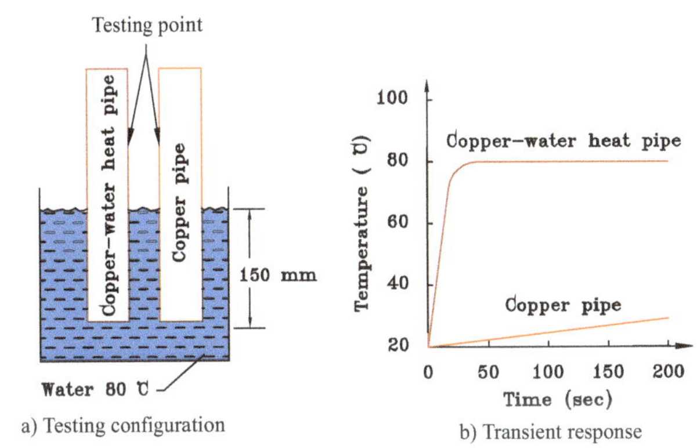

Fig. 24 Transient response of a typical copper-water heat pipe compared to a copper pipe of the same dimensions.

Three of the important design parameters that are used to compare heat pipes and other two-phase thermal devices are the thermal resistance, the axial heat transport capacity (W-m) and the maximum radial heat flux at evaporator (W/cm²). Table 5 presents typical experimental operating values reported in literature for a number of heat pipes tested under various operating conditions. The operating temperature, wick design, wall material and dimensions of the heat pipes tested are listed in this table. Also, the axial heat transport rate (W), axial heat transport capability (W-m) and the radial heat flux (W/cm²) are given. The radial heat flux or axial heat transport values reported may not necessarily be the limiting values.

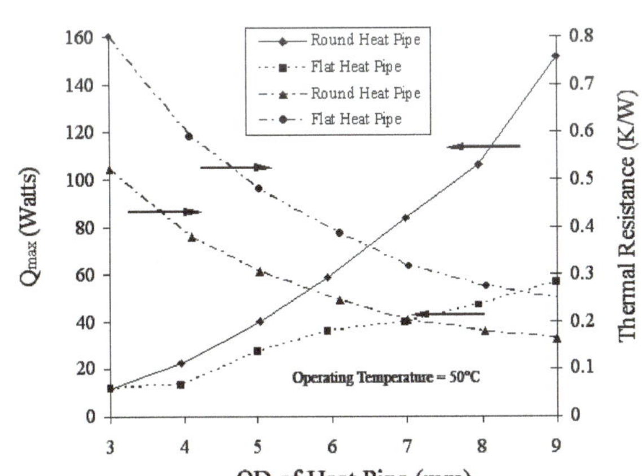

Fig. 25 Maximum heat transfer rate and thermal resistance of several commercially available copper-water heat pipes with fine fiber wicks for horizontal operation with L=300 mm, L_e=50 mm and L_c = 250 mm.

Table 5 Some typical experimental heat pipe results

Working Fluid	Operating Temperature	Wick Design	Wall Material	Dimensions	Axial Heat Transport (W)	Axial Heat Transport Capability (W-m)	Radial Heat Flux (W/cm^2)	Reference
Methane	-140°C	Circumferential mesh wick and arterial wick	Stainless steel	L_t = 0.4 m L_e = 3 cm L_c = 11.5 cm D_i = 7 mm	12	3.93	1.82	Rosler et al. (1987)
Ammonia	25°C	Monogroove	Aluminum	L_t = 5.5 m L_e = 0.91 m L_c = 0.91 m D_v = 13.4 mm D_l = 6.32 mm slot width = 0.381 mm	700	3200	7.26	Alario et al. (1982)
Water	50°C	Axial grooves	Copper	Outer pipe same as above; Inner pipe: D_o = 29.7 mm D_i = 25.4 mm	1300	940	Inner pipe: 2.32 Outer pipe: 1.47	Faghri and Thomas (1989) (annular heat pipe)
Water	50, 70, 80°C	3 wraps of a copper wire screen mesh with a wire diameter of 0.11 mm and 3937 strands per meter	Copper	D_o = 19.05 mm D_i = 15.75 mm L_t = 355.6 mm L_e = 101.6 mm L_c = 152.4 mm	20-180	4.6-41	0.33-2.96	Kempers et al. (2008)
Water	60-95°C	Micro capillary grooves (trapezoidal and rectangular)	Copper	L_t = 120.65 mm W = 13.41 mm L_e = 15.6 mm L_c = 34.4 mm L_a = 70 mm	130 (horizontal) 170 (vertical)	15.6 20.4	90 (horizontal) 150 (vertical)	Hopkins et al. (1999)
Water	75°C	Circumferential copper screen (50 mesh)	Copper	L = 1.0 m L_e = (4)@6.4 cm L_c = 30 cm D_v = 25.4 mm D_i = 22 mm	392	283	1.92	Faghri and Buchko (1991) (multiple evaporators)
Water	100°C	Axial grooves	Copper with rectangular cross section	L_t = 120 mm H = 2 mm W = 7 mm L_e = 20 mm L_c = 20 mm	70	8.4	35	Plesch et al. (1991)
Water	100°C	Copper wick, 150 mesh	Copper	D_o = 19.1 mm D_i = 17.3 mm L_t = 610 mm L_e = 393 mm L_c = 170 mm	570	187	2.4	El-Genk and Huang (1993)
Water		Axial rectangular grooves	Copper	L_t = 82 mm	40		20	Cao et al. (1997)
Water	100 °C	Axial rectangular grooves	Copper	L_t = 120 mm W = 7 mm H = 3 mm	50 (horizontal) 70 (vertical)		25 (horizontal) 35 (vertical)	Gao et al. (2000)
Water	160°C	Double-wall artery	Copper	L_t = 1.2 m L_e = 0.2 m L_c = 0.2 m D_v = 13.4 mm D_o = 22.2 mm	900	900	6.45	Faghri et al. (1984)
Therminol VP-1	300-400°C	Thermosyphon	316-stainless steel	D_o = 15.8 mm D_i = 6 mm L_t = 209 mm L_e = 45 mm L_c = 100 mm	35-70	4.8-9.6	1.57-3.13	Jouhara and Robinson (2009)
Sulfur-Iodine 5% wt	350°C	Thermosyphon	Mild steel	D_o = 25 mm D_i = 20 mm L_t = 1000 mm L_e = 400 mm L_c = 550 mm	390	204.8	1.24	Trovole and Raine (1995)

Working Fluid	Operating Temperature	Wick Design	Wall Material	Dimensions	Axial Heat Transport (W)	Axial Heat Transport Capability (W-m)	Radial Heat Flux (W/cm^2)	Reference
Sodium	430-790°C	Circumferential stainless steel screen (100 mesh)	Stainless steel	L_t = 1.0 m L_e = (4)@5.3 cm L_c = 29.2 cm D_o = 26.7 mm D_i = 21.5 mm	1309	979	6.64	Faghri et al. (1991a) (multiple evaporators)
Sodium	650°C	Nickel sintered powder metal wick and an artery	Stainless steel	D_o = 21.3 mm D_i = - L_t = 460 mm L_e = 100 mm L_c = 160 mm $L_{c,inactive}$ = 100 mm	1400	322	21	Yamawaki et al. (1998)
Sodium	800°C	Double-wall artery	Stainless steel	L_t = 2.0 m L_e = 25 cm L_c = 91 cm D_o = 2.22 cm	758	1076	4.35	Ponnappan (1989)
Mercury	630°C	Thermosyphon	316L stainless steel	D_o = 25.4 mm D_i = 21 mm L_t = 1000 mm L_e = 200 mm L_c = 640 mm	1922	1115	12	Vieira da Cunha and Mantelli (2009)
NaK (with Argon as NCG)	700°C	304 stainless wick	304L stainless steel	D_o = 19.1 mm D_i = - L_t = 600 mm L_e = 114 mm $L_{c,1}$ = 73 mm $L_{c,2}$ = 66 mm $L_{c,inactive}$ = 50 mm	250	57	3.65	Anderson and Tarau (2008)
Lithium	1227°C (1500 K)	Free-floating Mo-41wt.%Re 400-mesh screen wick (0.41 mm thick) A 0.37 mm annular gap separates the wick from the wall	molybdenum	D_o = 19.1 mm D_i = 16.06 mm L_t = 1800 mm L_e = 300 mm L_c = 1470 mm	4000	3660	22.2	Tournier and El-Genk (2003)

Thermal Transformer

The heat pipe can be effectively used as a thermal transformer to link energy sources and sinks having different heat fluxes. Energy can be added at a high heat flux rate to the heat pipe over a small surface area and removed over a larger surface area at a lower heat flux, or vice versa. Thermal flux transformation ratios as large as 15 to 1 can be achieved using heat pipes.

Power Flattening

A heat pipe that provides power flattening can be easily designed. A constant output heat flux at the condenser can be maintained for variations in the input heat flux in the evaporator section.

Temperature Control

By actively determining the conditions at the condenser (sink), a heat pipe can be designed to keep a nearly constant temperature at the evaporator even though the rate of heat input to evaporator varies. This is the basis for the variable conductance heat pipe which is discussed by Faghri (1995). The gas-loaded heat pipe, in which a noncondensible gas and a gas reservoir are constructed with a conventional heat pipe, is the simplest means to actively determine the conditions at the condenser.

Freedom of Design

Since the basic mechanisms of a heat pipe are the capillary pumping of a wick and the evaporation and condensation of a working fluid, heat pipe design is only restricted by the need to provide for these mechanisms. With this in mind, heat pipes can be designed so that they fit various shapes and configurations. These geometries range from simple cylindrical or flat heat pipes, to curved plate heat pipes for solar collectors and leading edge heat pipes.

Maintenance-Free

Since heat pipes are a closed systems and do not need any external electrical or mechanical drives, they can operate over long periods of time without maintenance.

10. HEAT PIPE ANALYSIS AND SIMULATION

Background

Numerical and analytical simulation of heat pipes has progressed significantly in the last several decades. The state of the art has been advanced in steady state, continuum transient, and frozen start up simulation for high-, intermediate-, and low-temperature heat pipes of conventional and nonconventional geometries.

In determining the heat capacity transmitted through a heat pipe or heat pipe performance characteristics, it is necessary to know the liquid and vapor pressure losses in the separate segments. The thermal fluid phenomena occurring within a heat pipe can be divided into four basic categories: (1) vapor flow in the core region, (2) liquid flow in the wick, (3) interaction between the liquid and vapor flows, and (4) heat conduction in the wall. Most of the detailed analytical and numerical research on heat pipes has been done on the vapor core region and wall heat conduction, since the liquid flow is difficult to describe with an exact theoretical model. Because of the presence of the wick structure, the analysis of categories (2) and (3) requires some empirical information that must be obtained from experimentation. Furthermore, the analyses of categories (2) and (3) are basically similar for heat pipes of various shapes. In contrast, the dynamics of vapor flow are related to the geometry and boundary specifications for nonconventional heat pipes.

The interaction between different regions can be best illustrated by following the real heat pipe operating condition, as shown by the flow chart in Fig. 26. Consider a heat pipe with an initial operating condition. If the heat load from the heat source to the heat pipe evaporator surface is increased at $t \geq 0$ to a higher level, the temperature of the container wall is increased accordingly, and more heat is transferred from the container to the heat pipe wick structure. In this case, the heat pipe container wall is interacting with the heat pipe wick structure. Meanwhile, the state of the liquid-vapor interface may change due to the change in liquid pressure and temperature. The interface evaporation mass flux is also increased because more heat is transferred from the wick structure to the interface. A higher mass flux at the interface will increase the vapor velocity, and result in a change in the vapor working condition. In the above process, the liquid flow in the wick structure is interacting with vapor flow via the liquid-vapor interface. At the same time, more liquid mass may condense onto the liquid-vapor interface in the condenser section due to the higher mass flow rate from the evaporator section. This will subsequently change the interface state. The interaction between the vapor flow and the liquid-vapor interface in this case can be called feedback. These feedback interactions will also occur between the liquid-vapor interface and wick structure, and between the wick structure and the container.

In Fig. 26, regions can interact only with regions directly above or below. The container region, for example, cannot interact with the vapor region directly. It can only interact with the vapor region via the wick structure region and the liquid-vapor interface. This is due to the physical locations of the different regions, which prescribe the relation between different regions. In general, all of the regions in the flow chart should be solved as a conjugate problem. However, in the real application, approximations are often introduced due to the complex nature of the problem. Some regions in the flow chart may be neglected or combined with other regions. Regions can also be decoupled from other regions. For instance, if we neglect the liquid flow in the wick structure, and solve the region as a heat conduction problem, the region associated with the wick structure may be combined with the region associated with the heat conduction in the container. If we are only interested in the vapor flow in the heat pipe, we can neglect the three preceding regions, and directly go to the fourth region associated with vapor flow. In some applications, the liquid flow in the wick structure is of interest. Therefore, the wick region can be decoupled from other regions, and solved as a one region problem with appropriate assumptions.

For the convenience of analysis, the frozen start up process of a high-temperature heat pipe or frozen low-temperate heat pipe can be divided into several periods based on experimental evidence.

1. The working fluid in the wick is frozen at the ambient temperature. For a heat pipe without any noncondensible gas present in the vapor space, the vapor core is completely evacuated before start up.
2. Heat input is started in the evaporator section. Heat is conducted through the pipe wall and wick structure, and melting of the working fluid begins at the wall-wick interface. Because the liquid-solid interface has not reached the wick-vapor boundary, no evaporation takes place and the vapor space is still evacuated.
3. The working fluid is completely melted in the evaporator section only, and liquid is evaporated at the liquid-vapor interface. The vapor flow in the rarefied or free molecular condition due to the low vapor pressure. Some of the working fluid in the wick is still frozen in the adiabatic and condenser sections.
4. Continuum vapor flow is established in the evaporator section as the vapor pressure increases due to evaporation. In the remainder of the heat pipe, however, the vapor is still in the rarefied condition.
5. The working fluid is completely melted and continuum flow is established over the entire heat pipe length. The heat pipe temperature increases steadily until the steady state is reached.
6. The heat pipe is operating under steady state operation.

Heat Source ↓ | Heat Sink ↑

Heat Conduction in Heat Pipe Container

Variables	*Phenomena*	*Governing Equations*
Temperature	Conjugate Effects	Heat Conductions
Heat Input	Radiation or Convection B.C.'s	Stefan-Boltzmann Law
Heat Output	Geometry Effects	Fourier Law

Liquid Flow and Heat Transfer in Wick Structure

Variables	*Phenomena*	*Governing Equations*
Temperature	Phase Change	Continuity Equation
Velocities	Porous Media Flow	Modified Navier-Stokes Equations
Pressure Drop	Interfacial Shear Stress	Averaged Energy Equation
Gravity	Liquid Entrainment	
	Boiling Limitation	

Liquid-Vapor Interface

Variables	*Phenomena*	*Governing Equations*
Contact Angles	Capillary Pressure	Interfacial Mass Balance
Interfacial Curvature	Disjoining Pressure	Interfacial Momentum Balance
Interface Position	Evaporation	Interfacial Energy Balance
Interfacial Mass Fluxes	Condensation	Clapeyron or Kelvin Equation

Vapor Flow in Heat Pipe Core

Variables	*Phenomena*	*Governing Equations*
Velocities	Compressibility Effects	Continuity Equation
Pressure Drop	Continuum & Rarefied Flows	Momentum Equations
Density	Mass Diffusion	Energy Equation
Temperature	Geometry Effects	Mass Diffusion Equation
Concentration	Sonic Limitation	Equation of State

Fig. 26 Flow chart for heat pipe operation and interaction between different regions.

The parameters limiting heat transport in conventional heat pipes due to vapor flow are the capillary, sonic, and entrainment limitations. The sonic limitation is not greatly influenced by any aspect of the wick structure. Therefore, the choking phenomena should be of primary interest in the vapor flow analysis due to the fact that the vapor velocity becomes significant compared with the sonic velocity. Furthermore, vapor flow analyses are required to predict the capillary limitation. In each concentration, there are several levels of approximation, ranging from a simple one-dimensional vapor flow analysis to a complete three-dimensional analysis, considering the conjugate nature of the wall and wick. Important modeling considerations include vapor compressibility, heat pipe geometry, and heat input distributions.

These methods cover various approximations and approaches, such as compressible vs. incompressible, analytical and closed-form solutions vs. numerical analyses, as well as one-dimensional, two-dimensional, and three-dimensional modeling considerations in each region. Faghri (2012) reviewed the advances in analysis and simulation of the different types of heat pipes under various operating conditions, also presented in this section.

This section summarizes the developments in modeling and discusses important results for various types of heat pipes under various operating conditions in the following order (Faghri, 2012):

1 – Steady state (period 6 above)
2 – Transient (period 5 above)
3 – Frozen start up (periods 1 through 6 above)
4 – Axially grooved heat pipes
5 – Thermosyphons
6 – Rotating heat pipes
7 – Loop heat pipes

8 – Capillary pump loop heat pipes
9 – Micro and miniature heat pipes
10 – Pulsating heat pipes
11 – Gas-loaded heat pipes

Steady-State

A steady state operational performance prediction is of significant value in the design of heat pipes. Faghri (1986) modeled the steady-state two-dimensional incompressible vapor flow in an annular heat pipe. In this study, the boundary-layer approximations were used to reduce the fully elliptic conservation equations to the partially parabolic Navier-Stokes equations, which were solved using a fully implicit, marching finite-difference scheme.

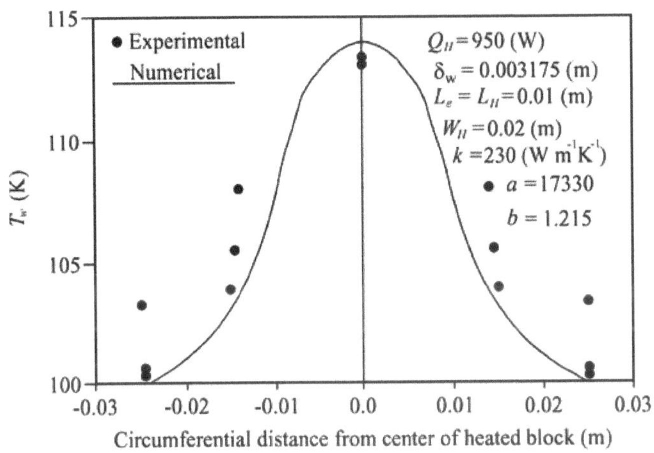

Fig. 27 Comparison of numerical results of Cao et al. (1989) with experimental data of Rosenfeld (1987).

This work was extended by Faghri and Parvani (1988) and Faghri (1989), where vapor compressibility was considered and the complete elliptic Navier-Stokes equations were solved. One- and two-dimensional models were developed for the vapor flow in an annular heat pipe, and the assumptions used by Faghri (1986) were verified. A major contribution of this work is the derivation of analytic expressions for the axial pressure drop within the annular heat pipe. These relations can also be simplified to describe a conventional cylindrical heat pipe and be used directly in the design process of heat pipes.

Block-heated heat pipes are often used as heat transfer mechanisms for energy conservation systems and electronic cooling. The fundamental difference between these systems and conventional heat pipe systems is the nonuniform block heating at the evaporator, which influences the temperature distributions and vapor flow patterns throughout the heat pipe. Cao et al. (1989) proposed a simplified model of this process by defining two separate governing equations for the evaporator section: one for the pipe wall under the heated block and one for the remaining unheated pipe wall. In this method, a power-law relationship was used for the boiling heat flux. The block-heated simulation was performed for the heat pipe studied experimentally by Rosenfeld (1987), which used a narrow line heated on a low-temperature heat pipe. The circumferential wall temperature profile predicted by Cao et al. (1989) is shown in Fig. 27, with good agreement with the experimental data. Cao et al. (1989) determined that the evaporative heat transfer coefficient has a strong influence on the wall temperature profile. This is because with a large evaporation coefficient, the majority of the input heat is removed through the phase change of the working fluid. With a lower evaporation coefficient, more heat is conducted circumferentially through the pipe wall, smoothing the temperature profile.

The importance of the effect of conjugate heat transfer within the heat pipe wall and wick was shown by Chen and Faghri (1990) with a model of the two-dimensional compressible vapor flow coupled to the two-dimensional conduction model in the wall and wick. This model solves the complete elliptic Navier-Stokes equations in the vapor with a finite control volume iterative technique. The simulation was performed for the high-temperature sodium/stainless-steel heat pipe studied in Ivanovskii et al. (1982) with the results shown in Fig. 28. They used three different methods: compressible elliptic, incompressible elliptic and compressible parabolic. Both the compressible elliptic and compressible parabolic solutions gave reasonable vapor temperature distributions, while the incompressible elliptic case shows the importance of including vapor compressibility.

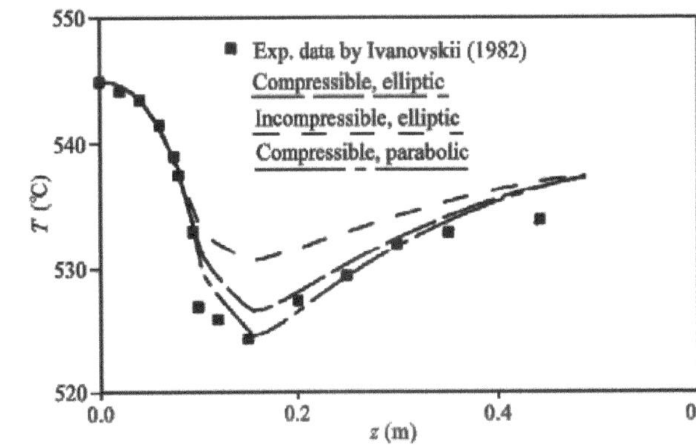

Fig. 28 The axial interface temperature profile along the sodium heat pipe with Q = 560 W, R_v = 0.007 m, L_e = 0.1 m, L_a = 0.05 m, k_l = 66.2 W/m^2-K, k_s = 19.0 W/m^2-K, δ_l = 0.0005 m, δ_w = 0.001 m. (Chen and Faghri, 1990)

This methodology was extended to model a low-temperature, multiple-evaporator heat pipe by Faghri and Buchko (1991). The heat pipe was simulated using a two-dimensional compressible formulation for the vapor flow, and a two-dimensional conduction model in the wall. However, Faghri and Buchko (1991) included the two-dimensional effects of liquid flow in the wick using volume-averaged velocities in the porous media. The conservation equations were solved using an elliptic finite control volume iterative scheme with the results as shown in Fig. 29. The experimental data corresponded to a water-copper multiple-evaporator heat pipe. One case was performed with only evaporator 1 active and a heat input of 97 W. Another had evaporators 1 and 2 active, with heat inputs of 99 W and 98 W, respectively.

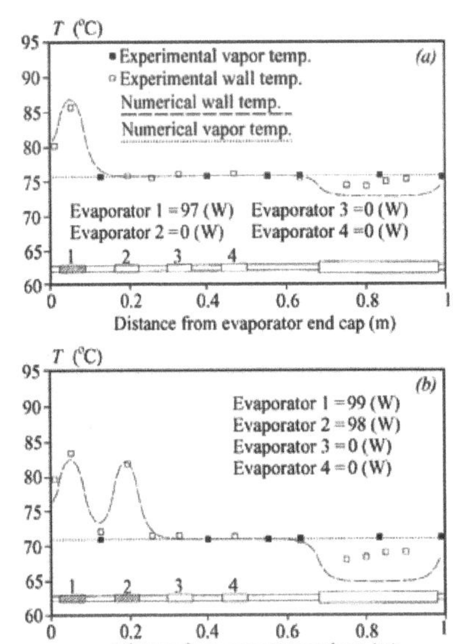

Fig. 29 Heat pipe wall and vapor temperature versus axial location for (a) single evaporator; (b) two evaporators (Faghri and Buchko, 1991).

The block heating phenomena studied by Cao et al. (1989) was further investigated by Schmalhofer and Faghri (1993a; 1993b). The complete three-dimensional compressible Navier-Stokes equations were solved in the vapor, and the wall and wick were simulated with a three-dimensional conduction model. The results of this simulation were compared with the experimental data of Schmalhofer and Faghri (1993a) for a water-copper heat pipe with good agreement (Schmalhofer and Faghri (1993b)).

An analytical and numerical study was carried out by Zhu and Vafai (1998b) for the steady incompressible vapor and liquid flow in flat plate heat pipes. The pseudo-three-dimensional analytical model employed the boundary layer approximation to describe the vapor flow conditions. The three-dimensional effects are discussed and the results show that a three-dimensional analysis is necessary if the vapor channel width-to-height ratio is less than 2.5.

Lefevre and Lallemand (2006) developed an analytical solution for a 2D hydrodynamic model for both liquid and vapor coupled to 3D heat conduction for the wall to simulate the steady state operation of a flat miniature heat pipe. The proposed model is also capable to predict the maximum heat transfer capability for a miniature heat pipe with multiple heat sources.

A detailed, three-dimensional steady-state model was developed by Xiao and Faghri (2008) to analyze the thermal hydrodynamic behaviors of flat heat pipes and vapor chambers without empirical correlations. The model accounts for the heat conduction in the wall, fluid flow in the vapor chambers and porous wicks, and the coupled heat and mass transfer at the liquid-vapor interface. The flat heat pipes with and without vertical wick columns in the vapor channel (Fig. 30) were investigated in the model by Xiao and Faghri (2008).

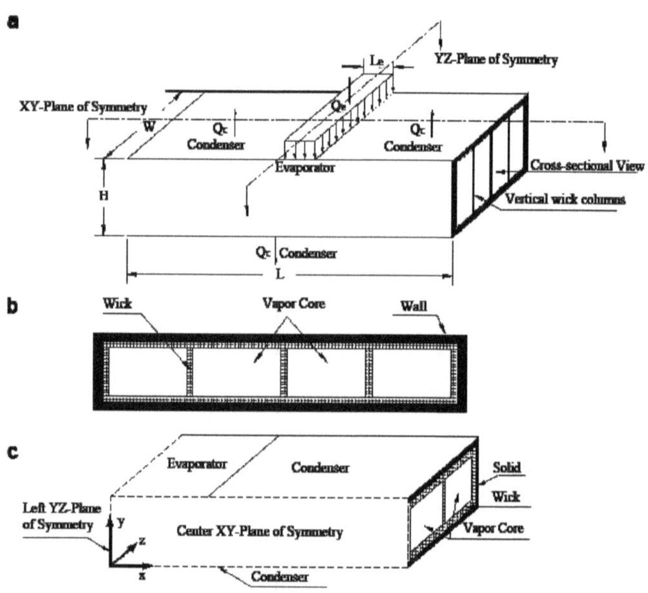

Fig. 30 Schematic diagram of the flat heat pipe: (a) geometry of the heat pipe; (b) cross-sectional view in YZ plane; and (c) computational domain (Xiao and Faghri, 2008).

Parametric effects, including the evaporative heat input and size on the thermal and hydrodynamic behavior in the heat pipes, were presented. The results showed that, the vertical wick columns in the vapor core can improve the thermal and hydrodynamic performance of the heat pipes, including thermal resistance, capillary limit, wall temperature, pressure drop, and fluid velocities due to the enhancement of the fluid/heat mechanism form the bottom condenser to the top evaporator. The results also predicted that higher evaporative heat input improves the thermal and hydrodynamic performance of the heat pipe, and shortening the size of heat pipe degrades the thermal performance of the heat pipe.

Figure 31 shows a comparison of the numerical results by Xiao and Faghri (2008) with the experimental data (Wang and Vafai, 2000) for the temperature distributions at the surface of the top and bottom walls of the flat heat pipe. The deviation in the condenser section in the top wall is mainly due to the variation of the convective heat transfer coefficient dependent of different surface temperature in the experiment, which is assumed to be constant in the numerical model.

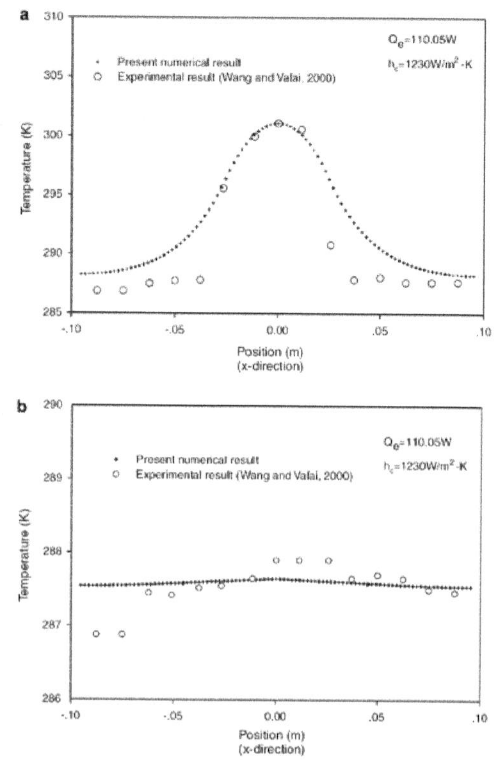

Fig. 31 Verification of numerical results for temperature distribution along a flat heat pipe with vertical wick columns at (a) the top wall; (b) the bottom wall at Z = W/2 (Xiao and Faghri, 2008).

Aghvami and Faghri (2011) presented a steady-state analytical thermal-fluid model, including the wall and both liquid and vapor flows for flat heat pipes and vapor chambers with different heating and cooling configurations. The parametric investigations showed that the assumption of uniform evaporation and condensation in the axial direction is valid only if the wall thermal conductivity is small. Shabgard and Faghri (2011) extended the above model to cylindrical heat pipes with multiple evaporator sections. The two-dimensional heat conduction in the heat pipe's wall was coupled with the liquid flow in the wick and the vapor hydrodynamics. It was found that neglecting the axial heat conduction through the wall can result in overestimated pressure drops up to 10% depending on the heat pipe specifications.

Transient

A first-order transient model of the vapor flow in a heat pipe was proposed by Jang et al. (1991). This model simulated heat pipe operation with one-dimensional compressible vapor flow in a porous pipe, accounting for laminar and turbulent skin friction. The results of this model are compared with the experimental data of Bowman (1987) in Fig. 32, where the transient vapor pressure at three locations along the heat pipe is shown. As in a conventional heat pipe, pressure recovery occurs in the condenser section can be seen. A major advantage of this method is the ability to capture sharp pressure gradients, such as those that would occur in supersonic flow or across shock waves.

A comprehensive transient heat pipe analysis is presented by Cao and Faghri (1990). This two-dimensional model accounts for vapor compressibility and couples the vapor flow with heat conduction in the wall and wick. For high-temperature heat pipes, it was determined that the conduction model is sufficient to describe heat transfer in the wick. This methodology allows heat pipe simulation for pulsed heat inputs with either a convective or radiative boundary condition at the outer pipe wall of the condenser. The transient vapor temperature profile for a pulsed heat input of Q = 623 W to 770 W with the convective boundary condition is shown in Fig. 33(a). The operating temperature of the heat pipe increases with time after the heat pulse. This is due to the coupling of the vapor to the wall, because the outer pipe wall temperature in the condenser must increase to reject the additional heat. The vapor temperature with the radiative boundary condition is shown in Fig. 33(b). Faghri et al. (1991b) adapted the numerical model of Cao and Faghri (1990) to a high-temperature heat pipe with multiple heat sources and sinks. Numerical results for continuum transient and steady-state operations with multiple heat sources were compared with the experimental results of Faghri et al. (1991a) and found to be in good agreement.

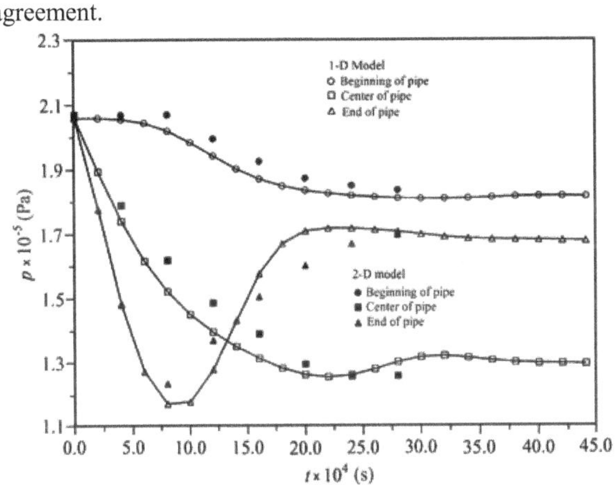

Fig. 32 Comparison of numerical results of Jang et al. (1991) with the experimental results of Bowman (1987) (Jang *et al.*, 1991).

A transient model capable of simulating nonconventional geometries was proposed by Cao and Faghri (1991) and Cao and Faghri (1993a). This method solves the one-dimensional compressible vapor flow coupled to the two- or three-dimensional heat conduction in the wall and wick. The one-dimensional vapor flow is easier to solve than the two-dimensional formulation, but friction coefficient information is required for both laminar and turbulent flow regimes. For complex geometries, such as a wing leading edge or spacecraft nosecap, a multidimensional formulation is used in the wall and wick. Conduction in the wall and wick is coupled to the vapor flow through a conjugate heat transfer solution technique for the elliptic conservation equations in the vapor. In this analysis, the liquid flow in the wick was assumed to be decoupled from heat pipe operation to determine the capillary limit for nonconventional geometries. This simulation was used to investigate the use of heat pipes for cooling the leading edge of reentry vehicles and hypersonic aircraft. In this case, a line heat source was applied at the leading edge stagnation point of the wing to simulate aerodynamic heating. The results of this simulation are shown in Fig. 34, where the extreme temperature gradient in the pipe wall at the leading edge can be seen. This is due to the severe heat flux (q = 75 x 106 W/m^2) at this point, which represents actual hypersonic heat loads.

A transient heat pipe model was developed by Zuo and Faghri (1997) using a transient two-dimensional wall and wick heat conduction and a quasi-steady-state one-dimensional vapor flow. The assumption of quasi-steady-state vapor was based on the fact that the vapor dynamics have a much smaller timescale than the wall conduction based on dimensional analysis. The transient two-dimensional wall heat conduction was solved by a boundary element method (BEM). A control volume finite difference method (FDM) was employed to solve the vapor flow continuity and momentum equations. Two iterative "estimate-correction" processes and a new pressure-correction method were incorporated in the vapor solution procedure. Significant reduction of the computation time was obtained by using this hybrid FDM/BEM. Comparisons with previous experimental and numerical results validated the assumption of a quasi-steady-state vapor and the solution methods.

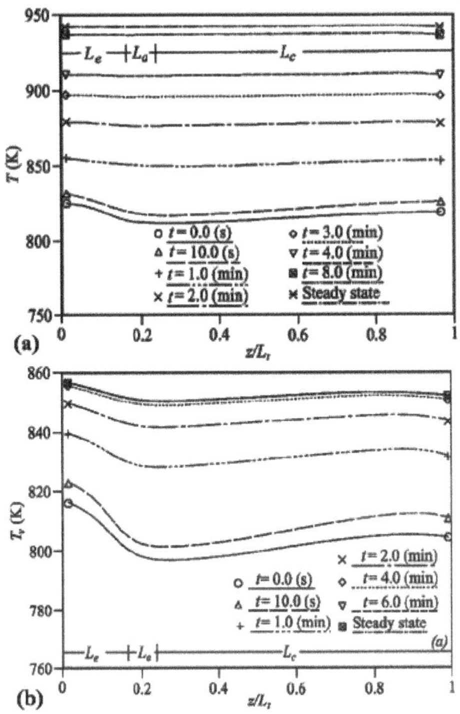

Fig. 33 Centerline vapor temperature for transient response to heat input pulse: (a) convective boundary condition; (b) radiative boundary condition (Cao and Faghri, 1990).

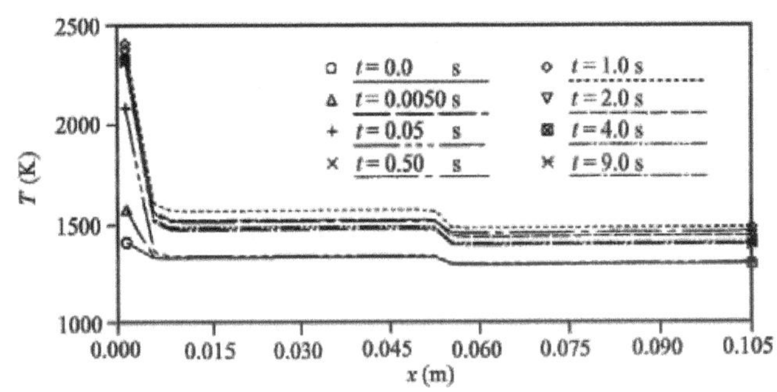

Fig. 34 Leading edge heat pipe outer wall temperature distribution (Cao and Faghri, 1993a).

An accurate, but simple lumped model using the "network" model was developed for transient heat pipe analysis by Zuo and Faghri (1998). The heat pipe consists of a number of components with different thermal resistances and dynamic responses in the network model. Governing equations of the transient heat pipe behavior were significantly simplified to a set of first-order, linear, ordinary differential equations. Figure 35 compares the transient network model with experimental results. For this specific case, the computational time required by the network model is about 1 minute on a standard desktop PC compared to several hours of CPU needed for a full two-dimensional numerical model. Zhu and Vafai (1998a) developed an analytical solution for quasi-steady and incompressible vapor flow with transient one-dimensional heat conduction in the wall and wick for an asymmetrical flat plate heat pipe.

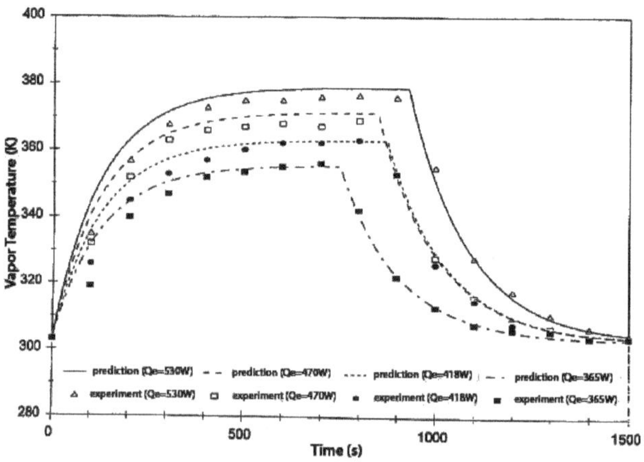

Fig. 35 Comparisons of predicted and measured transient vapor temperatures (Zuo and Faghri, 1998).

A more detailed two-dimensional transient numerical analysis of flat and cylindrical screen wick heat pipes was performed with no empirical correlations by Rice and Faghri (2007), while including the flow in the wick. Single and multiple heat sources were used as well as constant, convective, and radiative heat sinks. The numerical model by Rice and Faghri (2007) does not fix the internal pressure references by a point, but allows it to rise and fall based on the physics of the problem. Accordingly, the capillary pressure needed in the wick to drive the flow was obtained for various heating configurations and powers. These capillary pressures, in conjunction with an analysis that predicts the maximum capillary pressure for a given heating load, were used to determine the dry-out limitations of a heat pipe.

Special considerations need to be accounted for when considering the interface between the wick and the vapor region because the vapor interacts with the liquid in the wick's pores and the solid wick itself. Figure 36 is a schematic of the vapor/wick interfacial region.

Transient models with couplings between the various layers (wall, wick, and vapor) have successfully predicted the performance of heat pipes under all operating conditions for porous wick structures by neglecting the curvature effect at the liquid vapor interface. As heat pipe size or wick thickness decreases in size in the micro ranges, mainly for high thermal conductivity wicks, details of transport at the liquid vapor interface may become important in determining the capillary limit. Several efforts (Tournier and El-Genk (1994), Rice and Faghri (2007), and Ranjan et al. (2011)) have been made to include the effect of varying the radius of the curvature of the liquid meniscus formed in the wick pores for a screen mesh wick in the full numerical simulation of heat pipes for predicting the capillary limit.

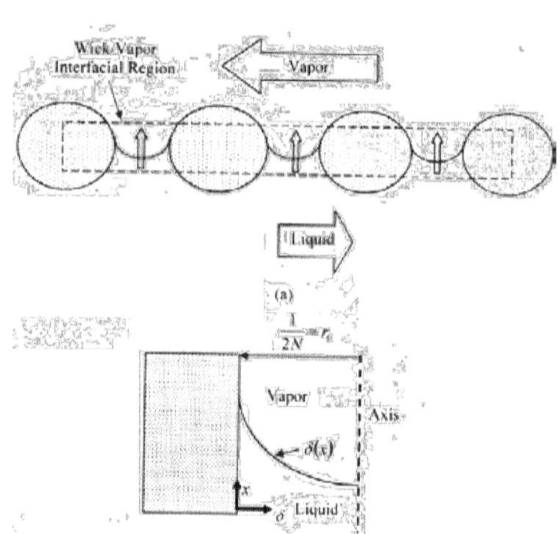

Fig. 36 Schematic of (a) a wick/liquid/vapor interfacial region and (b) an idealized single pore (Rice and Faghri, 2007).

Frozen Start Up

In some applications, heat pipes are required to transfer energy with the working fluid initially in the solid phase. As heat transfer progresses, the frozen working fluid melts, and the heat pipe undergoes a transition to normal operation. To fully understand frozen start up, significant efforts have been made toward numerical simulation of this process.
Early frozen start up research classified the frozen start up process into six periods, as discussed earlier (Jang *et al.*, 1990). These stages describe the condition of the working fluid as solid, mushy, or liquid, and the condition of the vapor flow as free molecular, partially continuum, continuum transient, and continuum steady. In the mathematical formulation by Jang et al. (1990), the heat transfer through the free molecular vapor flow was neglected and the frozen start up was solved in the wick and wall for the first two stages.

During the early start up period, the vapor density in the heat pipe vapor space is extremely low and partly loses its continuum characteristics. This condition is referred to as rarefied vapor flow. During the early stage of start up, when the melting interface has reached the wick-vapor interface, the liquid-vapor interface cannot be considered to be adiabatic. Otherwise, there would be no molecular vapor accumulation in the vapor space and the vapor would never reach the continuum flow state. The rarefied vapor flow was simulated by a self-diffusion model (Cao and Faghri, 1993b), where self-diffusion refers to the interdiffusion of particles of the same species due to the density gradient. This method has the advantage of modeling the heat transfer through the rarefied vapor flow as coupled to the phase change of the working fluid in the wick. However, this model is only valid for the early stages of frozen start up.

A complete numerical simulation of frozen start up was completed by Cao and Faghri (1993c) by combining the rarefied self-diffusion model of Cao and Faghri (1993b) with the continuum transient model of Cao and Faghri (1990). The model completely described frozen start up, including the effects of conjugate heat transfer within the wall. Heat transfer and fluid flow in the wick coupled to the vapor flow were simulated. This methodology was applied to simulate the multiple-evaporator high-temperature sodium/stainless-steel heat pipe studied by Faghri et al. (1991a) shown in Fig. 37(a) and the single evaporator sodium heat pipe of Ponnappan (1990) as shown in Fig. 37(b). The comparison with the experimental data is excellent, and the location and progression of the hot front is closely simulated as a function of time.

As mentioned previously, it may not always be possible to initiate frozen start up of a heat pipe. To address this difficulty, Cao and Faghri (1992) derived a closed-form analytical expression to describe heat pipe operation from frozen start up to continuum steady state. Also, a criterion was derived for the frozen start up limitation. The analytical model was derived under the assumption that the temperature profile at any time can be represented by three linear temperature approximations: one each in the hot and cold section of the heat pipe and one across the hot-cold front. The results of the model are shown in Fig. 38(a) when compared with the experimental frozen start up profiles of Faghri et al. (1991a), and in Fig. 38(b) when compared with the data of Ponnappan (1990).

It was determined that the frozen start up of a high-temperature heat pipe depends on several factors, such as the difference between the melting temperature of the working fluid and the ambient temperature, the liquid density of the working fluid, the porosity of the wick, and the physical dimensions of the heat pipe. These parameters were combined into a nondimensional parameter, which is a measure of the frozen start up ability of the heat pipe. This frozen start up limit (FSL) was validated using different experimental cases. Determination of the frozen start up limitation is one of the most significant contributions to frozen start up heat pipe analysis in recent years. The limit provides a fundamental design tool when considering heat pipe operation from a frozen state (Cao and Faghri, 1992).

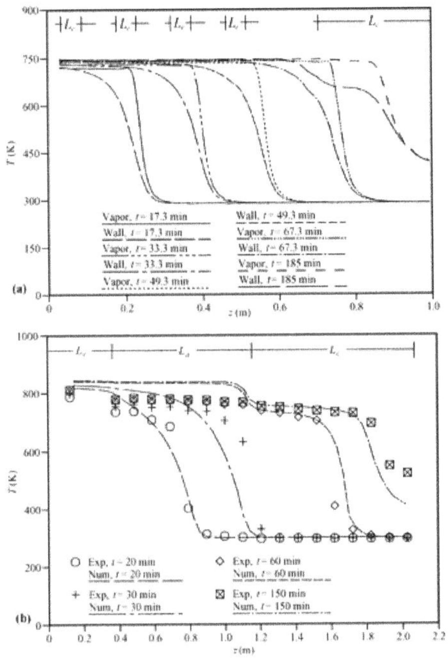

Fig. 37 Wall temperature prediction for frozen start up by Cao and Faghri (1993c) compared with the experimental data of (a) Faghri et al. (1991a); (b) Ponnappan (1990) (Cao and Faghri, 1993c).

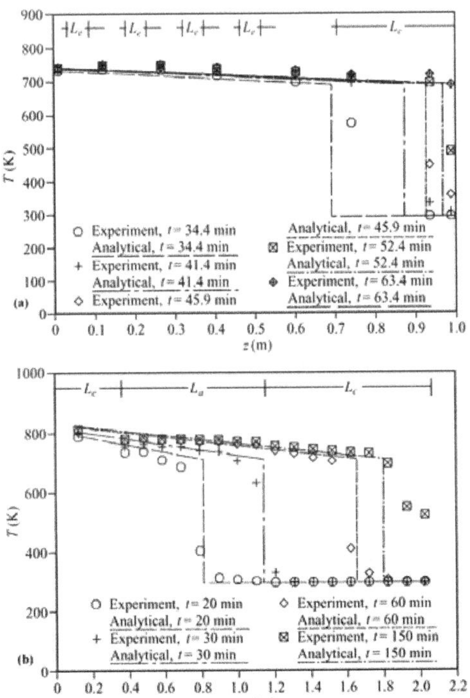

Fig. 38 Analytical wall temperature prediction for frozen start up by Cao and Faghri (1992) compared with experimental data of a (a) Faghri et al. (1991a); (b) Ponnappan (1990) (Cao and Faghri, 1992).

The aforementioned 2D models were quite successful and accurately predicted the frozen startup of low and high temperature heat pipes under various operating conditions by including both 2D free molecular and continuum vapor core regimes and freezing and melting phenomena by assuming flat liquid-vapor interface. Other frozen startup simulation efforts (Hall et al. (1994) and Tournier and El-Genk (1996)) were made by assuming 1-D vapor flow regimes, while accounting for the local interfacial radius of the curvature of the liquid meniscus.

Axially Grooved Heat Pipes

In the modeling of axially-grooved heat pipes (AGHP), the fluid circulation should be considered along with the heat and mass transfer processes during evaporation and condensation. Khrustalev and Faghri (1995a) have developed a mathematical model for low-temperature axially grooved heat pipes. The Khrustalev and Faghri (1995b) model accounts for the heat transfer in the microfilm region, which is of significant importance. Khrustalev and Faghri (1995a) modeled the heat transfer through the liquid film and the fin between the grooves (Fig. 39) in both the evaporator and condenser with respect to the disjoining pressure, interfacial thermal resistance and surface roughness, which was not accounted for by previous investigators. In this model, heat transfer processes in the heat pipe container and working fluid were considered to be one-dimensional in the radial direction, thereby neglecting the axial heat conduction. It was also assumed that no puddle flow occurred in the AGHP, no part of the condenser was blocked by the liquid working fluid, the liquid was distributed uniformly between the grooves, and the fluid flow along the axis was related to the capillary potential gradient, as described by the main radius of curvature of the liquid in a groove.

The numerical model is an iterative mathematical procedure which involved the following problems:

1. Heat transfer in the evaporating film on a rough surface.
2. Heat transfer in the condensate film on the fin top surface.
3. Heat conduction in a metallic fin and liquid meniscus.
4. Fluid circulation in the AGHP.

The first three problems have been presented in detail in by Khrustalev and Faghri (1995b). These three interconnected problems are included in the mathematical procedure along with the fluid circulation equations presented by Khrustalev and Faghri (1995a).

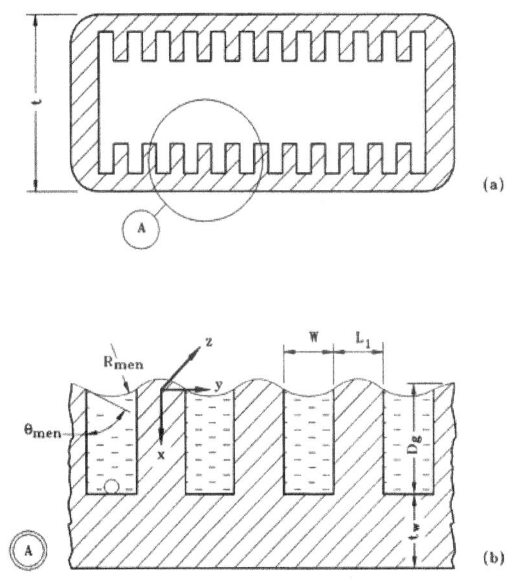

Fig. 39 Flat axially-grooved heat pipe cross-sections.

Khrustalev and Faghri (1995a) compared their numerical results with the experimental data by Schlitt et al. (1974), who studied an AGHP with the following geometry: $L_t = 0.914$ m, $L_c = 0.152$ m, $0.15 \leq L_e \leq 0.343$ m, $W = 0.610$ mm, $D_g = 1.02$ mm, $L = 0.430$ mm, $R_v = 4.43$ mm, $R_o = 7.95$ mm, $\gamma = 0$, $N = 27$, and $t_t = 0$. The working fluids were ammonia and ethane, and the container material thermal conductivity was assumed to be $k_w = 170$ W/(m-K). The longitudinal distribution of the meniscus contact angle, which is influenced by the inclination angle and the fluid pressure variation (Δp), is shown in Fig. 40(a) for ammonia (L_e = 0.343 m, T_v = 250 K). For positive values of the inclination angle φ (when the condenser end is elevated) the points of minimum and maximum liquid surface curvature (the so-called "dry" and "wet" points) were shifted from the ends of the heat pipe towards the adiabatic section. As a result, the meniscus contact angle distributions in the heat loaded sections were even more uniform than those in the horizontal case. Note that the meniscus angles in the condenser differ very slightly from the maximum value, and their values in the evaporator can be almost the same as that in the condenser for positive φ. The corresponding longitudinal distributions of the local heat transfer coefficients in the evaporator and condenser and also wall and vapor temperature variations are shown in Figs. 40(b) and 40(c). The local heat transfer coefficient in the middle of the condenser was about two times larger than that at the entrance or end cap, which resulted in the external wall temperature variation shown in Fig. 40(c). In the evaporator section, the variation of the local heat transfer coefficient was weaker (for heat loads which are not close to the maximum), so the wall temperature profile was very smooth. The temperature drop along the vapor flow was less than 0.01 K.

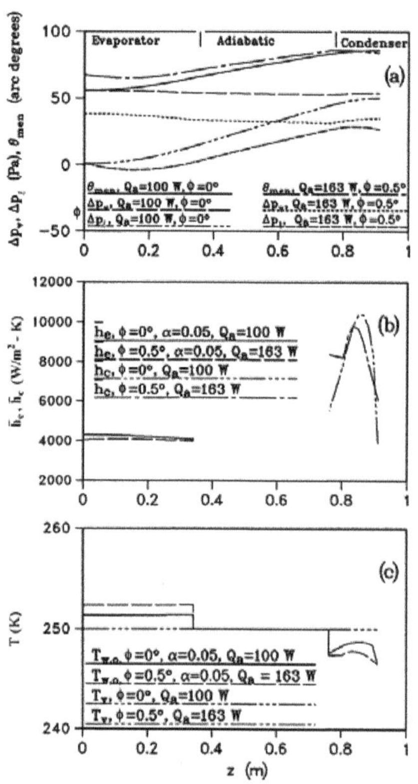

Fig. 40 Performance characteristics of the ammonia-Al heat pipe (T_v = 250 K): (a) Meniscus contact angle and fluid pressure; (b) Local heat transfer coefficients; (c) Wall and vapor temperatures (Khrustalev and Faghri, 1995a).

Advances in modeling of enhanced flat miniature heat pipes with capillary grooves were reviewed by Faghri and Khrustalev (1997). Lefèvre et al., (2010) extended the model by approximating the temperature along the heat pipe using different thermal resistances in their nodal model. The axial heat conduction cannot be neglected when the heat pipe wall is thick and/or the operating temperature is in the intermediate or high range. Do et al. (2008) extended the model by Khrustalev and Faghri (1995a; 1995b) to include the axial heat conduction along the heat pipe wall. The wall temperature profile and maximum heat transport rate obtained from the model developed by Do et al. (2008) are compared in Figure 41 with the experimental data of Hopkins et al. (1999) for a flat miniature heat pipe with a micro rectangular axial grooved wick structure.

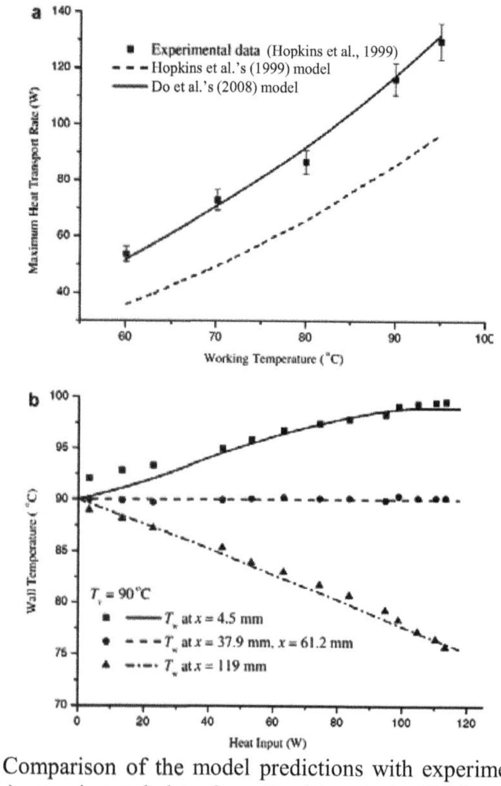

Fig. 41 Comparison of the model predictions with experimental data ((symbols) experimental data from Hopkins et al. (1999) and (lines) numerical simulation results from the Do et al. (2008) model): (a) maximum heat transport rate, and (b) wall temperature. (Adopted from Do et al. (2008))

Thermosyphons

When applying the Nusselt theory to a closed, two-phase thermosyphon, a variable vapor condensation rate must be considered. This approach was taken by Spendel (1984), who modeled a conventional thermosyphon condenser section with a two-dimensional incompressible formulation in the vapor space and an application of the Nusselt theory for the falling liquid film. With this model, Spendel (1984) performed a limited parametric study on the importance of including the interfacial shear stress and vapor pressure drop in determining the falling film thickness and the resulting Nusselt number for the thermosyphon. Based on these effects, the local Nusselt number in the condenser of a thermosyphon can vary as much as 60% from the approximate formulation derived by Nusselt.

An examination of conventional and concentric annular thermosyphons was performed by Faghri et al. (1989a), by which an improved flooding limit formulation was determined. Furthermore, in this study, the effects of the empirically obtained interfacial shear stress of the counterflowing vapor on the falling liquid film were considered in the condenser section alone.

Harley and Faghri (1994b) coupled a general quasi-steady Nusselt-type solution of the falling film with the complete two-dimensional vapor solution to simulate the transient two-dimensional behavior of thermosyphons with variable properties. This model is significantly different from the previous models in that it simulates the entire thermosyphon, rather than only the condenser section. A two-dimensional transient formulation for the vapor is coupled with unsteady heat conduction in the pipe wall. Furthermore, the quasi-steady falling condensate film in the condenser, adiabatic, and evaporator sections are simulated by accounting for the variable vapor condensation rate, interfacial shear stress, and vapor pressure drop. The numerical results of Harley and Faghri (1994b) were compared with the experimental data of Mingwei et al. (1991).

The transient axial variations of the outer wall and centerline vapor temperatures are shown in Fig. 42(b), where the agreement with the steady-state experimental data of Mingwei et al. (1991) is quite good. The transient centerline axial vapor velocities for this low-temperature case are small and the axial pressure drop is negligible. The low axial vapor velocity is due in part to both the relatively high vapor density and latent heat of water and the low heat flux input to the evaporator. Obviously, compressibility is not a factor for the low-temperature thermosyphon because the maximum Mach number is much less than 0.3. The negligible axial pressure variation is typical of low-temperature two-phase heat transfer devices due to the low vapor velocities.

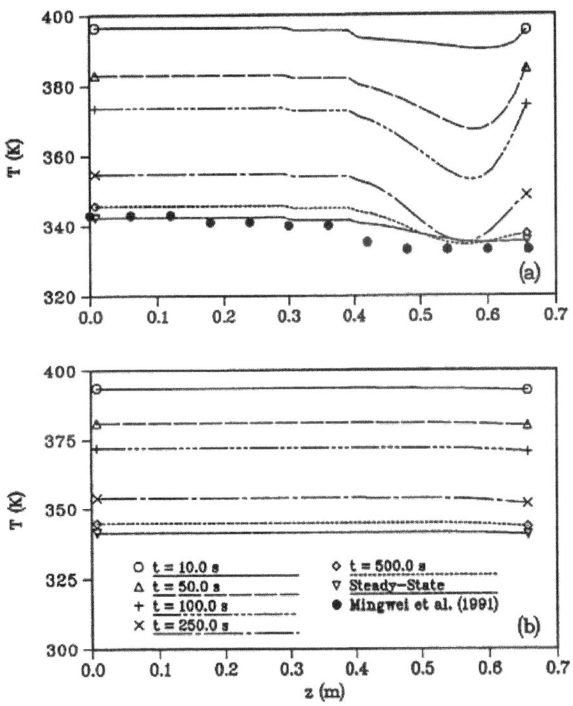

Fig. 42 Transient temperature profiles for the low-temperature thermosyphon experimentally studied by Mingwei et al. (1991): (a) transient outer wall temperature profiles; (b) transient centerline vapor temperature profiles (Harley and Faghri, 1994b).

The condensation process is affected by the presence of a small amount of noncondensible gas in a thermosyphon. During normal operation, the noncondensible gas is swept to the condenser by the vapor, which partially blocks the condenser section. This blockage reduces the area available for the heat transfer. Gas-loaded thermosyphons were studied by Hijikata et al. (1984), where a steady two-dimensional model that neglected heat conduction through the wall and falling liquid film was developed. However, since only the condenser section of the thermosyphon was modeled, a complex system of boundary conditions with several different empirical parameters was required. The behavior of the noncondensible gas was described by a modified two-dimensional steady-state species equation.

A more complete model of the condenser section of a gas-loaded thermosyphon was developed by Kobayashi and Matsumoto (1987). This two-dimensional steady-state model accounts for diffusion across the vapor-gas interface but neglects the falling film and axial heat conduction through the pipe wall.

Peterson and Tien (1989) extended the steady-state model of Hijikata et al. (1984) by accounting for the effect of heat conduction in the pipe wall. Due to the integral simplification of the governing equations, however, this model can only be used in cases where heat conduction in the wall is either dominant or negligible. Furthermore, this formulation only models the condenser section where all properties were assumed to be constant.

Harley and Faghri (1994a) coupled a general quasi-steady Nusselt-type solution of the falling film with the complete two-dimensional vapor-gas formulation to simulate the transient two-dimensional behavior of gas-loaded thermosyphons with variable properties. This formulation is significantly different from the previous models in that it simulates the entire thermosyphon rather than only the condenser section, using a two-dimensional transient formulation for the vapor and gas that is coupled to unsteady heat conduction in the pipe wall. Furthermore, this formulation models the quasi-steady falling condensate film in the condenser, adiabatic, and evaporator sections that accounts for the variable condensation rate, interfacial shear stress, and vapor pressure drop.

Zuo and Gunnerson (1995) presented a numerical model of inclined thermosyphon performance. Liquid-vapor interfacial shear stress and the effects of working fluid inventory at various inclination angles were included in the model. They analyzed limiting mechanisms of the dryout and flooding and demonstrated that the model is capable of predicting the performance of an inclined thermosyphon. A one-dimensional steady-state mathematical model describing natural circulation two-phase flow in a thermosyphon with a tube separator was developed by Lin and Faghri (1997b). Void fraction distributions along the thermosyphon were obtained under various operating conditions. Liquid fill ratios for steady flow were suggested. Lin and Faghri (1998b) numerically studied the hydrodynamic stability of natural circulation two-phase flow in a high performance thermosyphon with tube separator. The simulation results showed that the operating temperature, the heat rate and thermosyphon inclination angle have significant influence on the flow instability.

El-Genk and Saber (1999) developed a one-dimensional, steady-state model to determine the operation envelopes of closed, two-phase thermosyphons in terms of dimensions, type, vapor temperature of working fluid, and power throughput. The thermosyphon operation-envelope was an enclosure with three critical boundaries, related to dryout, boiling, and flooding limits. The calculations showed that an increase in the thermosyphon diameter, evaporator length, or vapor temperature expanded the operation-envelope, while an increase in the length of either the condenser or the adiabatic section only slightly changed the envelope's upper and lower boundaries. Pan (2001) presented a condensation model for a two-phase, closed thermosyphon by considering the interfacial shear stress due to the mass transfer and interfacial velocity. The relative velocity ratio and the momentum transfer factor greatly affect the condensation heat transfer in the thermosyphon. A sub-flooding limit was proposed to capture the interaction between the condensation and evaporation in the thermosyphon.

Jiao et al. (2008) developed a model to investigate the effect of the filling ratio on the distribution of the liquid film and liquid pool. The total heat transfer rate of the liquid pool, including natural convection and nucleate boiling, was calculated by combining their effective areas and heat transfer coefficients. The correlation for effective area was obtained based on experimental results. The range for the filling ratio was proposed for the steady and effective operation of the thermosyphon based on analysis and comparison. The effects of heat input, operating pressure, and geometries of the thermosyphon on the range of the filling ratio were also discussed. Jiao et al. (2012) further developed the model presented by Jiao et al. (2008) to investigate the effect of the filling ratio on dryout, flooding and boiling limits. The experiments were conducted using nitrogen as working fluid and the experimental results were compared with the calculations. The maximum filling ratio was introduced, beyond which the heat transfer performance could be declined due to the accumulation of the liquid in condenser end. The sensitivity of the operation range to the operating pressure and geometries were also analyzed.

Rotating Heat Pipes

The rotating heat pipe concept was first proposed by Gray (1969), who showed that such a rotating heat pipe can transfer significantly more heat than a similar stationary heat pipe.

Faghri et al. (1993) performed a rotating heat pipe vapor flow analysis by using a two-dimensional axisymmetric formulation to determine the influence of the rotation rate on the vapor pressure drop and interfacial shear stresses. They found that pressure drop and shear stress correlations for conventional heat pipes do not apply in rotating heat pipes as the vapor flow profiles are significantly different from the patterns seen in stationary heat pipes. However, that model did not consider heat transfer in the pipe wall or the counterflowing liquid film. Harley and Faghri (1995) developed a detailed transient numerical simulation of rotating heat pipes, including the thin liquid condensable film, vapor flow, and heat conduction in the pipe wall.

The complete rotating heat pipe simulation was done by Harley and Faghri (1995) for the heat pipe experimentally studied by Daniels and Al-Juimaily (1975) for a heat input of Q = 800 W and a rotational rate of ω = 600 rpm. The dimensions of this heat pipe were L_t = 0.325 m, L_e = 0.05 m, L_a = 0.123 m, L_c = 0.152 m, R_v = 0.02125 m, R_o = 0.02650 m, and α = 2 deg. The heat pipe wall was copper and the working fluid was Freon-113. A true comparison with actual data was impossible due to the assumption for the outer heat transfer coefficient. An approximate heat transfer coefficient (600 W/m^2-K) was determined from an examination of the wall temperature profile given by Daniels and Al-Juimaily (1975). Because both the wall temperature and total heat input were given, the assumption of an ambient temperature is needed to obtain the external convective heat transfer coefficient.

The axial variation of the outer wall temperature is shown in Fig. 43(a). This assumed heat transfer coefficient resulted in the isothermal operation of the rotating heat pipe, which can be seen in Fig. 43(b), where the centerline vapor temperature is nearly constant along the length of the rotating heat pipe. Daniels and Al-Jumaily (1975) determined the vapor temperature with two pressure transducers located on the centerline at the two end caps, where the vapor was assumed to be saturated. As can be seen in Fig. 43(b), the assumption of a saturated vapor is valid because the vapor velocity is low enough to neglect compressibility effects.

Lin and Faghri (1997a) investigated the flow behavior and the related heat transfer characteristics of stratified flow in axially rotating heat pipes with cylindrical and stepped wall configurations using theoretical and semi-empirical models. The predicted results and experimental data showed a good agreement. Lin and Faghri (1997c) presented a mathematical model of the hydrodynamic performance of a rotating miniature heat pipe (RMHP) with a grooved inner wall surface. Influences of operating temperature, rotational speed, and liquid-vapor interfacial shear stress on the maximum performance and optimum liquid fill amount were discussed. Pressure drops of the axial liquid flow and vapor flow were demonstrated.

The region of hysteretic annular flow in rotating stepped wall heat pipes was experimentally determined by Lin and Faghri (1998a). A model for predicting the condensation heat transfer coefficient was also proposed and the theoretical results were compared with experimental data. Lin and Faghri (1999) developed a theoretical model that describes the evaporating film flow in a rotating miniature heat pipe with an axial triangular grooved internal surface. The theory of thin liquid film vaporization heat transfer was used to predict the evaporation heat transfer in the micro region. The effects of disjoining pressure, surface tension, and centrifugal force on the flow are discussed. It was concluded that the influence of rotational speed on the evaporation heat transfer in the micro region can be neglected. Therefore, the heat transfer analysis in the micro region at zero rotational speed can be applied to the case of rotational operation. Cao (2010) developed numerical and analytical analysis for miniature high temperature rotating heat pipes for applications in gas turbine cooling.

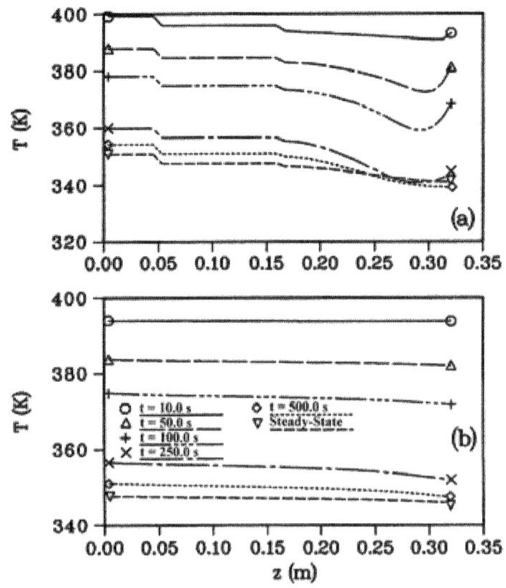

Fig. 43 Numerical simulation of the low-temperature rotating heat pipe: (a) transient outer pipe wall temperatures; (b) transient centerline vapor temperatures (Harley and Faghri, 1995).

All of the aforementioned analyses focused on a pure vapor flow with no noncondensible gases in the rotating heat pipe core. Harley and Faghri (2000) accounted for the noncondensible gas, the vapor pressure drop, the thin liquid film, the interfacial shear stress, and the heat conduction in the pipe wall of a rotating heat pipe without the need for specifying condenser inlet conditions. The problem is solved as a conjugate coupled heat transfer problem. A parametric study was also presented to determine the influence of rotational speed, heat input and output, and the mass of noncondensible gas on a rotating heat pipe.

Loop Heat Pipes

Various numerical and analytical models have been developed in order to understand the fundamental mechanisms of thermal and hydrodynamic couplings between LHP components, as well as to quantitatively predict the LHP operational characteristics. Most of the models are based on energy and pressure balances written for each component of the LHP. Some examples of steady-state LHP modeling studies are Maydanik et al. (1994), Kaya and Hoang (1999), Hoang and Kaya (1999), Kaya and Ku (1999), Muraoka et al. (2001), Kaya and Ku (2003), Hamdan (2003), Chuang (2003), Furukawa (2006), Kaya and Goldak (2006), and Launay et al. (2007c; 2008). Transient LHP modeling efforts include Cullimore and Bauman (2000), Hoang and Ku (2003), Launay et al. (2007a; 2007b), Kaya et al. (2008), Chernysheva and Maydanik (2008), and Khrustalev (2010).

LHP transient modeling of two-phase systems with capillary evaporators using a Thermal Desktop was pioneered by Cullimore and Bauman (2000) and extended by Khrustalev (2010) to upgraded loop heat pipe systems with complex radiators, multiple components, and varying environmental conditions. Launay et al. (2008) proposed separate closed-form solutions for two distinct LHP operating modes: the variable conductance mode (VCM), and the fixed conductance mode (FCM). These modes are exhibited on an experimentally observed LHP operating curve, depicting the operating temperature (usually represented by the compensation chamber temperature or evaporator temperature) versus the heat input. These models are also linked to the distribution of the condenser heat transfer area: one part is used for condensation, and the other is for liquid subcooling.

In terms of the closed-form solutions, the LHP operating temperature can be easily calculated as a function of the heat input. In Figs. 44 and 45, predictions by full numerical simulation and simplified closed-form solutions, are compared with experimental measurements for two LHP designs (Chuang, 2003; Boo and Chung, 2004) using acetone and ammonia as the working fluids, respectively. It appears that the relative difference between the full numerical simulation and closed-form solutions peaks at the transition between the VCM and FCM; however, the maximum difference is less than 15%. The general agreement of the modeling predictions with the experimental data validates both the model and closed-form solutions as useful tools for possible LHP designs. A recent review of the fundamentals, operation, and design of loop heat pipes was given by Ambirajan et al. (2012).

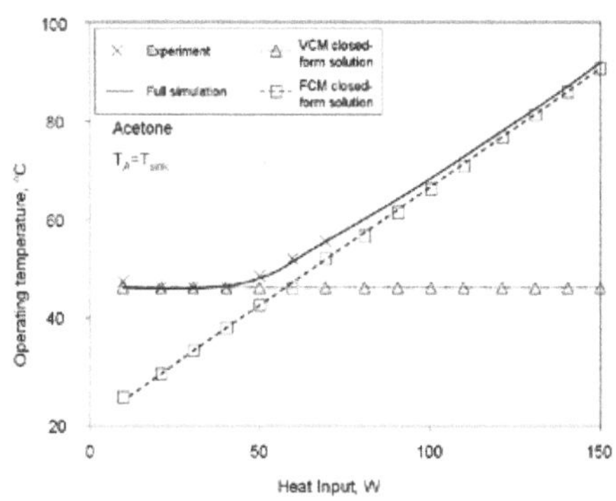

Fig. 44 Comparison of LHP modeling predictions (Launay *et al.*, 2008) with experimental results (Boo and Chung, 2004) for acetone as the working fluid (Launay et al., 2008).

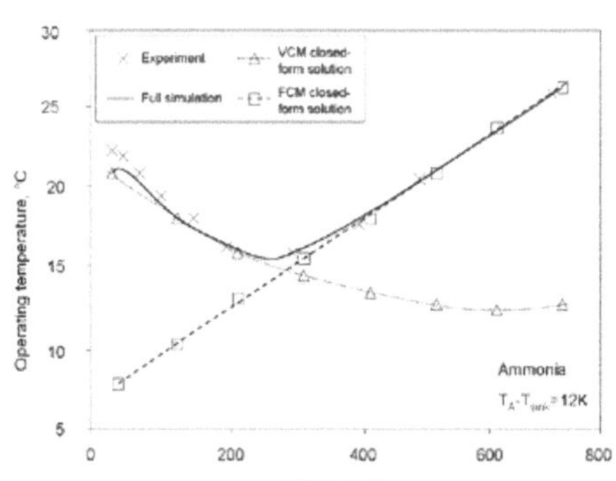

Fig. 45 Comparison of LHP modeling predictions (Launay *et al.*, 2008) with experimental results (Chuang, 2003) for ammonia as the working fluid (Launay et al., 2008).

Capillary Pumped Loop Heat Pipes

There are three approaches related to the analysis of CPLs. The first approach is to examine the heat transfer without including the fluid flow analysis in the wick or vapor regions of the evaporator (Kiper *et al.*, 1990). Since an analytical approach was used for solving the energy equation instead of a numerical approach, approximations were made in the solution procedure. It should be noted that the fluid flow problem is of significant importance in the evaporator and no meaningful conclusions can be made without a combined fluid flow and heat transfer analysis. This approach also included a lumped analysis of the evaporator and subcooler, in which an exponential temperature profile was assumed in the analysis. A second approach included a detailed and accurate analysis of the evaporator component for steady and transient performance, based on solving the complete differential forms of the momentum and energy equations (Cao and Faghri, 1994b; 1994c). This methodology is the most appropriate technique for analyzing the CPL system. The third approach is related to an overall system simulation using existing semi-empirical correlations under 1 g conditions to calculate the pressure drop and heat transfer coefficients in different parts of the CPL (Kroliczek *et al.*, 1984; Ku *et al.*, 1986a; 1986b; 1987a; 1987b; Chalmers *et al.*, 1988; Ku *et al.*, 1988; Benner *et al.*, 1989; Schweickart and Buchko, 1991; Ku, 1993). Some analytical studies concerning the complex transient phenomena of startup have been conducted (Cullimore, 1991).

Micro Heat Pipes

Khrustalev and Faghri (1994) developed a detailed mathematical model which allows the examination of the heat and mass transfer processes in a micro heat pipe (MHP). The model gave the distribution of the liquid in a micro heat pipe and its thermal characteristics as a function of the liquid charge and the applied heat load. The liquid flow in the triangular-shaped corners of a micro heat pipe with a polygonal cross section is considered by accounting for the variation of the curvature of the free liquid surface and the interfacial shear stresses due to liquid-vapor interaction. The predicted results are compared to existing experimental data. The importance of the liquid fill, minimum wetting contact angle, and the shear stresses at the liquid-vapor interface in predicting the maximum heat transfer capacity and thermal resistance of the micro heat pipe is demonstrated.

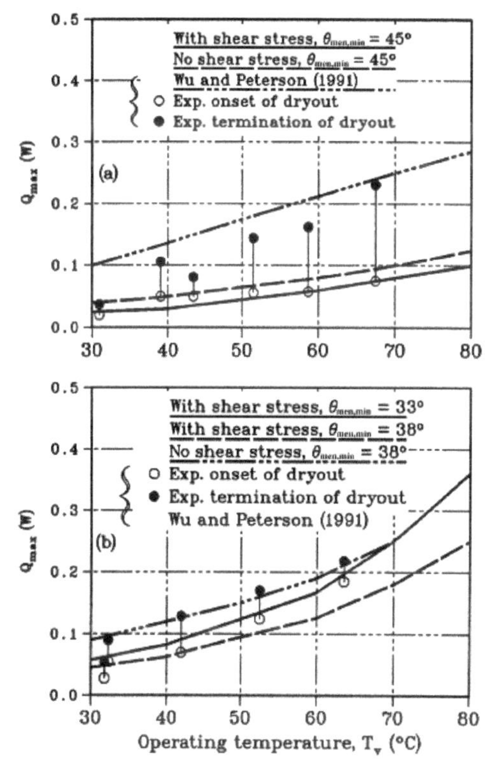

Fig. 46 Maximum heat transfer versus operating temperature: (a) Copper-water MHP; (b) Silver-water MHP (Khrustalev and Faghri, 1994).

The comparison of the numerical results (Khrustalev and Faghri, 1994) and the experimental data reported by Wu and Peterson (1991) for the maximum heat transfer capacity are shown in Fig. 46 for micro heat pipes with copper and silver casings. While the data for the minimum contact angles are contradictory and can be influenced by many physical factors, the numerical results are presented for $\theta_{men,min}$= 33°, 38° and 45°. The experimental results of the onset of dryout and the present numerical prediction show a good agreement. Neglecting the shear stress at the free surface of the liquid due to liquid-vapor frictional interaction ("no shear stress" in Fig. 46) can lead to an overestimation of the maximum heat transfer capacity. The shear stress at the free liquid surface influences the liquid distribution along the heat pipe, which can result in an increase of the liquid locking of the condenser end, compared to the case of neglecting this shear stress.

Sartre et al. (2000) and Suman and Kumar (2005) extended Khrustalev and Faghri's model (1994) to include heat conduction in the wall to simulate micro heat pipes of polygonal shapes. Wang and Peterson (2002) and Launay et al. (2004b) have improved Khrustalev and Faghri's model (1994) to simulate arrays of micro heat pipes made of several alluminim wires bounded between two aluminum sheets.

Pulsating Heat Pipes

Since slug flow is the primary flow pattern in PHPs, most existing efforts on modeling have focused on slug flow. Miyazaki and Akachi (1996) proposed a simple analytical model of self-exciting oscillation based on an oscillating feature observed during experimentation. The reciprocal excitation of pressure oscillation due to changes in the heat transfer rate caused by the oscillation of the void fraction was investigated. Oscillation of the void fraction was out of phase with the pressure oscillation by $-\pi/2$. This model indicates that an optimal charge ratio exists for a particular PHP. If the charge ratio is too high, the PHP will experience a gradual pressure increase followed by a sudden drop. Insufficient charging will cause chaotic pressure fluctuation. However, proper charging will generate a symmetrical pressure wave.

Miyazaki and Akachi (1998) derived the wave equation of pressure oscillation in the PHP based on the self-excited oscillation, in which the reciprocal excitation between the pressure oscillation and void fraction was accounted for. Miyazaki and Arikawa (1999) investigated the oscillatory flow in the PHP and then measured the wave velocity, which fairly agreed with the predictions of Miyazaki and Akachi's model (1998). Hosoda et al. (1999) reported a simplified numerical model of a PHP, in which temperature and pressure were calculated by solving the momentum and energy equations for two-dimensional, two-phase flow. However, the thin liquid film that surrounds a vapor plug on the tube wall and the friction between the tube and the working fluid were neglected. Experimental results were used as initial conditions for the model. The numerical results for pressure in the PHP were higher than the experimental results; however, the model showed that the propagation of vapor plugs induced fluid flow in the capillary tubes.

Zuo et al. (Zuo *et al.*, 1999; 2001) attempted to model the PHP by comparing it to an equivalent single spring-mass-damper system, with the parameters of the system affected by heat transfer. Zuo et al. (2001) showed that the spring stiffness increases with increasing time, and therefore, the amplitude of oscillation must decrease with increasing time; this is in contradiction with steady oscillations observed in PHP operation. Wong et al. (1999) modeled an open-loop PHP by considering it as a multiple spring-mass-damper system with the flow modeled under adiabatic conditions for the entire PHP. A sudden pressure pulse was applied to simulate local heat input into a vapor plug.

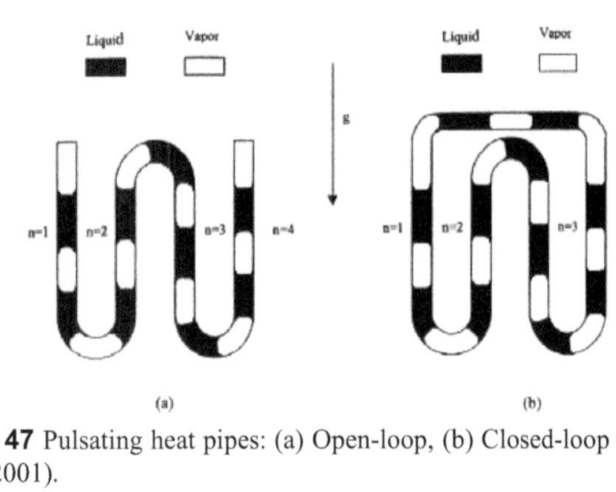

Fig. 47 Pulsating heat pipes: (a) Open-loop, (b) Closed-loop (Shafii *et al.*, 2001).

Shafii et al. (2001) developed a theoretical model to simulate the behavior of liquid slugs and vapor plugs in both closed- and open-loop PHPs with two turns (see Fig. 47). The model solved for the pressure, temperature, plug position and heat transfer rates. The most significant conclusion was that the majority of the heat transfer (95%) is due to sensible heat; not due to the latent heat of vaporization. Latent heat served only to drive the oscillating flow. Sakulchangsatjatai et al. (2004) applied Shafii et al.'s model (2001) to model closed-end and closed-loop PHPs as an oscillating two-phase heat and mass transfer in a straight pipe, while neglecting the thin liquid film between the vapor plug and the pipe wall.

Zhang et al. (2002) analytically investigated oscillatory flow in a U-shaped miniature channel-a building block of PHPs. A significant difference between this model and other mathematical models is the nondimensionalizing of the governing equations. Flow in the tube was described by two dimensionless parameters, the non-dimensional temperature difference and the evaporation and condensation heat transfer coefficients. It was found that both the initial displacement of the liquid slug and gravity had no effect on the amplitude and angular frequency of the oscillation. In addition, the amplitude and frequency of oscillation were increased by increasing the dimensionless temperature difference. The amplitude and frequency of oscillation were correlated to the heat transfer coefficients and temperature difference.

Zhang and Faghri (2003) investigated oscillatory flow in a closed-end pulsating heat pipe with an arbitrary number of turns (see Fig. 48). The results showed that for a PHP with few turns (less than 6), the amplitude and frequency of oscillation were independent of the number of turns. The motion of the vapor plugs was identical for odd numbered plugs once a steady state had been reached. Even numbered plugs also exhibited identical motion. Odd and even numbered plugs had the same amplitude, but were out of phase by a factor of π. As the number of turns increased above 6, the odd and even numbered plugs no longer showed identical oscillation. Each plug lagged slightly behind the next. Each plug remained out of phase by π, however. (see Fig. 49).

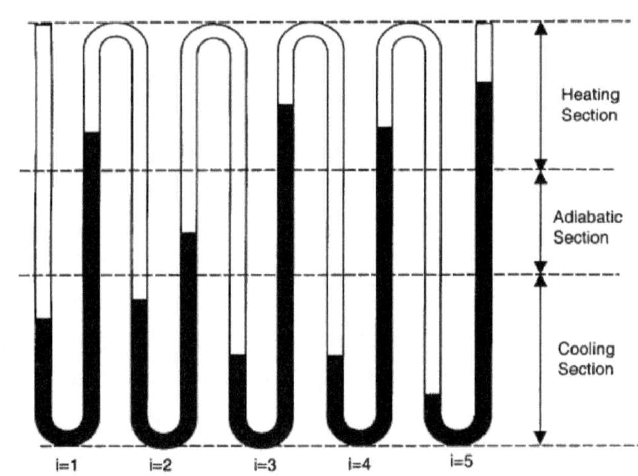

Fig. 48 Open-loop PHP with arbitrary turns (N=5) (Zhang and Faghri, 2003).

Dobson and Harms (1999) investigated a PHP with two open ends. The open ends were parallel and pointed in the same direction. These ends were submerged in water, while the evaporator section was coiled and attached to a float so that it remained out of the water. The evaporator was heated and the oscillatory fluid motion produced a net thrust. A numerical solution of the energy equation and the equation of motion for a vapor plug was presented to predict the plug's temperature, position and velocity. Oscillatory motion in the PHP generated a net average thrust of 0.0027N. Heat transfer due to sensible heat was not taken into account. Dobson (2004; 2005) proposed the use of open-ended PHP in conjunction with two check valves to pump water; however, the maximum attainable mass flow rates are on the order of mg/s, which is hardly enough to irrigate fields. An improved model for liquid slug oscillation that considered pressure difference, friction, gravity, and surface tension was also presented.

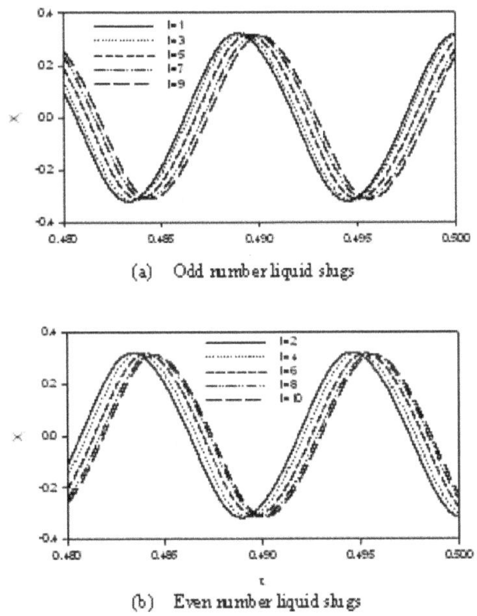

Fig. 49 Displacement of liquid slugs (N=10) (Zhang *et al.*, 2002).

Zhang and Faghri (2002) proposed models for heat transfer in the evaporator and condenser sections of a PHP with one open end by analyzing thin film evaporation and condensation. Heat transfer in the evaporator is the sum of evaporative heat transfer in the thin liquid film and at the meniscus. Heat transfer in the condenser is similarly calculated, and sensible heat transfer to the liquid slug is also considered. It is found that the overall heat transfer is dominated by the exchange of sensible heat, not by the exchange of latent heat. Shafii et al. (2002) further developed their earlier numerical model (Shafii *et al.*, 2001) by including an analysis of the evaporation and condensation heat transfer in the thin liquid film separating the liquid and vapor plugs. Both open- and closed-loop PHPs were considered, and each displayed similar results. As shown in Fig. 50, the total heat transfer is due mainly to the exchange of sensible heat (~95%). Total heat transfer slightly increases as the surface tension of the working fluid increases. The total heat transfer significantly decreases with decreased heating section wall temperature. Increasing the diameter of the tube results in a higher total heat transfer.

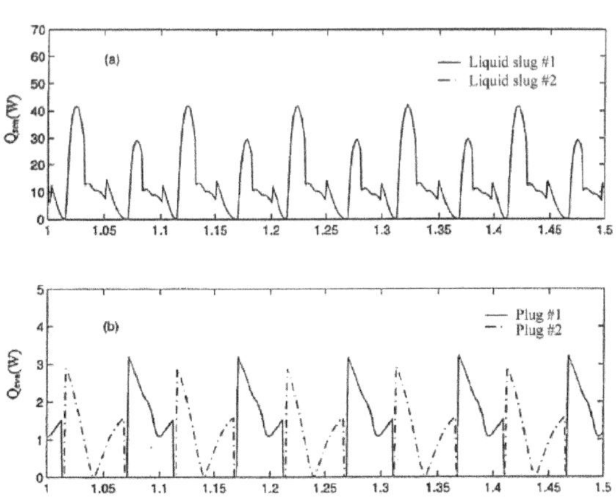

Fig. 50 Heat transfer rate: (a) sensible heat; (b) evaporative heat (Shafii *et al.*, 2002).

Liang and Ma (2004) presented a mathematical model describing the oscillation characteristics of slug flow in a capillary tube. In addition to the modeling of oscillating motion, numerical results indicated that the isentropic bulk modulus generates stronger oscillations than the isothermal bulk modulus. While it demonstrated that the capillary pipe diameter, bubble size, length and the number of unit cells determine the oscillation, the capillary force, gravitational force, and initial pressure distribution of the working fluid significantly affects the frequency and amplitude of oscillating motion in the capillary tube. Ma et al. (2002; 2006) analytically described the oscillating motion by performing force balances of the thermally driven, capillary, frictional, and elastic restoring forces on a liquid slug. Pressure differences between the evaporator and the condenser were related to the temperature difference between the evaporator and the condenser by the Clapeyron-Clausius equation. The temperature difference between the evaporator and the condenser of the PHP was utilized as a driving force of the oscillating motion. Ma et al. (2006) concluded that oscillating motion depends on the charge ratio, total characteristic length, diameter, temperature difference between the evaporation and condenser sections, and the working fluid and operating temperature. When compared to experimental results, the mathematical model under-predicted the temperature difference between the evaporator and condenser when compared to experimental results (Ma *et al.*, 2002).

Holley and Faghri (2005) presented a numerical model for a PHP with a sintered copper capillary wick and with flow channels that have different diameters. The effects of the varying channel diameters, inclination angles, and the number of parallel channels were presented. When a channel was of a smaller diameter, it induced circulation of the fluid, which in turn increased the heat load capability of the PHP. The modeled PHP performed better in bottom heat mode (smaller temperature differential) than in top heat mode. Varying the mean Nusselt number had little effect on the PHP performance. As the number of parallel channels increased, its heat load capability increased, and the PHPs sensitivity to gravity decreased.

Khandekar et al. (2002) used an Artificial Neural Network (ANN) to predict the PHP performance. The ANN was of the fully connected feed forward configuration, and was trained using 52 sets of experimental data from a closed-loop PHP. The ANN was fed the heat input and fill ratio of each data set and then calculated the effective thermal resistance of the PHP. The ANN model predicted the thermal performance for this type of PHP but neglected many parameters that affect PHP performance, including tube diameter, number of parallel channels, length of the PHP, inclination angle, and properties of the working fluid. If the ANN had more input nodes with which to consider these parameters, it would be a more effective model. Even then it would require considerably well-organized experimental data for the ANN to be effective.

Khandekar and Gupta (2007) modeled heat transfer in a radiator plate with an embedded PHP using a commercial package FLUENT. However, oscillatory flow and heat transfer of the PHP were not modeled in this investigation. The contribution of the PHP on the heat transfer in the radiator plate was considered using an effective thermal conductivity obtained from the experiment. Zhang and Faghri (2008) made a detailed review of pulsating heat pipe analysis. Khandekar et al. (2010) presented the experimental and theoretical methodologies to predict hydrodynamic characteristics in unidirectional two-phase Taylor bubble flows and recommended directions for further research on modeling of pulsating heat pipes.

Gas-Loaded Heat Pipes

Noncondensible gas-buffered heat pipes (NCHPs) offer isothermal operation for varying heat loads by changing the overall thermal resistance of the heat pipe. As the heat load increases, the vapor temperature and total pressure increase in the heat pipe. This increase in total pressure compresses the noncondensible gas in the condenser, thereby increasing the surface area available for heat transfer. This maintains a constant heat flux and temperature. NCHP operation can also be examined through the partial vapor pressure. As the heat input increases, the total pressure increases. However, because the total pressure is comprised of both the partial vapor and partial gas pressures, the increase in total pressure is accounted for by an increase in the partial gas pressure, maintaining a constant vapor pressure, and thus a constant vapor temperature.

Noncondensible gas-buffered heat pipes have been modeled previously with several different levels of approximation. The classic flat-front analytical model of Marcus and Fleishman (1970) neglected all diffusion across the vapor-gas interface. Later studies modeled one-dimensional steady diffusion (Edwards and Marcus, 1972) and one-dimensional transient diffusion (Shukla, 1981). Rohani and Tien (1973) studied the steady-state two-dimensional diffusion process in a noncondensible gas-buffered heat pipe. The importance of modeling radial diffusion in a NCHP was demonstrated where the noncondensible gas tends to accumulate at the liquid-vapor interface, which retards vapor condensation. However, Rohani and Tien (1973) neglected the effect of conjugate heat transfer through the wall as well as transient effects. The importance of the transient response of a NCHP was shown by Shukla (1981), as transient thermal overshoots were discovered. This study, however, did not include the effects of radial heat conduction in the wall or wick or the radial diffusion effects of the noncondensible gas.

Harley and Faghri (1994c) studied the two-dimensional, transient NCHP operation, including the effects of conjugate heat transfer through the wall, and is modeled through a solution of the general differential conservation equations. This procedure was used to simulate the high-temperature heat pipe experimentally studied both with and without noncondensible gases. The major advantage over previous models was that this model treats the noncondensible gas as a separate entity that is described by mass transport phenomena. Additionally, the energy transport through the wall is coupled to the transient operation of the heat pipe through the use of a conjugate solution technique. The complete behavior of the heat pipe, along with the location and two-dimensional shape of the noncondensible gas front, are modeled from the initial continuum-flow, liquid-state start up to steady-state conditions. The proposed model by Harley and Faghri (1994c) predicted the existing experimental data for the operation of high-temperature heat pipes with and without noncondensible gases.
The gas-loaded heat pipe experimentally studied by Ponnappan (1989) was simulated (Case 2), with results shown in Figs. 51 and 52. This case has a higher radiative emissivity than Case 1, which results in a decrease in the thermal resistance at the outer condenser surface. In the experiment performed by Ponnappan (1989), the emissivity was increased when the noncondensible gas was added, so that a near-constant operationg temperature (compared with Case 1) could be maintained. In Fig. 51, the wall and vapor temperatures decreased significantly in the condenser section due to the presence of the noncondensible gas. The gas density increased in the condenser during transient operation, as shown in Fig. 52(a). The steady-state wall temperature is in good agreement with the data by Ponnappan (1989).

A transient lumped heat pipe formulation for conventional heat pipes is presented, and the lumped analytical solutions for different boundary conditions at the evaporator and condenser are given by Faghri and Harley (1994). For high temperature heat pipes with a radiative boundary condition at the condenser, a nonlinear ordinary differential equation is solved. In an attempt to reduce computational demands, a transient lumped conductive model was developed for noncondensible gas-loaded heat pipes. The lumped flat-front transient model was extended by accounting for axial heat conduction across the sharp vapor-gas interface. The analytical solutions for conventional and gas-loaded heat pipes were compared with the corresponding numerical results of the full two-dimensional conservation equations and experimental data, showing a good agreement.

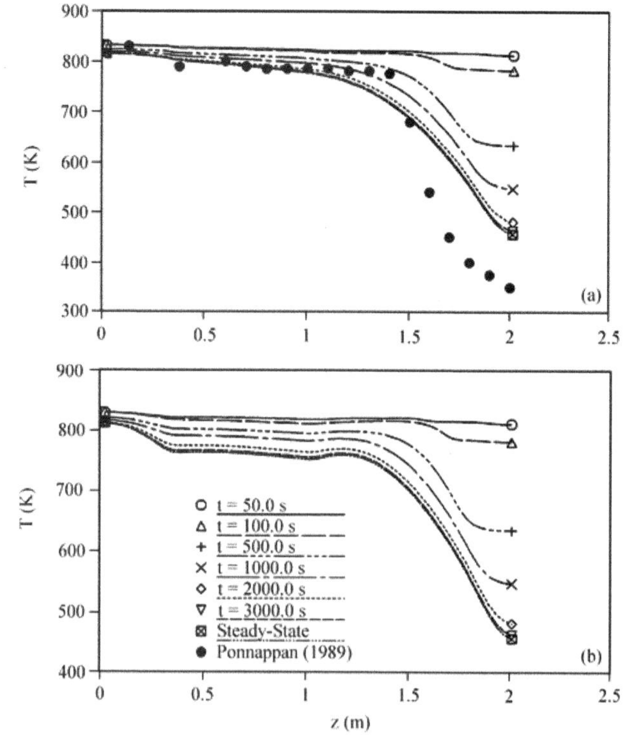

Fig. 51 Temperature profiles for the gas-loaded heat pipe with Q_{in} = 258 W (Case 2): (a) transient wall temperature profile; (b) transient centerline temperature profile (Harley and Faghri, 1994c).

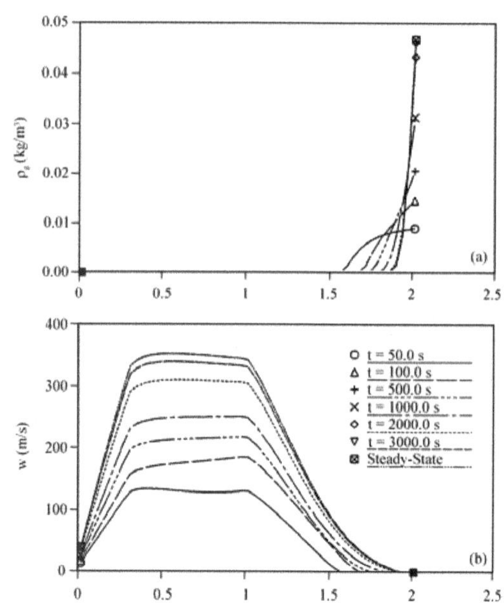

Fig. 52 Vapor-gas dynamics for the gas-loaded heat pipe with Q_{in} = 451 W (Case 2): (a) transient centerline gas density profile; (b) transient centerline axial velocity profiles (Harley and Faghri, 1994c).

11. HEAT PIPE APPLICATIONS

Heat pipes have been applied in many ways since their introduction in 1964 (Vasiliev, 2005; Mochizuki *et al.*, 2011). Depending on their intended use, heat pipes can operate over a temperature range from 4.0 to 3000 K. In all cases, their applications can be divided into three main categories: separation of heat source and sink, temperature equalization, and temperature control. Due to their extremely high thermal conductivity, heat pipes can efficiently transport heat from a concentrated source to a remotely mounted sink. This property can enable dense packing of electronics, for example, without undue regard for heat sink space requirements. Another benefit of the high thermal conductivity is the ability to provide an accurate method of temperature equalization. For example, a heat pipe mounted between two opposing faces of an orbiting platform will enable both faces to maintain constant with equal temperatures, thus minimizing thermal stresses. The temperature control is a result of the capability of heat pipes to transport large quantities of heat very rapidly. This feature enables a source of varying flux to be kept at a constant temperature as long as the heat flux extremes are within the operating range of the heat pipe.

Electronic and Electrical Equipment Cooling

Miniaturization of electronic components is accompanied by increased demands on heat dissipation systems due to the increased density of the components. For example, the digital computer has evolved from a massive system that filled an entire room to a unit which can be stored in a small briefcase. However, the overheating problems associated with the dense packing of heat-generating integrated circuit chips used in the computer (CPU and GPU Cooling) have escalated dramatically. Since the reliability of these and other types of electronic components is sensitive to their operating temperature, steps have been taken to improve heat dissipation by using heat pipes. Other applications to electronic cooling have included rectifiers, thyristors, transistors, traveling wave collectors, audio and RF amplifiers and high density semiconductor packages.

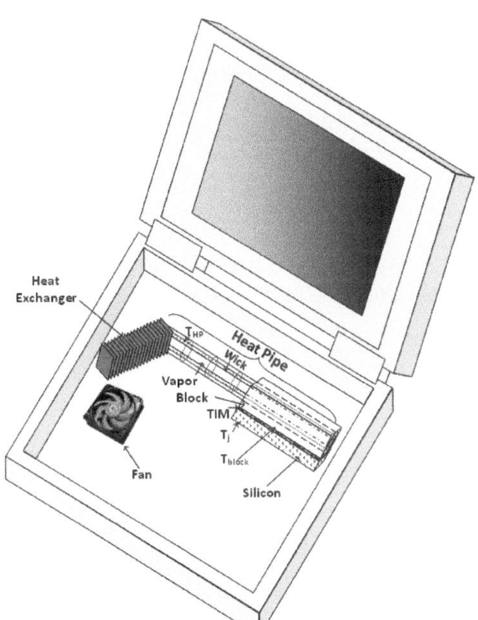

Fig. 53 Remote heat exchanger using a heat pipe embedded in a block for a laptop computer.

After the introduction of the Pentium processor in 1993, the processor performance and power consumption trend has significantly increased annually. In the year 2000, the heat flux was approximately 10-15 W/cm^2, eventually reaching 120-150 W/cm^2 in 2010. The average power consumption for laptop/notebook computers is currently between 25-50 W, while desktops and servers consume between 80 and 150 W. Regardless of components, power level, or type of computer/processor, as electronics are packed into smaller volumes, it is important to optimize the thermal management components with higher cooling rates.

Presently, almost all laptop/notebook computers use a heat pipe remote heat exchanger (RHE), where a heat pipe is used to transfer heat from the processor to the heat exchanger. Heat pipes have been an important part of laptop/notebook computer cooling systems and will remain so for the foreseeable future. Figure 53 shows a simplified typical layout of a heat pipe remote heat exchanger for laptop/notebook computers, using the heat pipe embedded inside a metal block (CPU case) over the silicon (die). For simplicity, a single heat sink is shown in Fig. 53 to illustrate the concept of thermal resistance optimization for personal computers.

The overall thermal resistance of this system is presented by the following equation:

$$Q = Heat\ dissipation\ (W) = \frac{T_j - T_\infty - T_{sys}}{R_{j-\infty}}$$

where

$R_{j-\infty} = R_{Si} + R_{TIM} + R_e + R_c + R_a + R_{HT-\infty}$
T_j = Junction temperature [°C]
T_∞ = Ambient temperature [°C]
T_{sys} = Impact of other energy sources in the system [°C]
$R_{j-\infty}$ = Thermal resistance of the CPU die to ambient [°C/W]
R_{Si} = Thermal resistance of silicon (microprocessor package) [°C/W]
R_{TIM} = Thermal resistance through thermal interface material (TIM) [°C/W]
R_e, R_c, R_a = The thermal resistances for the evaporator, condenser, and adiabatic section of the heat pipe, respectively.
$R_{HT-\infty}$ = Thermal resistance of heat exchanger to ambient [°C/W]

The thermal resistances of both the evaporator and condenser sections of the heat pipe are much higher than the adiabatic sections. Figure 54 (Machiroutu *et al.*, 2006) shows a simple pie chart with fractional distributions of various thermal resistances for a heat pipe remote heat exchanger technology for laptop computers. Machiroutu et al. (2006) also showed that the overall thermal resistances presented in Fig. 54 can be reduced by 17% by reducing the thermal resistances of the evaporator and condenser sections.

Mochizuki et al. (2007) presented the various typical CPU thermal resistances versus heat dissipation (Fig. 55) by assuming a junction temperature of 100°C, outside ambient temperature of 35°C, and a system temperature rise of 10°C. Mochizuki et al. (2006) presented a number of potential future cooling technologies for higher power cooling chips, including vapor chambers and loop heat pipes (LHPs) for desktop computers and servers. Figure 56 (Mochizuki *et al.*, 2006) fins are soldered directly to the vapor chamber and proposed to replace the solid copper metal interface heat spreaders (IHS) with a two-phase micro-channel vapor chamber. Wu et al. (2011) proposed a cold energy storage system using wickless thermosyphons for cooling large scale data centers (8800 kW).

Annular heat pipes are proposed for use in the fusing units of copy machines by some copy machine manufacturers as heat drums for fast thermal response and uniform fusing (Jalilvand *et al.*, 2006). The proposed annular heat pipe is composed of two unequal diameters, where the evaporation and condensation is separated in the radial direction so that the heat is transferred in the radial direction.

While miniaturization is not a significant factor in large-scale electrical equipment, the application of heat pipes to this area is important due to the increase in efficiency when the components operate at lower temperatures (Oslejsek and Polasek, 1976; Giessler *et al.*, 1987; Momose *et al.*, 1987; Alonso and Perez, 1990).

Electric motors have been designed which incorporate heat pipes around the periphery of the rotor for cooling during operation. Another design replaced the solid motor shaft with an on-axis rotating heat pipe with an internal taper for cooling. These designs removed heat from the interior of the electric motor so that the bearings operated at a lower temperature, which increased efficiency. Also, since the resistance of the electric windings decreases with temperature, less power was required to maintain a particular load. Similarly, circuit breakers and electric transformers designed with heat pipe cooling benefit from the decrease in internal temperature.

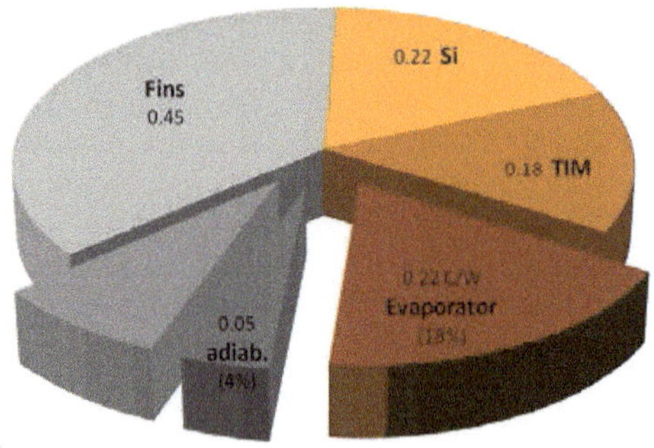

Fig. 54 Pie chart showing component thermal resistances of a heat pipe remote heat exchanger for a laptop PC (Machiroutu *et al.*, 2006).

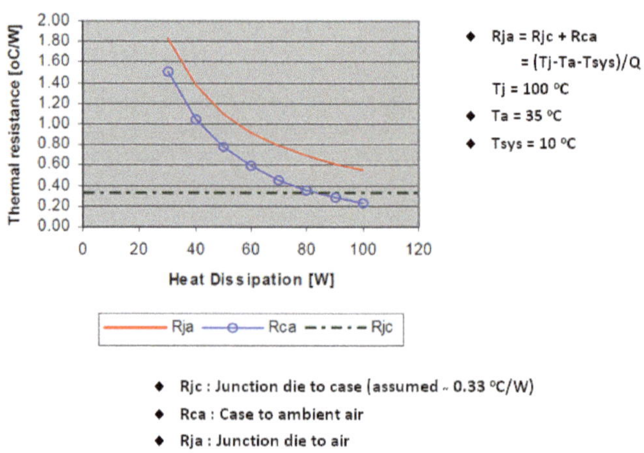

Fig. 55 Various typical CPU thermal resistances versus heat dissipation (Mochizuki *et al.*, 2007).

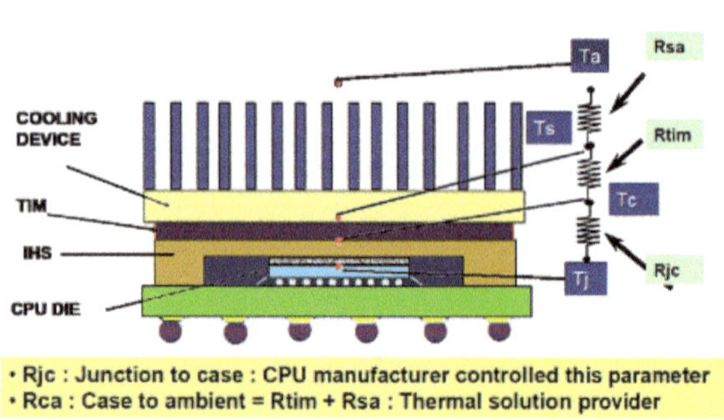

Fig. 56 Thermal solution integration for desktop and servers (Mochizuki *et al.*, 2006).

Underground power transmission cables that are placed close to hot water or steam pipes are sometimes subjected to limitations in the maximum permissible current due to thermal interference. Therefore, it is advantageous to place maintenance-free cooling systems using heat pipes at these locations. Several heat pipes are placed at the hot spot along the direction of the cable, which are bent so that the condenser sections are above ground or in the soil away from the heated section (Iwata *et al.*, 1984). The two methods of removing heat, either by natural convection to the atmosphere or by conduction to the soil, provide sufficient cooling so that higher current loads are capable of being provided to the user.

Energy Systems

With home heating costs increasing, more attention has been focused on the use of heat pipes to collect solar energy (Bienert, 1973; Roberts, 1978). A relatively simple design incorporates a bank of inclined thermosyphons exposed to the south side of a residence. Solar energy is absorbed and transported into the living space, where it is convected to the interior air or stored in a water tank for later use. During the night, the thermosyphons essentially act as thermal diodes, since the only way heat can be transferred from the interior to the outside is by axial conduction through the pipe walls. A similar design can be used for desalinating sea water using solar energy. In this application, however, a heat pipe would be positioned at the focal point of a trough-shaped parabolic reflector in order to generate the high temperatures and heat fluxes necessary for desalination.

Singh et al. (2011) presented the design and characteristics of various energy conservation systems and renewable energy systems utilizing heat pipes as the thermal control mechanism. A wide range of energy systems including data center cooling, agricultural products cold storage, bakery waste heat recovery and automotive dashboard cooling were discussed. It was argued that zero emission and economical advantages can be achieved by using thermosyphon and capillary pumped loop.

Concentrating photovoltaic (PV) systems use low-cost optical systems such as the Fresnel lens, a mini-reflecting mirror that can concentrate solar intensity from 200 to 1000 suns. The concentrated solar energy delivered from the solar cell is from 20 to $100W/cm^2$. Part of the energy is directly converted to electricity, while the remainder is removed as waste heat. Heat pipe cooling systems (Fig. 57) were developed to passively remove the high heat flux at the PV cell and reject it to the ambient by natural convection (Akbarzadeh and Wadowski, 1996; Gi and Maezawa, 2006; Anderson *et al.*, 2008).

Heat transfer devices and methodologies for novel thermal energy storage systems (TES) for various applications including concentrating solar power (CSP) generation systems using latent heat phase change materials (PCMs) are in great demand in the energy field. However, latent heat thermal energy storage PCMs, despite their great potential, suffer from low thermal conductivity. Faghri (1990; 1991) invented two methods (Figs. 58 and 59) to significantly increase the thermal conductivity of PCMs by embedding micro or conventional heat pipes in the PCM for applications in thermal energy storage systems and heat exchangers.

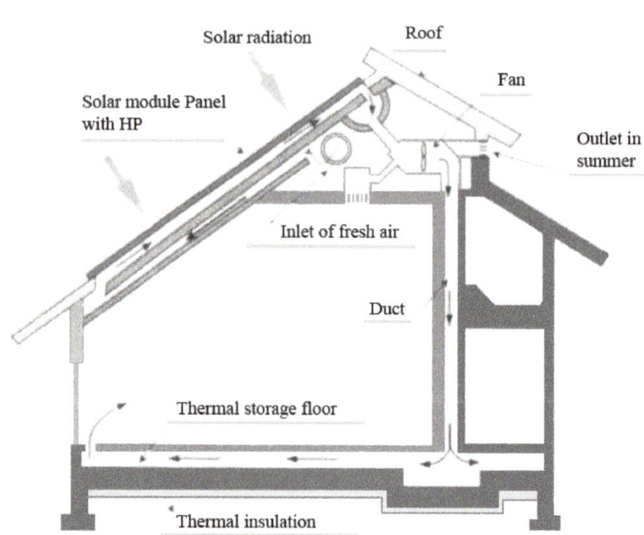

Fig. 57 Solar house which has an energy collection mechanism (Gi and Maezawa, 2006).

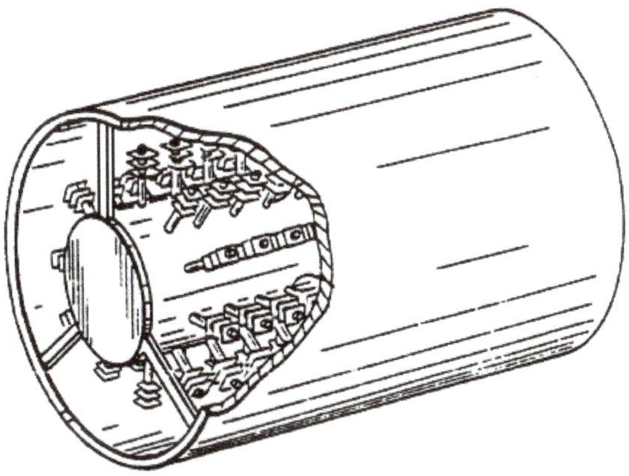

Fig. 58 Thermal energy storage system (Faghri, 1990).

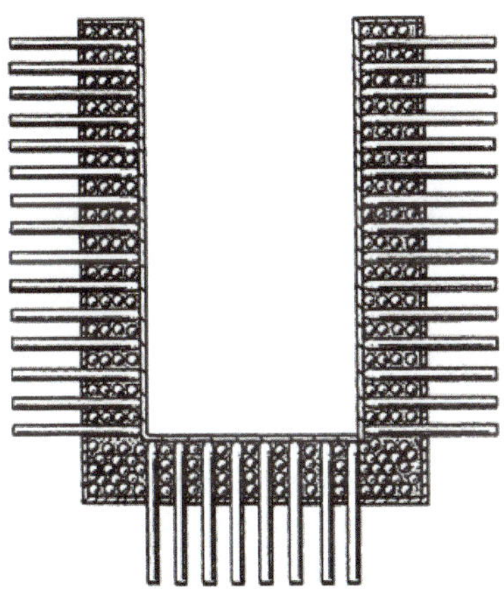

Fig. 59 Micro heat pipe exchanger (Faghri, 1991).

The effect of heat pipes to improve the performance of latent heat thermal energy storage systems has been investigated experimentally and theoretically (Shabgard *et al.*, 2010; Robak *et al.*, 2011; Nithyanandam and Pitchumani, 2012). The experimental results of Robak et al. (2011) showed 60% and more than 100% increase in the heat transfer rates during melting and solidification, respectively, compared to the fin-assisted storage systems as well as non-fin, non-heat pipe configurations. Shabgard et al. (2010) developed a thermal network modeling approach for computationally-efficient simulation of such heat pipe-assisted latent heat thermal energy storage systems. The modeling results also showed substantial heat transfer enhancements during charging and discharging of latent heat thermal energy storage systems equipped with heat pipes. The developed thermal network model was later extended to simulate a large scale heat pipe-assisted cascaded latent heat thermal energy storage system for CSP applications (2012).

Fuel Cells are versatile energy conversion devices with numerous potential applications: large electrical plants, stationary electricity generation, vehicle propulsion and small portable power (Faghri and Guo, 2005; 2008). In recent years, proton exchange membrane fuel cells (PEMFCs), direct methanol fuel cells (DMFCs), solid oxide fuel cells (SOFCs) and molten carbonate fuel cells (MCFCs) have become the major types of fuel cell technologies that draw the most attention for commercialization. Heat pipe technology is beneficial for fuel cell development in two main ways: using heat pipes in fuel cells as thermal management components and using the heat pipe concept in the fuel cell systems to achieve passive high-effective fluids-thermal management. Most fuel cell systems produce electrical energy at high efficiencies that may range from 40% to 60% based on the lower heating value (LHV) of the fuel. For example, a fuel cell operating at 1.0 kW and 50% efficiency generates 1.0 kW of waste heat. This heat may be dissipated by convection, conduction, radiation or phase change. The heat generated in a fuel cell stack may be dumped to the atmosphere, but often, it is used in other system components that require heat. In some cases, the heat is used to run a thermodynamic cycle for additional power generation. Heat pipes can be utilized in fuel cell systems for thermal management purposes which allow for effective use of the fuel cell byproduct, heat, leading to a substantial increase in heat transfer and overall system efficiency (Faghri and Guo, 2005; 2008).

The matter of heat generation in fuel cell stacks presents challenges for thermal management. Stacks operating at 40% to 60% efficiency generate heat at the same rate to more than twice the rate of electric generation. Due to changes in mass concentration, temperature gradients, and in some cases, phase change throughout the stack, the heat generation is not uniform. This non-uniform heat generation further increases thermal gradients in the stack. Increasing the mobility of the heat is a challenge that, if met, leads to three main benefits: (1) The risk of stack failure due to overheating is reduced, (2) The stack operates more closely to its design temperature, resulting in better power density and efficiency, and (3) The heat can be reused, perhaps for reactant preheating, prevaporization, combined cycle operation, or cogeneration. For example, the heat pipe embedded with a bipolar plate (Faghri and Guo, 2008) is an innovative approach that would increase heat transfer in fuel cell stacks while requiring significantly smaller thermal gradients and much smaller volumes and weights than alternative methods. As conceptualized in Fig. 60, a bipolar plate is fabricated with holes, into which micro heat pipes are inserted and bonded. The micro heat pipes that are shown in Fig. 60 are embedded in an axial and transverse direction.

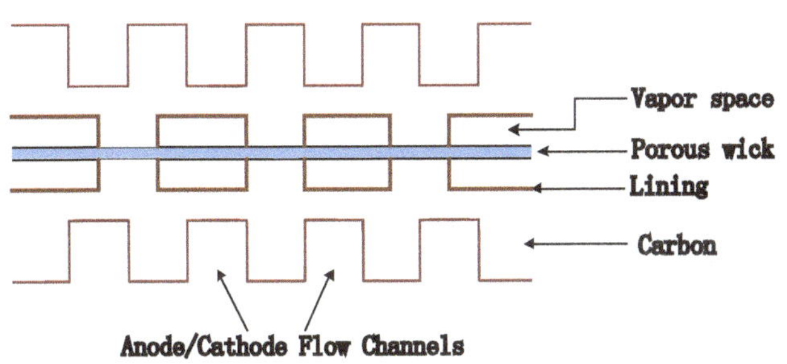

Fig. 60 Bipolar plate integrated heat pipe (Faghri and Guo, 2008).

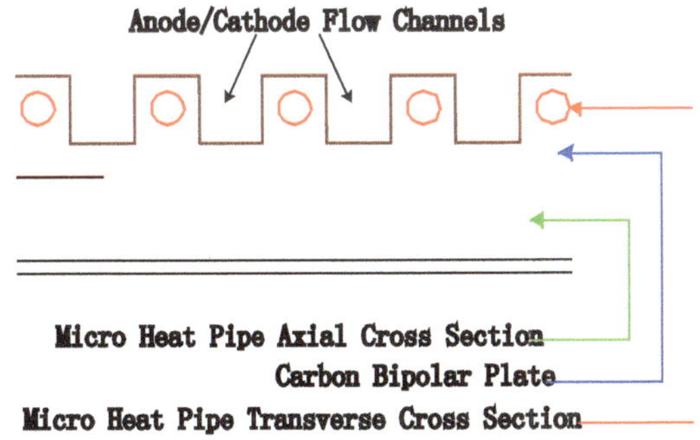

Fig. 61 Micro heat pipe embedded bipolar plate (Faghri and Guo, 2008).

Another method for thermal control in the fuel cell stack is presented with the integrated bipolar plate flat heat pipe (Faghri and Guo, 2008). For example, as conceptualized in Fig. 61, a carbon bipolar plate contains a flat heat pipe. A challenge for developing this component is sealing the heat pipe. Due to the permeability of the carbon, two halves of the bipolar plates are coated with a liner--silver activated brazing alloy (ABA). Also in the enclosure is a porous wick of metal foam, felt, or porous carbon. The two halves are then sealed together, perhaps with brazing in an inert gas. The structure of the interior is designed to allow good electrical conductivity, thermal conductivity to the wick, and structural support, while the working fluid is typically at a pressure different than the surroundings.

Passive DMFC technology uses various capillary approaches to manage methanol and water without the need for a complex micro-fluidic subsystem (Fig. 62) (Guo and Faghri, 2006a; 2006b; Faghri and Guo, 2005; 2009). At the core of this new technology is a unique passive system that uses the heat pipe concept for fuel delivery. Furthermore, the fuel cell is designed for both passive water management and effective carbon dioxide removal. The passive components that are critical to the fuel cell design are the fuel delivery, and air-breathing and water recirculation systems. The passive fuel delivery system stores pure methanol, which can be mixed with water in situ without the use of a pumping system, and can be passively supplied to the fuel cell anode at an optimal concentration. Since water is needed in the anode for the methanol reaction to occur, the water created in the cathode can be passively supplied to the anode. This water recirculation, in conjunction with passive methanol fuel delivery, can dramatically extend the operation time of the fuel cell per refueling. The passive mass transfer concept (wick structure) developed in heat pipe technology is an effective approach for mass transfer management in various fuel cell technologies. The proposed DMFC technologies developed were operated passively, without moving parts, which resulted in a highly reliable system. Due to their significantly longer charging life, passive miniature DMFC system are seriously being considered for replacing the battery for applications such as cell phones, digital cameras and laptops.

Current dish/Stirling systems use directly illuminated receiver design in which the concentrated solar radiation is absorbed by tubular receiver which also serve as the heater tubes of the Stirling engine. Replacing the directly illuminated receiver design with spherical heat pipe receivers improves the performance by about 20% (Andraka *et al.*, 1996). Two primary reasons contribute to this improvement; (a) in the heat pipe receiver the temperature difference between the output gas temperature and the receiver peak temperature is much less than the directly illuminated receiver due to nearly isothermal operation of the heat pipe. This allows higher operating temperatures resulting in higher conversion efficiencies, and (b) in heat pipe receiver heat is transferred to the Stirling engine by condensation on the entire surface of the heater tubes, whereas in directly illuminated receivers only one side of the tubes is irradiated. Hence the "dead volume" of the engine is decreased and higher compression ratios can be achieved which result in greater system efficiency (Andraka *et al.*, 2012).

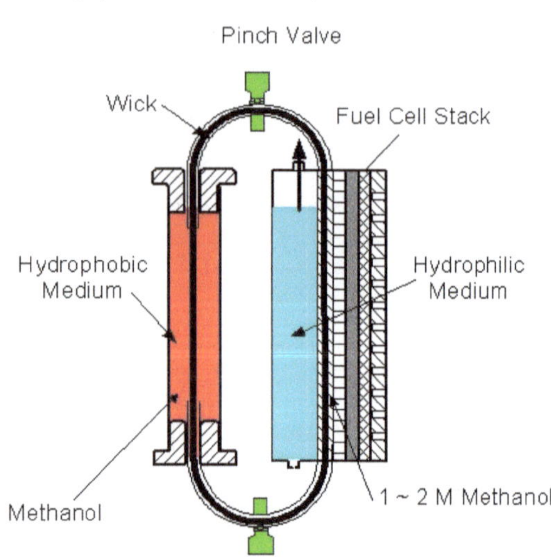

Fig. 62 Passive direction methanol fuel cell using heat pipe concept (Faghri and Guo, 2008).

Aerospace and Avionics

Heat pipes are very attractive components in the area of spacecraft cooling and temperature stabilization due to their low weight penalty, zero maintenance, and reliability. Structural isothermalization is an important problem with respect to orbiting astronomy experiments due to the possible warpage from solar heating. During orbit, an observatory is fixed on a single point such as a star. Therefore, one side of the spacecraft will be subjected to intense solar radiation, while the other is exposed to deep space. Heat pipes have been used to transport the heat from the side irradiated by the sun to the cold side in order to equalize the temperature of the structure. Heat pipes are also being used to dissipate heat generated by electronic components in satellites.

Early experiments of heat pipes for aerospace applications were conducted in sounding rockets which provided six to eight minutes of 0-g conditions. In 1974, ten separate heat pipe experiments were flown in the International Heat Pipe Experiment (McIntosh *et al.*, 1976). Also in 1974, heat pipe experiments were conducted aboard the Applications Technology Satellite-6 (Kirkpatrick and Brennan, 1976), in which an ammonia heat pipe with a spiral artery wick was used as a thermal diode. With the use of the space shuttle, flight testing of prototype heat pipe designs continued at a much larger scale. In 1983, on the eighth space shuttle flight, a 6-ft. monogroove heat pipe with Freon 21 as the working fluid was flight tested as a heat pipe radiator (Rankin, 1984). The Space Station Heat Pipe Advanced Radiator Element, which consisted of a 50-ft. long high capacity monogroove heat pipe encased in a radiator panel, was flown on the space shuttle in 1989 (Brown *et al.*, 1990), and during a 1991 shuttle flight, two heat pipe radiator panels were separately flight tested (Brown *et al.*, 1991). Heat pipe thermal buses were proposed which facilitate a connection between heat-generating components and external radiators (Morgownik and Savage, 1987; Amidieu *et al.*, 1987; Peck and Fleischman, 1987). The components may be designed with a clamping device which can be directly attached to the heat pipe thermal bus at various points in the spacecraft. In 1992, two different axially grooved oxygen heat pipes were tested in a Hitchhiker Canister experiment that was flown aboard the Shuttle Discovery (STS-53) by NASA and the Air Force to determine startup behavior and transport capabilities in micro gravity (Brennan *et al.*, 1993).

An advanced capillary structure which combined re-entrant and a large number of micro grooves for the heat pipe evaporator was investigated in microgravity conditions during the 2005 FOTON-M2 mission of the European space agency (Schulze *et al.*, 2007). Swanson (2007) presented the NASA thermal technical challenges & opportunities for the new age of space exploration with emphasis on heat pipes and two phase thermal loops.

A heat pipe laser mirror has been fabricated in order to test the feasibility of this technology compared to water-cooled or uncooled mirrors for high power lasers (Barthelemy *et al.*, 1978). Presently, uncooled mirrors are limited to only a few seconds of use prior to distortion from thermal overheating. Water-cooled mirrors have longer service times, but are subjected to high internal pressures causing distortion, which must be removed by polishing under pressure. A copper-water heat pipe mirror was constructed, which was not affected by the problems associated with conventional mirrors due to the heat pipe action. The experiments conducted used a 10 kW carbon dioxide laser, and it was found that heat pipes can be used with success if the heat pipe is sufficiently preheated.

Thermal diodes have been proposed for use in cooling low-temperature sensors, such as an infrared detector in low subsolar earth orbits (Williams, 1978). This type of heat pipe was proposed due to its characteristic of being able to cool the instrument during normal operation, but effectively insulating it when exposed to an external heat flux. One type of thermal diode uses a liquid reservoir at the evaporator end of the heat pipe, which does not communicate with the wick structure. During normal operation, the reservoir is empty. If the condenser is subjected to an external heat flux, however, the working fluid condenses in the reservoir, causing the wick to dry out. This results in the heat pipe becoming an insulator, because heat can only be conducted axially through the thin pipe wall. Heat pipes have also been qualified and/or used for thermal control applications in avionic systems including aircrafts with more electric architectures.

Radioisotope Stirling systems are proposed to replace the radioisotope thermoelectric generating systems as a long-lasting electricity generation solution in space missions due to their higher efficiency (Thieme and Schreiber, 2003). In the current radioisotope Stirling systems if the Stilrling engine stops, the heat removal from the system would be ceased and the insulation will be spoiled to prevent damage to the clad fuel, but the mission will also be ended. Alkali-metal variable conductance heat pipes are proposed and tested to allow multiple stops and restarts of the Stirling engine (Tarau and Anderson, 2010). In the proposed design, the evaporator of the heat pipe is connected to the heat generation module. During the normal operation, the heat is transferred from the heat generation unit to the heater head of the Stirling engine by evaporation and condensation of the sodium working fluid. When the Stirling engine stops, the temperature and pressure of the heat pipe working fluid increases. The higher pressure inside the heat pipe compresses the non-condensable gas and opens up a radiator through which the heat is dissipated and the system temperature stabilizes. Once the Stirling engine restarts, the temperature and pressure drop and the radiator is covered by the non-condensable gas to prevent unwanted heat rejection.

Heat Exchangers and Heat Pumps

Increases in the cost of energy have promoted the use of heat pipe technology in industrial applications. Due to their high heat transfer capabilities with no external power requirements, heat pipes are being used in heat exchangers for various applications. In the power industry, heat pipe heat exchangers are used as primary air heaters on new and retrofit boilers. The major advantages of heat pipe heat exchangers compared to conventional heat exchangers are that they are nearly isothermal and can be built with better seals to reduce leakage. Heat pipe air heaters should also be cheaper than conventional tubular heat exchangers, as they are smaller and can be shipped in a small number of modules. Heat pipe heat exchangers can serve as compact waste heat recovery systems which require no power, a low pressure drop and are easy to install on existing lines. Heat pipe heat exchangers can be categorized into gas-gas, gas-liquid, and liquid-liquid type heat units. Among these three, gas-gas heat pipe heat exchangers have the widest application in industry. A gas-gas heat pipe exchanger consists of a group of externally finned heat pipes which reclaim waste heat (Hassan and Accensi, 1973; Holmes and Field, 1986). These units eliminate cross-contamination due to the solid wall between the hot and cold gas streams. Also, the heat pipe design is totally reversible (heat can be transferred in either direction). Gas-gas energy recovery units typically fall into three categories: heat recovery in air-conditioning systems (low temperatures), recovery of excess process heat for space heating (moderate temperature), and recovery of waste heat from high temperature exhaust streams for reuse in the process (preheating of combustion air, for example). The units for these applications vary in size and construction depending on the specific application, but many commercial models are now available that implement this heat pipe design.

Gas-liquid heat pipe exchangers are less commonly available than gas-gas models due to the fact that the present design of waste heat boilers is very efficient. In the past, exhaust heat from boilers was simply dispersed to the atmosphere. Figure 63 shows schematics for waste heat recovery with liquid-gas and gas-gas heat exchangers. Faghri (1993a) invented an innovative design for a centrifugal heat pipe vapor absorption heat pump (Fig. 64). The heat pipes in this heat pump system are disk-shaped, with one face partially or completely being the evaporator and the opposite face partially or completely being the condenser (Faghri, 1994). The wick is designed such that the centrifugal force aids in the delivery of the condensate to the evaporator. This design will significantly improve the heat and mass transfer characteristics of the rotating components of the vapor-absorption heat pump by increasing the heat pumping capacity that can be packaged in a given value. This results in a more efficient and compact vapor-absorption heat pump system.

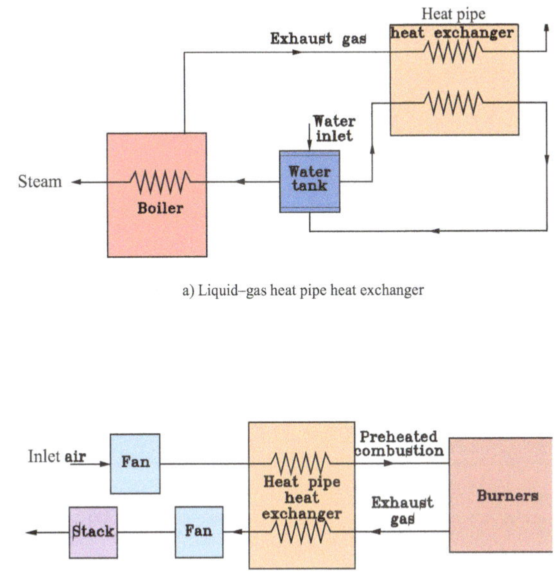

a) Liquid–gas heat pipe heat exchanger

b) Gas–gas heat pipe heat exchanger

Fig. 63 Waste heat recovery using heat pipe heat exchangers.

Gas Turbine Engines and the Automotive Industry

The temperature limitation is one of the most crucial limiting factors related to the efficiency of a gas turbine aircraft engine or power gas turbine. An increased turbine inlet temperature decreases both the specific fuel and air consumption, while increasing efficiency. This desire for a high turbine inlet temperature, however, is often in conflict with materials available to withstand the high temperature. As a result, innovative cooling systems for hot-gas-path components are required. Among the hot-gas-path components, the first-stage rotor blade and nozzle guide vane require the most challenging cooling consideration. In addition to improving energy utilization efficiency, effective cooling could also drastically improve the reliability of high-speed rotating components. It has been observed that the creep life of turbine blades is reduced by half with every 10 to 15°C rise in metal temperature. Therefore, the temperature of the turbine blade must be kept within certain tolerable limits.

The primary cooling technology in use today for turbine blades and nozzle guide vanes (NGVs) is the film cooling technology. Since 1995, a number of efforts have been initiated to utilize the concepts of miniature, radially rotating, high-temperature heat pipes, for gas turbine blades and disk cooling. This also includes a series of experimental and numerical studies (Cao, 1996; 1997; 2010; Zuo *et al.*, 1998). High temperature heat pipe cooling is a promising cooling technology for gas turbine hot components, such as first stage rotor blades, nozzle guide vanes, and rotor disks. This has the potential to significantly reduce the temperature of these hot components, allowing for a much higher gas turbine inlet temperature, while also reducing the consumption of high-pressure compressor air.

A Stirling engine heated by sodium heat pipes has been constructed and tested (Alleau *et al.*, 1984; 1987; Meijer and Khalili, 1990). This engine operates at highest efficiency when the thermal energy is supplied to it at a very constant temperature and high heat fluxes. Helium is heated to increase its pressure, which is used to drive pistons and a crankshaft for the generation of power from thermal energy. The sodium heat pipes deliver heat from a molten salt heat storage system to the gaseous helium which is used to drive the engine. The coupling of a Stirling cycle engine with sodium heat pipes can also be used in the direct conversion of solar energy.

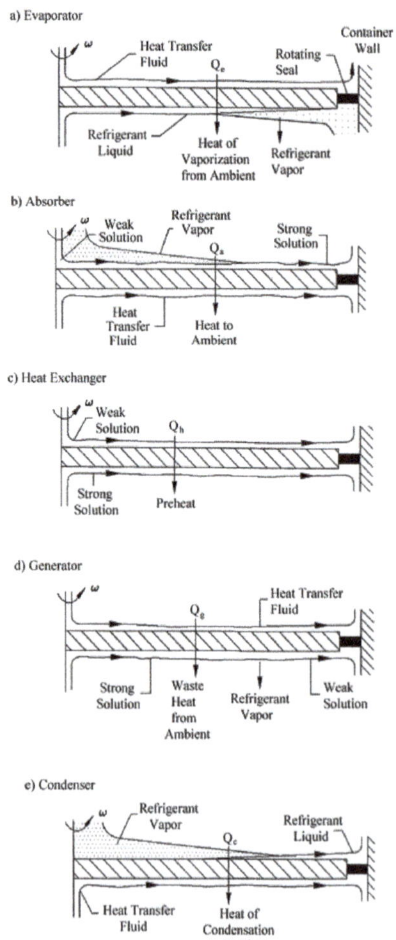

Fig. 64 Schematic of centrifugal heat pipe heat pump vapor-absorption cycle (Faghri, 1993a).

Nguyen et al. (1992) proposed a thermosyphon Rankine engine for power generation using solar and other available waste heat energy sources. The basis of the engine (Fig. 65) is the modification of a heat pipe to incorporate a turbine and thereby, making the system into a Rankine cycle engine to convert thermal energy to electrical energy.

Another potential application of heat pipes has been toward the cooling of vehicle brake systems (Maezawa *et al.*, 1981). Normally, the brake pads and rotors of a conventional automobile brake system suffer greatly from the effects of elevated temperatures. Thermal cycling induces stresses and causes increased amounts of corrosion, which shorten the working lives of the rotors.

High temperatures at the rotor-pad interfaces cause embrittlement of the brake pad material, which also reduces the service life. Rotating and stationary heat pipes have been proposed to cool both the rotor and brake pads, which would result in lower, more uniform temperatures. The use of this development could also increase the service life of brake pads so that 'lifetime' pads, lasting the entire life of the automobile, may become a reality. Vibration/shock-tolerant capillary two-phase loops are developed to meet the cooling requirements of the high performance electronics of future military vehicles which undergo intense shocks during operation (Tang and Park, 2008). The shock tolerant feature is achieved by utilizing an in-situ sintered wick which also reduces the manufacturing costs.

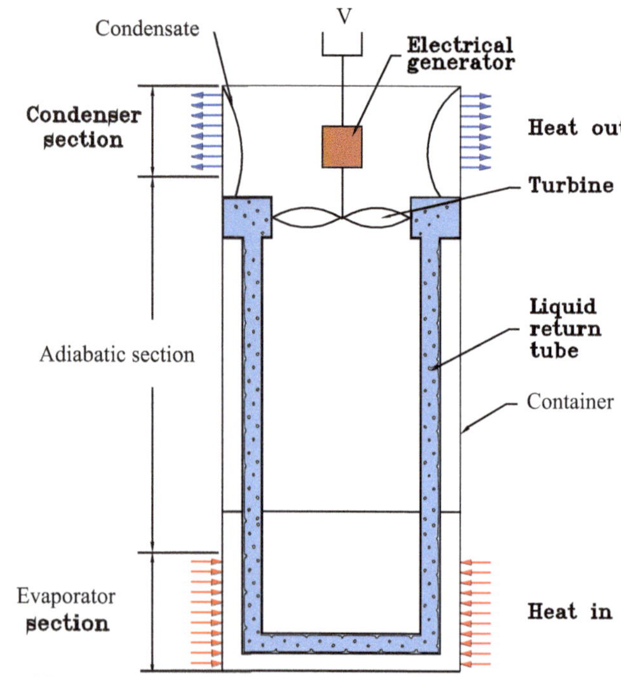

Fig. 65 Experimental thermosyphon Rankine engine (Nguyen *et al.*, 1992).

Production Tools

An important application of heat pipes is in the field of die casting and injection molding (Winship, 1974; Reay, 1977). The most obvious use of heat pipes in this field is the removal of heat during the solidification process; however, heat pipes are also useful in minimizing thermal shock in the dies. Die casting involves introducing a material in molten form into a closed die. The material is cooled until it solidifies, the part is removed, and the process repeated. An important consideration is the time required for the material to cool into solid form, so most dies are water cooled. It is often difficult to cool inaccessible parts of the die, however, so heat pipes are used to cool these sections. These heat pipes can also be used to preheat the die to assure the continuous flow of molten material due to the reversible nature of heat pipes.

Medicine and Human Body Temperature Control

Another application is of heat pipes is related to human physiology. A surgical probe incorporating a cryogenic heat pipe is being used to destroy tumors in the human body (Basiulis, 1976). This type of surgery, where the tissue is frozen instead rather than irradiated, is beneficial because the surrounding tissue sustains practically no damage. Also, the surgery results in very little bleeding or pain. The cryoprobe is a hand-held device with a reservoir of liquid nitrogen and a 12-inch heat pipe extension, which is maintained at approximately 77 K for about one-half hour. Fletcher and Peterson (1997) invented a catheter using micro heat pipes that provide precise temperature control for treating diseased tissue.

Another application related to human physiology concerns the control of body temperature (Faghri *et al.*, 1989b; Faghri, 1993b). In regions or occupations where humans are exposed to extreme temperatures, such as workers in polar regions or foundry workers, adverse health effects from these environments are evident. Frostbite on the extremities in cold regions and heat exhaustion in warm climates are very serious problems which must be handled with extreme care. These problems can be avoided with the use of gloves, socks, and suits in which heat pipes are placed in order to transfer heat either to or from parts of the body (Fig. 66). In cold climates, heat pipes could bring heat from the torso to the extremities such as the fingers, toes, and ears to prevent frostbite. Figure 67 shows a conceptual design for cold weather handwear with heat pipes, where body heat is transferred from the forearm to the fingers. In very hot environments, such as those experienced by fire fighters, a cold suit employing heat pipes could be developed which would be lighter and less bulky than the suits presently worn. This type of suit would also be beneficial in the respect that the wearer would be kept cooler, resulting in more time available for extracting people from a burning building, for instance.

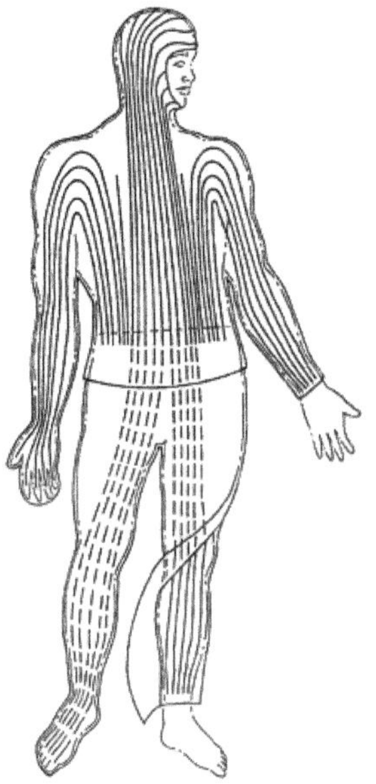

Fig. 66 Temperature regulation system for the human body using heat pipes (Faghri, 1993b).

Similar garments or blankets using heat pipes (Faghri, 1993b) could also be applied to medical patients whose own body temperature regulation system is impaired or not functioning. Patients with spinal cord injuries or people with Multiple Sclerosis often become overheated during hot weather because of the inability of the nervous system to control their body temperature. A heat pipe suit equipped with an external heat exchanger could significantly improve the quality of life of these patients by allowing them to be more exposed to the elements. Likewise, patients who are bedridden often become overly warm or cold. In some cases, a blanket with a single-phase fluid loop is placed on the patient for heating or cooling. Since this type of blanket is quite heavy due to the fluid loop, a heat pipe blanket has been proposed, which would decrease the weight and improve the heat transfer and temperature uniformity.

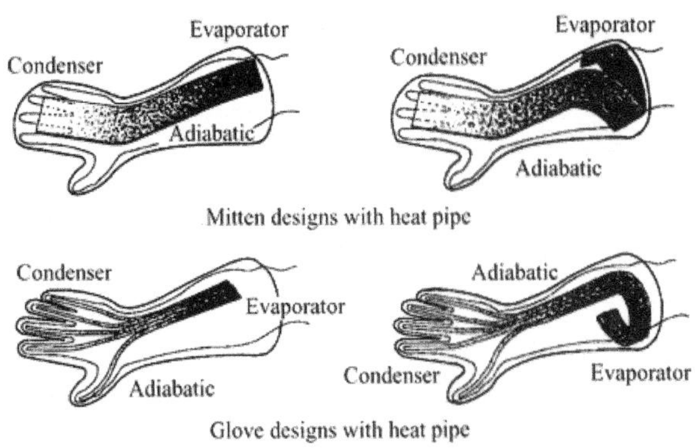

Fig. 67 Models of cold weather handware with heat pipes (Faghri *et al.*, 1989b).

It has been found experimentally that when the temperature of the cauterizing forceps used in surgical operations increase above 80°C, the tissue under operation tends to stick to the forceps. Small-diameter heat pipes are employed to precisely control the forceps temperature by removing heat from the forceps tip as well as the tissue (Bilski and Broadbent, 2010). The heat generated during cauterization evaporates the working fluid of the heat pipe. The vapor moves towards the cold section of the heat pipe where it dissipates its thermal energy content and condenses. Capillary action then returns the condensed liquid to the forceps tip. This technology represents the smallest known mass-produced heat pipe assembly (with a diameter of 2.34 mm) and makes it possible to passively control the temperature of forceps tips with diameters ranging from 0.5 to 4 mm.

Ovens and Furnaces

One of the first applications of heat pipes (actually, two-phase closed thermosyphons) was in baking ovens (Anonymous, 1867). Normal flame-heated baking ovens warmed the firebrick in the firebox, which then conducted the heat to the baking chamber so that the baked food was not contaminated by the combustion products. With the advent of the heat pipe, heat was transferred to the baking chamber by the evaporation and condensation of the working fluid within the array of tubes. This arrangement was not only more efficient, saving up to 25 percent of the fuel normally needed, but also resulted in a more uniform oven temperature. A similar concept is the gas-fired restaurant griddle, which is a flat plate heat pipe separating the gas fire from the cooking surface (Basiulis, 1976). This provides a surface with an extremely uniform temperature, even with heavy food loading. It also has the benefit of a fast warm-up time from a cold start, and the efficiency is comparable to the above-mentioned baking oven. Other food processing equipment has been designed using heat pipe technology such as braisers, kettles, saucepans, and deep fat fryers (Lamp, 1978).
High temperature furnaces have been developed for heat-treating, sintering, and other applications (Finlay *et al.*, 1976; Brost *et al.*, 1990). These furnaces are actually annular heat pipes with liquid-metal working fluids. The isothermal behavior of the interior of the annular heat pipe makes it ideal for use as a furnace. The temperature gradient across the length is very small, which is always a significant problem with conventional furnaces. Due to the operation of the heat pipe, only a small portion of the outer surface of the annular heat pipe needs to be heated. Upon insertion of a cold charge, the area in contact with the cold charge becomes the condenser. Annular heat pipe furnaces are also being used in the calibration of temperature measurement devices (Chengsheng *et al.*, 1984; Bassani *et al.*, 1990). Thermocouples, RTD's and thermistors have been calibrated using a variable conductance heat pipe furnace; the temperature of which was measured with a traceable platinum RTD. Optical pyrometers, which measure surface temperature from the emitted radiation, have also been calibrated with heat pipe blackbody sources.

Permafrost Stabilization

Permafrost stabilization is an important matter in designing the foundations of buildings or other structures in arctic regions. During the summer months, heat is conducted into the permafrost, which tends to melt the water trapped in the soil. Normally, this thawing would allow the foundation to sink or shift, thereby damaging the building. To avoid this problem, engineers responsible for the Trans-Alaskan pipeline placed heat pipes around the pilings of the pipeline supports (Fig. 68) (Waters, 1976). The heat from the soil adjacent to the pilings was dissipated to the ambient by the heat pipes, which maintained a permafrost bulb around the pilings throughout the year.

Mashiko et al. (1989) developed an artificial permafrost storage facility in Hokkaido, Japan (Fig. 69), where the evaporator of the heat pipe is buried underground and the condenser is exposed to the ambient. The heat pipe transfers the ground heat to the ambient in winter which freezes the soil. When the temperature rises in the spring, the heat is not transmitted from the ambient to the ground (thermal diode), so the frozen soil is kept throughout the year. The main feature of this system is that no power is required, and the temperature is kept constant for a long period of time.

Manufacturing

High temperature heat pipes have been proposed for use in the manufacturing of glass bottles (Brost *et al.*, 1973). The glass bottle forming procedure starts by periodically dipping a steel piston into a steel form filled with molten glass. This forms a hollow glass tube, which is later blown into its final shape. The initial glass temperature is around 1100°C and the surface temperature of the piston needs to be kept around 600°C. At higher piston temperatures, the glass will stick to the piston, and at lower temperatures, the glass viscosity increases, causing insufficient deformation during the forming process. Insufficient deformation is the cause of thin-walled bottles which contribute to the waste rate. A stainless-steel/potassium heat pipe was proposed and tested, and it was found that the heat pipe could be kept nearly isothermal. This resulted in a higher dipping frequency and a reduced amount of glass bottle waste.

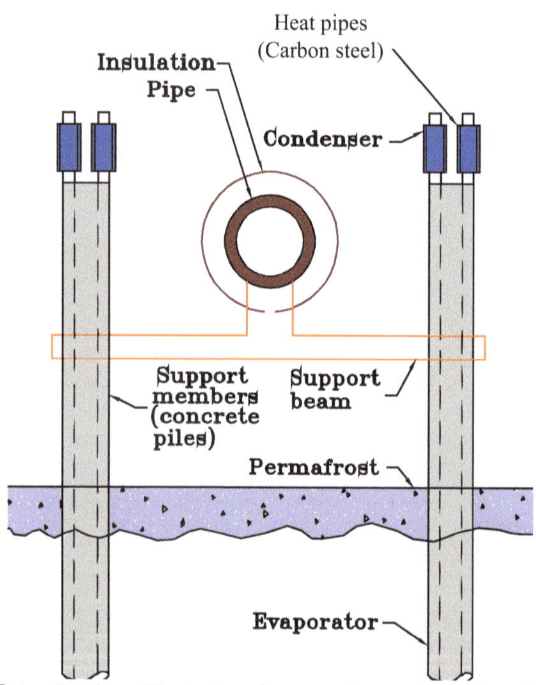

Fig. 68 Application of heat pipes for permafrost stabilization of structural supports on the Trans-Alaskan pipeline.

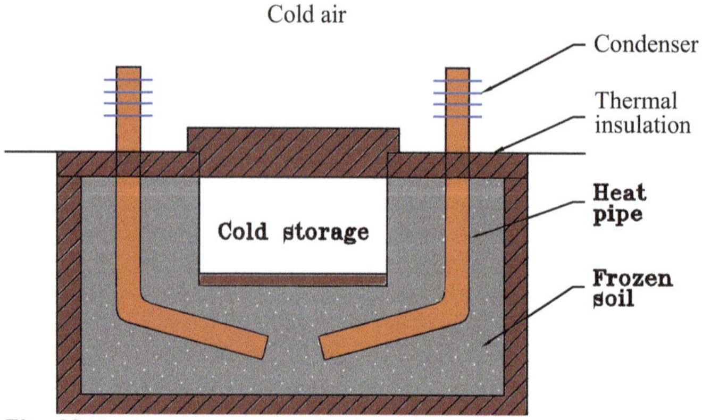

Fig. 69 Cold energy storage using heat pipes.

Transportation Systems and Deicing

Several innovations using heat pipe technology have been proposed which could improve the safety and reliability of transportation systems. Heat pipes have been used to melt the ice and snow on roadways, bridges, and aircraft runways by transporting geothermal heat stored in the ground to the pavement (Suelau *et al.*, 1976; Bartsch *et al.*, 1987). Shiraishi et al. (1992) presented several newly developed snow removal and deicing methods using heat pipes such as prevention of snow damage to support wires for telephone poles, a snow melting system for pavement, and a large scale snow melting system for roads in Japan. Corrugated, long heat pipes are utilized for these applications.

People working on ships at sea during the winter months are plagued by the problems associated with icing. Anything exposed to the elements, such as the decks and handrails, are constantly icing over during the winter. The usual method of removing the ice is to use an axe or hatchet, which is extremely dangerous when the seas are rough. The potential for severe cuts or falling overboard is always present. To solve this problem, heat pipes have been incorporated into the decks and handrails to transport waste heat from the engine (Matsuda *et al.*, 1981). This heat constantly melts the ice before large deposits are formed. Another similar deicing problem addressed by the use of heat pipes concerns navigation buoys (Larkin and Dubuc, 1976). Off the east coast of Canada, navigation buoys are frequently subjected to ocean spray and freezing temperatures, which results in the buoys capsizing due to the weight of the ice. Researchers have constructed a prototype buoy which was, in essence, a large ammonia thermosyphon heated from the sea. It was experimentally determined that the superstructure of the thermosyphon was kept free from ice, but that the auxiliary float seriously iced up.

12. CONCLUSIONS

An overview of various types of heat pipes is presented in this review. This includes heat pipe fundamentals, operations, heat transport limitations, and full simulation. This review provides a self contained document to design and simulate various types of heat pipes under different operating conditions. Numerical and analytical analyses of heat pipes have progressed significantly over the last several decades. State-of-the-art modeling is capable of predicting thermal performance under various operating conditions for conventional heat pipes. In particular, advances related to the full simulation of heat pipes under steady state, continuum transient and frozen startup operation for conventional heat pipes have been very successful, despite complex multiphase and multidomain transport phenomena in heat pipes. In general, it has been shown that heat pipe simulations must include conjugate heat transfer with the wall, wick and vapor, since these affect both the transient and steady state operating conditions. More fundamental works are needed to better understand the physical phenomena of pulsating heat pipes and loop heat pipes. An accurate simulation of liquid/vapor interface, including multiphase phenomena in various wicks, is important to accurately predict the heat transport limitation of heat pipes.

Acknowledgment: Copyright permission from ASME

REFERENCES

Anonymous, 1867, "The Paris Exhibition – Perkins' Portable Oven," *The Engineer,* p. 519.

Aghvami, M., and Faghri, A., 2011, "Analysis of Flat Heat Pipes with various Heating and Cooling Configurations," *Applied Thermal Engineering,* **31**(14-15), 2645-2655. http://dx.doi.org/10.1016/j.applthermaleng.2011.04.034

Akbarzadeh, A., and Wadowski, T., 1996, "Heat Pipe-Based Cooling Systems for Photovoltaic Cells Under Concentrated Solar Radiation," *Applied Thermal Engineering,* **16**(1), 81-87. http://dx.doi.org/10.1016/1359-4311(95)00012-3

Alario, J., Haslett, R., and Kosson, R., 1981, "The Monogroove High Performance Heat Pipe," AIAA-81-1156, *AIAA 16th Thermophysics Conference,* Palo Alto, CA.

Alario, J., Haslett, R., and Kosson, R., 1982, "The Monogroove High Performance Heat Pipe," *Progress in Astronautics and Aeronautics,* **83**, 305-324.

Alario, J. P., 1984, "Monogrove Heat Pipe Radiator Shuttle Flight Experiment: Design, Analysis, and Testing," SAE-84-0950, *Proceedings of the 14th Intersociety Conference on Environmental systems,* San Diego, CA.

Alleau, T., Bricard, A., Chabanne, J., Kermorgant, H., and de Lallee, J., 1984, "Sodium Heat Pipes for Thermal Energy Supply of a Stirling Engine," *Proceedings of the 5th International Heat Pipe Conference,* Tsukuba, Japan, 54-58.

Alleau, T., Bricard, A., and Thouvenin, A., 1987, "Stirling Engine Coupled with a Sodium Boiler," *Proceedings of the 6th International Heat Pipe Conference,* Grenoble, France, 748-752.

Alonso, G., and Perez, R., 1990, "Cooling of Power Transformers with Two-Phase Thermosyphons," *Proceedings of the 7th International Heat Pipe Conference,* Minsk, USSR.

Ambirajan, A., Adoni, A. A., Vaidya, J. S., Rajendran, A. A., Kumar, D., and Dutta, P., 2012, "Loop Heat Pipes: A Review of Fundamentals, Operation, and Design," *Heat Transfer Engineering,* **33**(4-5), 387-405. http://dx.doi.org/10.1080/01457632.2012.614148

Amidieu, M., Moscheti, B., and Tatry, M., 1987, "Develpement of a Space Deployable Radiator Using Heat Pipes," *Proceedings of the 6th International Heat Pipe Conference,* Grenoble, France, 380-385.

Anderson, W. G., Dussinger, P. M., Sarraf, D. B., and Tamanna, S., 2008, "Heat Pipe Cooling of Concentrating Photovoltaic Cells," *33rd IEEE Photovoltaic Specialists Conference,* San Diego, CA. http://dx.doi.org/10.1109/PVSC.2008.4922577

Anderson, W. G., Bonner, R. W., Dussinger, P. M., Hartenstine, J. R., Sarraf, D. B., and Locci, I. E., 2007, "Intermediate temperature fluids life tests - Experiments," *Collection of Technical Papers - 5th International Energy Conversion Engineering Conference,* **2**, 926-941.

Anderson, W. G., and Tarau, C., 2008, "Variable conductance heat pipes for radioisotope stirling systems," *Proceedings of STAIF 2008,* **969**, 679-688. http://dx.doi.org/10.1063/1.2845031

Andraka, C. E., Rawlinson, K. S., Moss, T. A., Adkins, D. R., Moreno, J. B., Gallup, D. R., Cordeiro, P. G., and Johansson, S., 1996, "Solar Heat Pipe Testing of the Stirling Thermal Motors 4-120 Stirling Engine," SAND96-1424C, *Proceedings of the 31st Intersociety Energy Conversion Engineering Conference,* **2**, 1295-1300. http://dx.doi.org/10.1109/IECEC.1996.553903

Andraka, C. E., Rawlinson, K. S., and Siegel, N. P., 2012, "Technical Feasibility of Storage on Large Dish Stirling Systems," Sandia Report SAND2012-8352.

Barthelemy, R., Jacobson, D., and Rabe, D., 1978, "Heat Pipe Mirrors for High Power Lasers," *Proceedings of the 3rd International Heat Pipe Conference,* Paper No. 78-391, Palo Alto, CA.

Bartsch, G., Butow, E., and Schroeder-Richter, D., 1987, "Deicing of Road Surfaces Employing Heat Pipes Installed in the Ground-Water Layer," *Proceedings of the 6th International Heat Pipe Conference,* Grenoble, France, 715-720.

Basiulis, A., 1976, "Follow-up on Heat Pipe Applications," *Proceedings of the 2nd International Heat Pipe Conference,* Bologna, Italy, 473-480.

Bassani, C., Lighthart, J., Sciamanna, G., Marcarino, P., and Fernicola, V., 1990, "Gas Controlled Heat Pipes for Primary Temperature Measurements," *Proceedings of the 7th International Heat Pipe Conference,* Minsk, USSR.

Benner, S., Costello, F., and Ku, J., 1989, "SINFAC Simulation of a High-Power Hybrid CPL," AIAA-89-0316, *Proceedings of the 27th Aerospace Sciences Meeting,* Reno, NV.

Bienert, W., 1973, "Heat Pipes for Solar Energy Collectors," Paper No. 12-1, *Proceedings of the 1st International Heat Pipe Conference,* Stuttgart, Federal Republic of Germany.

Bilski, W. J., and Broadbent, J., 2010, "Feel the Heat: Thermal Design Trends in Medical Devices," *MD+DI,* **32**(6).

Boo, J. H., and Chung, W. B., 2004, "Thermal Performance of a Small-Scale Loop Heat Pipe with PP wick," *Proceedings of the 13th International Heat Pipe Conference,* Shanghai, China, 259-264.

Bowman, W. J., 1987, Simulated Heat Pipe Vapor Dynamics, Ph.D. Dissertation, Air Force Institute of Technology, Dayton, OH.

Brennan, P. J., Thienel, L., Swanson, T., and Morgan, M., 1993, "Flight Data for the Cryogenic Heat Pipe (CRYOHP) Experiment," AIAA-93-2735.

Brost, O., Groll, M., and Mack, H., 1990, "High-Temperature Lithium Heat Pipe Furnace for Space Applications," *Proceedings of the 7th International Heat Pipe Conference,* Minsk, USSR.

Brost, O., Groll, M., Neuer, G., and Schubert, K., 1973, "Industrial Applications of Alkali-Metal Heat Pipes," Paper No. 11-3, *Proceedings of the 1st International Heat Pipe Conference,* Stuttgart, Federal Republic of Germany.

Brown, R., Gustafson, E., Gisondo, F., and Hutchison, M., 1990, "Performance Evaluation of the Grumman Prototype Space Erectable Radiator System," AIAA Paper No. 90-1766.

Brown, R., Kosson, R., and Ungar, E., 1991, "Design of the SHARE II Monogroove Heat Pipe," AIAA 91-1359, *Proceedings of AIAA 26th Thermophysics Conference,* Honolulu, HI.

Cao, Y., 1996, "Rotating Micro/Miniature Heat Pipes for Turbine Blade Cooling Applications," *AFOSR Contractor and Grantee Meeting on Turbulence and Internal Flows,* Atlanta, GA.

Cao, Y., 1997, "A Feasibility Study of Turbine Disk Cooling by Employing Radially Rotating Heat Pipes," Final Report for Summer Faculty Research Program, Air Force Research Lab, Turbine Engine Division.

Cao, Y., 2010, "Miniature High-Temperature Rotating Heat Pipes and their Applications in Gas Turbine Cooling," *Frontiers in Heat Pipes (FHP),* **1**, 023002.
http://dx.doi.org/10.5098/fhp.v1.2.3002

Cao, Y., and Faghri, A., 1990, "Transient Two-Dimensional Compressible Analysis for High-Temperature Heat Pipes with Pulsed Heat Input," *Numerical Heat Transfer; Part A: Applications,* **18**(4), 483-502.
http://dx.doi.org/10.1080/10407789008944804

Cao, Y., and Faghri, A., 1991, "Transient Multidimensional Analysis of Nonconventional Heat Pipes with Uniform and Nonuniform Heat Distributions," *Journal of Heat Transfer,* **113**(4), 995-1002.
http://dx.doi.org/10.1115/1.2911233

Cao, Y., and Faghri, A., 1992, "Closed-Form Analytical Solutions of High-Temperature Heat Pipe Startup and Frozen Startup Limitation," *Journal of Heat Transfer,* **114**(4), 1028-1035.
http://dx.doi.org/10.1115/1.2911873

Cao, Y., and Faghri, A., 1993a, "Conjugate Modeling of High-Temperature Nosecap and Wing Leading-Edge Heat Pipes," *Journal of Heat Transfer,* **115**(3), 819-822.
http://dx.doi.org/10.1115/1.2910765

Cao, Y., and Faghri, A., 1993b, "Simulation of the Early Startup Period of High-Temperature Heat Pipes from the Frozen State by a Rarefied Vapor Self-Diffusion Model," *Journal of Heat Transfer,* **115**(1), 239-246.
http://dx.doi.org/10.1115/1.2910655

Cao, Y., and Faghri, A., 1993c, "A Numerical Analysis of High-Temperature Heat Pipe Startup from the Frozen State," *Journal of Heat Transfer,* **115**(1), 247-254.
http://dx.doi.org/10.1115/1.2910657

Cao, Y., and Faghri, A., 1994a, "Micro/Miniature Heat Pipes and Operating Limitations," *Journal of Enhanced Heat Transfer,* **1**(3), 265-274.

Cao, Y., and Faghri, A., 1994b, "Analytical Solutions of Flow and Heat Transfer in a Porous Structure with Partial Heating and Evaporation on the Upper Surface," *International Journal of Heat and Mass Transfer,* **37**(10), 1525-1533.
http://dx.doi.org/10.1016/0017-9310(94)90154-6

Cao, Y., and Faghri, A., 1994c, "Conjugate Analysis of a Flat-Plate Type Evaporator for Capillary Pumped Loops with Three-Dimensional Vapor Flow in the Groove," *International Journal of Heat and Mass Transfer,* **37**(3), 401-409.
http://dx.doi.org/10.1016/0017-9310(94)90074-4

Cao, Y., Faghri, A., and Mahefkey, E. T., 1989, "The Thermal Performance of Heat Pipes with Localized Heat Input," *International Journal of Heat and Mass Transfer,* **32**(7), 1279-1287.
http://dx.doi.org/10.1016/0017-9310(89)90028-8

Cao, Y., Gao, M., Beam, J. E., and Donovan, B., 1997, "Experiments and Analyses of Flat Miniature Heat Pipes," *Journal of Thermophysics and Heat Transfer,* **11**(2), 158-164.
http://dx.doi.org/10.2514/2.6247

Chalmers, D. R., Fredley, J., Ku, J., and Kroliczek, E. J., 1988, "Design of a Two-Phase Capillary Pumped Flight Experiment," SAE 88-1086, *Proceedings of 18th Intersociety Conference on Environmental Systems,* San Francisco, CA.

Chen, M. M., and Faghri, A., 1990, "An Analysis of the Vapor Flow and the Heat Conduction through the Liquid-Wick and Pipe Wall in a Heat Pipe with Single Or Multiple Heat Sources," *International Journal of Heat and Mass Transfer,* **33**(9), 1945-1955.
http://dx.doi.org/10.1016/0017-9310(90)90226-K

Chengsheng, H., Tinghan, L., Yaopu, W., and Zengqi, H., 1984, "Design and Performance of Low Temperature Heat Pipe Blackbody," *Proceedings of the 5th International Heat Pipe Conference,* Tsukuba, Japan, 74-81.

Chernysheva, M. A., and Maydanik, Y. F., 2008, "Numerical Simulation of Transient Heat and Mass Transfer in a Cylindrical Evaporator of a Loop Heat Pipe," *International Journal of Heat and Mass Transfer,* **51**(17-18), 4204-4215.
http://dx.doi.org/10.1016/j.ijheatmasstransfer.2007.12.021

Chuang, P. A., 2003, "An Improved Steady-State Model of Loop Heat Pipe Based on Experimental and Theoretical Analyses," Ph.D. Thesis, Pennsylvania State University, State College, PA.

Cotter, T. P., 1965, "Theory of Heat Pipes," Los Alamos Scientific Laboratory Report No. LA-3246-MS.

Cotter, T. P., 1984, "Principles and Prospects for Micro Heat Pipes," *Proceedings of 5th International Heat Pipe Conference,* Tsukuba, Japan, 328-335.

Cullimore, B., 1991, "Startup Transient in Capillary Pumped Loops," AIAA-91-1374, Honolulu, HI.

Cullimore, B., and Bauman, J., 2000, "Steady State and Transient Loop Heat Pipe Modeling," SAE 2000-01-2316, *Proceedings of the 34th International Conference on Environmental Systems (ICES),* Toulouse, France.

Daniels, T. C., and Al-Jumaily, F. K., 1975, "Investigations of the Factors Affecting the Performance of a Rotating Heat Pipe," *International Journal of Heat and Mass Transfer,* **18**(7-8), 961-973.
http://dx.doi.org/10.1016/0017-9310(75)90190-8

Do, K. H., Kim, S. J., and Garimella, S. V., 2008, "A Mathematical Model for Analyzing the Thermal Characteristics of a Flat Micro Heat Pipe with a Grooved Wick," *International Journal of Heat and Mass Transfer,* **51**(19-20), 4637-4650.
http://dx.doi.org/10.1016/j.ijheatmasstransfer.2008.02.039

Dobson, R. T., 2004, "Theoretical and Experimental Modelling of an Open Oscillatory Heat Pipe Including Gravity," *International Journal of Thermal Sciences,* **43**(2), 113-119.
http://dx.doi.org/10.1016/j.ijthermalsci.2003.05.003

Dobson, R. T., 2005, "An Open Oscillatory Heat Pipe Water Pump," *Applied Thermal Engineering,* **25**(4), 603-621.
http://dx.doi.org/10.1016/j.applthermaleng.2004.07.005

Dobson, R. T., and Harms, T. M., 1999, "Lumped Parameter Analysis of Closed and Open Oscillatory Heat Pipes," *Proceedings of the 11th International Heat Pipe Conference,* Tokyo, Japan, 12-16.

Edwards, D. K., and Marcus, B. D., 1972, "Heat and Mass Transfer in the Vicinity of the Vapor-Gas Front in a Gas-Loaded Heat Pipe,"

Journal of Heat Transfer, Transactions ASME, **94**(2), 155-162. http://dx.doi.org/10.1115/1.3449887

El-Genk, M. S., and Huang, L., 1993, "An Experimental Investigation of the Transient Response of a Water Heat Pipe," *International Journal of Heat and Mass Transfer,* **36**(15), 3823-3830. http://dx.doi.org/10.1016/0017-9310(93)90062-B

El-Genk, M. S., and Saber, H. H., 1999, "Determination of Operation Envelopes for Closed, Two-Phase Thermosyphons," *International Journal of Heat and Mass Transfer,* **42**(5), 889-903. http://dx.doi.org/10.1016/S0017-9310(98)00212-9

Faghri, A., 1989, "Performance Characteristics of a Concentric Annular Heat Pipe: Part II-Vapor Flow Analysis," *Journal of Heat Transfer,* **111**(4), 851-857. http://dx.doi.org/10.1115/1.3250796

Faghri, A., 1990, "Thermal Energy Storage Heat Exchangers," U.S. Patent No. 4976308.

Faghri, A., 1991, "Micro Heat Pipe Energy Storage Systems," U.S. Patent No. 5000252.

Faghri, A., 1993a, "Centrifugal Heat Pipe Vapor Absorption Heat Pump," U.S. Patent No. 5201196.

Faghri, A., 1993b, "Temperature Regulation System for the Human Body using Heat Pipes," U.S. Patent No. 5269369.

Faghri, A., 1994, "Centrifugal Heat Pipe System," U.S. Patent No. 5297619.

Faghri, A., 1995, Heat Pipe Science and Technology, 1st ed., Taylor & Francis, Washington, D.C.

Faghri, A., and Buchko, M., 1991, "Experimental and Numerical Analysis of Low-Temperature Heat Pipes with Multiple Heat Sources," *Journal of Heat Transfer,* **113**(3), 728-734. http://dx.doi.org/10.1115/1.2910624

Faghri, A., Buchko, M., and Cao, Y., 1991a, "A Study of High-Temperature Heat Pipes with Multiple Heat Sources and Sinks: Part I - Experimental Methodology and Frozen Startup Profiles," *Journal of Heat Transfer,* **113**(4), 1003-1009. http://dx.doi.org/10.1115/1.2911193

Faghri, A., Buchko, M., and Cao, Y., 1991b, "A Study of High-Temperature Heat Pipes with Multiple Heat Sources and Sinks: Part II - Analysis of Continuum Transient and Steady- State Experimental Data with Numerical Predictions," *Transactions of the ASME Journal of Heat Transfer,* **113**(4), 1010-1016. http://dx.doi.org/10.1115/1.2911194

Faghri, A., Chen, M. M., and Morgan, M., 1989a, "Heat Transfer Characteristics in Two-Phase Closed Conventional and Concentric Annular Thermosyphons," *Journal of Heat Transfer,* **111**(3), 611-618. http://dx.doi.org/10.1115/1.3250726

Faghri, A., Gogineni, S., and Thomas, S., 1993, "Vapor Flow Analysis of an Axially Rotating Heat Pipe," *International Journal of Heat and Mass Transfer,* **36**(9), 2293-2303. http://dx.doi.org/10.1016/S0017-9310(05)80114-0

Faghri, A., and Guo, Z., 2009, "Vapor Feed Fuel Cells with a Passive Thermal-Fluids Management System," U.S. Patent No. 7625649B1.

Faghri, A., and Parvani, S., 1988, "Numerical Analysis of Laminar Flow in a Double-Walled Annular Heat Pipe," *Journal of Thermophysics and Heat Transfer,* **2**(3), 165-171. http://dx.doi.org/10.2514/3.81

Faghri, A., Reynolds, D., and Faghri, P., 1989b, "Heat Pipes for Hands," *Mechanical Engineering,* **111**(6), 72-75.

Faghri, A., Stewart, R. J., and Rainey, C. L., 1984, "Axial Variation of Local Heat Flux Along the Condenser Section of a Double Wall Artery High Capacity Heat Pipe," *Proceedings of the 5th International Heat Pipe Conference,* Tsukuba, Japan, 13-17.

Faghri, A., 2012, "Review and Advances in Heat Pipe Science and Technology," *Journal of Heat Transfer,* **134**(12), 123001. http://dx.doi.org/10.1115/1.4007407

Faghri, A., and Guo, Z., 2005, "Challenges and Opportunities of Thermal Management Issues Related to Fuel Cell Technology and Modeling," *International Journal of Heat and Mass Transfer,* **48**(19-20), 3891-3920. http://dx.doi.org/10.1016/j.ijheatmasstransfer.2005.04.014

Faghri, A., and Harley, C., 1994, "Transient Lumped Heat Pipe Analyses," *Heat Recovery Systems and CHP,* **14**(4), 351-363. http://dx.doi.org/10.1016/0890-4332(94)90039-6

Faghri, A., and Khrustalev, D., 1997, "Advances in Modeling of Enhanced Flat Miniature Heat Pipes with Capillary Grooves," *Journal of Enhanced Heat Transfer,* **4**(2), 99-109.

Faghri, A., and Thomas, S., 1989, "Performance Characteristics of a Concentric Annular Heat Pipe: Part I-Experimental Prediction and Analysis of the Capillary Limit," *Journal of Heat Transfer,* **111**(1-4), 844-850. http://dx.doi.org/10.1115/1.3250795

Faghri, A., 1986, "Vapor Flow Analysis in a Double-Walled Concentric Heat Pipe," *Numerical Heat Transfer,* **10**(6), 583-595.

Faghri, A., and Guo, Z., 2008, "Integration of Heat Pipe into Fuel Cell Technology," *Heat Transfer Engineering,* **29**(3), 232-238. http://dx.doi.org/10.1080/01457630701755902

Finlay, I., Cree, D., and Blundell, D., 1976, "Performance of a Prototype Isothermal Oven for Use with an 'X' Band Microwave Noise Standard," *Proceedings of the 2nd International Heat Pipe Conference,* Bologna, Italy, 545-554.

Fletcher, L. S., and Peterson, G. P., 1997, U.S. Patent No. 5591162.

Furukawa, M., 2006, "Model-Based Method of Theoretical Design Analysis of a Loop Heat Pipe," *Journal of Thermophysics and Heat Transfer,* **20**(1), 111-121. http://dx.doi.org/10.2514/1.14675

Gao, M., Cao, Y., Beam, J. E., and Donovan, B., 2000, "Structural Optimization of Axially Grooved Flat Miniature Heat Pipes," *Journal of Enhanced Heat Transfer,* **7**(6), 361-369.

Gaugler, R., 1944, "Heat Transfer Device," U.S. Patent No. 2350348.

Gi, K., and Maezawa, S., 2006, "Study on Solar Cell Cooling by Heat Pipe," *Proceedings of the 8th International Heat Pipe Symposium,* Kumamoto, Japan, 295-299.

Giessler, F., Sattler, P., and Thoren, F., 1987, "Heat Pipe Cooling of Electrical Machines," *Proceedings of the 6th International Heat Pipe Conference,* Grenoble, France, 557-564.

Gray, V. H., 1969, "The Rotating Heat Pipe—A Wickless, Hollow Shaft for Transferring High Heat Fluxes," *Proceedings of ASME/AIChE Heat Transfer Conference,* Minneapolis, MN, 1-5.

Grover, G., 1966, "Evaporation-Condensation Heat Transfer Device," U.S. Patent No. 3229759.

Grover, G. M., Cotter, T. P., and Erickson, G. F., 1964, "Structures of very High Thermal Conductance," *Journal of Applied Physics,* **35**(6), 1990-1991.
http://dx.doi.org/10.1063/1.1713792

Guo, Z., and Faghri, A., 2006a, "Development of Planar Air Breathing Direct Methanol Fuel Cell Stacks," *Journal of Power Sources,* **160**(2), 1183-1194.
http://dx.doi.org/10.1016/j.jpowsour.2006.03.045

Guo, Z., and Faghri, A., 2006b, "Miniature DMFCs with Passive Thermal-Fluids Management System," *Journal of Power Sources,* **160**(2), 1142-1155.
http://dx.doi.org/10.1016/j.jpowsour.2006.03.013

Hall, M. L., Merrigan, M. A., and Reid, R. S., 1994, "Status Report on the THROHPUT Transient Heat Pipe Modeling Code," AIP-CP-301, *Proceedings of the 11th Symposium on Space Nuclear Power and Propulsion,* American Institute of Physics, New York, NY, 965-970.

Hamdan, M. O., 2003, "Loop Heat Pipe (LHP) Modeling and Development by Utilizing Coherent Porous Silicon (CPS) Wicks," Ph.D. Thesis, University of Cincinnati, OH.

Harley, C., and Faghri, A., 1994a, "Transient Gas-Loaded Thermosyphon Analysis," *Proceedings of the 10th International Heat Transfer Conference,* Brighton, England.

Harley, C., and Faghri, A., 2000, "Transient Gas-Loaded Rotating Heat Pipes," *Proceedings of the 15th National and 4th ISHMT/ASME Heat and Mass Transfer Conference,* Pune, India.

Harley, C., and Faghri, A., 1994b, "Complete Transient Two-Dimensional Analysis of Two-Phase Closed Thermosyphons Including the Falling Condensate Film," *Journal of Heat Transfer,* **116**(2), 418-426.
http://dx.doi.org/10.1115/1.2911414

Harley, C., and Faghri, A., 1994c, "Transient Two-Dimensional Gas-Loaded Heat Pipe Analysis," *Journal of Heat Transfer,* **116**(3), 716-723.
http://dx.doi.org/10.1115/1.2910927

Harley, C., and Faghri, A., 1995, "Two-Dimensional Rotating Heat Pipe Analysis," *Journal of Heat Transfer,* **117**(1), 202-208.
http://dx.doi.org/10.1115/1.2822304

Hassan, H., and Accensi, A., 1973, "Spacecraft Application of Low Temperature Heat Pipes," Paper No. 9-1, *Proceedings of the 1st International Heat Pipe Conference,* Stuttgart, Federal Republic of Germany.

Hijikata, K., Chen, S. J., and Tien, C. L., 1984, "Noncondensible Gas Effect on Condensation in a Two-Phase Closed Thermosyphon," *International Journal of Heat and Mass Transfer,* **27**(8), 1319-1325.
http://dx.doi.org/10.1016/0017-9310(84)90059-0

Hoang, T. T., and Kaya, T., 1999, "Mathematical Modeling of Loop Heat Pipe with Two-Phase Pressure Drop," AIAA Paper No. 1999-3448.

Hoang, T., and Ku, J., 2003, "Transient Modeling of Loop Heat Pipes," AIAA Paper No. 2003-6082.

Holley, B., and Faghri, A., 2005, "Analysis of Pulsating Heat Pipe with Capillary Wick and Varying Channel Diameter," *International Journal of Heat and Mass Transfer,* **48**(13), 2635-2651.
http://dx.doi.org/10.1016/j.ijheatmasstransfer.2005.01.013

Holmes, H. R., and Field, A. R., 1986, "The Gas-Tolerant High-Capacity Tapered Artery Heat Pipe," AIAA-86-2343, *Proceedings of AIAA/ASME 4th Joint Thermophysics and Heat Transfer Conference,* Boston, MA.

Hopkins, R., Faghri, A., and Khrustalev, D., 1999, "Flat Miniature Heat Pipes with Micro Capillary Grooves," *Journal of Heat Transfer,* **121**(1), 102-109.
http://dx.doi.org/10.1115/1.2825922

Hosoda, M., Nishio, S., and Shirakashi, R., 1999, "Meandering Closed Loop Heat Transport Tube (Propagation Phenomena of Vapor Plug)," AJTE99-6306, *Proceedings of the 5th ASME/JSME Joint Thermal Engineering Conference,* San Diego, CA, USA, 1-6.

Ivanovskii, M. N., Sorokin, V. P., and Yagodkin, I. V., 1982, The Physical Principles of Heat Pipes, Clarendon Press, Oxford.

Iwata, Z., Sakuma, S., and Koga, S., 1984, "Heat Pipes for Local Cooling of Underground Power Cables," *Proceedings of the 5th International Heat Pipe Conference,* Tsukuba, Japan, 153-155.

Jalilvand, A., Katsuta, M., Saito, K., Toyonaga, J., and Mochizuki, M., 2006, "The study of the thermal performance of concentric annular heat pipe with the application in the heat roll of fusing unit of copy machine," *Annals of the Assembly for International Heat Transfer Conference 13.*
http://dx.doi.org/10.1615/IHTC13.p22.60

Jang, J. H., Faghri, A., and Chang, W. S., 1991, "Analysis of the One-Dimensional Transient Compressible Vapor Flow in Heat Pipes," *International Journal of Heat and Mass Transfer,* **34**(8), 2029-2037.
http://dx.doi.org/10.1016/0017-9310(91)90214-Y

Jang, J. H., Faghri, A., Chang, W. S., and Mahefkey, E. T., 1990, "Mathematical Modeling and Analysis of Heat Pipe Start-Up from the Frozen State," *Journal of Heat Transfer,* **112**(3), 586-594.
http://dx.doi.org/10.1115/1.2910427

Jiao, B., Qiu, L. M., Gan, Z. H., and Zhang, X. B., 2012, "Determination of the Operation Range of a Vertical Two-Phase Closed Thermosyphon," *Heat and Mass Transfer,* **48**(6), 1043-1055.
http://dx.doi.org/10.1007/s00231-011-0954-x

Jiao, B., Qiu, L. M., Zhang, X. B., and Zhang, Y., 2008, "Investigation on the Effect of Filling Ratio on the Steady-State Heat Transfer Performance of a Vertical Two-Phase Closed Thermosyphon," *Applied Thermal Engineering,* **28**(11-12), 1417-1426.
http://dx.doi.org/10.1016/j.applthermaleng.2007.09.009

Jouhara, H., and Robinson, A. J., 2009, "An Experimental Study of Small-Diameter Wickless Heat Pipes Operating in the Temperature Range 200c to 450c," *Heat Transfer Engineering,* **30**(13), 1041-1048. http://dx.doi.org/10.1080/01457630902921113

Kaya, T., and Ku, J., 1999, "A Parametric Study of Performance Characteristics of Loop Heat Pipes," SAE Paper No. 2001-01-2317.

Kaya, T., and Goldak, J., 2006, "Numerical Analysis of Heat and Mass Transfer in the Capillary Structure of a Loop Heat Pipe," *International Journal of Heat and Mass Transfer,* **49**(17-18), 3211-3220. http://dx.doi.org/10.1016/j.ijheatmasstransfer.2006.01.028

Kaya, T., and Hoang, T. T., 1999, "Mathematical Modeling of Loop Heat Pipes and Experimental Validation," *Journal of Thermophysics and Heat Transfer,* **13**(3), 314-320. http://dx.doi.org/10.2514/2.6461

Kaya, T., and Ku, J., 2003, "Thermal Operational Characteristics of a Small-Loop Heat Pipe," *Journal of Thermophysics and Heat Transfer,* **17**(4), 464-470. http://dx.doi.org/10.2514/2.6805

Kaya, T., Pérez, R., Gregori, C., and Torres, A., 2008, "Numerical Simulation of Transient Operation of Loop Heat Pipes," *Applied Thermal Engineering,* **28**(8-9), 967-974. http://dx.doi.org/10.1016/j.applthermaleng.2007.06.037

Kempers, R., Robinson, A. J., Ewing, D., and Ching, C. Y., 2008, "Characterization of Evaporator and Condenser Thermal Resistances of a Screen Mesh Wicked Heat Pipe," *International Journal of Heat and Mass Transfer,* **51**(25-26), 6039-6046. http://dx.doi.org/10.1016/j.ijheatmasstransfer.2008.04.001

Khalkhali, H., Faghri, A., and Zuo, Z. J., 1999, "Entropy Generation in a Heat Pipe System," *Applied Thermal Engineering,* **19**(10), 1027-1043. http://dx.doi.org/10.1016/S1359-4311(98)00089-1

Khandekar, S., and Gupta, A., 2007, "Embedded Pulsating Heat Pipe Radiators," *Proceedings of the 14th International Heat Pipe Conference,* Florianopolis, Brazil, 22-27.

Khandekar, S., Panigrahi, P. K., Lefèvre, F., and Bonjour, J., 2010, "Local Hydrodynamics of Flow in a Pulsating Heat Pipe: A Review," *Frontiers in Heat Pipes (FHP),* **1**, 023003. http://dx.doi.org/10.5098/fhp.v1.2.3002

Khandekar, S., Schneider, M., Schäfer, P., Kulenovic, R., and Groll, M., 2002, "Thermofluid Dynamic Study of Flat-Plate Closed-Loop Pulsating Heat Pipes," *Microscale Thermophysical Engineering,* **6**(4), 303-317. http://dx.doi.org/10.1080/10893950290098340

Khrustalev, D., 2010, "Advances in transient modeling of loop heat pipe systems with multiple components," *AIP Conference Proceedings,* **1208**, 55-67.

Khrustalev, D., and Faghri, A., 1995a, "Thermal Characteristics of Conventional and Flat Miniature Axially-Grooved Heat Pipes," *Journal of Heat Transfer,* **117**(4), 1048-1054. http://dx.doi.org/10.1115/1.2836280

Khrustalev, D., and Faghri, A., 1994, "Thermal Analysis of a Micro Heat Pipe," *Journal of Heat Transfer,* **116**(1), 189-198. http://dx.doi.org/10.1115/1.2910855

Khrustalev, D., and Faghri, A., 1995b, "Heat Transfer during Evaporation on Capillary-Grooved Structures of Heat Pipes," *Journal of Heat Transfer,* **117**(3), 740-747. http://dx.doi.org/10.1115/1.2822638

Khrustalev, D., and Faghri, A., 1995c, "Heat Transfer in the Inverted Meniscus Type Evaporator at High Heat Fluxes," *International Journal of Heat and Mass Transfer,* **38**(16), 3091-3101. http://dx.doi.org/10.1016/0017-9310(95)00003-R

Khrustalev, D., and Faghri, A., 1996a, "Enhanced Flat Miniature Axially Grooved Heat Pipe," *Journal of Heat Transfer,* **118**(1), 261-264. http://dx.doi.org/10.1115/1.2824057

Khrustalev, D., and Faghri, A., 1996b, "Estimation of the Maximum Heat Flux in the Inverted Meniscus Type Evaporator of a Flat Miniature Heat Pipe," *International Journal of Heat and Mass Transfer,* **39**(9), 1899-1909. http://dx.doi.org/10.1016/0017-9310(95)00270-7

Kiper, A. M., Feric, G., Anjum, M., and Swanson, T. D., 1990, "Transient Analysis of a Capillary Pumped Loop Heat Pipe," AIAA-90-1685, *Proceedings of AIAA/ASME 5th Joint Thermophysics and Heat Transfer Conference,* Seattle, WA.

Kirkpatrick, J. P., and Brennan, P. J., 1976, "Long Term Performance of the Advanced Thermal Control Experiment," *Proceedings of the 2nd International Heat Pipe Conference,* Bolgna, Italy, 629-646.

Kobayashi, Y., and Matsumoto, T., 1987, "Vapor Condensation in the Presence of Non-Condensible Gas in the Gravity Assisted Thermosyphon," *Proceedings of 6th International Heat Pipe Conference,* Grenoble, France.

Kroliczek, E. J., Ku, J., and Ollendorf, S., 1984, "Design, Development, and Test of a Capillary Pump Heat Pipe," AIAA-84-1720, *Proceedings of AIAA 19th Thermophysics Conference,* Snowmass, CO.

Ku, J., 1993, "Capillary Pump Loop for the Systems of Thermal Regulation of Spacecraft," *Proceedings of ASME National Heat Transfer Conference,* Atlanta, GA.

Ku, J., Kroliczek, E. J., Butler, D., Schweickart, R. B., and McIntosh, R., 1986a, "Capillary Pumped Loop GAS and Hitchhiker Flight Experiments," AIAA-86-1249, *Proceedings of AIAA/ASME 4th Joint Thermophysics and Heat Transfer Conference,* Boston, MA.

Ku, J., Kroliczek, E. J., McCabe, M. E., and Benner, S. M., 1988, "A High Power Spacecraft Thermal Management System," AIAA-88-2702, *Proceedings of AIAA Thermophysics, Plasmadynamics, and Lasers Conference,* San Antonio, TX.

Ku, J., Kroliczek, E. J., and McIntosh, R., 1987a, "Analytical Modelling of the Capillary Pumped Loop," *Proceedings of the 6th International Heat Pipe Conference,* Grenoble, France.

Ku, J., Kroliczek, E. J., and McIntosh, R., 1987b, "Capillary Pumped Loop Technology Development," *Proceedings of the 6th International Heat Pipe Conference,* Grenoble, France.

Ku, J., Kroliczek, E. J., Taylor, W. J., and McIntosh, R., 1986b, "Functional and Performance Tests of Two Capillary Pumped Loop Engineering Models," AIAA-86-1248, *Proceedings of AIAA/ASME 4th Joint Thermophysics and Heat Transfer Conference,* Boston, MA.

Lamp, T., 1978, "Development of a Heat Pipe Pan Assembly for a Gas-Fired Commercial Tilting Braiser," Paper No. 78-399, *Proceedings of the 3rd International Heat Pipe Conference,* Palo Alto, CA.

Larkin, B., and Dubuc, S., 1976, "Self De-icing Navigation Buoys Using Heat Pipes," *Proceedings of the 2nd International Heat Pipe Conference,* Bologna, Italy, 529-536.

Launay, S., Sartre, V., and Bonjour, J., 2007a, "Effects of Fluid Thermophysical Properties on Loop Heat Pipe operation," *Proceedings of the 14th International Heat Pipe Conference,* Florianopolis, Brazil.

Launay, S., Platel, V., Dutour, S., and Joly, J. L., 2007b, "Transient Modeling of Loop Heat Pipes for the Oscillating Behavior Study," *Journal of Thermophysics and Heat Transfer,* **21**(3), 487-495. http://dx.doi.org/10.2514/1.26854

Launay, S., Sartre, V., and Bonjour, J., 2007c, "Parametric Analysis of Loop Heat Pipe Operation: A Literature Review," *International Journal of Thermal Sciences,* **46**(7), 621-636. http://dx.doi.org/10.1016/j.ijthermalsci.2006.11.007

Launay, S., Sartre, V., and Bonjour, J., 2008, "Analytical Model for Characterization of Loop Heat Pipes," *Journal of Thermophysics and Heat Transfer,* **22**(4), 623-631. http://dx.doi.org/10.2514/1.37439

Launay, S., Sartre, V., and Lallemand, M., 2004a, "Experimental Study on Silicon Micro-Heat Pipe Arrays," *Applied Thermal Engineering,* **24**(2-3), 233-243. http://dx.doi.org/10.1016/j.applthermaleng.2003.08.003

Launay, S., Sartre, V., Mantelli, M. H., De Paiva, K. V., and Lallemand, M., 2004b, "Investigation of a Wire Plate Micro Heat Pipe Array," *International Journal of Thermal Sciences,* **43**(5), 499-507. http://dx.doi.org/10.1016/j.ijthermalsci.2003.10.006

Le Berre, M., Launay, S., Sartre, V., and Lallemand, M., 2003, "Fabrication and Experimental Investigation of Silicon Micro Heat Pipes for Cooling Electronics," *Journal of Micromechanics and Microengineering,* **13**(3), 436-441. http://dx.doi.org/10.1088/0960-1317/13/3/313

Lefèvre, F., Lips, S., and Bonjour, J., 2010, "Investigation of Evaporation and Condensation Processes Specific to Grooved Flat Heat Pipes," *Frontiers in Heat Pipes (FHP),* **1**, 023001. http://dx.doi.org/10.5098/fhp.v1.2.3001

Lefèvre, F., and Lallemand, M., 2006, "Coupled Thermal and Hydrodynamic Models of Flat Micro Heat Pipes for the Cooling of Multiple Electronic Components," *International Journal of Heat and Mass Transfer,* **49**(7-8), 1375-1383. http://dx.doi.org/10.1016/j.ijheatmasstransfer.2005.10.001

Liang, S. B., and Ma, H. B., 2004, "Oscillating Motions of Slug Flow in Capillary Tubes," *International Communications in Heat and Mass Transfer,* **31**(3), 365-375. http://dx.doi.org/10.1016/j.icheatmasstransfer.2004.02.007

Lin, L. C., and Faghri, A., 1998a, "Condensation in Rotating Stepped Wall Heat Pipes with Hysteretic Annular Flow," *Journal of Thermophysics and Heat Transfer,* **12**(1), 94-99. http://dx.doi.org/10.2514/2.6307

Lin, L., and Faghri, A., 1997a, "Heat Transfer Analysis of Stratified Flow in Rotating Heat Pipes with Cylindrical and Stepped Walls," *International Journal of Heat and Mass Transfer,* **40**(18), 4393-4404. http://dx.doi.org/10.1016/S0017-9310(97)00060-4

Lin, L., and Faghri, A., 1997b, "Steady-State Performance in a Thermosyphon with Tube Separator," *Applied Thermal Engineering,* **17**(7), 667-679. http://dx.doi.org/10.1016/S1359-4311(96)00084-1

Lin, L., and Faghri, A., 1997c, "Steady-State Performance of a Rotating Miniature Heat Pipe," *Journal of Thermophysics and Heat Transfer,* **11**(4), 513-518. http://dx.doi.org/10.2514/2.6292

Lin, L., and Faghri, A., 1998b, "An Analysis of Two-Phase Flow Stability in a Thermosyphon with Tube Separator," *Applied Thermal Engineering,* **18**(6), 441-455. http://dx.doi.org/10.1016/S1359-4311(97)00046-X

Lin, L., and Faghri, A., 1999, "Heat Transfer in Micro Region of a Rotating Miniature Heat Pipe," *International Journal of Heat and Mass Transfer,* **42**(8), 1363-1369. http://dx.doi.org/10.1016/S0017-9310(98)00270-1

Ma, H. B., Hanlon, M. A., and Chen, C. L., 2006, "An Investigation of Oscillating Motions in a Miniature Pulsating Heat Pipe," *Microfluidics and Nanofluidics,* **2**(2), 171-179. http://dx.doi.org/10.1007/s10404-005-0061-8

Ma, H. B., Maschmann, M. R., and Liang, S. B., 2002, "Heat Transport Capability in a Pulsating Heat Pipe," AIAA 2002-2765.

Machiroutu, S., Kluge, B., Kuroda, M., and Pokharna, H., 2006, *Proceedings of the 8th International Heat Pipe Symposium,* Kumamoto, Japan, 46-50.

Maezawa, S., Suzuki, Y., and Tsuchida, A., 1981, "Heat Transfer Characteristics of Disk-Shaped Rotating, Wickless Heat Pipe," *Proceedings of the 4th International Heat Pipe Conference,* London, UK, 725-734.

Marcus, B. D., and Fleishman, G. L., 1970, "Steady State and Transient Performance of Hot Reservoir Gas-Controlled Heat Pipes," ASME Paper No. 70-HT/SpT-11.

Mashiko, K., Okiai, R., Mochizuki, M., Ryokai, K., Tsuchiya, F., and Fukuda, M., 1989, "Development of an Artificial Permafrost Storage Using Heat Pipes," Paper No. 89-HT-15, *ASME/AIChE National Heat Transfer Conference,* Philadelphia, PA.

Matsuda, S., Miskolczy, G., Okihara, M., Okihara, T., Kanamori, M., and Hamada, N., 1981, "Test of a Horizontal Heat Pipe Deicing Panel for Use on Marine Vessels," *Proceedings of the 4th International Heat Pipe Conference,* London, UK, 3-10.

Maydanik, Y. F., 2005, "Loop Heat Pipes," *Applied Thermal Engineering,* **25**(5-6), 635-657. http://dx.doi.org/10.1016/j.applthermaleng.2004.07.010

Maydanik, Y. F., Fershtater, Y. G., and Solodovnik, N., 1994, "Loop Heat Pipes: Design, Investigation, Prospects of Use in Aerospace Technics," SAE Paper 941185.

Maydanik, Y. F., Ferchtater, Y. G., and Goncharov, K. A., 1991, "Capillary Pump Loop for the systems of Thermal Regulation of Spacecraft," *4th European Symposium on Space Environment and Control Systems, ESA SP-324.* Florence, Italy.

McIntosh, R., Ollendorf, S., and Harwell, W., 1976, "The International Heat Pipe Experiment," *Proceedings of the 2nd International Heat Pipe Conference,* Bologna, Italy, 589-592.

Meijer, R., and Khalili, K., 1990, "Design and Testing of a Heat Pipe Gas Combustion System for the STM4-120 Stirling Engine," *Proceedings of the 7th International Heat Pipe Conference,* Minsk, USSR.

Mingwei, T., Li, C., Hongji, Z., and Qingming, T., 1991, "Start-Up Behavior of the Frozen Two-Phase Closed Steel-Water Thermosyphons," *Proceedings of the 8th International Heat Pipe Conference.*

Miyazaki, Y., and Akachi, H., 1996, "Heat Transfer Characteristics of Looped Capillary Heat Pipe," *Proceedings of the 5th International Heat Pipe Symposium,* Melbourne, Australia, 378-383.

Miyazaki, Y., and Akachi, H., 1998, "Self Excited Oscillation of Slug Flow in a Micro Channel," *Proceedings of the 3rd International Conference on Multiphase Flow,* Lyon, France.

Miyazaki, Y., and Arikawa, M., 1999, "Oscillatory Flow in the Oscillating Heat Pipe," *Proceedings of the 11th International Heat Pipe Conference,* Tokyo, Japan, 131-136.

Mochizuki, M., Nguyen, T., Saito, Y., Kiyooka, F., and Wuttijumnong, V., 2007, "Advanced micro-channel vapor chamber for cooling high power processors," *Proceedings of the ASME InterPack Conference, IPACK 2007,* Vancouver, Canada, 695-702.

Mochizuki, M., Saito, Y., Kiyooka, F., and Nguyen, T., 2006, "High Power Cooling Chips by Heat Pipes and Advanced Heat Spreader," *8th International Heat Pipe Symposium,* Kumamoto, Japan, 214-221.

Mochizuki, M., Nguyen, T., Mashiko, K., Saito, Y., Nguyen, T., and Wuttijumnong, V., 2011, "A Review of Heat Pipe Applications Including New Opportunities," *Frontiers in Heat Pipes (FHP),* **2**, 013001.
http://dx.doi.org/10.5098/fhp.v2.1.3001

Momose, T., Ishimaru, H., Murase, T., Tanaka, S., and Ishida, S., 1987, "Application of Heat Pipe to Electron-Position Collider TRISTAN," *Proceedings of the 6th International Heat Pipe Conference,* Grenoble, France, 798-799.

Morgownik, A., and Savage, C., 1987, "Design Aspects of a Deployable 10 kW Heat Pipe Radiator," *Proceedings of the 6th International Heat Pipe Conference,* Grenoble, France, 351-356.

Muraoka, I., Ramos, F. M., and Vlassov, V. V., 2001, "Analysis of the Operational Characteristics and Limits of a Loop Heat Pipe with Porous Element in the Condenser," *International Journal of Heat and Mass Transfer,* **44**(12), 2287-2297.
http://dx.doi.org/10.1016/S0017-9310(00)00259-3

Nguyen, T., Johnson, P., Akbarzadeh, A., Gibson, K., and Mochizuki, M., 1992, "Design, Manufacture and Testing of a Closed Cycle Thermosyphon Rankine Engine," *Proceedings of the 8th International Heat Pipe Conference,* Beijing, China.

Nithyanandam, K., and Pitchumani, R., 2012, "Computational studies on a latent thermal energy storage system with integral heat pipes for concentrating solar power," *Applied Energy,* (in press).
http://dx.doi.org/10.1016/j.apenergy.2012.09.056

Oslejsek, O., and Polasek, F., 1976, "Cooling of Electrical Machines by Heat Pipes," *Proceedings of the 2nd International Heat Pipe Conference,* Bologna, Italy, 503-514.

Pan, Y., 2001, "Condensation Heat Transfer Characteristics and Concept of Sub-Flooding Limit in a Two-Phase Closed Thermosyphon," *International Communications in Heat and Mass Transfer,* **28**(3), 311-322.
http://dx.doi.org/10.1016/S0735-1933(01)00237-8

Peck, S., and Fleischman, G., 1987, "Lightweight Heat Pipe Panels for Space Radiators," *Proceedings of the 6th International Heat Pipe Conference,* Grenoble, France, 362-367.

Perkins, J., 1836, UK Patent No. 7059.

Peterson, G. P., 1992, "Overview of Micro Heat Pipe Research and Development," *Applied Mechanics Reviews,* **45**(5), 175-189.
http://dx.doi.org/10.1115/1.3119755

Peterson, G. P., and Ma, H. B., 1999, "Temperature Response of Heat Transport in a Micro Heat Pipe," *Journal of Heat Transfer,* **121**(2), 438-445.
http://dx.doi.org/10.1115/1.2825997

Peterson, P. F., and Tien, C. L., 1989, "Numerical and Analytical Solutions for Two-Dimensional Gas Distribution in Gas-Loaded Heat Pipes," *Journal of Heat Transfer,* **111**(3), 598-604.
http://dx.doi.org/10.1115/1.3250724

Plesch, D., Bier, W., Seidel, D., and Schubert, K., 1991, "Miniature Heat Pipes for Heat Removal from Microelectronic Circuits," *Proceedings of the ASME Winter Annual Meeting,* Atlanta, GA.

Ponnappan, R., 1989, "Studies on the Startup Transients and Performance of a Gas Loaded Sodium Heat Pipe," Technical Report, WRDC-TR-89-2046.

Ponnappan, R., 1990, "Comparison of Vacuum and Gas Loaded Mode Performances of a LMHP," AIAA-90-1755, *Proceedings of AIAA/ASME 5th Joint Thermophysics and Heat Transfer Conference,* Seattle, WA.

Ranjan, R., Murthy, J. Y., Garimella, S. V., and Vadakkan, U., 2011, "A Numerical Model for Transport in Flat Heat Pipes Considering Wick Microstructure Effects," *International Journal of Heat and Mass Transfer,* **54**(1-3), 153-168.
http://dx.doi.org/10.1016/j.ijheatmasstransfer.2010.09.057

Rankin, J. G., 1984, "Integration and Flight Demonstration of a High Capacity Monogroove Heat Pipe Radiator," AIAA-84-1716, *Proceedings of the AIAA 19th Thermopyhsics Conference,* Snowmass, CO.

Reay, D., 1977, "Heat Pipes – A New Diecasting Aid," *Proceedings of the 1st National Diecasters Conference,* Birmingham Exhibition Centre, UK.

Rice, J., and Faghri, A., 2007, "Analysis of Porous Wick Heat Pipes, Including Capillary Dry-Out Limitations," *AIAA Journal of Thermophysics and Heat Transfer,* **21**(3), 475-486.
http://dx.doi.org/10.2514/1.24809

Robak, C. W., Bergman, T. L., and Faghri, A., 2011, "Enhancement of Latent Heat Energy Storage using Embedded Heat Pipes," *International*

Journal of Heat and Mass Transfer, **54**(15-16), 3476-3484. http://dx.doi.org/10.1016/j.ijheatmasstransfer.2011.03.038

Roberts, C., 1978, "A Non-tracking Moderately Focusing Heat Pipe Solar Collectors," Paper No. 78-402, *Proceedings of the 3rd International Heat Pipe Conference,* Palo Alto, CA.

Rohani, A. R., and Tien, C. L., 1973, "Steady Two-Dimensional Heat and Mass Transfer in the Vapor-Gas Region of a Gas-Loaded Heat Pipe," *Journal of Heat Transfer,* **95**(3), 377-382. http://dx.doi.org/10.1115/1.3450067

Rosenfeld, J. H., 1987, "Modeling of Heat Transfer into a Heat Pipe for a Localized Heat Input Zone," *AIChE Symposium Series,* **83**, 71-76.

Rosler, S., Groll, M., Supper, W., and Konev, S., 1987, "Analysis and Experimental Investigation of a Cryogenic Methane Heat Pipe," *Proceedings of the 6th International Heat Pipe Conference,* Grenoble, France, 219-222.

Sakulchangsatjatai, P., Terdtoon, P., Wongratanaphisan, T., Kamonpet, P., and Murakami, M., 2004, "Operation Modeling of Closed-End and Closed-Loop Oscillating Heat Pipes at Normal Operating Condition," *Applied Thermal Engineering,* **24**(7), 995-1008. http://dx.doi.org/10.1016/j.applthermaleng.2003.11.006

Sartre, V., Zaghdoudi, M. C., and Lallemand, M., 2000, "Effect of Interfacial Phenomena on Evaporative Heat Transfer in Micro Heat Pipes," *International Journal of Thermal Sciences,* **39**(4), 498-504. http://dx.doi.org/10.1016/S1290-0729(00)00205-2

Schlitt, K. R., Kirkpatrick, J. P., and Brennan, P. J., 1974, "Parametric Performance of Extruded Axial Grooved Heat Pipes from 100 K to 300 K," AIAA-74-724, *Proceedings of AIAA/ASME Thermophysics and Heat Transfer Conference.*

Schmalhofer, J., and Faghri, A., 1993a, "A Study of Circumferentially-Heated and Block-Heated Heat Pipes-I. Experimental Analysis and Generalized Analytical Prediction of Capillary Limits," *International Journal of Heat and Mass Transfer,* **36**(1), 201-212. http://dx.doi.org/10.1016/0017-9310(93)80080-E

Schmalhofer, J., and Faghri, A., 1993b, "A Study of Circumferentially-Heated and Block-Heated Heat Pipes-II. Three-Dimensional Numerical Modeling as a Conjugate Problem," *International Journal of Heat and Mass Transfer,* **36**(1), 213-226. http://dx.doi.org/10.1016/0017-9310(93)80081-5

Schulze, T., Sodtke, C., Stephan, P., and Gambaryan-Roisman, T., 2007, "Performance of Heat Pipe Evaporation for Space Applications with Combined Re-Entrant and Micro Grooves," *Proceedings of the 14th International Heat Pipe Conference,* Florianopolis, Brazil.

Schweickart, R. B., and Buchko, M. T., 1991, "Development and Test Results of a Two-Phase Reservoir for Thermal Transport Systems used in Micro-Gravity," *SAE (Society of Automotive Engineers) Transactions,* **100**, 1887-1899.

Shabgard, H., Bergman, T. L., Sharifi, N., and Faghri, A., 2010, "High Temperature Latent Heat Thermal Energy Storage using Heat Pipes," *International Journal of Heat and Mass Transfer,* **53**(15-16), 2979-2988. http://dx.doi.org/10.1016/j.ijheatmasstransfer.2010.03.035

Shabgard, H., and Faghri, A., 2011, "Performance Characteristics of Cylindrical Heat Pipes with Multiple Heat Sources," *Applied Thermal Engineering,* **31**(16), 3410-3419. http://dx.doi.org/10.1016/j.applthermaleng.2011.06.026

Shabgard, H., Robak, C. W., Bergman, T. L., and Faghri, A., 2012, "Heat Transfer and Exergy Analysis of Cascaded Latent Heat Storage with Gravity-Assisted Heat Pipes for Concentrating Solar Power Applications," *Solar Energy,* **86**(3), 816-830. http://dx.doi.org/10.1016/j.solener.2011.12.008

Shafii, M. B., Faghri, A., and Zhang, Y., 2001, "Thermal Modeling of Unlooped and Looped Pulsating Heat Pipes," *Journal of Heat Transfer,* **123**(6), 1159-1172. http://dx.doi.org/10.1115/1.1409266

Shafii, M. B., Faghri, A., and Zhang, Y., 2002, "Analysis of Heat Transfer in Unlooped and Looped Pulsating Heat Pipes," *International Journal of Numerical Methods for Heat and Fluid Flow,* **12**(5), 585-609. http://dx.doi.org/10.1108/09615530210434304

Shiraishi, M., Mochizuki, M., Mashiko, K., Ito, M., Sughiharu, S., Yamagishi, Y., and Watanabe, F., 1992, "Snow Removal and Deicing Using Flexible Long Heat Pipes," *Symposium on Snow and Snow Related Problems,* International Glaciological Society.

Shukla, K. N., 1981, "Transient Response of a Gas-Controlled Heat Pipe," *AIAA Journal,* **19**(8), 1063-1070. http://dx.doi.org/10.2514/3.7842

Singh, R., Mochizuki, M., Nguyen, T., and Akbarzadeh, A., 2011, "Applications of Heat Pipes in Energy Conservation and Renewable Energy Based Systems," *Frontiers in Heat Pipes (FHP),* **2**, 033003. http://dx.doi.org/10.5098/fhp.v2.3.3003

Spendel, T., 1984, "Laminar Film Condensation Heat Transfer in Closed Two-Phase Thermosyphons," *Proceedings of 5th International Heat Pipe Conference,* Tsukuba, Japan.

Stenger, F. J., 1966, "Experimental Feasibility Study of Water-Filled Capillary Pumped Heat Transfer Loops," NASA X-1310, Nasa LeRC Report.

Suelau, H., Kroliczek, E., and Brinkman, C., 1976, "Application of Heat Pipes to Deicing Systems," *Proceedings of the 2nd International Heat Pipe Conference,* Bologna, Italy, 515-528.

Suman, B., and Kumar, P., 2005, "An Analytical Model for Fluid Flow and Heat Transfer in a Micro-Heat Pipe of Polygonal Shape," *International Journal of Heat and Mass Transfer,* **48**(21-22), 4498-4509. http://dx.doi.org/10.1016/j.ijheatmasstransfer.2005.05.001

Swanson, T. D., 2007, "Thermal Control Technologies for the New Age of Space Exploration," *Proceedings of the 14th International Heat Pipe Conference,* Florianopolis, Brazil.

Tang, X., and Park, C., 2008, "Vibration/Shock-Tolerant Capillary Two-Phase Loop Technology for Vehicle Thermal Control," *Proceedings of the ASME Summer Heat Transfer Conference,* Jacksonville, FL, 531-537. http://dx.doi.org/10.1115/HT2008-56349

Tarau, C., and Anderson, W. G., 2010, "Sodium Variable Conductance Heat Pipe for Radioisotope Stirling Systems - Design and Experimental Results," *IECEC 2010, 8th Annual International Energy Conversion Engineering Conference,* Nashville, TN.

Thieme, L. G., and Schreiber, J. G., 2003, "NASA GRC Stirling Technology Development Overview," *AIP Conference Proceedings,* **654**, 613-620.
http://dx.doi.org/10.1063/1.1541346

Tournier, J. M., and El-Genk, M. S., 1994, "A Heat Pipe Transient Analysis Model," *International Journal of Heat and Mass Transfer,* **37**(5), 753-762.
http://dx.doi.org/10.1016/0017-9310(94)90113-9

Tournier, J. M., and El-Genk, M. S., 1996, "A Vapor Flow Model for Analysis of Liquid-Metal Heat Pipe Startup from a Frozen State," *International Journal of Heat and Mass Transfer,* **39**(18), 3767-3780.
http://dx.doi.org/10.1016/0017-9310(96)00066-X

Tournier, J. M., and El-Genk, M. S., 2003, "Startup of a Horizontal Lithium-Molybdenum Heat Pipe from a Frozen State," *International Journal of Heat and Mass Transfer,* **46**(4), 671-685.
http://dx.doi.org/10.1016/S0017-9310(02)00324-1

Trefethen, L., 1962, "On the Surface Tension Pumping of Liquids or a Possible Role of the Candlewick in Space Exploration," G.E. Tech. Info., Serial No. 615 D114.

Trovole, H. P., and Raine, J. K., 1995, "The Design and Heat Pipe Tests for a Line Focus Solar Stirling Domestic Generation System," *Proceedings of the Institution of Mechanical Engineers, Part A: Journal of Power and Energy,* **209**(1), 27-36.

Vasiliev, L. L., 2005, "Heat Pipes in Modern Heat Exchangers," *Applied Thermal Engineering,* **25**(1), 1-19.
http://dx.doi.org/10.1016/j.applthermaleng.2003.12.004

Vieira da Cunha, A. F., and Mantelli, M. H., 2009, "Analytical and Experimental Analysis of a High Temperature Mercury Thermosyphon," *Journal of Heat Transfer,* **131**(9), 1-7.
http://dx.doi.org/10.1115/1.3089551

Wang, Y. X., and Peterson, G. P., 2002, "Analysis of Wire-Bonded Micro Heat Pipe Arrays," *Journal of Thermophysics and Heat Transfer,* **16**(3), 346-355.
http://dx.doi.org/10.2514/2.6711

Wang, Y., and Vafai, K., 2000, "An Experimental Investigation of the Transient Characteristics on a Flat-Plate Heat Pipe during Startup and Shutdown Operations," *Journal of Heat Transfer,* **122**(3), 525-535.
http://dx.doi.org/10.1115/1.1287725

Waters, E. D., 1976, "Heat Pipes for the Trans-Alaskan Pipeline," *Proceedings of the 2nd International Heat Pipe Conference,* Bolonga, Italy, 803-814.

Williams, R., 1978, "Investigation of a Cryogenic Thermal Diode," Paper No. 78-417, *Proceedings of the 3rd International Heat Pipe Conference,* Palo Alto, CA.

Winship, J., 1974, "Diecasting Sharpens its Edge," *American Machinist,* **118**, 77-78.

Wong, T. N., Tong, B. Y., Lim, S. M., and Ooi, K. T., 1999, "Theoretical Modeling of Pulsating Heat Pipe," *Proceedings of the 11th International Heat Pipe Conference,* Tokyo, Japan, 159-163.

Wu, X. P., Mochizuki, M., Mashiko, K., Nguyen, T., Nguyen, T., Wuttijumnong, V., Cabusao, G., Singh, R., and Akbarzadeh, A., 2011, "Cold Energy Storage Systems using Heat Pipe Technology for Cooling Data Centers," *Frontiers in Heat Pipes (FHP),* **2**, 013005.
http://dx.doi.org/10.5098/fhp.v2.1.3005

Wu, D., and Peterson, G. P., 1991, "Investigation of the Transient Characteristics of a Micro Heat Pipe," *Journal of Thermophysics and Heat Transfer,* **5**(2), 129-134.
http://dx.doi.org/10.2514/3.239

Xiao, B., and Faghri, A., 2008, "A Three-Dimensional Thermal-Fluid Analysis of Flat Heat Pipes," *International Journal of Heat and Mass Transfer,* **51**(11-12), 3113-3126.
http://dx.doi.org/10.1016/j.ijheatmasstransfer.2007.08.023

Yamawaki, S., Yoshida, T., Taki, M., and Mimura, F., 1998, "Fundamental Heat Transfer Experiments of Heat Pipes for Turbine Cooling," *Journal of Engineering for Gas Turbines and Power,* **120**(3), 580-587.
http://dx.doi.org/10.1115/1.2818186

Zhang, Y., and Faghri, A., 2002, "Heat Transfer in a Pulsating Heat Pipe with Open End," *International Journal of Heat and Mass Transfer,* **45**(4), 755-764.
http://dx.doi.org/10.1016/S0017-9310(01)00203-4

Zhang, Y., and Faghri, A., 2003, "Oscillatory Flow in Pulsating Heat Pipes with Arbitrary Numbers of Turns," *Journal of Thermophysics and Heat Transfer,* **17**(3), 340-347.
http://dx.doi.org/10.2514/2.6791

Zhang, Y., Faghri, A., and Shafii, M. B., 2002, "Analysis of Liquid-Vapor Pulsating Flow in a U-Shaped Miniature Tube," *International Journal of Heat and Mass Transfer,* **45**(12), 2501-2508.
http://dx.doi.org/10.1016/S0017-9310(01)00348-9

Zhang, Y., and Faghri, A., 2008, "Advances and Unsolved Issues in Pulsating Heat Pipes," *Heat Transfer Engineering,* **29**(1), 20-44.
http://dx.doi.org/10.1080/01457630701677114

Zhu, N., and Vafai, K., 1998a, "Analytical Modeling of the Startup Characteristics of Asymmetrical Flat-Plate and Disk-Shaped Heat Pipes," *International Journal of Heat and Mass Transfer,* **41**(17), 2619-2637.
http://dx.doi.org/10.1016/S0017-9310(97)00325-6

Zhu, N., and Vafai, K., 1998b, "Vapor and Liquid Flow in an Asymmetrical Flat Plate Heat Pipe: A Three-Dimensional Analytical and Numerical Investigation," *International Journal of Heat and Mass Transfer,* **41**(1), 159-174.
http://dx.doi.org/10.1016/S0017-9310(97)00075-6

Zuo, Z. J., North, M. T., and Ray, L., 1999, "Combined Pulsating and Capillary Heat Pipe Mechanism for Cooling of High Heat Flux Electronics," *Proceedings of ASME Heat Transfer Device Conference,* Nashville, TN, 2237-2243.

Zuo, Z. J., and Faghri, A., 1997, "Boundary Element Approach to Transient Heat Pipe Analysis," *Numerical Heat Transfer; Part A: Applications,* **32**(3), 205-220.
http://dx.doi.org/10.1080/10407789708913888

Zuo, Z. J., and Faghri, A., 1998, "A Network Thermodynamic Analysis of the Heat Pipe," *International Journal of Heat and Mass Transfer,* **41**(11), 1473-1484.
http://dx.doi.org/10.1016/S0017-9310(97)00220-2

Zuo, Z. J., Faghri, A., and Langston, L., 1998, "Numerical Analysis of Heat Pipe Turbine Vane Cooling," *Journal of Engineering for Gas Turbines and Power,* **120**(4), 735-743. http://dx.doi.org/10.1115/1.2818461

Zuo, Z. J., and Gunnerson, F. S., 1995, "Heat Transfer Analysis of an Inclined Two-Phase Closed Thermosyphon," *Journal of Heat Transfer,* **117**(4), 1073-1075. http://dx.doi.org/10.1115/1.2836287

Zuo, Z. J., North, M. T., and Wert, K. L., 2001, "High Heat Flux Heat Pipe Mechanism for Cooling of Electronics," *IEEE Transactions on Components and Packaging Technologies,* **24**(2), 220-225. http://dx.doi.org/10.1109/6144.926386

HEAT TRANSFER BETWEEN SOLID SURFACES IN CONTACT HAVING A RELATIVELY LARGE GAP (EXTERNAL THERMAL RESISTANCE OF A HEAT PIPE)

Yasushi Koito* Masakazu Tsurumoto, Fumiya Hatano, Toshio Tomimura

Department of Advanced Mechanical Systems, Kumamoto University, Kumamoto-shi, Kumamoto, 860-8555, Japan

ABSTRACT

Experimental and numerical analyses are conducted to investigate the heat transfer characteristics of two rectangular rods in contact, where a relatively large gap is present. Sixteen types of wedge-shaped gaps having different sizes are investigated in the experiment. Using the thermal resistance network method, the temperature distribution is obtained numerically and the discussion is made on the heat transfer characteristics around the gap. It is confirmed that the constriction/spreading of heat flow inside the rods greatly affects the thermal resistance of the gap. Fairly good agreement is obtained between the experimental and the numerical results.

Keywords: *Electronics cooling, Thermal management, Thermal contact resistance, Wedge-shaped gap, Constriction/spreading heat flow.*

1. INTRODUCTION

A heat pipe has been widely used in electronics cooling (Faghri, 2014; Mochizuki *et al*., 2011). In this application, since the heat pipe is in contact with a heat source and a heat sink, not only internal thermal resistances of the heat pipe but also external thermal resistances, namely, thermal contact resistances of the heat pipe with the heat source and the heat sink are important factors for thermal management. Thermal interface material such as thermal grease is used to reduce the thermal contact resistance; however, it cannot be neglected if a relatively large gap is present between surfaces.

Concerning the measurement of contact heat transfer between two cylindrical specimens, the effect of curvature of contact surfaces on the accuracy of the measurement was theoretically investigated by Madhusudana (2000). Rostami *et al*. (2001) conducted the numerical analysis on the contact heat transfer between two solids. Between the contact surfaces, a gap having a shape of conical and that of circular were considered, and the thermal resistances were studied. Wahid (2003) numerically analyzed the heat flow of a cylinder having a conical gap. The numerical results were shown changing the angle of the gap and the thermal conductivity of gas in the gap. Kumar *et al.* (2004) presented a theoretical model to predict the contact heat transfer between curvilinear surfaces. Both a macro contact area corresponding to curvilinear surfaces in contact and a micro contact area for surface roughness were modeled. It was shown that the prediction agreed closely with those measured in the experiments. Bahrami *et al*. (2004) developed an approximate analytical model for determining a thermal contact resistance of nonconforming random rough contacts in the presence of an interstitial gas. Macro/micro gaps and macro/micro contacts were considered in modeling. Good agreement was shown between the calculated results by the model and the experimental data. Bahrami *et al*. (2006) also published a review article concerning the thermal contact resistance models for nonconforming rough surfaces.

To the authors' knowledge, the number of studies concerning the contact heat transfer between solid surfaces having a relatively large gap is very small. Moreover, sufficient information has not been published on the effect of thermal grease applied to the gap. In the present study, therefore, a relatively large gap having a shape of wedge is made between solid surfaces, and the heat transfer characteristics are investigated experimentally. The thermal grease is used, and the width and the height of the wedge-shaped gap are changed. The numerical analysis is also conducted and the discussion is made on the heat transfer characteristics around the gap.

2. EXPERIMENTAL METHOD

As shown in Fig. 1, a wedge-shaped gap is made between two rectangular brass rods; the bottom surface of the upper rod is partially wedge-shaped while the top surface of the bottom rod is entirely flat. The dimensions of the rods are 25 mm × 25 mm × 60 mm (length), and the thermal conductivity is 108 W/(m·K). Sixteen types of wedge-shaped gaps categorized into two groups (G1, G2) are prepared as shown in Table 1, where a and b are the width and the height of the gap, respectively. In addition, the measurement is also conducted for the two rods in contact where the wedge-shaped gap is not made. This measurement is categorized as G0 in this study.

Figure 2 shows a schematic of a measuring apparatus. This is basically the same apparatus as used in the previous study (Tomimura *et al*., 2011). As shown in the figure, the lower rod is placed on the cooling device, where cooling water is supplied from a thermostatic bath. The ceramic heater having the dimensions of 25 mm × 25 mm × 1 mm (thickness) is mounted on the top surface of the upper rod. A silicone grease having the thermal conductivity of 1.0 W/(m·K) is used to the gap. By using a lever and weights, a prescribed nominal contact

* *Corresponding author. Email: koito@gpo.kumamoto-u.ac.jp*

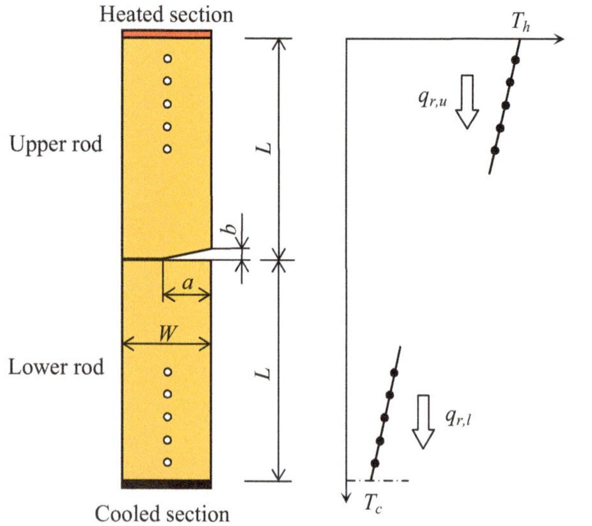

Fig. 1 Wedge-shaped gap between two rods and temperature distribution (Open circles represent the measuring points of temperatures)

Table 1 Specification of wedge-shaped gaps

Gap	a [mm]	b [μm]
G1-1	13.1	52
G1-2	13.3	108
G1-3	12.9	207
G1-4	12.8	310
G1-5	12.7	405
G1-6	12.7	610
G1-7	12.7	817
G1-8	12.9	1041
G2-1	17.7	48
G2-2	18.4	104
G2-3	18.3	211
G2-4	18.0	302
G2-5	17.6	401
G2-6	17.6	605
G2-7	17.7	820
G2-8	17.8	1027

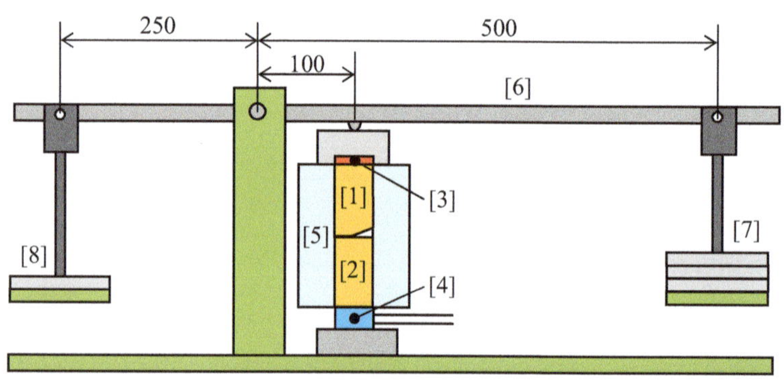

[1] Upper rod [4] Cooling device [7] Weight
[2] Lower rod [5] Thermal insulator [8] Balance weight
[3] Ceramic heater [6] Lever

Fig. 2 Schematic diagram of experimental apparatus

pressure is applied to the rods. The rods are thermally insulated using a polystyrene foam. For temperature measurement, five small holes are drilled on each rod, and T-type thermocouples of 0.5 mm in diameter are inserted by applying a thermally conductive silver paste. The measuring points of the temperatures are also shown in Fig. 1.

In experiment, the top surface of the upper rod is heated by the heater while the bottom surface of the lower rod is cooled by the cooling device, and the temperature distribution at steady state is obtained by the thermocouples. The heat input and the cooling temperature are 20.5 W – 26.0 W and 10 °C – 14 °C, respectively. Although the contact surface area between the upper and the lower rod varies with the width of the gap, a, the nominal contact pressure applied on the rods is kept at 1.6 MPa throughout the experiment.

A schematic of a steady-state temperature distribution inside the rods is also shown in Fig. 1. By extrapolating the measured temperature as shown by the solid lines in the figure, the temperatures at the top surface of the upper rod, T_h, and the bottom surface of the lower rod, T_c, are calculated, and then the thermal resistance of the wedge-shaped gap, R_g, is obtained by

$$R_g = R_t - 2R_r \tag{1}$$

where R_t is the total thermal resistance between the heated and the cooled section of the rods, and R_r the thermal resistance of each rod. R_t and R_r are expressed respectively as

$$R_t = \frac{T_h - T_c}{Q} \tag{2}$$

$$R_r = \frac{L}{\lambda_r W^2} \tag{3}$$

where L, W^2, and λ_r are the length (= 60 mm), the cross-sectional area (= 25 mm × 25 mm), and the thermal conductivity (= 108 W/(m·K)) of the rod, respectively. When one-dimensional heat flow is assumed in the rods, the heat fluxes in the upper rod, $q_{r,u}$, and the lower rod, $q_{r,l}$, are

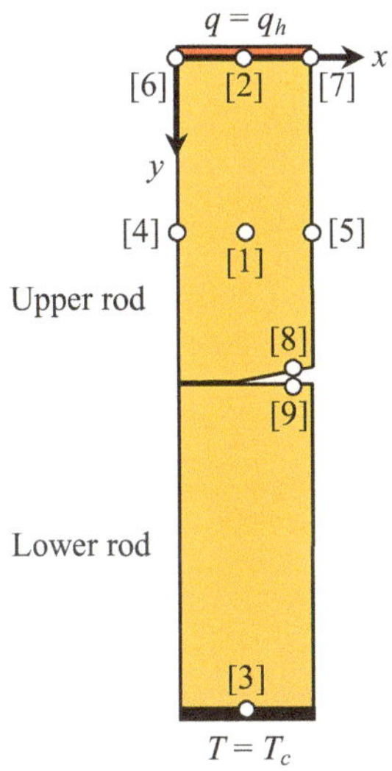

Fig. 3 Computational domain and coordinate system

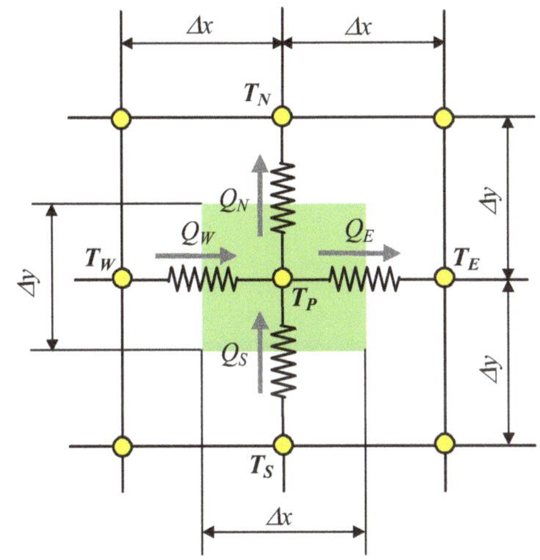

Fig. 4 Thermal resistance network for inner point [1]

calculated by using Fourier's law, and then the heat transfer rate, Q, is obtained by

$$Q = \left(\frac{q_{r,u} + q_{r,l}}{2} \right) W^2 \tag{4}$$

3. NUMERICAL ANALYSIS

The computational domain is shown in Fig. 3 with the coordinate system. The heat transfer characteristics of the upper rod, the lower rod and the wedge-shaped gap filled with the thermal grease are investigated in the x–y coordinate system. Using the thermal resistance network method, the steady-state two-dimensional heat conduction is calculated numerically under the boundary conditions of a uniform heat flux at the top surface of the upper rod (heated section), q_h, and a uniform temperature at the bottom surface of the lower rod (cooled section), T_c. The adiabatic boundary condition is exposed for the left and the right side of the computational domain.

As mentioned above, the thermal resistance network method is used to obtain the temperature distribution inside the computational domain. The followings are the representatives of discretized equations concerning the temperatures of a nodal point P indicated by [1] – [9] in Fig. 3.

For inner nodal points of the computational domain, the thermal resistance network is described as shown in Fig. 4. The temperature T_P at the inner nodal point P indicated by [1] is given by

Point [1]: $$T_P = \frac{\frac{\Delta y}{\Delta x}(T_W + T_E) + \frac{\Delta x}{\Delta y}(T_S + T_N)}{2\left(\frac{\Delta y}{\Delta x} + \frac{\Delta x}{\Delta y} \right)} \tag{5}$$

where T_W, T_E, T_S, and T_N are the temperatures at the neighboring nodal points W, E, S, and N, respectively. Δx and Δy are the x-direction and the y-direction distance between two adjacent nodal points. The derivation of this equation is described in Appendix A.

For the nodal points at the heated section, the cooled section, the left and the right side of the adiabatic section, which are indicated by [2] – [5] in Fig. 3, the temperatures T_P are given respectively by

Point [2]: $$T_P = \frac{\lambda_r \left\{ \frac{\Delta y}{2\Delta x}(T_W + T_E) + \frac{\Delta x}{\Delta y} T_S \right\} + q_h \Delta x}{\lambda_r \left(\frac{\Delta y}{\Delta x} + \frac{\Delta x}{\Delta y} \right)} \tag{6}$$

Point [3]: $$T_P = T_c \tag{7}$$

Point [4]: $$T_P = \frac{\frac{\Delta x}{2\Delta y}(T_S + T_N) + \frac{\Delta y}{\Delta x} T_E}{\frac{\Delta y}{\Delta x} + \frac{\Delta x}{\Delta y}} \tag{8}$$

Point [5]: $$T_P = \frac{\frac{\Delta x}{2\Delta y}(T_S + T_N) + \frac{\Delta y}{\Delta x} T_W}{\frac{\Delta y}{\Delta x} + \frac{\Delta x}{\Delta y}} \tag{9}$$

The following two equations are applied at the left (point [6]) and the right (point [7]) corner of the heated section:

Point [6]: $$T_P = \frac{\lambda_r \left(\frac{\Delta y}{\Delta x} T_E + \frac{\Delta x}{\Delta y} T_S \right) + q_h \Delta x}{\lambda_r \left(\frac{\Delta y}{\Delta x} + \frac{\Delta x}{\Delta y} \right)} \tag{10}$$

Point [7]: $$T_P = \frac{\lambda_r \left(\frac{\Delta y}{\Delta x} T_W + \frac{\Delta x}{\Delta y} T_S \right) + q_h \Delta x}{\lambda_r \left(\frac{\Delta y}{\Delta x} + \frac{\Delta x}{\Delta y} \right)} \tag{11}$$

The thermal resistance network for the wedge-shaped gap is shown in Fig. 5, where x_a is the x-direction distance from the left corner of the gap. The nodal point P in this network corresponds to point [8] in Fig.

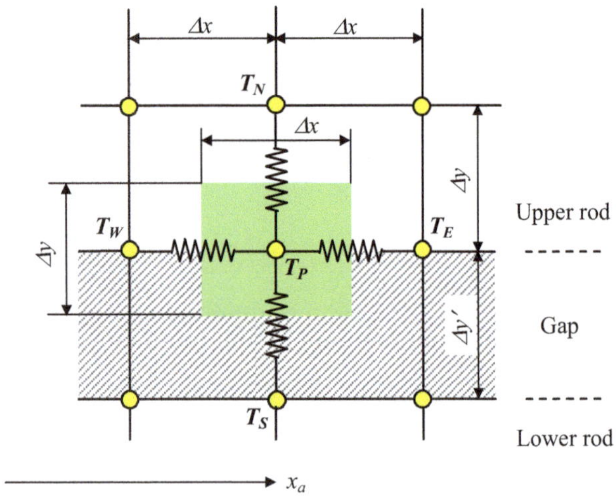

Fig. 5 Thermal resistance network for interfacial point [8]

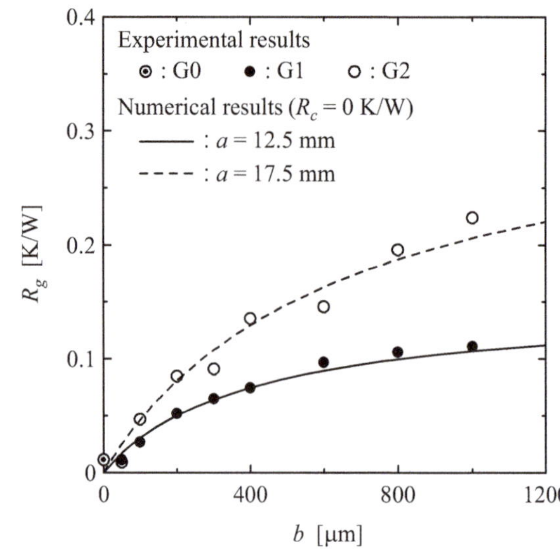

Fig. 6 Comparison between experimental and numerical results

3. Inside the wedge-shaped gap, the y-direction distance between two adjacent nodal points, $\Delta y'$, is calculated by

$$\Delta y' = x_a \frac{b}{a} \tag{12}$$

It is noted here that $\Delta y'$ is not constant but increases with x_a. The temperature T_P at the interfacial point P between the upper rod and the gap indicated by [8] is given by

Point [8]:

$$T_P = \frac{\dfrac{\lambda_r \Delta y + \lambda_g \Delta y'}{2\Delta x}(T_W + T_E) + \dfrac{\lambda_g \Delta x}{\Delta y'} T_S + \dfrac{\lambda_r \Delta x}{\Delta y} T_N}{\lambda_r\left(\dfrac{\Delta y}{\Delta x} + \dfrac{\Delta x}{\Delta y}\right) + \lambda_g\left(\dfrac{\Delta y'}{\Delta x} + \dfrac{\Delta x}{\Delta y'}\right)} \tag{13}$$

In the same way, the temperature T_P at the interfacial point P between the lower rod and the gap indicated by [9] is given by

Point [9]:

$$T_P = \frac{\dfrac{\lambda_r \Delta y + \lambda_g \Delta y'}{2\Delta x}(T_W + T_E) + \dfrac{\lambda_r \Delta x}{\Delta y} T_S + \dfrac{\lambda_g \Delta x}{\Delta y'} T_N}{\lambda_r\left(\dfrac{\Delta y}{\Delta x} + \dfrac{\Delta x}{\Delta y}\right) + \lambda_g\left(\dfrac{\Delta y'}{\Delta x} + \dfrac{\Delta x}{\Delta y'}\right)} \tag{14}$$

Using an Excel spreadsheet, the calculation is conducted and the numerical result of the temperature distribution inside the computational domain is obtained.

4. RESULTS AND DISCUSSION

4.1 Comparison between Experimental and Numerical Results

Concerning the relation between the thermal resistance of the wedge-shaped gap, R_g, and the height of the gap, b, the experimental results for gap G1 and G2 are shown in Fig. 6 by symbols. The experimental result is also shown for the two rods in contact where the wedge-shaped gap is not made, namely, $b = 0$. This measurement is indicated by G0 in this figure. As expected, R_g increases with b because the size of the wedge-shaped gap becomes larger. The numerical results are moreover shown in this figure by lines. Because the experimental result of the thermal resistance at $b = 0$ is very small, the thermal contact resistance between the upper and the lower rod, R_c, is neglected ($R_c = 0$) and the numerical results are obtained. In spite of the neglect of R_c, fairly good agreement is found between the experimental and the numerical results, confirming the validity of the present numerical analysis.

4.2 Heat Transfer Analysis

By using the numerical model, the detailed discussion is made on the heat transfer characteristics of the wedge-shaped gap.

Heat transfer characteristics of rods and gap: The temperature distributions inside the upper rod, the lower rod, and the wedge-shaped gap are shown in Fig. 7 for (a) b = 200 μm, (b) b = 400 μm, and (c) b = 800 μm. Because the thermal conductivity of the silicone grease used in the gap (λ_g = 1.0 W/(m·K)) is much smaller than that of the rod (λ_r = 108 W/(m·K)), the constriction and the spreading of heat flow are observed near the gap inside the upper and the lower rod, respectively. Moreover, the constriction and the spreading of heat flow become conspicuous with the increase in b. The five open circles on each rod correspond to the measuring points where the thermocouples are placed in the experiment. It is found that the measuring points are not affected by the constriction and the spreading of heat flow inside the rods. Therefore, in the present experiment, the heat fluxes in the upper rod, $q_{r,u}$, and the lower rod, $q_{r,l}$, can be obtained accurately by using the measured temperature and Fourier's law under the assumption that the heat flow is one-dimensional.

When the one-dimensional heat flow in y-direction inside the upper rod, the lower rod, and the wedge-shaped gap is assumed, the thermal resistance of the gap, R_g', is theoretically given by

$$R_g' = \left\{\frac{\lambda_r W(W-a)}{2L} + \frac{\lambda_r \lambda_g}{\lambda_r - \lambda_g}\frac{Wa}{b}\ln\left(\frac{\lambda_r - \lambda_g}{\lambda_g}\frac{b}{2L} + 1\right)\right\}^{-1} - 2R_r \tag{15}$$

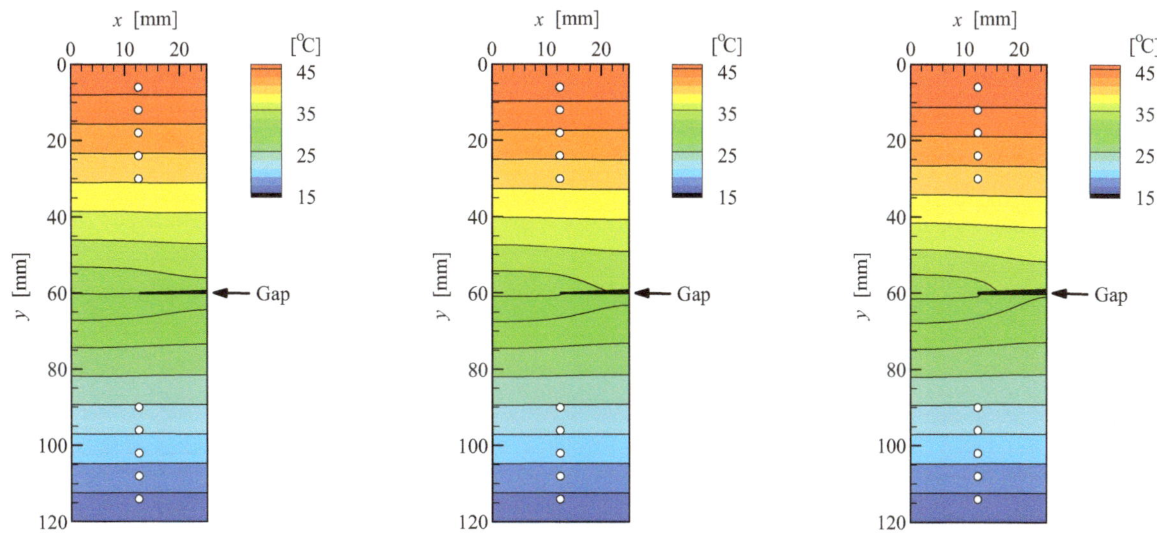

(a) $b = 200$ μm (b) $b = 400$ μm (c) $b = 800$ μm

Fig. 7 Temperature distribution inside rods and gap at $a = 12.5$ mm
(Open circles represent the measuring points of temperatures in the experiment)

where the thermal contact resistance between the upper and the lower rod, R_c, is neglected ($R_c = 0$). The derivation of this equation is described in Appendix B. It is noted here that the constriction and the spreading of heat flow inside the upper and the lower rod are not considered in Eq. (15). The numerical results, R_g, obtained by using the two-dimensional thermal resistance network method are compared in Fig. 8 with the theoretical results, R_g', by Eq. (15). It is observed that R_g is much lower than R_g', implying that the thermal resistance of the gap is strongly affected by the occurrence of the constriction and the spreading of heat flow inside the rods.

Effects of thermal contact resistance and contact area: Changing the thermal contact resistance between the upper and the lower rod, R_c, the relation between the thermal resistance of the gap, R_g, and the height of the gap, b, is calculated as shown in Fig. 9. In this calculation, the thermal contact resistance, R_c, is added in the thermal resistance network between the bottom surface of the upper rod and the top surface of the lower rod. Furthermore, the relation between R_g and the width of the gap, a, is also obtained as shown in Fig. 10. In this calculation, the thermal contact resistance, R_c, is neglected ($R_c = 0$) and the numerical results are obtained for $b = 200$ μm, $b = 400$ μm, and $b = 800$ μm. The dependence of R_g on R_c and a confirms that the thermal contact resistance and the contact area of the unshaped surface are also important factors dominating the heat transfer between the surfaces in contact having the gap. This is owing to the fact that a relatively large amount of heat is transferred through the unshaped contact surfaces of the upper and the lower rod.

Effect of thermal grease: The numerical analysis is moreover conducted when the thermal grease is not used inside the wedge-shaped gap. In this case, the air having the thermal conductivity of 0.026 W/(m·K) is considered inside the gap. Concerning the relation between the thermal resistance, R_g, and the height, b, the numerical results with and without the thermal grease are shown in Fig. 11, where the thermal contact resistance, R_c, is neglected ($R_c = 0$). As expected, R_g without the thermal grease is much larger than that with the grease. For the case of gap without the thermal grease, it is found that R_g increases steeply from $b = 0$ and thereafter increases very slightly with b. For the case of gap with the thermal grease, on the other hand, R_g increases gradually from $b = 0$. The ratio of R_g with the thermal

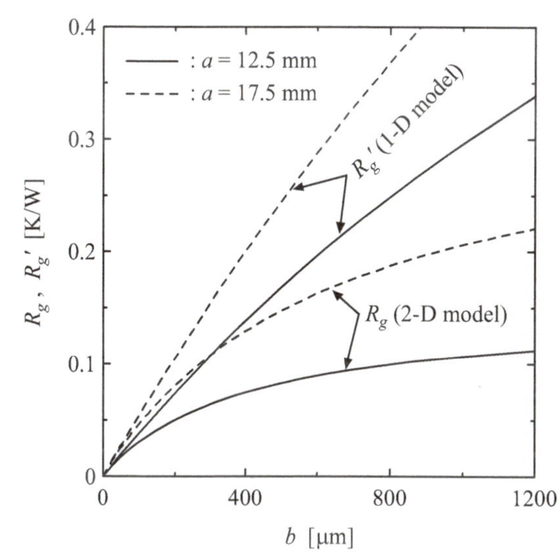

Fig. 8 Comparison between numerical results by two-dimensional model and analytical results by Eq. (15)

grease to R_g without the grease is denoted by ξ, and also shown in Fig. 11. It is confirmed that the thermal grease is most effective for $b = 0$, namely, a flat surface without the gap; the effect of the thermal grease is reduced with the increase in b.

5. CONCLUSIONS

Experimental and numerical studies are conducted on the heat transfer characteristics of the two rods in contact having a relatively large wedge-shaped gap. The findings of this study can be summarized as follows:

- The thermal resistance network method is used in the present numerical analysis. Since the numerical results agree well

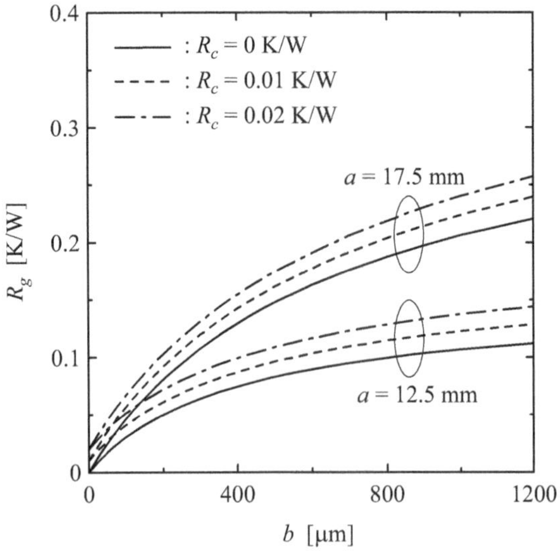

Fig. 9 Effect of thermal contact resistance on thermal resistance of wedge-shaped gap

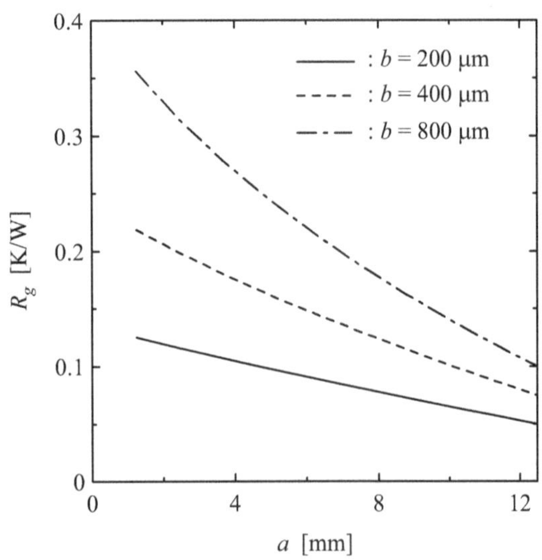

Fig. 10 Effect of contact surface area on thermal resistance of wedge-shaped gap

with the experimental ones, the validity of the numerical analysis is confirmed.

- The constriction and the spreading of heat flow inside the upper and the lower rod are visualized. The thermal resistance of the gap is strongly affected by the occurrence of the constriction and the spreading of heat flow inside the rods.
- The thermal contact resistance and the contact area of the unshaped surface are important factors dominating the heat transfer between the surfaces in contact having the gap.
- The thermal grease is most effective for a flat surface without a relatively large gap; the effect of the thermal grease is reduced with the increase in height of the gap.

NOMENCLATURE

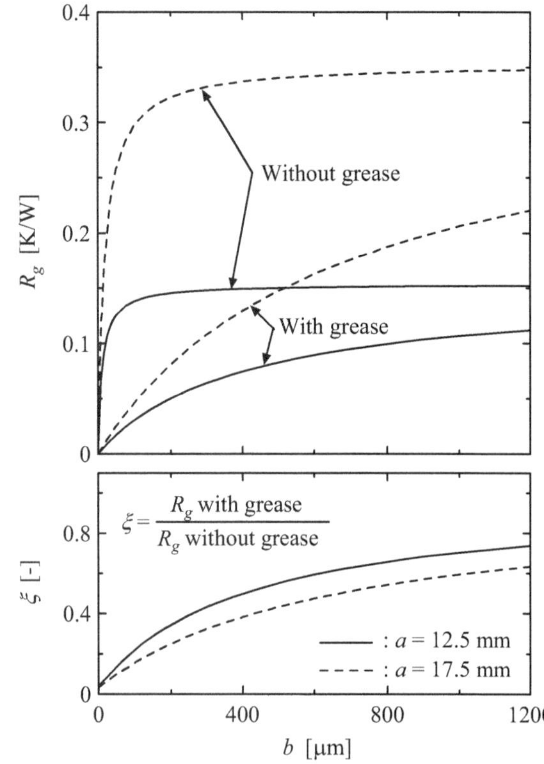

Fig. 11 Thermal resistance of wedge-shaped gap with and without thermal grease

a	width of gap (mm, m)
b	height of gap (μm, m)
L	length of rod (mm, m)
Q	heat transfer rate (W)
q	heat flux (W/m^2)
R	thermal resistance (K/W)
R_c	thermal contact resistance (K/W)
R_g'	thermal resistance calculated by Eq. (15) (K/W)
R_t	total thermal resistance (K/W)
T	temperature (°C)
W	width of rod (mm, m)
x	x coordinate (mm, m)
x_a	x-direction distance from left corner of gap (mm, m)
y	y coordinate (mm, m)

Greek Symbols

Δx	x-direction distance between adjacent nodal points (mm, m)
Δy	y-direction distance between adjacent nodal points (mm, m)
$\Delta y'$	y-direction distance between adjacent nodal points calculated by Eq. (12) (mm, m)
λ	thermal conductivity (W/(m·K))
ξ	ratio of R_g with grease to R_g without grease (-)

Subscripts

c	cooled section
E	neighboring nodal point on east side
eff	effective
g	gap
h	heated section
l	lower
N	neighboring nodal point on north side
P	central nodal point

r	rod
S	neighboring nodal point on south side
u	upper
W	neighboring nodal point on west side

REFERENCES

Bahrami, M., Yovanovich, M. M., and Culham, J. R., 2004, "Thermal Joint Resistances of Nonconforming Rough Surfaces with Gas-Filled Gaps," *Journal of Thermophysics and Heat Transfer*, **18**(3), 326-332.

Bahrami, M., Culham, J. R., Yavanovich, M. M., and Schneider, G. E., 2006, "Review of Thermal Joint Resistance Models for Nonconforming Rough Surfaces," *Applied Mechanics Reviews*, **59**(1), 1-12. http://dx.doi.org/10.1115/1.2110231

Faghri, A., 2014, "Heat Pipes: Review, Opportunities and Challenges," *Frontiers in Heat Pipes*, **5**(1), 013001. http://dx.doi.org/10.5098/fhp.5.1

Kumar, S. S., Abilash, P. M., and Ramamurthi, K., 2004, "Thermal Contact Conductance for Cylindrical and Spherical Contacts," *Heat and Mass Transfer*, **40**(9), 679-688. http://dx.doi.org/10.1007/s00231-003-0433-0

Madhusudana, C. V., 2000, "Accuracy in Thermal Contact Conductance Experiments – The Effect of Heat Losses to the Surroundings," *International Communications in Heat and Mass Transfer*, **27**(6) 877-891. http://dx.doi.org/10.1016/S0735-1933(00)00168-8

Mochizuki, M., Nguyen, Th., Mashiko, K., Saito, Y., Nguyen, Ti., and Wuttijumnong, V., 2011, "A Review of Heat Pipe Application Including New Opportunities," *Frontiers in Heat Pipes*, **2**(1), 013001. http://dx.doi.org/10.5098/fhp.v2.1.3001

Rostami, A. A., Hassan, A. Y., and Lim, P. C., 2001, "Parametric Study of Thermal Constriction Resistance," *Heat and Mass Transfer*, **37**(1), 5-10. http://dx.doi.org/10.1007/s002310000087

Tomimura, T., Shiotsu, Y., Koito, Y., Ishizuka, M., and Hatakeyama, T., 2011, "Evaluation of Effective Thermal Conductivity of Multilayer Printed Circuit Board," AJTEC2011-44232, *Proceedings of the ASME/JSME 2011 8th Thermal Engineering Joint Conference*, Honolulu, Hawaii.

Wahid, S. M. S., 2003, "Numerical Analysis of Heat Flow in Contact Heat Transfer," *International Journal of Heat and Mass Transfer*, **46**(24), 4751-4754. http://dx.doi.org/10.1016/S0017-9310(03)00320-X

APPENDIX A: DERIVATION OF EQ. (5)

In Fig. 4, the energy balance at the nodal point P is expressed as

$$Q_W + Q_S = Q_E + Q_N \tag{A-1}$$

where Q_W, Q_S, Q_E, Q_N are heat transfer rates at the nodal points W, S, E, N, respectively. By using Fourier's law, they are calculated respectively by

$$Q_W = \Delta y\, \lambda \frac{T_W - T_P}{\Delta x} \tag{A-2}$$

$$Q_S = \Delta x\, \lambda \frac{T_S - T_P}{\Delta y} \tag{A-3}$$

$$Q_E = \Delta y\, \lambda \frac{T_P - T_E}{\Delta x} \tag{A-4}$$

$$Q_N = \Delta x\, \lambda \frac{T_P - T_N}{\Delta y} \tag{A-5}$$

where λ is the thermal conductivity. Substitution of Eqs. (A-2) to (A-5) into Eq. (A-1) yields Eq. (5).

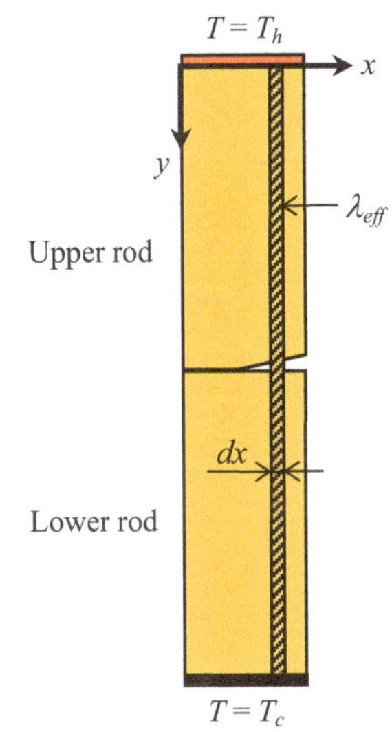

Fig. B-1 Analytical system for one-dimensional calculation

APPENDIX B: DERIVATION OF EQ. (15)

As shown in Fig. B-1, an element having a thickness of dx is considered. When one-dimensional heat flow in y-direction is assumed, the effective thermal conductivity of the element, λ_{eff}, is obtained from

$$\frac{2L}{\lambda_{eff}} = \frac{L - x_a b/a}{\lambda_r} + \frac{x_a b/a}{\lambda_g} + \frac{L}{\lambda_r} \tag{B-1}$$

By using λ_{eff}, the heat transfer rate from the heated to the cooled section, Q, is calculated as

$$\begin{aligned} Q &= W\left(\int_0^{W-a} \lambda_r \frac{T_h - T_c}{2L} dx + \int_{W-a}^{W} \lambda_{eff} \frac{T_h - T_c}{2L} dx \right) \\ &= \left\{ \frac{\lambda_r W(W-a)}{2L} + \frac{\lambda_r \lambda_g}{\lambda_r - \lambda_g} \frac{Wa}{b} \ln\left(\frac{\lambda_r - \lambda_g}{\lambda_g} \frac{b}{2L} + 1 \right) \right\} (T_h - T_c) \end{aligned} \tag{B-2}$$

and therefore the total thermal resistance from the heated to the cooled section , R_t, is obtained as

$$R_t = \frac{T_h - T_c}{Q} = \left\{ \frac{\lambda_r W(W-a)}{2L} + \frac{\lambda_r \lambda_g}{\lambda_r - \lambda_g} \frac{Wa}{b} \ln\left(\frac{\lambda_r - \lambda_g}{\lambda_g} \frac{b}{2L} + 1 \right) \right\}^{-1} \tag{B-3}$$

Substitution of Eq. (B-3) into Eq. (1) yields Eq. (15).

NUMERICAL SIMULATION ON FLOW AND HEAT TRANSFER IN OSCILLATING HEAT PIPES

S.F. Wang[a,*], Z.R. Lin[a], Z.Y. Lee[b], and L.W. Zhang[b]

[a]*Key Laboratory of Enhanced Heat Transfer and Energy Conservation of the Ministry of Education, South China University of Technology, Guangzhou City, Guangdong Province, 510640, China*
[b] *Novark Technology Inc, Shenzhen City, Guangdong Province, 518172, China*

ABSTRACT

In order to predict the heat transport capability of oscillating heat pipes (OHPs), a mathematical and physical model of OHP was built to simulate the process of flow and heat transfer in vertical bottom heating mode. Water was used as working fluid. Mixture model in FLUENT was used for two-phase flow numerical simulation. The result showed that the numerical simulation was successful to reproduce the behavior of the internal flow of OHP, including vapor generation in evaporation section, oscillation phenomena caused by the pressure difference and heat transfer due to oscillation. Comparing with the experimental tests, the simulation results agreed with the experimental records fairly well.

Keywords: *oscillating heat pipes (OHPs), numerical simulation, heat transfer, flow, predicting correlation*

1. INTRODUCTION

An oscillating heat pipe (OHP) is a two-phase flow device used for transferring heat without any moving mechanical parts (Akachi and Polášek, 1996; Polášek and Zelko, 1997). It consists of tubes/channels of capillary dimensions arranged in a serpentine manner and joined end to end as shown in Fig.1. It is predicted as one of the most promising solution for higher heat dissipation compact cooling. The heat transfer characteristics of OHPs have become a hot research. However, most studies focused on the experimental investigations. Only a few researches paid attention to the numerical simulations. Zuo (Zuo *et al.*, 2001) simulated the operation of vapor and liquid slugs in OHP by the single spring-mass-damper system, which was far away from the experimental results. Wong (Wong *et al.*, 1999) predicted the direction of vapor and liquid slugs by the multiple spring-mass-damper system, but without considering the effect of heat exchange. Dobson (Dobson, 2004) and Faghri *et al.* (Shafii *et al.*, 2002; Zhang *et al.*, 2002) established the mass, momentum and energy conservation equations of OHP by the control volume method. Khandekar (Khandekar *et al.*, 2002) proposed an artificial neutral network to predict the thermal performance of OHP, which based on an amount of experimental data. In fact, there was still no an effective numerical simulation to predict the thermal performance of OHPs successfully.

With the aim of exploring potential applications of OHP technology, it is very important to predict the thermal performance of OHP by the numerical simulation, rather than by a large number of experiments. In this work, a mathematical and physical model of OHP was built to simulate the process of flow and heat transfer. The simulation results were compared with the experimental results.

2. EXPERIMENTAL APPARATUS AND PROCEDURE

In the experiments, an experimental apparatus of OHPs were established for thermal performance test to study the flow and heat transfer. The four-turn OHP was selected as a typical shape. Pure water was used as working fluid. The operational orientation was vertical bottom heating mode.

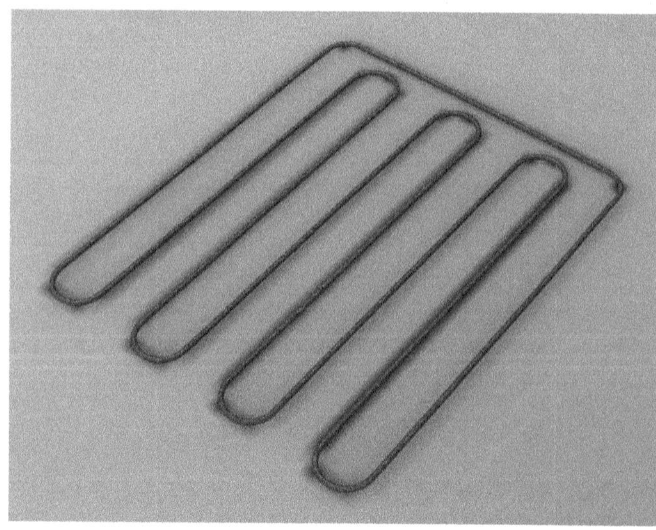

Fig. 1 Prototype of OHP

In the thermal performance test, two types of copper tubes with outer diameter (D_o) of 2.5 and 3mm, inner diameter (D_i) of 1.3 and 1.8 mm were used as manufacturing material to be bent into the OHPs respectively. The heat transfer length (L) of 200 mm were adopted for comparative experiments. In the liquid filling process, internal channels of copper tubes were first exhausted and the working fluid was filled fully into the tube under the pressure difference. Liquid filling ratio was controlled around 50±5% by the second vacuum, which discharged the excess liquid. The experimental apparatus of OHPs was shown in Fig. 2. The nickel chrome electric wires were wound around the copper tube,

* *Corresponding author*

which was wrapped in thermal insulation adhesive plaster, as evaporation section. They were connected to the transformer, which supplied heating power by adjusting the current and voltage. The length of evaporation section was 20 mm, the same as the condensation section; and the condensation section was cooled by water (25±0.05°C). The OMEGA K-type thermocouples were installed to measure the wall temperature at different positions of OHPs. The detailed location of thermocouples was shown in Fig. 2.The pulse of temperature was measured to reflect the internal working fluid oscillation indirectly.

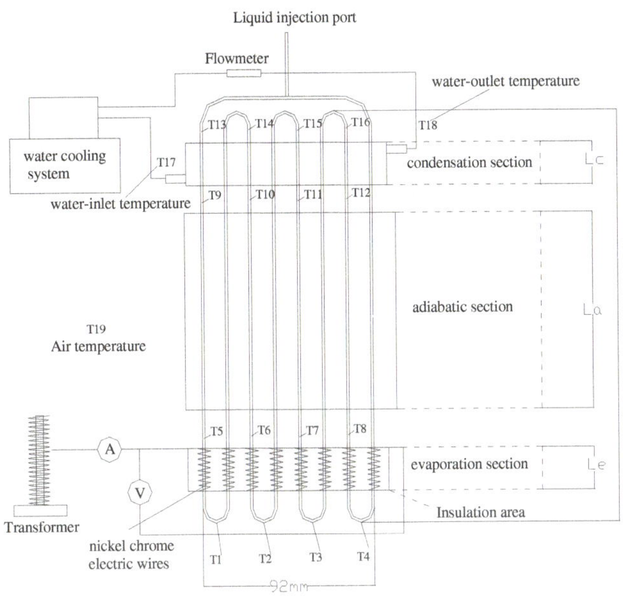

Lc: the length of condensation section; La: the length of adiabatic section; Le: the length of evaporation section;

Fig. 2 Experimental apparatus of OHPs

All tests were conducted at an ambient temperature of 25±1°C. Base on the application requirements, when the average temperature of evaporation section was over 100°C, the experiment would be stopped. The thermal resistance of OHPs was equal to temperature difference between condensation section and evaporation section divided by heating power. $R_{OHP}=(T_e-T_c)/Q_h$ (R_{OHP}: thermal resistance of OHP; T_e: average temperature of T_1, T_2, T_3 and T_4 approximately instead of evaporation section; T_c: average temperature of T_{13}, T_{14}, T_{15} and T_{16} approximately instead of condensation section; Q_h: heating power).

3. DESCRIPTION OF SIMULATION

3.1 Physical model

The 2D physical model was built to simulate the internal flow and heat transfer in OHPs, shown in Fig.3. Pure water was used as working fluid. The internal flow of OHP was mainly vapor-liquid two phase flow. It was assumed that the density of liquid phase was incompressible, and the vapor phase was compressible. The phase and density changed with the temperature changed. The geometric model and mesh diagram were shown in Fig.3. The design length of evaporation section was 20mm, and the length of condensation section was 20mm.

3.2 Definite conditions

As the OHP was operated in a negative pressure, we assumed that the water vaporization temperature is 50°C accord to the experimental condition. It meant that the liquid phase of working fluid evaporated over 50°C and the vapor phase condensed below 50°C. The heat flux in evaporation section depended on the power input. The heat transfer coefficient in condensation section was set according to the actual situation. The insulation section was set to insulation boundary condition.

In the cases, mixture model was used for flow and heat transfer simulation. The user defined function was added to calculate mass and energy transport, in order to achieve heat transfer process of evaporation and condensation.

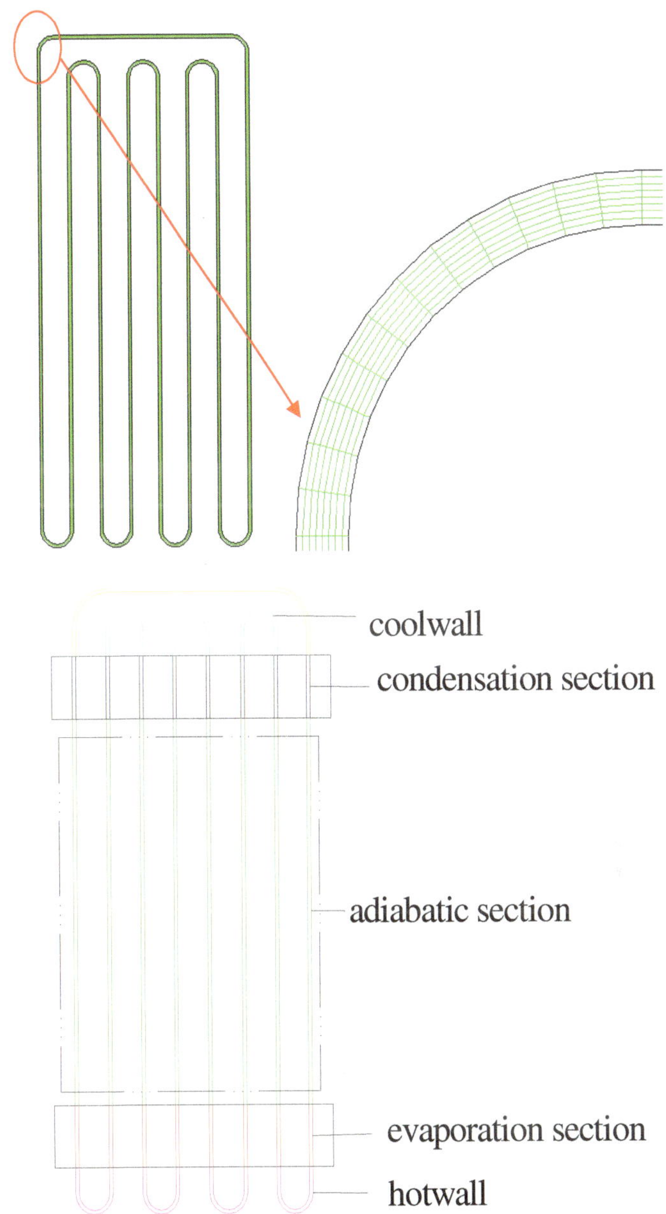

Fig. 3 Geometric model and mesh diagram

4. SIMULATION ON START-UP PROCESS

In the simulation case, the four-turn OHP with inner diameter (D_i) of 1.3 and heat transfer length of 200 mm was selected as 2D model for comparison. The heating power was 24W. As shown in Fig.4, the start-up of OHP had been recorded, which went through the process of heat transfer. From 1s to 5s, the temperature in evaporation section increased due to power input. When the time up to 7s, the evaporation section reached a certain temperature, the generation of bubbles began. The region of high temperature in evaporation section was expanding from 7s to 9s. As the temperature increased, the bubbles jetted across the evaporation section to the condensation section. High temperature

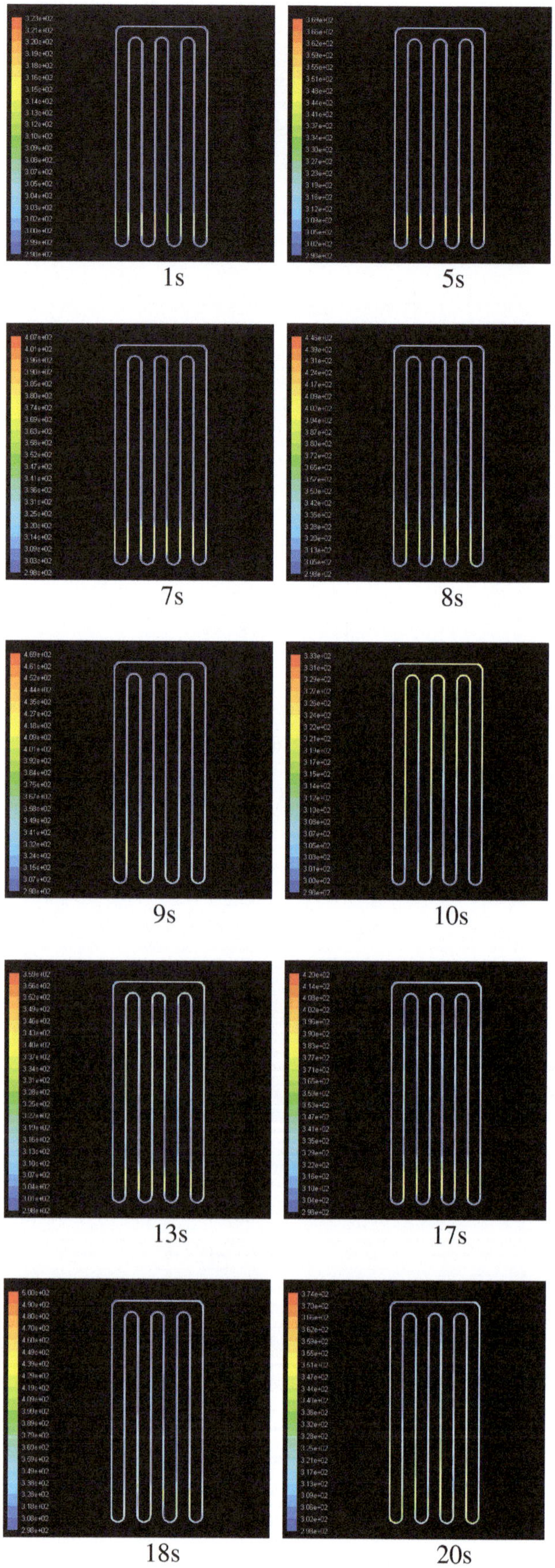

Fig. 4 Contours of temperature change in OHP start-up process

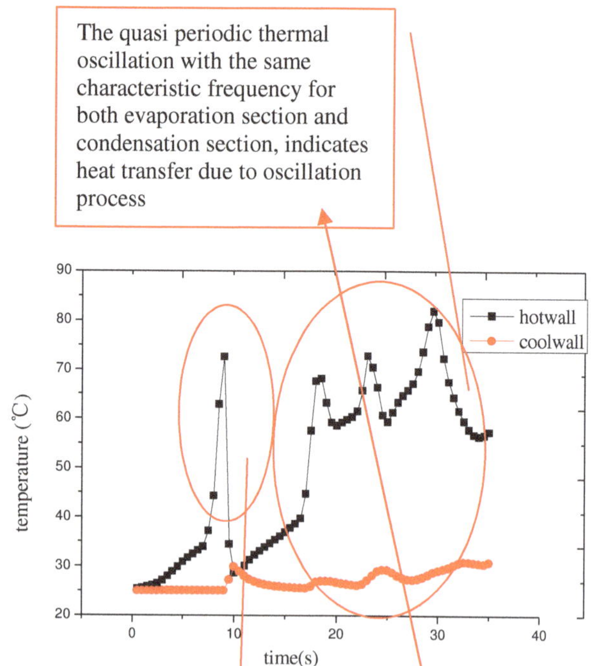

Fig. 5 Temperature monitors record for OHP startup process simulation (heating power=24W)

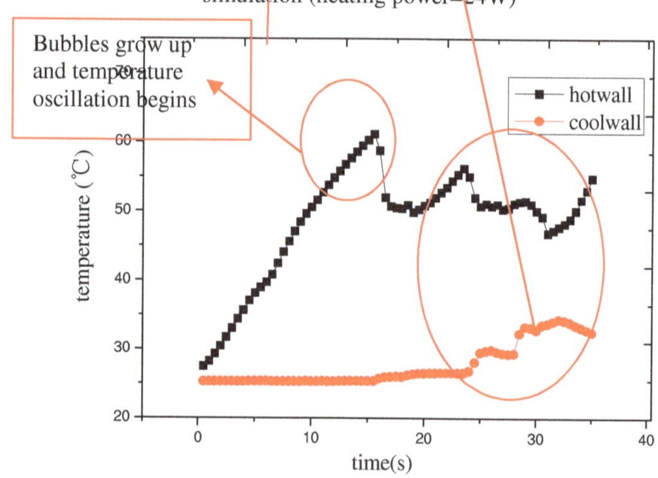

Fig. 6 Temperature record for OHP startup test (heating power=24W)

appeared in the condensation section, while the evaporation section was cooled down in 10s. After that, the vapor slugs (bubbles) and liquid slugs (the liquid region between bubbles) oscillation occurred and made the heat transfer from evaporation section to condensation section. The temperature distribution of OHP became uniform. In 13s, vapor slugs (bubbles) condensed into liquid in the condensation section. With the help of gravity and capillary forces, the condensed liquid returned back to the evaporation section. At the same time, the temperature in evaporation section increased from 10s to 17s, which reserved energy for the next oscillation cycle. In 18s to 20s, the new cycle began. The evaporation section received the energy and reached a certain temperature. The condensed liquid evaporated and generated the bubbles again. As a result, the heat transfer process of OHP mainly depended on this vapor and liquid slug oscillation, which occurred due to pressure difference between evaporation and condensation.

Therefore, the numerical simulation was successful to reproduce the startup process of OHP, including vapor generation in evaporation section, oscillation phenomena caused by the pressure difference and heat transfer due to oscillation process. They matched with the experimental observation. As shown in Fig.5, the temperature change in

OHP startup process (the total time is 35 seconds) was recorded. It was found that when the hotwall (marked in Fig.3) reaches a certain temperature (about 75°C), the temperature oscillation began. The oscillation amplitude and temperature difference between evaporation and condensation section was decreasing with the startup of OHP. The quasi periodic thermal oscillation with the same characteristic frequency appeared in both evaporation section and condensation section, which indicated heat transfer due to oscillation process. Comparing to the experimental records in Fig.6, the trend of temperature change in simulation was consistent with experimental results.

5. NUMERICAL SIMULATION COMPARED WITH EXPERIMENTAL RESULTS

5.1 Comparison of temperature curves

In the case, the OHP with inner diameter of 1.3mm and heat transfer length of 200mm was selected for comparison between simulation and experimental results. Fig.7 was the temperature record for OHP performance test in 24W; while Fig. 8 was the temperature record for OHP 2D simulations in 24W. As can been seen in Fig.8, the oscillation range in temperature was consistent with the experimental record. The average temperature of the experimental record in evaporation section was about 57°C; while condensation section's was about 32°C.The average temperature of 2D simulation record in evaporation section was about 63°C; while condensation section's was about 28°C.

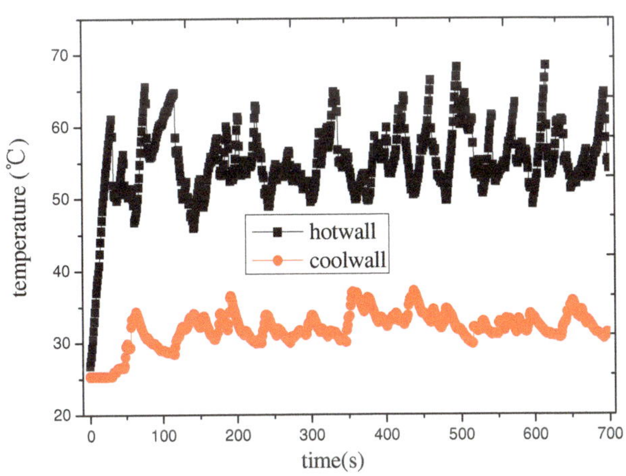

Fig. 7 Temperature record for OHP performance test in 24W (D_i=1.3mm, D_o =2.5mm, L=200mm)

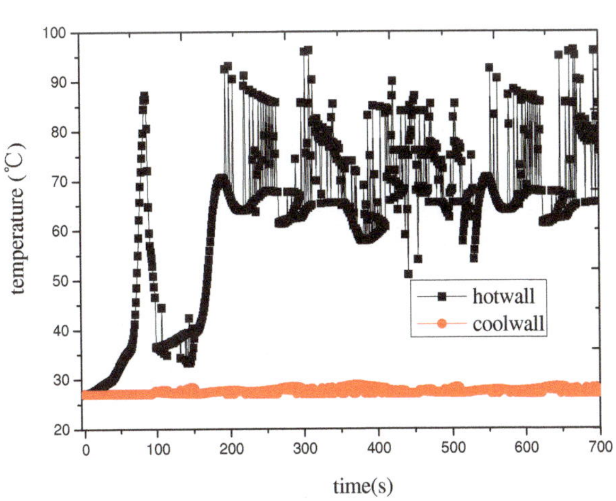

Fig. 8 Temperature record for OHP 2D simulation in 24W (D_i=1.3mm, D_o =2.5mm, L=200mm)

The simulation temperature was slightly higher than the experimental record, but it represented the simulation model was successful to predict the OHP performance and operation. Furthermore, it was noticed that the simulation record needed some time to develop into the stable oscillation state after the startup process. In contrast, the experimental record reached the stable oscillation state earlier. Therefore, the longer simulation, the more accurate results can be obtained, closer to the real operation.

5.2 Comparison of thermal resistance

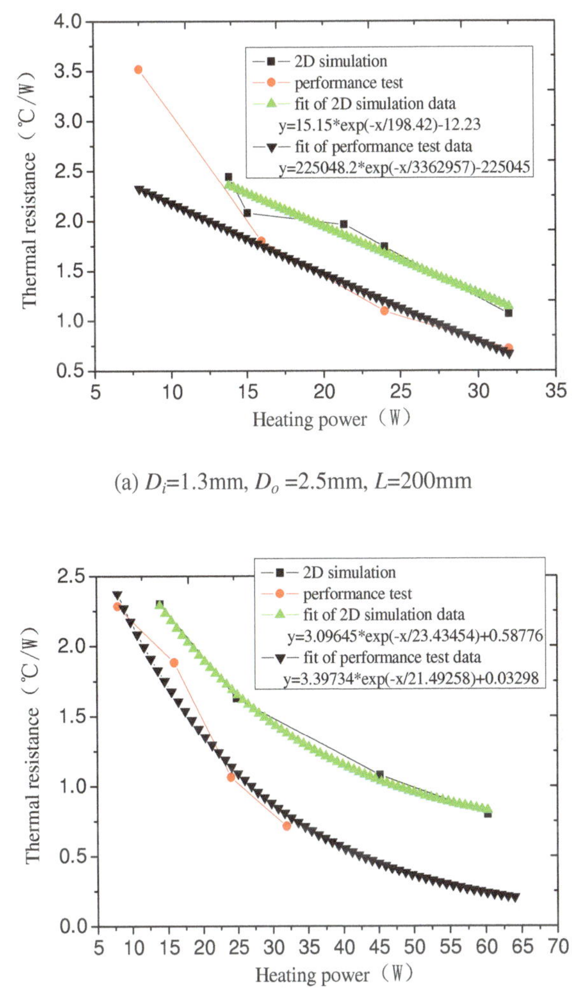

(a) D_i=1.3mm, D_o =2.5mm, L=200mm

(b) D_i=1.8mm, D_o =3mm, L=200mm

Fig. 9 Comparison of thermal resistance among 2D simulation and performance test

After the section discussed above, it proved that 2D mathematical model could successfully simulate the internal flow of an OHP. In this section, several cases with different parameters were selected for comparison in different operating conditions. As can be seen in Fig.9, the thermal resistance of 2D model simulation was higher than the test data. From the comparison of fitting equations between 2D simulation and performance tests, the gap kept stable. This may be due to the condition difference between the calculations and experiments.

In conclusion, it was believed that 2D model was considered as the good solution for OHP simulation. In order to fix the error of the simulation results, a fitting formula to consider the design parameters of OHP need to create for predicting the thermal performance of OHPs.

5.3 Correlation prediction of OHPs

For reducing the error of the simulation results, it needed a fitting formula to fix the gap between the simulation and the performance test. Considering the past prediction method mentioned by Khandekar (Khandekar *et al.*, 2003), Katpradit (Katpradit *et al.*, 2005), Charoensawan (Charoensawan *et al.*, 2007), Arslan and Ozdemir (Arslan and Ozdemir, 2008), Sakulchangsatjatai (Sakulchangsatjatai *et al.*, 2004), the dominating dimensionless parameters, including D_i/L, Ja, Bo and $1+(\rho_v/\rho_l)^{0.25}$, and some operating parameters including ΔT, Q were used to fix a correlation for predicting the heat transport capability of OHPs.

D_i/L is the ratio of the inner diameter to that of the heat transfer length of the tube and it represents the geometry of the OHPs. The jakob number ($Ja=C_p\Delta T/h_{fg}$) is the ratio of sensible heat to that of the latent heat of the working fluid. The bond number ($Bo=D_i[g(\rho_l-\rho_v)/\sigma]^{0.5}$) is the ratio of the buoyancy force to that of the surface tension force of the working fluid. The Wallis number ($Wa=1+(\rho_v/\rho_l)^{0.25}$) can be used to explain the flooding phenomenon that influences dry-out to occur at the evaporation section. ΔT is the temperature difference between evaporation section and condensation section from the simulation results. Q is the power input.

According to the dimensionless method, the 2D simulation prediction of thermal resistance in vertical bottom heating mode was:

$$R_{2D}=0.02669\times\Delta T^{1.28803}\times Q^{-1.28803}\times[Di/L]^{-0.72677}\times[Cp\Delta T/h_{fg}]^{-0.57583}\times\left[Di[g(\rho_l-\rho_v)/\sigma]^{0.5}\right]^{0.50135}\times\left[1+(p_v/p_l)^{0.25}\right]^{-11.66237}$$

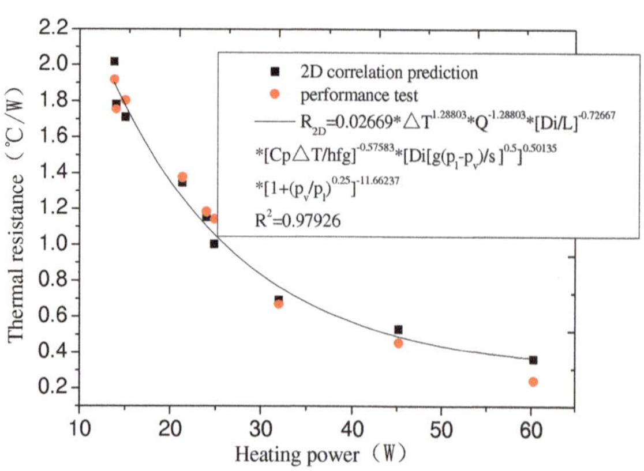

Fig. 10 Comparison of measurement and prediction

As can be seen in Fig.10, R-square of 2D simulation prediction was 0.98175, which represents high reliability for the prediction fitting formula. The correlation prediction agrees with the experimental results fairly well. In conclusion, the operation of OHPs can be predicted by inner diameter, heat transfer length, heating power input and temperature difference required. Instead, OHPs can be designed according to the actual conditions.

6. CONCLUSIONS

In order to predict the heat transport capability of OHPs, a mathematical and physical model of OHP was built to simulate the internal flow and heat transfer in vertical bottom heating mode. Water was used as working fluid. Mixture model in FLUENT was used for two-phase flow numerical simulation. Conclusions of the studies can be summarized as follows:

(1) The numerical simulation was successful to reproduce the startup process of OHP, including vapor generation in evaporation section, oscillation phenomena caused by the pressure difference and heat transfer due to oscillation. The quasi periodic thermal oscillation with the same characteristic frequency for evaporation section and condensation section, indicated heat transfer phenomenon between these two sections. Comparing with the experimental observations, the simulation results agreed with the experimental results fairly well.

(2) According to the comparison of temperature curves record and thermal resistance between 2D model and performance tests, it was found that mixture model was successful to simulate the OHP operation, which more consistent with the experimental data.

(3) In order to reduce the error of the simulation results, the dimensionless method was used to create a fitting formula to fix the gap between the simulation and the performance test. The simulation correlation could predict the heat transport capability of OHPs successfully.

The 2D simulation prediction of thermal resistance in vertical bottom heating mode was:

$$R_{2D}=0.02669\times\Delta T^{1.28803}\times Q^{-1.28803}\times[Di/L]^{-0.72677}\times[Cp\Delta T/h_{fg}]^{-0.57583}\times\left[Di[g(\rho_l-\rho_v)/\sigma]^{0.5}\right]^{0.50135}\times\left[1+(p_v/p_l)^{0.25}\right]^{-11.66237}$$

ACKNOWLEDGEMENTS

The authors deeply appreciate the financial support offered by National Natural Science Foundation of China (Granted No.51146009), International Cooperation and Exchange Program from the Ministry of Science and Technology of China (Grant No. 2011DFA60290), and Novark Technology INC.

NOMENCLATURE

Bo	Bond number ($=D_i[g(\rho_l-\rho_v)/\sigma]^{0.5}$)
C_p	Constant specific heat (J/kg·K)
D_i	Inner diameter (mm)
D_o	Outer diameter (mm)
g	Gravitational acceleration (m/s^2)
h_{fg}	Latent heat of vaporization (J/kg)
ΔT	Temperature difference between outlet and inlet ($=T_{18}-T_{17}$)
Ja	Jakob Number ($=C_p\Delta T/h_{fg}$)
L	Heat transfer length (mm)
L_e	Length of evaporation section (mm)
L_c	Length of condensation section (mm)
L_a	Length of adiabatic section (mm)
S	Heat transfer area (m^2)
Wa	Wallis Number ($1+(\rho_v/\rho_l)^{0.25}$)
Q	Power input (W)
R	Thermal resistance of OHP (□/W)
Greek Symbols	
ρ	density (kg/m^3)
σ	Surface tension (N/m)
Subscripts	
c	Condensation section
e	Evaporation section
l	Liquid
v	Vapor

REFERENCES

Akachi H., Polášek F., 1996, "Pulsating heat pipes," *5th International Heat Pipe Symposium*, Melbourne, Australia, 17-20.

Arslan G., Ozdemir M., 2008, "Correlation to predict heat transfer of an oscillating loop heat pipe consisting of three interconnected columns," *Energy Conversion and Management*, 49(8), 2337-2344.

Charoensawan P., Terdtoon P., 2007, "Thermal Performance Correlation of Horizontal Closed-Loop Oscillating Heat Pipes." *9th Electronics Packaging Technology Conference*, 906-909.

Dobson R. T., 2004, "Theoretical and experimental modeling of an open oscillatory heat pipe including gravity," *International Journal of Thermal Sciences*, 43(2), 113-119.

Katpradit T., Wongratanaphisan T., Terdtoon, P. Kamonpet P., Polchai A., Akbarzadeh A., 2005, "Correlation to predict heat transfer characteristics of a closed end oscillating heat pipe at critical state," *Applied Thermal Engineering*, 25(14-15), 2138-2151.

Khandekar S., Cui X., Groll M., 2002, "Thermal Performance Modeling of Pulsating Heat Pipes by Artifical Neutral Network," *11th International Heat Pipe Conference*, Moscow, Preprint V2, B2, 1-5

Khandekar S., Charoensawan P., Groll M., Terdtoon P., 2003, "Closed loop pulsating heat pipes: Part B. Visualization and semi-empirical modeling ," *Applied Thermal Engineering*, 2003, 23(16), 2021-2033.

Polášek F., Zelko M., 1997, "Thermal control of electronic components by heat pipes and thermosyphons; A historical overview," *10th International Heat Pipe Conference*, Stuttgart, Germany, 21–25.

Sakulchangsatjatai P., Terdtoon P., Wongratanaphisan T., Kamonpeta P., Murakami M., 2004, "Operation modeling of closed-end and closed-loop oscillating heat pipes at normal operating condition," *Applied* Thermal Engineering, 24(7), 995-1008.

Shafii M. B., Faghri A., Zhang Y. W., 2002, "Analysis of Heat Transfer in Unlooped and Looped Pulsating Heat pipes," *International Journal of Numerical Methods for Heat &Fluid Flow*, 12 (3), 585-609.

Wong T., Tong B. Y., Lim S. M., Ooi K.T., 1999, "Theoretical Modeling of Pulsating Heat Pipe," *11th International Heat Pipe Conference*, Tokyo, Japan, 159-163.

Zuo Z. J., North M. T., Wert K. L., 2001, "High Heat Flux Heat Pipes Mechanism for Cooling of Electronics," *IEEE Transactions on Components and Packing Technologies*, vol.24, no.2, 220-225.

Zhang Y. W., Faghri A., Shafii M. B., 2002, "Analysis of Liquid-vapor Pulsating Flow in a U-shaped Miniature Tube," *International Journal of Heat and Mass Transfer*, 45 (12), 2501-2508.

17

LAMINATION AND CHARACTERIZATION OF A POLYETHYLENE-TEREPHTHALATE FLEXIBLE MICRO HEAT PIPE

Guan-Wei Wu, Sih-Li Chen, Wen-Pin Shih*

[a] *Department of Mechanical Engineering, National Taiwan University, Taipei,10617, Taiwan*

ABSTRACT

A polyethylene-terephthalate (PET) flexible micro heat pipe (FMHP) for heat dissipation has been fabricated and tested. The structure of flexible heat pipe consists of PET, Cu tape, Cu mesh and rubber. These materials are all flexible and inexpensive. Therefore, the heat pipe can bend and the wick structure will not collapse. Besides, the fabrication process can reduce both the material cost and machining time. Experiment results reveal that the minimum thermal resistance in this FMHP is 0.157 °C/W.

Keywords: *polyethylene-terephthalate, flexible, micro heat pipe*

1. INTRODUCTION

For developing high-performance portable appliances with light weight and slim body, consumer electronics will require higher density mounting. Therefore, heat dissipation becomes an extremely important issue for electronics production. Heat pipe is a two-phase heat transfer exchanger with a very high thermal conductivity. The working fluid is filled into the heap pipe in vacuum. When the working fluid absorbs the heat from the heat source, it evaporates at the boiling point and the vapor delivers to the cooling end. Then the vapor releases heat and condenses to liquid. The wick supplies capillary force, facilitating the liquid to return to the heat source again. This two-phase thermal cycle needs minimal temperature gradients by enthalpy difference between phase changes (Yau and Ahmadzadehtalatapeh, 2010).

Conventionally, heat pipe are manufactured by metals such as aluminum and copper (Take and Webb, 2001; Lim *et al.*, 2008). Many reports focused on micro heat pipe fabrication using silicon wafer for patterning hundred-micrometer structures (Berre *et al.*, 2003). Although such materials could against high pressure difference, they have low or even no flexibility and are expensive (Sobhan *et al.*, 2007). In recent years, a novel fabrication method has been applied to produce microchannel or package material in the flexible heat pipe (Oshman *et al.*, 2009; Lin *et al.*, 2009; Tanaka *et al.*, 2009).

In this paper, the flexible polyethylene-terephthalate (PET) plastic film is used to replace the metal as the package material of the heat pipe. A relatively simpler approach to fabricate flexible micro heat pipe (FMHP) by lamination is also presented.

2. STRUCTURE DESIGN OF MICRO HEAT PIPE

The structure of flexible heat pipe consists of PET, Cu tape, Cu mesh and rubber, as shown in Fig.1. The width and length of the presented heat pipe are 20 mm and 110 mm, respectively (Fig. 2). The total thickness of the heat pipe is 1.8 mm. The capillary structure is made of Cu mesh, allowing the condensed liquid to go to the evaporator (liquid line). The gap between the upper and lower Cu mesh is supported by rubber, and the vapor will pass through this channel gap to the condenser (vapor line).

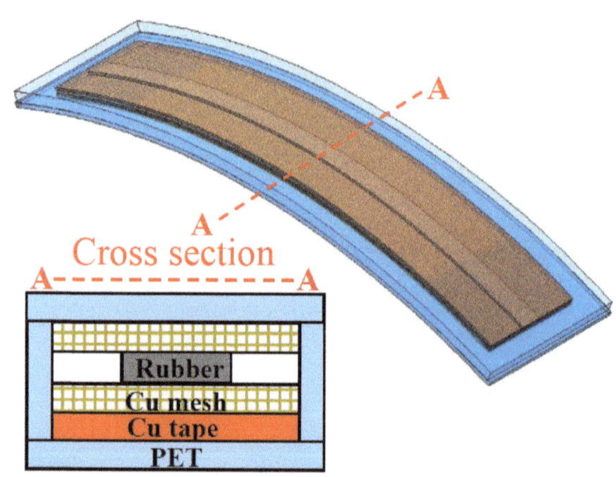

Fig. 1 Structure of the flexible heat pipe

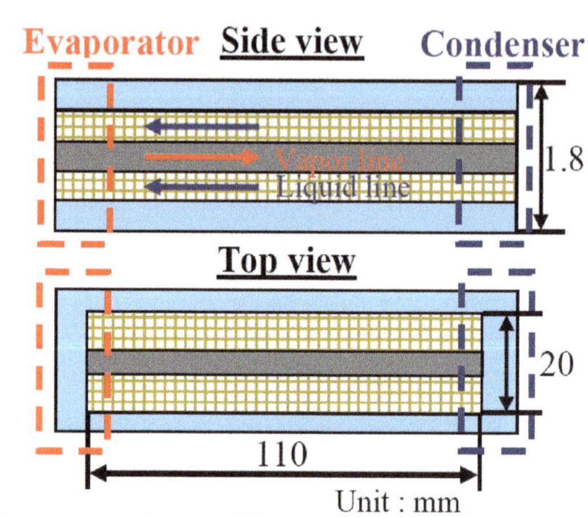

Fig. 2 Dimensions of the flexible heat pipe

* *Corresponding author. Email:wpshih@ntu.edu.tw*

3. FABRICATION PROCESS

To fabricate the FMHP, the PET is firstly cut into two areas and attached to a Cu tape, as shown in Fig. 3(a). The function of these two bare areas for the Cu tape is to contact the heat source and heat sink directly. Then the Cu mesh, rubber and another Cu mesh are attached layer by layer, as shown in Figs. 3(b) and (c). The PET is covered on the upper Cu mesh. Finally, a hot laminator is used for bonding the top and the bottom PET layers, as shown in Fig. 3(d).

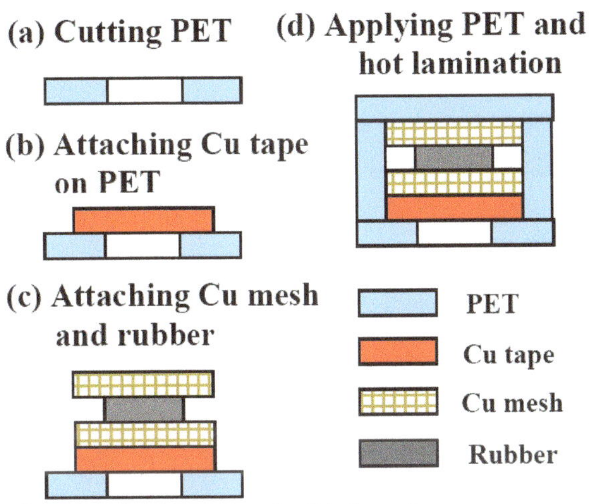

Fig. 3 Process for fabricating flexible heat pipe

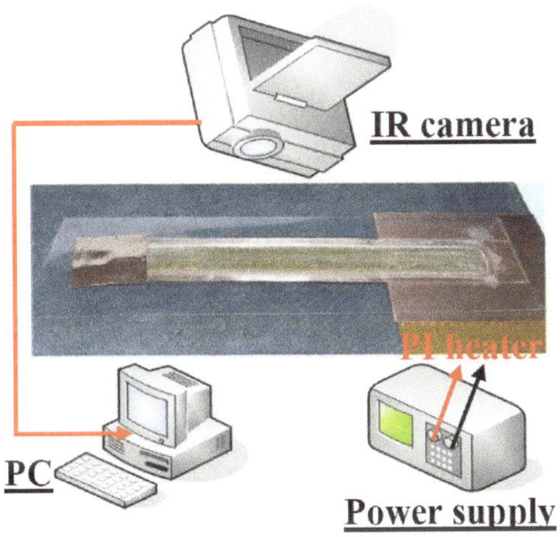

Fig. 4 Experiment set-up

Table 1 Specifications of the FMHP.

Packaging material	PET
Wick structure	Cu mesh
Working fluid	Methanol
Evaporator	PI heater
Condenser	Natural convection
Heat pipe width (w)	2 cm
Heat pipe height (h)	1.8 mm
Length between evaporator and condenser (L_{eff})	7.5 cm

4. EXPERIMENT SET-UP AND PROCEDURE

An infrared (IR) camera, NEC TVS-500EX, is used to measure the thermal resistance of the heat pipe. The temperature measurement range and wavelength are -40-500 °C and 8-14 μm, respectively. A power supply provides the power to the copper/polyimide heater as the heating source to drive the working fluid inside heat pipe, as shown in Fig. 4. The condensation section is cooled using air by natural convection and the ambient temperature is kept at 25 °C.

5. RESULTS OF EXPERIMENT

5.1 Thermal Resistance Test at Horizontal Orientation

When the working fluid, methanol, absorbs the heat until reaching its boiling point, it evaporates and goes to the condenser along the vapor line, as shown in Fig. 5. The IR image of the working heat pipe with different input powers at 36.9% filling ratio is shown in Fig. 6. For the input power larger than 4.7 W, the heat will be delivered from the evaporator to the condenser with increasing input power.

The measured IR image of the FMHP is shown in Fig. 7. The red and black numbers are the highest and lowest temperature in the square area, respectively. The thermal resistance, R_t, is calculated by

$$R_t = \frac{T_e - T_c}{W} \tag{1}$$

where T_e(°C) the evaporator temperature, T_c(°C) the condenser temperature, and W(Watts) the input power.

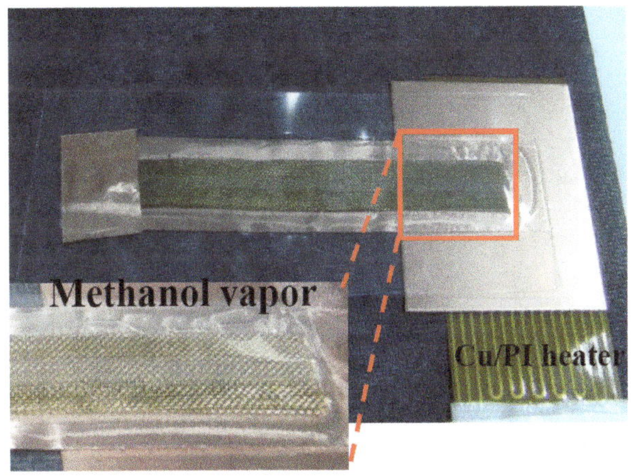

Fig. 5 Methanol vapor in the fabricated micro heat pipe

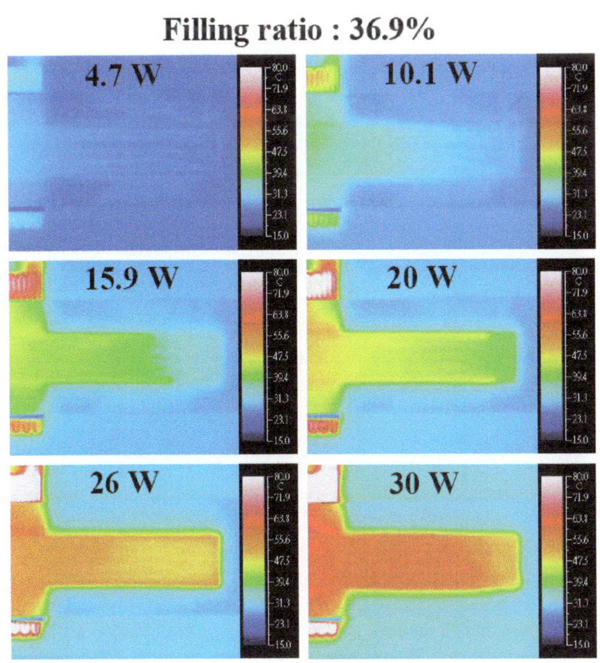

Fig. 6 IR image of FMHP at different input powers

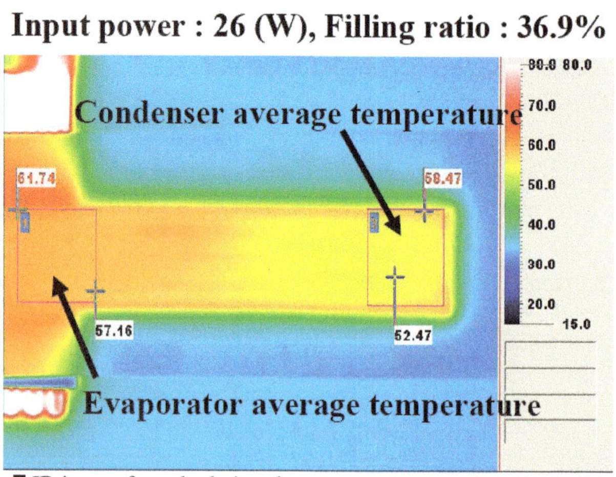

Fig. 7 IR image for calculating the average temperature of evaporator and condenser

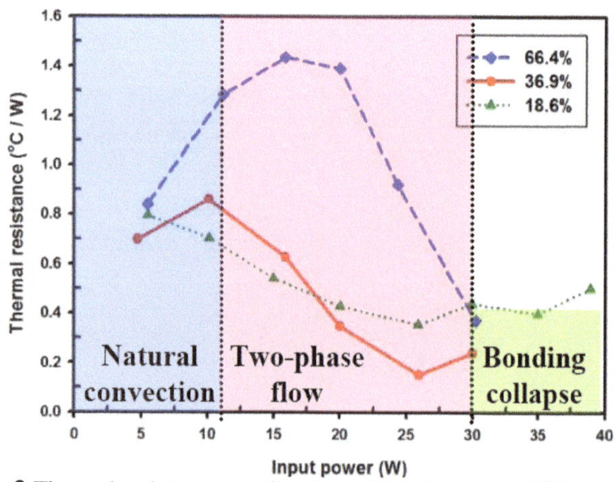

Fig. 8 Thermal resistance as a function of input power at different filling ratios

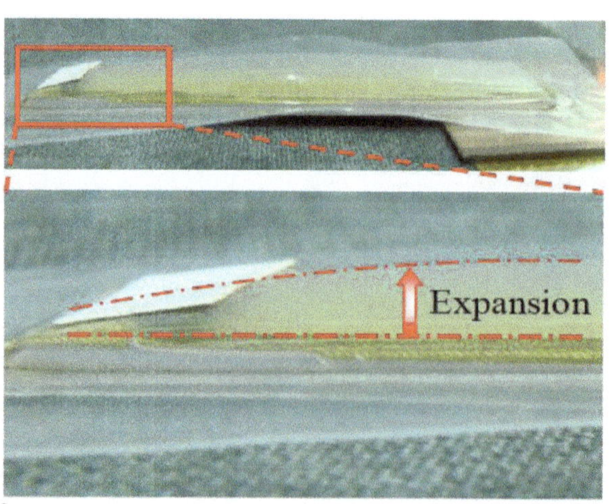

Fig. 9 Delamination in FMHP

The minimum thermal resistance of FMHP occurs at 26 W. The average temperatures in the evaporator and the condenser are 58.94 °C and 54.86 °C, respectively when the input power is 26 W. Therefore, the thermal resistance of the fabricated heat pipe is 0.157 °C/W, which is suitable for practical applications.

The results of the experiment on measuring the thermal resistance with variation input powers at 18.6%, 36.9% and 66.4% filling ratios are shown in Fig. 8. Three operating conditions could be observed in the FMHP. When the input power is smaller than 10 W, the methanol is just under natural convection in liquid phase. If the input power is between 10-30 W, a thermal cycle of two-phase flow is generated inside the heat pipe. Under this condition, the heat will rapidly transfer from the evaporator to the condenser. However, when the input power is larger than 30 W, the PET will be delaminated because of high methanol vapor pressure at 36.9% and 66.4% filling ratios, as shown in Fig. 9.

At low filling ratio, though the methanol vapor has lower flow resistance and higher drive pressure, a little working fluid takes participation in the heat exchange at the evaporator, the heat is delivered finitely to the condenser. At high filling ratio, more heat could be taken to the condenser but the flow resistance is higher and the drive pressure is lower (Lips *et al.*, 2010). In this experiment, the optimal filling ratio is 36.9% for the highest thermal conductivity.

The effective thermal conductivity, K_{eff}, is (El-Nasr and S. M. El-Haggar, 1996; Dorf, 2004)

$$K_{eff} = \frac{Q_{in} L_{eff}}{A_{eff}(T_e - T_c)} \tag{2}$$

where Q_{in} the power input, A_{eff} the cross-sectional area of the heat pipe, L_{eff} the effective length between the evaporator and condenser, T_e the average temperature of the evaporator, and T_c the average temperature of the condenser. Fig. 10 is the effective thermal conductivity as a function of the input power at different filling ratios.

The maximum effective thermal conductivity of FMHP occurs at 26 W. The effective area and length is 3.6×10^{-5} m and 0.075 m, respectively. The temperature difference between the evaporator and the condenser is 4.08 K. Therefore, the effective thermal conductivity is 13261 W/m*K.

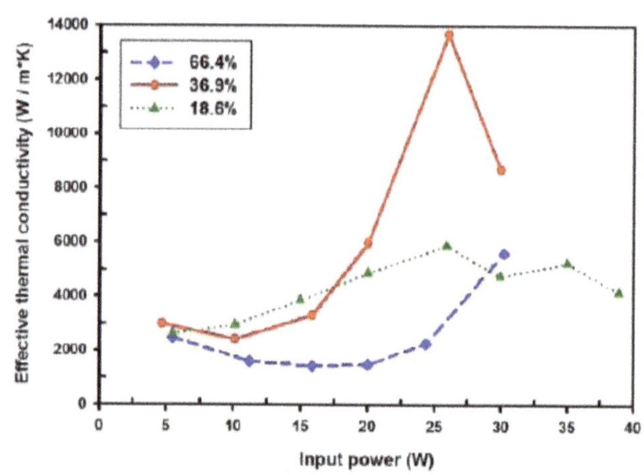

Fig. 10 Effective thermal conductivity as a function of the input power at different filling ratios

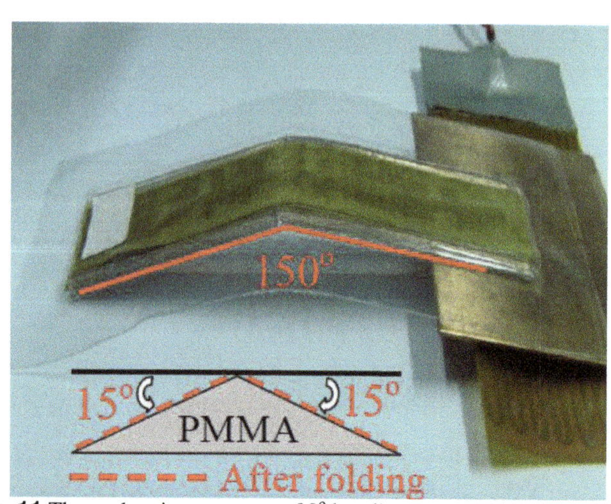

Fig. 11 Thermal resistance test at 30° bending angle

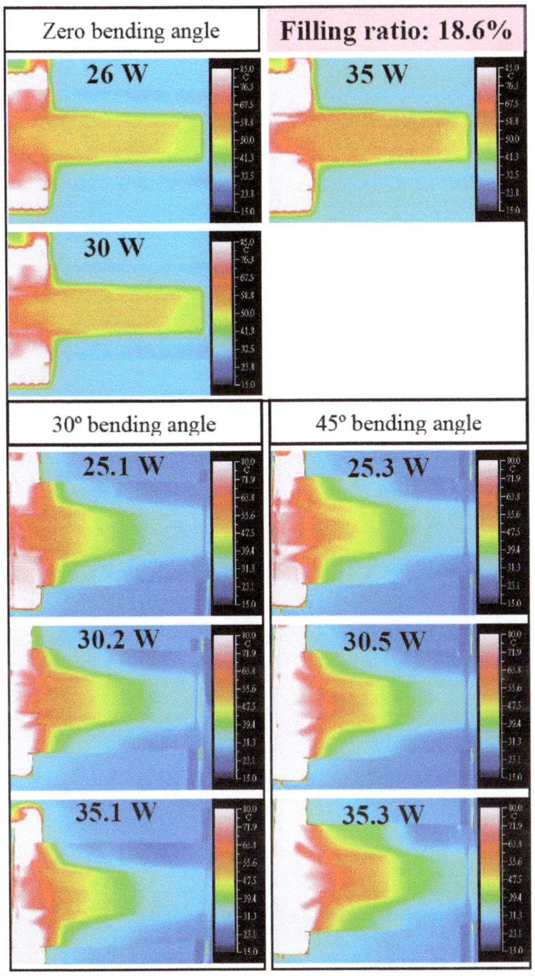

Fig. 12 IR images with different input powers and bending angles

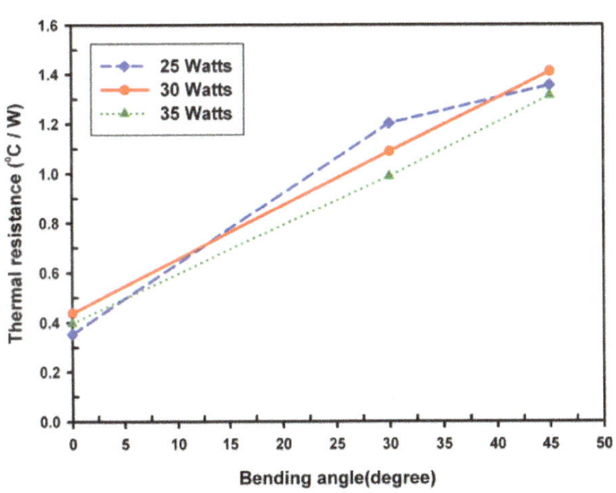

Fig. 13 Thermal resistance with various bending angles at different input powers

5.2 Effect of Different Bending Angles on Thermal Resistance

The thermal resistance of the bent heat pipe is measured and compared with the straight one. The photograph of the thermal resistance test at the 30° bending angle is shown in Fig. 11. A poly(methyl methacrylate) (PMMA) mold with triangular form is used as the supporting structure. The folding line of the MFHP is along the tip of the triangular, and the bending angle is symmetric on both sides.

The IR images of the evaporator for measuring the thermal resistance at different bending angles of 0° (horizontal), 30° and 45° are shown in Fig. 12. The heat is accumulated in the evaporator region when the bending angle increases from 0° to 30°. When the heat pipe is bent, the cross-sectional area will reduce at the bending section. The methanol vapor will be blocked on the evaporator region and can not be easily delivered to the condenser. Although the thermal resistance can be slightly reduced by increasing the input power, it is still high in general. Accordingly, the thermal resistance will increase with higher bending angles, as shown in Fig. 13. For even larger bending angles ($>45^o$) and input powers (>45 W), the high vapor pressure will cause the PET to delaminate.

6. CONCLUSIONS

A FMHP using PET films has been fabricated in this work. For packaging the heat pipe, a laminator is used to bond the PET films. It is a rapid and low cost fabrication process for manufacturing heat pipe. The steady-state thermal resistance of straight and bent FMHP has been analyzed and discussed. The optimal filling ratio and input power are 36.9% and 25 W, respectively for the highest thermal conductivity. The fabricated FMHP shows low thermal resistance (0.157 $^oC/W$) and high effective thermal conductivity (13261 W/m*K).

ACKNOWLEDGEMENTS

This work has been sponsored by the National Science Council, Taiwan with the contract number: 98-2221-E-002-126-MY3.

REFERENCES

Berre, M.L., Launay, S., Sartre, V., and Lallemand, M., 2003, "Fabrication and Experimental Investigation of Silicon Micro Heat Pipes for Cooling Electronics," *Journal of Micromechanics and Microengineering*, **13**(436), 436-441.

Dorf, R.C., 2004, *The Engineering Handbook*, 2nd ed., CRC Press.

El-Nasr, A.A. and El-Haggar, S.M., 1996, "Effective Thermal Conductivity of Heat Pipes," *Heat and Mass Transfer*, **32**(1-2), 97-101.

Lim, H.T., Kim, S.H., Im, H.D., Oh, K.H., and Jeong, S.H., 2008, "Fabrication and Evaluation of a Copper Flat Micro Heat Pipe Working under Adverse-Gravity Orientation," *Journal of Micromechanics and Microengineering*, **18**(10), 105013.

Lin, Y.H., Kang, S.W., and Wu, T.Y., 2009, "Fabrication of Polydimethylsiloxane (PDMS) Pulsating Heat Pipe," *Applied Thermal Engineering*, **29**(2-3), 573-580.

Lips, S., Lefèvre, F., and Bonjour, J., 2010, "Combined Effects of the Filling Ratio and the Vapour Space Thickness on the Performance of a Flat Plate Heat Pipe," *International Journal of Heat and Mass Transfer*, **53**(4), pp. 694-702.

Oshman, C.J., Shi, B., Li, C., Yang, R.G., Lee, Y.C., and Brigh, V.M., 2009, "Fabrication and Testing of a Flat Polymer Micro Heat Pipe," *Proceedings of Transducers'09*, Denver, CO, USA.

Sobhan, C.B., Rag, R.L., and Peterson, G.P., 2007, "A Review and Comparative Study of the Investigations on Micro Heat Pipes," *International Journal of Energy Research*, **31**(6-7), pp. 664-688.

Take, K. and Webb, R.L., 2001, "Thermal Performance of Integrated Plate Heat Pipe with a Heat Spreader," *Journal of Electronic Packaging*, **123**(3), 189-195.

Tanaka, K. Abe, Y., Nakagawa, M., Piccolo, C., and Savinoe, R., 2009, "Low-Gravity Experiments of Lightweight Flexible Heat Pipe Panels with Self-Rewetting Fluids," *Interdisciplinary Transport Phenomena: Fluid, Thermal, Biological, Materials and Space Science*, **1161**, 554-561.

Yau, Y.H. and Ahmadzadehtalatapeh, M., 2010, "A Review on the Application of Horizontal Heat Pipe Heat Exchangers in Air Conditioning Systems in the Tropics," *Applied Thermal Engineering*, **30**(2-3), 77-84.

EFFECT OF FILLING RATIO ON THERMAL CHARACTERISTICS OF WIRE-MESH HEAT PIPE USING COPPER OXIDE NANOFLUID

R. Manimaran*, K. Palaniradja, N. Alagumurthi

Department of Mechanical Engineering, Pondicherry Engineering College, Puducherry, India

ABSTRACT

An experimental study was carried out to investigate the thermal characteristics of heat pipe using nanofluid and Deionized water (DI water) is presented in this work. The CuO/water nanofluid is used as one of the working fluid in experimental heat pipe with concentration of 1.0% wt .The test section of the heat pipe is made of copper tube with outer diameter 22 mm, inner diameter 20.8 mm and length 600 mm. The heat pipe is tested with (DI water) and nanofluid respectively. This study focused on the effects of heat input, fill ratio and angle of inclination on the thermal efficiency and thermal resistance of heat pipe. The experimental results indicate that the thermal efficiency increases when nanoparticles are added with the DI water (nanofluid) and also the heat pipe which uses nanofluid as working fluid shows lower thermal resistance value when compared to the heat pipe which uses DI water alone.

Keywords: *heat pipe, angle of inclination, nanofluid, efficiency*

1. INTRODUCTION

Effective thermal management of modern electronic system and the miniaturization results in generation of more heat flux and this leads to the generation of higher temperature within the system, thus affecting the performance of the system. Therefore, effective cooling devices are needed to keep the systems operating at their reliable working temperature levels. Hence the concept of traditional old passive cooling does not meet the requirement of effective cooling whereas other technology like liquid cooling satisfies the required cooling needs of the electronic components, especially when Gaugler (1944) introduced the concept of heat pipe which can provide effective heat transfer widely used in the thermal management of electronic systems. Their ability to transport the heat over a substantial distance at low temperature drops has made them a much reliable and efficient cooling device.

Heat pipes are common thermal control components on satellites, where temperature differences exist in different sides of the spacecraft due to solar radiation or internal dissipation of electronic devices and surfaces oriented towards the deep space. On the other hand, heat pipes are used as passive exchangers in electronics cooling, air conditioning, propulsion, power generation, energy recovering, chemical engineering and other applications. Grover et al. (1964) carried out Research and development activities that are generally concerned with new heat pipes development with improved thermal performances, i.e. higher thermal conductance, higher heat transport capacity and larger operating temperature range.

Choi (1995) introduced the concept of dispersing the nanoparticle into the fluids which showed enhanced thermal conductivity compared to the fluids containing millimeter and micrometer sized particles. Y. Xuan and Li (2003) and You et al. (2003) studied the convective heat transfer characteristics of nanofluids and showed the enhancement of critical heat flux of nanofluids and pool boiling characteristics when compared with pure water. Y.Xuan and Q. Li, (2000) developed a model for preparing the various nanofluids by directly mixing various nanoparticles with base fluids which increase the thermal conductivity of nanofluids. Sidi et al. (2005) investigated, by numerical simulation, the hydrodynamic and thermal characteristics of a laminar forced convection flow of nanofluids inside a straight heated tube and showed that the heat transfer enhancement increases considerably with an augmentation of the flow Reynolds number. Wang and Mujumdar (2007) has presented a detailed review on the heat transfer application of nanofluids and also given direction for future research directions.

Esen (2004) studied experimentally about the working of solar cooking system by using vacuum tube collectors with heat pipes containing refrigerant as working fluid. Vasiliev (2008) carried out a detailed study about the application of micro and miniature heat pipes in modern heat exchangers for cooling electronic components and fuel cells cooling.Heris.et al.(2006) investigated the laminar flow convective heat transfer through circular tube for Cuo and Al_2O_3 oxide nanoparticles and showed the highest heat transfer enhancement for higher volume fraction. Jang (2007) studied the effects of various parameters that affected the thermal conductivity of nanofluids and also its influence on the particle volume fraction of the nanoparticles. Abreu and Colle (2004) considered the effect of evaporator length, filling ratio of working fluid, cooling temperature and heat flux to verify the overall thermal resistance of two phase thermosyphon experimentally and also provided the information for design of solar domestic hot water system.

Kang et al. (2006) used silver nanofluid to investigate the thermal performance of a grooved circular heat pipe and showed that the thermal resistance is decreased when compared with pure water. Do et al. (2010) conducted experiment on the screen mesh wicked heat pipes using Al_2O_3 nanofluids and showed that the heat pipe performance has been increased up to 40%. A two –dimensional analysis by Shafahi et al.(2010) is presented to study the thermal performance of a cylindrical heat pipe utilizing nanofluids and they found that the nanoparticles within the liquid enhance the thermal performance of the heat pipe by reducing the thermal resistance. Yang et al. (2008) showed that CuO nanofluid can improve the thermal performance of the heat pipe and also presented that an optimal mass concentration of 1.0 wt% is enough

* Corresponding author : email: manimaranpec@gmail.com.

to achieve the maximum heat transfer enhancement.Teng et al (2010) conducted experiment and found that the optimal thermal efficiency occurred when the charge amount was 60%, also the added nanoparticles at the concentration of 1.0 wt.%, showed the optimum value of thermal efficiency can be enhanced by 16.8% when compared with based fluid. Han et al. (2011) tested hybrid working fluid system and found that hybrid nanofluids were not much effective comparing with the pure nanoparticle nanofluid system. Senthilkumar et al. (2011) showed experimentally that the aqueous solution n-Hexanol based copper nanofluids enhances the thermal efficiency of the heat pipe and the effect of nanoparticles in the aqueous solution of n-Hexanol has a great effect on the reduction of the thermal resistance than that of DI water and copper nanofluid loaded heat pipes.

The aim of the present work is to investigate, experimentally the thermal performance of the heat pipe by using DI water and nanofluid as the working fluids. The experimental parameters considered here are heat input, filling ratio and angle of inclination.

2. EXPERIMENTAL SETUP AND PROCEDURE

2.1 Preparation and characterization of nanofluids

Single step synthesis via wet chemical method has been used for preparing copper oxide nanofluid. Zhu et al. (2011) studied that the nanofluids of different morphology can be obtained by changing the synthesis parameters In this procedure all the chemicals including $CuCl_2$, NaOH, and HCL acid were of analytical grade and purchased from Aldrich chemicals and no purification process has been carried out. In this method the preparation of nanoparticles has been carried out along with the nanofluids. In this preparation process 100 ml of DI water is taken in a 500 ml of beaker. To this 0.51 gm of $CuCl_2$ was added along with one gram of NaOH pellets. The reaction mixture was stirred under magnetic stirring for about one hour along with heating. The colour of the solution changes from blue to black after the reaction and then the mixture was cooled to room temperature.

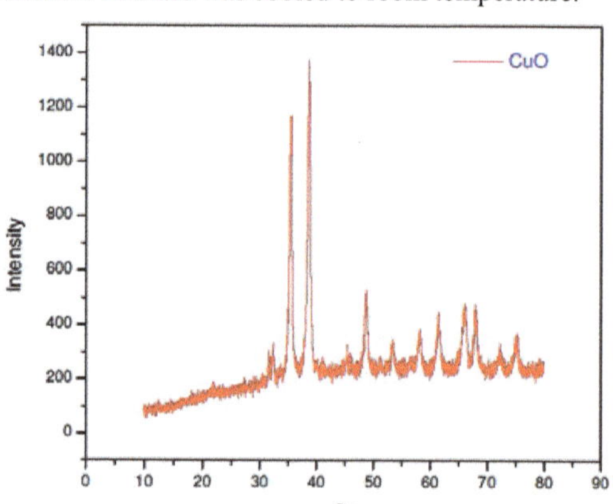

Fig. 1 X-ray diffractogram for the copper oxide nanoparticles

The PH value of so formed copper oxide wet precipitate is neutralized by adding droplets of hydrochloric acid (HCL).Then the copper oxide wet precipitate is washed with DI water to remove the impurity ions present in the solution. Copper oxide nanofluids were obtained by dispersing the wet precipitate into the required amount of DI water under ultrasonic vibration for about four hours to have uniform dispersion of copper oxide nanoparticles.

Characterization of nanoparticles has been carried out by using XRD and SEM.Fig.1 Shows the XRD pattern of the typical sample. All the peaks on the XRD can be indexed to that of monoclinic CuO according to the literature (JCPDS, File No 01-080-1916).The average crystal size is 20 nm which is calculated using Debye-Scherrer formula. Fig.2 and 3 shows the SEM image of CuO nanoparticle and dispersion of nanoparticles in DI water respectively.

Fig. 2 SEM image for the copper oxide nanoparticle sample

Fig. 3 SEM image for the copper oxide nanofluid

2.2 Experimental setup

A schematic diagram of the experimental setup is shown in Fig.4.In this research work the water and Cuo nanofluids are used as working fluids and the average size of the nanoparticle used in this study is 20 nm .The concentration of the nanoparticle used in this study is 1% by weight. The body of the heat pipe is made out of copper with the length of 600 mm, outside diameter 22 mm and the inside diameter 20.8 mm.Two layers of stainless steel screen mesh wick (number 160) is layered on the inner diameter of the heat pipe. The heat pipe consists of three sections namely evaporator section 150 mm, adiabatic section 300 mm and condenser section of 150 mm in length. An electric heater was applied on the evaporator region to provide the required heat supply and a wattmeter with the required power range and a variac has been integrated into the electric heater circuit to evaluate the exact power supplied. The evaporator and condenser regions were insulated using the glass wool (0.33W/m K) to prevent the heat loss from these regions. The condensation length was cooled by circulating water to maintain a constant temperature of 30°C ± 1°C at the inlet section and the mass flow rate is kept at constant value by using flow meter with uncertainty ±0.01L/min.Twelve K-type thermocouples with uncertainty ± 0.1°C were calibrated against quartz thermometer and soldered on the surface of the heat pipe to measure the wall temperature at different locations of the heat pipe including four at the evaporator region, three at the adiabatic region, three at condenser region and the remaining two at the inlet and outlet of the cooling water circuit. The vacuum pump model DOA-P707-AA and 1/8 HP is used for evacuating the heat pipe.

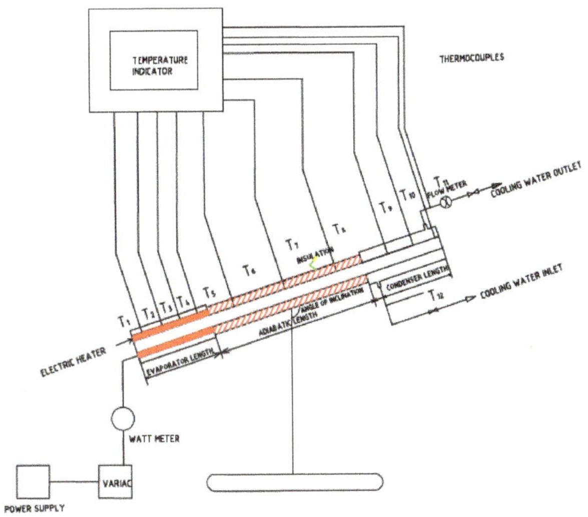

Fig. 4 Experimental setup

2.3 Experimental procedure

Two identical heat pipes of same dimensions has been used to carry out the experiment i.e. one for DI water and another for nanofluid.For each new charging the same heat pipe is used. Before filling up the working fluid, the heat pipe is evacuated using the vacuum pump to remove the dissolved gases and after evacuation the heat pipe is initially filled with 12.75ml (25% fill ratio) of the working fluid. The evaporator region is heated using the power supply to the required value with the help of auto transformer. The temperature rise was monitored for steady state condition and the temperatures are recorded at various locations of the heat pipe using the selector switch. Then the experiment is repeated for different fill ratios (25%, 50%, 75% and 100%) and angle of inclination (0^0,30^0,60^0 and 90^0).Filling ratio means the percentage of the evaporator section volume that is filled by the working fluids.

3. RESULTS AND DISCUSSION

3.1 Effect of fill ratio on the thermal efficiency of the heat pipe

Figures 5 and 6 indicate the variation of filling ratio in the evaporator region on the thermal efficiency of the heat pipe under various angles of inclinations for both DI water and nanofluids. Choi (1995) studied that the addition of nanoparticles to the base fluid will leads to the increase of thermal conductivity of the working fluid which lead to the increase in heat transfer capability. The heat pipe thermal efficiency can be calculated from the ratio of cooling capacity of water flowing at the condenser section to the heat supplied at the evaporator section.

In general the efficiency of the heat pipe increases as the fill ratio of the evaporator region increases due to the reason that fluid absorption capacity greatly depends on the filling ratio of the working fluid and more space for the working fluid. The results indicates that the efficiency reaches a maximum value for a fill ratio of 75% and gradually decreases when the fill ratio is increased for both the working fluids due to the reason that higher fill ratio leads to lower space for vapour in the condenser area which leads to the decrease of heat transfer rate of the working fluid at the evaporator section. At 75% fill ratio there was about 14.3% increase in efficiency occurs for copper oxide nanofluid heat pipe when compared with DI water heat pipe.

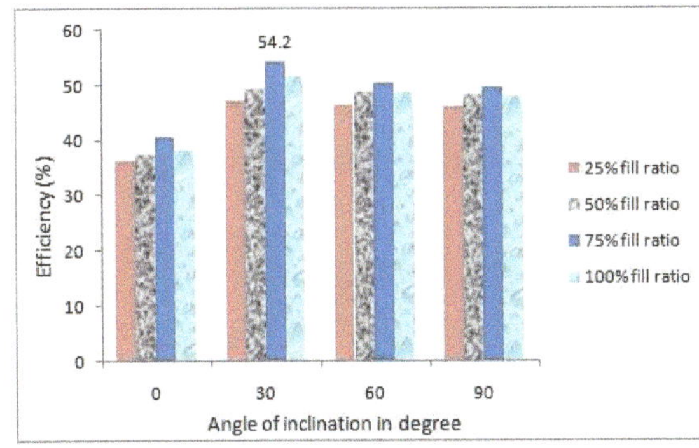

Fig. 5 Efficiency vs. angle of inclination of DI water for various fill ratio

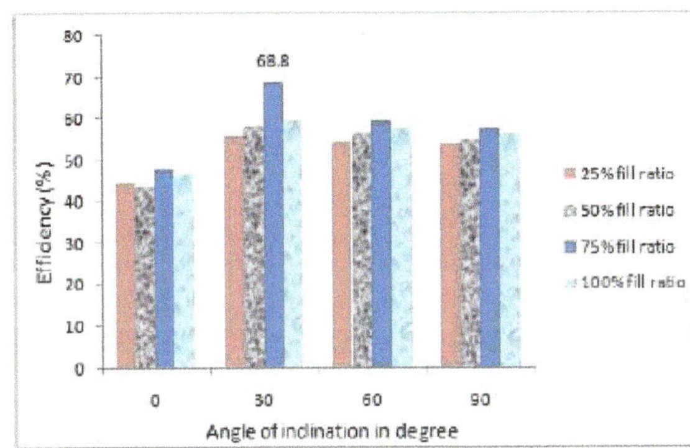

Fig.6 Efficiency vs. angle of inclination of nanofluids for various fill ratio

3.2 Influence of fill ratio on the thermal resistance of heat pipe

The thermal resistance (R) of the heat pipe can be calculated by the relation

$$R = \frac{Te - Tc}{Q} \tag{1}$$

where T_e and T_c are the evaporator and condenser surface temperature of the heat pipe respectively and Q is the heat supplied at the evaporator region of the heat pipe. Figures 7-10 shows the variation of thermal resistance for two different working fluids that occur at different heat pipe inclination and fill ratio.

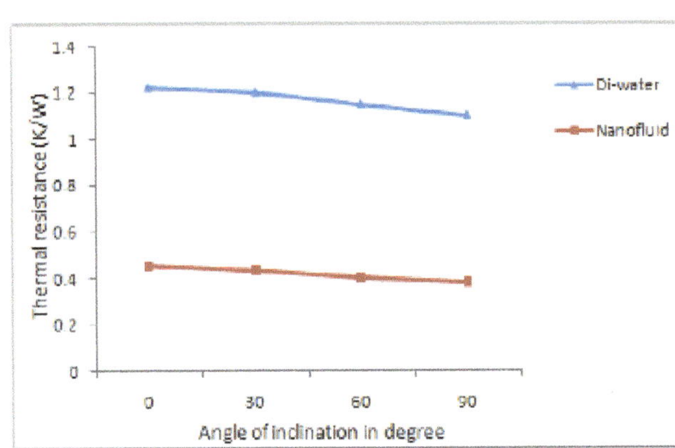

Fig. 7 Thermal resistance distributions for different angle of inclination at 25% fill ratio

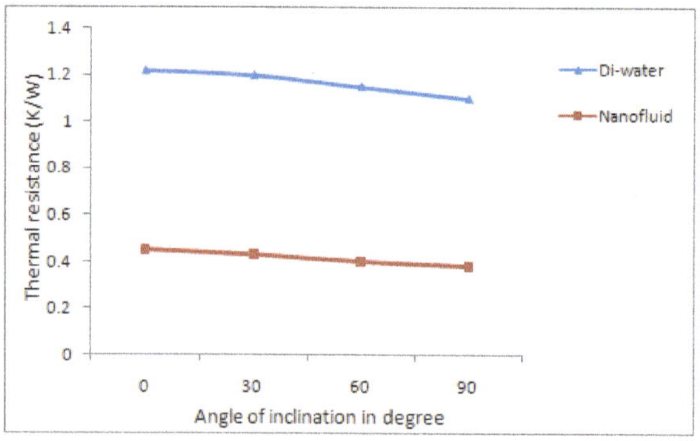

Fig. 8 Thermal resistance distributions for different angle of inclination at 50% fill ratio

The graphs indicate that the thermal resistance of the heat pipe decreases when the inclination angle increases for both the working fluids. Also the heat pipe which uses nanofluid as working fluid shows lower thermal resistance when compared to that of the heat pipe which uses DI water alone. An optimum thermal resistance value of 0.33 K/W is obtained when the heat pipe is operated under 75% by using nanofluid as a working fluid and the maximum decrease of thermal resistance was about 62 % when compared to DI water. The suspension of nanoparticle in the base fluid increases the thermal conductivity of working fluid. Wen and Y. Ding, (2005) and Patel et al. (2005) studied the formation of molecular nanolayer on the surface of nanoparticle occurs due to the absorption of liquid molecules by the nanoparticles which has higher thermal conductivity compared to the base fluid (DI water).

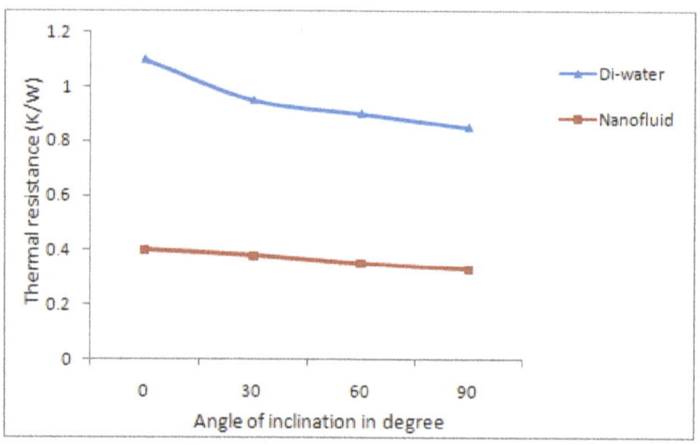

Fig. 9 Thermal resistance distributions for different angle of inclination at 75% fill ratio

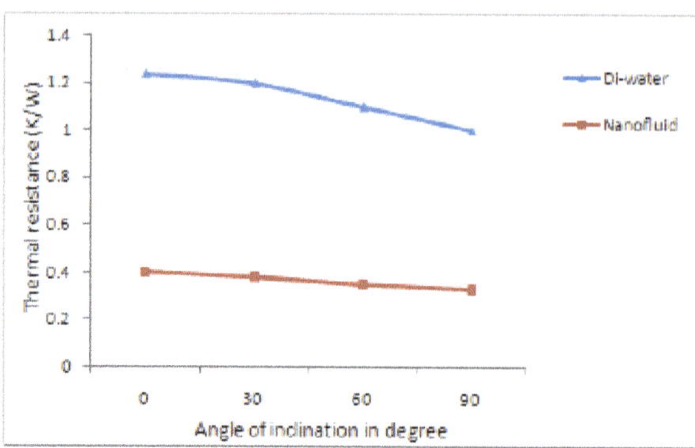

Fig. 10 Thermal resistance distributions for different angle of inclination at 100% fill ratio

CONCLUSION

The experiment has been conducted on the heat pipe for various parameters such as angle of inclination and fill ratio. The thermal efficiency for DI water and nanofluid increases when the fill ratio increases and the maximum efficiency is obtained for 75% fill ratio and for an angle of 30°.Comparatively the heat pipe which uses nanofluid as the working fluid shows higher efficiency than the heat pipe using DI water. As far as thermal resistance is concerned the value decreases when the angle of inclination increases and optimum thermal resistance is obtained at 75% fill ratio when the heat pipe is operated by using nanofluid as working fluid with maximum decrease of 62% when compared with heat pipe using DI water alone. From the above results it can be said that nanofluids have higher potential for heat transfer enhancement and can be used for practical heat transfer application process.

ACKNOWLEDGMENT

The authors thank Dr.S.Sendhilnathan, Assistant Professor, Anna University of technology pattukotai campus and Dr.R.Senthilkumar, Assistant Professor, Annamalai University for their help to carry out this work

REFERENCES

Gaugler, R. S, 1944, "Heat Transfer Devices," U.S. patent 2,350,348

Grover and G. M., Cotter, T.P. Erikson, G.F, 1964, "Structures of Very High Thermal Conductivity," J. *Appl. Phys,* 35, 1190–1191.

Choi ,U.S, 1995, "Enhancing Thermal Conductivity of Fluids with Nanoparticles," ASME *FED* 231, 99–103.

Xuan, Y, and Li, Q., 2003, "Investigation on Convective Heat Transfer and Flow Features of Nanofluids," *ASME Journal of Heat Transfer.*125, 151–155.

You, S.M., Kim, J.H., and Kim, K.H., 2003, "Effect of Nanoparticles on Critical Heat Flux of Water in Pool Boiling Heat Transfer," *Applied Physics Letters,* 83, 3374–3376.

Xuan, Y., and Li, Q., 2000, "Heat Transfer Enhancement of Nanofluids," *International Journal of Heat and Fluid Flow,* 21, 58–64.

Maiga, S., Palm,S.,Nguyen,C.,Roy,G.,Galanis,N.,2005,"Heat Transfer Enhancement by Using Nanofluids in Forced Convection Flows," *International Journal of Heat and Fluid Flow,*26,530–546.

Wang, X, Qi., and Mujumdar, A.S., 2007, "Heat Transfer Characteristics of Nanofluids: A Review," *International Journal of Thermal Sciences,* 46, 1-19.

Esen, M., 2004, "Thermal Performance of a Solar Cooker Integrated Vacuum-Tube Collector with Heat Pipes Containing Different Refrigerants," *Solar Energy,* 76, pp. 751–757.

Vasiliev, L.L., 2008, "Micro and Miniature Heat Pipes-Electronic Component Coolers," *Applied Thermal Engineering,* 28, 266–273.

Heris, S.Z., Etemad, S.G., and Esfahany, M.N., 2006, "Experimental Investigation of Oxide Nanofluids Laminar Flow Convective Heat Transfer," *International Communications in Heat and Mass Transfer,*33, 529–535.

Jang, S.P., 2007, "Effect of Various Parameters on Nanofluid Thermal Conductivity". *Journal of Heat Transfer,* 129, 617-623.

Abreu, S. L., and Colle, S., 2004, "An Experimental Study of Two-Phase Closed Thermosyphons for Compact Solar Domestic Hot-Water Systems," *Solar Energy,* 76, 141.

Kang, S.W., Wei, W.C., Tsai, S.H., and Yang, S.Y 2006, "Experimental Investigation of Silver Nano-fluid on Heat Pipe Thermal Performance," *Applied Thermal Engineering*. 26, 2377–2382.

Do, H., Ha, K. H.J., and Jang, S.P., 2010, "Thermal Resistance Of Screen Mesh Wick Heat Pipe Using The Water-based Al_2O_3 Nanofluids." *International Journal of Heat and Mass Transfer,* 25, 5888-5894.

Shafahi, M., Bianco, V., Vafai, K., and Manca,O.,2010, "An Investigation of The Thermal Performance of Cylindrical Heat Pipes Using Nanofluids," Int. *J. Heat Mass Transfer*, 53, 376–383.

Yang, X.F., Liu, Z.H., and Zhao, J., 2008, "Heat Transfer Performance of A Horizontal Microgrooved Heat Pipe Using CuO Nanofluids." *J Micromech Microeng,* 18,035038.

Teng, T.P., Hsu, H.G., Mo, H.E., and Chen, C.C., 2010, 'Thermal Efficiency of Heat Pipe With Alumina Nanofluid," *Journal of Alloys and Compounds.* 504S, S380–S384.

Han, W. S., and Rhi, S.H., 2011, 'Thermal Characteristics of Grooved Heat Pipe with Hybrid Nanofluids," *Thermal science,* 15,195-206.

Senthilkumar,R.,Vaidyanathan,S.,andSivaraman,B.,2011,"Experimental Analysis of Cylindrical Heat Pipe using Copper Nanofluid with an Aqueous Solution of n-Hexanol," *Frontiers in Heat Pipes,* 2, 033004,

Zhu, H., Han, D., Meng, Z., Wu, D., and Zhang, C., 2011, "Preparation and Thermal Conductivity of CuO Nanofluid via a Wet Chemical Method," *Nanoscale Research Letters*.6, 181.

Wen, D., and Ding, Y., 2005, "Effect of Particle Migration on Heat Transfer in Suspensions of Nanoparticles Flowing through Minichannels," *Microfluidic.Nanofluidic*.1, 183-189.

Patel, H.E., Sundararajan, T., Pradeep, T., Dasgupta, A., Dasgupta N., and Das, S.K., 2005, "A Micro-convection Model For Thermal Conductivity Of Nanofluids," *PRAMANAJ.Phys*.65, 863-869.

THE EFFECT OF THICKNESS AND PERMEABILITY OF WICK STRUCTURE ON L-SHAPE HEAT PIPE PERFORMANCE USING DIFFERENT WORKING FLUIDS

Mohamed H. A. Elnaggar

Engineering Department, Palestine Technical College, Deir EL-Balah, Gaza Strip, Palestine.

ABSTRACT

This paper presents a two-dimensional FEM simulation of working fluid behavior inside L-shape heat pipe to predict wall temperature and liquid pressure under natural and forced convection modes by using ANSYS-FLOTRAN. This work aimed to numerically study the effect of thickness and permeability of wick structure using different working fluids on L-shape heat pipe performance, where water and methanol were used as working fluids. The heat inputs for natural and forced convection modes comprised 20 W and 35 W respectively whereas the heat pipe wick thicknesses were taken as 0.5 mm and 0.75 mm. The results showed that at forced convection the use of water as working fluid with 0.5 mm sintered copper wick recorded the smallest temperature difference (9.8 °C) while methanol with 0.75 mm screen mesh wick recorded the largest temperature difference (28.27 °C).

Keywords: *L-shape heat pipe; ANSYS-FLOTRAN; Wick structure; Working fluid; Liquid pressure drop.*

1. INTRODUCTION

Effective cooling of electronic components is important for the successful functioning and high reliability of electronic devices. The rapid developments in microprocessors necessitate enhanced processing power to ensure faster operations. According to (Shao and Zhang 2011), "Local hot spot heat fluxes within the CPU are exceeding $100W/cm^2$ causing the CPU temperature rise, malfunction and failure of CPU without a very effective heat transfer device." Consequently, the requirement for cooling techniques to dissipate the associated heat is of utmost importance. Moreover, the present standard metallic heat sinks are obsolete in many ways and are not sufficient to address the increased cooling needs that are sought by many of today's electronic devices. Hence, it is highly desirable to explore high-performance cooling devices, especially for CPU cooling. Heat pipe has been identified and proved as one of the viable and promising options to achieved this purpose, due to its simple structure, flexibility and in particular, high efficiency.

Substantial numerical and experimental works have been reported on the application of heat pipes for electronic cooling. Reviews by (Faghri 1994), (Groll et al. 1998) and (Vasiliev 2005) provided detailed information on this topic. In addition, the main theoretical and numerical studies on heat pipes with capillary grooved structures were investigated by (Khrustalev and Faghri 1995a; b), (Faghri and Khrustalev 1997), (Khrustalev and Faghri 1999), (Zhang and Faghri 2001), (Jacolot et al. 2008), (Do et al. 2008), (Do and Jang 2010).

Two dimensional models that analyze the liquid and vapor flow and wall temperature along horizontal cylindrical heat pipe are developed by (Mistry et al. 2010), (Nouri-Borujerdi and Layeghi 2004), (Mahjoub and Mahtabroshan 2008). As well, some researchers investigated the flat heat pipes experimentally and numerically such as (Wang and Vafai 2000), (Maziuk et al. 2001), (Carbajal et al. 2007).
The differences in the shape of heat pipe affect its performance as the behaviors of the fluid and the wick structure inside the heat pipe play important role in the transmission of heat. However, the previous works on numerical characterization by taking into account the behavior of the working fluid considered only conventional horizontal cylindrical heat pipe.

In the present study, a two-dimensional FEM simulation of working fluid behavior inside L-shape heat pipe was performed to predict liquid pressure and the wall temperature. The complexity of the simulation of the present heat pipe lies in the fact that the evaporator is at x direction while the condenser is at y direction. The simulations were done by using ANSYS-FLOTRAN software and were validated by experimental results. The main objective of this work is to numerically study the effect of thickness and permeability of wick structure using different working fluids on L-shape heat pipe performance as water and methanol were used as working fluids.

2. MATHEMATICAL FORMULATION

The finned flat heat pipe under investigation is shown in Fig. 1. This finned heat pipe serves to cool the CPU of a modern notebook such as ASPIRE4720Z with Intel Pentium dual-core processor T2390, 1.86GHz and memory up to 4 GB DDR2. In this system, the heat pipe is used to transfer heat from CPU to the fins in a remote location, usually in the sides or corners of the notebook PC. The finned flat heat pipe consists of copper base plate of size 30 mm × 30 mm and condenser section of 72 mm long, with 50 aluminum rectangular fins of 20 mm × 10 mm size. The system is supported by radial fan with dimensions of 73 mm × 73 mm × 10 mm (width ×length × thickness). The specifications of the heat pipe are shown in Fig. 1 and summarized in Table 1.
In this study, the adapted mathematical model was based on the following assumptions: vapor and liquid flows are assumed to be in steady state, laminar and incompressible; the vapor is treated as ideal gas; there is no heat generation due to phase change and chemical reaction in the system; the liquid and vapor phases are coupled at the liquid–vapor interface, and the vapor injection and suction at the liquid vapor interface are uniform; the physical properties are constant.

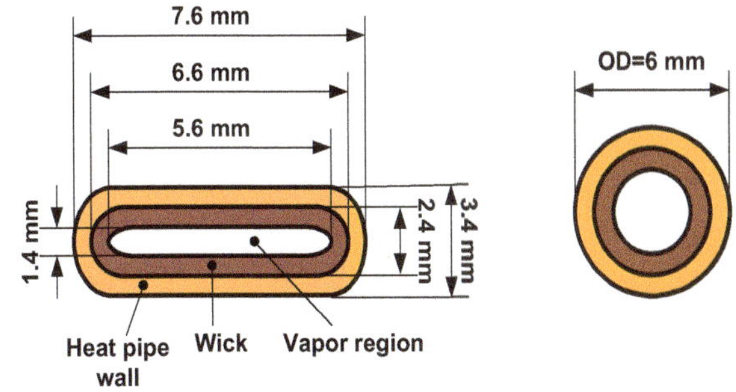

(a) Cross-section of flat heat pipe and equivalent (original) diameter (OD)

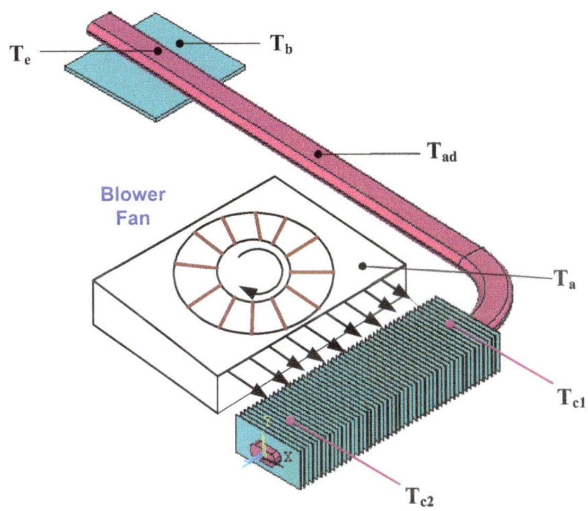

(b) Thermocouples locations

Fig.1. Finned flat heat pipe (a) Heat pipe cross-section (b) Thermocouples locations

Table 1 Specifications of finned L-shape heat pipe

Characteristic	*Dimensions/Material*
Evaporator length L_e	30 mm
Condenser Length L_c	72 mm
Adiabatic length L_{ad}	110 mm
Flat heat pipe	7.6 mm ×3.4 mm (width × thickness)
Original diameter (OD)	6 mm
Heat pipe Container thickness	0.5 mm
Wick thickness	0.5 mm
Working fluid	Water
Wick structure	Sintered copper powder
Material of pipe	Copper

The resulting governing equations are listed as follows:

Vapor region

Continuity: $$\frac{\partial u_v}{\partial x}+\frac{\partial v_v}{\partial y}=0 \quad (1)$$

where u and v are components of velocity in x and y directions, respectively.

Momentum:

$$\rho_v\left(u_v\frac{\partial u_v}{\partial x}+v_v\frac{\partial u_v}{\partial y}\right)=-\frac{\partial p}{\partial x}+\mu_v\left(\frac{\partial^2 u_v}{\partial x^2}+\frac{\partial^2 u_v}{\partial y^2}\right) \quad (2)$$

$$\rho_v\left(u_v\frac{\partial v_v}{\partial x}+v_v\frac{\partial v_v}{\partial y}\right)=-\frac{\partial p}{\partial y}+\rho g+\mu_v\left(\frac{\partial^2 v_v}{\partial x^2}+\frac{\partial^2 v_v}{\partial y^2}\right) \quad (3)$$

Energy:

$$\rho_v c_p\left(u_v\frac{\partial T}{\partial x}+v_v\frac{\partial T}{\partial y}\right)=k_v\left(\frac{\partial^2 T}{\partial x^2}+\frac{\partial^2 T}{\partial y^2}\right) \quad (4)$$

where, g is the acceleration of gravity, ρ_v vapor density, μ_v is the effective viscosity of vapor for laminar case is merely the dynamic viscosity, C_p specific heat and k_v is thermal conductivity of vapor.

Liquid wick region:

Continuity: $$\frac{\partial u_l}{\partial x}+\frac{\partial v_l}{\partial y}=0 \quad (5)$$

where, u and v are components of velocity in x and y directions, respectively

Momentum:

$$\rho_l\left(u_l\frac{\partial u_l}{\partial x}+v_l\frac{\partial u_l}{\partial y}\right)=-\frac{\partial P_l}{\partial x}+\mu_l\left(\frac{\partial^2 u_l}{\partial x^2}+\frac{\partial^2 u_l}{\partial y^2}\right)+R_x \quad (6)$$

$$\rho_l\left(u_l\frac{\partial v_l}{\partial x}+v_l\frac{\partial v_l}{\partial y}\right)=-\frac{\partial P_l}{\partial y}+\rho_l g+\mu_l\left(\frac{\partial^2 v_l}{\partial x^2}+\frac{\partial^2 v_l}{\partial y^2}\right)+R_y \quad (7)$$

R_x and R_y are distributed resistance components in x and y directions, respectively. A distributed resistance is a convenient way to approximate the effect of porous media. Any fluid element with a distributed resistance will have a real constant set number greater than 1 (ANSYS Release 11.0 documentation 2008). The total liquid pressure gradients along x and y directions are given by (ANSYS Release 11.0 documentation 2008):

$$\frac{\partial P_l}{\partial x_{resistance}}=\{-K\rho u_l|u_l|+\frac{f}{D_h}\rho u_l|u_l|+C\mu u_l\} \quad (8)$$

$$\frac{\partial P_l}{\partial y_{resistance}}=\{-K\rho v_l|v_l|+\frac{f}{D_h}\rho v_l|v_l|+C\mu v_l\} \quad (9)$$

where, K is the localized head loss, ρ is the density, μ is the viscosity, f is the friction coefficient and C is the FLOTRAN permeability expressed in m^{-2}, which is inverse of the physical permeability (ANSYS Release 11.0 documentation 2008).

Energy:

$$\rho_l c_{p,l}\left(u_l\frac{\partial T_l}{\partial x}+v_l\frac{\partial T_l}{\partial y}\right)=k_e\left(\frac{\partial^2 T_l}{\partial x^2}+\frac{\partial^2 T_l}{\partial y^2}\right)+Q_v \quad (10)$$

where, g, ρ, μ, C_p, k_e and Q_v are gravitational acceleration, density, dynamic viscosity, specific heat, effective thermal conductivity for liquid-wick structure and volumetric heat flux respectively. Subscripts v and l refers to vapor and liquid regions respectively. k_e is the effective thermal conductivity of the liquid-wick structure for sintered powder wick, as expressed by (Peterson 1994):

$$k_e=\frac{k_l[(2k_l+k_w)-2(1-\varphi)(k_l-k_w)]}{(2k_l+k_w)+(1-\varphi)(k_l-k_w)} \quad (11)$$

For screen mesh wick, k_e calculated from (Peterson 1994):

$$k_e=\frac{k_l[(k_l+k_w)-(1-\varphi)(k_l-k_w)]}{(k_l+k_w)+(1-\varphi)(k_l-k_w)} \quad (12)$$

where, φ is porosity, k_l and k_w are thermal conductivity of liquid and wick material respectively.
The steady state thermal conductivity equation to predict the wall temperature is:

$$k_s\left(\frac{\partial^2 T_s}{\partial x^2}+\frac{\partial^2 T_s}{\partial y^2}\right)=0 \tag{13}$$

where, k_s is solid thermal conductivity and T_s is solid (wall) temperature.

Boundary conditions:

Fig. 2 illustrates the detailed boundary conditions of the computational domain.

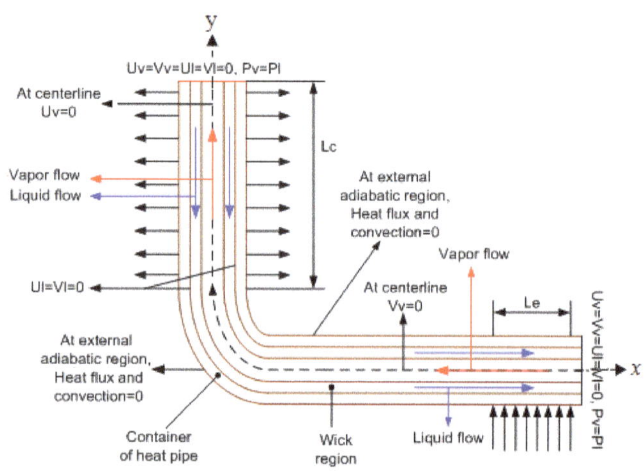

Fig. 2. Illustration of boundary conditions

At both ends of the heat pipe $u_v = v_v = u_l = v_l = 0$, and $P_v = P_l$

At the centerline of evaporator section $v_v = 0$, $\frac{\partial u_v}{\partial y} = 0$ and $\frac{\partial T}{\partial y} = 0$

At the centerline of condenser section $u_v = 0$, $\frac{\partial v_v}{\partial x} = 0$ and $\frac{\partial T}{\partial x} = 0$

For the solid (container) – liquid (wick) interface $u_l = v_l = 0$

At the adiabatic section, $\rho_v v_v = \rho_l v_l = 0$

The continuity of mass fluxes in *y* direction at the vapor liquid interface yields

$\rho_v v_v = \rho_l v_l = -\rho_v v_l$ where v_l is the vapor injection velocity expressed as (Nouri-Borujerdi and Layeghi 2004):

$$v_1 = \frac{Q_h}{2\rho_v \pi R_v L_e h_{fg}} \tag{14}$$

Similarly, the continuity of mass fluxes in *x* direction at the vapor liquid interface yields

$\rho_v u_v = \rho_l u_l = \rho_v u_l$ where u_l is the vapor suction velocity as given by (Nouri-Borujerdi and Layeghi 2004):

$$u_1 = \frac{Q_h}{2\rho_v \pi R_v L_c h_{fg}} \tag{15}$$

Where, L_e and L_c are lengths of the evaporator and condenser respectively, and h_{fg} is the latent heat of the working fluid.

The interface temperature (T_{int}) is calculated by the Clausius-Clapeyron equation, assuming the saturation temperature (T_0) and vapor pressure (P_0) at the liquid-vapor interface (Kaya and Goldak 2007):

$$T_{int} = \frac{1}{\frac{1}{T_0} - \frac{R}{h_{fg}}\ln\left(\frac{P_v}{P_0}\right)} \tag{16}$$

For the solid-liquid interface

At the evaporator part $K_e \frac{\partial T_l}{\partial y} = k_s \frac{\partial T_s}{\partial y}$

At the condenser part $K_e \frac{\partial T_l}{\partial x} = k_s \frac{\partial T_s}{\partial x}$

At the external heat pipe wall
$$\begin{cases} \text{Evaporator} \quad k_s \dfrac{\partial T}{\partial y} = q_e \\ Adiabatic \quad \dfrac{\partial T}{\partial y} = 0 \; and \; \dfrac{\partial T}{\partial x} = 0 \\ Condenser \quad -k_s \dfrac{\partial T}{\partial x} = h(T_s - T_a) \end{cases}$$

where, h is convection heat transfer coefficient, and T_s and T_a are wall surface and ambient temperatures respectively.

3. MESHING AND NUMERICAL SOLUTION

The modeling approach was adapted to the works of (Zhu and Vafai 1999), (Kaya and Goldak 2007) and (Nouri-Borujerdi and Layeghi 2004), taking into account the amendments to suit the L-shape heat pipe and the simulations were done by using ANSYS-FLOTRAN. The heat from the base was transferred to the heat pipe and then to the fins and from fins to the surrounding air. In the present two-dimensional numerical simulation of the flat heat pipe an obstacle was that the top view section was broad (7.6 mm) while the side view section was thin (3.4 mm) as shown in Fig. 1(a), and this prompted to calculate the equivalent diameter of the heat pipe. Accordingly, by assuming the flat heat pipe as cylindrical heat pipe of equivalent perimeter, the equivalent diameter of the flat heat pipe was obtained as 6 mm.

The computational domain is divided into two estranged regions: 1) vapor region, and 2) liquid-wick structure and heat pipe wall. The simulation is carried out by using the element FLUID141. The vapor region is meshed with 4900 fine triangular elements, whereas the liquid-wick and solid regions are given relatively very finer meshing, forming 52456 triangular elements. Preconditioned Conjugate Gradient (PCG) was used as iterative solvers. The convergence criterion was taken as 1e-007 were the convergence criterion is a value below which the convergence monitors for each active degree of freedom must drop to cause the time step to terminate (ANSYS Release 11.0 documentation 2008). For solving the set of vapor matrix equations, the Conjugate Residual Method was used; however when solving equations of solid and liquid simultaneously, the non-symmetric matrix equations appeared, and to overcome this phenomenon the Preconditioned Biconjugate Gradient Stabilized Method (BiCGSTAB) was selected.

4. RESULTS AND DISCUSSION

4.1 Natural Convection

In the natural convection mode, the simulation was performed with heat input of 20 W. At natural convection, the physical properties of the working fluid were taken at 120 °C as summarized in Table 2.
The wick structure consisted of sintered copper powder wick which has fine pore size causing low permeability (1.17×10^{-1}1 m2) (Mwaba et al. 2006). The effective thermal conductivity of liquid-wick structure is 3.69 W/mK, which was calculated from Eq. (11).

The wall temperature results were compared with experimental data to verify the accuracy of the simulation results at the same operation conditions.

Table 2 Physical properties of water and vapor at 120 °C (Faghri 1995)

	Water	*Vapor*
Density kg/m^3	945	1.12
Viscosity kg/m.s	2.6×10^{-4}	1.34×10^{-5}
Thermal conductivity W/mK	0.682	0.0234
Latent heat J/kg	2.2×10^{6}	-
Surface tension N/m	0.055	-
Pressure N/m^2	-	$2.02\text{x}10^{5}$

As shown in Fig. 3, the highest temperature reaches to 126.74 °C, located in the evaporator and then falls relatively in the adiabatic section to 120 °C and the temperature stay uniform in this region and then the temperature drops again to 114.94 °C in the condenser section. It is worth noting that the simulation temperatures including evaporator temperature (T_e), adiabatic section temperature (T_{ad}), condenser temperatures (T_{c1}) and (T_{c1}) are 125.19 °C, 120.0°C, 119.45 °C, and 114.87 °C respectively which are close to the experimental results except in the adiabatic section temperature (T_{ad}) which is less satisfactory due to the assumption of no heat loss in the adiabatic section. Moreover, as the experiment is isolated by glass tunnel, the adiabatic temperature affected by the removal heat. The good agreement indicates the validity of the present simulation methodology.

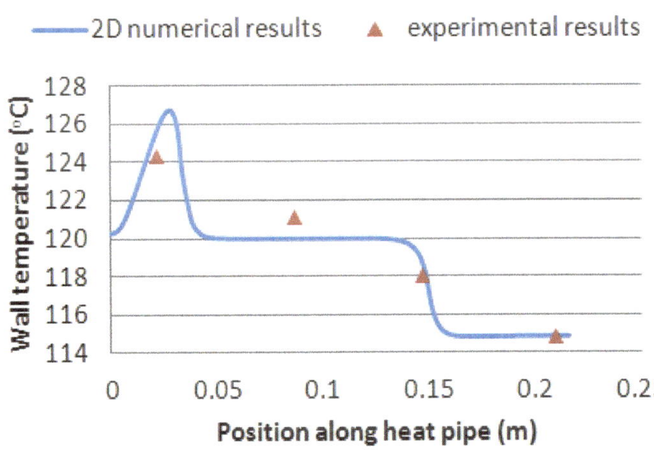

Fig. 3. Wall temperature distribution along the heat pipe at Q=20 W in natural convection and comparison with the experimental results.

Methanol as working fluid

A comparative study was made by replacing water with methanol as working fluid. In the natural convection mode, the simulation was performed with heat input of 20 W. At Q=20 W and natural convection, the physical properties of the working fluid were taken at 120 °C as summarized in Table 3.

Table 3 Physical properties of liquid and vapor of methanol at 120 °C

	Liquid	*Vapor*
Density kg/m^3	696	7.7
Viscosity kg/m.s	2.0×10^{-4}	1.28×10^{-5}
Thermal conductivity W/mK	0.196	0.0022
Latent heat J/kg	0.95×10^{6}	-
Surface tension N/m	0.0135	-
Pressure N/m^2	-	$6.41\text{x}10^{5}$

The use of screen mesh wick with porosity 0.62 and permeability1.93×10-10 (Mwaba et al. 2006) and water as working fluid resulted in value of effective thermal conductivity of water-wick structure 1.42 W/mK according to Eq. (12). As noted in Table 3, the thermal conductivity of methanol is small (0.196 W/mK) compared with water (0.682 W/mK) at the same operating temperature. This leads to low effective thermal conductivity of Liquid-wick region 0.445 W/mK. In order to know the influence of the wick thickness on the thermal performance of L-shape heat pipe in the present simulation, the wick thicknesses were taken as 0.5 mm to 0.75 mm for using both water and methanol as working fluids.

Fig. 4 shows the wall temperature distribution of the heat pipe in natural convection with reference to change in working fluid and thicknesses of screen mesh wick ($K=1.93\times10^{-10}$ m^2) at Q=20 W. It is observed that use of water as working fluid leads to small temperature difference between evaporator and condenser compared to that with methanol as working fluid. This is attributed to the fact that the thermal conductivity of water is higher than that of methanol, which leads to the increase of effective thermal conductivity of water-wick region. The effect of wick thickness is also observed in Fig. 4; the temperature difference of the heat pipe with wick thickness 0.5 mm is less than that with wick thickness 0.75 mm. Water with 0.5 mm screen mesh wick recorded the smallest temperature difference (24.56 °C) while methanol with 0.75 mm screen mesh wick recorded the largest (54 °C). It is also observed that the temperature distribution at the adiabatic region is uniform and constant at about 120 °C for all the cases.

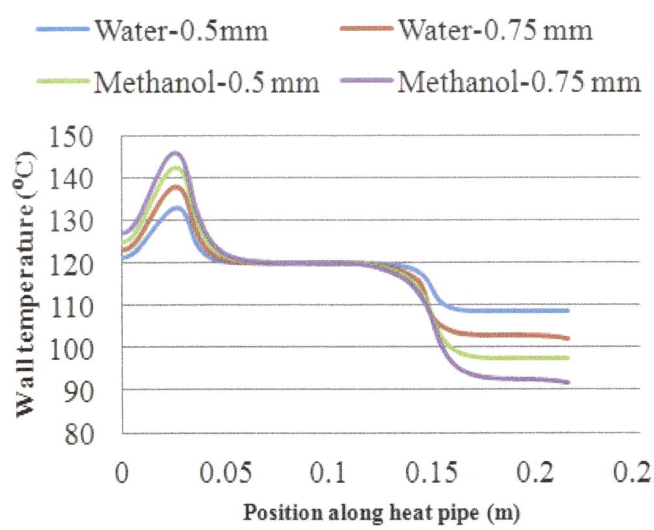

Fig. 4. Wall temperature distribution of heat pipe in natural convection with reference to variation of working fluid and thicknesses of screen mesh wick ($K=1.93\times10^{-10}$ m^2) at Q=20 W.

The comparison of heat pipes with screen mesh wick and different working fluids and wick thicknesses, is shown in Fig. 5. With increasing wick thickness, the flow area increases and it causes decrease in pressure drop in the liquid-wick region. It is observed that the maximum liquid pressure drop (692.43 Pa) is obtained for methanol with 0.5 mm wick thickness while the minimum pressure drop (338.94 Pa) is obtained for water with 0.75mm wick thickness. Although the liquid pressure drop of methanol is large owing to its higher viscosity, the pressure drop of water with wick thickness of 0.5 mm (501.81 Pa) is greater than that of methanol with wick thickness of 0.75 mm (468.37 Pa). This indicates that the effect of thickness on the pressure drop is more significant than the effect of the type of liquids used.

4.2 Forced Convection

In this mode, the applied heat is 35 W with coolant airflow rate of 6.5m3/h. As shown in Fig. 6, the highest temperature (52.94 °C) is located at the evaporator section. The adiabatic area between the evaporator and condenser has uniform temperature distribution about 47 °C, and it begins to decline at the condenser region to reach to 43.05°C towards the end of the condenser section.

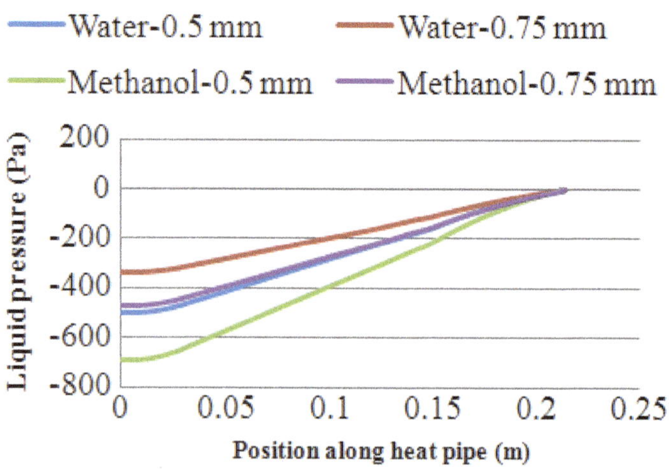

Fig. 5. Liquid pressure distribution along the heat pipe at Q=20 W in natural convection with various wick thicknesses (Screen mesh $K=1.93\times10^{-10}$ m^2).

Fig. 6 shows the wall temperature distribution along the heat pipe at coolant airflow of 6.5m3/h. The predict temperatures Te, Tad, Tc1 and Tcl are observed to be 51.52 °C, 47.00°C, 46.23 °C, and 43.06 °C respectively; this decline in temperatures is obviously due to the enhanced cooling by forced convection. These values of temperatures are in good agreement with experimental results except evaporator temperature (T_e) . This is due to (T_e) being affected by cooling air flow which decreases it.

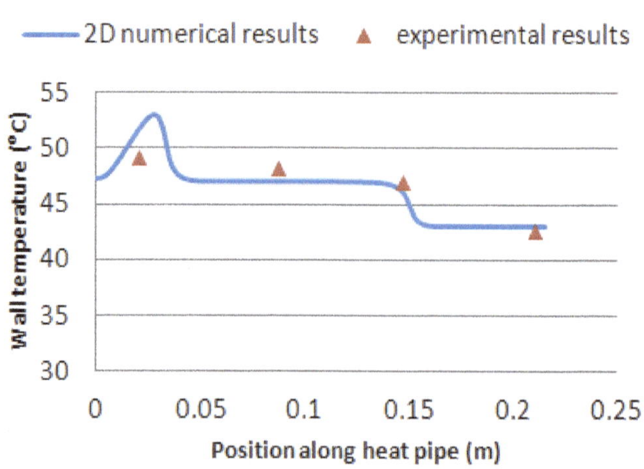

Fig. 6. Predicted temperature distribution along heat pipe at Q=35 W with coolant airflow rate of 6.5 m^3/h and comparison with the experimental results.

In order to show the effect of parameters and compare their influences on the performance of heat pipe under forced convection, the parameters are collected in Fig. 7. It is observed that the use of sintered copper powder wick leads to small temperature difference between evaporator and condenser compared to screen mesh wick as the sintered copper wick has low permeability 1.17×10-11 m^2, which causes increase of effective thermal conductivity of wick region. Additionally, the use of water as working fluid leads to small temperature difference compared to that with methanol as working fluid. This is because the thermal conductivity of water is higher than that of methanol, which increases the effective thermal conductivity of water-wick region compared to the methanol-wick region. The effect of wick thickness is also observed; the temperature difference of the heat pipe with wick thickness 0.5 mm is less than that of heat pipe with wick thickness 0.75 mm. The water with 0.5 mm sintered copper wick recorded the smallest temperature difference (9.8 °C) while methanol with 0.75 mm screen mesh wick recorded the largest temperature difference (28.27 °C). It is also observed that there is a rapprochement between the temperature distribution of water with 0.75 mm screen mesh wick and methanol with 0.5 screen mesh wick. Additionally, it is observed that the temperature distribution at the adiabatic region is uniform and constant about 47 °C for all cases. Methanol with copper powder wick was excluded because it results in exceedingly high pressure in the liquid-wick region.

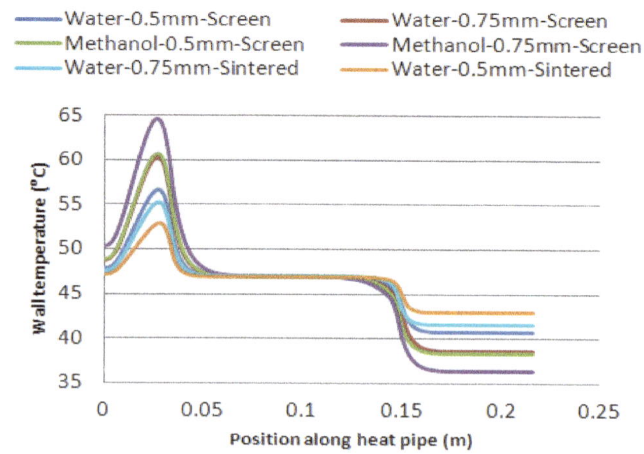

Fig. 7. Wall temperature distribution along the heat pipe at Q=35W and forced convection ($Q^*=6.5$ m^3/h) with various working fluid and wick types and thicknesses.

In the forced convection, heat input of 35 W is used, with increasing heat input, the flow rate of liquid increases, so it causes of increasing the pressure drop in liquid region. Fig. 8 shows the liquid pressure distribution along the heat pipe at Q=35 W under forced convection ($Q^*=6.5$ m^3/h) with various sintered copper wick thicknesses. It is observed that liquid pressure drop is decreased from 6563 Pa to 4219 Pa when the wick thickness is increased from 0.5 mm to 0.75 mm, as increasing the thickness of the wick leads to increase in the liquid flow area through the wick, which causes reduction in liquid pressure drop liquid-wick region.

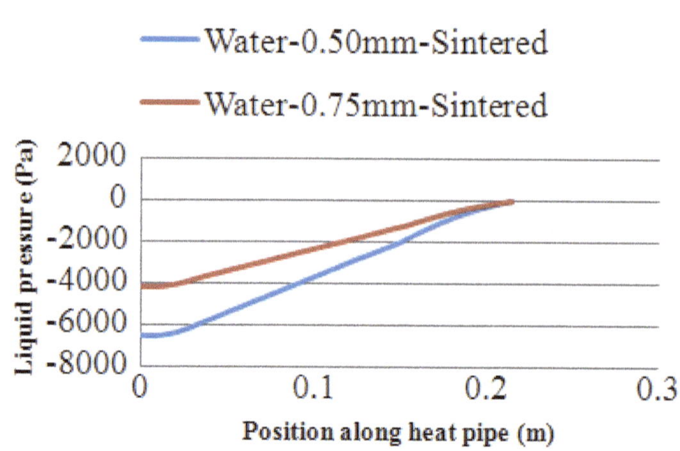

Fig. 8. Predict liquid (Water) pressure distribution at Q=35 W in forced convection ($Q^*=6.5$ m^3/h) of sintered copper powder wick with various thicknesses.

Fig. 9 clearly shows the comparison of heat pipes with screen mesh wicks of difference working fluids and wick thicknesses. It is observed that the maximum liquid pressure drop (1789 Pa) is obtained when methanol with 0.5 mm wick thickness is used, while the minimum pressure drop (633 Pa) is obtained when water with 0.75mm wick thickness is used. Although the liquid pressure drop of methanol is large due to its higher viscosity, the pressure drop of methanol with wick thickness of 0.75 mm (1083 Pa) is less than the pressure of water with wick thickness of 0.5 mm (1129 Pa). This is because, the increased thickness of the wick leads to increase in the liquid flow area through the wick, which causes reduction in liquid pressure drop at the liquid-wick region.

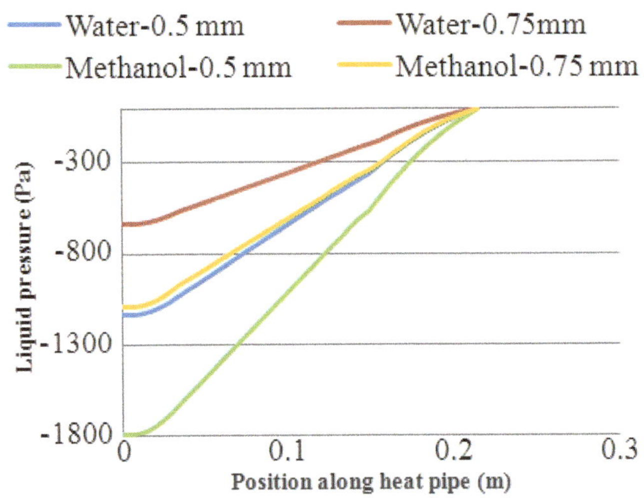

Fig. 9 Liquid pressure distribution along the heat pipe at Q=35W at Q^{*}=6.5 m^3/h with various working fluid and wick thickness at Screen mesh

5. CONCLUSION

A two-dimensional FEM simulation of working fluid behavior inside L-shape heat pipe was performed by using ANSYS-FLOTRAN software to study the effect of thickness and permeability of wick structure using water and methanol as different working fluids on L-shape heat pipe performance.

The study concluded with the following key points:

- The increase in thickness of the wick caused a decrease in liquid pressure drop in the liquid-wick region while the use of methanol as working fluid increased the liquid pressure.
- The use of sintered copper powder wick resulted in small temperature difference between the evaporator and condenser compared to screen mesh wick.
- Although good performance of heat pipe was achieved by employing water with 0.5 mm thickness of sintered copper powder, the unfavorable high pressure resulted in low heat transport capability at the liquid–wick region.

REFERENCES

ANSYS Release 11.0 documentation, Cartographer, 2008.

Carbajal, G., Sobhan, C. B., Peterson, G. P., Queheillalt, D. T., and Wadley, H. N. G., 2007, "A quasi-3D analysis of the thermal performance of a flat heat pipe," *International Journal of Heat and Mass Transfer*, 50(21-22), 4286-4296.

Do, K. H., and Jang, S. P., 2010, "Effect of nanofluids on the thermal performance of a flat micro heat pipe with a rectangular grooved wick," *International Journal of Heat and Mass Transfer*, 53(9-10), 2183-2192.

Do, K. H., Kim, S. J., and Garimella, S. V., 2008, "A mathematical model for analyzing the thermal characteristics of a flat micro heat pipe with a grooved wick," *International Journal of Heat and Mass Transfer*, 51(19-20), 4637-4650.

Faghri, A., 1994, "Recent advances in the numerical analysis of heat pipes," *Computational Mechanics*, 14(5), 480-491.

Faghri, A.,1995, *Heat Pipe Science and Technology*, Taylor & Francis Group.

Faghri, A., and Khrustalev, D., 1997, "Advances in Modeling of Enhanced Flat Miniature Heat Pipes with Capillary Grooves, " *Journal of Enhanced Heat Transfer*, 4(2), 99-109.

Groll, M., Schneider, M., Sartre, V., Chaker Zaghdoudi, M., and Lallemand, M., 1998, "Thermal control of electronic equipment by heat pipes," *Revue Générale de Thermique*, 37(5), 323-352.

Jacolot, R., Romestant, C., Ayel, V., Bertin, Y., and Hugon, J., 2008, "Visualization and Experimental Analysis of Fluid Behavior in Grooved Heat Pipes," *Experimental Heat Transfer*, 21(3), 188 - 205.

Kaya, T., and Goldak, J., 2007, "Three-dimensional numerical analysis of heat and mass transfer in heat pipes," *Heat and Mass Transfer*, 43(8), 775-785.

Khrustalev, D., and Faghri, A., 1995a, "Heat Transfer During Evaporation on Capillary-Grooved Structures of Heat Pipes," *Journal of Heat Transfer*, 117(3), 740-747.

Khrustalev, D., and Faghri, A., 1995b, "Thermal Characteristics of Conventional and Flat Miniature Axially Grooved Heat Pipes," *Journal of Heat Transfer*, 117(4), 1048-1054.

Khrustalev, D., and Faghri, A, 1999, "Coupled Liquid and Vapor Flow in Miniature Passages With Micro Grooves," *Journal of Heat Transfer*, 121(3), 729-733.

Mahjoub, S., and Mahtabroshan, A., 2008, "Numerical Simulation of a Conventional Heat Pipe," *Proceedings Of World Academy Of Science, Engineering And Technology*, 29, 117-122.

Maziuk, V., Kulakov, A., Rabetsky, M., Vasiliev, L., and Vukovic, M., 2001, "Miniature heat-pipe thermal performance prediction tool – software development," *Applied Thermal Engineering*, 21(5), 559-571.

Mistry, P. R., Thakkar, F. M., De, S., and DasGupta, S., 2010, "Experimental Validation of a Two-Dimensional Model of the Transient and Steady-State Characteristics of a Wicked Heat Pipe," *Experimental Heat Transfer*, 23(4), 333 - 348.

Mwaba, M. G., Huang, X., and Gu, J., 2006, "Influence of wick characteristics on heat pipe performance, " International Journal of Energy Research, 30(7), 489-499.

Nouri-Borujerdi, A., and Layeghi, M., 2004, "A Numerical Analysis of Vapor Flow in Concentric Annular Heat Pipes," *Journal of Fluids Engineering*, 126(3), 442-448.

Peterson, G. P., 1994, *An Introduction to heat pipes: Modeling, Testing, and Applications*, New York: John Wiley & Sons, INC.

Shao, W., and Zhang, Y., 2011, "Effects of Capillary and Gravitational Forces on Performance of An Oscillating Heat Pipe," *Frontiers in Heat Pipes (FHP)*, 2, 023003.

Vasiliev, L. L., 2005, "Heat pipes in modern heat exchangers," *Applied Thermal Engineering*, 25(1), 1-19.

Wang, Y., and Vafai, K., 2000, "An experimental investigation of the thermal performance of an asymmetrical flat plate heat pipe," *International Journal of Heat and Mass Transfer*, 43(15), 2657-2668.

Zhang, Y., and Faghri, A., 2001, "Numerical Simulation Of Condensation On A Capillary Grooved Structure," *Numerical Heat Transfer, Part A: Applications*, 39(3), 227-243.

Zhu, N., and Vafai, K., 1999, "Analysis of cylindrical heat pipes incorporating the effects of liquid-vapor coupling and non-Darcian transport--a closed form solution" *International Journal of Heat and Mass Transfer*, 42(18), 3405-3418.

THEORETICAL STUDY AND REVIEW ON THE OPERATIONAL LIMITATIONS DUE TO VAPOUR FLOW IN HEAT PIPES

Rémi Bertossi, Cyril Romestant, Vincent Ayel*, Yves Bertin

Institut Pprime, CNRS - ENSMA - Université de Poitiers, UPR 3346
Département Fluide, Thermique et Combustion
1, avenue Clément Ader 86961 Futuroscope Chasseneuil Cedex, FRANCE

ABSTRACT

Vapour flow in heat pipes is subject to two operating limitations that can restrict the performances of such technologies. The first one is the viscous limitation: it occurs when the operating temperature is closed to the triple point and the pressure slope on the saturation curve, related to the temperature glide, is not large enough to counterbalance the viscous pressure losses of the vapour flow. The second one is the sonic limitation: it can appear when the vapour velocity becomes high so that it reaches the speed of sound. This paper presents a detailed description of the physical phenomena which take place in these two types of limitations. Some analytical expressions of maximal heat flux induced by the latter are also proposed.

Keywords: *viscous limitation, sonic limitation, analytical study, vapour, heat-pipes*

1. INTRODUCTION

Heat pipes working are very dependent on their operational limitations. Indeed, maximal performances are limited by the operating conditions of the device. Five types of limitations are classically defined. Fig. 1 represents the operating domain of heat pipes as a function of the different limitations that can come into play (Faghri, 1995). The latter are notably very dependent on the thermophysical properties of the working fluid; for example, for working temperatures closed to the triple point or to the critical one, the performances can be seriously deteriorated. Besides, limitations are also directly linked to the characteristics of the fluid flow.

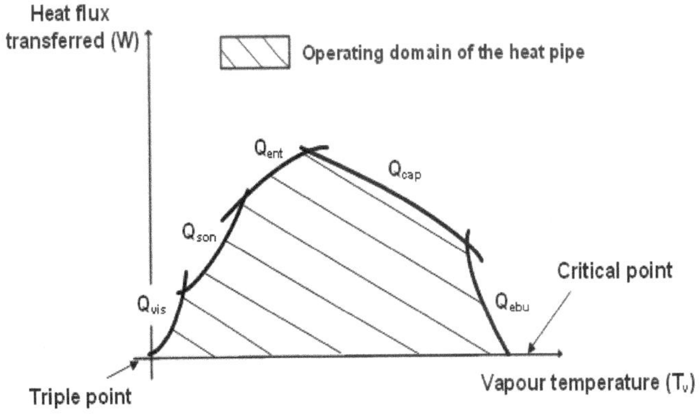

Fig. 1 Operating limitations interfering in the heat pipe working

The operating limitations can be divided into two main categories: the first ones linked to the liquid flow (Q_{ent}, Q_{cap}, Q_{boil} on Fig. 1) and the other ones linked to the vapour flow (Q_{vis}, Q_{son} on Fig. 1). For the liquid flow, one has:

- The entrainment limitation, Q_{ent}, that can occur when the vapour velocity is very high so that liquid particles can be pulled out from the liquid flow which is consequently altered and not able to carry out properly heat flow from the condenser into the evaporator.
- The capillary limitation, Q_{cap}, that is linked to the maximal pumping capacity of the heat pipe that counterbalances the pressure losses both in vapour and liquid flows. The latter depends on the heat pipe filling, the porous characteristics, the working fluid and the operating conditions. The capillary limitation represents the main limitation for capillary pumping heat pipes. It is mainly relative to the liquid flow because the viscous pressure losses are generally more important in the liquid flow than in the vapour flow.
- the boiling limitation, Q_{boil}, that can be attained when the superheat at the wall becomes too important in the evaporator, then triggering nucleation at the liquid wall and altering liquid flow.

The two other limitations, Q_{vis} and Q_{son}, strictly concern the vapour flow. They correspond to the viscous and the sonic limitations, respectively. The first one occurs for flows whose temperatures are much closed to the triple point (Fig. 1). The second one corresponds to vapour flow velocities which have attained the speed of sound; both are generally preponderant for low density of the vapour phase. In macro heat pipes, the vapour flow is not directly linked to the liquid flow. Indeed, it is generally admitted that the variations of the liquid film thickness along the heat pipe do not affect the section of the vapour phase. Consequently, determining the maximal transfer capacity of heat pipes, for specific operational limitations, only taking vapour flow into account, is quite possible.

* *Corresponding author. Email: vincent.ayel@ensma.fr*

This paper focuses on these two types of limitations concerning the vapour flow and presents a description of the physical phenomena interfering in this flow for different operating conditions. Analytical expressions of maximal heat flows induced by these limitations are also proposed. Note that all fluids thermophysical properties have been taken from the NIST Scientific and Technical Database (except for the Sodium).

2. VISCOUS LIMITATION

2.1 Theoretical analysis

When the operating temperature is close to the triple point, the maximal pressure gradient of vapour flow between the evaporator and the condenser in heat pipes tends to greatly decrease; this behaviour can be explained thanks to Clausius-Clapeyron equation (Eq. (1)), due to the decrease of the pressure slope dP_v/dT_v as the saturated temperature decreases:

$$dP_v = \frac{\rho_v h_{lv}}{T_v} dT_v \tag{1}$$

This relation can be used here because the vapour is assumed to be at saturation state all along the heat pipe. So, when vapour density is very low (typically, inferior to 10^{-2} kg.m^{-3}), a given temperature gradient generates a very low pressure gradient. To illustrate this phenomenon, Table 1 gathers some typical numerical values of the derivatives of the saturation curve as functions of temperatures, for four classical working fluids.

Table 1 Numerical values of the derivative of the saturation curve $(dP_v\ dT_v)$ for different fluids.

Fluid	T_v (K)	$\rho_v h_{lv}/T_v$ (Pa.K^{-1})	*Fluid*	T_v (K)	$\rho_v h_{lv}/T_v$ (Pa.K^{-1})
Water	273.16	44.4	Sodium	600	**0.16**
	278.15	60.8		700	2.45
	283.15	82.2		800	17.3
	293.15	144.7		900	75.7
Mercury	273.15	**2.7x10^{-3}**	Lithium	500	**0.81x10^{-9}**
	283.15	**6.5x10^{-3}**		600	**0.27x10^{-6}**
	313.15	**0.065**		700	**1.58x10^{-3}**
	373.15	1.96		800	**0.032**
Triple Point: Water 273.16 K, Sodium 371 K, Mercury 234.3K, Lithium 453.7 K					

In Table 1, values written in bold characters correspond to very low values of the derivative of the saturation curve and so, to situations where the viscous limitation can preferentially occur. Water density at temperatures close to the triple point is low, but remains high enough to ensure a significant pressure gradient. On the contrary, for liquid metals, the value is so little that the pressure drop is not sufficient to counterbalance viscous pressure losses and the viscous limitation can be quickly attained. Consequently, heat transfer capability of liquid metal heat pipes can be highly deteriorated at temperatures closed to the triple point.

In order to completely determine the flow pressure drop, Navier-Stokes equations should be completely solved since the vapour flow is *a priori* at least two-dimensional because of parietal injection of heat flux.

As far as we know, as these complicated numerical calculations are still not performed in the literature, an approximated solution, taking into account the main physical phenomena, can be developed in order to have an acceptable order of magnitude of the viscous limitation.

In a first extent, a numerical approximated solution is presented. Starting from this approximation, an analytical approximated solution is then proposed to easily obtain an evaluation of the maximal heat flux corresponding to this viscous limitation. Note that, in this study, the vapour flow characteristics are considered not to be those of a rarefied gas, considering flow regimes with Knudsen number much lower than unity.

2.2 Approximated numerical solution

In this section, one considers heat pipe with a circular section and with lengths of evaporator, adiabatic zone and condenser important compared to the inner diameter of the vapour section (L_i D_v 10 with i - e, a or c, for evaporator, adiabatic zone and condenser, respectively). Besides, boundary conditions are assumed to be symmetrical all around the heat pipe. For these reasons, pressure is here considered only as a function of the abscissa z; radial gradients are neglected (Fig. 2). As viscous limitation is defined as regards to relative low pressure gradient and given the cylindrical geometry of the problem, the hypothesis of a Poiseuille-type laminar and incompressible flow is adopted here.

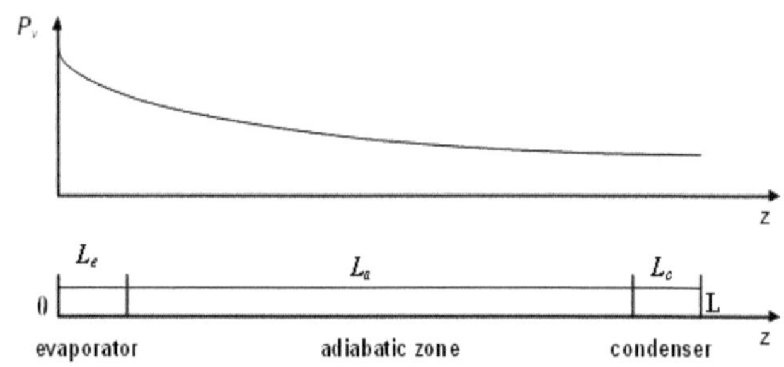

Fig. 2 Evolution of the vapour pressure field along the heat pipe.

In this case, the pressure losses along the heat pipe are given by the following relation:

$$\frac{dP_v(z)}{dz} = -\frac{32\mu_v(z)}{D_v^2}\overline{U}_v(z) \tag{2}$$

with $\overline{U}_v(z)$ the average vapour velocity in a given section of the heat pipe and D_v the diameter of the vapour section, assumed here to be constant.

Let us notice that the vapour dynamic viscosity $\mu_v(z)$ is dependent on both pressure and temperature, so it is constant in a cross-section of the heat pipe but varies along the vapour flow, *i.e.* along abscissa z. Besides, mass and heat flux carried out by the vapour flow in a section of the heat pipe are linked by the classical relation (3):

$$Q(z) = \dot{m}(z)h_{lv} \tag{3}$$

Here, the latent heat of phase change h_{lv} is considered constant because, for temperatures closed to the triple point, it does not vary much. The mass flow in a section is obviously given by the following equation:

$$\dot{m}(z) = \rho_v(z)\overline{U}_v(z)S_v \tag{4}$$

Like viscosity, vapour density is assumed constant in a cross-section and is evaluated at saturated temperature defined by the pressure in the considered section. In Eq. (4), S_v corresponds to the transversal cross-section of the vapour flow ($S_v = \pi r_v^2$). Then, the average vapour velocity in a given section can be obtained by:

$$\overline{U}_v(z) = \frac{Q(z)}{h_{lv}\pi r_v^2 \rho_v(z)} \tag{5}$$

Combining equations (2) and (5), one obtains the final expression for vapour viscous pressure losses:

$$\frac{dP_v(z)}{dz} = -\frac{8}{\pi r_v^4 h_{lv}} \frac{\mu_v(z)Q(z)}{\rho_v(z)} \tag{6}$$

Q(z) can be easily calculated along the heat pipe integrating the parietal heat flux density both at evaporator and condenser. Besides, if the vapour is considered at saturation state all along the heat pipe, the dynamic viscosity and the density can be expressed only as functions of pressure. But, as seen before, the pressure is a function of *z*; so, it is possible to integrate Eq. (6) in order to obtain the pressure profile.

The viscous limitation Q_{vis} is chosen as the **maximal heat flux** that can be carried out by the vapour flow for a given temperature, taken in the middle of the adiabatic zone. The latter temperature corresponds to the reference operating conditions in heat pipes. The boundary condition used for this limitation is the saturation pressure at the triple point at the end of the condenser (at $z = L$ on Fig. 2), which corresponds to the maximal pressure gradient between the adiabatic zone and the extremity of the condenser. Consequently, for a given heat flux density at the condenser, the pressure profile can be calculated, and, particularly, the pressure in the middle of the adiabatic zone can be determined. **The saturated temperature corresponding to this pressure is used to define the viscous limitation point.** So, by varying the condensation heat flux, the curve $Q_{vis}(T_v)$ can be drawn and the viscous limitation characterized.

This numerical solution imposes the resolution of Eq. (6) but the value of Q_{vis} cannot be directly obtained. That is why, in the following section, thanks to simplifying hypotheses, one proposes here an analytical expression for the viscous limitation Q_{vis}.

2.3 Approximated analytical solution

To obtain an analytical expression of the viscous limitation Q_{vis}, Busse (1973) has made a strong hypothesis, also used by Faghri (1995): the vapour flow is assumed to be isothermal. This hypothesis seems to be arguable given the evolution of the saturation curve (Table 1). However, if the latter is retained, in first approximation, ideal gas law gives, between two cross-sections:

$$\frac{P_v(z)}{\rho_v(z)} = \frac{P_{v,0}}{\rho_{v,0}} = r_g T_{v,0} \tag{7}$$

where the subscript "0" refers to the section at the evaporator cap ($z = 0$ on Fig. 2).

If variations of the vapour dynamic viscosity can be neglected compared to those of vapour density along the heat pipe, with equations (6) and (7), one obtains the following equation with separate variables:

$$P_v(z)\frac{dP_v(z)}{dz} = -\frac{8\mu_v}{\pi r_v^4 h_{lv}} \frac{P_{v,0}Q(z)}{\rho_{v,0}} \tag{8}$$

Integrating relation (8) between $z = 0$ and L, the following equation is obtained:

$$P_{v,0}^2 - P_{v,L}^2 = \frac{16\mu_v}{\pi r_v^4 h_{lv}} \frac{P_{v,0}}{\rho_{v,0}} \int_0^L Q(z)dz \tag{9}$$

Let us define Q_{vis} as the maximal value of *Q(z)*, i.e., for the overall heat flux transferred along the heat pipe (in the adiabatic zone), the effective length of the heat pipe can be defined as:

$$L_{eff} = \frac{1}{Q_{vis}} \int_0^L Q(z)dz \tag{10}$$

Then, Eq. (9) gives:

$$Q_{vis} = \frac{\pi r_v^4 h_{lv}}{16\mu_v L_{eff}} \left(1 - \frac{P_{v,L}^2}{P_{v,0}^2}\right) \rho_{v,0} P_{v,0} \tag{11}$$

This expression does not involve the adiabatic temperature or pressure, but the conditions at both extremities of the heat pipe (at $z = 0$ and $z = L$). In accordance with the preceding analysis led in section 2.2, Eq. (9) can also be integrated between the centre of the adiabatic zone (at $z_{adia} = L_e + L_a/2$) and the end of the condenser (at $z = L$). Indeed, the upstream flow is out of interest given the fact that the reference temperature is defined in the adiabatic zone, given as reference temperature and pressure. Consequently, one obtains:

$$Q_{vis} = \frac{\pi r_v^4 h_{lv}}{16\mu_v L'_{eff}} \left(1 - \frac{P_{v,L}^2}{P_{v,z_{adia}}^2}\right) \rho_{v,z_{adia}} P_{v,z_{adia}} \tag{12}$$

with: $L'_{eff} = \frac{1}{Q_{vis}} \int_{z_{adia}}^{L} Q(z)dz$ and $z_{adia} = L_e + L_a/2$.

It has been said that the flow characteristics at the end of the condenser (at $z = L$) are equal to those at the triple point TP. Finally, the viscous limitation is expressed as:

$$Q_{vis} = \frac{\pi r_v^4 h_{lv,z_{adia}}}{16\mu_{v,z_{adia}} L'_{eff}} \left(1 - \frac{P_{v,TP}^2}{P_{v,z_{adia}}^2}\right) \rho_{v,z_{adia}} P_{v,z_{adia}} \tag{13}$$

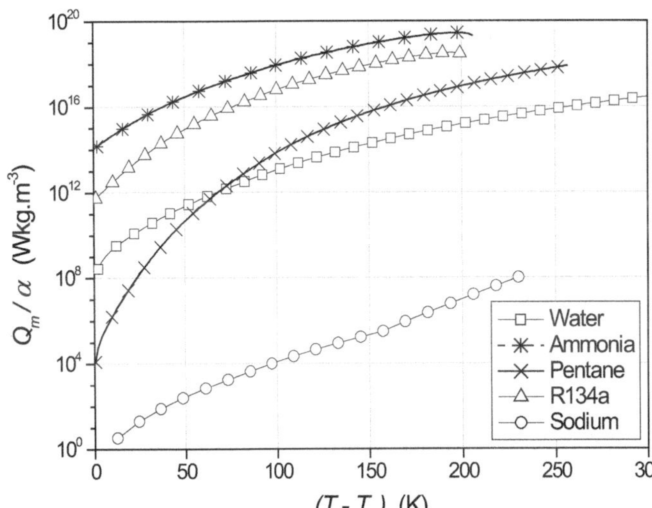

Fig. 3 Evolution of Q_{vis}/α as a function of $(T - T_{TP})$ for five fluids.

This formulation gives a very simple expression of the viscous limitation. However, given the hypothesis of an isothermal flow, it should be taken with some caution. Let define α as a constant geometrical parameter in Eq. (13) : $\alpha = \pi r_v^4/16L'_{eff}$. On Fig. 3, the evolution of Q_{vis}/α is represented as a function of $(T - T_{TP})$ for five working fluids, T_{TP} being the temperature at the triple point. One can notice that the evolutions are quite different according to the fluids, with several orders of magnitude differentiating the ones to the others. As vapour metals are most often subject to this kind of limit, sodium is obviously here the fluid most affected by the viscous limit. The value of

Q_{vis}/a is at least four orders of magnitude lower than the four other fluids! Except sodium, pentane seems to be the less performing fluid for temperatures close to the triple point. Besides, as seen previously, for all fluids, the viscous limitation is minimal for the lowest temperatures.

3. SONIC LIMITATION

3.1 Theoretical analysis

Sonic limitation analytical expression

On Fig. 1, the sonic limitation appears generally after the viscous limitation with increasing temperatures, close to the triple point. This study mainly concerns the metal fluid heat pipes with very low vapour density. It occurs when vapour density is low so that its velocity reaches the speed of sound. A shock wave can then be generated and block the vapour flow, thereby preventing transmission of the heat flux from the evaporator into the condenser. Compressibility effects and pressure losses have also to be taken into account. In this paper, elementary physical hypotheses will be used to study the different possible vapour flow behaviours; a model for vapour flow will subsequently be adopted and used to obtain a simple expression of sonic limitation. Quantitative comparisons with experimental data will also be performed.

As already presented by Bertossi *et al.* (2011), let us assume the vapour flow in a cylindrical heat pipe to be 1D; the maximal heat flux, attained when vapour velocity reaches the speed of sound ($U_s = \sqrt{\gamma r_g T_v}$), can be expressed as:

$$Q_{son} = \rho_v U_s \frac{\pi}{4} D_v^2 h_{lv} \tag{14}$$

The reference temperature $T_{v,a}$ being taken in the adiabatic zone "a", one easily obtains the following expression:

$$Q_{son} = \frac{\pi}{4} D_v^2 h_{lv,a} \rho_{v,a} \sqrt{\gamma r_g T_{v,a}} \tag{15}$$

In Eq. (15), in order to obtain the thermodynamic state in the adiabatic zone "a", one needs to use three equations for determining the three unknown parameters: $T_{v,a}$, $P_{v,a}$ and $\rho_{v,a}$. First of all, in most studies concerning heat pipes, vapour is assumed to be an ideal gas (Deverall *et al.*, 1970; Faghri, 1995; Peterson, 1994). Consequently, for a reversible and adiabatic flow, one may suppose here that the vapour phase, at the evaporator, follows the behaviour of an ideal gas: $P_v/\rho_v = r_g T_v$.

Then, as regards the variations in pressure between the beginning of the evaporator and the adiabatic zone, one may consider a momentum balance between the two parts. Differentiating mass flow rate ($\dot{m}_v = \rho_v \overline{U}_v S_v$) gives:

$$\frac{d\dot{m}_v}{\dot{m}_v} = \frac{d\rho_v}{\rho_v} + \frac{d\overline{U}_v}{\overline{U}_v} + \frac{dS_v}{S_v} \tag{16}$$

Vapour density ρ_v and velocity $\overline{U}_v$ are classically linked by a thermodynamic relation. Consequently, evolution of the couple (ρ_v, $\overline{U}_v$) is a function of the variations of the section and of the mass flow rate. Considering evolution of (ρ_v, $\overline{U}_v$) with respect to Eq. (16), an increase of relative mass flow induces an equivalent decrease of the section. So, vapour flow in constant section of a pipe with matter injection ($dS_v = 0$, $d\dot{m}_v > 0$) shows exactly the same influence than in case of constant mass flow rate in convergent nozzle ($dS_v < 0$, $d\dot{m}_v = 0$). This analogy is illustrated on Fig. 4. In the same way, vapour flow with condensation and no variation of section is comparable to constant mass flow with increasing section (Deverall *et al.*, 1970).

For isentropic ideal gas flow, Candel (1990) has shown that, when the speed of sound is reached in a nozzle throat, speed and mass flow rate attain a maximal value. In the same way, in heat pipes, the sonic limitation corresponds to the heat flux transferred when maximal mass flow is attained at the end of evaporator, leading us to consider that it is when the vapour flow reaches the speed of sound.

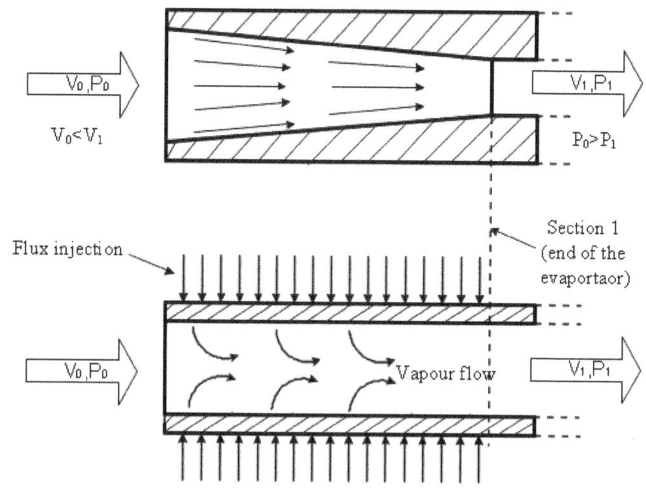

Fig. 4 Analogy of incompressible vapour flow in a heat pipe with convergent nozzle.

For a 1D vapour flow, neglecting the viscous forces, a momentum balance for the whole evaporator (with a constant section) gives: $P_{v,0} = P_{v,1} + \rho_{v,1} V_{v,1}^2$. In the previous equation, "0" refers to the entrance of the evaporator ($z = 0$) with a velocity equal to zero, and "1" refers to the "sonic throat" with maximal velocity (at the inlet of the adiabatic zone). Expressing vapour velocity as a function of Mach number *Ma* and considering ideal gas law, the momentum balance gives:

$$\frac{P_{v,0}}{P_{v,1}} = 1 + \gamma Ma_1^2 \tag{17}$$

If sonic limitation occurs when $V_{v,1}$ equals the speed of sound at the end of the evaporator ($Ma_1 = 1$), one has:

$$\frac{P_{v,0}}{P_{v,1}} = 1 + \gamma \tag{18}$$

Finally, the last equation needed to solve the system comes from the assumption that one will take concerning the temperature evolution in the vapour flow. In the following section, one will consider an isothermal flow, an isentropic flow and, finally, a flow following the saturation state.

Experimental review on sonic limitation

Unfortunately, only few experimental studies have been laid on sonic limitation. In a previous experimental work performed by Gagneux (1979) on a sodium heat pipe, one can notice in Fig. 5 that the reference temperature has to be defined precisely when ploting operating limitation curves. Indeed, Fig. 5 shows two separated curves obtained with same experimental data: one obtained with T_A as the reference temperature (adiabatic temperature), while the other is obtained with T_0 as the reference temperature (temperature at the beginning of the evaporator). The location of the thermal sensors for T_0 and T_A are shown on the scheme of Fig. 5.

Taking these observations as a starting point, it appears that the definition of the adiabatic temperature $T_{v,a}$ taken as reference temperature in Eq. (15) is very important for the sonic limitation evaluation. So, two classical approaches, considering an isothermal, or an isentropic, vapour flow, will be presented and discussed in the following part, whereas a new approach, considering the vapour phase at saturation state all along the heat pipe flow, will be also considered and compared to the two other approaches.

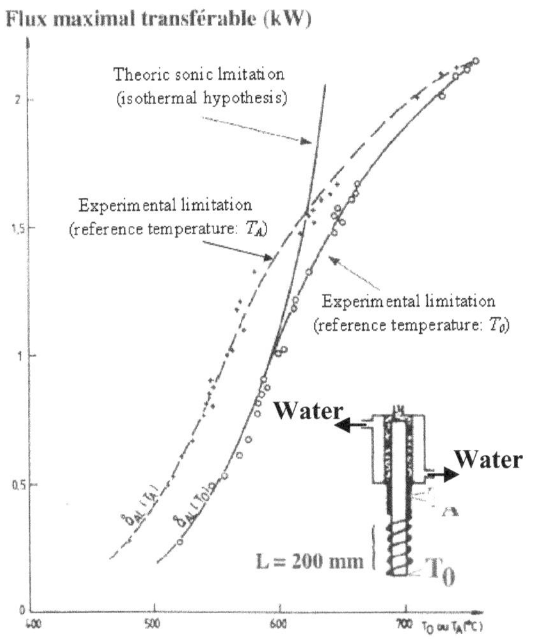

Fig. 5 Experimental results of sonic limitation for a sodium heat pipe (Gagneux, 1979).

3.2 Review and discussion on existing approaches

In Eq. (15), if the major difficulty lies in the values to be assigned to $h_{lv,a}$, $\rho_{v,a}$ and $T_{v,a}$ in the adiabatic zone, two classical approaches have been used involving isothermal and isentropic evolutions of the vapour flow between the evaporator and the adiabatic zone.

Isothermal model

Busse (1973) has studied an isothermal vapour flow ($T_{v,0}/T_{v,1} = 1$), considering the vapour as an ideal gas (see "Isothermal evolution" on Fig. 6). Using Eq. (15), with such a hypothesis, he found the following expression for the sonic limitation:

$$Q_{son,isot} = \frac{\pi}{4} D_v^2 h_{lv,a} \rho_{v,a} \sqrt{\gamma r_g T_{v,0}} \tag{19}$$

where "0" refers to the entrance to the evaporator, and $T_{v,a} = T_{v,0}$. Assuming that the vapour is at saturation state at the entrance of the evaporator ($T_{v,0} = T_{sat}(P_{v,0})$), that the evolution of the pressure along the evaporator is considered as isentropic[2] and given by Eq. (18), and using the ideal gas law, one finally obtains:

$$Q_{son,isot} = \frac{\pi}{4} D_v^2 h_{lv,a} \frac{P_{v,0}}{(1+\gamma)} \sqrt{\frac{\gamma}{r_g T_{v,0}}} \tag{20}$$

Isentropic model

In isentropic models found in the literature (Deverall *et al.*, 1970; Faghri, 1995; Peterson, 1994), the pressure evolution along the evaporator is given again by Eq. (18) and the vapour is still considered as an ideal gas (see "isentropic evolution" on Fig. 6). For an isentropic flow, an energy balance for the whole evaporator gives:

$$\frac{T_{v,0}}{T_{v,1}} = 1 + \frac{\gamma - 1}{2} Ma_1^2 \tag{21}$$

And, if $Ma_1 = 1$, one has:

$$\frac{T_{v,0}}{T_{v,1}} = \frac{\gamma + 1}{2} \tag{22}$$

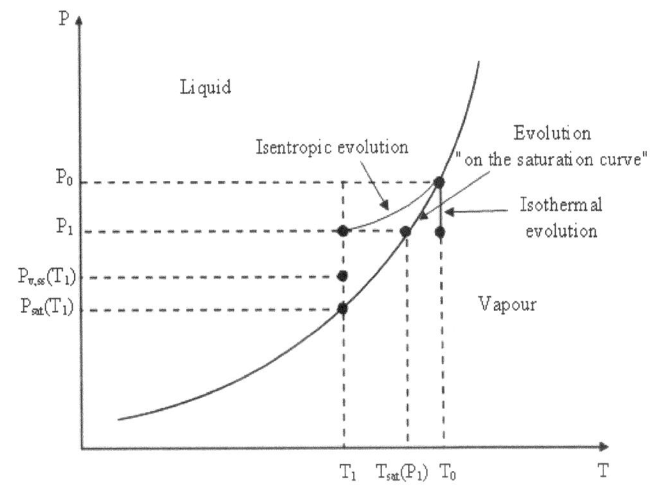

Fig. 6 Evolutions of the vapour flow along the saturation curve.

Using the general expression of Eq. (15) and the ideal gas law, the following expression can be obtained:

$$Q_{son,isen} = \frac{\pi}{4} D_v^2 h_{lv,a} \frac{P_{v,a}}{\sqrt{T_{v,a}}} \sqrt{\frac{\gamma}{r_g}} \tag{23}$$

Assuming that the vapour is at saturation state at the entrance of the evaporator, and using relations (18) and (22) for isentropic evolutions of both pressure and temperature, one finds:

$$Q_{son,isen} = \frac{\pi}{4} D_v^2 h_{lv,a} \frac{P_{v,0}}{(1+\gamma)\sqrt{T_{v,0}}} \sqrt{\frac{1+\gamma}{2}} \sqrt{\frac{\gamma}{r_g}} \tag{24}$$

where "0" refers to the entrance of the evaporator at saturation state ($T_{v,0} = T_{sat}(P_{v,0})$).

Discussion on the two models

The isothermal model's weakness derives from the fact that temperature evolution is assumed to be constant, even though it is generally admitted that along heat pipes, vapour temperature may evolve to some extent. Due to the thermophysical properties of the fluids, such evolution may modify the results obtained with the model, which has consequently been deemed somewhat unrealistic; that is one reason why alternative approaches have been developed.

Concerning the isentropic model, if the hypothesis of an ideal gas flow in a convergent nozzle is taken classically into account, the vapour phase cannot be considered at saturation state all along the heat pipe. Moreover, on the phase diagram, an isentropic evolution from point "0" in the saturation curve necessarily leads to a working point in the throat "1" situated in the liquid phase (cf. Fig. 6). As a result, the hypothesis of an isentropic flow with an operating point at saturation conditions likewise appears to be unrealistic, unless the vapour phase remains in a metastable state.

[2] Considering the evolution of the pressure along the evaporator as isentropic permits to easily compare the models the ones to the others (*i.e.* for the same pressure drop).

Let us imagine the vapour phase to be at supersaturated state ($P_{v,ss}$ on Fig. 6). Carey (1992) has shown that while a supersaturation state is conceivable for the vapour; a limitation exists beyond which droplet condensation can be triggered. Based on kinetic analysis of the droplets growing in the vapour phase and comparing the theory with experimental results, Carey obtained the following relation:

$$\frac{P_{v,ss}(T_v)}{P_{sat}(T_v)} = e^{\left(\frac{E^*\sqrt{-\ln J^*}}{2\sqrt{E^*}+\left(-\ln J^*\right)^{3/2}}\right)} \tag{25}$$

with $E^* = \dfrac{16\pi\sigma^3}{3k_B\rho_l r_g^2 T_v^3}$ and $J^* = J\dfrac{M\rho_l}{N_A}\sqrt{\dfrac{\pi M}{2\sigma N_A}}\left(\dfrac{r_g T_v}{P_{sat}(T_v)}\right)$

where $J = 10^6\,m^{-3}s^{-1}$ is the flux of droplets number per unit of volume, obtained through experimental results (Carey, 1992).

Eq. (5) determines the maximal supersaturation pressure $P_{v,ss}$ that can be admitted in the vapour phase for a given temperature T_v. Table 2 gathers numerical applications that locate the reference points of the vapour flow for Mach number equal to 1 at the end of the evaporator, taking an initial operating point (P_0, T_0) at saturation state. The different characteristic points are represented on Fig. 6, assuming that the vapour phase reaches the speed of sound at the throat: considering an isentropic evolution, the pressure at the throat P_1 always highly exceeds the maximal pressure of supersaturation $P_{v,ss}(T_1)$. As a result, **appearance of droplets in the vapour phase inevitably occurs when the flow velocity at the end of the evaporator is sonic for an isentropic flow.**

Table 2 Numerical applications for the evolution of the vapour phase at the evaporator.

	Fluid	Water		Methanol	Sodium
Initial point "0"	T_0	40	60	0	427
	$P_0 = P_{sat}(T_0)$	7370	19920	4050	95.1
Isentropic evolution	T_1 (Eq. (22))	-12	4.5	-45.5	310
	P_1 (Eq. (18))	3100	8300	1690	40
Saturated evolution	$T_1 = Tsat(P_1)$	23	42	-13	377
	P_1 (Eq. (18))	-	840	127	4
Supersatur a-tion state	$-ln(J^*)$	-	55.8	52.4	42.0
	E^*	-	103	30.7	3.05
	$P_{v,ss}(T_1)$	-	3150	270	5

Temperatures in °C, Pressures in Pa, J^* and E^* dimensionless

Nevertheless, Levy (1968), considering a diphasic vapour flow model (vapour phase containing liquid droplets), gave results very close to those obtained by Busse's isothermal model (Eq. (20)). So, the use of this complex model seems to be somewhat inappropriate. This idea is supported by the lack of experimental data on the subject. Moreover, vapour in heat pipes is supposed to be totally dried and droplets can lead to deterioration of heat transport capacity of the heat pipe. Consequently, the vapour cannot remain at supersaturation state. This hypothesis is confirmed by the liquid vapour interface which is necessarily close to the saturation state since it represents a zone for the condensation.

Thus, the vapour flow can no longer easily be considered neither isothermal nor as isentropic. Given this conclusion, vapour flow behaviour has been modified in the following part, and a new model has been built into the study.

3.3 Sonic limitation considering saturated vapour flow

In this case, it is assumed that vapour flow follows the saturation curve of the working fluid: it is supposed to be saturated and dry all along the heat pipe. From this hypothesis, a new expression of the sonic limitation may be proposed.

New expression for the sonic limitation

First, as already said, evolution of the thermophysical properties P_v, T_v and ρ_v on the saturation curve does not follow the ideal gas law. However, the gaz constant $r_g = P(T_{sat})/\rho(T_{sat})T_{sat}$ does not significantly vary for temperatures close to the triple point, and r_g can be approximated, in a first approach, as a constant. Fig. 7 confronts the evolution of the ratio between r_g and the specific gas constant ($r_{specific} = R/M_{vap}$), as a function of ($T - T_{TP}$). The curves are drawn with regard to six fluids: water, methanol, ammonia, pentane, R134a and sodium. For the latter, the data close to the triple point are missing so the curves have been obtained for temperatures relatively far from the triple point. For methanol, water and pentane, the ratio $r_g/r_{specific}$ is particularly close to unity up until ($T - T_{TP}$) reaches 100 K, and decreases for greater temperature differences. For ammonia and R134a, the ratio remains close to unity up until ($T - T_{TP}$) reaches 50 K and decreases for greater temperature differences. These are the fluids that have the highest density at these temperature levels. For sodium, one can suppose that in cases where ($T - T_{TP}$) is inferior to 220 K, the ratio remains close to 1.

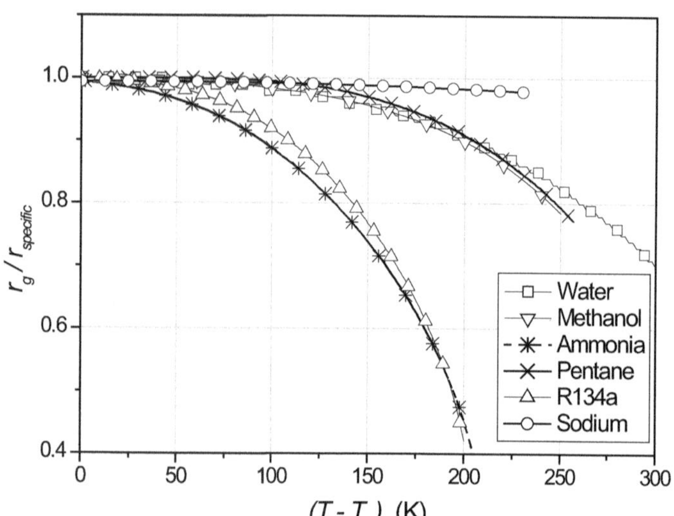

Fig. 7 Evolution of the ratio $r_g/r_{specific}$ as a function of (T-T_{TP}) for six fluids.

For the six fluids, the approximation $r_g/r_{specific} \approx 1$ is no longer valid when temperature increases on the saturation curve because of the pressure increase and on account of the irreversible phenomena associated with molecular shocks. If the vapour is assumed to be at saturation state all along the heat pipe, integrating the Clausius-Clapeyron equation between sections "1" and "0" on Fig. 6, and using the ideal gas law ($r_g = P(T_{sat})/\rho(T_{sat})T_{sat}$), Eq. (26) is obtained:

$$\frac{P_{v,0}}{P_{v,1}} = e^{\frac{h_{lv}}{r_g}\left(\frac{1}{T_{v,1}}-\frac{1}{T_{v,0}}\right)} \tag{26}$$

Starting from this, replacing "1" by "a", using Eq. (15) and the previous approximation concerning r_g, one finally obtains:

$$Q_{son,sat} = \frac{\pi}{4} D_v^2 h_{lv,a} \frac{P_{v,a}}{\sqrt{T_{v,a}}} \sqrt{\frac{\gamma}{r_g}} \tag{27}$$

As in Eq. (20), to have the same pressure drop between the two evaporator's extremities for the three models (isothermal, isentropic and saturated vapour), the pressure evolution along the evaporator is given by Eq. (18). Combined with Eq. (26), one finally obtains:

$$Q_{son,sat} = \frac{\pi}{4} D_v^2 h_{lv,a} \frac{P_{v,0}}{(1+\gamma)} \sqrt{\frac{\gamma}{r_g T_{v,0}}} \frac{1}{\sqrt{1 - r_g / h_{lv,a}\, T_{v,a}\, ln(1+\gamma)}} \tag{28}$$

where "0" and "a" are evaluated at the saturation state according to Eq. (26).

Comparison of the different expressions for the sonic limitation

The ratio between the sonic limitation in the case of isentropic flow ($Q_{son,isen}$, Eq. (24)) and in the case of saturated vapour flow ($Q_{son,sat}$, Eq. (28)) is given by the following relation:

$$Q_{son,sat} / Q_{son,isen} = \sqrt{\frac{2}{(1+\gamma)\left(1 - \frac{r_g T_{v,a}}{h_{lv,a}} ln(1+\gamma)\right)}} \tag{29}$$

In Eq. (29), the ratio $h_{lv,a,sat}/h_{lv,a,isen}$ is set close to unity because, for temperatures close to the triple point, the latent heat of phase change does not significantly vary. Given this principle, Fig. 8 shows that the ratio $Q_{son,sat}/Q_{son,isen}$ is close to 1 for temperatures near the triple point: in general, it is included in the interval **[0.9;1]** for temperature differences lower than 100 K, except for the sodium where it is lower than 0.9, suggesting that this fluid is more sensitive to the isentropicity of the evolution than the others.

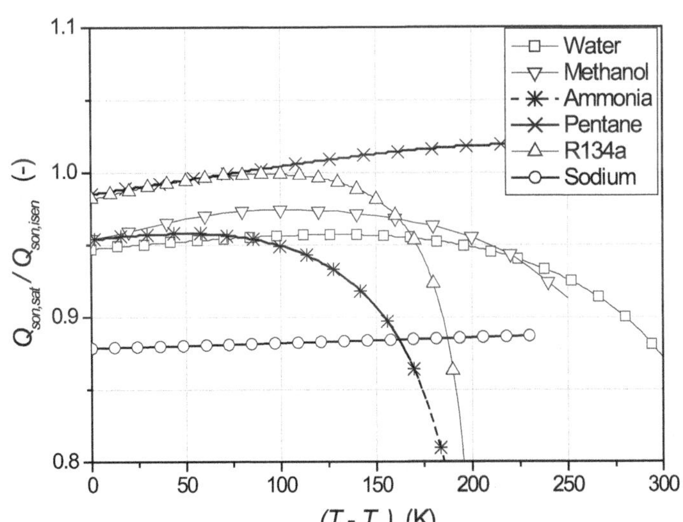

Fig. 8 Evolution of $Q_{son,sat}/Q_{son,isen}$ as a function of $(T\text{-}T_{TP})$.

In the same way, the ratio between the sonic limitation in the case of isothermal flow ($Q_{son,isot}$, Eq. (20)) and in the case of saturated vapour flow ($Q_{son,sat}$, Eq. (28)) is given by:

$$Q_{son,sat} / Q_{son,isot} = \left(1 - r_g T_{v,a}\, ln(1+\gamma) / h_{lv,a}\right)^{-1/2} \tag{30}$$

The latter ration is typically included between **1** and **1.05** for temperatures near the triple point, as can be seen on Fig. 9. As a consequence, seeing the curves of Fig. 8 and 9, the sonic limitation determined with saturated vapour flow is situated between the isentropic model and the isothermal one.

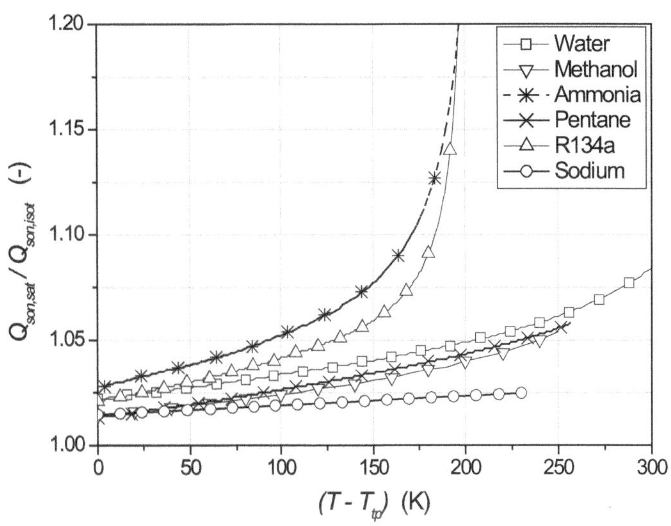

Fig. 9 Evolution of $Q_{son,sat}/Q_{son,isot}$ as a function of $(T\text{-}T_{TP})$.

From these conclusions, it would be wise to compare the model results to the experimental ones obtained by Gagneux (1979) on a sodium heat pipe. Fig. 5 shows that the experimental results have a good agreement with Busse's isothermal model (Eq. (20)). Unfortunately, the measurements are not accurate enough to differentiate one of the three models to the others. Thus, for the instant, the approach considering a vapour evolution on the saturation curve has no major advantages compared to the isothermal model in view of the low accuracy obtained by experimental results; but, at least, the present model is physically quite more rigorous than the two other ones. In all cases, the actual sonic limitation, like that considering a saturation flow, is included between those considering the first two hypotheses, both bounding the vapour flow regime.

3.4 Sonic limitation considering 2D vapour flow

Busse (1973) has also treated the case of a two-dimensional, laminar and isothermal vapour flow. If the hypothesis of an isothermal flow is arguable, it still seems to be acceptable according to the previous section. In order to neglect radial pressure gradient, Busse has considered a symmetrical heat flux injection all around the heat pipe and a large evaporator length compared to the vapour diameter. Besides, the viscous losses are neglected. A momentum balance on the whole evaporator gives the relation:

$$\overline{P}_{v,0} - \overline{P}_{v,1} = \overline{\rho_{v,1} V_{v,1}^2} \tag{31}$$

Yet, the term $\overline{V_{v,1}^2}$ cannot be directly evaluated with the data of the hydrodynamic problem because only the average velocity can be calculated as a function of the mass flow. Busse's study consisted in determining the ratio:

$$A_s = \frac{\overline{V_{v,1}^2}}{\overline{V}_{v,1}^2} \tag{32}$$

(30)

Considering a cosine velocity profile and a maximal velocity at the end of the evaporator, Busse found the value of A_s equal to 1.11. If the vapour phase is considered at saturation state all along the evaporator, with Busse's results, one obtains:

$$Q_{son} = \frac{\pi}{4\sqrt{1.11}} D_v^2 h_{lv,a} \rho_{v,a} \sqrt{\gamma r_g T_{v,a}} \quad (33)$$

Eq. (33) gives values of the sonic limitation that are 5l lower than those given by Eq. (28) corresponding to 1D saturated flow; the difference is small but not negligible.

Finally, DeMichele and Davis (1972) have performed numerical calculations for a two-dimensional laminar compressible flow considering the vapour as an ideal gas. The flow has been assumed to be isentropic, without viscous friction neither heat exchanges. They have obtained a large amount of results for several liquid metals. However, the latter results remain very close to the 1D model: the values obtained are superior of about 10l to experimental results, which is consistent with the present model (Eq. 28) which gives results only 5l lower than isentropic model (Fig. 8).

3.5 Sonic limitation considering viscous pressure losses

Up to now, viscous pressure losses have not been taken into account in the different models of sonic limitation presented above. However, for heat pipes with long adiabatic zone, the pressure losses of the vapour flow become important due to the high values of its velocity (close to the speed of sound). Considering the latter pressure losses, the sonic limitation can be this time defined as the maximal transfer capability for a flow having a sonic velocity at the end of the adiabatic zone, *i.e.* where the vapour pressure is the lowest. Indeed when the vapour velocity is high enough, the inertial effects can cause an increase of the vapour pressure in the condensation zone (Romestant, 2000).

However, when the sonic limitation is determined using the temperature at the centre of the adiabatic zone as a reference (as it is the case in our model), the upstream vapour behaviour do not influence the sonic limitation. Only the pressure losses between the middle of the adiabatic zone and the condenser inlet are at the origin of a decrease of the sonic limitation. Consequently, using the adiabatic temperature as a reference, the present model is less sensitive to the viscous pressure losses.

Besides, Dobran (1989) led a study on the sonic limitation on a 1D flow of an ideal gas taking into account viscous losses both in evaporator and adiabatic zone. This model highlighted the apparition of a sonic velocity and the influence of the viscous losses for the determination of the sonic limitation. Yet, Dobran determined the sonic limitation using temperature at the evaporator inlet as a reference, which makes increase the difference with a model which does not take into account viscous losses. This last observation is an additional argument to choose the adiabatic temperature as a reference to determine the sonic limitation.

3.6 Vapour radius determination: application to sonic limitation

In the case of the sonic limitation, as seen before, the condenser does not take part in the calculations and the liquid film thickness is consequently assumed to be constant all along the adiabatic zone. Besides, neglecting the shear stresses causes only a little error on the sonic limitation.

4. CONCLUSIONS

In this paper, two heat pipes operating limitations, linked to the vapour flow, have been studied. The first one, viscous limitation, can be calculated assuming a laminar vapour flow. An analytical approach has been developed to obtain a first order solution. However, a numerical integration is possible to improve the precision of the calculations.

The second one, sonic limitation, has been defined showing that the classical approaches contain some hypotheses that are physically not acceptable even if the latter lead to results that remain close the one to the other and to experimental data. An original approach has been proposed: a more realistic hypothesis formulated on the hypothesis that the vapour flow remains at saturation state and leading to a simple expression of the sonic limitation.

Finally, let us notice that the transition between the two limits still represents an entire domain to investigate. Indeed, given the hypotheses of a pure viscous flow on one hand and of pure inertial flow, without viscous losses on the other hand, the existence of a transition zone where the physical phenomena are both present is predictable. Then, numerical calculations would be compulsory, this type of calculations only concerning a given type of heat pipe.

NOMENCLATURE

D	Diameter (m)
g	acceleration of gravity (m s^2)
h_{lv}	latent heat of phase change (J kg)
L	heat pipe length (m)
M	Molar mass (kg.mol^{-1})
Ma	Mach number (-)
$\dot{m}$	mass flow (kg s)
P	pression (Pa)
q	heat flux density (W.m^{-2})
Q	heat flux (W)
r	radius (m)
r_g	specific gas constant (J kg·K)
S	surface (m^2)
T	temperature (K)
U	velocity along x-axis (m s)
V	velocity (m s)
y,z	coordinate (m)

Greek Symbols

α	geometric parameter (m^3)
γ	isentropic coefficient (-)
δ	film thickness (m)
μ	dynamic viscosity (Pa.s)
ρ	density (kg m^3)
τ	shear stess (N m^2)

Subscripts

a	adiabatic
c	condenser
cap	capillary
e	evaporator
eb	ebullition
eff	effective
ent	entrainement
isen	isentropic
l	liquid
sat	saturation
son	sonic
specific	specific
ss	supersaturation
v	vapour
vis	viscous
w	wall
0	entrance of the evaporator
1	inlet of the adiabatic zone

REFERENCES

Bertossi, R. , Romestant, C., Ayel, V., Bertin, Y. 2010, "A new expression for sonic limitation in heat pipes", Proceedings of *VIII*

Minsk International Seminar "Heat Pipes, Heat Pumps, Refrigerators, Power Sources", 405-412.

Busse, C.A, 1973, "Theory of the ultimate heat transfer limit of cylindrical heat pipes", *International Journal of Heat and Mass Transfer*, **16**, 169-185.

Candel, S., 1990, *Mécanique des Fluides*, Dunod Universiț .

De Michele, D.W., Davis, M.V., 1972, "Vapor transport limits of liquid metal heat pipes", Nuclear Technology, **15**, 366-383.

Deverall, J.E., Kemme, J.E., Florschuetz, L.W., 1970, "Sonic limitation and startup problems of heat pipes," *Los Alamos scientific laboratory, report LA-4518*.

Dobran, F., 1989, "Supressions of the sonic heat transfer limit in high temperature heat pipes", *Journal of Heat Transfer*, **111**, 605-610.

Faghri, F.P., A., 1995, *Heat pipe science and technology*, Taylor et Francis.

Gagneux, P., 1979, *Contribution à l'étude des caloducs à sodium*, PhD Thesis, Universiț de Poitiers, LET, ENSMA.

Levy, E.K., 1968, "Theorical investigation of heat pipes operating at low vapor pressures", *ASME Aviation Space Conference*, **15**, 671-676.

NIST Scientific and Technical Database, National Institute of standards and Technology,

Peterson, G.P., 1994, *Heat Pipes: modeling testing and applications*, John Wiley & Sons, Inc.

Romestant, C., 2000, *Etudes théoriques et expérimentales de caloducs et de thermosiphons soumis à de fortes accélerations*, PhD Thesis, Universiț de Poitiers, LET, ENSMA.

THE COMPACT STEAM TURBINE –A NOVEL APPROACH TO SMALL SCALE COMBINED HEAT AND POWER GENERATION

Markus Kley[a], Rüdiger Kretschmer[b], Yorck Leschber[b], Ralf Rieger[b,*]

[a] *Hochschule Aalen, Beethovenstr. 1, 73430 Aalen, Germany*
[b] *DHBC e.V., Schlossberg 12, 67366 Weingarten, Germany*

ABSTRACT

Current power plant technology needs to be simplified in order to be useful for small-scale systems. The small-scale turbine which represents an integrated small power plant originally invented at the Hochschule Darmstadt addresses this demand. Essentially, it is a very special application of a rotating heat pipe. It consists of rotating parts, thus reversing the traditional arrangement in a turbine. The device is divided by a plate into an evaporation and condensation chamber. This construction allows higher pressure differences so that a RHP can operate as a central part of a small power plant.
Keywords: *Rotating Heat Pipe, Small-Scale Turbine, Micro CHP, solar thermal power.*

1. INTRODUCTION

It was pointed out that small-scale CHP systems need to be simplified in order to be competitive with medium or large-scale systems (Dong et al., 2009). A novel approach is the Kompakte Dampf Turbine (KDT, meaning compact steam turbine), originally invented at the Hochschule Darmstadt, which addresses this demand (Heddrich et al., 2004). In this article, the compact steam turbine is examined in relation to a rotating heat pipe (RHP). The KDT can serve in a cogeneration plant (micro-CHP) to convert thermal energy into electricity, the remaining energy being used for heating. The KDT is about 30 cm in diameter and about 30 cm in height. With an additional external power generator it represents a complete power plant, not only a turbine. One big advantage of the KDT is its simplicity. For its construction only very few parts are necessary (Fig.1), namely an evaporator, a turbine which consists of a rotating cylinder with a rotating dividing plate and nozzles on it plus an optional blade wheel system. Through the rotational force the turbine also acts as a pump.

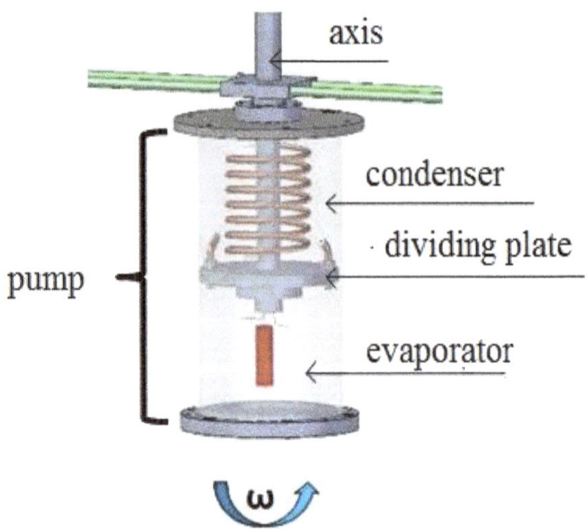

Fig. 1 Sketch of the KDT.

2. THE NEWTONIAN BUCKET AS THE UNDERLYING MECHANICAL PRINCIPLE

One of the crucial principles of the KDT is the "Newtonian Bucket". This term describes a cylinder partially filled with a liquid and set in rotation. Depending on the speed of rotation ω, the free surface of the liquid is at first parabolic rapidly assuming the shape of an annular tube. The pressure on the inner wall is the sum of atmospheric pressure and the pressure due to the centrifugal force of the rotating liquid (Fig. 2).

The shape of the surface (eqs. (1) and (2)) is derived in many textbooks, for example Sigloch (2004).

$$F_g = mg \qquad F_c = mx\omega^2 \tag{1}$$

$$\frac{dy}{dx} = \frac{F_c}{F_g} = \frac{x}{g}\omega^2 \quad => y = \frac{\omega^2}{2g}x^2 \tag{2}$$

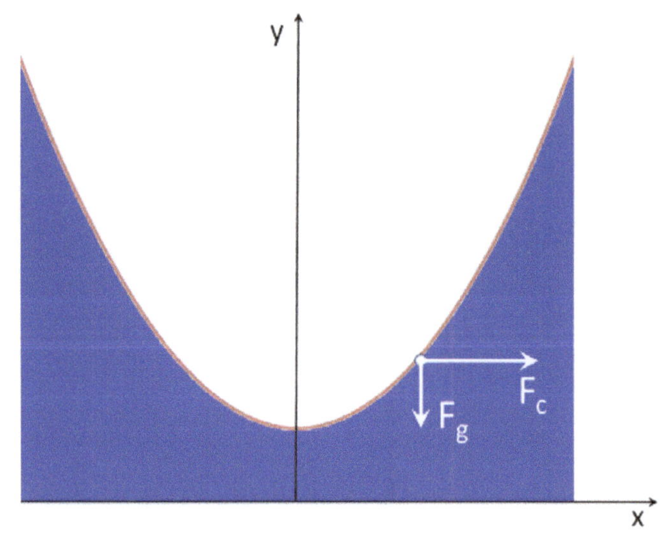

Fig. 2 The principle of "Newtonian bucket".

**Corresponding author, Email: ralf.rieger@dhbc.de*

3. ROTATING HEAT PIPE

As far as the heat transfer properties of the KDT are concerned the KDT is a variation of the rotating heat pipe (RHP). One attribute of the RHP are the large heat transfer coefficients. The RHP is a variation of the heat pipe: a heat transfer device that has been developed since the 1960s which is shown in Fig. 3. Gray (1969) formulated the initial proposal of a RHP and an overview is given by Reay and Kew (2006).

On the evaporator side the heat evaporates the liquid. The pressure on the high temperature side is higher than on the low temperature side resulting in a flow of the vapor to the condenser. The condensed liquid is returned to the evaporator side either by gravity (thermosyphon) or by the capillary (heat pipe) forces.

Since the vapor and liquid fluid reside in the same space, there are no large temperature and pressure differences in this RHP. Therefore the rotating heat pipe is not designed as thermal engine, where large pressure differences are necessary, despite some earlier considerations by Nguyen (T. Nguyen et al., 1999).

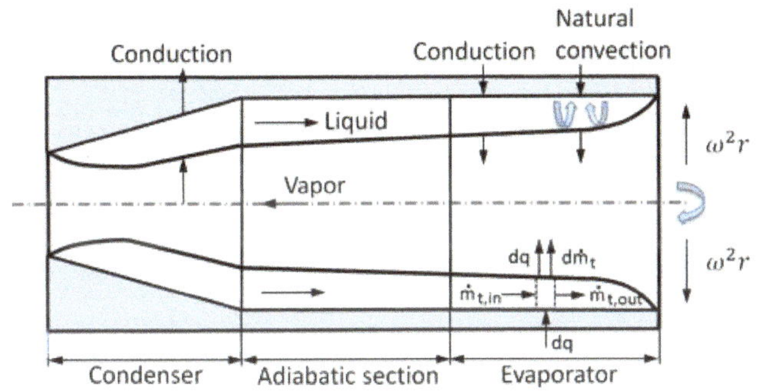

Fig. 3 Processes in the rotating heat pipe. Description is from Song (Song *et al.,* 2003).

In a RHP the condensed fluid is returned as a result of the centrifugal forces. Sometimes the condenser end of an RHP has an inclination (taper angle). This form increases the centrifugal forces, i.e. improves the flow back to the evaporator.

4. FROM RHP TO KDT

The step from RHP to KDT is the introduction of a dividing plate. One can view the KDT as a rotating cylinder partially filled with liquid, divided by a plate. Between the edge of the plate and the inner wall of the cylinder there is a small gap. The "Newtonian Bucket" effect pushes the liquid up the cylinder wall, covering the gap. The liquid then seals the gap which has a width s, effectively insulating the upper and the lower chamber from each other (Fig. 4).

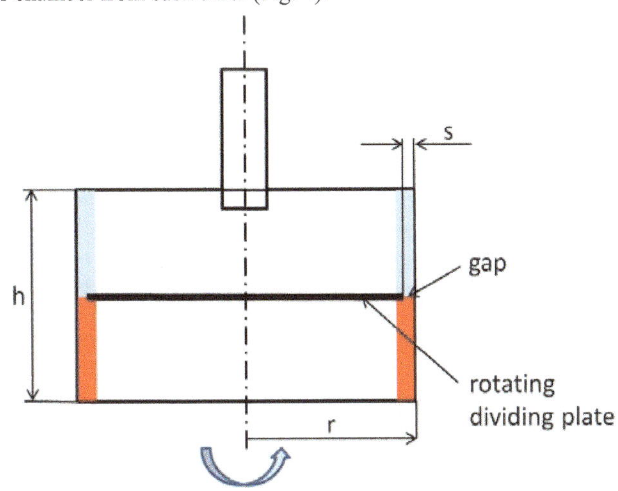

Fig. 4 Sketch of a fast rotating cylinder partially filled with liquid.

The total pressure on the inner cylinder wall is the sum of the gas pressure and the centrifugal pressure from the rotating liquid. If the gas pressure in the lower chamber is increased then there will be a new equilibrium. Part of the liquid is pushed through the gap in the upper chamber. The total pressure on the cylinder wall must be equal: In the lower chamber the gas pressure (p*) is increased and the centrifugal pressure lowered. In the upper chamber the gas pressure (p) remains the same and the centrifugal pressure increases (since there is more liquid there), as shown in Fig. 5.

The rotating dividing plate has an important function: It effectively seals the chambers in collaboration with the liquid and therefore different gas pressures can be maintained in the chambers.

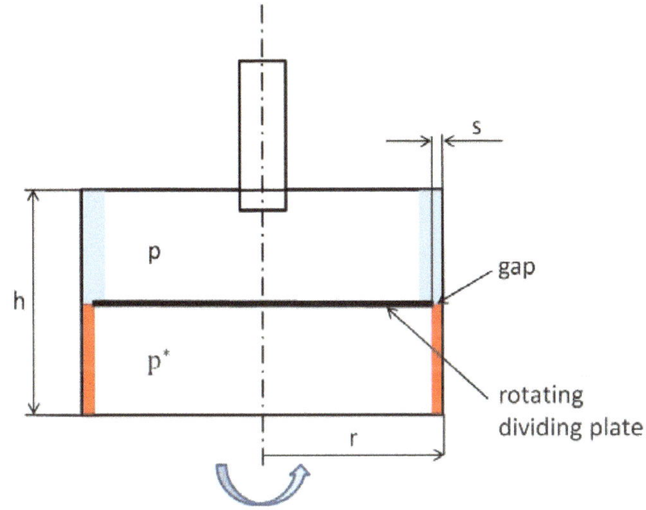

Fig. 5 Same as Fig.4 with different pressure in evaporator and condenser.

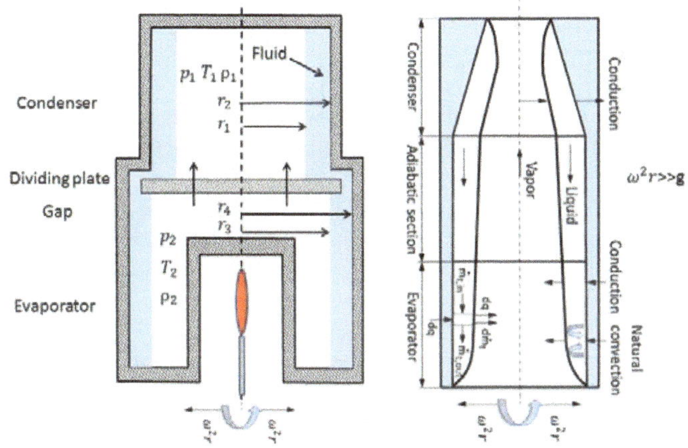

Fig. 6 Comparison between KDT and Rotating Heat Pipe.

In Fig. 6 a comparison between KDT and RHP is given. Here p_1 is the gas pressure in the upper chamber, p_2 in the lower chamber. The radius r_1 is the distance of the fluid surface from the center in the upper chamber, r_3 in the lower chamber. r_2 and r_4 are the radii of the upper chamber and lower chamber. T_1 is temperature of the condensate, T_2 temperature of the vapor. The right side of Fig. 6 is from Song et al (2003).

If one now introduces holes in the partitioning plate allowing for vapor transfer from one chamber to another, one can see the similarity between the RHP and the KDT. The lower chamber serves as evaporator and the upper chamber as condenser.

In a traditional RHP as shown in Fig. 3 only relatively small pressure gradients between evaporator and condenser are possible.

These are typically a few Kelvin and a few hundred Pascal (Song et. al., 2003). In contrast, the KDT can support temperature differences of several hundred Kelvin and pressure gradients of several MPa. The reason for this difference is that the dividing plate together with the rotating fluid effectively seals the two chambers from each other.

In the KDT the nozzles serve as the first stage of a propulsion system which can be enhanced by suitable turbine blades. There is of cause a crucial difference between the KDT and an RHP: The KDT will generate power not merely transport heat.

5. CLAUSIUS RANKINE CYCLE FOR THE KDT

In principle the KDT can use any working fluid. However, for developing the first prototypes we concentrate on dry Organic Rankine Cycle fluids (ORC), like pentane, toluene or butylbenzene which have a number of thermodynamic advantages at low temperatures. The visualization of the Organic Rankine Cycle is given in Fig. 7.

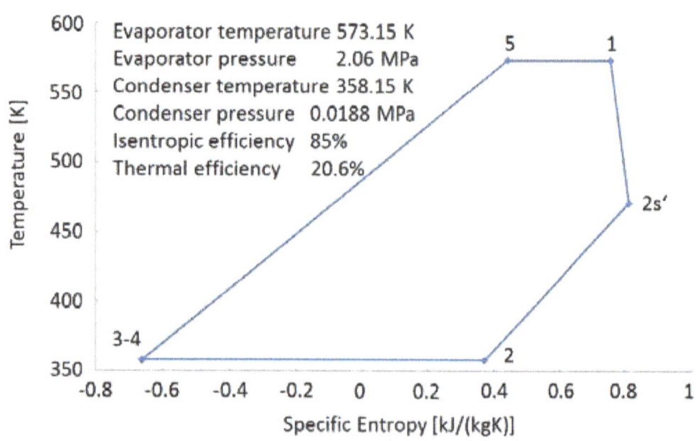

Fig. 7 T-S diagram for Organic Rankine Cycle (p-xylene) for typical temperatures of a CHP.

The Calculation is based on Fluidprop (Colonna *et al.*, 2004). To simplify the description reference is given to the study by Liu where a simple ideal ORC is evaluated (Liu et al., 2004). The main five processes include the increase of pressure by pumping (3→4 rotating KDT with angular velocity ω), heating by evaporator (4→5, e.g. light input in Fig. 8), evaporating (5→1), providing work output via expander (1→2s', nozzles in Fig. 8) , and cooling by condenser (2s'→2, cooler in Fig 8) and condensing (2→3, cooler in Fig 8).

The new aspect is that this cycle takes place in a modified RHP with high pressure difference, so that the weak performance described by Nguyen et al., (1999) should be overcome.

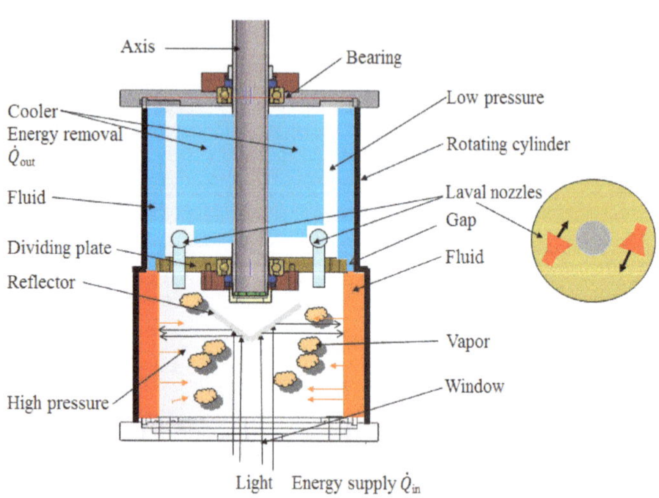

Fig. 8 The Organic Rankine Cycle in the KDT.

In Fig. 8 the complete KDT is shown. Heating, evaporation, expansion, condensation and pumping is embedded in a rotating, cylindrical device. Energy for heating the working fluid can be provided via conventional fuel like gas or via regenerative energy like solar heating.

A calculation has been carried out for p-xylene and is shown in Table 1 and plotted in Fig 7. The calculation is based on Fluidprop (P. Colonna and T.P. van der Stelt, 2004). More accurate results are from Lai et al., (2011) which result in a thermal efficiency of 20.4%. For this rough estimation and discussion of the KDT principle a high accuracy is not needed.

In Table 1 the calculations for Fig. 7 are presented. The calculations are normalized to 20kW. The mass flows of the ORC fluids under consideration are in the neighborhood of 30-50 g/s whereas water would only yield < 10g/s due to its large latent heat.

Table 1: Calculations for Fig. 7

State	*Spec. Enthalpy [kJ/kg]*	*Spec. Entropy [kJ/(kgK)]*	*T[K]*	*Pressure [MPa]*
1	406.1	0.7554	573.15	2.06
2s'	260.7	0.8107	470.83	0.019
2	77.9	0.3694	358.15	0.019
3	-290.7	-0.6598	358.15	0.019
4	-289.0	-0.6623	358.15	2.06
5	224.3	0.4382	573.15	2.06

6. EQUILIBRIUM AND PUMPING OF THE KDT

As described earlier the radii of the fluid surfaces in the condenser and evaporator are a result of the balancing from gas and centrifugal pressures. The details are shown in Fig. 6.

For a steady state in this special RHP we write down a simple equation for the total pressures in the condenser and the evaporator.

$$p_{a1} + \int_{p(r1)}^{p(r4)} dp = p_{a2} + \int_{p(r3)}^{p(r4)} dp \tag{3}$$

Using $dp = \rho\omega^2 r dr$ results in:

$$p_{a1} + \int_{r1}^{r4} \rho_1\, \omega^2 r dr = p_{a2} + \int_{r3}^{r4} \rho_2\, \omega^2 r dr \tag{4}$$

Integrating and rearranging

$$p_{a2} - p_{a1} = \frac{1}{2}\rho_1\omega^2(r_4^2 - r_1^2) - \frac{1}{2}\rho_2\omega^2(r_4^2 - r_3^2) \tag{5}$$

Normalizing to ρ_1 describes a specific enthalpy difference.

$$\frac{p_{a2}-p_{a1}}{\rho_1} = \frac{1}{2}\omega^2(r_4^2 - r_1^2) - \frac{\rho_2}{2\rho_1}\omega^2(r_4^2 - r_3^2) \tag{6}$$

Alternatively the equations can be derived from Gerber (1975).

In a power plant the Clausius Rankine Cycle has a state change which is correlated to some relatively small pumping power. The state change from 3→4 must be regarded (see Table 1). The increase in enthalpy results in needed pumping energy. The specific enthalpy increase Δh_{ts} for p-xylene is about 1700 J/kg. This enthalpy increase is relatively small and therefore not visible in a typical T-S diagram like Fig. 7. Introducing Δh_{ts} into (6) results in:

$$\frac{p_{a2}-p_{a1}}{\rho_1} + \Delta h_{ts} = \frac{1}{2}\omega^2(r_4^2 - r_1^2) - \frac{\rho_2}{2\rho_1}\omega^2(r_4^2 - r_3^2) \tag{7}$$

At a calculated mass flow of 0.027 kg/s for p-xylene at $\dot{Q}_{in}$ of 20 kW delivers a pumping power of 46 W (see Table 2). The rotating system provides this small amount of power because a typical efficiency for small ORC turbines is about 10% (Mikielewicz et al., 2009; Belozerov et al., 2012). This corresponds to about 2.5% of the desired shaft power of about 2 kW. Solving equation (7) for r_3 results in:

$$r_3^2 = \frac{2(p_{a2}-p_{a1})}{\omega^2 \rho_2} + \frac{\rho_1}{\omega^2 \rho_2}\Delta h_{ts} - \frac{\rho_1}{\rho_2}(r_4^2 - r_1^2) + r_4^2 \quad (8)$$

Equation (8) shows that the vapor pressures (p_{a1}, p_{a2}) lead to different radii r_3 (and similarly r_1) of the liquid ring tubes in the evaporator and the condenser. Differences in the vapor pressure are compensated by corresponding geodesic pressures so that overall there is a pressure balance.

Table 2: Calculated fluid properties according to equation (8).

Fluid	*Angular velocity ω[rad/s]*	*Δp =p₂-p₁ [MPa]*	*ρ [kg/m³] state 3*	*ρ [kg/m³] state 4*	*Enthalpy difference Δh₃₋₄ [J/kg]*
p-xylene	1000	2.04	786.79	475.25	1700
water	1000	0.74	968.69	969.02	330

Fluid	*r₃ [m]*	*Pumping power [Watt]*	*Power for acceleration [Watt]*	*Mass-flow [kg/s]*	*Thermal efficiency [%]*
p-xylene	0.1432	46	277	0.0270	20.6
water	0.1500	2	71	0.0063	20.6

Table 2 shows a listing of pressures, radii and densities for p-xylene and water. The heating takes place in the evaporator. Calculations were done with Fluidprop (Colonna et al. 2004). Here the pressures, radii and densities are listed. Exemplary insertion of values into (8) delivers results for p-xylene and water: the radii of KDT were set at ω=1000 rad/s to r_1= 0.120m and r_2= 0.142m, r_4=0.150m. The assumption was to have equal thermal efficiencies.

In the static cooler it is necessary that the condensed working fluid is accelerated by the dividing plate. The required power for accelerating the fluid is indicated in Table 2 in the column "Power for acceleration". The required power is much less for water than it is for p-xylene as working fluid.

In a KDT the power for pumping provided by the rotating system is very low compared to the expected electrical power output. Under operational conditions a dynamical equilibrium is reached. If radii and rotational speed are suitably chosen then for given densities a solution at small pumping power is found where KDT simultaneously pumps and seals.

7. CONCLUSIONS

Previously the KDT was presented only as low-budget power source for micro-CHP. In this article it was shown that the working principle of the KDT is closely related to that of a RHP. This similarity allows us to exploit the research previously done on RHP. An example for typically required pumping power in the rotating system is given. The invention of Heddrich et al. (2003) is an approach to make the KDT as a special RHP feasible for use as a highly integrated and simple in construction Micro CHP for use with solar, regenerative (e.g. biofuels) and conventional energy sources.

The KDT is essentially a fully integrated power plant with numerous advantages: Very few parts, no valves, a liquid seal between low maintenance costs. The final goal is to develop a small scale robust integrated power plant which could be produced in a serial production for a small amount of money.

NOMENCLATURE

c_p	specific heat (J/kg·K)
h	latent heat of phase change (J/kg)
k	thermal conductivity (W/m·K)
p	pressure (kg/m·s²)
t	time (s)
T	temperature (K)
m	mass (kg)
r	radius (m)
x	x-coordinate (m)
y	y-coordinate (m)
$\dot{Q}_{in}$	energy supply
$\dot{Q}_{out}$	energy removal

Greek Symbols

ρ	density (kg/m³)
ω	angular velocity (rad/s)

Subscripts

g	gravitational component
c	centrifugal component

REFERENCES

Belozerov, A., Kley, M., Leschber, Y., Rieger, R., Shatalov, I., Wiedmann, M., 2012, "A New Integrated Power Plant with a Small-Scale Turbine for the Organic Rankine Cycle," RUDN Bulletin 2012 No. 3, pp. 104-113.

Colonna P., van der Stelt, T.P, 2004, FluidProp: a Program for the Estimation of Thermo Physical Properties of Fluids, Energy Technology Section, Delft University of Technology.

Dong L., Liu, H., and Riffat, S., 2009, "Development of Small-Scale and Micro-Scale Biomass-Fuelled CHP Systems," Appl. Thermal Eng., **29**, 2119 – 2126.

Gerber, N, 1975, Rigidly Rotating Liquids in Closed Partially-Filled Cylindrical Cavities, Ballistic Research Laboratories Report No. 2462.

Gray, V.H., 1969, "The Rotating Heat Pipe – A Wickless, Hollow Shaft for Transferring Heat Fluxes," Proc. ASME/AIChE Heat Transfer Conf., Minneapolis, pp. 1-5, ASME Paper No. 69-HT-19.

Heddrich, W., Lautner, H., Hein, L., Losch, M., 2004, Thermal Power Machine for Converting Thermal Energy into Mechanical Energy and Use Thereof, Patent DE10315746, 2004.

Lai, N. A., Wendland, M., Fischer, J., 2011, "Working Fluids for High-Temperature Organic Rankine Cycles," *Energy,* **36**, 199-211.

Liu, B.-T., Chien, K.H., Wang, C.C., 2004, "Effect of Working Fluids on Organic Rankine Cycle for Waste Heat Recovery," *Energy*, **29**, 1207–1217.

Mikielewicz, J., Piwowarski, M., Kosowski, K., 2009, "Design Analysis of Turbines for Co-Generating Micro-Power Plant Working according to Dance with Organic Rankine's Cycle," *Polish Maritime Research*, **16**, Special issue, 34-38.

Nguyen, T., Mochizuki, M., Mashiko, M., Sauciuc, I., Akbarzadeh, A., Johnson, P., Kusaba, S., Suzuki, H., 2009, "Heat Pipe Turbine for Production of Electrical Power from Renewable Sources," Proceedings of 5th ASME/JSME Joint Thermal Engineering Conference, March 15-19, 1999, San Diego, California

Reay, D. A., Kew, P. A., Heat Pipes, Amsterdam, Heidelberg, 2006, 5. ed.; XVII

Sigloch, H., 2004, Technische Fluidmechanik, Springer, 5.Auflage.

Song, F., Ewing, D., Ching, C. Y., 2003, "Fluid Flow and Heat Transfer Model for High-Speed Rotating Heat Pipes," *International Journal of Heat and Mass Transfer*, **46**, 4393–4401.

HEAT PIPE HEAT EXCHANGER FOR HIGH TEMPERATURE NUCLEAR REACTOR TECHNOLOGY

R.T Dobson*, R Laubscher

Department of Mechanical Engineering, University of Stellenbosch, Stellenbosch
*Private Bag X1, MATIELAND **7602**, South Africa*

ABSTRACT

When a high temperature nuclear reactor is used to supply process heat, nuclear regulators require an intermediate heat transfer loop between the primary reactor coolant stream and the secondary process heat stream. A sodium-charged heat pipe heat exchanger design is presented that circumvents the need for an intermediate coolant loop. This is done by physically separating the reactor coolant and secondary coolant by two pipe walls, and a vapour section and a liquid section. This tritium diffusion resistant design for the re-heater, preheater, steam generator and super-heater heat supply system of a 232 MW-electrical pebble bed modular reactor heat source for a superheated Rankine power cycle is presented. A theoretical model was then used to produce preliminary design sizes (tube sizes, tube lengths, working fluid, etc.) for a sodium-charged heat exchanger to be used with a high temperature gas-cooled reactor.

Keywords: Heat pipe heat exchanger; sodium working fluid, tritium, high temperature, nuclear reactor.

1. INTRODUCTION

About 17 per cent of the world's electricity is produced by nuclear reactors, as compared with the 4 per cent produced by so-called renewable energy sources. The renewable sources have the disadvantage that their supply is not consistent and by and large depend on the weather. Nuclear power on the other hand has the disadvantage of having a poor public acceptance, but at this time is the only commercially viable non-fossil fuel source of energy. In an attempt to ensure an appropriate roadmap for future nuclear technology development a set of guide lines have been drawn up by the so-called Generation IV Nuclear Forum (US DOE, 2002). These guide lines may be summarised as being to:

- Ensure sustainability, focusing on waste management and resource utilization
- Ensure economic competiveness by reducing costs and financial risks
- Ensure and enhance the safe and reliable goals, improved accident management and minimization of consequences, investment protection and reduced need for off-site emergency response.
- Ensure proliferation resistance and physical protection by implementing means for controlling and securing nuclear material and nuclear facilities.

Together with these guidelines six nuclear power reactors systems were identified as the most attractive for further research and development, namely the Gas-Cooled Fast Reactor, Lead-Cooled Fast Reactor, Molten Salt Reactor, Sodium-Cooled Fast Reactor, Supercritical-Water-Cooled Reactor and the Very-High-Temperature Reactor Systems. An important characteristic of all these reactor systems is that they will operate at much higher temperatures than present-day pressurised and boiling water nuclear power reactors.

Higher temperature heat sources are becoming an ever-increasing imperative in the process industry for the production of various plastics, fertilisers, hydrogen generation, coal-to-liquid fuel and process heat. Current high temperature reactor technology is capable of producing reactor coolant temperatures in excess of 950 °C. At these temperatures however, tritium which is a radioactive contaminant found in the reactor coolant stream, is able to contaminant the secondary stream by diffusing through the steel retaining walls in the heat exchanger between the reactor coolant and secondary coolant. [Tritium (H3) is a radioactive isotope of hydrogen and has a half-life for beta-decay of 246 days.] Further if equipment is contaminated with tritium, good nuclear-maintenance practice requires that 10 half-lives (about eight years) must elapse before ordinary maintenance operations are carried out. To circumvent the possibility of this tritium contamination of the secondary stream, regulatory authorities require that the system designer consider the incorporation of an intermediate heat exchange system as shown in Fig. 1a (IAEA, 1996). Figure 1b, on the other hand, shows the piping layout incorporating a heat pipe heat exchanger (HPHE), thereby obviating the need for an additional pump and pipe work and high pressure heat exchanger shell, but even adding an extra tritium diffusion barrier at, ostensibly, no extra cost. A drawing of such a heat pipe heat exchanger (HPHE) is shown in Fig 2. It consists of an enclosure That had been evacuated prior to charging it with a working fluid of which the lower half is filled with liquid and the upper part is its vapour. The hot and high pressure stream passes through tubes that are submerged in the liquid pool; and as this is where boiling occurs it may also be termed the evaporator section. In the condenser section the vapour condenses on the tubes in which the colder, but also high pressure, stream flows. In this section of the container the vapour condenses on the colder tubes; and hence is also termed the condenser section. The condensate collects along the lowest portion of the tube and then then falls, under the influence of gravity, as droplets back into the liquid pool. In this way latent heat is transfer with essentially no resistance to flow or pressure loss, reminiscent of the working

Corresponding author email: rtd@sun.ac.za

fluid flow pattern in a closed two-phase thermosyphon-type (natural circulation) heat pipe.

Another advantage of this HPHE design is that the working fluid container need not be a pressure vessel; as the high pressure streams are conveniently contained in the small-diameter tubes. By an appropriate choice of the working fluid the pressure in the HPHE container may be designed to operate at more-or-less atmospheric pressure and thereby eliminating the need for any large-diameter high-pressure vessels. Should any tritium diffuse through the higher temperature tubes, or for that matter any non-condensable gasses (NCGs) find their way into the working fluid the lighter gasses would find their way to the highest point in the container. Here the NCG may be conveniently extracted.

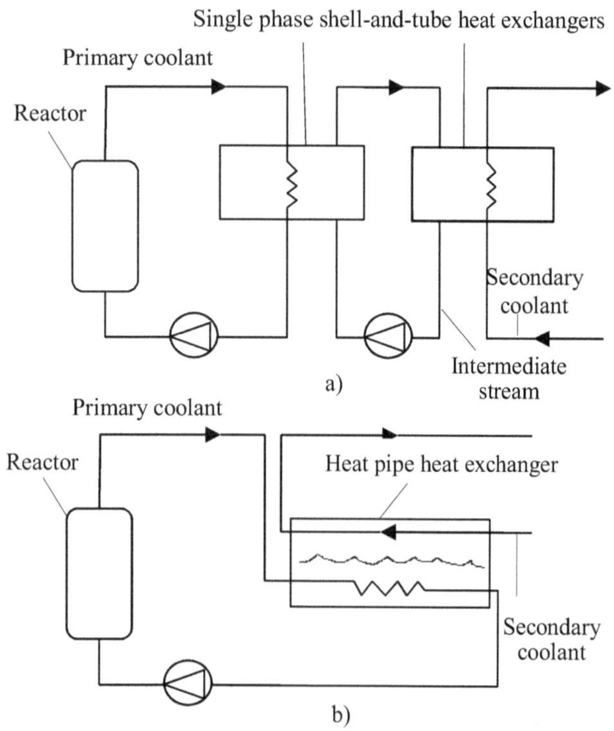

Fig. 1 Current regulatory power loop with intermediate heat exchange circuit (a), and heat pipe heat exchanger (b).

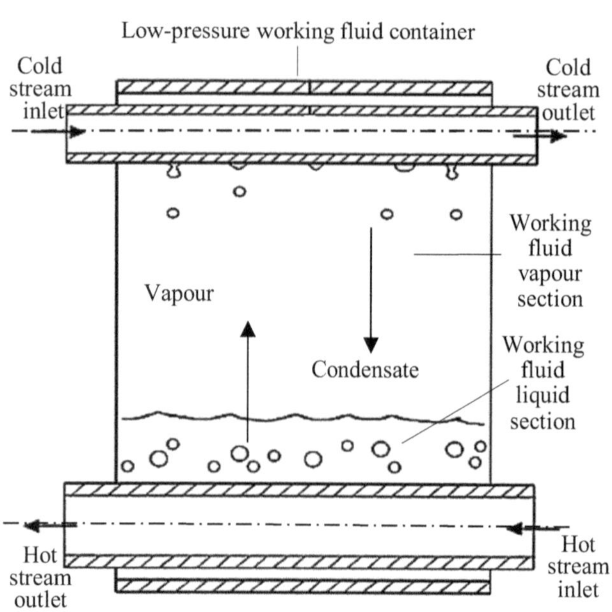

Fig. 2 Sectional view of a conceptual heat pipe heat exchanger in which the the hot and cold stream stream tubes pass through the working fluid container (only one evaporator and one only condenser tube are shown).

The objectives of this paper may now be summarised as being to:

- Present a design concept of a specially designed heat pipe-type heat exchanger to completely eliminate the intermediate heat transfer circuit including the high-pressure heat exchanger, pipework and pump.
- Present the basic heat design features of the preheater, super-heater and re-heater heat exchangers for an approximately 232 MWelectrical pebble bed modular reactor (PBMR) steam Rankine cycle electrical power generating unit.

To meet these objectives background design theory including working fluid choice, boiling and condensation thermal coefficients and the operating states are given in section 2.A summary of the simulation model is given in section 3. The design results sizes are given in section 4 and finally in section 5 conclusions are drawn.

Table 1 Liquid metals used in heat pipes (Mills, 1995).

Working fluid	Melting point, at 1 atm °C	Boiling point at 1 atm °C	Latent heat MJ/kg	Liquid density (kg/m^3)	Useful range (°C)
Sodium	98	883	3.9	900	600 - 1200
Lithium	180	1340	19.5	514	1000 - 1800
Lead	327	1746	0.9	10530	650 - 900
Potassium	64	759	2.0	814	500 - 1000

2. BACKGROUND THEORY

2.1 Working Fluid

High temperature nuclear reactor applications would typically require operating temperatures of between 650 and 950 °C. From table 1 it is seen that sodium's useful working range is between 600 and 1 200 °C and hence would be the working fluid of choice. Further at these temperatures sodium's saturated vapour pressure will be 15 kPa and 960 kPa, respectively. These pressure are not too low or too high and thus the need for special high-vacuum equipment and specially designed high pressure vessels is eliminated (Faghri, 1995). Sodium is also compatible with typical high temperature steels.

2.2 Boiling Heat Transfer Coefficient

The boiling heat transfer coefficient is given by Shah (1992) as

$$h_b = C\dot{q}^{0.7}\mathrm{Pr}^{m} \quad (1)$$

where Pr < 0.001, C = 13.7 and m = 0.22, Pr > 0.001, C = 6.9 and m = 0.12. This correlation is for pool boing for liquid metals at pressures between 0.3 and 1000 kPa, reduced pressures of between 4.4 x 10^{-6} and 2.4 x 10^{-2} and heat fluxes of between 12 and 2 800 kW/m^2.

2.3 Condensation Heat Transfer Coefficient

The laminar film condensation heat transfer coefficient is determined theoretically by using the force balance and energy balance on a liquid element within the film around the condenser tube, and not empirically using generalisations of data through correlations, as in the case of nucleate pool boiling (Mills, 1995). The heat transfer coefficient for a laminar condensate film is dependent on the thermal conductivity of the fluid medium and the film thickness that is formed on the condenser tube and is given as

$$h_c = \frac{k_l}{\delta} \tag{2}$$

where $\delta = \left(\frac{3}{4}\frac{\rho_l v_l^2}{(\rho_l-\rho_v)g}\frac{\mathrm{Re}}{\sin\phi}\right)^{1/3}$

and

$$\mathrm{Re} = \left(\frac{4}{3}\frac{2k_l(T_{sat}-T_s)d}{\mu_l h_{fg}}\left[\frac{4}{3}\frac{(\rho_l-\rho_v)g}{\rho_l v_l^2}\right]^{\frac{1}{3}}(2.5872)\right)$$

Integrating along the pipe to determine the average heat transfer is given by (Mill, 1995) as

$$\bar{h}_c = 0.728\frac{(\rho_l-\rho_v)gh_{fg}k_l^3}{Nv_l D(T_{sat}-T_s)} \tag{3}$$

2.4 Operating States

Figure 3 gives a schematic layout drawing of a high temperature heat source for a 232 MW-electrical superheated Rankine cycle electrical power generating unit for a helium heating stream leaving the reactor at 750 °C; with the quantitative values of the working point states given in table 2. This layout is based on the steam supply requirements for a conceptual design for a coal-to-liquid fuel processing plant (Botha, 2013). In Fig. 3 the conventional shell and tube preheater, superheater and reheater have been replaced by HPHEs, similar to the design shown in Fig. 2. The reactor is essentially a pebble bed modular reactor and the power unit is assumed to be a superheated Rankine cycle. [To improve power generation thermal efficiency a supercritical Brayton cycle with inlet pressures of between 20 and 30 MPa and inlet temperature 950 °C

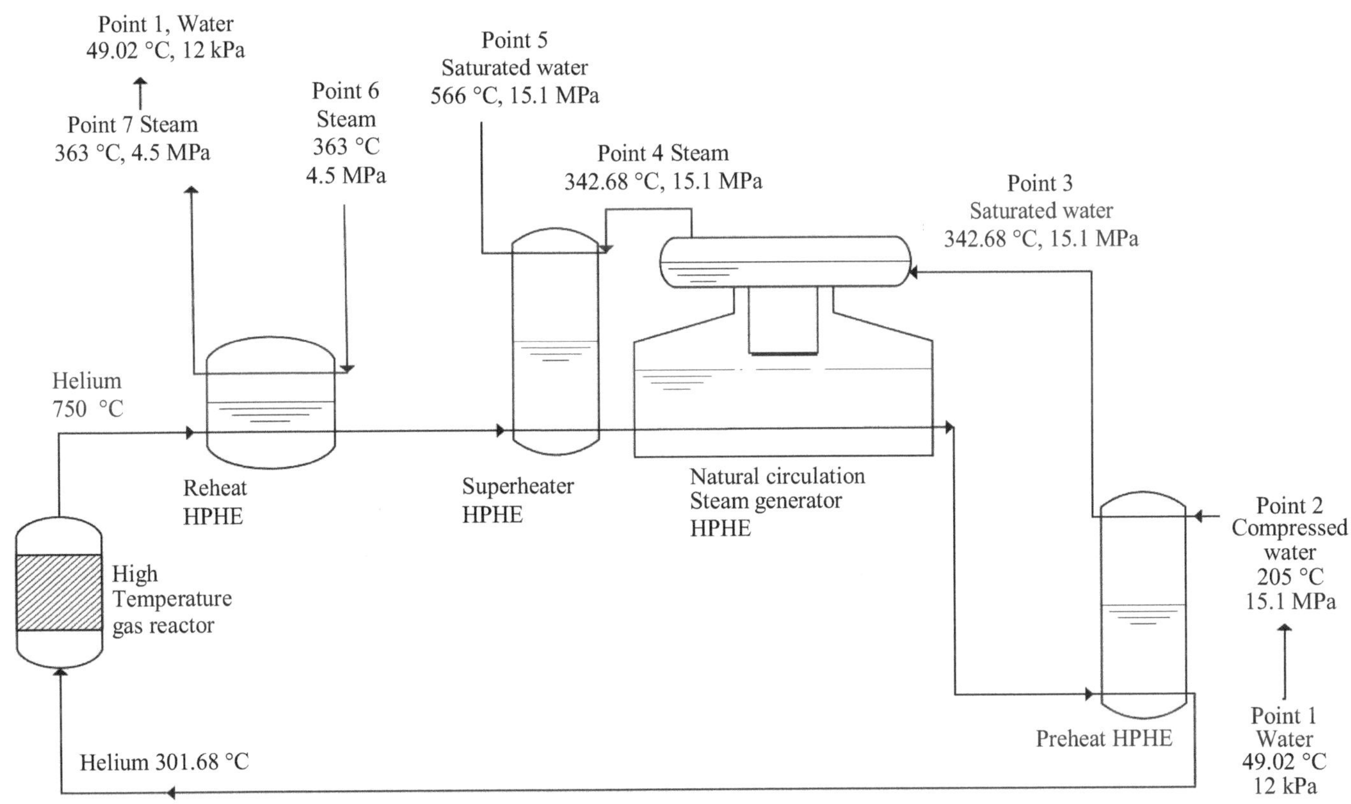

Fig. 3 Primary loop and heat exchanger layout of single high temperature gas-cooled reactor system.

Table 2 Thermodynamic properties at each point indicated in figure 3.

Point	T, °C	P, MPa	s, kJ/kg.K	h, kJ/kg	x, %
1	49.074	0.012	0.69148	205.462	< 0
2	205	15.1	2.356275	880.3575	< 0
3	342.68	15.1	3.6848	1610.3	0.01%
4	342.68	15.1	5.3108	2610.8	100.00%
5	566	15.1	6.56933	3491.856	> 100
6	364.085	4.5	6.56933	3116.487	> 100
7	566	4.5	7.217364	3592.53	> 100
8	49.074	0.012	7.217364	2307.832	88.20%

Table 3 Inlet/outlet temperatures for each of the required heat exchangers and the required heat transfer rate from the helium.

Preheater HPHE		
$\dot{Q}_{ht}$	49.5	MW
$T_{steam,in}$	205	°C
$T_{steam,out}$	342.68	°C
Superheater HPHE		
$\dot{Q}_{ht}$	59.65	MW
$T_{steam,in}$	342.68	°C
$T_{steam,out}$	566	°C
Reheat HPHE		
$\dot{Q}_{ht}$	32.121	MW
$T_{steam,in}$	364.08	°C
$T_{steam,out}$	566	°C

Table 4 Design specification for preheater heat pipe heat

Preheater HPHE		
$\dot{m}_{steam}$	67.745	kg/s
$\dot{m}_{helium}$	89.30133	kg/s
N_{tubes}	200	number
P_{HPHE}	0.04987	bar
OD_{COND}	0.08	m
OD_{EVAP}	0.2	m
$T_{steam,in}$	364.085	°C
$T_{steam,out}$	564.837	°C
$T_{helium,in}$	750	°C
$T_{helium,out}$	679.962	°C
$\dot{Q}_{in}$	32.638	MW
L_{tube}	14	m

Table 5 Design specification for the superheat heat pipe heat exchanger

Super heater HPHE		
$\dot{m}_{steam}$	67.745	kg/s
$\dot{m}_{helium}$	89.30133	kg/s
N_{tubes}	320	number
P_{HPHE}	0.02035	bar
OD_{COND}	0.08	m
OD_{EVAP}	0.2	m
$T_{steam,in}$	342.696	°C
$T_{steam,out}$	563.9329	°C
$T_{helium,in}$	679.6595	°C
$T_{helium,out}$	553.798	°C
$\dot{Q}_{in}$	58.871	MW
L_{tube}	25	m

Table 6 Design specification for reheat heat pipe heat exchanger

Reheater HPHE		
$\dot{m}_{steam}$	67.745	kg/s
$\dot{m}_{helium}$	89.30133	kg/s
N_{tubes}	400	number
P_{HPHE}	0.03162	bar
OD_{COND}	0.08	m
OD_{EVAP}	0.2	m
$T_{steam,in}$	205.04	°C
$T_{steam,out}$	340.377	°C
$T_{helium,in}$	407.883	°C
$T_{helium,out}$	300.583	°C
$\dot{Q}_{in}$	50.237	MW
L_{tube}	35	m

could have been considered. At present, but with somewhat lower inlet temperatures of between 550 and 700 °C, there are some 600 fossil fuelled supercritical power stations around the world (Zhao Zongrang, 2005). In the context in which this paper is applied, namely a coal-to-liquid (CTL) fuel plant, a superheated Rankine (rather than a supercritical cycle) was considers appropriate. In Botha's (2013) study a thermally-less efficient plant was considered from a nuclear reactor proof-of-concept point of view, wherein overall plant efficiency but mainly reliability, carbon footprint, pollution and process heat and financial risk were taken into account.]

The superheated Rankine cycle pressure-enthalpy (P-h) and temperature-entropy (T-s) diagrams are given in Fig. 4 and Fig. 5. By expanding the steam in two stages and reheating the working fluid between the turbine stages; this increases the thermal efficiency and ensures desirable moisture content at final turbine stage (Cengel, 2006). Table 3 summarises the design requirements for the three heat pipe heat exchangers shown in Fig. 3.

3. THEORETICAL MODELLING

The HPHE shown in Fig. 2 consists of a housing or container that is charged with the working fluid, and a hot stream section and cold stream section and can be discretised into a number vertical elements, 1, 2, 3, 4 ……i…..N-1, to N, as shown in Fig. 6. Each vertical element, in turn, is subdivided into three one-dimensional control volumes, the hot stream, the two-phase working fluid and the cold stream as shown in Fig. 7.

Consider the energy equation, neglecting potential and kinetic energy and as given by

$$\frac{\Delta E}{\Delta t} = (\dot{m}h)_{in} - (\dot{m}h)_{out} + \dot{Q}_{in} - \dot{Q}_{out} \qquad (4)$$

Equation 1 may now be applied to each of the three control volumes in each of the i = 1 to N elements shown in Fig. 7. However before the conservation energy can be applied a number of additional assumptions are needed, namely that the vapour is assumed to flow uniformly upwards, and the condensates drops vertically downwards in each element, there is no interfacial thermal resistance between the liquid and vapour; in each vertical element the liquid and vapour are at the same (saturated) temperature, and that although the temperature of the heating and cooling fluids change the fluid pressure does not drop along the length of the pipes.

Applying the conservation of energy, as given by equation 4, to each of the three control volumes in the ith element to get (Laubscher, 2013)

$$T_{i\,hot}^{t+\Delta t} = \left(\frac{\rho c_p T^t}{(\rho c_p)^{t+\Delta t}}\right)_{i\,hot} + \frac{\Delta t}{(\rho A \Delta x c_p)_{i\,hot}^{t+\Delta t}} \left[(\rho v A c_p T)_{i\,hot\,in} - (\rho v A c_p T)_{i\,hot\,out} - \dot{Q}_{i\,evap}\right]^t \qquad (5)$$

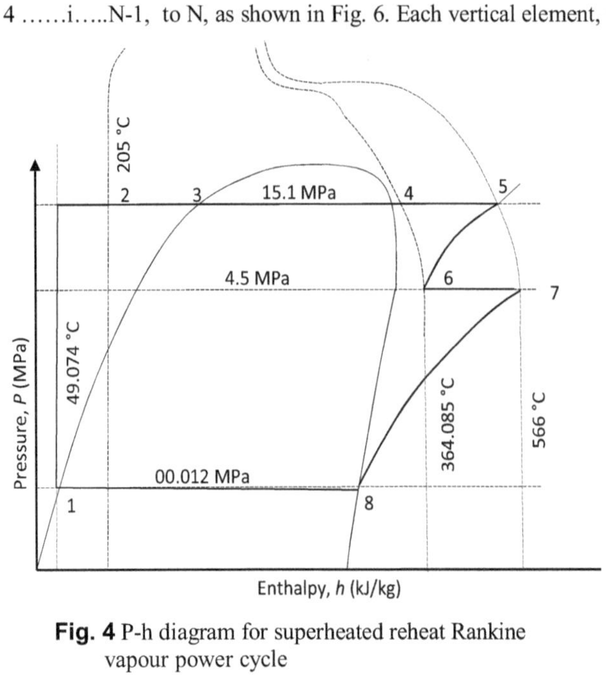

Fig. 4 P-h diagram for superheated reheat Rankine vapour power cycle

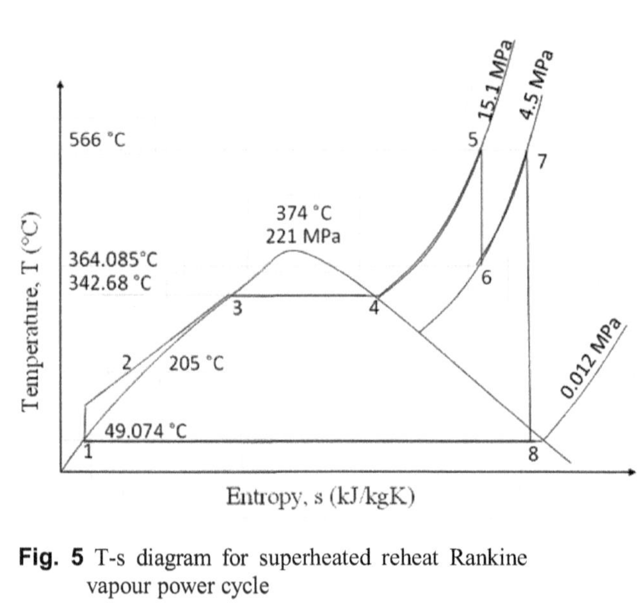

Fig. 5 T-s diagram for superheated reheat Rankine vapour power cycle

$$T^{t+\Delta t}_{\mathrm{i\,cold}} = \left(\frac{\rho c_p T^t}{(\rho c_\mathrm{p})^{t+\Delta t}}\right)_{\mathrm{i\,cold}} + \frac{\Delta t}{(\rho A \Delta x c_\mathrm{p})^{t+\Delta t}_{\mathrm{i\,cold}}}$$
$$\left[(\rho vAc_\mathrm{p}T)_{\mathrm{i\,cold\,in}} - (\rho vAc_\mathrm{p}T)_{\mathrm{i\,cold\,out}} - \dot{Q}_{\mathrm{i\,evap}}\right]^t \quad (6)$$

$$T^{t+\Delta t}_{\mathrm{i\,int\,fluid}} = \left(\frac{\rho c_\mathrm{v} T^t}{(\rho c_\mathrm{v})^{t+\Delta t}}\right)_{\mathrm{i\,int\,fluid}}$$
$$+\frac{\Delta t}{(\rho A \Delta x c_\mathrm{v})^{t+\Delta t}_{\mathrm{i\,int\,fluid}}}\left[\dot{Q}_{\mathrm{i\,evap}} - \dot{Q}_{\mathrm{i\,cond}} - \dot{Q}_{\mathrm{i\,loss}}\right]^t \quad (7)$$

Equations 5 to 7 have been presented in an explicit finite difference form suitable for numerical solution and are numerically stable provided the time step is less than $\Delta t < \Delta x/v$; namely provided that a fluid particle does not move a distance greater than the control volume

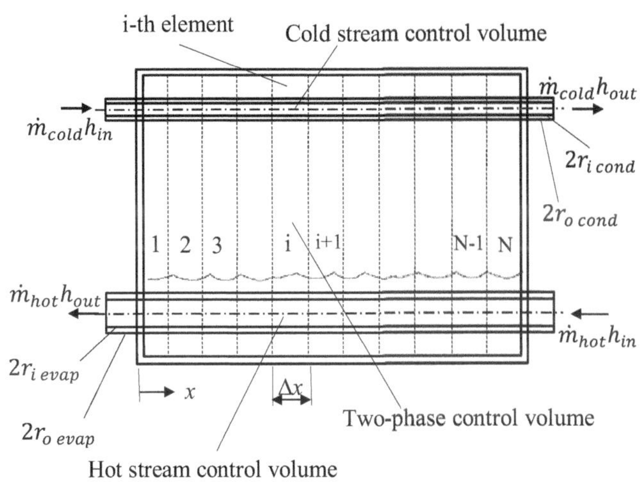

Fig. 6 Discretisation of heat exchanger into a series of elements, each element consisting of a hot heating, two-phase and a cold cooling control volume, as shown in figure 5.

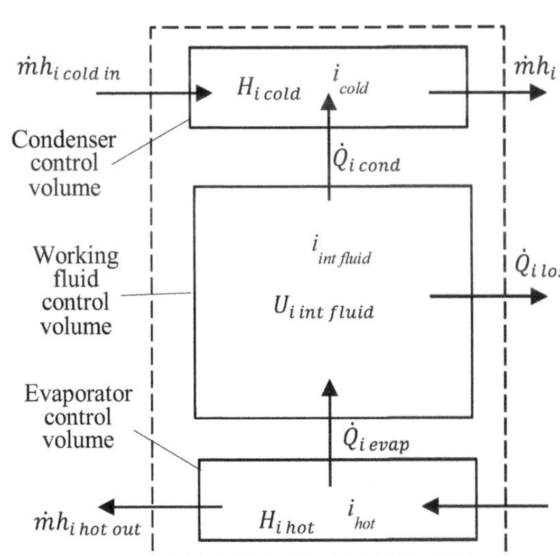

Fig. 7 Depiction of three control volumes constituting an i-th element along the heat pipe heat exchanger.

Δx. Note that the density and specific heat at time step $t + \Delta t$ are $\rho^{\,t+\Delta t} = \rho^{\,t} + (\Delta\rho/\Delta t)^{t-\Delta t/2}\Delta t$ and $c^{t+\Delta t} = c^t + (\Delta c/\Delta t)^{t-\Delta t/2}\Delta t$, respectively. The heat transfer rates $\dot{Q}_\mathrm{i}$ from the heating gas to the working fluid and from the working fluid to the cooling liquid are given by equations 5 to 7, respectively as

$$\dot{Q}_{\mathrm{i\,evap}} = \frac{\bar{T}_{\mathrm{i\,hot}} - T_{\mathrm{i\,int\,fluid}}}{\sum R_{\mathrm{evap}}} \quad (8)$$

$$\dot{Q}_{\mathrm{i\,cond}} = \frac{T_{\mathrm{i\,int\,fluid}} - \bar{T}_{\mathrm{i\,cold}}}{\sum R_{\mathrm{cond}}} \quad (9)$$

where $\bar{T}_{\mathrm{i\,hot}}$ and $\bar{T}_{\mathrm{i\,cold}}$ are the average hot stream and cold stream control volume temperatures, and assuming a linear temperature distribution along each control volume, the average heating fluid and cooling fluid temperatures are given, respectively as

$$\bar{T}_{\mathrm{i\,hot}} = (T_{\mathrm{i\,hot\,in}} + T_{\mathrm{i\,hot\,out}})/2 \quad (10)$$

and

$$\bar{T}_{\mathrm{i\,cold}} = (T_{\mathrm{i\,cold\,in}} + T_{\mathrm{i\,cold\,out}})/2 \quad (11)$$

$T_{\mathrm{i\,int\,fluid}}$ is the internal fluid control volume temperature and the thermal resistance for the evaporator $\sum R_{\mathrm{evap}}$ and the condenser $\sum R_{\mathrm{cond}}$ are given by

$$\sum R_{\mathrm{evap}} = \frac{1}{h_\mathrm{h}A_\mathrm{h}} + \frac{\ln(d_\mathrm{o}/d_\mathrm{i})}{2\pi k\Delta x} + \frac{1}{h_\mathrm{b}A_\mathrm{b}} \quad (12)$$

and

$$\sum R_{\mathrm{cond}} = \frac{1}{h_\mathrm{c}A_\mathrm{c}} + \frac{\ln(d_\mathrm{o}/d_\mathrm{i})}{2\pi k\Delta x} + \frac{1}{h_\mathrm{c}A_\mathrm{c}} \quad (13)$$

where heat transfer area $A = \pi d\Delta x$ and the internal pipe heat transfer coefficients in equations 12 and 13 may be calculated using the Gnielinski correlation and with the Petukhov friction factor (Mills, 1995) as

$$h_\mathrm{h} = \frac{k}{d}\frac{(f/8)(\mathrm{Re_d}-1000)\mathrm{Pr}}{1+12.7(f/8)^{1/2}(\mathrm{Pr}^{2/3}-1)} \quad (14)$$

and

$$f = (0.790\ln\mathrm{Re_D} - 1.64D)^{-0.2} \quad (15)$$

4. RESULTS

Solving equations 7, 8 and 9, for steady state conditions, and for the design parameters given in table 3, possible design specifications for the preheater, superheater and reheater are given in tables 4, 5 and 6, respectively. For the given heat transfer rate, inlet and outlet temperatures of the hot and cold streams the evaporator/condenser tube sizes, evaporator/condenser tube length through heat exchanger core and the number of tubes for each of the three HPHEs have been obtained for a reactor outlet temperature of 750 °C.

The variation of the heat transfer rate as a function of the diameter of the tubes, number of tubes and length of the tubes is shown in figure 8a, b, c and d, respectively for the superheater design given in table 5. The results are shown in Fig.8. For each of the graphs shown in Fig. 8 a single parameter, such as evaporator/condenser diameter, was varied while the others, tube length and tube amounts, was kept constant.

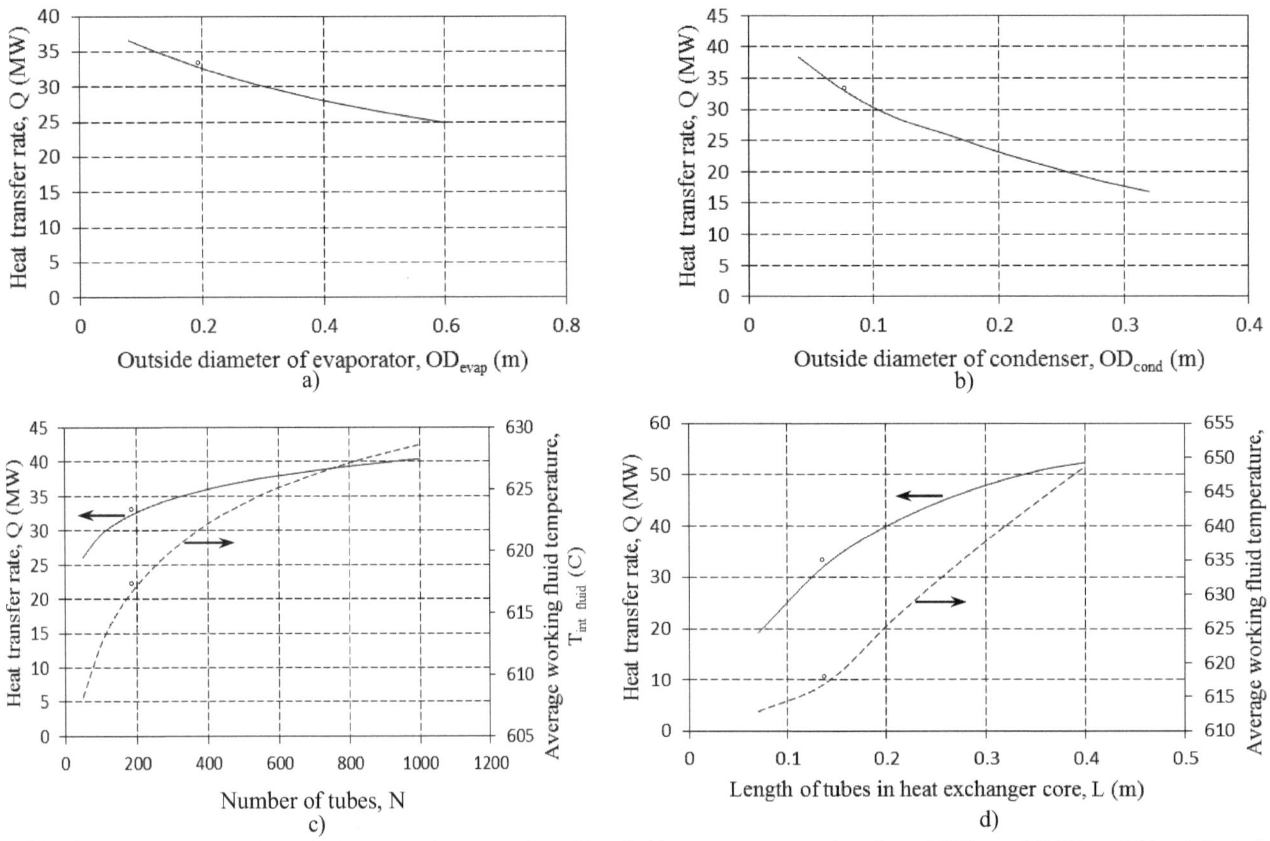

Fig. 8 Sensitivity analysis of the heat transfer rate and working fluid temperature as a function of different HPHE variables (The "•" represents base-case values as per table 5).

From Fig. 8a and Fig. 8b it is seen that if the diameter of the condenser or evaporator tubes are increased the heat transfer rate decreases. This is due to the fact that for the thermal resistance due to forced convection within the tubes is; $R_i \gg R_w, R_e$ and then $D \propto 1/Re$ and $Re \propto Nu \propto h$ and then $h \propto 1/R_i$. From Fig. 8c it is seen that as the number of evaporator and condenser tubes increase, the heat transfer rate increases. This is because the heat-transfer area increases which resulting in a greater heat transfer rate and increased working fluid temperature within the HPHE. Similarly for Fig. 8d as the HPHE as the tube length is increased so too does the heat transfer area increase.

5. CONCLUSIONS and RECOMMENDATIONS

- In Fig. 3 a natural circulation water-tube steam generator and steam drum is shown, but the drum can also be regarded as a type of two-phase heat pipe heat exchanger. The hot stream heats the working fluid which in turn heats the water-tubes in which the stems is generated; reminiscent of a typical two phase closed thermosyphon-type heat pipe. However, unlike the pre- super- and re-heaters, through which the water is force-circulated, the water flows as a two-phase closed natural circulation loop. In this case the energy and momentum are coupled and hence have to be solved simultaneous and this would require a more complex solution method than is given in section 3. As such the design of the steam generator has not been considered in this paper.
- Sodium would probably be the most suitable working fluid to use in the novel HPHE presented in this paper.
- In general, the Shah boiling correlation will most likely produce the most accurate heat transfer coefficient results.
- Thermodynamic principles and correlations available in the published literature were used to calculate the required heat transfer rates required at each heat exchanger section in the nuclear power plant cycle. In an actual sodium charged heat pipe heat exchanger (HPHE) additional experimental work will be needed to obtain the heat transfer coefficients
- The theoretical model was successfully used to produce a reasonable preliminary design for a heat pipe heat exchanger (HPHE) preheater, a superheated and a reheater (using the projected cycle state points as predicted for the Chinese high temperature nuclear reactor of pebble-bed modular design, also termed the HTR-PM).

NOMENCLATURE

A	area, m^2
c_p	specific heat, J/kg K
c_v	specific heat (constant volume), J/kg K
d	diameter, m
f	friction factor
g	gravity, m/s^2
H	total enthalpy, J
h	heat transfer coefficient, W/m^{2}K
h	specific enthalpy, K/kg
h_{fg}	latent heat of vaporization, J/kg
i	element/CV designation
k	thermal conductivity, W/mK
L	length, m
$\dot{m}$	mass flow rate, kg/s
N	number (of tubes)
P	pressure, Pa
P_r	reduced pressure $P_r = P/P_{crit}$
P_{crit}	critical pressure, Pa
Pr	Prandtl number, $Pr = c_P \mu_l / k$
$\dot{Q}$	heat transfer rate, W

$\dot{q}$	heat flux, W/m^2
Re	Reynolds number
r	radius, m
s	Specific entropy, J/kg
T	temperature, °C
t	time, s
U	total internal energy, J
u	fluid velocity, m/s

Greek symbols

δ	film thickness, m
ρ	density, kg/m^3
ν	kinematic viscosity, $\nu = \mu/\rho$
μ	dynamic viscosity kg/ms

Subscripts

avg	average
b	boiling
c	condensing
cold	cold stream
cond	condenser
crit	critical
d	diameter
e	External
evap	evaporator
hot	hot stream
ht	heat transfer
i	internal
in	into
int	internal
l	liquid
o	outside
out	out
s	surface
sat	saturation
v	vapour
w	wall

ACKNOWLEDGEMENTS

Financial assistance from the South African Heat Pipe Association and PBMR Pty LTD is great fully acknowledged

REFERENCES

Botha F., 2013, Simulation of a Syngas-from-Coal Production Plant coupled to a High Temperature Nuclear Reactor, *MSc Eng Thesis*, University of Stellenbosch, Stellenbosch, South Africa.

Faghri, A., 1995, *Heat Pipe Science and Technology*, Taylor and Francis.

IAEA (International Atomic Energy Agency), 1991, Safety Related Terms for Advanced Nuclear Plants, IAEA-TECDOC-626: Vienna.

IAEA (International Atomic Energy Agency), 1996, Progress in Design, Research and Development, and Testing of Safety Systems for Advanced Water Cooled Reactors, IAEA-TECDOC-872, Vienna.

Jousse, A., 2007, Tritium Transport in Very High temperature Reactors for Hydrogen Production, California: University of California at Berkeley.

Laubscher, R., 2013, Development Aspects of a High Temperature Heat Pipe Heat Exchanger for High Temperature Gas-Cooled Nuclear Reactor Systems, MSc Eng, Stellenbosch: University of Stellenbosch, South Africa.

Laubscher, R. and Dobson, R.T., 2013, "Boiling and Condensation Heat Transfer Coefficients for a Dowthem-A Charged Heat Pipe", 11th International Heat Pipe Symposium (11th IHPS), Beijing, June 9 – 12.

Laubser, R. and Dobson, R.T., 2013, "Theoretical and Experimental Modelling of a Heat Pipe Heat Exchanger for High Temperature Nuclear Reactor Technology", *Applied Thermal Engineering*, accepted for publication.
http://dx.doi.org/10.1016/j.applthermaleng.2013.06.063

Lamarsh, J. R., 2001, *Introduction to Nuclear Engineering*, Third edition. New Jersey: Prentice Hall.

Mills, A. F., 1995, *Heat Transfer*, 2nd Edition, Prentice Hall, New Jersey.

Sabharwall, G. P., 2008, Theoretical Design of Thermosyphon for Process Heat Transfer from NGNP to Hydrogen Plant.

Shah, M., 1992, A Survey of Experimental Heat Transfer Data for Nucleate Pool Boiling of Liquid Metals and a New Correlation. *International Journal of Heat and Fluid Flow*, **13**(4). http://dx.doi.org/10.1016/0142-727X(92)90007-V

Thome, J., 2006, *Boiling Heat Transfer on External Surfaces in Engineering Data Book III*, Wolverine Tube Inc. pp. 9-1 to 9-11.

Cengel, 2006. *Thermodynamics: An Engineering Approach*, McGraw Hill, New York.

US DOE, 2002, *A Technological Roadmap for Generation IV Nuclear Energy Systems*, Issued by U.S. DOE Nuclear Energy Research.

Advisory Committee and the Generation IV International Forum, 03-GA50034, December 2002.

Zhao Zongrang, 2005, Development and Application of Supercritical Coal-fired Units and CFB Boilers in China, Presentation, *Thermal Power Research Institute* (TPRI), Cebu, The Philippines, 26 January, 2005.

PREVENTION POSSIBILITY OF NUCLEAR POWER REACTOR MELTDOWN BY USE OF HEAT PIPES FOR PASSIVE COOLING OF SPENT FUEL

Masataka Mochizuki*, Thang Nguyen, Koichi Mashiko, Yuji Saito, Randeep Singh, Tien Nguyen, Vijit Wuttijumnong

Fujkura Ltd. 1-5-1, Kiba, Koto-Ku, Tokyo 135-8512, Japan

ABSTRACT

In Japan, following a major earthquake, a 15-metre tsunami disabled the power supply and cooling of three Fukushima Daiichi reactors, causing a nuclear accident on 11 March 2011. All three cores largely melted in the first three days. The accident was rated 7 on the INES (International Nuclear Event Scale) scale, due to high radioactive releases in the first few days. After two weeks the three reactors (units 1-3) were stable with water addition but no proper heat sink for removal of decay heat from fuel. By July they were being cooled with recycled water from the new treatment plant, and reactor temperatures had fallen to below 80°C at the end of October, and official 'cold shutdown condition' was announced in mid December. In this paper, the nuclear power plant No. 4 at Fukushima was chosen for the analysis - thermal power 2,381 MW; electrical power 784 MW; 658 spent fuel bundles (39,456 spent fuel rods); spent fuel water pool volume 1,400 tonnes. Detail analysis of various heat pipe design cases to determine the best design concept in terms of cooling power, construction and cost are presented. The best design when considering thermal safety margin and cost is the heat pipe cooling 0.9 MW; 1,662 heat pipe modules; water temperature will reach to peak 68 °C after 75 hours, and will be saturated at 50 °C after 2,000 hours. The estimated cost for complete heat pipe cooling system is about 2.16 Millions USD.

Keywords: *Boiling water reactor, Pressurized water reactor, Fast breeder reactor, Nuclear power, Spent fuel, Heat pipe, Loop heat pipe.*

1. INTRODUCTION

1.1 What Happened at Fukushima Nuclear Reactors?

The Great East Japan Earthquake of magnitude 9.0 on 11 March 2011 did considerable damage in the region, and the large tsunami it created caused very much more. The earthquake was centered 130 km offshore the city of Sendai in Miyagi prefecture on the eastern cost of Honshu Island (the main part of Japan), and was a rare and complex double quake giving a severe duration of about 3 minutes. Japan moved a few meters east and the local coastline subsided half a meter. The tsunami inundated about 560 sq km and resulted in a human death toll of over 19,000 and much damage to coastal ports and towns with over a million buildings destroyed or partly collapsed.

Eleven reactors at four nuclear power plants in the region were operating at the time and all shut down automatically when the quake hit. Subsequent inspection showed no significant damage to any from the earthquake. The reactors proved robust seismically, but vulnerable to the tsunami. Power, from grid or backup generators, was available to run the Residual Heat Removal (RHR) system cooling pumps at eight of the eleven units, and despite some problems they achieved 'cold shutdown' within about four days. The other three at Fukushima Daiichi lost power almost an hour after the quake, when the entire site was flooded by the 15-metre tsunami. This disabled 12 of 13 back-up generators on site and also the heat exchangers for dumping reactor waste heat and decay heat to the sea. The three units lost the ability to maintain proper reactor cooling and water circulation functions. Electrical switchgear was also disabled. Thereafter, many weeks of focused work centered on restoring heat removal from the reactors and coping with overheated spent fuel ponds.

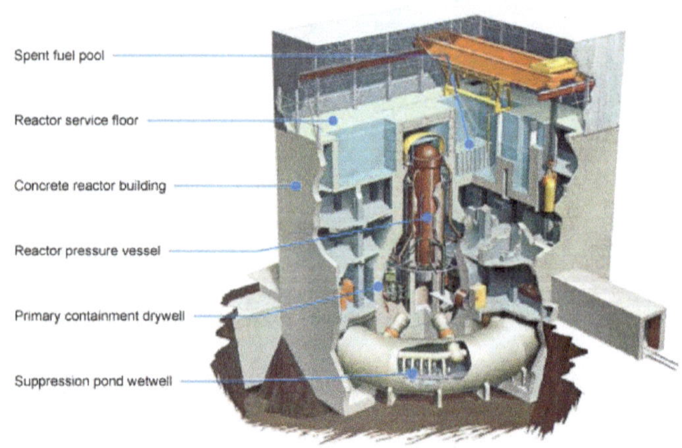

Fig. 1 Boiling water reactor (BWR) Mark I

The Fukushima Daiichi reactors are GE boiling water reactors (BWR) of an early (1960s) design supplied by GE, Toshiba and Hitachi, with what is known as a Mark I containment as shown in Fig. 1. Reactors 1-3 came into commercial operation 1971-75. Reactor power is 460 MWe for unit 1, 784 MWe for units 2-5, and 1100 MWe for unit 6.

It appears that no serious damage was done to the reactors by the earthquake, and the operating units 1-3 were automatically shut down in response to it, as designed. At the same time all six external power supply sources were lost due to earthquake damage, so the emergency diesel generators located in the basements of the turbine buildings started up. Initially cooling would have been maintained

* *Corresponding author. Email: mmotizuk@fujikura.co.jp*

through the main steam circuit bypassing the turbine and going through the condensers. Then, 41 minutes later the first tsunami wave hit, followed by a second 8 minutes later. These waves submerged and damaged the seawater pumps for both the main condenser circuits and the auxiliary cooling circuits, notably the Residual Heat Removal (RHR) cooling system. They also drowned the diesel generators and inundated the electrical switchgear and batteries, all located in the basements of the turbine buildings, therefore, there was a station blackout, and the reactors were isolated from their ultimate heat sink. All this put those reactors 1-3 in a dire situation and led the authorities to order, and subsequently extend, an evacuation while engineers worked to restore power and cooling. The 125-volt DC batteries for units 1 & 2 were flooded and failed, leaving them without instrumentation, control or lighting. Figure 2 shows a schematic of water level at nuclear reactor 1-4 site. It showed that the site was designed for expected tsunami level of only 5.7 m, while the actual tsunami was about 15 m.

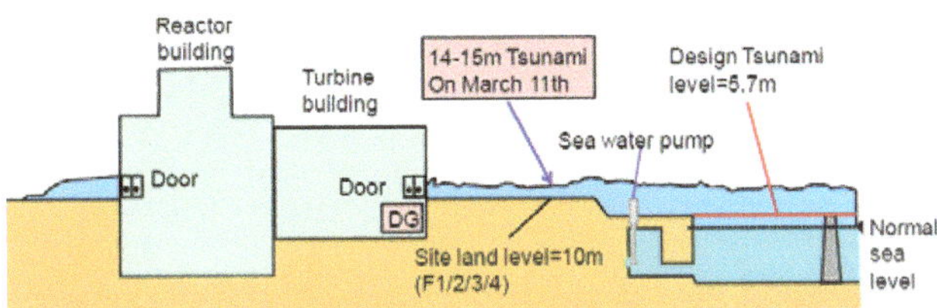

Fig. 2 Schematic of water level at nuclear reactor 1-4 site

When the power failed, at about one hour after shutdown of the fission reactions, the reactor cores would still be producing about 1.5% of their nominal thermal power, from fission product decay which are about 22 MW in unit 1 and 33 MW in units 2 & 3. Without heat removal by circulation to an outside heat exchanger, this produced a lot of steam in the reactor pressure vessels housing the cores, and this was released into the dry primary containment (PCV) through safety valves. Later this was accompanied by hydrogen, produced by the interaction of the fuel's very hot zirconium cladding with steam after the water level dropped. As pressure started to rise here, the steam was directed into the suppression chamber under the reactor, within the containment, but the internal temperature and pressure nevertheless rose quite rapidly. Water injection commenced, using the various systems provide for this and finally the Emergency Core Cooling System (ECCS). These systems progressively failed over three days, so water injection to the reactor pressure vessel (RPV) was with fire pumps, but this required the internal pressures to be relieved initially by venting into the suppression chamber/ wet well.

Inside unit 1, it is understood that the water level dropped to the top of the fuel about three hours after the scram and the bottom of the fuel 1.5 hours later. The temperature of the exposed fuel rose to some 2800°C so that the central part started to melt after a few hours and by 16 hours after the scram most of it had fallen into the water at the bottom of the RPV. There was a hydrogen explosion on the service floor of the building above unit 1 reactor containment, blowing off the roof and cladding on the top part of the building, after the hydrogen mixed with air and ignited due to the oxidation of the zirconium cladding at high temperatures in the presence of steam produces hydrogen exothermically, with this exacerbating the fuel decay heat problem. In unit 1 most of the core as corium comprised of melted fuel and control rods, was assumed to be in the bottom of the RPV, but later it appeared that it had mostly gone through the bottom of the RPV and eroded about 65 cm into the drywell concrete below in which is 2.6 m thick. This reduced the intensity of the heat and enabled the mass to solidify. Figure 3 shows the picture of nuclear reactor site damage for unit 1, 2, 3 and 4 before (picture above) and 13 days (picture below) after the accident. Tepco has said that the three reactors, with unit 4, are written off and will be decommissioned.

Fig. 3 Nuclear reactor 1-4 site before and after the accident

1.2 Nuclear Reactor Introduction

Nuclear power has the potential to support the electric energy needs of the growing population. Nuclear energy share in global electricity production is growing fast due to its high energy density, advanced reactor technology, low greenhouse gas emissions, ease of installation and plant expansion. In nuclear power plants, kinetic energy produced by the nuclear fission of the radioactive material (usually uranium-235 or plutonium-239) is converted to heat and thereby to useful electrical power. Nuclear fission provides very high density energy, for example one kg of U-235 can produce 3 million times of energy generated by equivalent mass of coal. Japan has total of 54 nuclear power reactors with total electric power capacity of 49 GW (30% of country's demand) and there is proposal for 19 new reactors with total capacity of 13 GW to be built in the near future.

Two most commonly used reactors in nuclear power plants are pressurized water reactor (PWR) and boiling water reactor (BWR). PWR pumps high pressure coolant (water) to the reactor core to extract energy from the nuclear fission reaction. The hot water is then passed through the steam generator where it heats up the secondary coolant and produce steam which is passed through turbine to generate electricity. Unlike PWR, in BWR the steam (~ 282 °C) is generated in the nuclear core that is directly used to drive the turbine. In case of an accident, BWR is more susceptible to radiation leak than PWR, due to direct utilization of the contaminated steam in the turbine located outside the primary containment. The BWR containment consists of drywell that houses reactor with related cooling system and wet well or suppression pool. The suppression pool contains water charge for core cooling during emergency reactor shutdown and for dumping excess heat (nuclear reaction control) during reactor operation. Figure 4 presents the schematic of the BWR based nuclear power plant, having reactor vessel with fuel and control rods assemblies, turbine and generator arrangement, sea water cooled condenser, suppression pool and, most importantly, electrically driven ECCS with pumps. Nuclear power plants normally use sea water for cooling purpose. In case of any malfunctioning, the nuclear reactors are automatically shut down by using control rod mechanism. After shutdown, the ECCS is required to transfer and dissipate the residual heat from the core and maintain reactor temperature within safer limits (< 100ºC). The ECCS, which is activated after reactor shut down, typically uses diesel generators to power number of pumps for spraying high pressure water on the hot core. If the active water cooling system stops due to loss of electrical power, then the reactor internal temperature and pressure will build-up due to steam formation from accumulated residual heat causing fuel meltdown (~ 1,800 ºC) and reactor vessel damage.

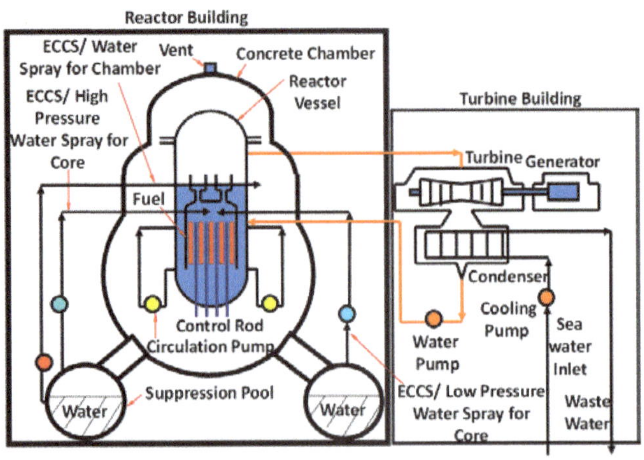

Fig. 4 Boiling water reactor with ECCS system

One of the worst nuclear accident at Fukushima nuclear power plant resulted from the failure of electric generators for ECCS that was caused by M9 earthquake and Tsunami triggered by it. The system described in Figure 4 is similar to the BWR used in Fukushima nuclear power plants. Active ECCS is not reliable due to their dependence on the electric power and therefore is prone to failure during adverse natural calamities. Mochizuki et al (2012) proposed a totally passive cooling system using heat pipe for ECCS. In this paper is a detail case study proposal of using heat pipe for cooling of spent fuel as shown in Fig. 5.

2. DESIGN CASES STUDIES

2.1 Conceptual Design Proposal of Cooling System for Spent Fuel Using Thermal Diode Heat Pipes

Table 1 shows the outline specification of nuclear power reactors in Fukushima facility in Japan. The nuclear power reactor No. 1 is the power plant that was damaged and caused nuclear leakage in 2011 natural disaster as mentioned in the introduction section. In this study, the nuclear reactor No. 4 has been chosen for detail thermal analysis and design proposal of cooling system for spent fuel using heat pipe. An example of the structure of the spent fuel unit is shown in Fig. 6.

The thermal specification for the nuclear power reactor No. 4 for the analysis in this paper is assumed as below:

- Thermal power: 2,381 MW.
- Electrical power: 784 MW.
- Fuel quantity: 34,956 units.
- Time of reactor shutdown from startup ~ 1.5 years.
- Fuels are remained in the reactor for about 1 month after shutdown, the fuels are then transferred to the water cool pool spent fuel.

Figure 7 (a) and (b) shows a front and plan view respectively of a schematic design concept of a cooling system for spent fuel using thermal diode heat pipes. The term thermal diode heat pipe means the heat flow only in one direction from the evaporator in the lower section, to the condenser in the upper section. The working liquid inside the heat pipe, after condensed in the condenser, is returned to the evaporator by gravity. The heat pipe evaporator section is submersed in the water pool of spent fuel. The spent fuel bundles of height approximate 4.5m totally submersed in the deep water pool of 10m in height and contained about 1400 tons of water. The heat pipe condenser length OF about 4m is exposed to ambient environment and it is cooled by natural air convection.

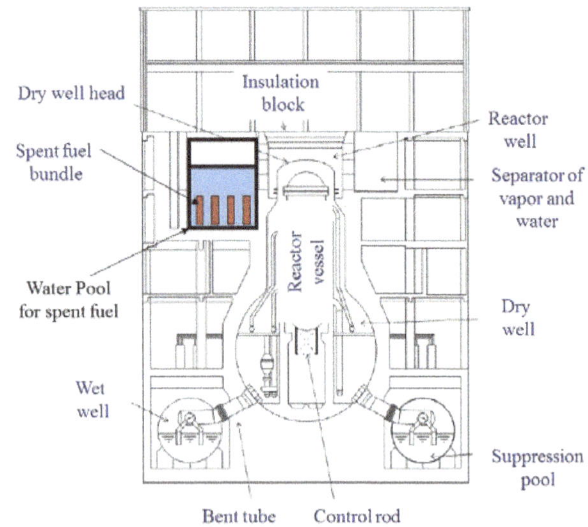

Fig. 5 Boiling water reactor with spent fuel cool pool

Table 1. Nuclear power reactor in Fukushima

	NO.1	NO.2	NO.3	NO.4	NO.5
Electric-Power (MW)	460	784			
Thermal Power (MW)	1,380	2,381			
Construction	Sept, 1967	May, 1969	Oct, 1970	Sept, 1972	Dec, 1971
Commercial Operation	March, 1971	July, 1974	March, 1976	Oct, 1978	April, 1978
Type of Nuclear Reactor		Boiling Water Reactor(BWR)			
Type of Reactor		Mark- I			
Contractor	GE	GE+Toshiba	Toshiba	Hitachi	Toshiba
Number of Fuel Bundle	400	548			
Length of Fuel Core(m)	4.35	4.47			
Number of Control Rod	97	137			
Reactor Vessel					
ID(m)	4.8	5.6			
Height(m)	20	22			
Weight(Ton)	440	500			
Concrete Chamber					
Height(m)	32	33		34	
OD of Top(m)	10	11			
OD of Bottom(m)	18	20			
Water Volume of Suppression Pool(Ton)	1,750	2,980			
Turbine					
Rotation(rpm)		1,500			
Inlet Temp. of vapor(℃)		282			
Vapor Pressure(Kg/cm².G)		66;8			
Fuel					
Type of Fuel		Oxide Uranium			
Weight(Ton)	69	94			

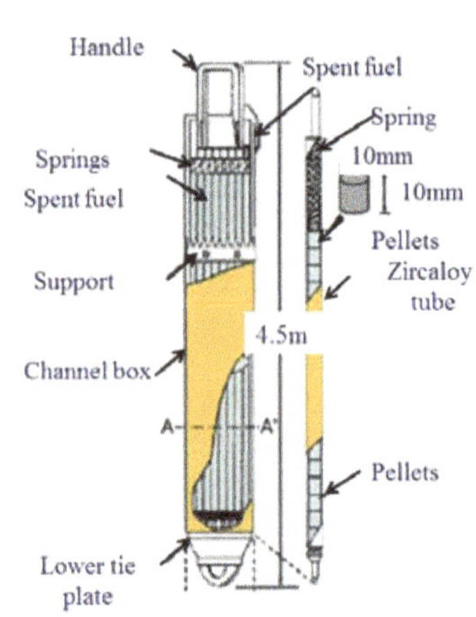

Fig. 6 Outline structure of the spent fuel

Figure 8 shows a concept and picture of the corrugated heat pipe. The corrugated heat pipe is chosen due to design robust and flexibility for integration into the system. The corrugated heat pipe is commercially available and had been using in snow melting of roads and pedestrian walkways in Japan. The heat pipe is made of stainless steel and working liquid is ammonium. The lengths of the evaporator and the condenser are about 6m and 4m respectively. For enhancement of air cooling, aluminum fins are mechanically insert in the condenser section.

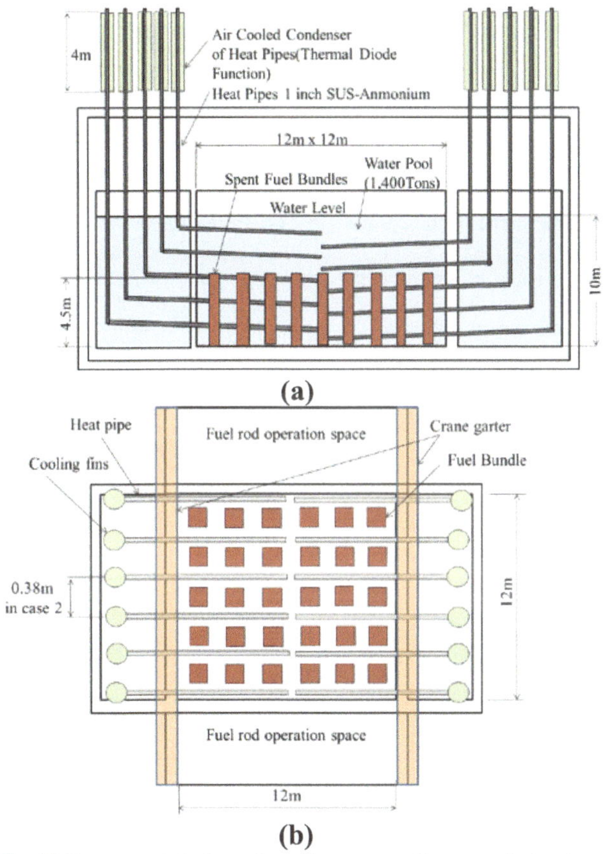

Fig. 7 Conceptual design of cooling system for spent fuel using thermal diode heat pipes (a) Front view; (b) Plan view

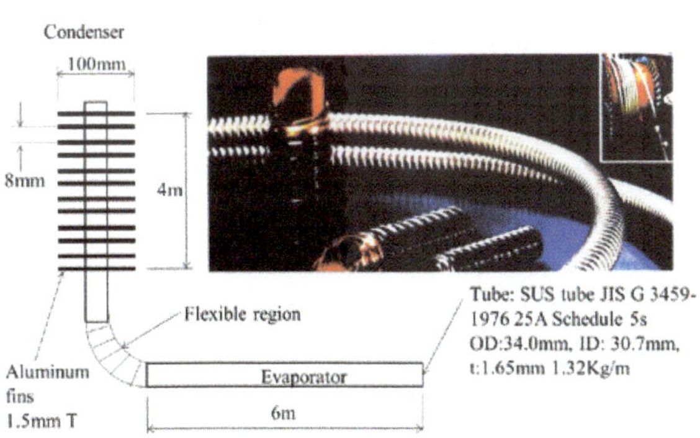

Fig. 8 Corrugated heat pipe

2.2 Thermal Design Analysis of Heat Pipes

The calculations are based on the reference (Mochizuki et al, 2012).

2.2.1 Heat Pipe Fin Air Side Heat Transfer Calculation

Table 2 shows an example of calculation of fin tube design. In brief, the calculation shows that with the air flow velocity of 1 m/s gives a reasonable fin to air heat transfer coefficient of about 12 W/m2K; air pressure drop 1.22 mmHg for the fin geometry of OD 0.1m, fin pitch 0.008m, fin thickness 0.0015m, and the fin efficiency 0.97.

Table 2. Heat Pipe Fin Air Side Heat Transfer Calculations

Property & Value	Symbol	Value	Unit
Thermal Conductivity of Air	λa	0.026	W/m.°C
OD of Tube	do	0.034	m
Viscosity of Air	μa	1.91.E-06	kg.s/m^2
OD of fin	Do-fin	0.10	m
ID of fin	Di-fin	0.034	m
Fin Pitch	Pf	0.008	m
Fin thickness	tf	0.0015	m
Fin heat transfer area per 1m	Ao	1.74	m^2/m
Bare tube heat transfer area per 1m	Ao'	0.0867	m^2/m
Specific Heat of Air	Cpa	1.005	kJ/kg.°C
Cooling air velocity	Vair	1	m/s
Density of Air	γa	1.129	kg/m^3
Max. Flow Rate of Air	Gmax	4,064	kg/m^2.hr
Heat Transfer Coefficient of Fin Tubes	αair	12	W/m^2.K
Pitch of tube on X direction	S1	0.12	m
Pitch of tube on Y direction	S2	0.12	m
f factor	f	1.06	Non
Number of tube rows	J	20	Non
Coefficient of Gravity	gc	1.27.E+08	kg.m/hr^2.kg
Pressure Drop of Air	Δair	1.217	mmAq
Thermal conductivity of fin (Al)	λfe	160	W/m.°C
Half of Fin thickness	yb	0.00075	m
Fin height	w	0.033	m
Coefficient of fins	Xf	0.29	Non
1/2 x Fin OD	xe	0.05	m
1/2 x Tube OD	xb	0.017	m
Coefficient of fin and tube height	z	2.94	Non
Fin efficiency	φf	0.97	Non

2.2.2 Heat Pipe Water Side Heat Transfer Calculation

Table 3 shows the calculation of heat transfer on the water side, assuming water temperature and heat pipe temperature of 50 °C and 40 °C respectively. With the heat pipes placed horizontal and natural convection heat transfer, the calculation results showed that the average heat transfer coefficient between heat pipe and cooling water is approximate 670 W/m^{2}K.

2.2.3 Heat Pipe Evaporation Heat Transfer Calculation

The evaporation heat transfer coefficient inside heat pipe is calculated using Imura correlation (Mochizuki et al, 2012). The boundary conditions and assumptions are given in the below. The evaporation heat transfer coefficient is calculated about 1,281 W/m^2.K as shown in Table 4.

- Working fluid ammonia at 40°C operating temperature.
- Heat pipe inclination 2 degrees.
- Evaporator length 6m.
- Heat transfer rate of heat pipe 0.257 kW.
- Heat flux at evaporator 445 W/m^2.
- Tube: SUS JIS G3459-1976, 25A, Schedule 5s, OD 34mm. ID 30.7mm, t1.65mm, 1.32 kg/m.

Table 3. Water Side Heat Transfer Calculations

Properties & Values	Symbol	Value	Unit
Prandtl number	Pr	3.57	Non
Gravity	g	9.8	m/s^2
Expansion coeeficient of water	β	0.00045	1/°C
OD of tube	do	0.034	m
Water temp.	tw	50	°C
Tube temp.	t∞	40	°C
Dynamic viscosity of water	νw	5.55.E-07	m^2/s
Grasphof number	Gr	5,627,161	
Average Nusselt number	Num	35.483	
Thermal conductivity of water	λw	0.6	W/m^2.°C
Average heat transfer coefficient	αw	670	W/m^2.°C

Table 4. Heat pipe Evaporator Heat Transfer Calculations

Property & Value	Symbol	Value	Unit
ID of tube	di	0.0307	m
Density of liquid	ρl	568.4	kg/m³
Density of vapor	ρv	14.48	kg/m³
Thermal conductivity liquid	λl	0.433	W/m.°C
Specific heat of liquid	Cpl	5020	J/kg.°C
Gravity	g	9.81	m/s²
Heat flux at evaporator	qe	445	W/m²
Latent heat of NH3	L	1,066,000	J/kg
Viscosity of NH3	μl	0.000114	Pa.s
χ factor	χ	1,669	Non
Inside pressure	Pin	1.87.E+06	Pa
Ambient pressure	Po	1.01.E+05	Pa
Boiling heat transfer coefficient	αe	1,281	W/m².°C

2.2.4 Heat Pipe Condensation Heat Transfer Calculation

The condensation heat transfer coefficient is calculated about 8857 $W/m^2.K$ as shown in Table 5.

Table 5. Heat pipe Condensation Heat Transfer Calculations

Properties & Values	Symbol	Value	Unit
Density of liquid	ρl	568.4	kg/m³
Thermal conductivity liquid	λl	0.433	W/m.°C
Gravity	g	9.81	m/s²
Inside diameter of Tube	di	0.0307	m
Length of condenser	Lc	4	m
Heat transfer area of condenser	Ac	0.386	m²
Heat transfer rate of heat pipe	Qhp	0.257	kW
Heat flux at condenser	qc	666.5	W/m²
Latent heat of NH3	L	1,066,000	J/kg
Viscosity of NH3	μl	0.000114	Pa.s
Condensation heat transfer coefficient	αc	8,857	W/m².°C

2.2.5 Summary of Overall Heat Pipe Heat Transfer Calculation

Table 6 shows a summary of overall heat pipe heat exchanger thermal performance.

Table 6. Summary of Overall Heat Pipe Heat Transfer Calculations

Properties & Values	Symbol	Value	Unit
Heat transfer coefficient at evaporator	αe	1,281	W/m².°C
ID of Tube	di	0.0307	m
Length of evaporator	Le	6	m
Heat transfer area at evaporator	Ae	0.578	m2
Heat transfer coefficient at water side	αw	670	W/m².°C
OD of Tube	do	0.034	m
Heat transfer area at water side	Aw	0.641	m²
Heat transfer coefficient at condenser	αc	8,857	W/m².°C
Length of condenser	Lc	4	m
Heat transfer area of inside condenser	Ac	0.386	m²
Heat transfer coefficient at fin tubes	αa	10.28	W/m².°C
OD of fin	Do-fin	0.1	m
ID of fin	Di-fin	0.034	m
Fin pitch	Pf	0.008	m
Fin thickness	tf	0.0015	m
Fin heat transfer area per 1m	Ao	1.736	m²/m
Bare tube heat transfer area per 1m	Ao*	0.087	m²/m
Heat transfer area at Condenser	Afin	7.290	m²
Fin efficiency	φf	0.97	Non
Total thermal resistance	Rhp	0.0177	°C/W

2.3 Design Cases Evaluation

2.3.1 Nuclear Decay Heat – No cooling case

The nuclear decay heat is calculated using equation (1) (Glasstone & Sesonske, 1967), where:

- P (t): Decay heat as function of time [W]
- P_o: Normal thermal power before shutdown [W]
- t: Time since reactor shutdown [s]
- t_s: Time of reactor shutdown from start-up [s]

$$\frac{P(t)}{P_o} = 0.066 \times \left[t^{-0.2} - (t_s + t)^{-0.2} \right] \qquad (1)$$

Figure 9 shows the nuclear decay heat and temperature rise of cooling water with time in case of no cooling, in which the water temperature increased over 100 °C just in one day. The boundary conditions and assumptions are stated in the below:

- Water cooling volume 1,400 tons at initial temperature 30 °C.
- Time of reactor shutdown from start-up: 1.5 years.
- Fuels are remained in the reactor for 1 month, and after transferred to water cooling pool.
- Decay heat at start of cooling = 3.6 MW.
- Heat dissipation at water surface is neglected.
- Ambient temperature 30 °C.

Figure 10 shows the cooling water temperature rise with time for heat pipe design case 1. The cooling water temperature increased to 100 °C in about 85 hours, then saturated to 70 °C after 2,000 hours. This case 1 seems small cooling capacity. Figure 11 shows the cooling water temperature rise with time for heat pipe design case 2. The cooling water temperature increased to 68 °C in about 75 hours, then saturated to 50 °C after 2,000 hours. This case 2 seems reasonable cooling capacity. Figure 12 shows the cooling water temperature rise with time for heat pipe design case 3. The cooling water temperature increased to 49 °C in about 40 hours, then saturated to 40 °C after 2,000 hours. This case 3 seems also reasonable cooling capacity. Figure 13 shows the cooling water temperature rise with time for heat pipe design case 4. The cooling water temperature increased to 39 °C in about 21 hours, then saturated to 35 °C after 1,500 hours. This case 4 seems is over design cooling capacity.

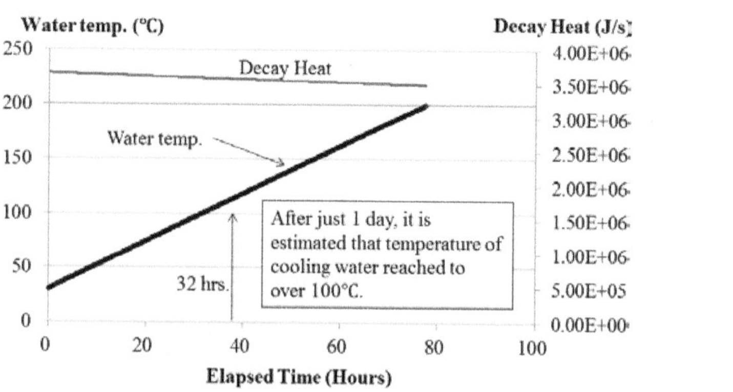

Fig. 9 Nuclear decay heat

2.3.2 Nuclear Decay Heat – Heat pipe cooling cases

Table 7 shows 4 cases of heat pipe design for evaluation of cooling performance and cost.

Table 7. Heat Pipe Design Cases

	Ratio of heat pipe cooling to total thermal power	Required thermal resistance [Rhp]	Quantity of heat pipes required
Case 1- Qhp = 0.45 MW	0.019%	2.19×10^{-5} °C/W	823
Case 2 – Qhp = 0.9 MW	0.038%	1.09×10^{-5} °C/W	1662
Case 3 – Qhp = 1.8 MW	0.076%	5.47×10^{-6} °C/W	3292
Case 4 – Qhp = 3.6 MW	0.151%	2.73×10^{-6} °C/W	6648

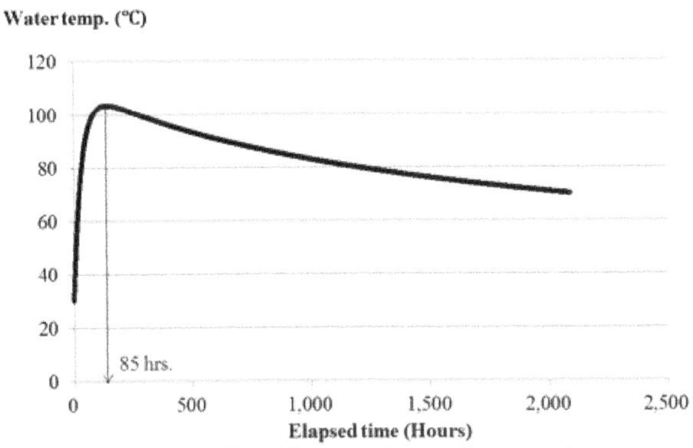

Fig. 10 Heat pipe design case 1

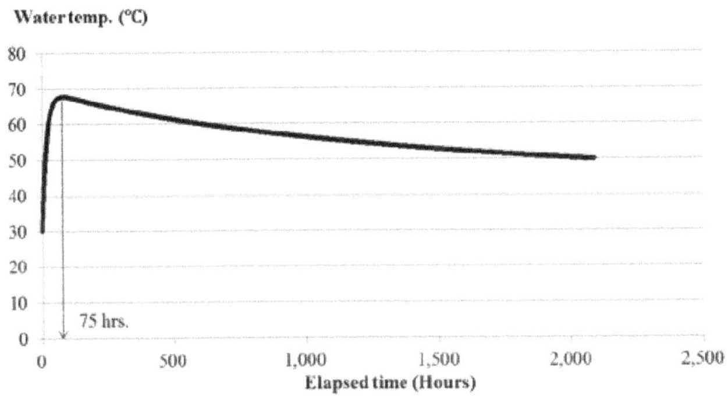

Fig. 11 Heat pipe design case 2

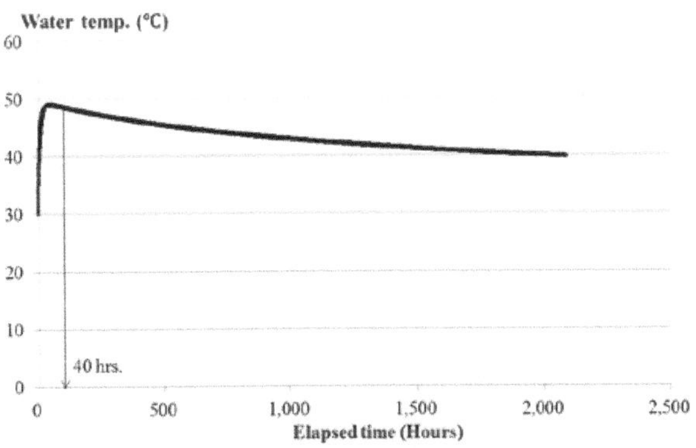

Fig. 12 Heat pipe design case 3

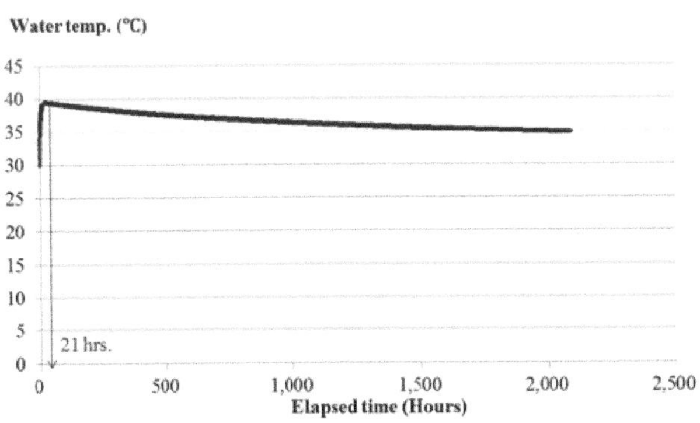

Fig. 13 Heat pipe design case

3. SUMMARY AND CONCLUDING REMARKS

1. The proposed heat pipe cooling system for cooling of spent fuel for Fukushima nuclear No. 4 is completely passive without any use of electrical power.
2. The general boundary conditions and assumptions for the thermal analysis are: Thermal power 2,381 MW; Electrical power 784 MW; Fuel quantity 34,956 units; Time of reactor shut down from started up 1.5 years; Fuels are remained in the reactor for about 1 month after shutdown, the fuels are then transferred to the water cool pool spent fuel; Decay heat at the start of cooling is 3.6 MW; Water volume in spent fuel pool is 1400 tons at initial temperature of 30 °C and ambient temperature 30 °C.
3. There are 4 cases of heat pipe design presented, with thermal performance and heat transfer capability comparison from under design to over design.
4. Heat pipe is a 1 inch diameter corrugated stainless steel and working liquid is ammonium. The heat pipe condenser is 4m long with 0.1m diameter aluminum fins, fin pitch 0.008m, fin thickness 0.0015m, and cooling is by natural air convection. The heat pipe evaporator is 6m long, horizontally placed at 2 degrees angle at the bottom of the spent fuel water cooling pool.
5. Without heat pipe cooling, the water temperature in the spent fuel pool will increased to 100 °C in 32 hours.
6. Heat pipe design case #1: Qhp = 0.45MW, Heat pipe quantity 823 pcs. After 85 hours, the water temperature increased to 100°C and it will be evaporate. This is under design cooling capacity.
7. Heat pipe design case #2: Qhp = 0.9MW, Heat pipe quantity 1,662 pcs. After 75 hours, the water temperature will reach to 68°C, then saturated to 50°C after 2000 hours. This design is very reasonable and economical design.
8. Heat pipe design case #3: Qhp = 1.8MW, Heat pipe quantity 3,292 pcs. After 40 hours, the water temperature will reach to 49°C, then saturated to 40°C after 2,000 hours. This design is reasonable with extra design safety.
9. Heat pipe design case #4: Qhp= 3.6MW, Heat pipe quantity 6,648pcs. After 21 hours, water temperature will reach to 39°C, then saturated to 35°C after 1500 hours. This is over design.
10. It is recommended to use design case #2 when considering reasonable design safety margin and economical cost.
11. The cost of heat pipes for case #2 is estimated about 1300 USD x 1666 pcs = 2.16 Million USD.

REFERENCES

Glasstone, S., Sesonske, A., *Nuclear Reactor Engineering*, 2nd Ed., Van Nostrand Reinhold, 1967.

Mochizuki, M., Singh, R., Nguyen, Th., Nguyen, Ti., Mashiko, k., Saito, Y., Wuttijumnong, V., Completely Passive Heat Pipe Based Emergency Core Cooling System For Nuclear Power Reactor, *16th International Heat Pipe Conference*, Lyon, France, May 20-24, 2012.

Yoshioka, R., Lino, K., *Technical Report Fukushima Accident Summary*, Association for the Study of Failure, Aug 19, 2011, Japan.

THE STUDY OF EVACUATED SOLAR COLLECTOR USING PULSATING HEAT PIPE

Jong Soo Kim[a], Soo Jung Ha[a], Sang Il Lee[a,*], Bang Ho Jeong[b]

[a] *Department of Refrigeration and Air-Conditioning Engineering, Pukyong National University, Nam-gu, Busan, 608-739, Republic of Korea*
[b] *Kyungunbeomyang Energy Co., LTD., Gangseo-gu, Busan, 618-280, Republic of Korea*

ABSTRACT

Solar heat energy, one of the typical renewable energy absorbs radiation energy from the sun and converts it into heat energy. Since it is sustainable and eco-friendly, it can be a good alternative to solve the problem of the depletion of fossil fuels. But it has low density. So, solar collector is very important to use more efficiently. From the viewpoint of low cost as well as short construction schedule for the manufacturing of heat exchanger, flexibility, and high performance, Pulsating Heat Pipe (PHP) draws special attention. Therefore, we use PHP for increasing efficiency of solar collector. In this study, we applied two types PHP to solar collector. One is U type, and other is Double U type. We made evaporating section of PHP by using a 'U' shape to prevent flooding which is the problem of conventional 'I' type heat pipe of solar collector. Condensing section of PHP was double tube which has the circular finned water tube inside for increasing heat transfer area. PHP was charged R-141b about 100% of evaporating section volume. And then, we compare the solar collector efficiency according to the evaporating section shape of the PHP on standard water flow rate. Consequently, heat collector efficiency of the U type solar collector is about 15% higher than Double U type solar collector. And both solar collector works reliably on the standard water flow rate.

Keywords: *Pulsating heat pipe, Evacuated solar collector, U type solar collector, Double U type solar collector*

1. INTRODUCTION

Solar energy, one of typical renewable energy absorbs radiation energy from the sun and converts it into heat energy. Since it is sustainable and eco-friendly, it can be a good alternative to solve the problem of the depletion of fossil fuels. But it has low density. So many studies about high efficiency solar collector have been carried out (Morrison et al., 2004; Dhariwal and Mirdha, 2005; Zambolin and Del Col, 2010). There are many types of the solar collector such as flat-plate collectors, stationary compound parabolic collectors and evacuated solar collectors. Among them, since conventional simple flat-plate solar collectors were developed for use in sunny and warm climates, their performance are greatly reduced when conditions become unfavorable during cold, cloudy, and windy days. However, evacuated solar collectors use a vacuum-sealed tube which reduces convection and conduction losses, so the collectors can operate at higher temperatures than flat-plate collectors. And their efficiency is higher at low incidence angles. This effect tends to give evacuated solar collectors an advantage over flat-plate collectors in terms of daylong performance.

Evacuated solar collectors use liquid-vapor phase change materials to transfer heat at high efficiency. These collectors feature a heat pipe (a highly efficient thermal conductor) placed inside a vacuum-sealed tube. The heat pipe contains a small amount of fluid that undergoes an evaporating-condensing cycle. In this cycle, solar heat evaporates the liquid and the vapor travels to the heat sink region, where it condenses and releases its latent heat. The condensed fluid returns to the solar collector and the process is repeated (Terpstra and Van Veen, 1987; Kalogirou, 2009).

Conventional heat pipe which applied to evacuated solar collector has very long 'I' type single tube evaporating section. And condensing section size is very small and located where block the water flow (Azard, 2008; Bin et al., 2013) as shown in Fig. 1. In the conventional type, since the working fluids inside evaporating section run to only one direction, when the sunshine is high, it became the cause of flooding lead to dryout. If flooding effect occurred, heat pipe performance was limited (Japan Association for Heat Pipe, 2001). Also the condensing section is very small, so its heat transfer area is not enough. And pressure drop is increased because it located where prevent the water flow (Kreith and Bohn, 2002; Cengel and Ghajar, 2012).

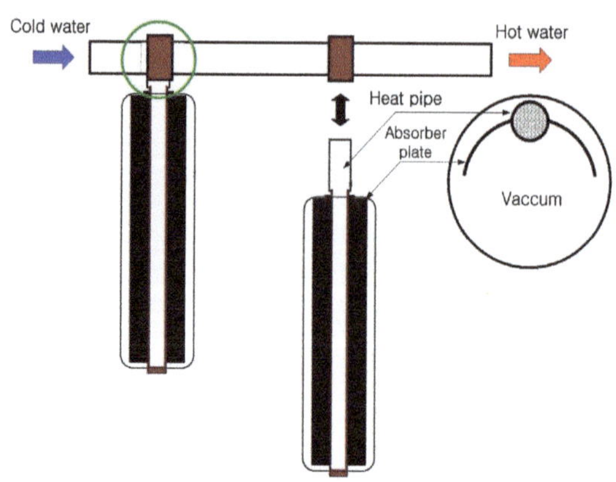

Fig. 1 Conventional 'I' type heat pipe solar collector

* *Corresponding author. Email: darksia77@pknu.ac.kr*

Our evacuated solar collector compensated the defect of conventional heat pipe solar collector. We used pulsating heat pipe (PHP) which has 'U' type evaporating section to prevent flooding by oscillating and irregular flow of vapor bubble. In the condensing section, water can flow straight inside the water tube without resistance by condensing section of PHP (compare the Fig. 1). Accordingly, pressure drop inside the water tube and water pump power are reduced. In addition, since the heat exchange is occurred on outer surface of water tube unlike conventional 'I' type heat pipe solar collector, heat transfer area of condensing section is increased by a fin that is attached to the outer surface of the water tube.

In this study, we applied two types PHP to solar collector. All of heat pipe have same condensing section shape, but evaporating section is different. One is the U type PHP and other is the Double U type PHP which has two U tube evaporating section. Using these solar collectors, we evaluate each solar collector performance on standard of water flow rate.

2. EXPERIMENTAL DEVICES AND METHODS

2.1 Pulsating heat pipe

A PHP is typically a meandering tube comprised of a serpentine channel with capillary dimension. Initially, the PHP is evacuated and then partially filled with the working fluid. Effects from surface tension cause the formation of liquid slugs interspersed with vapor bubbles. As heat is applied to the evaporating section, the working fluid begins to evaporate. This results in an increase of vapor pressure inside the tube which causes the bubbles in the evaporator zone to grow and pushes the liquid towards the condensing section. As the condensing section cools, the vapor pressure reduces and condensation of bubbles occurs. This process between the evaporating section and condensing section is continuous and results in an oscillating motion within the tube. Heat is transferred through latent heat of the vapor and through sensible heat transported by the liquid slugs (Reay and Kew, 2005). Operating procedure of PHP is shown in Fig. 2.

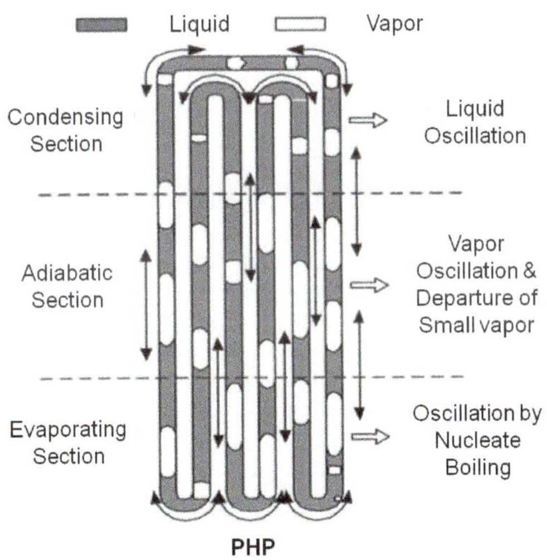

Fig. 2 Operating principle of PHP

2.2 Experimental Devices

Fig. 3 is a schematic diagram of experimental devices. Solar collectors were placed in outdoors where unimpeded area and installed on 35 degree inclination angle to the south with pyrheliometer. Heated water by solar collector was stored at the heat storage tank. And the pump with control valve and stop valve was installed for the flow control. T-type thermocouple was used to measure the inlet and outlet water temperature.

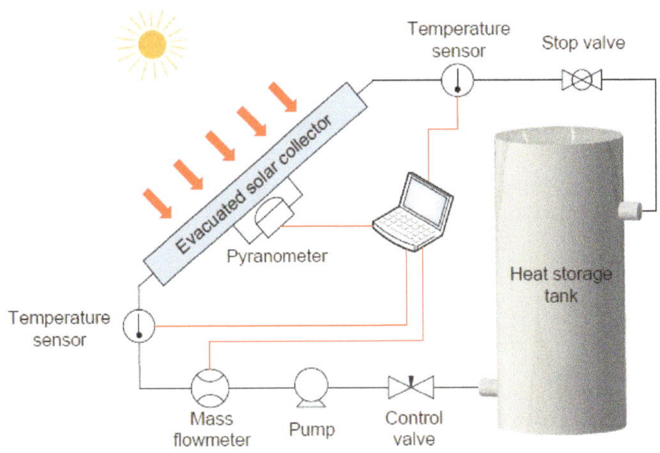

Fig. 3 Schematic diagram of solar collector system

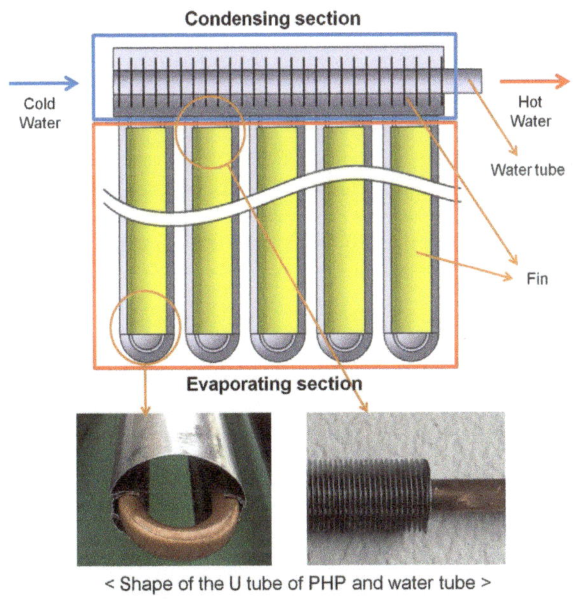

Fig. 4 Internal shape of U type solar collector

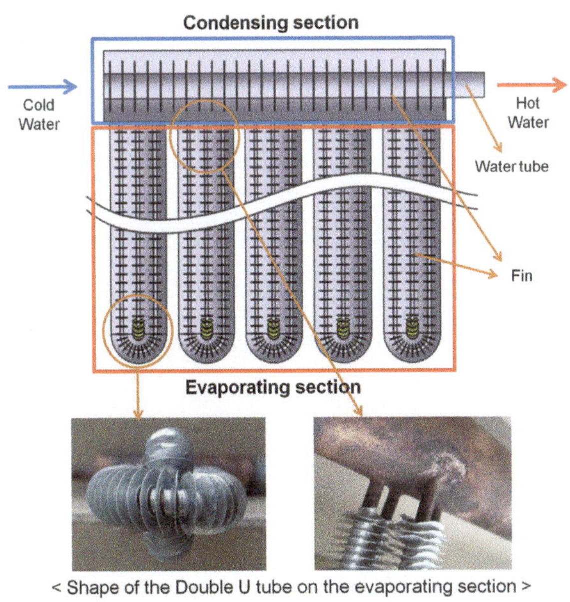

Fig. 5 Internal shape of Double U type solar collector

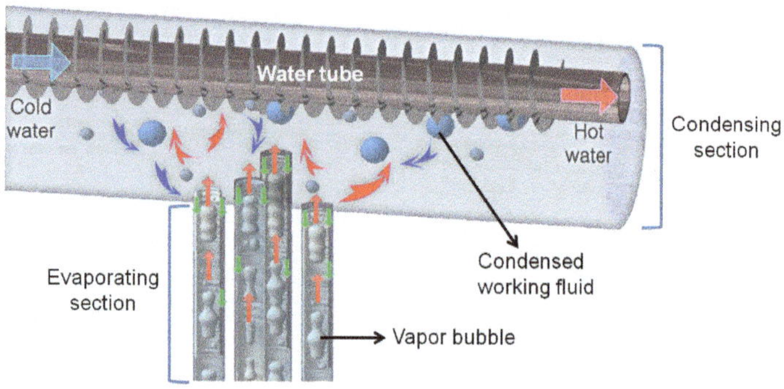

Fig. 6 Operating principle of PHP inside the solar collector

Table 1 Specification of the test equipments

Parameter		*Specification*	
Collecting Area(A_C)		U type	1.48 m^2
		Double U type	1.30 m^2
Storage Tank Volume		340 L	
Glass concentric evacuated tube		U type	58 mm
		Double U type	47 mm
Selective surface		Black chrome coating	
U type PHP *Working fluid: R-141b (2000cc)	Evaporating section	Length	1495 mm
		Pipe(O.D.)	7.94 mm
		Pipe(I.D.)	6.34 mm
		Fin Diameter	30.00 mm
	Condensing section	Length	1165 mm
		Pipe(O.D.)	15.88 mm
		Pipe(I.D.)	13.84 mm
		Fin Diameter	35.40 mm
		Fin pitch	2.00 mm
Double U type PHP *Working fluid: R-141b (1365cc)	Evaporating section	Length (1)	1560 mm
		Length (2)	1382 mm
		Pipe(O.D.)	4.76 mm
		Pipe(I.D.)	3.60 mm
		Fin Diameter	14.65 mm
	Condensing section	Length	1165 mm
		Pipe(O.D.)	15.88 mm
		Pipe(I.D.)	13.84 mm
		Fin pitch	35.40 mm
		Fin Diameter	2.00 mm

In this study, the condensing section of the each PHP is same, but U type solar collector has 21 U tubes and Double U type has 14 Double U tube as the evaporating section. Spirally-coiled circular aluminum fin was installed on evaporating section of Double U type PHP. Aluminum fin which transfer heat by conduction from inner surface of concentric evacuated tube to PHP was installed on evaporating section of U type PHP. And U type PHP was used more large U tube than Double U type PHP to complement decrease PHP performance due to reduction in numerals of U tube.

Condensing section of PHP was double tube which has the circular aluminum finned water tube inside. The detailed shape and specification of them were shown in Fig. 4, Fig. 5 and Table 1.

Vacuum of PHP could be reached to 0.04 Pa. Working fluid was R-141b which has high merit number and stability for container from the results of previous research by 100% of evaporating section volume (Lee et al., 2011). Therefore Double U type PHP was charged by 1365cc and U type PHP was charged by 2000cc.

Fig. 6 is operating principle of the solar collector using the PHP. If working fluid within heat pipe is vaporized in evaporating section, phase change occurred and vapor bubbles are generated. This vapor bubbles are forwarded to condensing section and heat exchange occur between water pipe and working fluid. After this heat exchange, condensed working fluid is moved to evaporating section. Working fluid transfers heat repeating this process.

2.2 Experimental Method

The experiment was conducted from 11 a.m. to 3 p.m. on sunny days. Water was supplied by 1 kg/min and 2 kg/min according to Standard of renewable energy facilities (1.2 kg/min per collector unit area) (Korea Energy Management Corporation (KEMCO), 2009). Data were collected every 2 seconds on the water inlet/outlet temperature, water flow rate and solar irradiance for calculating solar collector efficiency. And then every data were arranged with the time average.

2.3 Solar Collector Efficiency

The quantity of heat transferred to the water in the tank can be calculated by water temperature variation, taking into account the water flow rate and its specific heat:

$$Q_w = \dot{m} c_{p\cdot w}(T_{o\cdot w} - T_{i\cdot w}) \tag{1}$$

The cumulative collection efficiencies (η) can be expressed as a ratio of the amount of heat stored in the tank to the total amount of solar energy irradiation on the collector for the same period of time (Rittidech and Wannapakne, 2007):

$$\eta = \frac{Q_w}{A_C G_T} = \frac{\dot{m} c_{p\cdot w}(T_{o\cdot w} - T_{i\cdot w})}{A_C G_T} = \frac{\dot{m} c_{p\cdot w} \int \Delta T dt}{A_C \int G_T dt} \tag{2}$$

3. RESULTS AND DISCUSSION

3.1 Performance analysis according to shape of PHP

The experiment was conducted sunny day when the solar irradiance was more than 800 W/m^2 and water flow rate was 1kg/min. Fig. 7 and Fig. 8 are graphs of the water inlet/outlet temperature and solar irradiance during measuring period. Its average values are shown in Table 2.

When the solar irradiance was 903.7 W/m^2, Double U type solar collector was collected heat by 431.9 W/m^2 and U type solar collector was collected heat by 544.9 W/m^2. Consequently, heat collector efficiency of the U type solar collector is about 13% higher than Double U type solar collector.

Heat transfer area of evaporating section of PHP is similar to each other (Double U type: 4.27 m^2, U type: 4.53 m^2). But tube diameter of evaporating section of U type PHP is bigger than Double U type. Therefore, although the charging ratio is same to each other as 100% of evaporating section volume, quantities of working fluid are 1.4 times more than Double U type PHP. Under the same charging ratio, quantity of working fluid is one of factors contributing to determine performance of solar collector. As a result, U type solar collector which has more working fluid has high efficiency compare with Double U type solar collector.

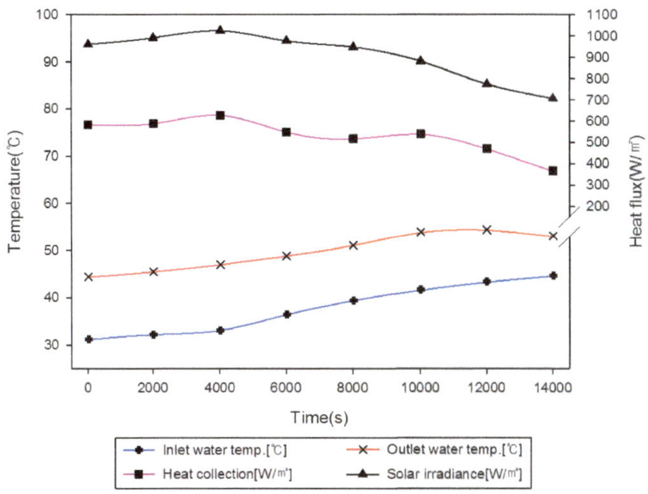

Fig. 7 Profile of water temperature and solar irradiance (U type solar collector, 1 kg/min)

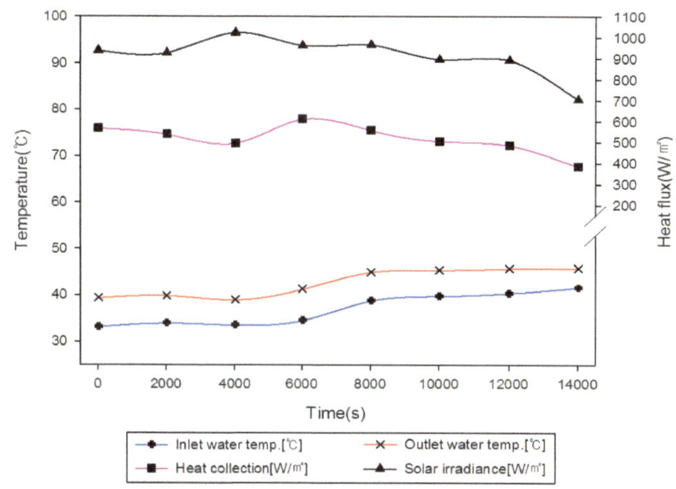

Fig. 9 Profile of water temperature and solar irradiance (U type solar collector, 2 kg/min)

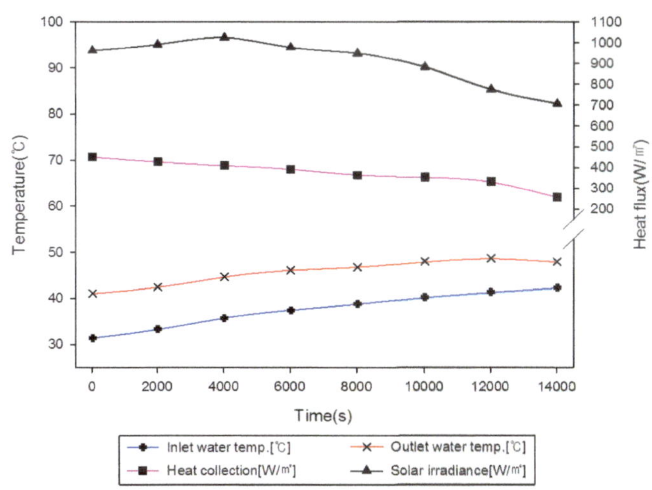

Fig. 8 Profile of water temperature and solar irradiance (Double U type solar collector, 1 kg/min)

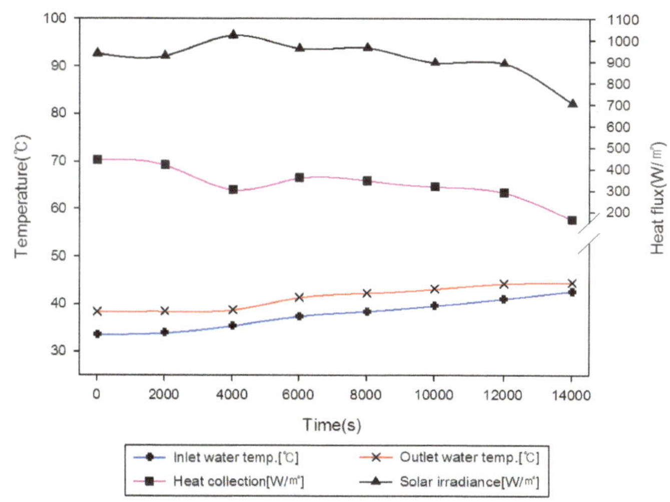

Fig. 10 Profile of water temperature and solar irradiance (Double U type solar collector, 2 kg/min)

Table 2 Test results of experiment (Water flow rate: 1kg/min)

Shape of PHP	*U type*	*Double U type*
Water flow rate (kg/min)	0.94	0.97
Water inlet temp. (℃)	37.9	37.8
Water outlet temp. (℃)	50.2	46.1
Ambient temp. (℃)	23.9	
Solar irradiance (W/m^2)	903.7	
Heat collection (W/m^2)	544.9	431.9
Efficiency (%)	60.3	47.8

Table 3 Test results of experiment (Water flow rate: 2kg/min)

Shape of PHP	*U type*	*Double U type*
Water flow rate (kg/min)	1.94	1.96
Water inlet temp. (℃)	37.2	37.8
Water outlet temp. (℃)	42.9	41.6
Ambient temp. (℃)	23.5	
Solar irradiance (W/m^2)	849.7	
Heat collection (W/m^2)	520.8	397.5
Efficiency (%)	61.3	46.8

3.2 Performance analysis according to water flow rate

Under the same experimental condition, we measured the performance of solar collector according to the water flow rate. Fig. 9 and Fig. 10 are presented water inlet/outlet temperature and solar irradiance variation when the water flow rate is 2 kg/min. Its average values are shown in Table 3.

When the solar irradiance was 849.7 W/m^2, Double U type solar collector was collected heat by 397.5 W/m^2 and U type solar collector was collected heat by 520.8 W/m^2. Consequently, heat collector efficiency of the U type solar collector is about 15% higher than Double U type solar collector. Compared with the results when the water flow rate is 1 kg/min, although temperature difference between inlet and outlet temperature is lower, solar collector efficiencies are almost equal. As a result, both solar collector works reliably on the standard water flow rate.

4. CONCLUSIONS

In this study, we applied two types PHP to solar collector and evaluate each solar collector performance on the standard water flow rate. The results of the experiment may be summarized as follows:

(i) From Table 3, when the solar irradiance was 849.7 W/m^2, Double U type solar collector collected heat by 397.5 W/m^2 and U type solar collector collected heat by 520.8 W/m^2. Consequently, heat collector efficiency of the U type solar collector is about 15% higher than Double U type solar collector.

(ii) Although both solar collectors have almost same heat transfer area of evaporating section of PHP and charging ratio, the quantities of working fluid of U type PHP are more than Double U type value. Therefore, U type solar collector shows much better performance compared with Double U type solar collector.

(iii) When the water flow rate was increased from 1 kg/min to 2 kg/min, water outlet temperature was decrease. But in each condition, solar collector efficiencies are almost equal in less than 1%. As a result, both solar collector works reliably on the standard water flow rate.

ACKNOWLEDGEMENTS

This work was supported by the solar heat collecting device using heat pipe of the Korea Institute of Energy Technology Evaluation and Planning (KETEP) grant funded by the Korea government Ministry of Knowledge Economy.
(No. 1677000201100200)

NOMENCLATURE

A_C	collecting area (m^2)
$c_{p\cdot w}$	specific heat (J/kg·K)
G_T	solar irradiance (W/m^2)
$\dot{m}$	water flow rate (kg/min)
$T_{i\cdot w}$	water inlet temperature (℃)
$T_{o\cdot w}$	water outlet temperature (℃)
Q_w	heat recovery amount of water (W)
η	solar collector efficiency (%)

REFERENCES

G. L. Morrison, I. Budihardjo and M. Behnia, 2004, "Water-in-glass evacuated tube solar water heaters", Solar energy, 76, 135-140.

S. R. Dhariwal and U. S. Mirdha, 2005, "Analytical Expressions for the Response of Flat-Plate Collector to Various Transient Conditions", *Energy Conversion and Management*, 46, 1809-1836.

E. Zambolin and D. Del Col, 2010, "Experimental analysis of thermal performance of flat plate and evacuated tube solar collectors in stationary standard and daily conditions", Solar Energy, 84, 1382-1396.

M. Terpstra and J. G. Van Veen, 1987, *Heat Pipes: Construction and Application*, Elsevier Applied Science.

S. A. Kalogirou, 2009, *Solar Energy Engineering*, Academic Press.

D. A. Reay and P. A. Kew, 2005, *Heat pipes*, 5th ed., Butterworth-Heinemann.

E. Azad, 2008, "Theoretical *and* experimental investigation of heat pipe solar collector", *Experimental Thermal and Fluid Science*, 32, 1666-1672.

Bin Du, Eric Hu and Mohan Kolhe, 2013, "An experimental platform for heat pipe solar collector testing", Renewable and Sustainable Energy Reviews, 17, 119-125.

Japan Association for Heat Pipe, 2001, *Practical Heat Pipe*, 2th ed., The Nikkan Kogyo Shimbun (in Japanese).

F. Kreith and M. S. Bohn, 2002, *Principle of Heat Transfer*, 6th ed., Thomson Learning.

Y. A. Cengel, and A. J. Ghajar, 2012, *Heat and Mass Transfer (Fundamentals and Applications)*, 4th ed., Mc Graw Hill.

Lee, S. I., Son, G. J., Kim, J. S., Ha, S. J., and Jung, B. H., 2011, "Performance Evaluation of the Evacuated Solar Collector using the Pulsating Heat Pipe", *Proceedings of International Heat Pipe Symposium*, 196-200.

Korea Energy Management Corporation (KEMCO), 2009, "Standard of Renewable Energy Facilities – Solar Collector" SH-101.

Rittidech. S., and Wannapakne. S., 2007, "Experimental Study of the Performance of a Solar Collector by Closed-End Oscillating Heat Pipe (CEOHP)", *Applied Thermal Engineering*, 27, 1978-1985.

AXIAL TEMPERATURE DISTRIBUTION AND PERFORMANCE OF R410A AND WATER FILLED THERMOSYPHON AT VARIOUS FILL RATIOS AND INCLINATIONS

K. S. Ong[a*], W. L. Tong[b], J. S. Gan[b] and N. Hisham[b]

[a] *Universiti Tunku Abdul Rahman, Jalan Universiti, 31900 Kampar, Malaysia*
[b] *School of Engineering, Monash University Sunway Campus, Jalan Lagoon Selatan, 46150 Bandar Sunway, Malaysia*

ABSTRACT

Axial wall temperature distribution of R410a and water filled thermosyphons at various fill ratios and inclination angles are experimentally determined. Power was supplied using resistance band heaters. A water cooled jacket surrounded the condenser section. Power input ranged from 100 – 830 W. Fill ratios of 0.93, 0.66, 0.4 and 0.25 were employed for the R410a pipe and 1.00, 0.75, 0.50 and 0.25 for the water filled pipe. Inclination angles were set at 30°, 50°, 70° and 90°. The results showed that the evaporator wall temperature was not uniform especially at high power input, low fill ratios and large inclinations.

Keywords: Heat pipe, R410a filled thermosyphon, water filled thermosyphon, inclined, fill ratio.

1. INTRODUCTION

Heat pipes are very effective heat transfer devices that are capable of transferring large amounts of heat through long distances without parasitic power for their operation. A two-phase closed thermosyphon is a wickless heat pipe. For simplicity and low cost, thermosyphons are preferred over wicked heat pipes. Thermosyphon heat exchangers have received considerable attention in industrial heat recovery and air conditioning systems and recently as solar collectors. In some air conditioners, the existing system ducting requires the equipment to fit in-line before and after the evaporator coil to act as pre-cool and reheat coils. In this instance, the heat pipe heat exchanger is required to be inclined due to space constraint. During evaporation in inclined thermosyphons, the liquid level in the evaporator section drops leading to liquid film thickness becoming non-axisymmetric. This is accentuated further as tilt angle increases and at low fill ratio. Hence the performance of the inclined single tube thermosyphon needs to be investigated in order to optimise the design of the inclined thermosyphon heat exchanger.

Hahne and Gross (1981) conducted an experimental investigation to observe the effect of inclination on the transport behaviour of a thermosyphon using a 2m long × 4cm ID steel pipe filled with R115. The evaporator section was electrically heated and the condenser water-cooled. Experiments were performed with inclinations from 0-90°. The quantity of fill was such that the heating zone was always flooded at all inclinations. Their results showed that the highest heat transport occurred at inclinations of around 50° to the horizontal.

Nguyen-Chi and Groll (1981) investigated the entrainment or flooding limit in a water filled 20mm OD × 2.5m long copper thermosyphon with fine circumferential grooves at its inner surface. Fill ratio was between 0.38–0.88, inclination angle from 10–80° from the horizontal and operating wall temperature between 20–80°C. The evaporator section was electrically heated and the condenser water-cooled. They derived an empirical correlation which allowed the prediction of the maximum performance of the inclined thermosyphon.

Negishi and Sawada (1983) experimentally visually observed the internal boiling and condensing phenomena inside a water and ethanol filled 330mm long × 13mm ID copper thermosyphon by fitting glass windows at both ends of the pipe. Heating and cooling were provided with hot and cold water jackets. Experiments were performed with fill ratio between 0.05–1.0, inclination angle between -10 to 90° to the horizontal and hot water temperature between 25– 85°C. Their results showed that it was necessary to have a fill ratio between 0.25–0.60 and inclination between 20 – 40° for a water filled pipe and 0.40–0.75 fill ratio and more than 5° inclination for ethanol filled pipe. They also determined the overall heat transfer coefficient to be between 2.4–3.0 × 10^3 W/m^2 K for the water filled pipe and between 0.9–1.1 × 10^3 W/m^2 K for ethanol.

Gurses and Cannistraro (1991) employed a finned water-cooled condenser wicked heat pipe for solar energy applications using 15mm ID × 18mm OD × 1930mm long copper pipe filled with 28cc of distilled water. They concluded that the performance depended upon inclination angle with the maximum in the vertical position. For effective solar operation, the inclination angle should be between 45–90°.

Zuo and Gunnerson (1995) presented a numerical model for a thermosyphon which considered the effects of dry-out and flooding in the upper portion of the evaporator section at low fill ratio for large diameter pipes (low aspect ratio) when inclined. Their predictions compared well with experimental data of other studies.

Shiraishi et al. (1996) correlated the critical transfer rate in an inclined thermosyphon by introducing a modified Kutateladze number. They employed six stainless tubes with 14 and 21mm ID of varying lengths to give aspect ratios of 11, 22 and 32. Fill liquids were

* *Corresponding author. Email: skong@utar.edu.my*

fluorinert (FC72), freon (R113), alternative freon (DF123), ethanol and water.

Terdtoon et al. (1996) investigated the effect of aspect ratio and Bond number on heat transfer characteristics of an inclined thermosyphon at normal operating condition using 9.5, 11.1 and 25.4mm ID copper tubes with R-22, ethanol and water at 0.80 fill ratio and inclination angles between 0–90°. Heating and cooling were provided with hot and cold water jackets at the evaporator and condenser sections. Evaporator temperature was maintained at 10°C. Various combinations of tube diameter and length resulted in aspect ratios of 5, 10, 20, 30 and 40. Their results showed that there was a range of optimum inclinations (40–70°) over which the thermosyphons exhibited better performances than when in a vertical position. The optimum inclination was dependent upon the aspect ratio.

Said and Akash (1999) showed that a wicked water filled heat pipe performed better than one without at inclinations of 30, 60 and 90°.

Payakaruk et al. (2000) determined the effects of Bond number, Froude number, Weber number and Kutaladze number on the heat transfer characteristics of inclined thermosyphon using 7.5, 11.1 and 25.4mm ID copper tubes. Fill liquids were R-22, R123, R134a, ethanol and water at fill ratios of 0.50, 0.80 and 1.0. Inclination angles were between 0–90° and aspect ratios from 5, 10, 20, 30 and 40. Heating and cooling were carried out using hot and cold water jackets at the evaporator and condenser sections. Saturation vapor temperature ranged from 0-30°C. Their results showed that the type of fill liquid affected the results but not the filling ratio.

Shalaby et al. (2000) experimentally investigated the heat transfer performance of a thermosyphon using a smooth 21mm ID × 1500mm long copper tube with R22 as working fluid. Electrical heating and cooling water jacket were used. Fill ratio was kept to between 0.30–1.0, inclination angle between 22.5–90° and power input varied from 100–300 W. They found that the optimum filling ratio was 0.50 and best inclination was 30°.

Nitipong et al. (2000) predicted the performance limit of a thermosyphon by assuming that dryout and flooding greatly influenced the performance limit. They correlated their predictions with experimental data from other researchers using R113 and R123 as working fluids to $\pm$ 25%.

Joudi and Witwat (2000) improved on the performance of a thermosyphon by adding a separator in the adiabatic section and adding on a three-layered wick in the evaporator section of three lengths of 20mm ID copper pipe filled with water. Evaporator and condenser lengths were kept constant at 10 and 150mm while adiabatic lengths varied at 100, 300 and 700mm. Electrical heating ranged from 5–32 kW/m^2. Condensing was by a cooling water jacket. Various fill ratios and inclinations were used. Their experiments showed that the presence of an adiabatic separator caused a marked improvement in all the heat pipes.

Loh et al. (2005) conducted a comparative study on heat pipe performance with three different wick structures (groove, mesh and sintered metal powder) in 4, 5 and 6mm OD pipes subjected to orientation angles from -90 to + 90° and at varying powers up to 30W. Their results showed that orientation had less impact on the sintered metal powder heat pipe compared to the others. Generally, performances depended upon pipe diameter, wick material and inclinations.

Noie (2005) investigated the heat transfer characteristics of a thermosyphon using a 25mm ID × 980mm long copper tube filled with water and fill ratios between 0.30 and 0.90. Heating was provided with electrical resistance wire and cooling by a water-cooled jacket. By varying the length of heating wire, aspect ratios were kept at 7.45, 9.80 and 11.8. He showed that the wall axial temperature distribution was almost uniform especially at the lower power inputs and that that maximum heat transfer rate at each aspect ratio occurred at different filling ratios.

Ong et al. (2008) investigated the heat transfer characteristics of a 32mm ID × 842mm long copper thermosyphon filled with R134a at fill ratio of 0.50 and 0.70 and R410a at fill ratio of 0.50 inclined at 45, 60, 75 and 90°. Heating was provided with electrical band resistance heaters and cooling by a water-cooled jacket. Power input varied from 100–500 W. They showed that the wall axial temperature distribution was more uniform at the lower power inputs. By assuming average wall temperatures at both evaporating and condenser sections, they determined the average overall thermal resistance to be 0.060 K/W for both cases.

Mozumder et. al. (2010) conducted experiments on a miniature 5mm ID × 150mm long heat pipe using water, methanol and acetone at 0.35, 0.55, 0.85 and 1.0 fill ratio and with power inputs up to 10W. They found that in the case of water and methanol filled pipes, fill ratio had minimum effect on performance while for the acetone filled pipe best performance was obtained at fill ratio of 1.0. They further concluded that fill ratios greater than 0.85 showed better results in general.

The performance of a water filled thermosyphon was investigated by Ong and Tong (2011) at fill ratios between 0.25 and 1.0 inclined at 30-90°. Power input varied from 300–830W using electrical band resistance heaters and a water-cooled jacket. They determined the average overall thermal resistance to be 0.017 K/W.

Not much work was found in the literature using R410a as fill liquid. The difference in performance between R134a and R410a filled thermosyphons was investigated by Ong (2011) at fill ratios from 0.5-0.7 and at inclinations from 45-90°. Water coolant flowrate was kept as high as possible to maintain a constant condenser wall temperature. It was shown that the overall thermal resistances for both the R134a and the R410a filled thermosyphons were about 0.060W/m^2 K.

From the literature review it could be concluded that the performance of a thermosyphon depended upon type of fill liquid, fill ratio, aspect ratio and power input. Past investigations have shown that while the wall temperatures in the evaporator and condenser sections are uniform in some cases they are non uniform in others. This is due to the very complex nature of the boiling process in the thermosyphon. As a result, the thermal resistance of the thermosyphon would be affected. The objective of this paper is to determine the axial wall temperature distribution and its effect on the overall thermal resistance of a R410a and a water filled thermosyphon at various fill ratios and inclinations.

2. EXPERIMENTAL INVESTIGATION

The thermosyphon shown in Figure 1 was fabricated from a 38mm O/D × 32mm I/D × 760mm long copper pipe. A 71mm I/D PVC pipe water cooled jacket arrangement with inlet and outlet hose connections surrounded the condenser section. Evaporator, adiabatic and condenser sections were 390, 90 and 327mm long, respectively. Aspect ratio (evaporator section length/pipe diameter) was about 12.2. Four electrical band heaters each capable of providing 250W was attached around the evaporator section. Power (P) to the band heaters was regulated using an auto-transformer and measured using laboratory standard voltmeter and ammeter. The thermosyphon was insulated all round with 50mm thick rockwool. Details of the thermosyphon and locations of type T thermocouples are shown in Figure 1.

The experimental set up is shown in Figure 2. Thermocouples were mechanically attached to the surface of the thermsyphon by binding wire. The thermocouples on the condenser wall section inside the water jacket were provided with water-proof tape over the binding wire in order to insulate them from the cooling effect of the water stream on the readings. Thermocouples were also attached to the inlet and outlet sections of the water jacket. Saturation temperature was measured with a type T thermocouple probe inserted about 150mm inside the thermosyphon. Saturation pressure was noted on a compound pressure gauge. The pressure gauge had a manufacturer's accuracy rating of 3% of full scale reading. It was only used to indicate and to give a visual warning if the thermosyphon was leaking. Coolant water was from mains supply and the flow rate was kept as high as possible, at about 0.0867 kg/s in order to maintain a near-uniform condenser temperature. It was measured with a float-in-glass type flow meter with an instrument quoted accuracy of $\pm 2\%$. There was no control over the coolant inlet temperature which varied between 27-29°C.

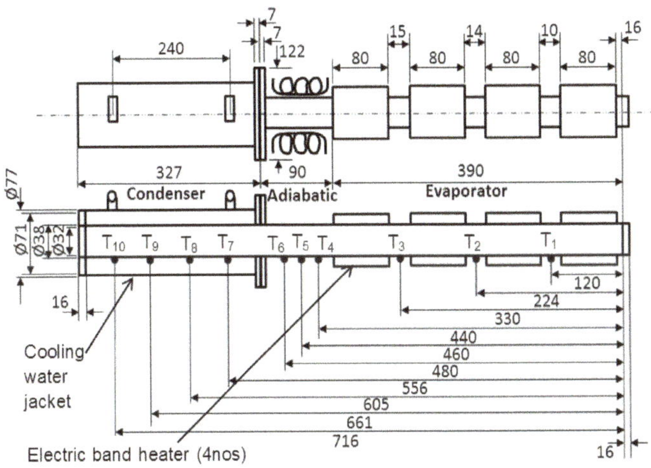

Fig. 1 Details of thermosyphon.

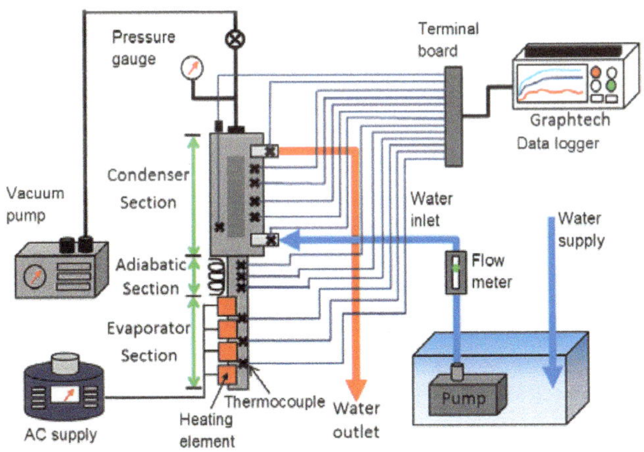

Fig. 2 Experimental set-up.

Experiments were conducted with the thermosyphon first filled with water and then with R410a. Input power was varied between 100 – 400W for the R410a filled pipe and 304–830W for the water filled pipe resulting in heat fluxes from 2.6–10.2 kW/m² and 7.8–21.2 kW/m², respectively. Fill ratios (FR) were kept at 0.93, 0.66, 0.40 and 0.25 for the R410a filled pipe and 1.00, 0.75, 0.50 and 0.25 for the water filled pipe. Fill ratio (FR) is defined as the ratio of the volume of fill liquid to the evaporator section volume. The thermosyphons were tested inclined at angles (θ) from 30, 50, 70 and 90° to the horizontal. Saturation temperature taken from standard saturation pressure-temperature tables were compared with measured values. The comparison was within ± 1°C. Each run was repeated 5 times in order to check for repeatability. It was found that the temperatures obtained were repeatable to within ± 1°C. Average values were then calculated from the 5 runs and utilised to plot the results.

3. EXPERIMENTAL RESULT

Since the coolant water mass flow rate was kept high, the temperature difference between coolant inlet and outlet temperatures was small, of the order of a few °C. Thus it was not possible to calculate accurately the heat rejected to the coolant water. In this paper, the thermosyphon was assumed perfectly insulated and power output was assumed equal to the electrical power input.

3.1 Axial Wall Temperature Distribution along R134a-filled Thermosyphon

Typical axial wall temperature distribution for the R410a filled thermosyphon at various power inputs and fill ratio of 0.93 are shown in Figure 3 for the vertical thermosyphon and in Figure 4 for the 30° inclined case. Results for fill ratio of 0.25 and 90° and 30° inclinations are shown in Figures 5 and 6, respectively. The results generally show that evaporator wall temperature increased with power input, as expected. The condenser wall temperatures also increased but to a lesser extent. Hence the temperature difference between evaporator and condenser section walls increased with power input. Contrary to what most researchers assume, the wall temperature distribution is not uniform especially at low fill ratios and large inclinations. Figure 3 for the vertical thermosyphon with high fill ratio of FR = 0.93 shows that the evaporator wall temperature was highest at the bottom and decreased gradually towards the adiabatic section. This was followed by a sharp drop in the adiabatic section and then a gradual decrease in the condenser section towards the top end of the pipe. For the same high fill ratio but at a greater inclination of 30°, Figure 4 shows that evaporator wall temperature first increased and then decreased along the pipe. This was more noticeable at high power input and low inclination.

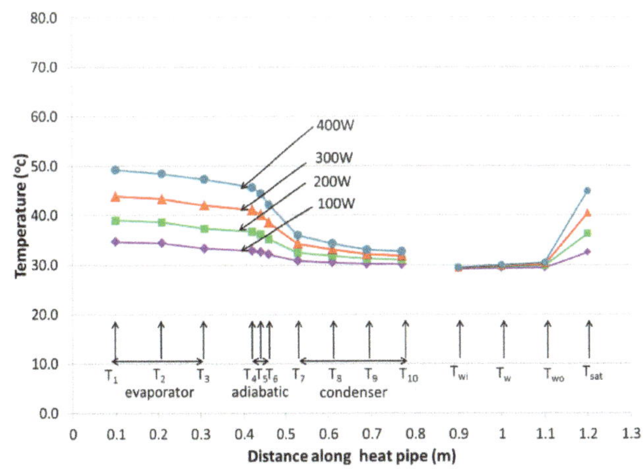

Fig. 3 Axial temperature distribution (R410a, FR = 0.93, $\theta = 90°$).

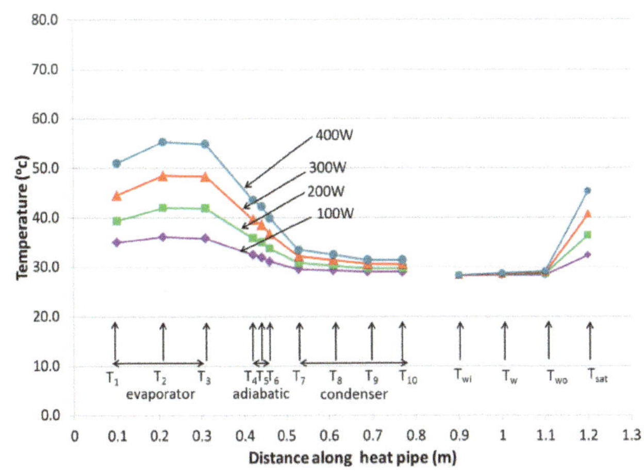

Fig. 4 Axial temperature distribution (R410a, FR = 0.93, $\theta = 30°$).

At low fill ratio, Figure 5 for the vertical thermosyphon show that the temperature declined from the bottom of the pipe towards the adiabatic and condenser sections especially at the high power inputs.

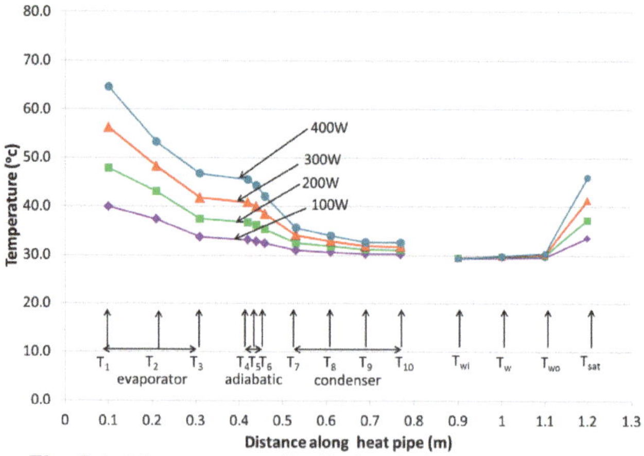

Fig. 5 Axial temperature distribution (R410a, FR = 0.25, θ = 90°).

Figure 6 show that at the low power input of 100W, the evaporator and condenser sections exhibited near uniform temperature distribution. The results show that dry out could have possibly occurred at the high temperature case normally associated with high power input, although it is not conclusive without visual observation. None-the-less, it would be quite correct to say that the axial wall temperature is uniform at low input power and for small fill ratios.

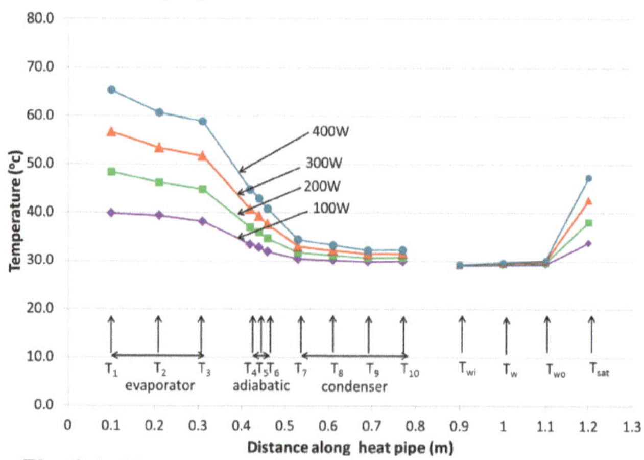

Fig. 6 Axial temperature distribution (R410a, FR = 0.25, θ = 30°).

As the thermosyphon is tilted from left to right, the liquid level on the left (top) and right (under) sides of the evaporator section of the pipe adjusts itself according to gravity. The thermocouples are mechanically attached to the left hand side of the evaporator section as shown in Figure 7. As the inclination increases, the liquid level rises towards the adiabatic section as illustrated in Figure 7. The immersion length and also the downward flowing liquid film thickness become non-axisymmetric and the heat transfer mechanism becomes more complicated. This makes the process of evaporation and condensation more complicated. Zuo and Gunnerson [5] considered this effect in their theoretical model. We would expect that as the inclination increases, dry out would more likely to occur on the right hand side of the thermosyphon.

3.2 Axial Wall Temperature Distribution along Water-filled Thermosyphon

Typical axial wall temperature distribution for the water filled thermosyphon are shown in Figure 8 for fill ratio of 1.00 and 90° inclination and in Figure 9 for fill ratio of 0.25 and 30° inclination. Similar trends are observed. The experiments were performed at power inputs greater than those for the R134a thermosyphon. The axial wall temperature distribution is non-uniform and decreased towards the top of the pipe. There was no temperature peak in the evaporator section as was noticed in the R134a filled pipe even though at the high power inputs. Dry out was noticed at the bottom of the pipe. At low fill ratio, Figure 9 shows that the highest temperature was observed to occur at the bottom of the pipe while the upper portion of the evaporator exhibited a near uniform temperature.

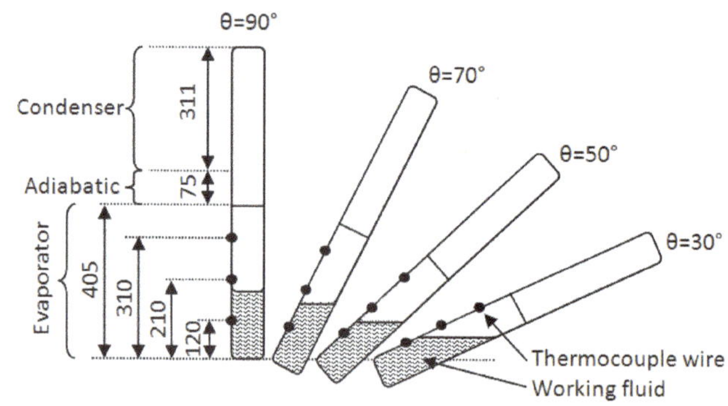

Fig. 7 Liquid level in thermosyphon with inclination (R410a, FR = 0.25).

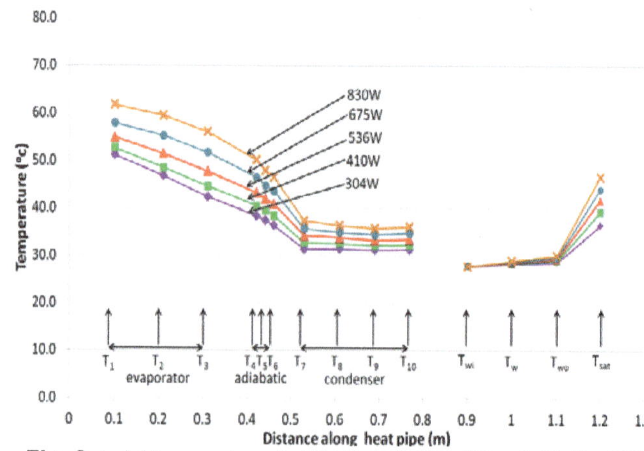

Fig. 8 Axial temperature distribution (Water, FR = 1.00, θ = 90°).

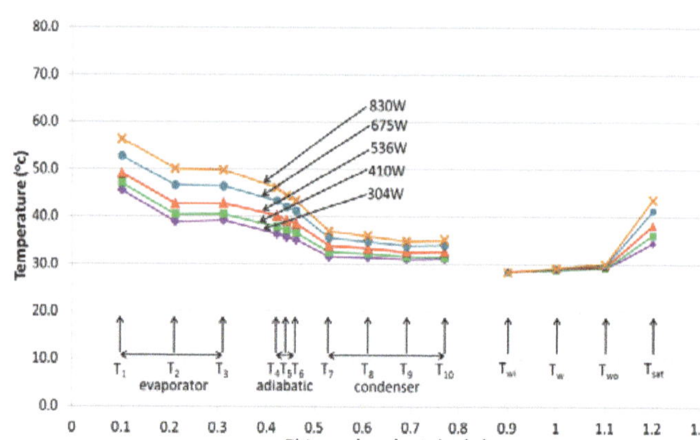

Fig. 9 Axial temperature distribution (Water, FR = 0.25, θ = 30°).

3.3 Performance of Thermosyphons

In order to compare the thermal performance of the thermosyphons at different fill ratios and inclinations, power output (P) was plotted against calculated mean operating temperature difference in Figures 10–13 for the R410a thermosyphon and in Figures 14–17 for the water filled unit. Since in some cases, the wall temperature were not uniform, a mean temperature difference (ΔT) was obtained by taking the difference between the average evaporator (T_e) and condenser temperature (T_c). As previously discussed, the axial wall temperature distribution depended upon power input, fill ratio and inclination. The calculated mean evaporator wall temperature was calculated by taking the average of the three thermocouples in the evaporator section. For the R410a-filled thermosyphon, the evaporator temperature varied from 1°C (Figure 3) to 9°C (Figure 4). For the water-filled thermosyphon, the evaporator temperature varied from 5°C (Figure 8) to 6°C (Figure 9). As a result of non-uniform wall evaporator and condenser wall temperatures, this can only give us an indication of how the thermosyphons behaved when compared with each other. Linear regression lines of $P = a\Delta T + c$ were then obtained and shown in Table 1. For the R410a thermosyphon, regression coefficients above 0.99 were obtained while it was above 0.95 for the water filled unit.

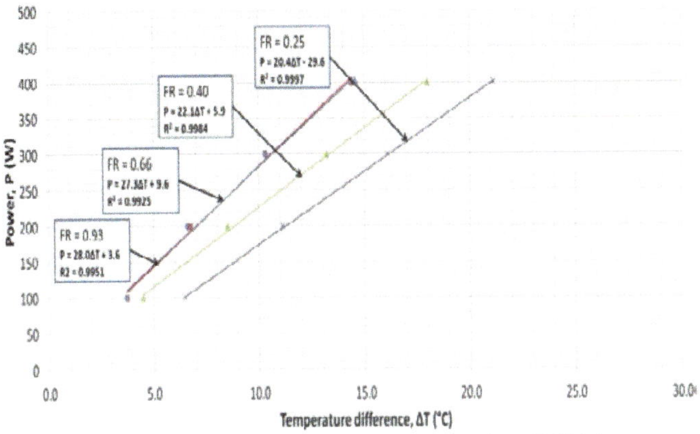

Fig. 10 Effect of fill ratio at 90° inclination (R410a).

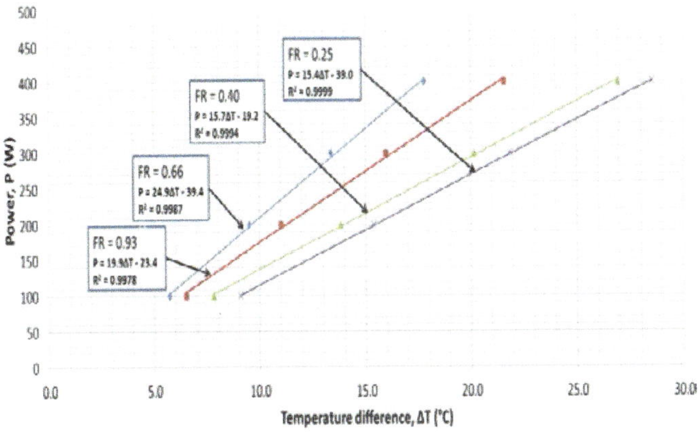

Fig. 11 Effect of fill ratio at 30° inclination (R410a).

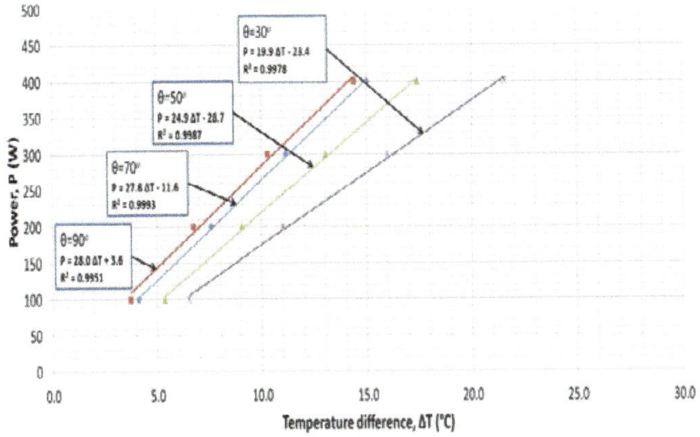

Fig. 12 Effect of inclination with FR = 0.93 (R410a).

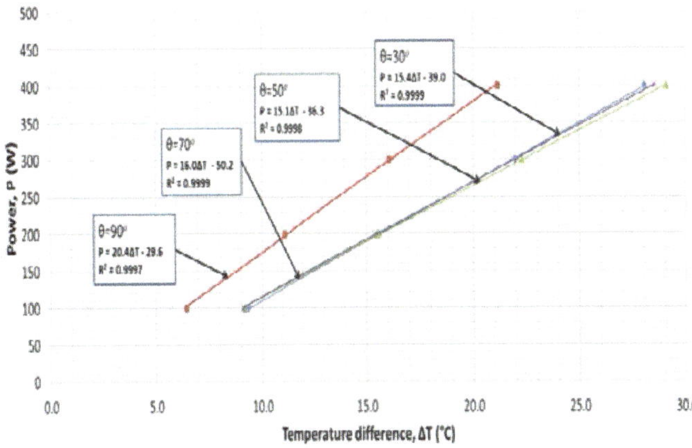

Fig. 13 Effect of inclination with FR = 0.25 (R410a).

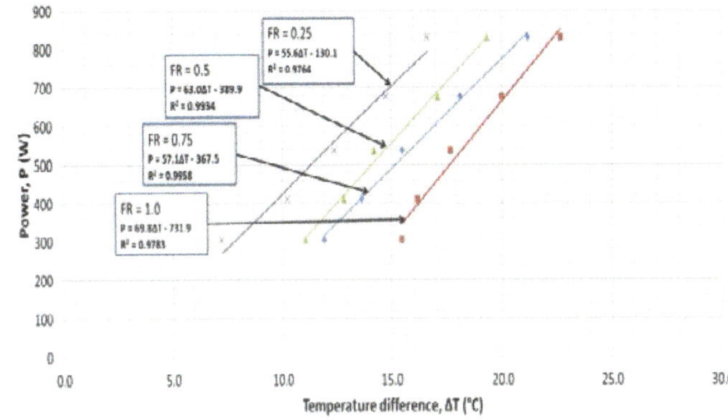

Fig. 14 Effect of fill ratio at 90° inclination (water).

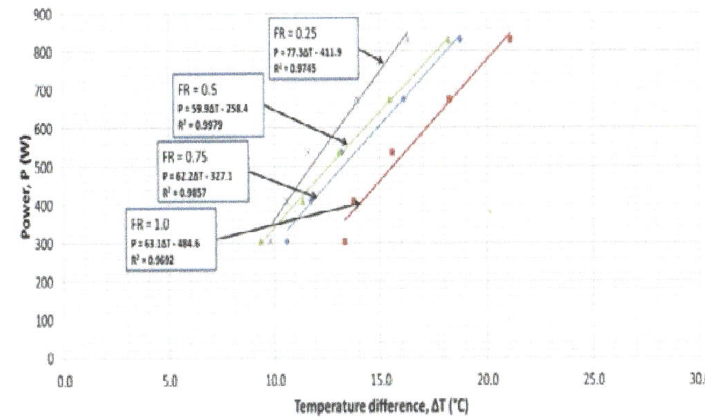

Fig. 15 Effect of fill ratio at 30° inclination (water).

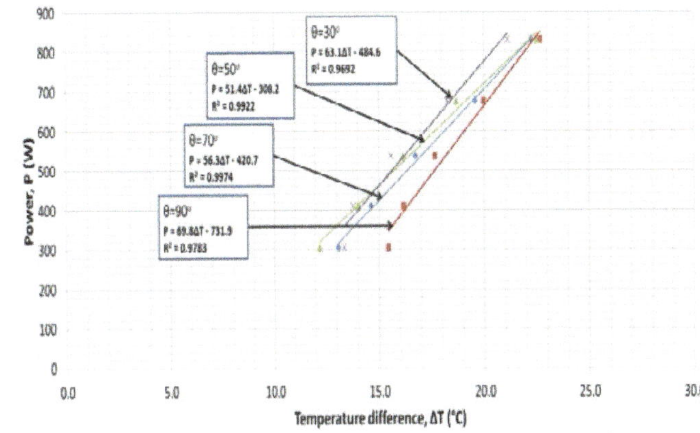

Fig. 16 Effect of inclination with FR = 1.00 (water).

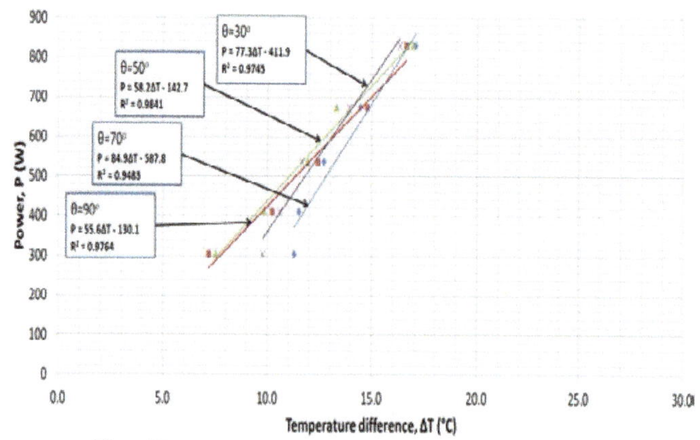

Fig. 17 Effect of inclination with FR = 0.25 (water).

Table 1 Linear regression coefficient from experimental results.

Inclination (degrees)	R410a (Fill ratio) 0.93			0.66			0.40			0.25		
	a	c	R^2	a	c	R^2	a	c	R^2	a	c	R^2
90	28.0	4	0.9951	27.3	-10	0.9984	22.1	6	0.9984	20.4	-30	0.9997
70	27.6	-12	0.9993	24.6	-19	0.9981	18.4	-37	0.9995	16.0	-50	0.9999
50	24.9	-29	0.9987	25.4	-37	0.9994	16.9	-13	0.9986	15.1	-36	0.9998
30	19.9	-23	0.9978	24.9	-39	0.9987	15.7	-19	0.9994	15.4	-39	0.9999

Inclination (degrees)	Water (Fill ratio) 1.00			0.75			0.50			0.25		
	a	c	R^2	a	c	R^2	a	c	R^2	a	c	R^2
90	69.5	-728	0.9810	63.3	-393	0.9924	57.3	-371	0.9954	73.8	-321	0.9618
70	55.9	-414	0.9966	68.2	-481	0.9739	61.6	-389	0.9984	84.0	-576	0.9470
50	51.4	-307	0.9924	72.1	-509	0.9464	66.1	-445	0.9934	58.3	-144	0.9824
30	63.3	-490	0.9690	60.0	-259	0.9989	62.4	-329	0.9871	77.7	-415	0.9723

The effect of fill ratio for the R410a pipe at 90° and 30° inclinations are shown in Figures 10 and 11. Figure 10 shows that the 0.93 and 0.66 fill ratio R140a thermosyphons performed equally well when operated in the vertical position. When tilted at 30°, Figure 11 indicates that the 0.66 fill ratio thermosyphon performed better than the 0.93 unit especially at high power input. The difference in operating temperature difference was only at 1.0°C at 100W and 3.5°C at 400 W. At 30° inclination and at 400W, the operating temperature difference was about 7°C.

The effect of inclination with fill ratios of 0.93 and 0.25 are shown in Figures 12 and 13. The results show that R410a thermosyphons performed better in the vertical position. Similar fill ratio and inclination effects are shown in Figures 14–17 for the water filled thermosyphon. However, in contrast with the R410a filled thermosyphon, the results showed that low fill ratio and inclined water filled pipes performed better than higher filled vertical ones.

4. CONCLUSIONS

The performance of R410a and water filled thermosyphons were investigated under various fill ratios and inclinations. The following conclusions were drawn:

- Axial wall temperature distribution was not uniform and depended upon power input, fill ratio and inclination as was noticed in some cases by previous investigators.
- Thermosyphon wall temperature increased with increasing input power.
- Dry out could have easily occurred in the case of high power inputs, low fill ratios and high inclinations.
- The R410a filled thermosyphon performed better in the vertical position at all fill ratios.
- The water filled thermosyphon performed better at low fill ratio and when inclined.

ACKNOWLEDGEMENTS

Thanks are extended to OYL R & D Sdn. Bhd. for their assistance in providing the R410a refrigerant and technical expertise.

REFERENCES

Gurses, A.C. and Cannistraro, L.T., 1991, "The Inclination Effect on the Performance of Water-filled Heat Pipes," Renewable Energy, 1, 667–674.

Hahne, E. and Gross, U., 1981, "The Influence of the Inclination Angle on the Performance of a Closed Two-phase Thermosyphon," Heat Recovery System, 1, 267–274.

Joudi, K. and Witwat, A.M., 2000, "Improvements of Gravity Assisted Wickless Heat Pipes," Energy Conversion and Management, 41, 2041-2061.

Loh, C.K., Harris, E. and Chou, D.J., 2005, "Comparative Study of Heat Pipes Performances in Different Orientations," 21st IEEE Semi-Therm Symposium.

Mozumder, A.K., Akon, A.F., Chowdury, M.S.H. and Banik, S.C., 2010, "Performance of Heat Pipe for Different Fluids and Fill Ratios," Journal of Mechanical Engineering, ME41, Trans. Mech. Eng. Div., Institution of Engineers, Bangladesh, 96-102.

Negeshi, K. and Sawada, T. 1983, "Heat Transfer Performance of an Inclined Two-phase Closed Thermosyphon," International Journal Heat Mass Transfer, 26, 1207–1213.

Nguyen-Chi, H. and Groll, M., 1981, " Entrainment or Flooding Limit in a Closed Two-phase Thermosyphon," Heat Recovery System, 1, 275–286.

Nitipong, S., Terdtoon, P., Tantakom, P. and Polchai, A., 2000, "A Performance Limit Model of an Inclined Two-phase Closed Thermosyphon," Proceedings 6IHPS, Thailand, 258-267.

Noie, S.H., 2005, "Heat Transfer Characteristics of a Two-phase Closed Thermosyphon," Applied Thermal Engineering, 25, 495-506.

Ong, K.S., Tung, A. and Tan, A.H., 2008, "Evaporating and Condensing Heat Transfer Coefficients in R134a and R410 Filled Thermosyphon at Low Temperatures," Proceedings 9IHPS, Malaysia, 107-115.

Ong, K.S. and Tong, W.L., 2011, "Inclination and Fill Ratio Effects on Water Filled Two-phase Closed Thermosyphon," Proceedings 10IHPS, Taiwan.

Ong, K.S., 2011, "Effects of Inclination and Fill Ratio on R134a and R410a Thermosyphon," Journal of Energy Heat and Mass Transfer, 33, 143-152.

Payakaruk, T., Terdtoon, P. and Ritthidech, S., 2000, "Correlations to Predict Heat Transfer Characteristics of an Inclined Closed Two-phase Thermosyphon at Normal Operating Conditions," Applied Thermal Engineering, 20, 781-790.

Said, S.A. and Akash, B.A., 1999, "Experimental Performance of a Heat Pipe," International Communication Heat Mass Transfer, 26, 679-684.

Shalaby, M.A., Araid, F.F., Sultan, G.I. and Awad, M.M., 2000, "Heat Transfer Performance of a Two-phase Closed Thermosyphon," Proceedings 6IHPS, Thailand, 269-278.

Shiraishi, M., Kim, Y.I., Murakami, M., and Terdtoon, P., 1996, "A Correlation for the Critical Heat Transfer Rate in an Inclined Two-phase Closed Thermosyphon," Proceedings 5IHPS, Australia, 248-254.

Terdtoon, P., Ritthide, S. and Shiraishi, M., 1996, "Effect of Aspect Ratio and Bond Number on Heat Transfer Rate in an Inclined Two-phase Closed Thermosyphon," Proceedings 5IHPS, Australia, 248-254.

Zuo, Z.J. and Gunnerson F.S., 1995, "Heat Transfer Analysis of an Inclined Two-phase Closed Thermosyphon," Journal Heat Transfer, 117, 1073-1075.

BOILING AND CONDENSATION HEAT TRANSFER COEFFICIENTS FOR A HEAT PIPE HEAT EXCHANGER

R. Laubscher, R.T. Dobson*

Department of Mechanical and Mechatronic Engineering, University of Stellenbosch, Stellenbosch
Private Bag X1, MATIELAND ***7602****, South Africa*

ABSTRACT

This paper reports on the experimental determination of the boiling and condensation heat transfer coefficients for the working fluid of a novel heat pipe heat exchanger at high operating temperatures. This technology is of importance in the high-temperature gas cooled nuclear power and process heat industries as it eliminates the probability of tritium finding its way into the product stream. The high temperature working fluid used to evaluate the heat pipe heat exchanger was Dowtherm-A, a special high temperature eutectic mixture of 26.5% diphenyl and 73.5% diphenyl oxide. A specially designed experimental heat pipe heat exchanger was built using a hot gas stream of up to 600 °C as the heating stream and cold water as the cooling stream. The basic features of the design are given. The experimental model is used as a proof of concept and to validate the theoretical heat transfer modeling analyses. Existing heat transfer coefficients correlations were considered; they include the Rohsenow, Stephan and Abdelsalam, Gorenflo, and Kutateladze correlations. None of these correlations gave satisfactory results and as such experimentally determined boiling and condensation correlations were determined and which gave reasonable results.

Keywords: Heat pipe heat exchanger, Dowtherm-A, heat transfer coefficients, boiling, condensation.

1. INTRODUCTION

Operating at fuel temperatures in excess of 950 °C, as most very high temperature gas-cooled nuclear reactors are expected to do, the formation of tritium in the reactor and its occurrence in the coolant stream is of concern. At these high temperatures tritium (H^3) has a particularly high diffusion rate through steel (Jousse, 2007) and may conceivably find its way into the process stream and ultimately into a consumer product. It has been suggested that this concern might be addressed by including an additional heat transfer loop between the primary reactor coolant loop and the process heat steam loop. Such an additional heat transfer loop would require an additional heat exchanger, additional pipework and circulation pump and a high temperature and pressure gas or low pressure but high temperature liquid metal as the working fluid. Because of the additional heat exchanger there will, needs be, also be an additional degradation of the available process stream temperature. A novel natural circulation two-phase thermosyphon-type heat pipe heat exchanger (HPHE) has been suggested by Dobson and Laubscher (2013). A concept drawing of such a heat transfer device is given in Fig. 1.

The concept HPHE shown in Fig. 1 consists essentially of a working fluid container more-or-less half charged with a liquid in thermal equilibrium with its vapour; and by a judicious choice of working fluid, at a relatively low pressure, say 0.5 to 1.5 kPa. The high pressure hot fluid stream is channeled through the evaporator section, whilst the stream being heated is channeled through high-pressure-withstanding tubes positioned in the vapour of the condenser section. The working fluid boils in the evaporator and condenses as a film on the colder tubes, collects and forms droplets which, under the influence of gravity, fall back into the liquid pool. In this way the latent heat of vaporization is transferred with essentially no temperature drop. [This heat transfer is thus reminiscent of a closed two-phase thermosyphon-type heat pipe.] The need for an extra heat exchanger, circulating pump and pipe work and high pressure working fluid container is thus also avoided.

Note that in Fig. 1 both the hot stream and the cold stream are shown as single tubes passing through the container. In an actual practical case there will be a large number of tubes and these tubes will not pass straight though the container. It is more likely that the hot and cold stream fluids will pass through a two-tube-pass heat exchange single shell (for the container) as in conventional tubular heat exchangers technology. The actual design would have to take careful cognisance of differential thermal expansion stresses, strains and deformation. The reason the concept has only two single tubes is for

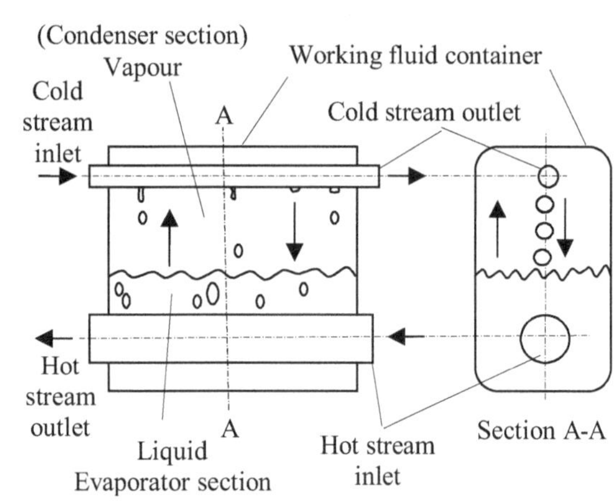

Fig. 1 Concept drawing of a novel heat pipe heat exchanger (HPHE).

----

* *Corresponding author email: rtd@sun.ac.za*

ease of experimental data capture and analysis, as single tube heating ("pool boiling") and condensing ("film-wise") geometries are well documented in the published literature.

Although the indicated HPHE working fluid for a 950 °C primary heating stream is sodium, it was deemed expedient from a safety and practical point of view to first conduct tests at an intermediate temperature. The highest non-toxic, non-flammable and commercially available working fluid for this purpose is Dowtherm-A. It is a eutectic mixture of 20.5% biphenyl and 70.5% diphenyl oxide, and is supplied by the Dow Chemical Company (1997). Its properties are given in Table 1 where it is seen that its boiling point at one atmosphere is 257.1 °C; and it is stored, shipped and suplied in conventional 210 litre (55 US gallons) standard steel drums (or so-called "barrels"). Unfortunately however the long-term stability of Dowtherm-A has as yet not been fully established and demonstrated.

A theoretical heat transfer model for a HPHE was developed (Laubscher, 2013), but the model then needed to be validated. To do this an experimental apparatus was designed, constructed and tested. To experimentally validate the theoretical model the boiling and condensation heat transfer coefficients for Dowtherm-A were needed. It thus follows that the objectives of this paper may be summarised as follows:

- Document and evaluate the existing and available boiling and condensation heat transfer coefficient correlations that could be applicable to Dowtherm-A.
- Observe the general flow behavior of the working fluid in the container by means of direct visual observation through windows in the sides of the experimental test apparatus. Also to check whether the vapour-liquid flow occurs in more-or-less vertical planes along the length of the tubes; as this was assumed to be the case in the theoretical model.
- Evaluate the applicability of the existing correlations for Dowtherm-A as working fluid, and if needs be generate a new and better correlation.

To meet these objectives a literature study was undertaken in section 2 in order to identify appropriate boiling and condensation correlations for inclusion into the theoretical model. The experimental work that was undertaken, in which a Dowtherm-A charged, HPHE was built and tested is described in section 3. Based on the experimental test results two new correlations were determined in section 4; one for the boiling and the other for the condensation heat transfer coefficients. An evaluation of the accuracy and applicability of the experimental and theoretical correlations was done and briefly discussed and conclusions are drawn in section 5.

Table 1 Thermal properties of Dowtherm-A (Dow Chemical Company, 1997).

Property	value	units
Freezing point	12	°C
Boiling point at 1 atmosphere	257.1	°C
Flash point	113	°C
Fire point	118	°C
Auto-ignition point	599	°C
Density at 20 °C	1056	kg/m^3
Critical temperature	497	°C
Critical pressure	313.4	kPa
Heat of combustion	30053	kJ/kg
Latent heat at 100 °C	345	kJ/kg
Dynamic viscosity at 25 °C	3.71	mPas
Thermal conductivity of vapour at 25 °C	0.0081	W/m°C
Thermal conductivity of liquid at 25 °C	0.1379	W/m°C
Saturated temperature at 50 kPa	226.67	°C

2. LITERATURE STUDY

When evaluating the thermal characteristics of a two-phase closed thermosyphon-type heat pipe it is important to consider the heat transfer coefficients between the internal working fluid and the wall surface in both the evaporator and the condenser sections (Pioro,1998), where the heat transfer coefficient is defined as

$$h = \frac{\dot{Q}}{A(T_{\text{surface}} - T_{\text{working fluid}})} \tag{1}$$

2.1 Boiling Heat Transfer Coefficient

For boiling, the heat transfer coefficient depends on the boiling regime, of which there are four: natural convection, nucleate, transition and film boiling. Of the four, nucleate boiling is the most desirable boiling regime in practice because high heat transfer rates can be achieved in this regime and also with a relatively small temperature difference between the heated surface and liquid (Cengel 2006). For nucleate boiling, the heat flux is the dominant factor influencing the heat transfer coefficient. The heat transfer rate is also influenced by the physical properties of the working fluid, which may even vary along the axial length of the heat exchanger. In general, nucleate boiling heat transfer coefficient correlations contain a wall heat flux, a temperature term to capture the fluid pressure and density and a fluid property dependent factor, such as the fluid Prandtl number, to capture the effect of thermal conductivity, surface tension and heat capacity (Pioro *et al.* 2004). Thome (2006) stated that experimental results for heat flux $\dot{q}$ and wall superheat ΔT typically fit to an exponential equation of one of the following forms, $\dot{q} \propto \Delta T^{\mathrm{n}}, h_{\mathrm{b}} \propto \Delta T^{\mathrm{n}}$ or $h_{\mathrm{b}} \propto \dot{q}^{\mathrm{n}}$, where n equals 3, 2 or 0.7, respectively.

The Rohsenow's (1962) boiling heat transfer correlation is the most widely used and can be used for any geometry, since it was found that the rate of heat transfer during nucleate boiling is essentially independent of the geometry and orientation of the heater surface; rather, it is more dependent on the bubble pumping. The Rohsenow correlation is

$$\frac{c_p \Delta T_{\mathrm{b}}}{h_{\mathrm{fg}}} = \mathrm{C_{sf}} \left[\frac{\dot{q}}{\mu_{\mathrm{l}} h_{\mathrm{fg}}} \sqrt{\frac{\sigma}{\mathrm{g}(\rho_{\mathrm{l}} - \rho_{\mathrm{v}})}}\right]^{0.33} \left(\frac{c_p \mu_l}{k_{\mathrm{l}}}\right)^{\mathrm{n}} \tag{2}$$

The results obtained using the Rohsenow correlation can be out by about 100% for the heat transfer rate, given a temperature difference between the saturated liquid and the heated surface (or the excess temperature); and given the heat transfer rate may be out by about 30% for the excess temperature. For Dowtherm A, the constant $\mathrm{C_{sf}}$ is 0.00695 and n is 1.72 (Jouhara, 2005).

Stephan and Abdelsalam (1980) formulated a correlation containing four groups of variables using regression techniques for water, refrigerants, organics and cryogens as

$$\frac{h_{\mathrm{b}} d_{\mathrm{bub}}}{k_{\mathrm{l}}} = 0.0546 \left[\left(\frac{\rho_{\mathrm{v}}}{\rho_{\mathrm{l}}}\right)^{0.5} \left(\frac{\dot{q} d_{\mathrm{bub}}}{k_{\mathrm{l}} T_{\mathrm{sat}}}\right)\right]^{0.67} \left[\frac{h_{\mathrm{fg}} d_{\mathrm{bub}}{}^2}{\mathrm{a}_{\mathrm{L}}^2}\right]^{0.248} \left[\frac{\rho_{\mathrm{l}} - \rho_{\mathrm{g}}}{\rho_{\mathrm{l}}}\right]^{-4.33} \tag{3}$$

where d_{bub} is the bubble departure diameter, and is given by

$$d_{\mathrm{bub}} = 0.0146(\beta) \left[\frac{2\sigma}{\mathrm{g}(\rho_{\mathrm{l}} - \rho_{\mathrm{g}})}\right]^{0.5} \tag{4}$$

where β is the contact angle that may be assigned a fixed value of 35°, irrespective of the fluid, and a_{L} is an an experimentally determined parameter. This correlation is for organic fluids and it could be expected that it would also generate accurate results for Dowtherm-A.

Gorenflo (1993) proposed a fluid-specific correlation and included the effect of surface roughness and reduced pressures and contains a working fluid specific and a reference heat transfer coefficient α_o, and is given by

$$h_b = \alpha_o F_{PF} \left(\frac{\dot{q}}{\dot{q}_o}\right)^{nf} \left(\frac{R_p}{r_{po}}\right)^{0.133} \tag{5}$$

where the pressure correction factor F_{PF} is given by

$$F_{PF} = 1.2P_r^{0.27} + 2.5P_r + \frac{P_r}{1-P_r} \tag{6}$$

and

$$nf = 0.9 - 0.3P_r^{0.3} \tag{7}$$

where R_p is the surface roughness and is set at 0.4 μm when unknown. This correlation is applicable for a reduced pressure range from about 0.0005 to 0.95, and is known to be applicable over a wide range of heat fluxes and pressures and according to Thome (2006) is also very reliable.

The Kutateladze et al. (1961, 1966) correlation is applicable to a wide range of conditions (fluid type, heater geometry and surface roughness) and is given by

$$\frac{h_b l_*}{k_l} = 0.44\left(\frac{1\times10^{-4}\dot{q}P}{g h_{fg}\rho_v \mu_l}\frac{\rho_l}{(\rho_l-\rho_v)}\right)^{0.7} Pr^{0.35} \tag{8}$$

2.2 Condensation Heat Transfer Coefficient

The laminar film condensation heat transfer coefficient is determined theoretically by using a force and energy balance over a liquid element within the condensate film flowing around the condenser tube, rather than by generalisations of experimental data (Mills, 1995). It is dependent on the thermal conductivity of the working fluid and the thickness of the condensate film on the condenser tube and is given by

$$h_c = \frac{k_l}{\delta} \tag{9}$$

The film thickness δ is a function of the Reynolds number, liquid and vapour densities, gravity and angle of location on the condenser wall ϕ and is given by

$$\delta = \left(\frac{3}{4}\frac{\rho_l v_l^2}{(\rho_l-\rho_v)g}\frac{Re}{\sin\phi}\right)^{1/3} \tag{10}$$

where the Reynolds number is given as

$$Re = 2.5872\frac{4}{3}\frac{2k_l(T_{sat}-T_s)d}{\mu_l h_{fg}}\left[\frac{4}{3}\frac{(\rho_l-\rho_v)g}{\rho_l v_l^2}\right]^{\frac{1}{3}} \tag{11}$$

Integrating around the periphery of the condenser tube, the average heat transfer coefficient for laminar film condensation of a horizontal tube can be derived, as

$$\bar{h}_c = 0.728\frac{(\rho_l-\rho_v)g h_{fg} k_l^3}{N v_l D(T_{sat}-T_s)} \tag{12}$$

3. EXPERIMENTAL WORK

An experimental heat pipe heat exchanger was designed, built and installed in the laboratory, as shown in Figure 2. As well as allowing for the experimental determination of the heat transfer rates and the boiling and condensation heat transfer coefficients, it also has windows in the front and backside thus allowing visual observation of internal working fluid behaviour. A hot stream flows through the evaporator pipe and heats the liquid and boiling occurs. The so-formed vapour fills the vapour section above the liquid. The vapour then condenses on the condenser tubes and heat is in turn then transferred from the condenser wall, through the tube wall and into the cold stream liquid.

The HPHE consists of a container for the working fluid, a 101.6 mm diameter evaporator tube and a 21.3 mm diameter condenser tube, all of 316L stainless steel, and two large glass windows. Also shown in Fig. 2 is the pipework, flow control valves, pressure relief valve, constant head water supply tank, working fluid sump and vacuum pump. (Not show is the continuous combustion chamber hot gas supply and the mild steel support structures.) K-type thermocouples were used for all the temperature measurements. The thermocouple junctions were positioned at four vertical-plane locations along the axial length of the pipes, 0.058, 0.248, 0.428 and 0.618 m from the hot stream inlet side. For each vertical plane three temperatures measurements were taken around the periphery of the heat exchange pipes, in the liquid next to and just above the evaporator pipe, and two above each other in the vapour space; giving a total of 44. Four thermocouples were used to measure the inlet and outlet temperatures of the heating gas and cooling water. The room temperature was also measured. Mass flow rate of the combustion gases was determined using a bell-mouth located at the gas-burner inlet to which was added the liquefied petroleum gas (LPG) fuel, as measured using an orifice-plate type flow meter. The mass flow rate of the cold water stream from the constant heat tank was determined using a measuring-cylinder and stop-watch.

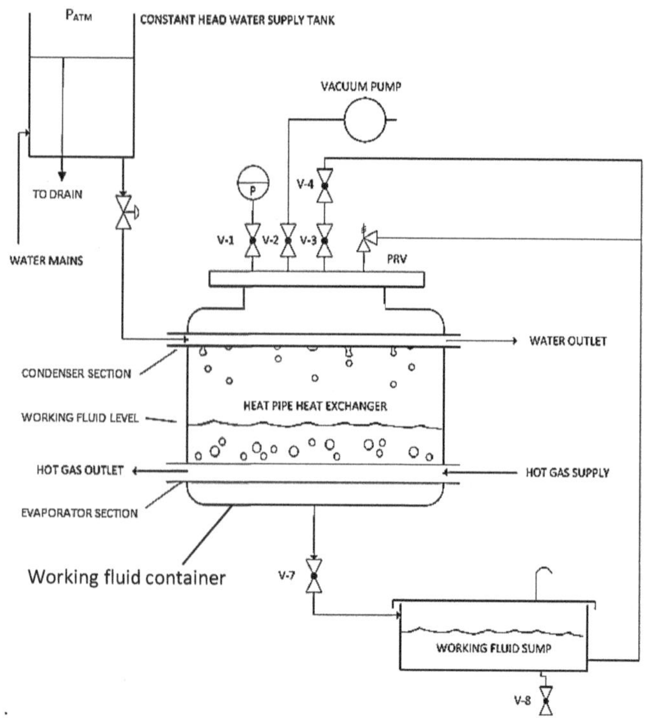

Fig. 2 Schematic layout of experimental setup. The dimensions of the working fluid container are (excluding the flanged instrumentation and observation platforms) 596 mm high, 671 mm wide and 516 mm deep. The hot gas tube diameter is 101.6 mm, whilst that of the cold water pipe in the vapour space is 21.8 mm.

4. RESULTS AND ANALYSIS

During experimentation visual inspection was done on the boiling and condensation processes by observing the evaporator and condenser through a glass window, which was incorporated into the experimental design. Figure 3 shows a photograph of the HPHE evaporator section whilst boiling. When observing the evaporator during operation the following is noted: the working fluid flows mainly radially away from the evaporator surface and the boiling frequency (bubble formation and separation) is observed to be almost constant. Figure 4 shows a photograph of the condenser section during operation. When observing the condensation process the following is noted: all along the length of the condenser droplets forms at the bottom of the condenser tube and when the droplet is large enough it detaches itself from the tube surface and falls into the working fluid pool. It is also noted that there is no droplet formation on the remaining periphery of the condenser tube. The more detailed boiling and condensation experimental procedure will be processed and discussed next.

4.1 Boiling Heat Transfer Coefficient

The experimental boiling heat transfer coefficient was determined by dividing the experimentally determined evaporator wall heat flux $\dot{q}_{in}$ with the difference between the experimentally measured difference between the saturated working fluid temperature and the average wall surface temperature $(\bar{T}_{s\,evap} - \bar{T}_l)$ as

$$h_b = \frac{\dot{q}_{in}}{(\bar{T}_{s\,evap} - \bar{T}_l)} \tag{13}$$

where

$$\dot{q}_{in} = \frac{\dot{Q}_{in}}{2\pi r_{evap\,o} L_{evap}} \tag{14}$$

and

$$\dot{Q}_{in} = \dot{m}_{hot} c_{p\,hot} (T_{hot\,inlet} - T_{hot\,outlet}) \tag{15}$$

where $\bar{T}_{s\,evap}$ is the average of all the evaporator surface thermocouple temperature readings and $\bar{T}_l$ is the average liquid pool temperature in the heat exchanger.

Figure 5 shows the experimentally determined boiling heat transfer coefficients as a function of the wall heat flux; also shown are the four correlations mentioned in the literature study. It is seen that the average experimentally determined heat transfer coefficient is about 5000 ±1000 W/m^{2}K and that it is relatively insensitive to the heat flux. All the correlations gleaned from the literature are all seen to under-predict the experimental values by roughly 3000 W/m^{2}K or about 70%. As this is a relatively large discrepancy an experimentally determined heat transfer coefficient correlation of form $h_b = C\dot{q}_{evap}^m Pr_r^n T_{sat}^z$ was considered. The liquid-surface temperature difference is captured by $\dot{q}_{evap}^m$, the ability of the heat to diffusion through the liquid by the Pr^n, and the fluid properties by T_{sat}^z. The constants C, m, n and z were determined using multivariable linear regression to give an experimentally determined boiling heat transfer coefficients of

$$h_b = 0.485 \times 10^{-4}\, \dot{q}_{evap}^{-0.07675} Pr^{0.686} T_{sat}^{0.9287} \tag{16}$$

Fig. 6 shows that equation 16 predicts the boiling heat transfer coefficients to within ±10% of the experimentally determined boiling heat transfer coefficients.

4.2 Condensation Heat Transfer Coefficient

The experimental condensation heat transfer coefficient was determined using

$$h_c = \frac{\dot{q}_{out}}{(\bar{T}_v - \bar{T}_{s,cond})} \tag{17}$$

where

$$\dot{q}_{out} = \frac{\dot{Q}_{out}}{2\pi r_{co} L_{cond}} \tag{18}$$

where

$$\dot{Q}_{out} = \dot{m}_{cold} c_{p\,cold} (T_{cold\,outlet} - T_{cold\,inlet}) \tag{19}$$

$\bar{T}_v$ is the average vapour temperature in the heat exchanger and $\bar{T}_{s\,cond}$ is the average condenser tube surface temperature. In figure 7 the condensation heat transfer coefficient as calculated by equation 18 is seen to be more or less constant for all the experimental value from 0 to 5000 W/m^{2}K. Assuming a correlation of the form $h_{cond} = CRe^m T_{sat}^n$ the constants C, m and n were also determined using multivariable linear regression to obtain a correlation with $R^2 = 0.927$ of the form

$$h_c = 0.01074 Re^{-2.667} T_{sat}^{4.3} \tag{20}$$

Figure 8 shows that equation 18 better predicts the condensation heat transfer rate than does the so-called Nusselt theory as given by equation 12, but with a relatively wide scatter band of ±70%.

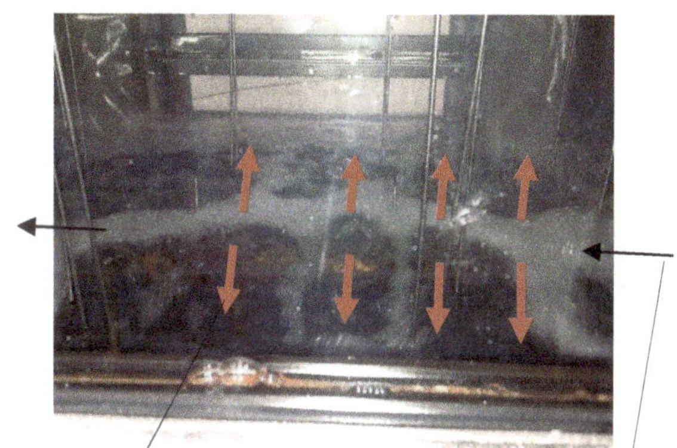

Fig. 3 Photograph of evaporator tube.

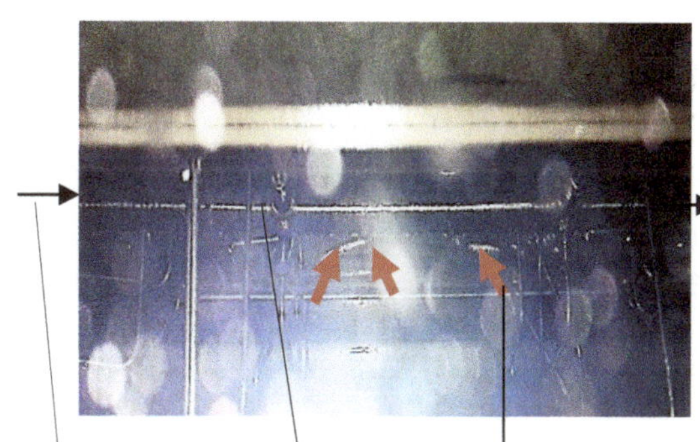

Fig. 4 Photograph of condensation process.

5. CONCLUSIONS

- The boiling heat transfer coefficient of the heat pipe heat exchanger using Dowtherm A was experimentally determined and it was found that the correlations ascribed to Rohsenow, Gorenflo, Stephan-Adelsalam, and Kutateladze and as given by equations 2, 3, 4 and 5, respectively were all some 70% too low. The experimentally determined correlation for boiling heat transfer coefficient as given by equation 16, however, captured the experimental values to within ±10%.

- The theoretically determined condensation heat transfer coefficient assuming film-wise condensation theory and as was calculated using equation 12 did not really correlate with the experimentally determined heat transfer coefficient values. As shown in Fig. 7, for experimentally determined values ranging from 1000 to 5000 W/m^2K, the theoretical value remained more-or-less constant at 1400 W/m^2K. From this it is concluded that traditional film-wise condensation theory should not be automatically be assumed to apply in the case of the condensation process as exhibited in the heat pipe heat exchanger.
- The experimentally determined condensation heat transfer coefficient correlation as given by equation 20 did indeed capture the average experimental values as seen in figure 8 but then only to within 70% of the experimental values.
- The large scatter of the experimental condensation heat transfer coefficient results as exhibited in Fig. 8 can be ascribed to the following affect: as the droplets form at the bottom of the condenser tube the film thickness increases and therefore the condensation heat transfer coefficient value drops. When the droplet detaches itself from the condenser surface the film thickness suddenly decreases and the condensation heat transfer coefficient value therefore rapidly increases. Also liquid tends to collect at the ends of the thermocouple probes measuring the vapour temperature.
- A large amount of carry-over of liquid out of the liquid pool (as the vapour escapes out of the liquid pool) was not noticed. However, as intimated in the objectives in the introduction, the does indeed tend to flow as vapour upwards and liquid downwards in more-or-less vertical planes along the length of the tubes with apparently a minimum of flow parallel to the direction of the heating and cooling tubes. The assumption uniform upwards and downwards flow in formulating the theoretical model (Dobson and Laubscher, 2013; Laubscher, 2013) was thereby indeed validated.

NOMENCLATURE

A	Area, m^2
a	Dimensional experimental variable
c_p	Specific heat, J/kg K
C_{sf}	Rohsenow constant
D	Diamter, m
D	Diamter, m

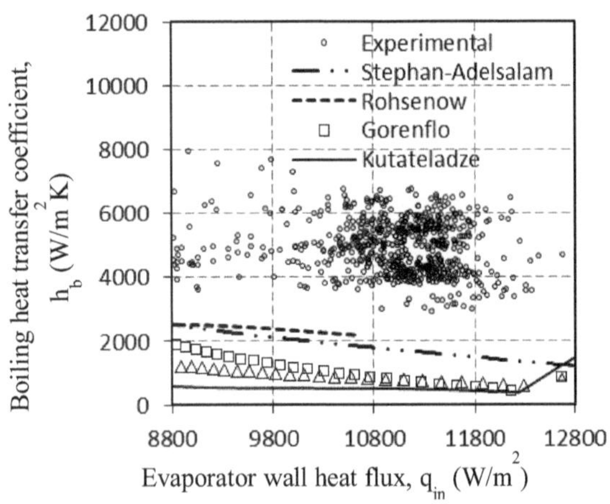

Fig. 5 Experimental and sundry boiling heat transfer coefficient correlations as a function of the evaporator wall heat flux.

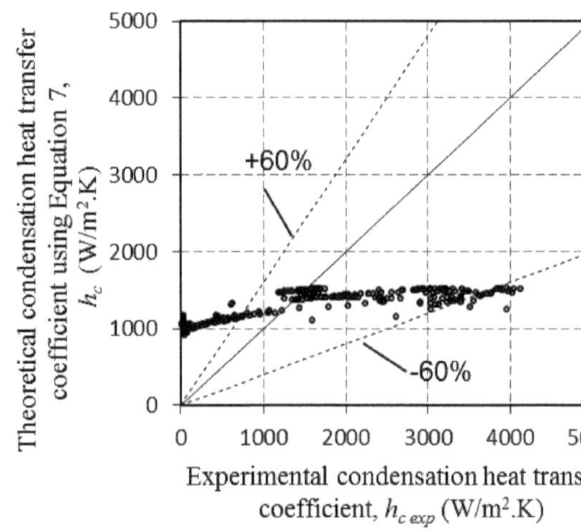

Fig. 7 Theoretical condensation heat transfer coefficient equation using Nusselt theory as a function of the experimental heat transfer coefficient results.

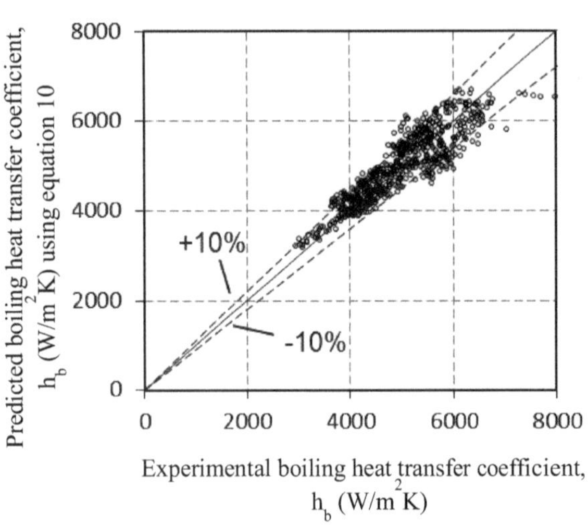

Fig. 6 Predicted boiling heat transfer coefficient (using equation 16) as a function the experimentally determined heat transfer coefficient.

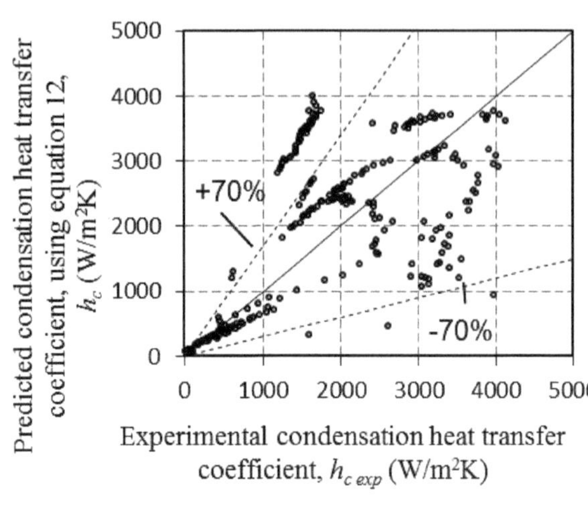

Fig. 8 Correlated (equation 19) condensation heat transfer coefficient equation as a function of the actual experimental heat transfer coefficient results.

F_{PF}	Factor
g	Gravity, m/s^2
h	Heat transfer coefficient, W/m^2K
h_{fg}	Latent heat of vapourisation, J/kg
k	Thermal conductivity, W/m K
L	Length, m
l_*	Characteristic length, m
$\dot{m}$	Mass flow rate, kg/s
nf	Gorenflo coefficient
N	number (of tubes)
P	Pressure, Pa
P_r	Reduced pressure
P_{crit}	Critical pressure, Pa
Pr	Prandtl number, $c_p\mu_l/k$
$\dot{Q}$	Heat transfer rate, W
$\dot{q}$	Heat flux, W/m^2
Re	Reynolds number, $Re = \rho VD/\mu$
R_p	Surface roughness, µm
r	Radius, m
T	Temperature, °C

Greek symbols

α	Heat transfer coefficient, W/m^2K
δ	(Film) thickness, m
ϕ	angle, ° or rad.
σ	Surface tension, N/m
ρ	Density, kg/m^3
μ	Viscosity kg/m.s
v	Kinematic viscosity $\nu = \mu/\rho$, m^2/s

Subscripts

avg	average
b	boiling
C	constant
c	condensing
cold	Cold stream
cond	Condenser
evap	Evaporator
hot	Hot stream
in	Into section
inlet	Inlet to HPHE
int	Internal
l	Liquid
o	Outside
out	Out of section
outlet	Out of HPHE
s	Surface
sat	Saturation
v	Vapour

REFERENCES

Cengel, Y., 2006, *Heat and Mass Transfer*, McGraw Hill, New York.

Dobson, R.T. and Laubscher, R., 2013, "Heat Pipe Heat Exchanger for High Temperature Nuclear Reactor Technology," *Proceedings of the 11th International Heat Pipe Symposium (11th IHPS)*, June 9-12, Beijing, China.

Dow Chemical Company, 1997, Dowtherm A Heat Transfer Fluid: Product technical data. [Online] Available at: http://msdssearch.dow.com/PublishedLiteratureDOWCOM/dh_0030/0901b803800303cd.pdf?filepath=/heattrans/pdfs/noreg/176-01337.pdf&fromPage=GetDoc

Gorenflo, D., 1993, *Pool Boiling*, VDI-Heat Atlas, Dusseldorf, Germany.

Jouhara, H., 2009, "An Experimental Study of Wickless Miniature Heat Pipes Operating in the Temperature Range 200 °C to 450 °C," *Heat Transfer Engineering*, **30**(13), 1041-1048.

Jousse, A., 2007, *Tritium Transport in Very High Temperature Reactors for Hydrogen Production*, Report Number UCBTH-07-005, University of California, 31 August.

Kutateladze, S.S and Borsnansky, V.M., 1966, "A Concise Encyclopaedia of Heat Transfer", Pergamon Press, New York.

Kutateladze, S.S, 1961, "Boiling Heat Transfer", *International Journal of Heat and Mass Transfer*, July, 31-45.

Laubscher, R. and Dobson, RT., 2013, "Theoretical and Experimental Modeling of a Novel Heat Pipe Heat Exchanger for High Temperature Nuclear Reactors Technology," *Applied Thermal Engineering*, In print.

Mills, A.F., *Heat Transfer*, Prentice Hall, New Jersey, 1995.

Mostinskii, I.L., 1963, "Application of the Rule of Corresponding States for Calculation of Heat Transfer and Critical Heat Flux," *Teploenergetika*, **4**, 66.

Pioro, I.L., 1998, "Experimental Evaluation of Constants for the Rohsenow Pool Boiling Correlation," *International Journal of Heat and Mass Transfer*, **29**, January, 2003-2013.

Pioro, I.L., Rohsenow, W. & Doerffer, SS., "Nucleate pool-boiling heat transfer II: assessment of prediction methods," *International Journal of Heat and Mass Transfer*, **23**, August 2004, pp. 5045-5057, 2004.

Rohsenow, W., "A Method of Correlating Heat Trasnfer Data for Surface Boiling of Liquids," Trans. ASME **74**, pp. 979-976, 1952.

Stephan, K. and Abdelsalam, M., 1980 "Heat Transfer Correlations for Natural Convection Boiling," *International Journal of Heat and Mass Transfer*, **23**(1), 73-87.

Thome, J., 2006, *Boiling Heat Transfer on External Surfaces*, Engineering Data Book III, Wolverine Tube Inc., pp. 9-1 and 9-11.

A REVIEW ON HEAT TRANSFER ENHANCEMENT STUDIES OF HEAT PIPES USING NANOFLUIDS

G. Kumaresan*, S. Venkatachalapathy

Department of Mechanical Engineering, National Institute of Technology, Tiruchirappalli, Tamilnadu, India, 620 015

ABSTRACT

Heat pipe is a special type of heat exchanger that transfers large amount of heat due to the effect of capillary action and phase change heat transfer principle. Recent development in the heat pipe includes high thermal conductivity fluids like nanofluids, sealed inside to extract the maximum heat. This paper reviews, influence of various factors such as heat pipe tilt angle, charged amount of working fluid, nanoparticles type, size, and mass/volume fraction and its effect on the improvement of thermal efficiency, heat transfer capacity and reduction in thermal resistance. The nanofluid preparation and the analysis of its thermal characteristics also have been reviewed.

Keywords: heat pipe, nanofluid, thermal resistance, capillary pressure, tilt angle

1. INTRODUCTION

In the emerging world, the field of electronics is one of the fast developing sciences and its contribution to the technology is rapidly growing day by day. During the end of 20th century, most of the electronic devices were larger in size and they had been adopted with fan or micro fin cooling system. These cooling methods occupied considerable volume and did not perform effectively whenever heat dissipation is high and this led to high component temperatures, which affect the performance of electronic devices. Due to the advancement in technology, compact devices were developed to dissipate large amount of heat and one such device is a heat pipe. The heat pipes are suitable devices for the cooling purpose and it was first introduced by Gaugler in 1942. Further developments were made by Groover in 1964 at Los Alamos scientific laboratories (Bejan and Kraus, 2003). The design was further modified and some parameters were changed to improve the performance. These are varying the wick structure (Naphon et al., 2009), base fluids (Senthilkumar et al., 2011), inclination angle (Kiatsiriroat et al., 2000; Naphon et al., 2008) operating pressure (Shafahi et al., 2010; Huminic et al., 2011), charged amount of working fluid (Liu et al., 2011; Mousa et al., 2011), dispersion of nanoparticles in the base fluid (Kiatsiriroat et al.,2000; Mousa et al.,2011), size of particles (Kang et al., 2006; Wang et al.,2010), kind of nanoparticles (Kang et al., 2006; Chen, 2010), mass/volume fraction of nanoparticles (Liu et al., 2011, Teng et al., 2010), heat input (Liu et al., 2011; Do et al., 2010) and geometry (Liu and Zhu., 2011) of heat pipe.

2. HEAT PIPE AND ITS LIMITATIONS

A heat pipe contains three different sections; an evaporator at one end, a condenser at other end and an adiabatic section in-between. Figure 1 shows the schematic arrangement of a heat pipe (Kreith and Bohn, 1997). Heat pipe is basically a sealed tube having a wick structure on the inner surface and filled with a fluid at saturated state. Evaporator is the place, where heat is absorbed by the fluid which creates temperature and thus density difference. In the condenser section, heat is rejected to the surrounding medium. The adiabatic section is externally covered with an insulation layer and it is just acting as a flow passage without any heat losses from the working fluid. The addition and removal of heat in the evaporator and condenser sections respectively, induces a pressure difference thus leading to vapor flow from evaporator to condenser. The liquid is retracted into the evaporator due to the capillary pressure in the wick structure and the process repeats.

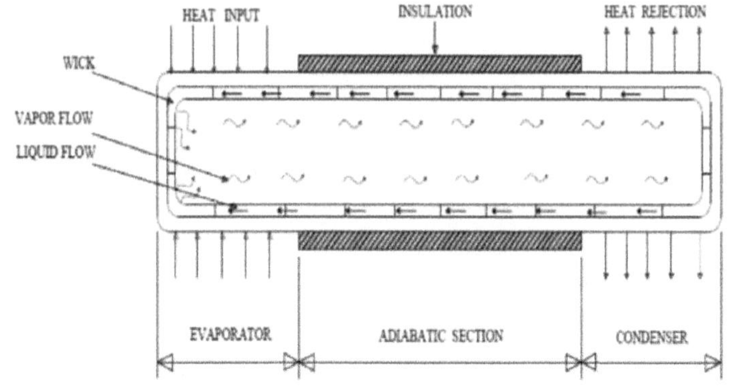

Fig. 1 Heat pipe

The maximum heat transport capacity of a heat pipe is influenced by two limitations; one that leads to heat pipe failure and the other that does not. Limitations that result in heat pipe failure are characterized by insufficient liquid flow to the evaporator for a given heat input, thus resulting in dry out of the evaporator section. The limits categorized under heat pipe failure are capillary limit, boiling limit and entrainment limit. However limitations not resulting in heat pipe failure do require that the heat pipe operate at an increased temperature for an increase in heat input. The three limits are viz. viscous limit, sonic limit and

*Corresponing author E mail : 411111053@nitt.edu

Condenser limit. Capillary pressure is the pressure difference created between the liquid–vapor interfaces that are essential for the energy transportation in the heat pipe. Sometimes, the driving force is insufficient to move the liquid from condenser to evaporator and evaporator dry out may take place, called capillary limit. An efficient heat pipe always maintains the maximum capillary pressure higher than the total pressure losses inside (Bejan and Kraus, 2003). This is given by the following equations.

$$(\Delta P_c)_{max} \geq \Delta P_{tot} \quad (1)$$

where, $\Delta P_{tot} = \Delta P_v + \Delta P_l + \Delta P_g + \Delta P_{ph}$ (2)

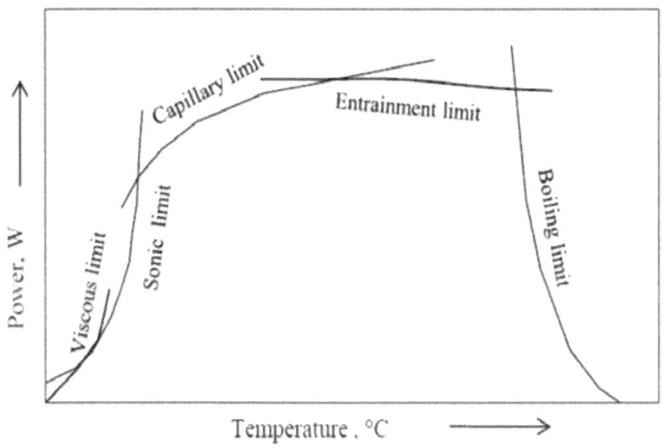

Fig. 2 Limitations of a heat pipe

When the applied heat flux in the evaporator leads to boiling, vapor bubbles are produced in the evaporator which may partially block the liquid flow coming from the condenser. This causes dry out condition in the evaporator, known as boiling limit. As the vapor passes in the counter flow direction to the liquid, high shear forces are developed. This entrains the liquid and resulting in insufficient liquid flow to the wick structure, known as entrainment limit. Operation of heat pipe at low temperatures creates low vapor pressure which may be insufficient to support the increased vapor flow. This condition is called viscous limit. Choking of heat pipe may occur due to low vapor densities and this is the sonic limit of heat pipe. Ideally, the applied heat flux in the evaporator should be equal to heat rejection from the condenser, which is controlled by convection and radiation to the surroundings and this is called condenser limit.

3. NANOFLUIDS AND ITS PREPARATION

In heat transfer applications, conventional fluids like water, oil, refrigerant, etc. are used in heat exchangers, IC engines, refrigerators and air conditioners. Heat transfer capability mainly depends and limited by the thermal conductivity of the working fluid. A method was introduced by Argonne laboratory in 1996 to raise the thermal conductivity of the conventional fluids. In this method nano–sized metallic and non–metallic particles having high thermal conductivity are dispersed in the base fluids (called nanofluids). Some of the commonly used metallic, non-metallic solids and the base fluids and their thermal conductivity values are listed in Table 1.

Thermal conductivity of a fluid can be improved by adding nanoparticles and thus preparation of nanofluid is important. Nanofluid preparation involves two methods: single step and two step method. The single step method is a process that combining the preparation of nanoparticles with the synthesis of nanofluids. Physical vapor deposition, liquid chemical method and chemical reduction method are some of the methods available to prepare the nanofluid by single step method. The fluid which is prepared by this method gives better stability and reduced agglomeration (collection of tiny particles to form a bulk mass that will settle more rapidly). But the single step method can be used only for low vapor pressure fluids. This method does not have a lengthy preparation process. Liu et al. (2006) prepared Cu/water nanofluid by chemical reduction method. Eastman et al. (2001) synthesized copper/ethylene glycol nanofluid through physical method. In the two step method, initially nano–scale sized metals, metal oxides, fiber particles and carbon nanotubes (CNT/NCT) are prepared. The dry powder is produced by various processes like chemical vapor condensation, mechanical alloying, etc. Thereafter, it is dispersed in the base fluids. The agglomeration is high in this method, because of its prolonged stages in the preparation. Xie et al. (2002) prepared aluminum oxide with three different base fluids viz. water, ethylene glycol (EG) and pump oil in two step method.

Table 1 Thermal conductivity of solids and liquids (Eastman et al., 1996)

Component	Material	Thermal Conductivity, W/m K
Metallic solids	Silver	429
	Copper	401
	Aluminum	237
Non-metallic solids	Diamond	3300
	Carbon nanotubes	3000
	Silicon	148
	Alumina (Al_2O_3)	40
Metallic liquids	Sodium @ 644K	72.3
Non-metallic liquids	(values at 300 K)	
	Water	0.613
	Ethylene glycol	0.253
	Engine oil	0.145

3.1 Thermal conductivity of nanofluids

Thermal conductivity is the primary property that influences the heat transport capacity of nanofluids. Most commonly used nanofluids and the percentage of increase in thermal conductivity values compared with the base fluids are presented in Table 2.

Table 2 Increase in thermal conductivity of nanofluid

Combinations	Thermal conductivity increase (%)	References
Cu/H_2O	78	Hwang et al.,2006
Cu/EG	40	Choi et al., 2001
Ag/H_2O	18	Pankaj et al., 2011
CNT/Poly oil	160	Xuan, and Li, 2000
NCT's/EG	30	Murshed et al., 2005
TiO_2/H_2O	30–33	Xinfang Li et al., 2007
Al_2O_3/H_2O	20	Xie et al., 2002

Several methods are available to estimate the thermal conductivity of nanofluids. Among them, Transient Hot Wire (THW) method is widely used by many researchers. Xie et al. (2002) in their work used Transient Hot Wire method to measure the thermal conductivity of Al_2O_3/H_2O nanofluid. Zhang et al. (2007) evaluated the thermal conductivity of Al_2O_3, TiO_2, CuO and CNT particles with water as base fluid by transient short hot wire technique. Murshed (2012) measured the effective thermal conductivity and thermal diffusivity of TiO_2, Al_2O_3 and Al nanoparticles with the varying volume fractions of 1–5% using a transient double hot wire technique. The effective thermal conductivity of a nanofluid can be calculated from the following equation (Li et al., 2009).

$$k_{eff} = \left[\frac{k_p + 2k_b + (k_p - k_b)\phi}{k_p + 2k_b - (k_p - k_b)\phi}\right] k_b \quad (3)$$

The thermal conductivity of nanofluids is found to increase with particle concentration and aspect ratio. Buongiorno et al. (2009) presented the thermal conductivity of different nanofluids in INPBE (International Nanofluid Property Benchmark Exercise). To validate the results, the experiments were conducted in more than two laboratories. For alumina nanoparticle with 1% volume fraction, the thermal conductivity ratio (The ratio of nanofluid thermal conductivity to the base fluid thermal conductivity) was 1.039±0.003. When the volume fraction increased to 3%, the thermal conductivity ratio increased to 1.121± 0.004. The reported thermal conductivity ratio was found to be varying from 1.003±0.008 to 1.204±0.010 for the type of nanofluids and range of volume fractions studied. Higher volume fractions and spherical shape particles gave good thermal conductivity enhancement.

3.2 Viscosity of nanofluids

Viscosity and temperature of any liquid is always interrelated with inverse proportionality. For the nanofluids also, the viscosity primarily depends on the temperature and the influence of particle volume fraction is also significant. If the viscosity is measured using capillary viscometer, the radius of capillary tube becomes an important parameter at higher volume fractions. Li et al. (2002) measured the viscosity of CuO/water nanofluid using capillary viscometer. Ding et al. (2006) investigated and proved that, there is direct proportionality relation between viscosity and particle volume fraction; inverse proportionality relation between viscosity and temperature. Singh et al. (2012) examined the dynamic viscosity and thermal conductivity of CNT–Ethylene glycol nanofluid with concentrations varied between 0.12–0.4 wt%. A rheometer was used to measure the viscosity with differentiating temperature limit of 25–60 °C, the result showed viscosity decreases with increases in temperature and rises with CNT concentrations.

There is no single equation to predict the viscosity of all nanofluids. This is because the different nanoparticles will have different properties and morphology. Various authors have developed equations to predict the viscosity of the nanofluids used in their study. However all formulas have been derived from the basic Einstein's (1906) equation,

$$\mu_{nf} = \mu_{bf}(1 + 2.5\phi) \quad (4)$$

For higher volume fractions, a modified form of this equation was developed by many researchers and Brinkman (1952) proposed the following relation,

$$\mu_{nf} = \mu_{bf}\frac{1}{(1-\phi)^{2.5}} \quad (5)$$

For isotropic suspension of spherical and rigid particles, Batchelor (1977) developed a relation to predict the viscosity,

$$\mu_{nf} = \mu_{bf}\,[1 + 2.5\phi] + 6.5\phi^2 \quad (6)$$

Venerus et al. (2009) used an equation to evaluate the viscosity of diluted suspension fluids.

$$\frac{\mu_{nf}}{\mu_{bf}} = 1 + [\mu]\phi + O(\phi^2) \quad (7)$$

where μ_o/μ_f is the ratio between zero shear rate viscosity of a diluted suspension and liquid phase viscosity and $[\mu]$ is the intrinsic viscosity, given by

$$[\mu] = lim_{\phi \to 0}\left[\frac{\frac{\mu_{nf}}{\mu_{bf}} - 1}{\phi}\right] \quad (8)$$

3.3 Stability of nanofluids

Nanofluid is a mixture of solid particles and a base fluid. Nanoparticles tend to aggregate with the time due to its high surface activity. The settling down of the particles creates obstruction to the flow velocity and clogging may occur particularly in microchannel flows. Sedimentation method is a simple and widely used one to find the stability of nanofluids. Hwang et al. (2006) measured the stability of various nanofluids such as multi–walled carbon nanotube (MWCNT), fullerene, copper oxide, silicon dioxide and silver nanoparticles in different base fluids like DI water, ethylene glycol, oil, silicon oil and poly-α-olefin oil by UV–vis spectrophotometer. The variation of supernatant particle concentration of nanofluid with sediment time was obtained by the measurement of absorption of nanofluids. Li et al. (2007) found that the stability of Cu/H_2O nanofluid was affected by pH value of water and the Cu particles concentrations.

Usually, the nanofluids taken in a transparent container and monitored for a certain time period and the changes are recorded by a digital camera to observe its stability. The fluids are prepared with the metal nanoparticles like Cu, Ag and Al, etc. which has poor stability. In those situations surfactants are mixed with nanofluids and thus the stability is improved. Murshed et al. (2005) used a mixture of Oleic acid and Cetyltrimethylammonium bromide (CTAB) surfactant with TiO_2 – DI water nanofluid to improve its stability. Nanoparticle concentrations, viscosity and pH value of base fluids affect the stability of nanofluids. Peng and Yu (2007) prepared and conducted the stability test on different nanofluids like CuO, Al_2O_3, Cu and Al dispersed in DI water by two step method. The results indicated that the nanoparticle concentration, viscosity and pH value of base fluids mainly affects the stability.

4. THERMAL PERFORMANCE OF HEAT PIPES

The thermal conductivity and heat transfer coefficient are the primary properties which play a vital role in the performance of heat pipes. The thermal conductivity of a heat pipe is very high, several hundred times compared with the best conducting metals of a same size. Kiatsiriroat et al. (2000) investigated the thermal performance of a thermosyphon with different mixture contents of TEG (Triethylene Glycol), Ethanol and water. They found that, the heat transfer rate varies with the TEG concentration in the mixture, because of the increase in critical heat flux compared with the ethanol-water mixture. The rate of heat transfer in a thermosyphon mainly depends on the TEG concentration present in the mixture and its flooding limit. Liu and Zhu (2011) studied the effect of CuO nanofluids on the heat transfer enhancement of a heat pipe with 0.5, 0.8, 1, 1.25, 1.5 and 2 wt.% concentrations. Peak performance was obtained at 1 % mass fraction and showed similar output for both evaporating and condensing heat transfer coefficients. They also found that the heat transfer coefficient was influenced by the operating pressure and best result was obtained at low pressures. The study was conducted with 7.45 kPa, 12.38 kPa and 19.97 kPa and better performance was obtained at 7.45 kPa pressures. Shafahi et al (2010) conducted a numerical study on the heat pipe performance analysis using Al_2O_3, CuO and TiO_2 nanofluids. The heat pipe reached maximum heat transport capacity at 5%, 7% and 15% for Al_2O_3, TiO_2 and CuO nanofluids respectively. When the concentrations exceeded critical level, the heat transport capacity was reduced. Hung et al. (2012) investigated the CuO–water nanofluid heat pipe with varying lengths of 0.3 m, 0.45 m and 0.6 m with a pipe diameter of 9.52 mm. In this study, the charged volume ratio (CVR), heating power and tilt angle were varied. The thermal conductivity was found to be increased with heat input. For the given heat input rates of 20–40 W, the maximum performance was attained at 40 W. The charged volume ratio was varied from 20–80%. The maximum thermal conductivity for a 0.3 m length heat pipe was achieved at a CVR of 20% with weight fraction, tilt angle, and heat input of 0.5 wt.%, 40°, and 40 W respectively. Compared with DI water heat pipe an increase of 22.7% in thermal conductivity was obtained. For 0.45 m length heat pipe, higher thermal conductivity was achieved at 40% of CVR, 1 % of weight fraction, 40° tilt angle and 40 W heating power. The increase in thermal conductivity was around 56.3% compared with DI water. The heat pipe thermal conductivity is calculated from the equation,

$$k_{HP} = \frac{Q}{A\,\Delta T} \tag{9}$$

4.1 Thermal efficiency

Thermal efficiency of a heat pipe is the ratio between the cooling load in the condenser (heat rejection) to the power supplied (heat input) in the evaporator. Naphon et al. (2008) in their investigation found that the thermal efficiency increased with the tilt angle of the heat pipe and the CVR of the working fluid. Introducing metallic nanoparticles with certain percentage of mass/volume fraction in the base fluids improved the heat pipe efficiency. Inclination angle of 60° results in maximum efficiency for DI water. When the DI water was replaced by alcohol with the fluid charged volume ratio remaining same at 66%, the inclination angle reduced to 45° for the maximum efficiency. For nanoparticles volume fraction of 0.1%, the rise in efficiency was 10.6% compared with that of deionized water. Senthilkumar et al. (2010) in their investigation found that the thermal efficiency decreased when the tilt angle exceeds 30° for DI water. For copper–water nanofluid and copper nanoparticles dispersed in the aqueous solution of n–Butanol, the deterioration in thermal efficiency was found for the tilt angles exceeding 45°. Noie et al. (2009) studied the thermal performance of a thermosyphon, filled with Al_2O_3–water nanofluid with volume fractions of 1%, 1.5%, 2%, 2.5% and 3%. The results clearly indicated that the efficiency increases with input power and the increase were large at lower input power and moderate at higher input powers. An efficiency gradient of 14.7% was obtained for the input power 48.4 – 97.1 W and much lower percentage gradient of 2.7% for 146.3 – 195.2 W with the volume fraction remaining same as 2%.

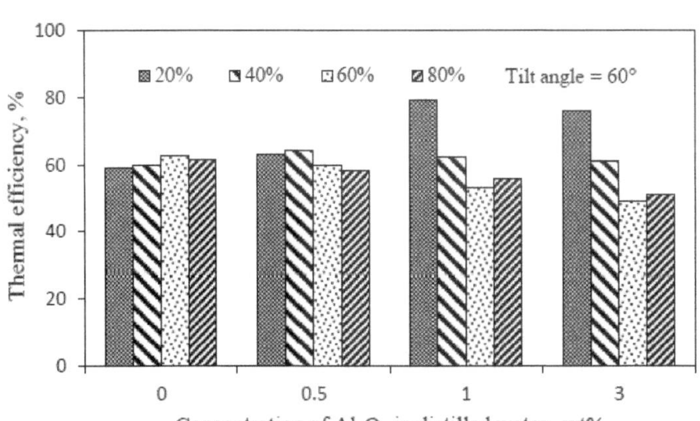

Fig. 3 Thermal efficiency comparison of Al_2O_3 nanofluid with different charged volumes (Teng et al., 2010)

Teng et al. (2010) studied the influence of CVR, tilt angle and weight fraction of Al_2O_3–water nanofluid on the performance of heat pipe. A tilt angle of 60° gave maximum efficiency for all the cases. Figure 3 shows the thermal efficiency variation for various concentrations of Al_2O_3 nanoparticles and different charged volume ratios. A CVR of 20% is preferable at higher concentrations, 1 and 3 wt. %.

4.2 Thermal resistance

Thermal resistance is a measurement that mainly affects the performance of heat pipes. Heat transport capacity depends on the temperature difference between the evaporator and the condenser ends. The nucleate boiling produces vapor bubbles which may block the liquid flow path and stop the heat transfer process. The addition of high thermal conductivity nanoparticles bombard the vapor bubbles and reduces its size. The thermal resistance of each part of the heat pipe is taken into account for calculating its overall thermal resistance. The mean operating temperature of a heat pipe which is required for estimating the transport limit is obtained from the resistance analogy. The overall thermal resistance (Bejan and Kraus, 2003) comprises of many resistances and the most important are saturated liquid–wick axial resistance (~10^{+4} °C/W), pipe wall axial resistance (~10^{+2} °C/W) and saturated liquid–wick resistances for the condenser and evaporator ends (~10^{+1} °C/W). Thermal resistance is defined as the ratio between the temperature difference in evaporator and condenser to the heat supplied.

$$R = \frac{T_e - T_c}{Q} \tag{10}$$

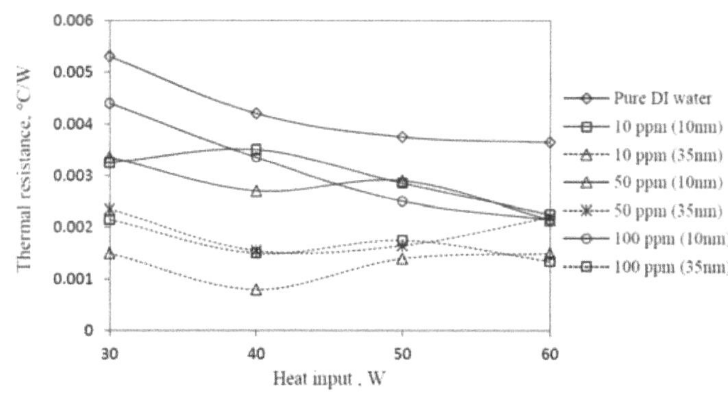

Fig. 4 Thermal resistance comparison of DI water and Ag nanofluids

Figure 4 shows the variation of thermal resistance with heat supplied for Ag nanofluid. The minimum thermal resistance was obtained for 10 ppm concentration and 35 nm size. Size of the nanoparticles affects the thermal resistance of heat pipe. Kang et al. (2006) in their study found that 10 nm size Ag nanoparticles with 1 ppm concentration and 50 W heat input reduced the thermal resistance by 52% compared with DI water. For the same heat pipe with 35 nm size particles and 10 ppm concentration, the thermal resistance was reduced by 81% compared with DI water at 40W input. Senthilkumar et al. (2011) used copper nanoparticle in two base fluids viz. DI water and n–Hexanol. The thermal resistance of copper nanofluid with aqueous solution of n–Hexanol was the lowest and does not vary much with heat input. Huminic, G. and Huminic, A., (2011) found that the thermal resistance reduced with increasing tilt angle in two phase closed thermosyphon using iron oxide nanofluid. The study conducted with different inclination angles, 30°, 45°, 60°, and 90°; iron oxide concentrations of 2.0% and 5.3%. Lowest thermal resistance was obtained at 90° inclination angle and 5.3% volume fraction.

Liu et al. (2011) studied the total heat resistance of a heat pipe. Comparison of pure DI water and 40 nm size Cu nanofluid with 1% volume concentration reduced the resistance. For the working pressure of 7.47 kPa and a heat input less than 85 W, the reduction in total resistance was around 60% with Cu nanofluid. When the heat input exceeded 85 W failure occurred due to dry out condition. Chen (2010) investigated the thermal resistance ratio, which increased with the concentration of the Ag nanoparticles. When the heat input was 40 W and the concentration being 5, 50 and 100 ppm, the ratio were respectively 0.51, 0.69 and 0.71. The thermal resistance ratio is given by,

$$\text{Thermal resistance ratio} = \frac{R_{water} - R_{nanofluid}}{R_{water}} \tag{11}$$

Wang et al. (2010) investigated the steady state operation and transient startup process of a CuO nanofluid heat pipe. Use of nanofluid not only improved the performance during steady operation condition but also reduces the startup time. The reduction in heat resistance was around 50% whereas the heat transport capacity was increased by 40% and the study was conducted with concentrations ranging from 0.5–2.0 wt.%. Huminic et al. (2011) used iron oxide nanoparticles with the concentrations of 2% and 5.3% in a thermosyphon heat pipe. The author reported that the heat transfer increases with increasing nanoparticles concentration and the tilt angle. The improvement in heat transfer was around 39% and 42% for the concentrations of 2% and 5.3% respectively compared with DI water. Four different sizes of gold nanoparticles, 8, 9.3, 15.6 and 21.3 nm were used by Tsai et al. (2004). The results were compared with DI water heat pipe and showed a reduction in heat resistance of 23%, 20%, 37% and 25% respectively. Wei et al. (2005) studied the thermal resistance of Ag nanofluid heat pipe with concentrations of 5, 10, and 15 ppm. A large reduction in heat resistance was observed, in the range of 30–70 % for 60, 80 and 100 W heat input compared with DI water.

Senthilkumar et al. (2010) studied the effect of tilt angle on thermal resistance using DI water, CuO–water nanofluid and CuO dispersed in n–Butanol. For all the heat inputs and tilts angles, lowest thermal resistance was obtained for n-Butanol with CuO as shown in Figure 5. Thermal resistance decreased with increasing heat input and a tilt angle of 60° gave the better performance.

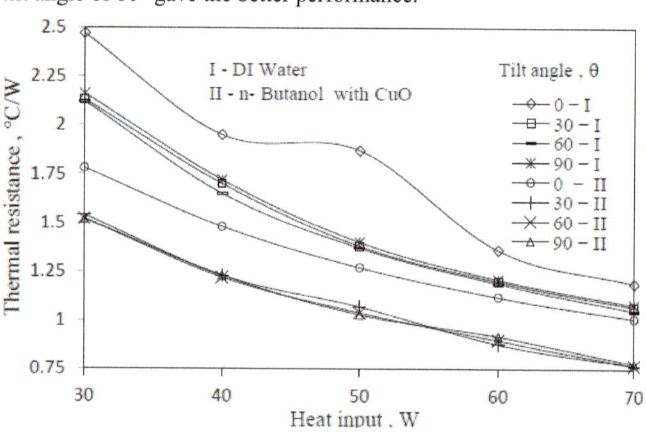

Fig. 5 Effect of tilt angle on DI water and nanofluid heat pipe

4.3 Temperature distribution

Wall temperature distribution plays an important role in the performance of heat pipes. The reducing temperature gradient between the evaporator and condenser section results in improved heat pipe performance. Liu et al. (2011) conducted a performance study on micro grooved heat pipes using different kinds of nanoparticles with water as base fluid. Thermal performance was improved with Cu and CuO nanoparticles while with SiO nanoparticles the performance deteriorated. At higher heat inputs, the performance with nanofluid is better whereas failure occurred for deionized water. Kang et al. (2009) used silver nanofluids of 10 and 35 nm, dispersed in deionized water with the concentrations of 1, 10 and 100 ppm. The heat input for the sintered heat pipe was varied from 30–70 W. For heat input exceeding 50W, dry out occurred with DI water while an increased performance was evident with silver nanofluid up to 70 W. Another study was conducted by the same authors (2006) using silver nanofluid of same size in grooved circular heat pipe. A drop in wall temperature of around 0.5°C was observed with concentration of 1 ppm. However the study was extended with different concentrations 10, 50 and 100 ppm. The wall temperature distribution decreased with increasing concentrations up to 50 ppm and a rise in temperature was observed beyond 50 ppm. Mousa (2011) used the following equation to find the average temperature of evaporator ($\overline{T}_e$) and condenser ($\overline{T}_c$) section. The relation is,

$$\Delta T = \overline{T}_e - \overline{T}_c \tag{12}$$

where,

$$\overline{T}_e = \frac{\sum_{i=1}^{i=N_{te}} T_{ei}}{N_e}, \overline{T}_c = \frac{\sum_{i=1}^{i=N_{tc}} T_{ci}}{N_c} \tag{13}$$

where N_{te} and N_{tc} are the number of thermocouples at the evaporator and condenser ends of the heat pipe.

Do et al. (2010) investigated the temperature distribution of a heat pipe with Al_2O_3 nanofluid for different heat inputs and two volume fractions, of 1% and 3%. The wall temperature was found to increase with heat input for all the cases whereas a reduction was observed with increasing concentrations. A drop in wall temperature of around 26.8°C was found at the evaporator end for 3% volume concentration of nanofluid compared with DI water. Moraveji and Razvarz (2012) used Al_2O_3 nanofluid in a sintered circular heat pipe having a bend of 90° in the adiabatic section. Enhancement in the heat pipe performance was obtained with increasing amount of Al_2O_3 dispersion. Lowest temperature was observed in the vicinity of heat pipe bend due to the impact of the vapor flow.

Figure 6 shows the temperature distribution along the length of a heat pipe with inclined grooves. The heat input and the tilt angle were respectively 80 W and 75°. DI water and CuO nanofluids were used and lower temperatures were obtained along the length of heat pipe for CuO nanofluid (Shafahi et al., 2010).

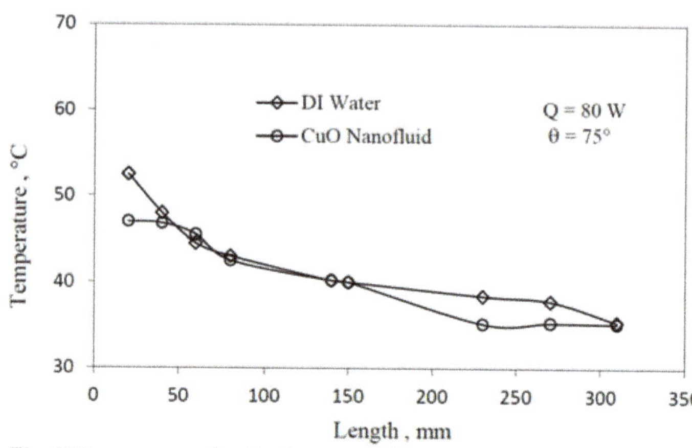

Fig. 6 Temperature distribution comparisons of CuO nanofluid and DI water

5. CONCLUSIONS

The review reports the use of conventional fluids and different nanofluids with varying mass/volume fractions in heat pipes. Nanoparticles like Ag, Au, Cu, CuO and Al_2O_3 were dispersed in various base fluids. Replacing the conventional fluid by nanofluid reduces the dry out problems and enhances the heat transfer capacity. Improvement in thermal efficiency and reduction in thermal resistance is witnessed with increasing mass/volume fraction of nanoparticles in base fluids. Orientation of the heat pipe also affects its thermal performance. The optimum performance is obtained at a tilt angle of around 60° for wick heat pipes and vertical position for thermosyphon heat pipes. Size of nanoparticles and its concentration has a strong influence on the temperature distribution. Effect of heat pipe geometry and its impact on heat transfer characteristics could be explored in future. Uses of hybrid nanofluids in heat pipes have also been deliberated.

NOMENCLATURE

A	area (m^2)
I	current (Ampere)
k	thermal conductivity (W/mK)
Q	heat supplied to the heat pipe (W)
R	thermal resistance (°C/W)
T	temperature (°C)
V	voltage (Volts)
ΔP	pressure drop (N/m^2)
ΔT	temperature difference (°C)

Greek symbols

θ	tilt angle (degree)
μ	viscosity (Ns/m^2)
ϕ	volume fraction

Subscript

b	base
bf	base fluid
c	condenser, capillary
e	evaporator
HP	heat pipe
g	body force/gravity
l	liquid
max	maximum
nf	nanofluid
ph	phase transition
p	particle
te	thermocouples at evaporator section
tc	thermocouples at condenser section
tot	total
v	vapor

REFERENCES

Batchelor, G.K., 1977, "The Effect of Brownian Motion on the Bulk Stress in a Suspension of Spherical Particles," *J. Fluid Mech.*, **83** (1) 97–117.

Bejan, A., and Kraus, A.D., 2003, *Heat Transfer Data Book*, John Wiley & Sons, Inc., 1181–121.

Brinkman, H.C., 1952, "The Viscosity of Concentrated Suspensions and Solution," *Journal of Chemical Physics,* **20**(4) 571–581.

Buongiorno, J., Venerus, D.C., Prabhat, N., McKrell, T., Townsend, J., Christianson, R., Tolmachev, Y.V., Keblinski, P., Hu, L-W., Alvarado, J.L., Bang, I.C., Bishnoi, S.W., Bonetti, M., Botz, F., Cecere, A., Chang, Y., Chen, G., Chen, H., Chung, S.J., Chyu, M.K., Das, S.K., Di Paola, R., Ding, Y., Dubois, F., Dzido, G., Eapen, J., Escher, W., Funfschilling, D., Galand, Q., Gao, J., Gharagozloo, P.E., Goodson, K.E., Gutierrez, J.G., Hong, H., Horton, M., Iorio, C.S., Jarzêbski, A.B., Jiang, Y., Jin, L.W., Kabelac, S., Kamath, A., Kedzierski, M.A., Kim, C., Kim, J.H., Kim, S., Kieng, L.G., Leong, K.C., Manna, I., Michel, B., Ni, R., Patel, H.E., Philip, J., Poulikakos, D., Reynaud, C., Savino, R., Singh, P.K., Song, P., Sundararajan, T., Timofeeva, E., Tritcak, T., Turanov, A.N., Van Vaerenbergh, S., Wen, D., Witharana, S., Yang, C.C., Yeh, W.H., Zhao, X.Z, Zhou, S.Q, 2009, "A Benchmark Study on the Thermal Conductivity of Nanofluids," *J. Appl. Phys.* **106,** 094312.

Chen, Y., 2010, "Experimental Study of Silver Nanofluid on Flat Heat Pipe Thermal Performance," *Journal of Marine Science and Technology,* **18**(5), 731–734.

Choi, S.U.S., Zhang, Z.G., Yu, W., Lockwood, F.E., and Grulke, E.A., 2001, "Anomalous Thermal Conductivity Enhancement in Nanotube Suspensions," *Applied Physics Letter,* **79**(14), 2252–2254.

Collier, J.G., and Thome, J.R., 1996, *Convective Boiling and Condensation*, 3rd ed., Oxford University Press, Oxford.

Ding, Y., Alias, H., Wen, D., and Williams, R.A., 2006, "Heat Transfer of Aqueous Suspensions of Carbon Nanotubes," *International Journal of Heat and Mass Transfer,* **49**, 240– 250.

Do, K.H., Ha, H.J., and Jang, S.P., 2010, "Thermal Resistance of Screen Mesh Wick Heat Pipes using the Water–based Al_2O_3 Nanofluids," *International Journal of Heat and Mass Transfer*, **53**(25–26), 5888–5894.

Eastman, J.A, Choi, S.U.S., Li, S., Thomson, L.J., and Lee, S.,1996, "Enhanced Thermal Conductivity through the Development of Nanofluids," Materials Science Division, Energy Technology Division 9700 S. Cass Avenue Argonne National laboratory Argonne, **457**(3).

Eastman, J.A., Choi, S.U.S., Li, S., Yu, W., and Thomson, L.J., 2001, "Anomalously Increased Effective Thermal Conductivities of Ethylene Glycol–Based Nanofluids Containing Copper Nanoparticles," *Applied Physics Letter,* **78**(6), 718–720.

Einstein, A., 1906, "A New Determination of Molecular Dimensions," *Annalen der physik*, **19**(4), 289–306.

El–Nasr, A.A, and El–Haggar, S.M., Effective Thermal Conductivity of Heat Pipes, 1996, *Heat and Mass Transfer, Springer Verlag,* **32**(1), 97–101.

Hwang, Y., Park, H.S., Lee, J.K., and Jung, W.H., 2006, "Thermal Conductivity and Lubrication Characteristics of Nanofluids," *Current Applied Physics,* **6**(1) 67–71.

Huminic, G., and Huminic, A., 2011, "Heat Transfer Characteristics of a Two–Phase Closed Thermosyphons using Nanofluids," *Experimental Thermal and Fluid Science*, **35**(3), 550–557.

Huminic, G., Huminic, A., Morjan, I., and Dumitrache, F., 2011, "Experimental Study of the Thermal Performance of Thermosyphon Heat Pipe using Iron Oxide Nanoparticles," *International Journal of Heat and Mass Transfer,* **54**(1–3), 656–661.

Hung, Y., Teng, T., and Lin, B., 2012, "Evaluation of the Thermal Performance of a Heat pipe using Alumina Nanofluids," *Experimental Thermal and Fluid Science,* **44**, 504–511.

Kang, S., Wei, W., Tsai, S., and Huang, C., 2009, "Experimental Investigation of Nanofluids on Sintered Heat Pipe Thermal Performance," *Applied Thermal Engineering,* **29**(5–6), 973–979.

Kang, S., Wei, W., Tsai, S., and Yang, S., 2006, "Experimental Investigation of Silver Nanofluid on Heat Pipe Thermal Performance," *Applied Thermal Engineering,* **26** (17–18), 2377–2382.

Kiatsiriroat, T., Nuntaphan, A., and Tiansuwan, J., 2000, "Thermal Performance Enhancement of Thermosyphon Heat Pipe with Binary Working Fluids," *Experimental Heat Transfer,* **13**(2), 137–152.

Kreith, F., and Bohn, M.S., 1997, *Principles of heat transfer*, 5th edition, PWS publishing company, 692–710.

Li, J.M., Li, Z.L., and Wang, B.X., 2002, "Experimental Viscosity Measurement of Copper Oxide Nanoparticles Suspensions," *Tsinghua Sci. Technology*, **7**(2), 198–201.

Liu, M., Lin, M.C., Huang, I., and Wang, C., 2005, "Enhancement of Thermal Conductivity with Carbon Nanotube for Nanofluids," *International communication in heat and mass transfer,* **32**(9), 1202–1210.

Liu, M., Lin, M.C., Tsai, C.Y., and Wang, C., 2006, "Enhancement of Thermal Conductivity with Cu for Nanofluid using Chemical Reduction Method," *International Journal of Heat Mass Transfer*, **49**(17–18), 3028–3033.

Liu, Z.H., and Zhu, Q.Z., 2011, "Application of Aqueous Nanofluids in a Horizontal Mesh Heat Pipe," *Energy Conversion and Management,* **52**(1), 292–300.

Li, X., Zhu, D., and Wang, X., 2007, "Evaluation on Dispersion Behavior of the Aqueous Copper Nano-Suspensions," *J. Colloid Interface Sci.*, **3**(10), 456–463.

Li, Y., Zhou, J., Tung, S., Schneider, E., and Xi, S., 2009, "A Review on Development of Nanofluid Preparation and Characterization," *Powder Technology,* **196**(2), 89–101.

Liu, Z., Li, Y., and Bao, R., 2011, "Compositive Effect of Nanoparticle Parameter on Thermal Performance of Cylindrical Micro–grooved Heat Pipe using Nanofluids," *International Journal of Thermal Sciences,* **50**(4), 558–568.

Liu, Z., Li, Y., and Bao, R., 2010, "Thermal Performance of Inclined Grooved Heat Pipes using Nanofluids," *International Journal of Thermal Sciences*, **49**(9), 1680 –1687.

Moraveji, M.K., and Razvarz, S., 2012, "Experimental Investigation of Aluminum Oxide Nanofluid on Heat Pipe Thermal Performance," *International Communication in Heat and Mass Transfer,* **39**(9), 1444–1448.

Mousa, M.G., 2011, "Effect of Nanofluid Concentration on the Performance of Circular Heat Pipe," *International Journal of Scientific & Engineering Research*, **2**(4), 2229–5518.

Murshed, S.M.S., 2012, "Simultaneous Measurement of Thermal Conductivity, Thermal Diffusivity, and Specific Heat of Nanofluids," *Heat Transfer Engineering,* **33**(8), 722–731.

Murshed, S.M.S., Leong, K.C., and Yang, C., 2005, "Enhanced Thermal Conductivity of TiO_2 Water Based Nanofluid," *International Journal of Thermal Sciences,* 44(4), 367–373.

Naphon, P., Assadamongkol, P., and Borirak, T., 2008, "Experimental Investigation of Titanium Nanofluids on the Heat Pipe Thermal Efficiency," *International Communications, in Heat and Mass Transfer,* **35**(10), 1316–1319.

Naphon, P., Thongkum, D., and Assadamongkol, P., 2009, "Heat Pipe Efficiency Enhancement with Refrigerant Nanoparticles Mixtures," *Energy Conversion and Management,* **50**(3), 772–776.

Noie, S.H., Heris, S.Z, Kahani, M., and Nowee, S.M., 2009, "Heat Transfer Enhancement using Al_2O_3–water Nanofluid in a Two–Phase Closed Thermosyphon," *International Journal of Heat and Fluid Flow,* **30**(4), 700–705.

Peng, X., Yu, X., 2007, "Influence Factors on Suspension Stability of Nanofluids," *J. Zhejiang Univ.: Engineering. Science,* **41,** 577–580.

Senthilkumar, R., Vaidyanathan, S., and Sivaraman B., 2011, "Experimental Analysis of Cylindrical Heat Pipe Using Copper Nanofluid with an Aqueous Solution of n–Hexanol," *Frontier in Heat Pipes,* **2**,033004.

Senthilkumar, R., Vaidyanathan, S., and Sivaraman B., 2010, "Performance Analysis of Heat Pipe Using Copper Nanofluid with Aqueous Solution of n–Butanol," *International Journal of Mechanical and Materials Engineering*, **1**(4), 251–256.

Shafahi, M., Bianco, V., Vafai, K., and Manca, O., 2010, "An Investigation of the Thermal Performance of Cylindrical Heat Pipes using Nanofluids," *International Journal of Heat and Mass Transfer,* **53**(1–3), 376–383.

Sharma, P., Baek, I., Cho, T., Park, S., and Lee, K.B., 2011, "Enhancement of Thermal Conductivity of Ethylene Glycol Based Silver Nanofluids," *Powder Technology,* **208**(1), 7–19.

Singh, N., Chand, G., and Kanagaraj, S., 2012, "Investigation of Thermal Conductivity and Viscosity of Carbon Nanotubes–Ethylene Glycol Nanofluids," *Heat Transfer Engineering,* **33**(9), 821–827.

Teng, T., Hsu, H., Mo, H., and Chen, C., 2010, "Thermal Efficiency of Heat Pipe with Alumina Nanofluid," *Journal of Alloys and Compounds,* **504**(1), S380–S384.

Tsai, C.Y., Chien, H.T., Ding, P.P., Chan, B., Luh, T.Y., and Chen, P.H., 2004, "Effect of Structural Character of Gold Nanoparticles in Nanofluid on Heat Pipe Thermal Performance," *Materials Letters*, **58**(9), 1461– 1465.

Venerus, D.C., Buongiorno, J., Christianson, R., Townsend, J., Bang, I.C., Chen, G., Chung, S.J., Chyu, M., Chen, H., Ding, Y., Dubois, F., Dzido, G., Funfschilling, D., Galand, Q., Gao, J., Hong, H., Horton, M., Hu, L., Iorio, C.S., Jarzebski, A.B., Jiang, Y., Kabelac, S., Kedzierski, M.A., Kim, C., Kim, J., Kim, S., McKrell, T., Ni., R., Philip., J., Prabhat, N., Song, P., Vaerenbergh, S.V., Wen, D., Witharana, S., Zhao, X., Zhou, S., 2010, "Viscosity Measurements on Colloidal Dispersions (Nanofluids) for Heat Transfer Applications," *Applied Rheology*, Vol. **20,** 44582.

Wang, G., Song, B., and Liu, Z., 2010, "Operation Characteristics of Cylindrical Miniature Grooved Heat Pipe using Aqueous CuO Nanofluids," *Experimental Thermal and Fluid Science,* **34**(8), 1415–1421.

Wei, W., Tsai, S., Yang, S., and Kang, S., 2005, "Effect of Nanofluid on Heat Pipe Thermal Performance," *Proc. 3rd IASME/WSEAS Int. Conf. on Heat Transfer, Thermal Engineering and Environment,* Corfu, Greece, 115–117.

Xie, H., Wang, J., Xi, T., Liu, Y., Ai, F., and Wu, Q., 2002, "Thermal Conductivity Enhancement of Suspensions Containing Nano sized Alumina Particle," *Journal of Applied Physics*, **9**(7), 4568 – 4572.

Xuan, Y., and Li, Q., 2000, "Heat Transfer Enhancement of Nanofluids," *International journal of Heat Mass Transfer,* **21**(1), 58–64.

Zhang, X., Gu, H., and Fujii, M., 2007, "Effective Thermal Conductivity and Thermal Diffusivity of Nanofluids Containing Spherical and Cylindrical Nanoparticles," *Experimental Thermal and Fluid Science,* **31**(6), 593–599.

SEPARATE-TYPE HEAT PIPE SOLAR RECEIVERS FOR CONCENTRATING SOLAR POWER

Yiding Cao*

Department of Mechanical and Materials Engineering, Florida International University, Miami, Florida 33174

ABSTRACT

This paper introduces separate-type heat pipe (STHP) based solar receiver systems that enable more efficient operation of concentrated solar power plants without relying on a heat transfer fluid. The solar receiver system may consist of a number of STHP modules that receive concentrated solar flux from a solar collector system, spread the high concentrated solar flux to a low heat flux level, and effectively transfer the received heat to the working fluid of a heat engine to enable a higher working temperature and higher plant efficiency. In general, the introduced STHP solar receiver has characteristics of high heat transfer capacity, high heat transfer coefficient in the evaporator to handle a high concentrated solar flux, non-condensable gas release mechanism, and lower costs. The STHP receiver in a solar plant may also integrate the hot cold tank based thermal energy storage system without using a heat transfer fluid.

Keywords: *High-temperature separate-type heat pipe (STHP), solar receiver, concentrated solar power plant*

1. INTRODUCTION

The concentrated solar power (CSP) generation system is a viable technology for conversion of solar energy into mechanical power and then into electricity through a generator. The solar power system generally includes a collector system that redirects or concentrates solar radiation from the sun onto a solar receiver system, which converts the received solar radiation into thermal energy. The thermal energy is then used as a heat source in a heat engine that converts the received thermal energy from the receiver system into mechanical power. Various CSP systems have been developed with additional advantages of integrating thermal energy storage (TES) mechanisms (Mancini et al., 1994; Stine and Diver, 1994; Lovegrove et al., 2007, Yang et al., 2009; Nithyanandam and Pitchumani, 2012a and 2012b; and Tao and Rayegan, 2011).

Because of economies of scale to compete with traditional fossil-fuel based utility industries, large-scale solar thermal power plants, such as solar tower and trough plants, have been gaining momentum in both governmental and private investments and developments (Tao and Rayegan, 2011). For large-scale CSP plants, a central solar receiver system is often employed in conjunction with a heat engine such as a steam turbine engine or gas turbine engine. Currently, a heat transfer fluid (HTF), such as a molten salt, is employed to receive solar energy and transport it from the receiver to the engine for the conversion of the received thermal energy to mechanical work. However, the maximum operating temperature of commonly used heat transfer fluids may be limited. At a high operating temperature, the heat transfer fluid may be chemically broken down. This limited operating temperature may inevitably affect the performance of the power plant as the power output and thermal efficiency of the related heat engine are highly dependent upon the temperature level of the heat source. Because of the relatively low operating temperature, the successful demonstrations of large-scale solar thermal power plants are primarily based on steam-turbine power plants that produce power at lower peak temperatures. In addition to the lower energy efficiency due to the limited peak working temperature of a heat transfer fluid, the amount of water consumption associated with wet cooling condenser may be unacceptable for the construction of a solar power plant in a desert. Alternative power systems, such as the advanced supercritical carbon dioxide power cycles, have been proposed (Ma, et al., 2012) to mitigate these problems.

The difficulties associated with the heat transfer fluid and water consumption of wet cooling may favor a gas turbine based power plant without relying on the use of a heat transfer fluid. First, without the constraint imposed by the heat transfer fluid, the gas turbine may work at an increased peak temperature to attain a substantially improved thermal efficiency as compared to the counterpart of a steam-turbine based power plant. Second, the gas turbine power plant may remove the need for wet cooling and be constructed in a location where water resource is limited. Additionally, the gas turbine power plant may incorporate compressor inter-cooling, reheat, and energy recovery from the exhaust stream, and could attain an overall thermal efficiency well above 50l even under a moderately high peak working temperature.

* Email: caoy@fiu.edu

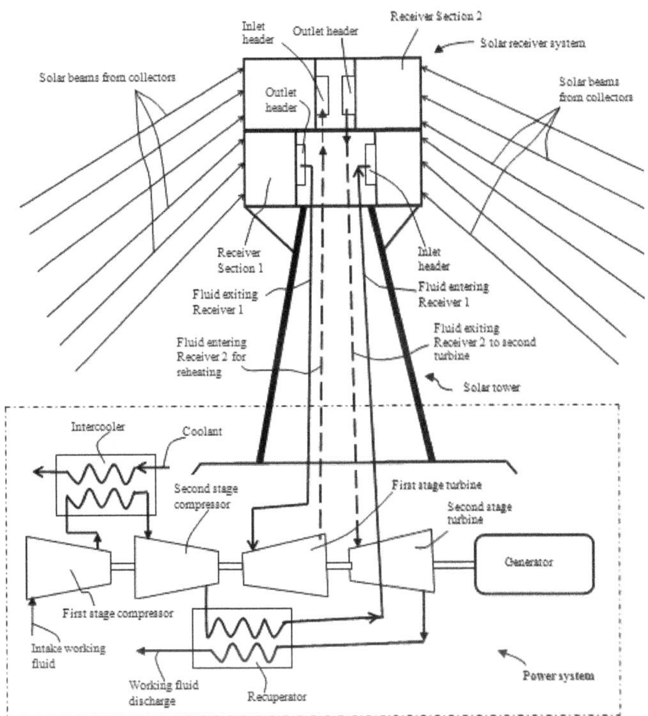

Fig. 1 Schematic of a solar tower power plant integrating compression intercooling, reheat, and recuperative techniques

Figure 1 illustrates schematically a concentrated solar tower power plant based on a gas turbine engine system. The power plant includes a central solar receiver that may be divided into two sections (sections 1 and 2), which are mounted on a solar tower. The receiver receives concentrated sunlight and coverts it into heat from a collector system (not shown) that redirects solar radiation from the sun onto the solar receiver. The gas turbine system includes compression intercooling, reheat, and recuperative techniques to increase its thermal efficiency. As illustrated of Figure 1, a working fluid enters a first stage compressor and exits the compressor at an outlet and then enters an intercooler. The working fluid flows through heat exchanger and transfers an amount of heat to a coolant stream. With a reduced temperature, the working fluid stream exits the intercooler and enters a second stage compressor. Upon exiting the second stage compressor, the further compressed stream enters a recuperator where it receives an amount of heat from an exhaust stream before the exhaust stream is discharged out of the recuperator. Having received the heat from the recuperator and raised its temperature, the working fluid stream exits the recuperator and flows up the tower and enters an inlet header of receiver 1. After acquiring an amount of heat from the receiver 1, the stream exits the receiver 1 through an outlet header and enters a first stage expander or turbine. The stream expands through the turbine and produces mechanical power. The produced mechanical power may be used to drive the compressors or to generate electricity through a generator. After expansion through the first stage turbine, the stream leaves the turbine and flows up the tower again into a second solar receiver through an inlet header, where it is reheated. The reheated stream exits the second receiver through an outlet header and flows down the tower and then enters a second stage turbine. The gas stream expands through the second turbine and produces mechanical power. After expansion through the second turbine, the stream exits the second turbine and becomes an exhaust stream. The exhaust stream then flows through the recuperator, transferring an amount of heat to the engine working fluid, and finally it is discharged out of the recuperator. For the gas turbine working in an open cycle, the exhaust stream is eventually released into the ambient. However, for closed cycles, such as a CO_2 or steam cycle, the discharged stream may return to the first stage compressor after a heat exchanger (not shown).

The solar tower power plant as illustrated above may represent one of the most attractive solar power generation systems from the view point of thermodynamics, as it can reach a thermal efficiency above 50l without requiring an extremely high peak working temperature (Moran, et al., 2009). However, two critical issues must be resolved before the system could be practically built. The first issue is that the receiver must be able to work at a high temperature and handle a high solar flux.

To attain a higher receiver efficiency of a solar receiver, the fraction of radiation loss from the receiver wall surface to the ambient should be reduced, particularly when the receiver surface temperature is high. The solar receiver efficiency may be approximately expressed by the following relation for a flat receiver:

$$\eta = 1 - \frac{\varepsilon_s \sigma (T_s^4 - T_{surr}^4)}{\alpha q_c''} \tag{1}$$

where ε_s is the emissivity of the receiver surface, σ is the Stefan-Boltzmann constant, T_s is the receiver surface temperature, T_{surr} is the ambient temperature, α is the absorptivity to the solar flux and q_c'' is the concentrated solar flux reaching the receiver surface. From the above equation, it may be seen that for given emissivity and absorptivity, a higher surface temperature may significantly increase the radiation loss, resulting in lower receiver efficiency. However, a higher surface temperature is generally desirable for a higher energy efficiency of the power plant. To maintain the receiver efficiency at an acceptably high level, the concentrated solar flux q_c'' must be increased. However, this increased solar flux (In some cases on the order of 50 to 100 W cm^2) in conjunction with the high working temperature may cause safety problem of the solar receiver for a power system of a gaseous working fluid. It is believed by the author of this paper that the high heat flux must be spread to a sufficiently low heat flux level before the received heat is transferred to the gaseous working fluid of the engine.

The second critical issue is the integration of thermal energy storage (TES) mechanisms with the power plant. The successful integration of the thermal energy storage in a heat transfer fluid based power plant is a bright spot among all solar power plants including PV and CSP power plants. The question is how the thermal energy storage mechanism may be integrated to power plant without the use of a heat transfer fluid.

2. SOLAR RECEIVERS BASED ON SEPARATE-TYPE HEAT PIPES

The heat pipe, particularly high-temperature heat pipe using a liquid metal as the working fluid, is a unique and simple heat transfer device that could work at a high temperature and handle a relatively high heat flux (Faghri, 1995; Cao and Faghri, 1991, 1992, and 1993; Ling and Cao, 2001; and Cao and Wang, 1995a and 1995b). It is a passive heat transfer device that could have an effective thermal conductance hundreds of times higher than the conductivity of copper. Therefore, intensive heat flux can be spread to a much larger surface area having a heat flux commensurate with that an engine working fluid could tolerate. Although a capillary effect may be employed to return the condensate from the condenser section to the evaporator section, the gravitational force should be employed in the present application to reduce the cost and increase the reliability. However, the use of traditional tubular heat pipes as a solar receiver may be impractical for large-scale power plants. In this case, hundreds or thousands of individual heat pipes may be needed, which could increase the power plant costs to an unacceptably high level. Additionally, the countercurrent liquid vapor flows may significantly limit the heat transfer capacity of the traditional heat pipes. For these reasons, the concept of low-temperature separate-type heat pipes (Jin and Chen, 1986 and Chen et al., 1986) that are used for industrial waste heat recovery may be borrowed and adapted for the present high-temperature solar applications (Cao, 2012a).

It is a common practice in a waste energy recovery field that a separate-type heat pipe (STHP) or gravity-assisted loop heat pipe may be employed as a large-scale heat exchanger unit (Jin and Chen, 1986 and Chen et al., 1986). Likewise, a receiver module based on high-temperature separate-type heat pipes (STHPs) or gravity-driven loop heat pipes may be employed, which may have the benefits of receiving and transferring a large amount heat at high temperatures with only a few heat-pipe modules required. In a particular situation, a receiver wall may constitute a single STHP module even for a relatively large power plant. Figure 2 schematically illustrates a proposed STHP based receiver module that includes a evaporator having a liquid drum at the bottom and a vapor drum at the top, and a finned condenser heat exchanger having a liquid drum at the bottom and a vapor drum at the top (Cao, 2012a). The vapor drum of the condenser is connected with the vapor drum of the evaporator through a vapor line while the liquid drum of the condenser is connected to the liquid drum of the evaporator through a liquid line to form a looped heat pipe system. The condenser heat exchanger is positioned at a sufficiently higher elevation as compared to the elevation of the evaporator to provide a sufficiently large gravitational driving force for the operation of the heat pipe.

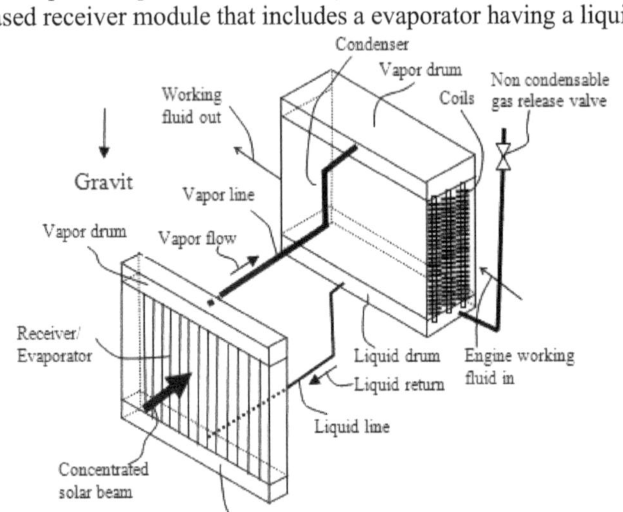

Fig. 2 Schematic of a high-temperature separate-type heat pipe module as a solar receiver unit (Cao, 2012a).

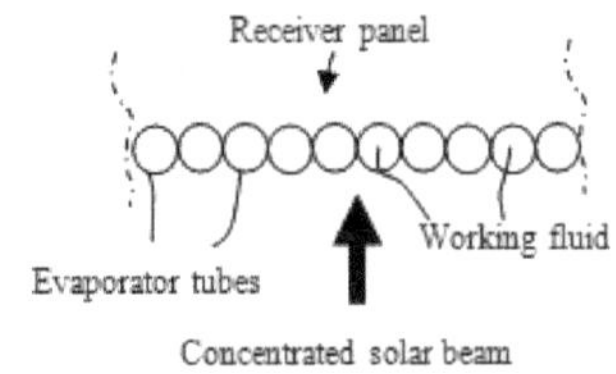

Fig. 3 Schematic of a receiver panel consisting of evaporator tubes between the liquid and vapor drums.

During the operation, the evaporator receives concentrated solar flux and converts it into heat that is then transferred to the working fluid inside the evaporator through boiling evaporation forced convection. The vapor generated inside the evaporator flows upward to the vapor drum and then through the vapor line into the vapor drum of the condenser, where it is redistributed to the interior spaces of finned tubes (the fins are optional depending on specific applications). The vapor flows downwardly inside the tubes and is condensed by transferring the heat to the working fluid of an engine system (not shown). The condensate is accumulated in the liquid drum and flows downwardly under the influence of gravity through the liquid line to the liquid drum of the evaporator, to supply liquid for boiling evaporation forced convection in the evaporator, and the aforementioned cycle is repeated.

Although the construction of the evaporator may have a number of options, for a large solar receiver, an evaporator that comprises a number of evaporator tubes between the vapor drum and the liquid drum may be employed. Figure 3 shows the cross-section of a tubes panel that forms the evaporator in Fig. 2. The panel consists of an array of tubes that are connected to the vapor drum at the top and to the liquid drum at the bottom and filled with an amount of heat pipe working fluid. For high-temperature operations, established working fluids are sodium that has a working temperature range of 900-1450 K and lithium that has a working temperature range of 1300-2000 K. These two working fluids have been employed and extensively tested in conventional heat pipes. The filling amount of sodium or lithium into the STHP is also very small as compared to the entire size of the STHP module, primarily a fraction of the interior volume of the evaporator.

Some of the biggest advantages of the separate-type heat pipe (STHP) system as compared to other types of heat pipes are large heat transfer loads and the flexibility of deploying the evaporator or condenser. In this case, the evaporator may be attached to a frame structure of a solar tower (not shown) and forms the outmost surface of the tower to receive concentrated solar flux from a collector system, while the condenser may be anchored to an inner structure of the tower (not shown) to facilitate the heat transfer from the condenser surface to the working fluid of a heat engine.

Although the evaporator of a single STHP module may provide a relatively large solar receiver wall and receive a large amount solar energy over 1000 kW, in some cases the size of a single evaporator may be still unable to accommodate the size of the collector or to meet the requirement for the heat load of the solar receiver system. For these reasons, a solar receiver wall in a receiver system may be constructed by a number of STHP solar receiver modules, as shown in Fig. 4 (Cao, 2012a). Individual

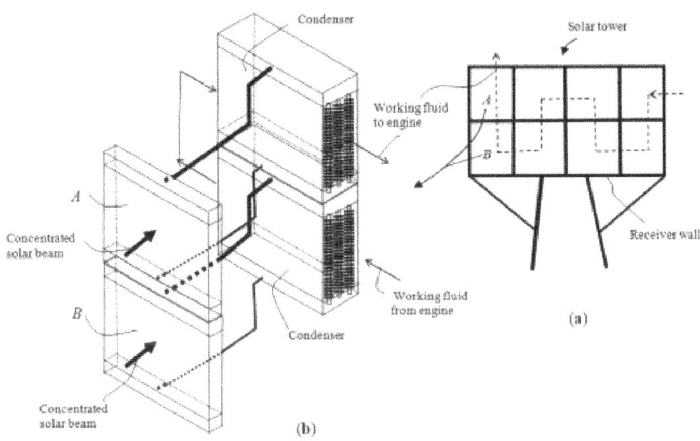

Fig. 4 Stacking arrangement of individual STHP modules (4a: A schematic frontal view of a receiver wall consisting of a number of STHP evaporators; 4b: Two STHP receiver modules with their evaporators being stacked one on top of the other)

Individual modules may be simply assembled side by side to form a wider receiver wall. Additionally, the evaporators of individual heat pipe modules may be stacked one on top of another, such as the stacking arrangement of evaporators *A* and *B* shown in Fig. 4b, to form a taller receiver wall. It may also be important that the gaps or inter-space in Fig. 4a and Fig. 4b between adjacent evaporators be minimized to form an integral receiver wall. Although the concentrated solar flux striking the receiver wall may be substantially uniform, the wall surface temperature may vary substantially, as the wall temperature is largely controlled by the engine working fluid temperature in the condenser. For this reason, the schematic serpentine arrangement of the engine working fluid in Fig. 4a could result in peak temperatures only over the last few heat pipe evaporators, which could substantially reduce the demand for high-temperature alloy materials for the construction of the receiver wall.

It should be noted that the vapor drum of an individual evaporator, as shown in Fig. 2, may be subjected to intensive concentrated solar flux once it is integrated with other evaporators to form a larger receiver wall, such as that in Fig. 4a. In this case, certain working fluid overcharging in the evaporator may be needed to avoid overheating the evaporator drum. However, this overcharging may result in a vapor-liquid two-phase mixture within the vapor drum, which may affect the heat transfer efficiency of the heat pipe system. To mitigate this problem, the vapor drum may be spaced back away from the evaporator and connected to the evaporator through a transition region as shown in Fig. 5 (Cao, 2012a). With this arrangement, the stacking of individual evaporators will not be affected and potential overheating of the vapor drum is avoided.

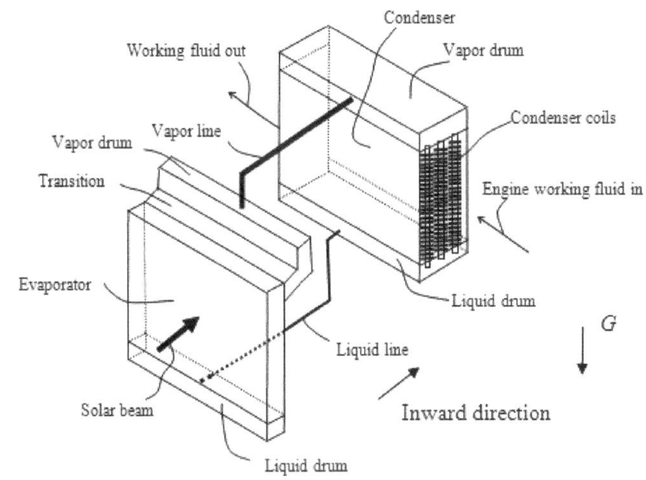

Fig. 5 A schematic view of a separate-type heat pipe receiver module with the vapor drum of the heat pipe evaporator being spaced away from the evaporator

The adoption of the separate-type heat pipe as the solar receiver has the following distinct advantages:

- **High heat transfer capacity**. In a conventional longitude heat pipe that combines the evaporator and condenser into a single pipe, the vapor flow and liquid return are countercurrent. When the heat transfer load is high accompanied by a high vapor velocity from the evaporator to condenser, this high vapor velocity may retard the condensate return from the condenser to the evaporator and cause a so-called entrainment limit. In a separate-type heat pipe, however, the vapor flow and liquid return are concurrent and the entrainment limit mentioned above no longer exists. As a result, the heat transfer capacity can be significantly increased. In low-temperature waste-heat recovery industrial applications using water as the working fluid, a heat transfer rate of more than 2000 kW per STHP module is not unusual.
- **High heat transfer coefficient in the evaporator.** In a STHP evaporator, the heat transfer is the combination of boiling and forced convection. While the boiling heat transfer coefficient may not be exceptionally high, having a similar magnitude to that of water pool boiling, the forced convection component may be significantly enhanced, which may be many times higher than the boiling component. The combined high heat transfer coefficient is essential to handle a high solar flux for increased receiver efficiency. Additionally, the heat transfer limitation using sodium as the working fluid is related to the critical heat flux of sodium pool boiling that may be above 200 W cm^2 (Noyes, 1963).

- **Non-condensable gas release valve.** One problem that may be encountered by a high-temperature heat pipe is the non-condensable gas generation inside the heat pipe. For a conventional heat pipe, the gas generation over time may reduce the heat transfer capability substantially. For a STHP, a non-condensable gas release valve may be installed, as shown in Fig. 2. The non-condensable gas generated may be periodically discharged out of the separate-type heat pipe, so that its performance can be maintained over a long period of time, which significantly reduces the maintenance costs.
- **Simple structure and no pumping power requirement**. The STHP has a very simple structure without the need for a wick structure. Additionally, the heat pipe is driven by gravity, which eliminates the need for a pumping system and related power consumption for the receiver. By using a tube panel structure as shown in Fig. 3, a receiver wall of 6×6 m^2 may be constructed by using a few STHP modules. The amount of the high-temperature metal used to construct the STHP receiver should be close to that required in a heat transfer fluid based solar receiver, because of similar receiver size and the heat transfer surface areas in the heat exchanger between the heat transfer fluid and the engine working fluid.
- **High heat flux spreading.** The condenser inside surface area may be 10 times higher than that of the evaporator. With the installation of the fins outside of the condenser tubes (Fig. 2), the condenser heat transfer surface area may be 100 times higher than that of the evaporator. Therefore, the 100 W cm^2 heat flux in the receiver may be spread to 1 W cm^2 in the condenser. With this low heat flux, the flow velocity of the engine working fluid in the condenser may be sufficiently low to minimize the loss of pressure head. This reduced pressure loss is essential to a gas turbine engine, as its thermal efficiency is directly related to the gas pressure before entering the turbine [Moran, et al., 2009]. The liquid-metal working fluid charge into the heat pipe is also very small, as it may be a fraction of the evaporator volume that is less than 10 percent of the entire heat pipe interior volume.

3. INTEGRATION OF THERMAL ENERGY STORAGE (TES) MECHANISM WITH THE SOLAR POWER PLANT

The above introduced STHP based solar receiver does not employ a heat transfer fluid. However, the hot cold tanks mechanism that has been successfully used in heat transfer fluid based solar power plants may also be employed for a solar power plant without the use of a heat transfer fluid. In this case, the working fluid of the heat engine may replace the functionalities of the heat transfer fluid to carry energy from the receiver to the storage tank or carry energy from the storage tank to the heat engine. Figure 6 illustrates schematically a solar tower power plant based on a simple gas turbine cycle and integrating a hot cold tank TES system. In this case, before entering the turbine, the hot engine working fluid that carries thermal energy

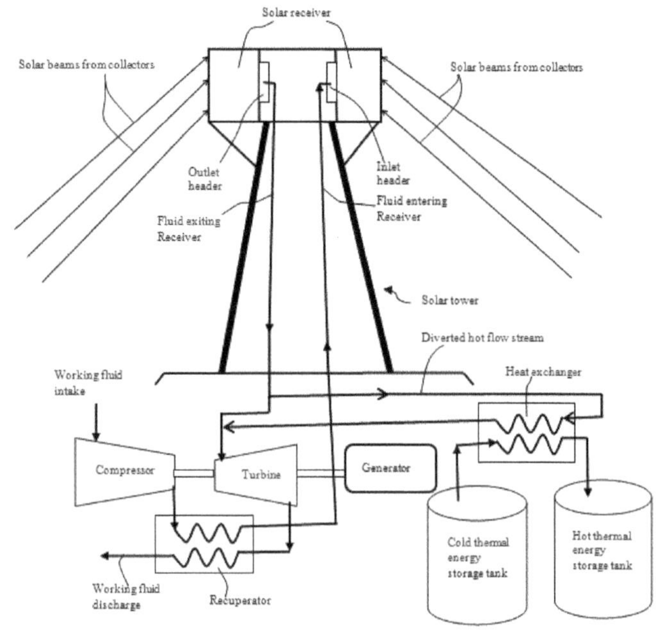

Fig. 6 Schematic illustration of the thermal charge process of a hot cold tank TES system integrated with a solar tower power plant without employing a heat transfer fluid (Cao, 2012a)

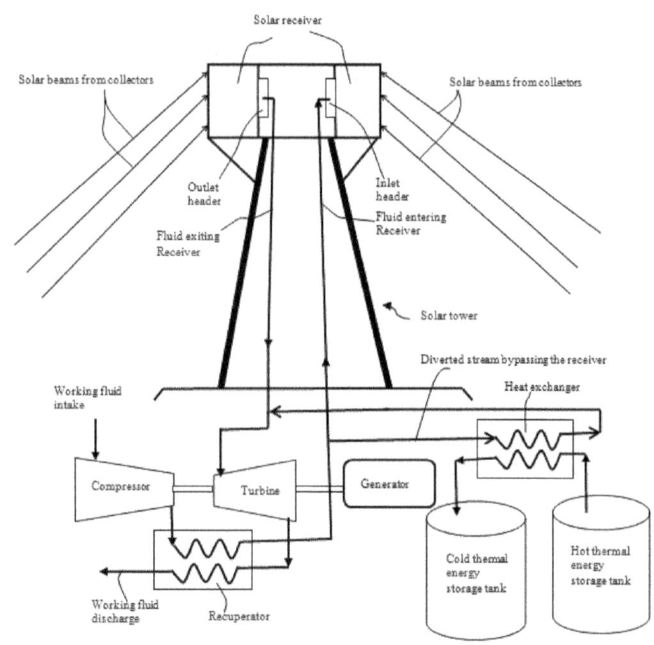

Fig. 7 Energy release of the hot cold tank TES system integrated with a solar tower power plant without employing a heat transfer fluid (Cao, 2012a)

from the solar receiver is partially or completely diverted to a heat exchanger associated with the hot and cold TES tanks. Through the heat exchanger, thermal energy is transferred from the hot engine working fluid to the TES material that is pumped from the cold TES material tank to the hot TES material tank to store the thermal energy therewithin. After passing an amount of thermal energy to the TES system, the working fluid is routed back to the turbine where some remained energy is converted into power. If the working fluid temperature at the exit of the turbine is too low, it may bypass the recuperator (not shown).

The above process is the thermal energy charge of the TES system. Figure 7 shows schematically the energy release process of the TES system. In this case, before entering solar receiver to acquire heat, the engine working fluid exiting the recuperator may be partially or completely diverted to the heat exchanger to receiver heat from the TES system, while the hot TES material is pumped from the hot tank to the heat exchanger. After transferring an amount of heat to the engine working fluid, the TES material with a reduced temperature returns back to the cold tank. At the same time, after acquiring an amount of heat from the TES system, the working fluid by-passes the solar receiver and enters the turbine to convert thermal energy into mechanical power. Alternatively, the engine working fluid exiting the heat exchanger may be routed back to the solar receiver (not shown). The above TES systems are based on the well-established hot/cold tank systems. Other options of TES systems as well as different heat pipe receiver configurations may also be possible (Cao, 2012a), which are not the topics of this paper.

4. EVALUATION OF THE HEAT TRANSFER COEFFICIENT IN THE EVAPORATOR OF THE STHP RECEIVER

As discussed earlier in this paper, for a STHP to be successfully employed as a solar receiver of a CSP system, it must be able to handle a high solar flux on the order of 50 to 100 W/cm^2, which would demand a very high heat transfer coefficient between the liquid metal working fluid and interior surface of the evaporator tube. Despite a very high thermal conductivity of the liquid metal, its boiling heat transfer coefficient is generally not exceptionally high, which may be a concern for high-temperature heat pipes to achieve a high flux. However, this difficulty may be resolved in a STHP, as the heat transfer in the evaporator of a STHP is the combination of boiling and forced convection.

The combined heat transfer coefficient for convective boiling in an upward vertical tube may be evaluated by the following correlation (Chen, 1963 and Faghri and Zhang, 2006):

$$h = Fh_l + Sh_b \tag{2}$$

where h_l and h_b are, respectively, the heat transfer coefficients for single-phase convection of the liquid and nucleate boiling. F and S are dynamic factors that modify the contribution of single-phase liquid convection and nucleate boiling, respectively, and may be evaluated by the following relations:

$$F = 2.35(0.213 + X_{tt}^{-1})^{0.736}, \text{ for } X_{tt}^{-1} > 1\text{; and } F = 1 \text{ for } X_{tt}^{-1} \le 1 \tag{3}$$

$$S = 1/(1 + 2.53\times10^{-6}\,\mathrm{Re}_{TP}^{1.17}) \tag{4}$$

where $\mathrm{Re}_{TP} = \mathrm{Re}_l F^{1.25}$ and X_{tt} is the Lockhart-Martinelli parameter defined by the following equation:

$$X_{tt} = \left(\frac{1-x}{x}\right)^{0.9}\left(\frac{\rho_v}{\rho_l}\right)^{0.5}\left(\frac{\mu_l}{\mu_v}\right)^{0.1} \tag{5}$$

The single-phase heat transfer coefficient for liquid may be calculated by using the following correlation for liquid metals with a constant heat flux boundary condition (Bergman et al., 2011):

$$Nu_D = 4.82 + 0.0185(\mathrm{Re}_D\,\mathrm{Pr})^{0.827} \tag{6}$$

The boiling heat transfer coefficient may be approximated by the following relation:

$$h_b = 4.0(q'')^{2/3} \tag{7}$$

Considering a sodium STHP solar receiver that has an overall receiver wall surface area of 6×6 m^2, the length of evaporator tube as shown in Fig. 2 and Fig. 3 is taken to be 6 m. In the following representative calculations, the tube outer diameter and inner diameter are taken to be 43 mm and 40 mm, respectively. A concentrated solar flux of 100 W/cm^2 and a working temperature of 1200 K are considered, and the average heat transfer coefficient with an average vapor quality of $\bar{x}_v = 0.5$ is calculated.

By assuming saturated conditions in both liquid and vapor drums, the mass flow rate of the working fluid in a single tube is:

$$\dot m = q''HD_o/h_{fg} = 10\times10^6\times6\times0.043/3.829\times10^6$$
$$= 0.0674 \quad \text{kg/s}$$

Then,

$$\mathrm{Re}_l = \frac{4\dot m(1-x_v)}{\pi D_i \mu_l} = \frac{4\times0.0674\times(1-0.5)}{\pi\times0.04\times1.514\times10^{-4}} = 7085.0$$

Therefore

$$h_l = \frac{k_l}{D_i}Nu_D = \frac{k_l}{D_i}(4.82 + 0.0185(\mathrm{Re}_D\,\mathrm{Pr})^{0.827})$$
$$= \frac{47}{0.04}(4.82 + 0.0185(7085\times0.00415)^{0.827})$$
$$= 6019.6 \quad \text{W/m}^2\text{-K}$$

According to Eq. (7),

$$h_b = 4.0\times(10^6)^{(2/3)} = 4.0\times(1\times10^6)^{(2/3)} = 40000 \quad \text{W/m}^2\text{-K}$$

According to Eq. (5),

$$X_{tt} = \left(\frac{1-x_v}{x_v}\right)^{0.9}\left(\frac{\rho_v}{\rho_l}\right)^{0.5}\left(\frac{\mu_l}{\mu_v}\right)^{0.1}$$
$$= \left(\frac{1-0.5}{0.5}\right)^{0.9}\left(\frac{0.379}{730}\right)^{0.5}\left(\frac{1.514\times10^{-4}}{0.2257\times10^{-4}}\right)^{0.1} = 0.0282$$

Therefore,

$$F = 2.35(0.213 + X_{tt}^{-1})^{0.736} = 2.35 \times (0.213 + (0.0282)^{-1})^{0.736}$$
$$= 32.63$$

Also,

$$\mathrm{Re}_{TP} = \mathrm{Re}_l F^{1.25} = 7085 \times 32.63^{1.25} = 5.53 \times 10^5$$

$$S = 1\ (1 + 2.53 \times 10^{-6} \times \mathrm{Re}_{TP}^{1.17}) = 1\ (1 + 2.53 \times 10^{-6} \times (5.53 \times 10^5)^{1.17})$$
$$- \ 0.07$$

Finally,

$$h = Fh_l + Sh_b = 32.63 \times 6019.6 + 0.071 \times 40000$$
$$= 1.96 \times 10^5 + 2840 = 1.99 \times 10^5 \quad \mathrm{W\ m^2\text{-}K}$$

According to the above result, the direct contribution of boiling to the overall heat transfer coefficient is very small. However, its agitation and enhancement on the forced liquid convective heat transfer has enabled a very high convective heat transfer coefficient. The temperature drop between inner surface of the evaporator tube and the heat pipe working fluid may be calculated as follows,

$$\Delta T = q''\ h = 10^6 / 1.99 \times 10^5 = 5.0^o C$$

The above result may indicate that the use of a STHP may be able to overcome the difficulty of relatively low heat flux of other high-temperature heat pipes and handle a high heat flux needed for CSP applications. It should be pointed out that the above temperature drop is only associated with that between the inner surface and the sodium working fluid, not including the temperature drop across the pipe wall that is related to the conductivity of the tube wall material. Depending upon a complete design of a STHP, other temperature drops over other sections of the system may occur, which is beyond scope of this paper. It should also be pointed out that the sodium filling in a SHTP is generally very small, corresponding to one third or one half of the inner volume of the evaporator, but its filling may significantly affect the performance of the STHP.

5. CONCLUSIONS

In this paper, the concept of low-temperature STHP in waste heat recovery industry is adapted for CSP solar receiver systems. The STHP based solar receiver has the advantages of high heat transfer capacity, high heat transfer coefficient in the evaporator to handle a high concentrated solar flux, non-condensable release mechanism, and lower costs. The STHP has a very simple structure, driven by gravity, and does not increase the use of high-temperature alloy materials. A preliminary calculation indicates that the heat transfer in the evaporator is exceptionally high due to the drastically enhanced convective heat transfer. This high heat transfer coefficient may allow a STHP solar receiver to handle a solar flux above 100 W cm². It is believed that the introduced STHP receiver is an enabling method for a gas turbine based solar power plant as the heat flux at the receiver must be spread to a much lower level for heat transfer to the gaseous engine working fluid.

Although the illustration of the STHP solar receiver is based on a solar tower power plant, the concept can also be employed for other solar power plants, including parabolic dish, Stirling engine, and Fresnel plants. The introduced STHP may also be employed for solar reactors of thermochemical storage systems (Cao, 2012b).

REFERENCES

Bergman, T.L, Lavine, A.S., Incropera, F.P., and Dewitt, D.P., 2011, *Fundamentals of Heat and Mass Transfer*, 7th Edition, John Wiley & Son.

Cao, Y. and Faghri, A., 1991, "Transient Multidimensional Analysis of Nonconventional Heat Pipes with Uniform and Nonuniform Heat Distributions," *ASME J. Heat Transfer*, Vol. 113, pp. 995-1002.

Cao, Y. and Faghri, A., 1992, "Closed-Form Analytical Solutions of High-Temperature Heat Pipe Startup and Frozen Startup Limitation," *ASME J. Heat Transfer*, Vol. 114, pp. 1028-1035.

Cao, Y. and Faghri, A., 1993, "A Numerical Analysis of High-Temperature Heat Pipe Startup from the Frozen State," *ASME J. Heat Transfer*, Vol. 115, pp. 247-254.

Cao, Y. and Wang, Q., 1995a, "Reciprocating Heat Pipes and Their Applications," *ASME J. of Heat Transfer*, Vol. 117, 1094-1096.

Cao, Y. and Wang, Q., 1995b, "Thermal Analysis of a Piston Cooling System with Reciprocating Heat Pipes," *Journal of Heat Transfer Engineering*, Vol. 16, pp. 50 - 57.

Cao, Y., 2012a, Heat Pipe Receivers for Concentrated Solar Power, US patent pending.

Cao, Y., 2012b, Heat Pipe Solar Thermochemical Reactor Units, US patent pending.

Chen, Y.G., Gao, M., and Xin, M.D., 1986, "Experiments of Heat Transfer Performance of Separate-Type Thermosyphon," *Int. Heat Pipe Symposium*, Osaka, Japan.

Chen, I.C., 1963, "A correlation for boiling heat transfer to saturated fluids in convective flow," ASME preprint 63-HT-34, Presented at the 6th National Heat Transfer Conference, Boston, MA.

Faghri, A., 1995, Heat Pipe Science and Technology, Taylor and Francis.

Faghri, A. and Zhang, Y., 2006, Transport Phenomena in Multiphase Systems, Elsevier, New York.

Jin, M.C. and Chen, Y.G, 1986, Heat Pipes and Heat Pipe Exchangers, Chongqing University Press, Chongqing, China.

Noyes, R.C., 1963, "An Experimental Study of Sodium Pool Boiling Heat Transfer," J. Heat Transfer -- May 1963 -- Volume 85, Issue 2, 125.

Ling, J. and Cao, Y., 2001, "Experimental Investigations of a Radially Rotating Miniature High-Temperature Heat Pipe for Turbine Blade Cooling," *ASME Journal of Heat Transfer*, Vol. 113, pp. 113-119.

Lovegrove, K., Taumoefolau, T., Paitoonsurikarn, S., Siangsukone, P., Burgess, G., Luzzi, A., Johnston, G., Becker, O., Joe, W., and Major, G., 2007, "Paraboloidal dish solar concentrators for multi-megawatt power generation," In: Solar World Congress 2007, Beijing, China.

Ma, D., Tuchi, C., Neises, T., and Wagner, M., 2012, "Thermodynamic Study of Advanced Supercritical Carbon Dioxide Power Cycles for High Performance Concentrating Solar Power Systems," *ASME 6th Conference on Energy Sustainability*, ES-FuelCell2012-91179.

Mancini, T.R., Kolb, G.J., and Chavez, J. M., 1994, "Solar thermal power today and tomorrow," *Mechanical Engineering* 1994; 116(8): 74-79.

Moran, M.J., Shapiro, H.N., and Boettner, D.B., 2009, *Fundamentals of Engineering Thermodynamics*, 7th edition, John Wiley & Sons.

Nithyanandam, K. and Pitchumani, R., 2012a, "Analysis of a Latent Thermocline Energy Storage System for Concentrating Solar Power Plants," *ASME 6th Conference on Energy Sustainability*, ES-FuelCell2012 □91389.

Nithyanandam, K. and Pitchumani, R., 2012b, "Numerical Analysis of Latent Thermal Energy Storage System with Embedded Thermosyphons," *ASME 6th Conference on Energy Sustainability*, ES-FuelCell2012 - 91416.

Noyes, R.C., 1963, "An Experimental Study of Sodium Pool Boiling Heat Transfer," J. Heat Transfer -- May 1963 -- Volume 85, Issue 2, 125.

Stine, W.B, and Diver, R.B., 1994, "A compendium of solar dish stirling technology," Sandia National Laboratories Report, SAND 93-7026.

Tao, Y. and Rayegan, R., 2011, "SOLAR ENERGY APPLICATIONS AND COMPARISONS," in Energy and Power Generation Handbook - ASME, Editor: K. R. Rao.

Yang, M., Yang, X., and Ding, J., 2009, "Heat transfer enhancement and performance of the molten salt receiver of a solar power tower." *Applied Energy*, 87(9): 2808-2811.

TEMPERATURE CONTROL WITH TWO PARALLEL SMALL LOOP HEAT PIPES FOR GLM PROGRAM

Dmitry Khrustalev[a,*], Chuck Stouffer[a], Jentung Ku[b], Jon Hamilton[c], Mark Anderson[c]

[a] *ATK Space, Beltsville, MD, 20705, USA*
[b] *NASA Goddard Space Flight Center, Greenbelt, MD, USA*
[c] *Lockheed Martin Space Systems, Palo Alto, CA, 94304, USA*

ABSTRACT

The concept of temperature control of an electronic component using a single Loop Heat Pipe (LHP) is well established for Aerospace applications. Using two LHPs is often desirable for redundancy/reliability reasons or for increasing the overall heat source-sink thermal conductance. This effort elaborates on temperature controlling operation of a thermal system that includes two small ammonia LHPs thermally coupled together at the evaporator end as well as at the condenser end and operating "in parallel". A transient model of the LHP system was developed on the Thermal Desktop™ platform to understand some fundamental details of such parallel operation of the two LHPs. Extensive thermal-vacuum testing was conducted with two thermally coupled LHPs operating simultaneously as well as with only one LHP operating at a time. This paper outlines the temperature control procedures for two LHPs operating simultaneously with widely varying sink temperatures. The test data obtained during the thermal-vacuum testing, with both LHPs running simultaneously in comparison with only one LHP operating at a time, are presented with detailed explanations.

Keywords: *Two-phase; temperature control; capillary wick.*

1. INTRODUCTION

Methods of controlling a single loop heat pipe (LHP) operating temperature were reviewed and summarized by Ku (2008). The most established approach is to maintain the reservoir temperature at a certain level (pre-selected set point temperature) by adding a relatively small amount of thermal energy to the reservoir utilizing a control heater in on/off mode with a feedback loop monitoring reservoir temperature. Alternatively, the control heater can be placed on the liquid return line as was reported by Bienert et al. (1999). Keeping the reservoir temperature at the set point level in the situation where the sink temperature varies in time while the heat dissipation from the heat source is constant, essentially eliminates temperature excursions on the component being cooled. This simple temperature control technique should be in principle applicable to two parallel LHPs, especially if they are not thermally coupled significantly. The extent to which the two LHP evaporators and condensers are thermally coupled together depends on the system design packaging. Possible transient heat flow rates between the two thermally coupled LHP evaporators, as well as between the two condensers, can result in the source temperature fluctuations unless operating temperatures of both loops are actively controlled. An extreme outcome of such thermal interaction can be only one LHP transporting the entire heat dissipated by the source accompanied by forcing the other LHP to become dormant. Since the two LHPs for GLM are strongly thermally coupled due to the system packaging, several aspects of operating the two LHPs simultaneously as well as one LHP at a time have been investigated for this particular application described below. The test data presented in this paper demonstrate that both LHPs can operate simultaneously in a stable and controlled manner.

2. GLM SYSTEM OVERVIEW, THERMAL REQUIREMENTS AND LHP CONFIGURATION

The Geostationary Lightning Mapper (GLM), shown in Figure 1, will locate and measure lightning in the Western hemisphere from geostationary orbit. The focal plane array (FPA) electronics interface with the LHP evaporators is only 4.5 cm x 8.8 cm and the expected constant heat dissipation is bounded by the range from 75 W to 105 W. The tight system component packaging around the LHP evaporators requires the use of 13 mm diameter evaporators and 10W startup heaters. The expected effective interface material heat transfer coefficients are 4650 W/(m^2-K) for the evaporator interface footprint and 1938 W/(m^2-K) for the condenser plate interface footprint of 90 mm x 530 mm. Operating temperature of the heat pipe-based radiator (not shown in Figure 1) interfaced with the LHP condenser plate, is anticipated to vary between -10°C and +12°C with some short-term excursions to +25°C with the rate of temperature change less than 1 K/minute. The LHP working fluid is anhydrous ammonia. The LHP primary evaporator wick is made of porous metal with pore radius of approximately 1.5 micrometers.

The main thermal requirement is to keep the FPA side of the LHP evaporator interface at 25°C ±2°C for any operating radiator temperature between -10°C and +12°C, including the time periods when the radiator operating temperature is transitioning between -10°C and

* *Corresponding author. Email: Dmitry.Khrustalev@ATK.com*

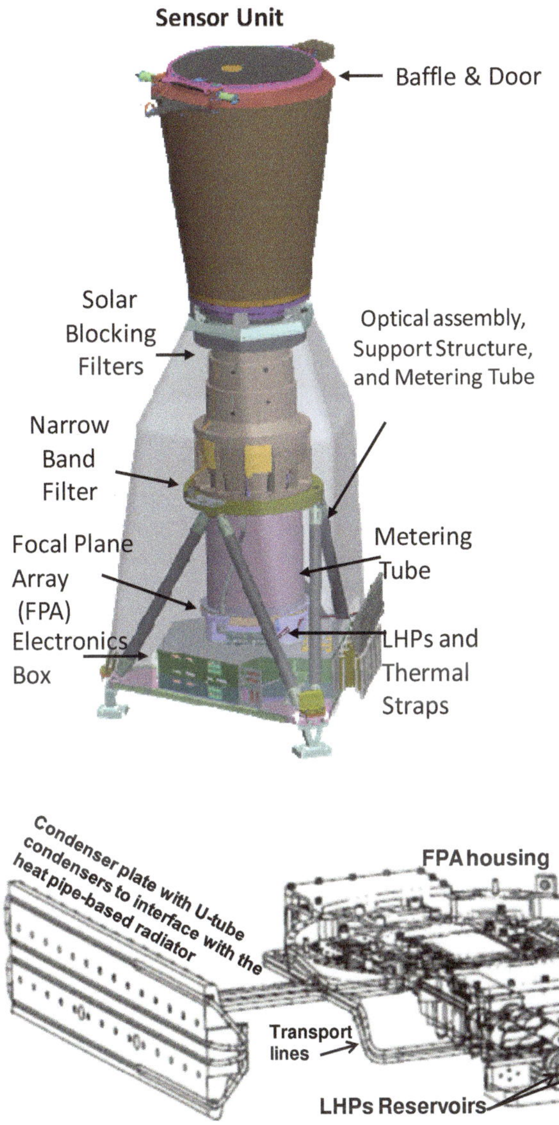

Fig. 1 View of the GLM instrument (Christian and Gheno, 2012) and FPA-LHP assembly packaging

+12°C. Such temperature control is required for only one LHP operating at a time as well as for both LHPs operating simultaneously.

The temperature controlling function of the LHPs is achieved by heat loading the LHP reservoir to maintain its temperature at a pre-determined level using the reservoir control heaters (6.5 W maximum) visible in Figure 2. This figure provides an image of the two LHPs coupled together and attached to a temporary support frame. The condenser plate is designed to be bolted to a heat pipe-based radiator with multiple bolts. The two U-tube condensers (3 mm outer diameter tubing) are soldered into the left and right sides of the condenser plate symmetrically, with the liquid return lines close to the plate central axis. The vapor inlets into the condensers are located at the outer sides of the condenser plate. The two evaporators are integrated into one aluminum block (45 mm x 88 mm x 17 mm) interfacing with the FPA surface through an interface material and thus are strongly coupled together thermally. The small diameter transport lines (3 mm outer diameter tubing) provide sufficient vibration isolation between the FPA and radiator. The two reservoirs are supported by a stainless steel bracket bolted to the FPA housing for structural integrity. This bracket also provides some thermal conductance between the FPA housing and the reservoirs. The reservoirs are shaped as cylinders approximately 125 mm long with outer diameters of 18 mm to fit into the allocated volume. Each evaporator has two startup heaters (10W each) situated on the back surface of the aluminum block interfacing with FPA (visible in Figure 5). Additional clarifications on the fluid flow schematics are provided in the next section describing thermal-fluid modeling conducted for the two coupled LHPs on the platform of Thermal Desktop™.

3. THERMAL-FLUID MODELING OF TWO COUPLED LHPS

An image of the Thermal Desktop™ (Sinda/Fluint) model for two GLM LHPs is shown in Figure 3. The model simulates transient circulation of the two-phase working fluid in the components of the two LHPs (condenser, evaporator, reservoir, and transport lines) and predicts the temperature distribution across the condenser plate. The model simultaneously solves the flow momentum, energy, and mass conservation equations for the two-phase and single-phase fluid flow for each fluid lump and path separately as well as for the two entire fluid submodels. It includes all three heat transfer mechanisms: convective, conductive (conduction in the condenser plate), and radiative (radiation from the environment to the LHP components), and

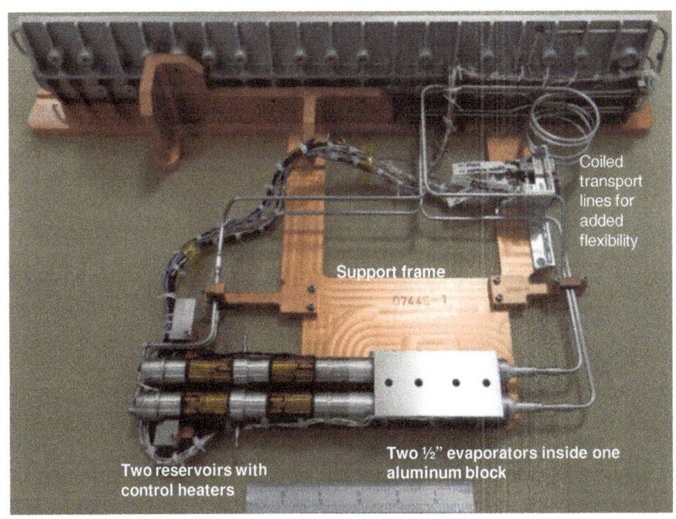

Fig. 2 Two GLM LHPs on a support frame

also accounts for the phase change heat transfer (evaporation and condensation). The modeling technique was documented in detail by Khrustalev (2010) and by Cullimore and Bauman (2000).

One important objective for this transient modeling was to simulate temperature controlling capability of the system utilizing control heaters on the reservoirs, and also to evaluate effect of the parasitic heating of the liquid return lines and reservoirs through radiation from the higher-temperature environment on the overall LHP thermal conductance. Another objective was to investigate if the two LHPs would be naturally sharing the common heat load (from FPA) with and without thermal coupling of the two reservoirs. Understanding gained from the modeling results and observations was useful for designing the LHPs. Figure 3 provides a fluid flow schematic and the temperature color map during steady state operation of the two LHPs with the total heat dissipation from the FPA of 105 W and the sink (radiator) temperature of 12°C. The predicted FPA interface temperature with the heat load of 105 W is 25.29°C and temperatures of the two reservoirs are within 0.11°C without any kind of control. Temperature of the liquid coming out of the condensers is 12.47°C and slightly increases to 13.04°C prior to flowing into the reservoir due to the radiation parasitic heating of the transport lines from the environment with temperature of 40°C.

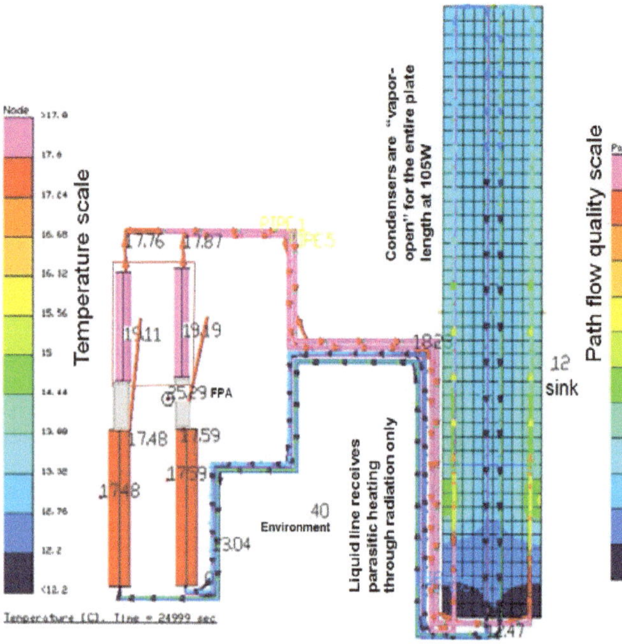

Fig. 3 Thermal Desktop™ model of two GLM LHPs

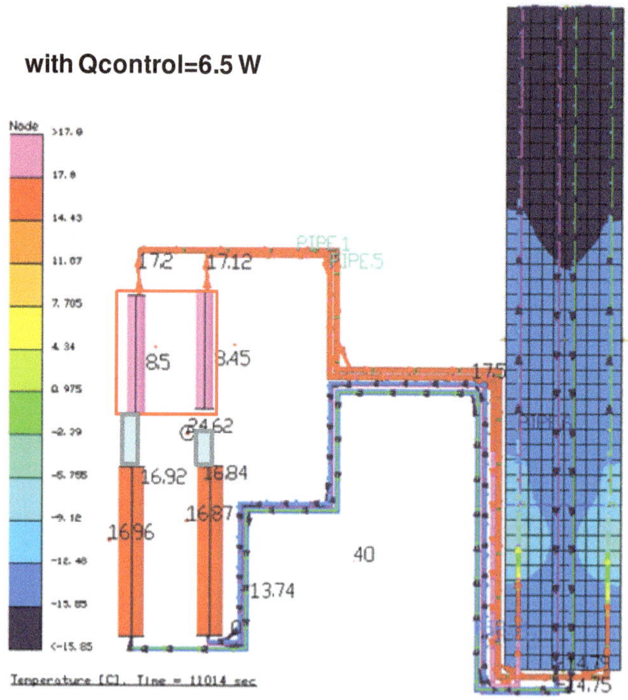

Fig. 4 Liquid blocking of the condensers in the temperature control mode with 6.5 W per reservoir and cold sink

The flow quality in the condensers and transport lines can be read using the color quality scale on the right of Figure 3: The outer lines of the condensers are not soldered to the condenser plate for a length of 50 mm and therefore contain vapor with quality of 1 (red color). The temperature distribution on the condenser plate can be observed using the color temperature scale on the left. Without applying control power to the reservoirs, the two condensers in Figure 3 contain two-phase fluid in the outer passes and single phase subcooled liquid on most length of the central lines of the condensers. Figure 4 presents a steady state with the sink temperature of -16°C and 6.5 W of the control power applied to each reservoir in order to maintain the FPA interface at 24.62°C. Since the temperature difference between the saturated vapor and the condenser plate is at the level of 30°C, the vapor condenses on a short length near the vapor lines entering the condenser plate and the rest of the condenser length is liquid-blocked. Having two-phase flow in the condenser on the entire length of the condenser plate at 105 W (see Figure 3) is perceived as a situation where the two LHPs are naturally sharing the common heat load. Maintaining an equal heat load sharing between the two LHPs with total heat loads of 75 W or lower, where the length of the two-phase flow is shorter than the condenser plate, will require the reservoirs to be controlled at the same set point temperature using control heaters. Generally, the heat load sharing can be achieved by keeping the reservoirs temperatures at the same pre-determined level by utilizing the control heaters for any heat load level on the evaporator interface. These statements are supported by the test data obtained during thermal-vacuum testing outlined in the following section.

1. THERMAL-VACUUM TEST SETUP

The test configuration for GLM LHPs is shown in Figure 5. The 5-kg aluminum block representing the FPA housing is held by four adjustable low-conductivity threaded rods with additional non-conductive supports, bolted to the test frame. Thermal interface gaskets were installed on the evaporator saddle and condenser plate operating interfaces. Six DC power supplies were used to power two control heaters, two startup heaters, cartridge heaters inside the 5-kg block, and also a cover heater shown in Figure 6. A data acquisition system (DAS) was recording readings from multiple thermocouples, thermistors, and heaters once per second. The DAS program allowed the control heaters to be turned on/off based on a pre-selected temperature sensor reading. To eliminate heat exchange with the vacuum chamber, the five surfaces of the 5-kg aluminum block surfaces were covered with several layers of multi-layer insulation (MLI). Also the entire test frame was additionally blanketed with hanging MLI blankets. To more accurately simulate the environmental heating through radiation to the reservoirs and transport lines, an aluminum heated cover, shown in Figure 6, was placed above the two reservoirs. The cover temperature was kept 25°C above the FPA-side pedestal (see Figure 7) temperature as soon as the cartridge heaters inside the 5-kg aluminum block were turned on. Locations of multiple type-T thermocouples and flight thermistors are shown in Figures 6 through 8. Thermocouples numbered form 62 to 70 (Figure 7) were used to calculate the average pedestal perimeter temperature, $\mathbf{T_{FPA}}$. Thermocouples with numbers from 18 to 21 and from 30 to 33 were used to calculate the average cold plate perimeter temperature $\mathbf{T_{sink}}$. The overall system conductance, **C** (W/K), including the LHP evaporator, condenser, and corresponding interfaces with the heat source and sink, is defined herein as follows

$$C=Q/(T_{FPA} - T_{sink}) \quad (1)$$

where Q is the electrical power dissipated by the cartridge heaters inside 5-kg aluminum block assuming an infinitesimal heat loss. An outline of the typical functional thermal-vacuum testing sequence is provided below.

5. TEST PROCEDURE SEQUENCE AND TEST DATA DISCUSSION

The main objectives of the thermal-vacuum testing included the following:

1. Evaluate and demonstrate the temperature controllability with the two GLM LHPs running simultaneously.

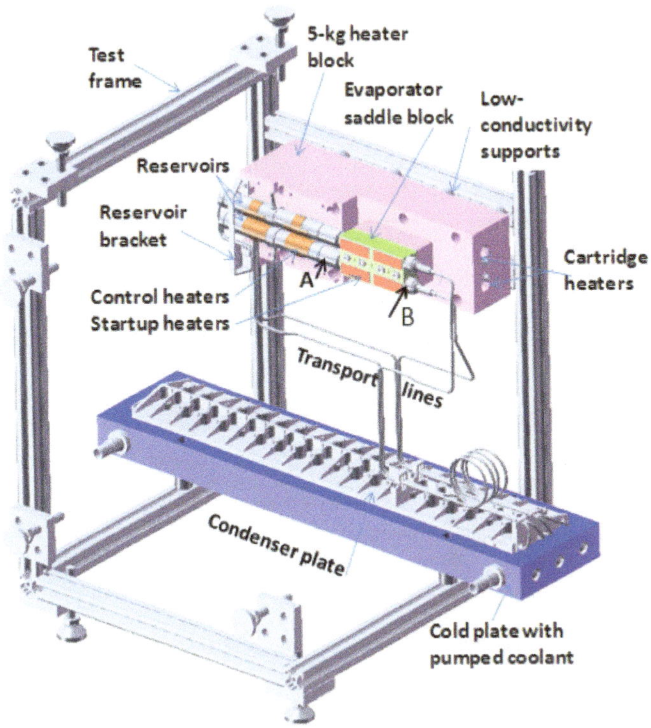

Fig. 5 Test setup configuration

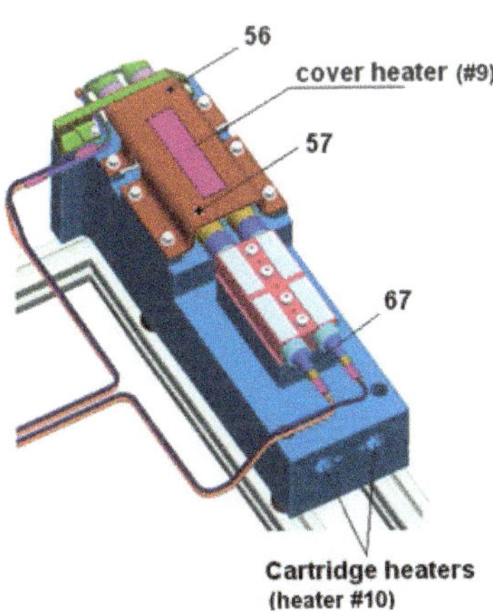

Fig. 6 Heated cover generating environmental heating to reservoirs

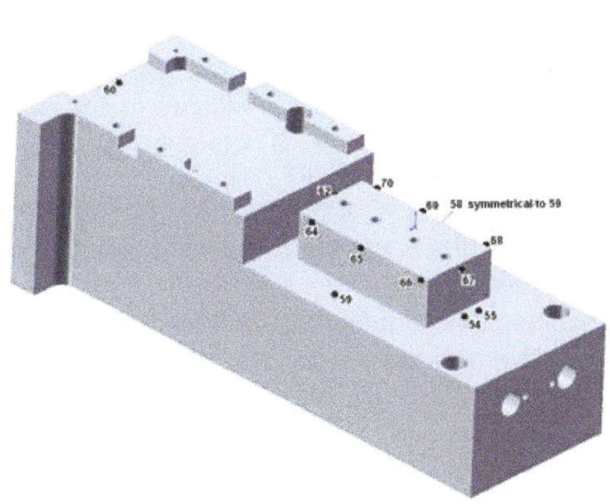

Fig. 7 Heated Thermocouple locations at the pedestal perimeter

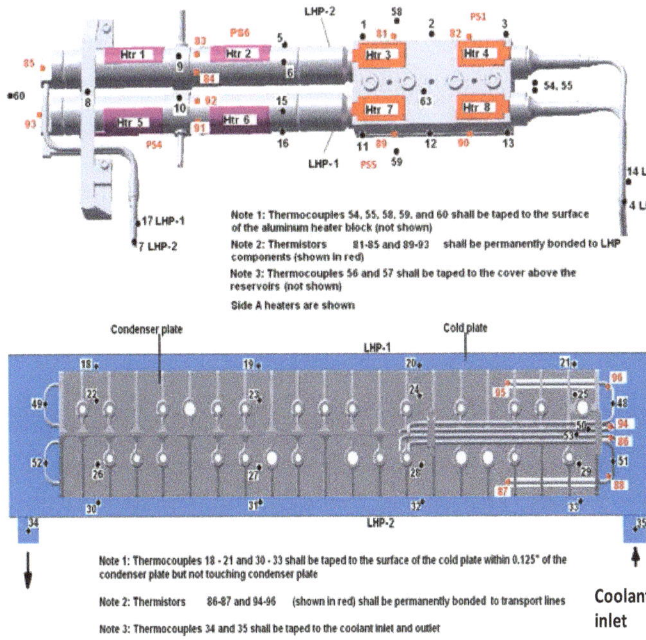

Fig. 8 Thermocouple/thermistor locations on the LHP components and cold plate with pumped coolant

2. Demonstrate the ability to maintain the FPA at 25°C±2°C with one LHP running alone or with two LHPs running simultaneously while the sink temperature varies between -10°C and +12°C.

3. Measure the overall system conductance for all applicable steady state cases.

The main procedure for temperature control with two LHPs operating simultaneously was Independent Reservoir Set Point control (IRSP) with the same temperature set point for both reservoirs.

From the multiplicity of thermal tests conducted for the GLM LHPs, two most representative "functional" tests are presented below by the plots for temperatures and heat loads versus time with explanations corresponding to different stages of the tests. While the "functional" test was repeated multiple times for three LHP orientations and different heat loads, test results from the orientation shown in Figure 5 and the heat load of 75 W on the 5-kg aluminum block are used for illustration in this paper.

5.1 Overview of Startup and Temperature Control Method

Startup of the LHP at low reservoir and evaporator temperature are more difficult for two reasons: 1) the high surface tension of the working fluid; and 2) the slope of the derivative dP/dT is smaller at lower temperatures. Past experience indicates that ammonia loops could start successfully for temperature at 15°C or higher. To simplify the test procedure, it was decided that the loop startup would not be initiated until temperatures of the evaporator and reservoir were raised to 15°C or above. Startup is conducted by turning on the two startup heaters on the evaporator block. Usually only one of the loops will start. The startup of the second loop is performed by raising the temperature of the reservoir of the operating loop several degrees so that the evaporator temperature will be higher than the reservoir temperature of the yet-to-start loop.

The following major stages of the "functional" test have been distinguished and are marked correspondingly in the graphics below:

Stage 1: Arriving at a repeatable conditions with the sink temperature of -35°C and pre-heated reservoirs at +15°C.

Stage 2: Startup of the first LHP with startup heaters and consequent heat loading of the 5-kg aluminum block at 75 W.

Stage 3: Gradual elevation of the sink temperature to +5°C and controlling the operating LHP.

Stage 4: Startup of the second LHP with startup heaters and controlling reservoirs with control heaters to maintain the FPA interface at +25°C.

Stage 5: Variation of the sink temperature from +5°C to -10°C, then to +5°C, +25°C, and back to +5°C while keeping the FPA interface at +25°C (up to 35°C for the hot sink at 25°C).

Stage 6: Shutting down circulation inside the LHPs by applying 5 W to the reservoirs.

5.2 Test Results with Both Loops Operating Simultaneously

The top (first) pane of the test data graph in Figure 9, as well as in the consecutive figures, shows temperatures of the LHPs components and sink varying versus time. The second pane monitors rate of the temperature change on the FPA interface. The 3rd pane is for the heat load on the 5-kg block (FPA). The 4th pane is for the heat loads generated by the two startup heaters and two control heaters on the two LHPs. Finally, the 5th pane plots the overall system conductance as defined by equation (1), specifically for steady state operation points.

Figure 9 covers the first four stages of the test and its description below is further divided into multiple steps for a higher resolution on time.

Step 1: Warming the block and evaporators. Since the evaporator and reservoir temperatures were below +15°C, startup heaters were turned on at 9:44 (one per LHP) at 10 W each, simultaneously with the reservoir heaters keeping reservoirs at the same temperature as evaporators (+1°C/-0°C).

Step 2: As soon as the LHP evaporators and the 5-kg aluminum block were at/above 15°C, all heaters were turned off at 10:34.

Step 3: Reservoir heating was done from 10:34 to 11:04 by utilizing heaters on both reservoirs to maintain their temperatures above the temperature of evaporators for about 30 minutes to yield repeatable conditions without fluid circulation in LHPs. All heaters were turned off at 11:04.

Step 4: Starting with 10 W on two startup heaters. One startup heater on each LHP was turned on at 11:06 with 10 W and fluid circulation in LHP2 started at 11:14 (see Figure 9) with the vapor line temperature at the entrance to the condenser increased to within 5°C of the reservoir temperature.

Step 5: 75 W (nominal FPA heat dissipation) was applied to the aluminum block at 11:16 and both startup heaters were turned off.

Step 6: The cover heater settings were adjusted to allow the cover near the two reservoirs to be heated to a temperature level 25°C higher than that on the aluminum pedestal (TC #67) in the auto-tracking mode.

Step 7: Preventing rapid cooling down. As soon as the reservoir temperature started to decrease due to the cold liquid entering it, the reservoir heater of the operating LHP2 was turned on to prevent a rapid cooling of the block/FPA. The heater power was adjusted to 3 W and then elevated back to 6 W to minimize the operating temperature variation. The aluminum pedestal cooling rate peaked at 1.1 K/minute at 11:17 and then reduced.

Step 8: The sink chiller set point temperature was changed to +5°C (from -41°C) with the temperature rate of change approximately 1.5 K/minute in order to simulate the sink temperature variation anticipated for the program. The temperature on the LHP evaporators gradually increased to 14°C due to the cold plate temperature increase.

Step 9: The LHP reservoir heater power was gradually reduced to zero by 11:50.

Step 10: Turning on LHP2 reservoir heater with the power level of 5 W at 11:55 AM and then powering the startup heater on LHP1 resulted in LHP1 startup, at which time the IRSP procedure was initiated at 12:08, keeping the FPA interface at 24°C.

Step 11: Power levels on both reservoir heaters were adjusted to 6 W at 12:30. The IRSP procedure with the reservoir set point at 18.4°C kept the 5-kg aluminum block temperature (measured on the aluminum pedestal) at approximately 24°C during that control heater power adjustment.

Continuation of the same test is shown in Figure 10.

Step 12: The cold plate (sink) temperature was varied from +5°C to -10°C, kept at -10°C for 40 minutes and then increased back to 5°C with the temperature change rate of approximately 1.5 K/minute. During this sink temperature variation the average pedestal (FPA) temperature stayed between 23.8°C and 24.2°C. The reservoir heaters duty cycles increased with the sink at -10°C, with corresponding time-averaged power levels of 2 W on LHP1 reservoir and 4 W on LHP2 reservoir.

Step 13: The cold plate temperature was further elevated to +25°C starting at 14:05 with the rate of 1.5 K/minute and then was kept at +25°C. During the time period from 14:37 to 15:00 the FPA heat load was turned off (for about one minute) and then quickly turned back on eight times without any significant temperature fluctuations on the LHP components. A steady state operation with both LHPs operating was achieved by 16:30 (See Figure 11). Although the IRSP procedure was effective all the time, the LHP1 reservoir heater duty cycle was virtually zero and about 1 W was continuously powered on the LHP2 reservoir heater. The highest aluminum pedestal (FPA) steady state average temperature was 33°C at 16:20, and the total FPA-sink conductance with both LHPs operating and 25°C sink temperature was around 9.5 W/K (see Figure 11 bottom pane).

Step 14: The sink temperature was then decreased to +5°C and LHP1 was shut down by 17:15 using the reservoir heater, while LHP2 was kept operating with the sink temperature again elevated to and being maintained at +25°C.

Step 15: LHP2 steady state operation was achieved at 17:52 with the highest aluminum pedestal (FPA) steady state average temperature of 33°C and corresponding LHP2 conductance of 9.5 W/K.

Step 16: The main heater (FPA) was turned off at 17:56 and one control heater on each reservoir was powered at 5 W, which stopped circulation in LHP2 within several minutes. The reservoir heater settings were adjusted to keep the reservoir temperature above that of the evaporator.

The overall source-to-sink conductance was 9.5 W/K for the sink temperature of 25°C with both LHPs operating simultaneously or only one LHP operating with the total heat load of 75 W. This observation is explained by noticing that with one LHP transporting only 37.5 W (out of 75 W total) its two-phase length in the condenser is smaller than the cold plate length (variable conductance mode) and the corresponding condenser-interface conductance is much lower than in the situation with the total heat load of 105 W where the entire cold plate is warmed up by the vapor condensing in the condenser, as shown in Figure 3.

Following this logic, two loop heat pipes operating simultaneously with the heat load of 105 W should have a higher overall source-to-sink conductance than only one LHP operating at a time. In fact, the test data in Figure 12 (for 105 W) do indicate that the overall conductance of the two LHPs was 12.5 W/K versus the conductance for only LHP1 operating of 9.5 W/K and for only LHP2 operating of 9.1 W/K. The test data in Figure 12 were obtained by adjusting the sink temperature in order to keep the FPA interface at 25°C±1°C with minimized control power applied to either of the two LHPs.

The highest sink temperature satisfying this condition was 16°C for the case of both LHPs operating simultaneously versus 14°C for only one LHP operating at a time. This difference is anticipated to become larger for higher heat loads, providing the motivation to run two LHPs in parallel for higher heat dissipations from FPA.

It is worth noting that turning on the second LHP at 11:00 was achieved easily by simply applying 5.5 W of the control power to the already running LHP1 reservoir in order to elevate its evaporator temperature and thus directing more of the thermal energy to flow from the 5-kg block to the LHP2 evaporator, starting the flow circulation in LHP2. The same principle was used in order to shut down the flow circulation in LHP1 at 12:45: applying 5.5 W to LHP1 reservoir for a longer time period did shut it down while LHP2 continued to operate. Also in support of the previous statement that lowering the heat load usually results in a lower LHP conductance, reducing the heat load from 105 W to 75 W at 14:15 (see Figure 12) led to the LHP2 conductance reduction from 9.1 W/K to 8.5 W/K.

5.3 Test Results with Only One Loop Operating

While the FPA temperature control with two LHPs operating simultaneously was demonstrated as presented above, utilizing only one LHP and keeping the second one dormant is discussed below. Figure 13 presents test data obtained with only LHP1 running during a similar complete "functional" test while applying approximately 1 W (time-averaged) to LHP2 reservoir to make sure it does not start the fluid circulation. This test with only one LHP functioning is discussed below.

Step 1: Warming the block and evaporators. Since the evaporator and reservoir temperatures prior to the test beginning were below +15°C, two startup heaters were turned on at 8:45 (one per LHP) at 10 W each simultaneously with the reservoir heaters keeping reservoirs at the same temperature with evaporators (+1°C/-0°C).

Step 2: As soon as the LHP evaporators and the 5-kg aluminum block were at/above 15°C, all heaters were turned off at 10:07.

Step 3: Reservoir heating was done from 10:07 to 10:37 by utilizing heaters on both reservoirs to maintain their temperatures above the temperature of evaporators for at least 30 minutes. All heaters were turned off at 10:37.

Step 4: Starting with 10 W on two startup heaters. One startup heater on each LHP was turned on at 10:38 with 10 W and fluid circulation in LHP1 started at 10:47 (see Figure 13) with the vapor line temperature at the entrance to the condenser sharply increased to within 5°C of the reservoir temperature.

Step 5: 75 W (nominal FPA power) was applied to the aluminum block and both startup heaters were turned off.

Step 6: The cover heater settings were adjusted to allow the cover near the two reservoirs to be heated to a temperature level 25°C higher than that on the aluminum pedestal (TC #67) in the auto-tracking mode.

Step 7: Preventing rapid cooling down. As soon as the LHP1 reservoir temperature started to decrease due to the cold liquid entering it, power was applied to the reservoir heater of the operating LHP1 to prevent a rapid cooling of the block/FPA. The aluminum pedestal cooling rate peaked at 1.5 K/minute at 10:45 and then reduced to +/- 0.5 K/minute for the rest of the test. The reservoir heater power level was then increased to 6 W.

Step 8: The chiller set point temperature was changed to 5°C (from -41°C) with the temperature rate of change approximately 1.5 K/minute. The temperature on the LHP evaporators gradually elevated to 11°C due to the cold plate temperature increase.

Step 9: The LHP1 reservoir heater power was gradually reduced to zero by 11:15.

Step 10: An operational steady state was achieved by 11:45 with LHP1 transporting 75 W to the cold plate at 4°C (5°C sink) with LHP1 conductance of 8.5 W/K.

Step 11: Power level on the LHP1 reservoir heater was adjusted to 6 W and the IRSP procedure was applied to LHP1 with the reservoir set point at 18°C in order to bring the 5-kg block and pedestal to 24.5°C.

Step 12: With the reservoir set point of 18°C, the steady state average pedestal (FPA) temperature was maintained at approximately 24.5°C at 12:30.

Step 13: The cold plate (and the "sink") temperature was varied from +5°C to -10°C, kept at -10°C for 40 minutes and then increased back to +5°C with the temperature change rate of approximately 1.5 K/minute. During this sink temperature variation the average pedestal (FPA) temperature stayed between 24°C and approximately 24.5°C. The reservoir heaters duty cycle was at its maximum with the sink at -10°C, with corresponding time-averaged power on the LHP1 reservoir of 6 W.

Step 14: The cold plate temperature was further elevated to +25°C starting at 13:45 with the rate of 1.5 K/minute and then was kept at +25°C to achieve a steady state with only LHP1 operating. The LHP1 reservoir heater duty cycle was zero at the steady state at 14:35. The highest aluminum pedestal (FPA) steady state average temperature was 33°C, which means the total FPA-sink conductance with LHP1 operating at 25°C sink temperature was 8 W/K.

Step 15: The sink temperature was then decreased to +5°C and the reservoir heater on LHP1 was powering with some duty cycle, keeping the LHP1 reservoir at 18°C (and with also about 1 W time-averaged heating on LHP2 reservoir).

Step 16: The main heater (FPA) was turned off at 15:03 and LHP1 reservoir heater was powered at 5 W (and heating on LHP2 reservoir increased to 3.5 W time-averaged), which stopped circulation in LHP1 within several minutes. The reservoir heater controller settings were adjusted to keep reservoir temperatures above those of the evaporators.

6 SUMMARY AND CONCLUSIONS

1. Test data demonstrated that two parallel LHPs thermally coupled on the evaporator end as well as on the condenser end, could operate simultaneously ("heat load sharing mode") and provide the temperature controlling function for the heat source during the sink temperature variation. Alternatively, only one LHP could be operated at a time in the temperature-control mode while the other LHP was kept "dormant" by applying a small heat load of approximately 1 W (time-averaged) to its reservoir.
2. The main method/procedure used to keep both LHPs operating simultaneously during the sink transients was Independent Reservoir Set Point Control (IRSP) where two reservoirs were kept at the same set point temperature by utilizing reservoir control heaters in the on/off mode.

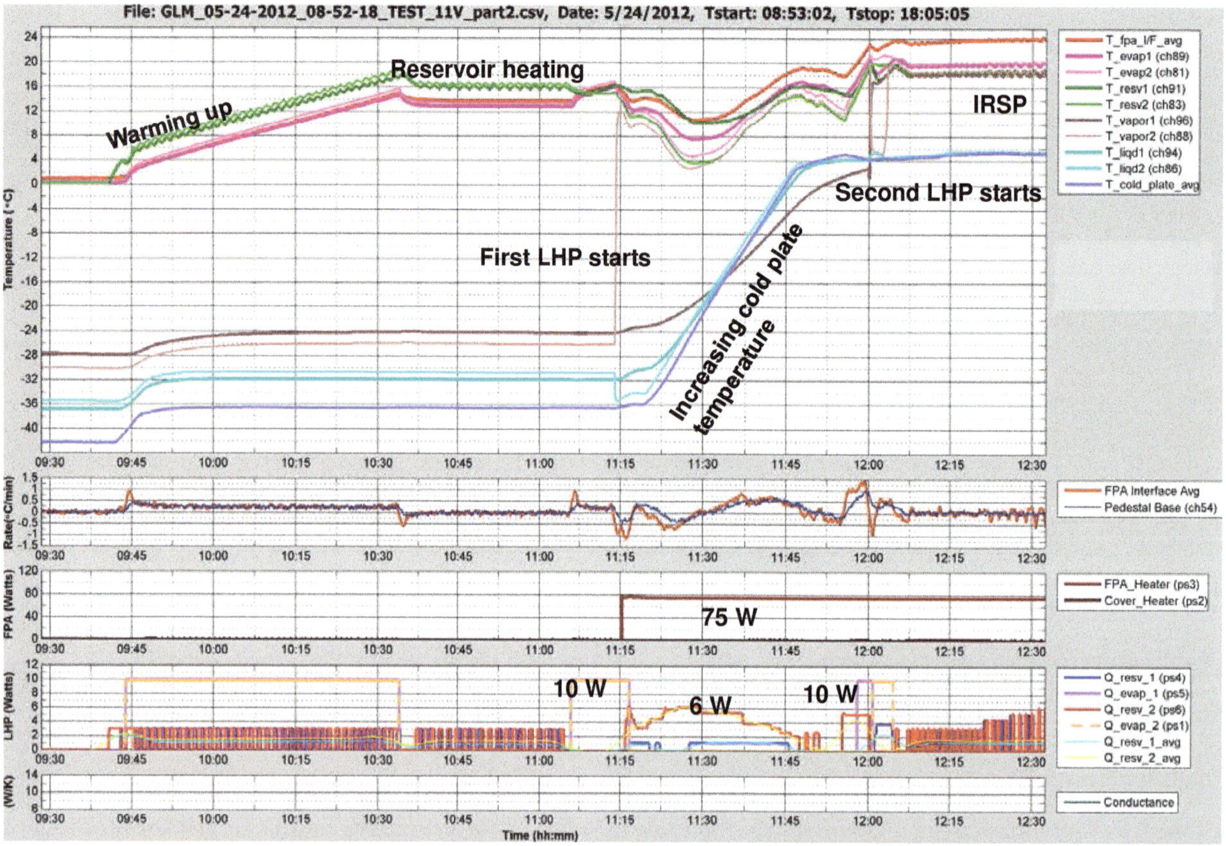

Fig. 9 Initial stages of the "functional" test: startup of the first and second LHPs

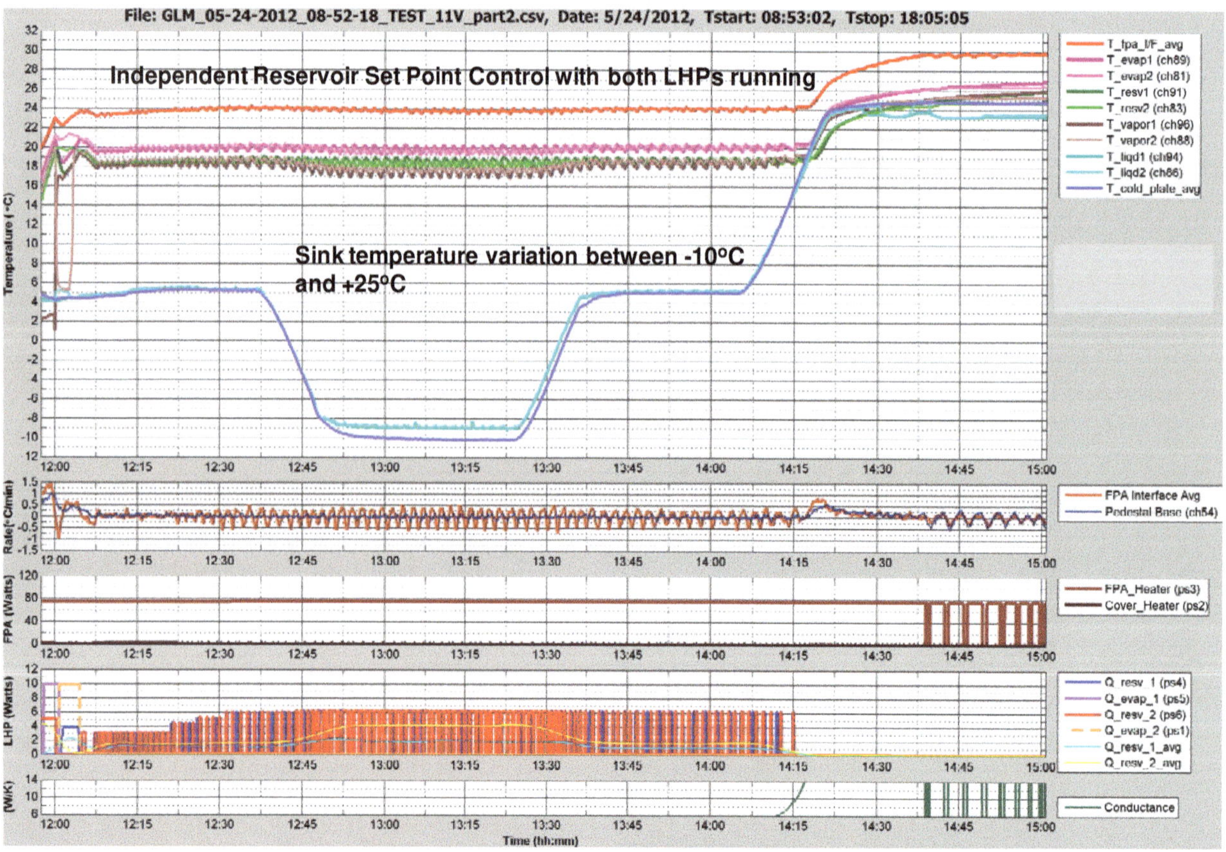

Fig. 10 Sink temperature variation with FPA interface maintained at +25°C

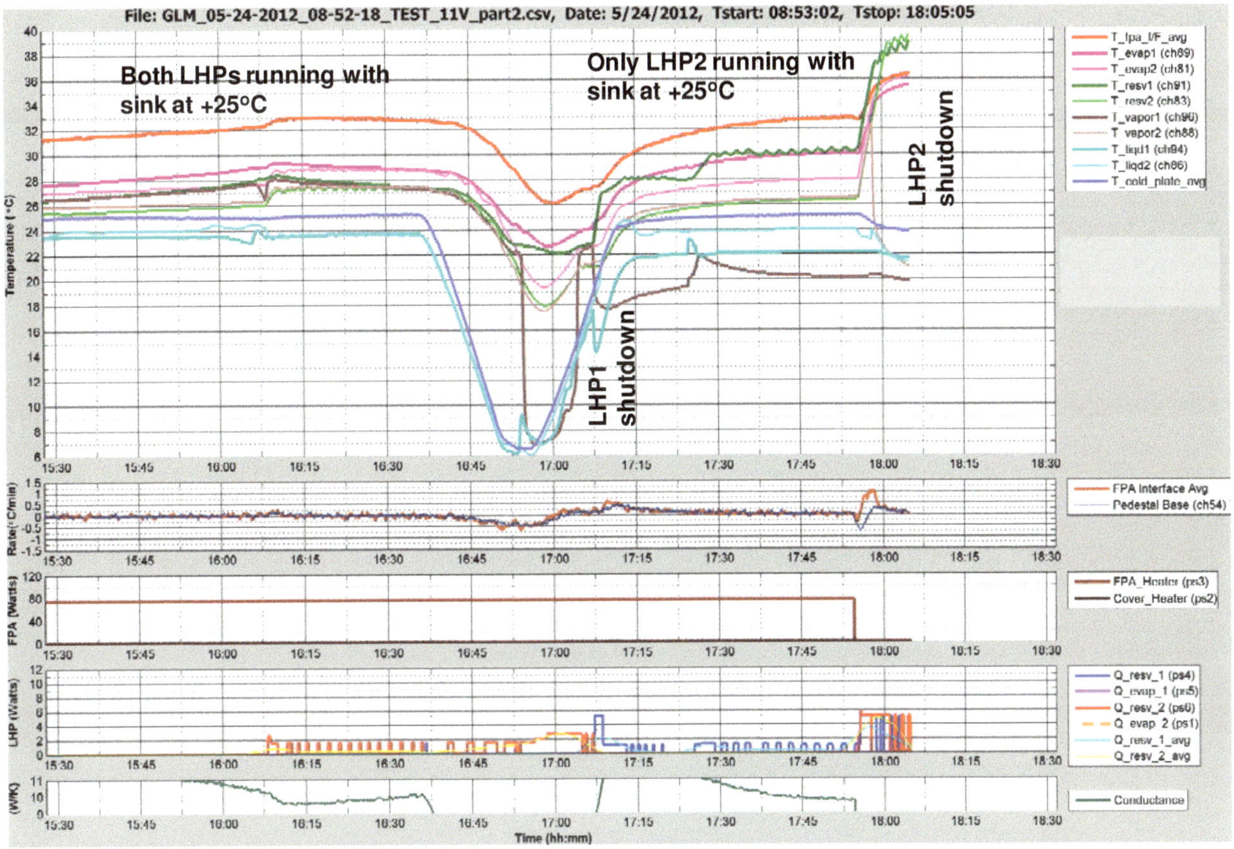

Fig. 11 Comparison of both LHPs operating with one LHP operating at the sink temperature of +25°C

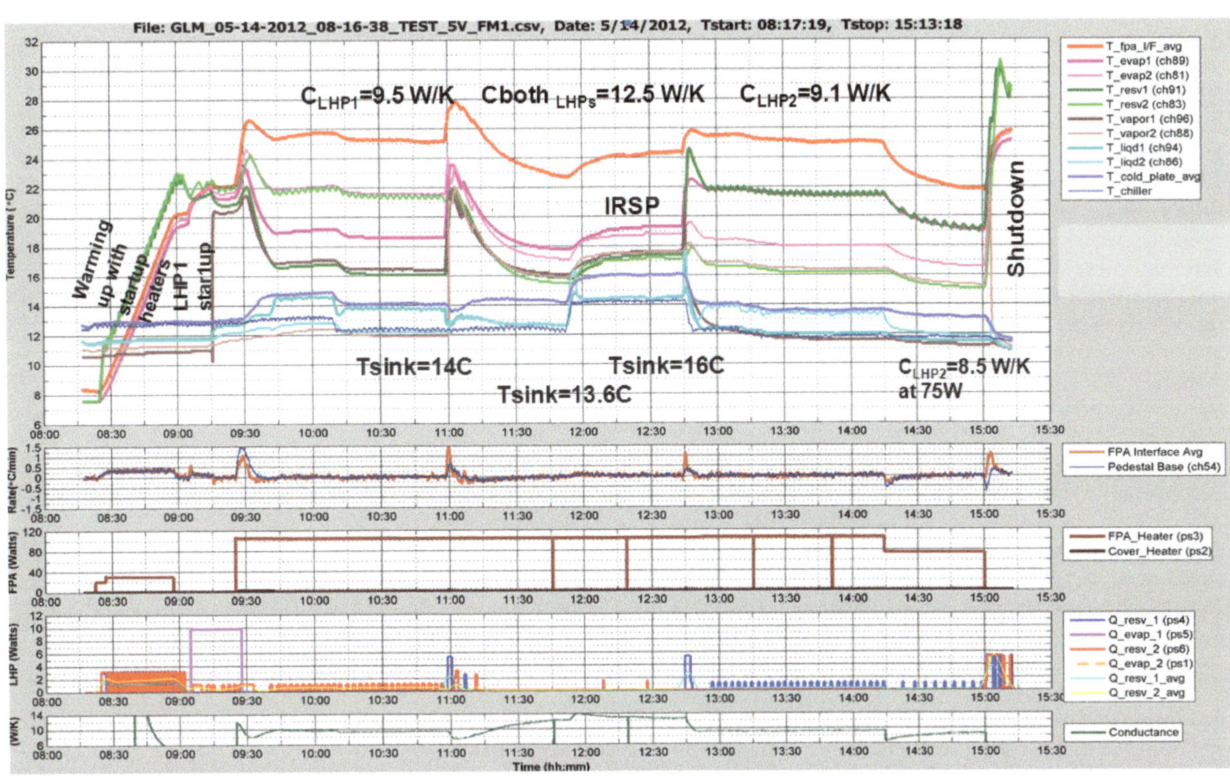

Fig. 12 Conductance comparison between both LHPs operating and only one LHP operating at 105 W

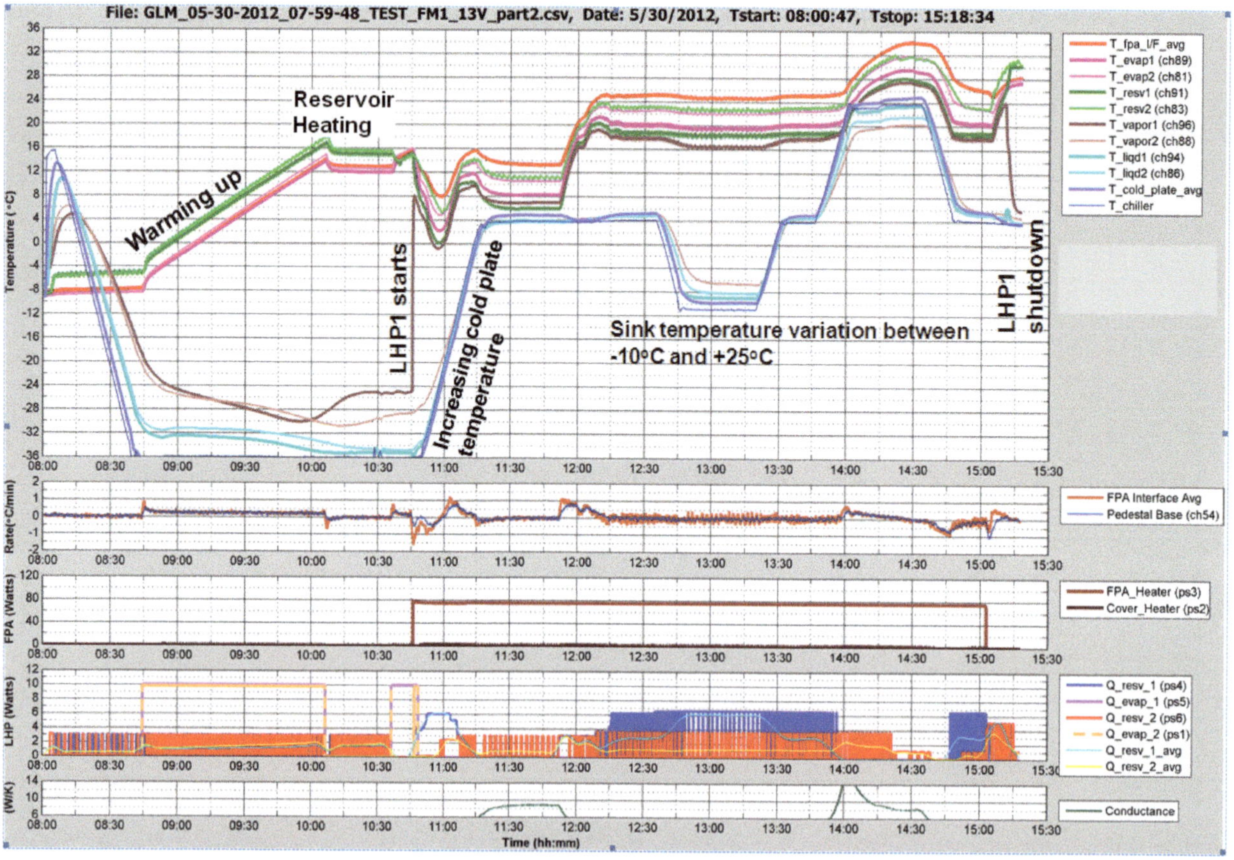

Fig. 13 Functional test with only LHP1 operating and controlling FPA temperature

3. The overall source-to-sink conductance (including the evaporator and condenser interfaces) was 12.5 W/K for the case of both LHPs operating simultaneously with the highest heat load of 105 W versus 9.5 W/K for only one LHP working. For the lowest heat load of 75 W though, there was no gain in the conductance for both LHPs operating versus only one LHP running at a time (~9.5 W/K for either case). This is explained by the naturally occurring liquid blocking of the condenser end, which is more pronounced for lower power levels and results in a non-uniform temperature of the condenser plate interfacing with the sink.
4. Starting the first LHP from the controlled conditions was repeatedly done (and with the success rate of 100%) by applying only 10 W per heater to the two startup heaters. Subsequently starting the second LHP with the first LHP already operating at the heat load of 75 W or 105 W was done multiple times by first applying a control heater power of several Watts to the operating LHP reservoir and then turning on the startup heater on the second LHP.
5. Keeping the FPA interface temperature at 25°C±1°C with the cold plate (sink) temperature of -10°C was achieved within the control power budget of 6.5 W per LHP, as had been predicted by the thermal analyses.
6. The LHP system developed for GLM can be operated with only one LHP running or both LHPs running on demand, which means that the LHP system can be considered as fully redundant while still offering an option to increase the overall source-to-sink conductance by turning on the second LHP in a hypothetical situation with a significantly elevated heat dissipation from the source.

ACKNOWLEDGEMENTS

Authors acknowledge NASA/NOAA funding and overall program guidance.

NOMENCLATURE

C	conductance (W/K)
Q	heat flow rate (W)
T	temperature (K)
t	time (seconds)
Subscripts	
avg	average
cond	condenser
ev	evaporator
res	reservoir
Acronyms	
FPA	Focal Plane Array
GLM	Geostationary Lightning Mapper
IRSP	Independent Reservoir Set Point control
LHP	Loop Heat Pipe
MLI	Multi-layer insulation

REFERENCES

Ku, J, 2008, "Methods of Controlling the Loop heat Pipe Operating Temperature," *Proceedings of the 33rd International Conference on Environmental Systems," (ICES)*, SAE Technical Paper 2008-01-1998, 2008.

Bienert, W., Krotiuk, W., and Nikitkin, M. 1999, "Thermal Control with Low Power, Miniature Loop Heat Pipes," *Proceedings of 29th International Conference on Environmental Systems, (ICES),* SAE Technical Paper 1999-01-2008, 1999.

Khrustalev, D., 2010, "Advances in Transient Modeling of Loop Heat Pipe Systems with Multiple Components," *Proceedings of Space, Propulsion & Energy Sciences International Forum (SPESIF) 2010,* Laurel, MD, Applied Physics Laboratory, March 18-20, 2010.

Cullimore, B., and Bauman, J., 2000, "Steady State and Transient Loop Heat Pipe Modeling," *Proceedings of the 34th International Conference on Environmental Systems," (ICES),* SAE 2000-01-2316.

Christian and Gheno, 2012, "Geostationary Lightning Mapper," American Meteorological Society (AMS) 92nd Annual Meeting, January 24, 2012.

EXPERIMENTAL INVESTIGATION OF IN SITU PRESSURE MEASUREMENT OF AN OSCILLATING HEAT PIPE

Fritz F. Laun and Brent S. Taft*

Air Force Research Labs, Albuquerque, New Mexico, 87108, USA

ABSTRACT

An investigation using in situ pressure measurements was conducted on an oscillating heat pipe (OHP) to better understand its chaotic nature and the relation between the frequency of oscillations and the thermal performance. The working fluid used was HPLC grade acetone with a 0.8 fill ratio by mass. Three different orientations were tested: top, horizontal, and bottom heating from 100 to 500 W in increments of 50 W. An aluminum water block was used with water at approximately 0°C with a flow rate of 0.9 l/min for the condenser section of the OHP. The condenser and evaporator section were each fitted with a pressure transducer mounted perpendicular to the OHP channel. The pressure and temperature measurements were compared. Complex pressure wave forms were observed and a natural frequency could not be determined at the high filling ratio used.

Keywords: pulsating heat pipe, PHP, OHP, two-phase, thermal spreader

1. INTRODUCTION

It is estimated that future satellite missions will require a heat-rejection capability measured in the 100's of W/cm^2, consistent with the general electronics trends of miniaturization and increasing power densities. Current generations of thermal spreaders available to deal with these large heat fluxes add significant weight, have a poor form factor, consume power, and can drastically increase the cost of manufacturing the final product. The passive thermal technologies available today are not able to deal with such high heat fluxes. However, a type of passive thermal system, the oscillating heat pipe (OHP), has the potential to successfully perform with high waste heat removal and minimal manufacturing cost.

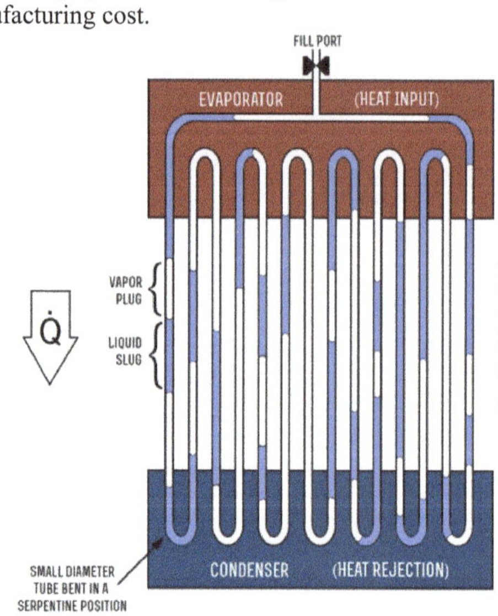

Fig. 1 Schematic of an oscillating heat pipe (Taft *et al.*, 2013)

With the development of the OHP in 1990, a great deal of research has gone into understanding its unique fluid flow characteristics (Akachi, 1990; Akachi 1993). However, electronics were less likely to be thermally limited and there was little need for a high heat flux thermal spreader. Recently attention has shifted to OHP research as electronics produce greater amounts of waste heat. The operating mechanisms of the OHP are still not well understood and further research must be done to create a viable thermal spreader.

The OHP, shown in Fig. 1, is a type of two-phase heat transfer device that is partially filled with a working fluid. The oscillatory motion of the working fluid efficiently transfers heat as a simply-formed, wickless heat pipe (Bhuwakietkumjohn and Rittidech, 2010). The two-phase oscillatory motion is induced by a pressure differential from the condenser (cold end) and the evaporator (hot end) that efficiently transfers heat. With adequate power, the temperature differential between the condenser and the evaporator can maintain an almost constant temperature. Typical OHPs exist as a single loop serpentine tubular design (T-OHP) or are machined on a flat-plate that is sealed with a cover sheet (FP-OHP). FP-OHPs have a better form factor and are able to deal with high heat fluxes (Borgenmeyer and Ma, 2007).

The primary operating mechanisms of classical heat pipes and OHPs differ substantially. OHP operation may be considered gravity independent at high power levels, while classic heat pipes are highly dependent on gravitational effects. Classic heat pipes require an intricate wick structure to promote capillary action to transport liquid back to the condenser while OHPs do not, greatly reducing the manufacturing cost (Dolgirev *et al.*, 2003). For high heat flux applications OHPs have a greater potential for successful operation than classic heat pipes (Zuo *et al.*, 2001). The heat transfer process in a classic heat pipe is primarily caused by latent heat transfer, while the OHP uses the phase change heat transfer to create the oscillatory motion, resulting in sensible heat that accounts for the bulk of heat transfer. These unique features make the OHP an ideal solution to future thermal cooling needs as power densities continue to increase. This paper investigates the pressure oscillations created inside the internal OHP channels due to the two-phase motion of the working fluid. A brief summary of existing literature is presented before discussing the experimental setup, results, and conclusion. It was found

* *Corresponding author, Email: afrl.rvsv@kirtland.af.mil*

that a dominate frequency of oscillations in the OHP did not exist and for a single temperature oscillation there are many pressure oscillations present.

2. BACKGROUND

Experimental and numerical investigations have been conducted to better characterize the pressure mechanism during OHP operation. Miyazaki and Akachi (1996) varied the heat flux and charging ratio to study the pressure oscillation characteristics of an OHP. Oscillation waves of long liquid slugs and vapor plugs were observed. An analytical model was then developed and it was determined that better heat transfer performance was achieved when frequency and amplitude of pressure oscillations was increased and decreased, respectively.

Kim et al. also varied the heat flux and charging ratio in an effort to better understand the pressure oscillation characteristics (Kim *et al.*, 2003). Experimental results were able to quantify a frequency of pressure oscillations between 0.1 Hz and 1.5 Hz at a charging ratio of 40% by volume. It was found that when the pressure waves were symmetric and sinusoidal the heat transfer performance was improved. However, at 20 and 80% fill rates the pressure wave forms were too complicated and thus a fundamental frequency could not be found and heat transfer performance was reduced. Suzuki (2000) studied the temperature oscillation characteristics of a copper OHP using water as a working fluid. The frequency of the temperature oscillations was found to be proportional to the effective thermal conductivity. A relation was developed between the effective thermal conductivity and the frequency of the temperature oscillations.

3. EXPERIMENTAL SETUP

A FP-OHP design, as shown in Fig. 2, was constructed as an embedded OHP in an aluminum structural panel. The OHP has a total of 20 turns and an overall channel length of 20 m. Traditionally an OHP may be manufactured in two separate pieces, later brazed together. However, ultrasonic consolidation (UC) was used to manufacture the OHP. UC is an additive/subtractive manufacturing technique that is ideally suited for the production of structurally integrated OHPs. UC allows for the creation of high strength materials with internal features that significantly exceed the strength to weight ratio of many current aerospace materials. A true metallurgic bond is created using UC deposition and CNC removal.

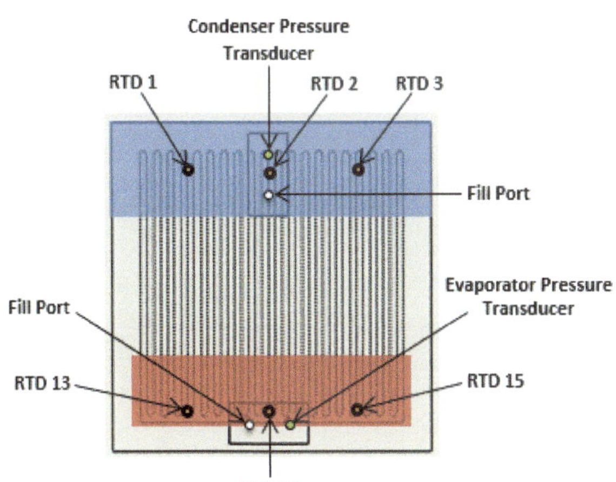

Fig. 2 Schematic of the FP-OHP design with the evaporator section highlighted in red and the condenser section in blue

In order to read pressure measurements inside the OHP, Swagelok VCR fittings and Omega pressure transducers were integrated into the panel in both the condenser and evaporator section. After leak testing using a helium leak detector to a leak rate of 10E-7 atm/cc/sec, a vacuum (3E10-3 Torr) was pulled on the sealed OHP and acetone (high performance liquid chromatography grade) was added with 0.8 fill ratio by mass (±0.01). The high fill ratio was used to help decrease orientation dependency, ensuring liquid was always present in the evaporator during startup.

The channel diameter used, 1.3 x 1.3 mm^2, adheres to the maximum allowable diameter for liquid to form vapor bubbles in a capillary tube when using acetone as a working fluid (Taft *et al.*, 2012):

$$D_{max} = 1.84\sqrt{\frac{\sigma}{g(\rho_l - \rho_v)}} \tag{1}$$

4. METHOD

The temperature differential of the OHP was measured using the experimental setup shown in Fig. 3. The system consisted of power supply, cooling bath, heater, aluminum water block, and data acquisition system (DAQ). Water was circulated through the aluminum water block at 0.9 l/min using a submergible pump in an ice bath. Heating and cooling were conducted on an area of 8 x 30 cm^2 and 15 x 30 cm^2, respectively. The adiabatic section was approximately 7.5 cm in length. Thermal paste (Artic Silver 5) was used as the thermal interface material between all contact surfaces. Four C-clamps were used with insulating pads to apply clamping pressure for the OHP to both the water block and the heater.

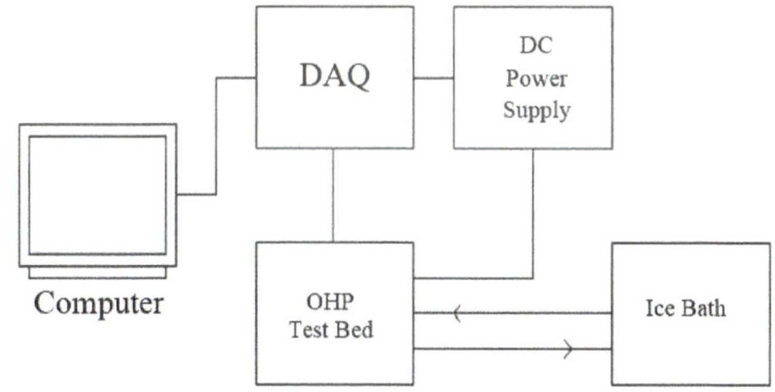

Fig. 3 Schematic of oscillating heat pipe experimental setup

A total of 16 thin film RTDs (100 Ohm) were placed at the surface of the OHP with an uncertainty of ±0.75°C and a response time of approximately 250 ms (90%). Temperature measurements were recorded at 10 Hz using a National Instruments data acquisition system with a customized Virtual Instrument (VI). The heat input was provided with a Lambda 150-10 power supply with a resolution of ±0.01 W. A test frame capable of horizontal and vertical orientations was used to support the OHP, heaters, and water block. High temperature foam insulation was used to insulate the system from the surrounding environment. Pressure measurements were recorded using two Omega pressure transducers, 0 - 0.2 and 0 - 3.4 MPa, in the condenser and evaporator section of the OHP. Pressure readings had an uncertainty of ±0.7 and 1.7 kPa for the condenser and evaporator section, respectively.

Heat input was increased from an initial 100 W in increments of 50 W to a maximum power output of 500 W. The power was measured using a voltage divider and current measurement card and recorded at 10 Hz. Each power level was held until a steady state condition was achieved and was then recorded for approximately five minutes. This procedure was performed for each orientation of the OHP three separate times to effectively produce a range of thermal resistances at each power level and orientation. This repetition helps to account for the variability in OHP performance resulting from the OHPs chaotic nature.

5. RESULTS AND DISCUSSION

The thermal resistance of the OHP was calculated by Eq. (2) over a period of five minutes of steady state conditions.

$$R_{OHP} = \frac{1}{3P}\left[\sum_{i=1}^{3} RTD_i - \sum_{i=13}^{15} RTD_i\right] \tag{2}$$

where $RTD_{1\text{-}3}$ and $RTD_{13\text{-}15}$ are located in the evaporator and condenser, respectively and P is the power input. The percent uncertainty in the calculated thermal resistance was calculated to be approximately 2%.

Oscillations began at a power level of 200 W, the condenser section data is shown in detail in Fig. 4. It can be seen that as the temperature readings from the RTDs increased, the pressure readings began to oscillate, indicating the movement and oscillations of liquid slugs and vapor plugs. As the temperature readings decreased, the pressure oscillations ceased as there was no longer enough energy to drive the oscillating motion. It can be seen that for one oscillation of an RTD there are many oscillations of liquid slugs and vapor plugs, creating the forced convection in the channel of the OHP that transfers sensible heat while phase change is the driving force, adding latent heat transfer.

The RTD measurements are damped due to the 1.6 mm wall between it and the OHP working fluid and are not a good indicator of actual oscillatory movement of liquid slugs and vapor plugs. The vapor plugs located in the condenser section are correlated with a decrease in the condenser temperature indicating a substantially decreased heat transfer compared to the oscillation of liquid slugs.

The theoretical saturation pressure of acetone may be compared to the observed pressure measurements at a given temperature. For comparison, at room temperature (~21°C) the OHP has a measured pressure of approximately 46.9 kPa in both the condenser and evaporator sections. The saturation pressure of acetone is 25.7 kPa at 21°C indicating that there may have been outgassing after the filling procedure. Air is more likely to be trapped in the pressure transducers as they are mounted above and perpendicular to the OHP channels. As the temperature of the acetone increased the difference between the theoretical and measured value of the pressure decreased. This may have been caused by the chaotic movement of the OHP dispersing any air pockets throughout the system. However, the OHP internal pressure at room temperature was consistent at 46.9 kPa for the entire duration of testing and several months later. It is most likely a common occurrence for OHPs to have a slight amount of air inside after filling, but simply cannot be reported in the majority of literature as pressure transducers are not commonly placed in OHPs after the filling process.

As the power input to the OHP increased the pressure drop caused by liquid slugs became more consistent and increased in frequency. This also corresponded to the frequency of temperature oscillations increasing with the increase of power. At every power level the condenser temperature always increased with the presence of liquid slugs and decreased when only vapor was present. The pressure difference between liquid slugs and vapor plugs at each power level remained fairly constant at approximately 24.1 kPa as the overall pressure increased due to the increased temperature.

The thermal resistance of the OHP decreased as the power level increased, as shown in **Error! Reference source not found.**. However, since three trials were used at each orientation and power level a range of thermal resistances was created. Overall the range of thermal resistance decreased as the power level increased as oscillations became more consistent. It was found that before oscillatory startup at low power levels (100 to 200 W) the bottom heating had better thermal performance as it created natural convection as a themosyphon. Once oscillations began (200 W) the orientation had little effect on thermal resistance of the OHP. This may have been caused by the high fill ratio (0.8 by mass) and there was too much liquid in the system, hindering performance.

A Fourier transformation was performed on the pressure oscillations in an attempt to find the natural frequency. The frequency of the pressure oscillations for the 200 and 500 W tests in the horizontal orientation are shown in

Fig. 6. It can been seen that there was no natural frequency found. This may be due to the OHP being over filled.

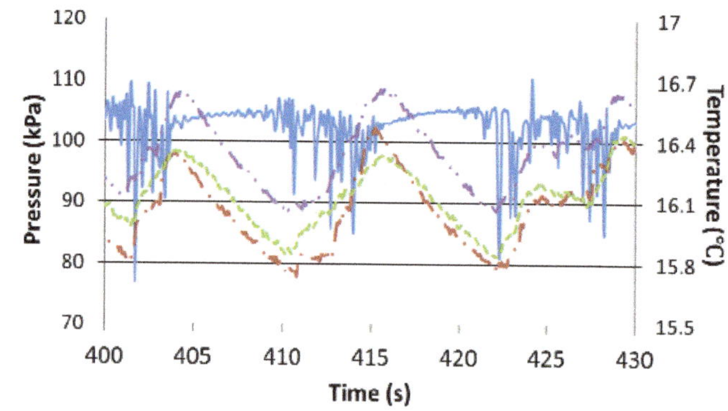

(a) q = 200 W

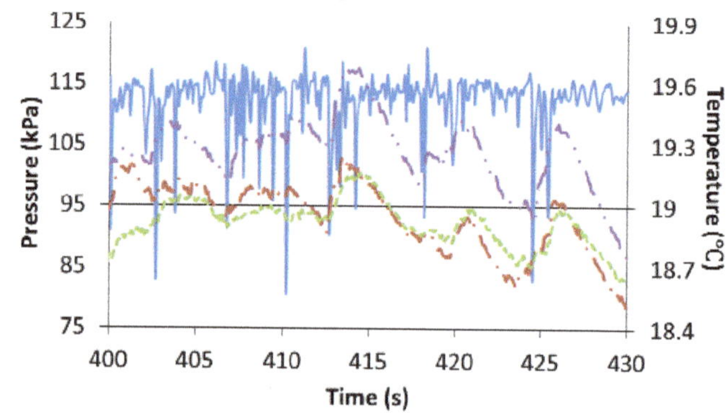

(b) q = 250 W

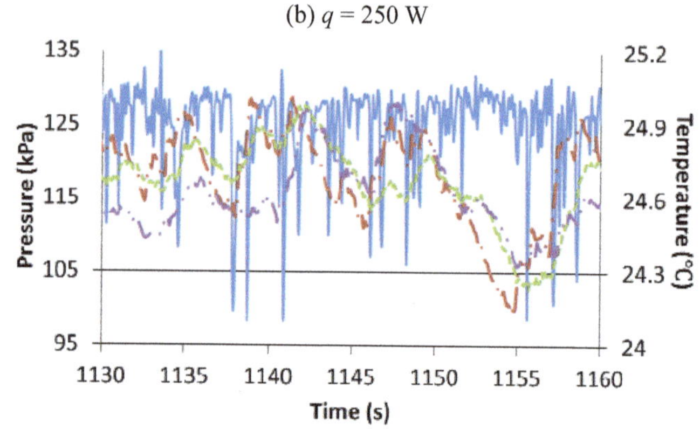

(c) q = 300 W

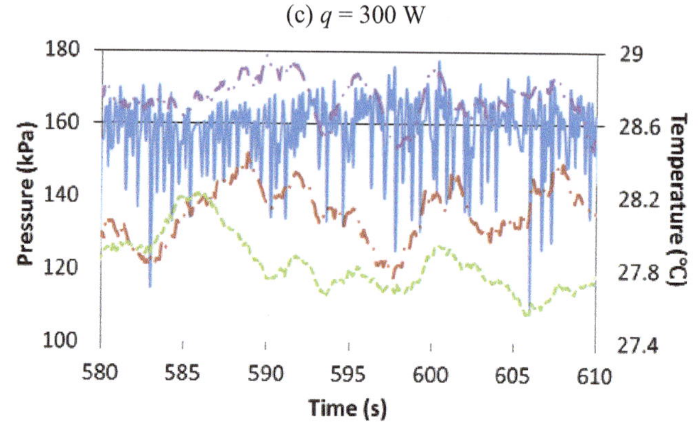

(d) q = 500 W

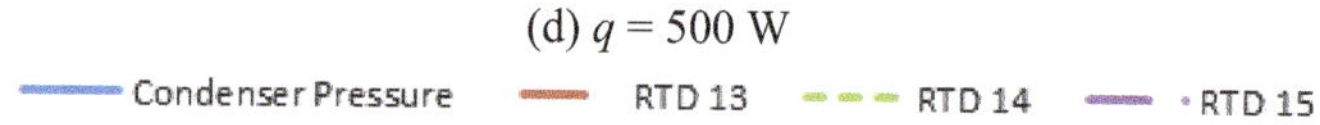

Fig. 4 Pressure and temperature measurements in the condenser section of the OHP at (a) 200 W, (b) 250 W, (c) 300 W, and (d) 500 W in the horizontal orientation

A phenomenon also observed by Kim et al., where it was found that at high fill ratios the OHP pressure wave forms were too complicated and a fundamental frequency could not be found (Kim *et al.*, 2003).

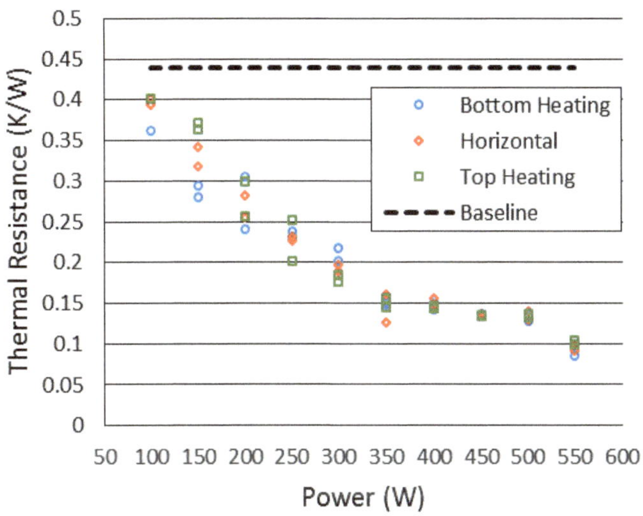

Fig. 5 Thermal resistance of the OHP for various power levels and orientations

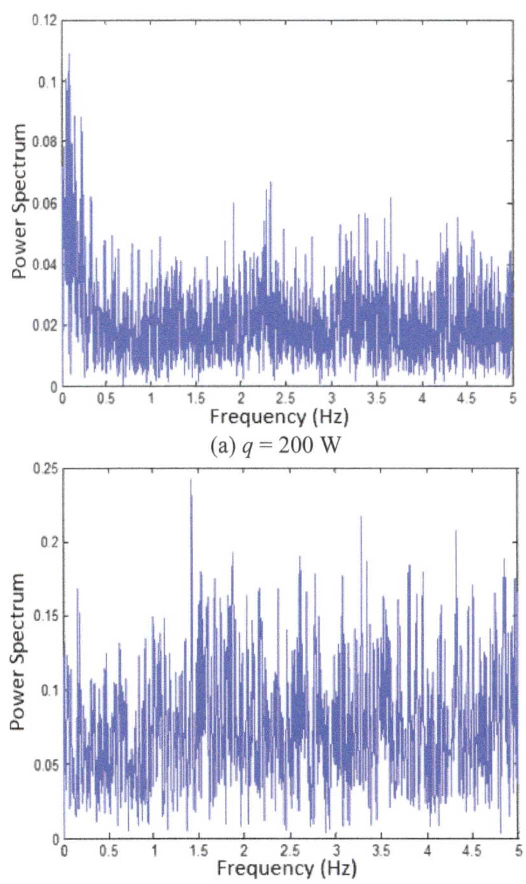

Fig. 6 Power spectra for the pressure oscillations of the OHP at (a) 200 W and (b) 500 W in the horizontal orientation.

6. CONCLUSIONS

It is important to find a working relationship between pressure and thermal performance in order to better quantify an optimum fill ratio for OHPs and better understand their chaotic nature. In this study, a definitive correlation between the frequency of pressure oscillations and the heat transfer performance of an OHP was not found. The pressure wave forms were too complicated and a natural frequency could not be found at a high filling ratio. It is worth pursuing in future research the effect of fill ratio on the frequency of pressure oscillations and the heat transfer performance of an OHP.

NOMENCLATURE

C	conductance (W/K)
P	power input (W)
P	pressure (Pa)
q	heat (W)
R	thermal resistance (K/W)
T	temperature (°C)

Greek Symbols

ρ	density (kg/m^3)
σ	surface tension (N/m)

Subscripts

l	liquid
OHP	oscillating heat pipe
v	vapor

REFERENCES

Akachi, H., 1990, "Structure of a Heat Pipe," U.S. Patent No. 4,921,041.

Akachi, H., 1993, "Structure of Micro-Heat Pipe," U.S. Patent No. 5219020.

Borgenmeyer, B., and Ma, H.B., 2007, "Experimental Investigation of Oscillating Motions in a Flat Plate Pulsating Heat Pipe," Journal of Thermophysics and Heat Transfer, 21(2), pp.405-409. http://dx.doi.org/10.2514/1.23263

Kim, J.S., Bui, N., Jung, H.S., and Lee, W.H., 2003, "The Study on Pressure Oscillation and Heat Transfer Characteristics of Oscillating Capillary Tube Heat Pipe," Journal of Mechanical Science and Technology, Vol. 17, No. 10, October 1, pp. 1533-1542.

Miyazaki, Y. and Akachi, H., 1996, "Heat Transfer Characteristics of Looped Capillary Heat Pipe," Proceeding of the 5th International Heat Pipe Symposium, Melbourne, pp. 378--383.

Bhuwakietkumjohn, N. and Rittidech, S., 2010, "Internal Flow Patterns on Heat Transfer Characteristics of a Closed-Loop Oscillating Heat-Pipe with Check Valves using Ethanol and a Silver Nano-Ethanol Mixture," Experimental Thermal and Fluid Science, Volume 34, Issue 8, Pages 1000-1007. http://dx.doi.org/10.1016/j.expthermflusci.2010.03.003

Suzuki, O., 2000-5, 'Heat transfer characteristics of a Bubble-Driven Capillary Heat Pipe,"37th National Heat Transfer Symposium of Japan, pp. 35--36.

Taft, B.S., 2013, "Non-Condensable Gases and Oscillating Heat Pipe Operation," Frontiers in Heat Pipes, Vol. 4, 013003 http://dx.doi.org/10.5098/fhp.v4.1.3003

Taft, B.S., Williams, A.D., and Drolen, B.L., 2012, "Review of Pulsating Heat Pipe Working Fluid Selection," Journal of Themophysics and Heat Transfer, Vol. 26, No. 4, pp. 651-656. http://dx.doi.org/10.2514/1.T3768

Dolgirev, Y.E., Gerasimov, Y.F., and Melkikh, A.V., 2003, "Theoretical and Experimental Study of Oscillating Heat Pipes with Few Turns," Journal of Engineering Physics and Thermophysics, Vol. 76, No. 5, September, pp. 996-1000. http://dx.doi.org/10.1023/B:JOEP.0000003212.51653.4a

Zuo, Z.J., North , M.T., and Wert, K.L., 2001, "High Heat Flux Heat Pipe Mechanism for Cooling Of Electronics," IEEE Transactions on Components and Packaging Technologies, Vol. 24, No. 2, June, pp. 220-225.
http://dx.doi.org/10.1109/6144.926386

AN EXPERIMENTAL PROCEDURE TO COMPARE THERMAL PERFORMANCE OF THERMOSYPHONS RAPIDLY

K. S. Ong* and Christopher Lim

Universiti Tunku Abdul Rahman, 31900 Kampar, Perak, Malaysia

ABSTRACT

There are various methods employed to determine the performance of heat pipes and two-phase closed thermosyphons. Perhaps the most common method involves measuring the rate of heat transfer from the evaporator to the condenser section with a water cooled jacket surrounding the latter. This entails higher costs and requires extended testing times. A method is proposed here to obtain comparative performance test data on various heat pipes rapidly and economically. The method does not require fabricating a water jacket. It merely involves measuring and comparing the transient temperature of a pre-heated container of hot water obtained first without the heat pipe and then with the heat pipe immersed into the container. The more efficient system will result in faster cooling of the container. The method is demonstrated using a number of different thermosyphons employing both water jackets and air cooled finned condensers. Their performances are compared in the paper.

Keywords: *Thermosyphon, comparative performance, rapid experimental procedure, air cooled condenser, water cooled condenser.*

1. INTRODUCTION

Heat pipes (HPs) are devices capable of transferring large quantities of heat effectively and efficiently. A HP typically consists of a sealed container lined with a wicking material. The container is evacuated and filled with just enough liquid to fully saturate the wick. A wickless HP is also known as a thermosyphon, shown in Fig.1. Heat applied to the evaporator section causes it to boil and vaporize, picking up latent heat of vaporization. The vapor then rises to the condenser section where it condenses. Here vapor gives up its latent heat of condensation. The condensate then travels back to the evaporator section of the heat pipe by gravity or capillary action. The performance of HPs depend on various factors such as type of fill liquid, fill ratio (FR), heat input at the evaporator section, inclination and physical dimensions of the pipe.

The performance of a HP could be evaluated by determining the amount of heat input at the evaporator section or by measuring the heat transferred at the condenser section to the coolant. At steady state, these two heat transfer rates should be equal if there is no heat loss from the pipe. In order to compare the performances of various heat pipes the heat transfer rates could be plotted against the temperature difference between the evaporator and condenser sections. This would give an indication of the thermal resistance of the pipe. A high thermal resistance would indicate a poor performing HP and vice versa.

Ong et al. (2014) investigated the performance of water and R410a filled thermosyphon at various power inputs (100 to 830 W), fill ratios (0.25 to 0.93) and inclinations (30° to 90°). The authors used a 38 mm OD x 807 mm long thermosyphon with a 71 mm OD cooling water jacket and electrical heating bands to supply heat to the evaporator section. By plotting input power against temperature difference between evaporator and condenser sections, found that the R410a filled thermosyphon performed better in the vertical position at all FR's and that the water filled thermosyphon performed better at low FR and when inclined. They showed that the evaporator wall temperature was not uniform especially at high power inputs, low fill ratios and large inclinations.

Shanthi and Velraj (2014) experimented with a two phase gravity assisted thermosyphon. The authors used a 12.5 mm ID x 300 mm long thermosyphon with a 200 mm long cooling water jacket and a 1000 W nichrome heating wire wound around the 75 mm long evaporator section. They measured the performance of their thermosyphon by plotting efficiency defined as ratio of power output over power input, versus power input and obtained values as high as 0.9. Anjakar and Yarasu (2012) determined the effect of condenser length on the performance of thermosyphon by using a 32 mm OD x 1000 mm long thermosyphon with a 450 mm long water cooled jacket and a 300 mm long nichrome heating element at the evaporator section. The authors determined the efficiency using the ratio of heat conducted away by the coolant water at the condenser section to the power input at the evaporator section and found values of 0.5 to 0.95. In theory the ratio should be equal to 1.0.

Buschmann and Franzke (2013) compared the performance of thermosyphons filled with deionised water (DI) and titanium oxide-based DI water and gold nanofluid. They showed that the nanofluids reduced thermal resistances up to 24 %. They also noted that nanoparticles depositing on the evaporator wall after prolonged heating operations resulted in an absence of nanoparticles being transported to the condenser section that eventually reduced the thermal transport capacity of the nanofluid and increased the overall thermal resistance. Huminic et al. (2011) experimented with thermosyphons filled with iron oxide nano particles and found that a concentration level of 5.3 % improved the thermal performance of their HP. Solomon et al. (2012) determined the thermal performance of a heat pipe with a nano particles-coated wick. They found that the thermal resistance reduced from 0.32 to 0.11 °C/W when subjected to power input ranging from 100 to 200 W, respectively.

Qu and Wang (2013) showed that the thermosyphon with a FR of 40 to 50 % resulted in faster response time and lowest overall thermal resistance. Shabgard et al. (2014) studied the thermal characteristics of

* *Corresponding author. Email: skong@utar.edu.my*

HPs with different FRs. They showed that when the HP is underfilled, dry out occurs and that overfilling causes the overall thermal resistance to increase. The authors suggested that HPs should be filled slightly more than the optimal FR to prevent the breakdown of liquid film.

Qu and Wang (2013) showed that high heat flux results in lower thermal resistance of the evaporator section. Wang (2009) investigated the transient thermal performance of a bent HP with internal grooved surface and showed that straight pipes responded faster to high heat flux compared to bent shape HPs. Wang also found heat pipes have response times varies from 50 to180 s depending on the pipe angle from 0 to 140°.

HPs behave differently when inclined at different angles. Wang (2009) noted that with the increase in inclination, the temperature difference between the evaporator and condenser section increased. Huminic and Huminic (2011) investigated the heat transfer characteristics of two-phase closed thermosyphons using nanofluids. The authors plotted thermal resistance versus inclination angle to determine the performance of their HP. They found that water has the highest thermal resistance of 0.05 K/W compared to iron oxide nano-particles.

In their investigation on using heat pipes to enhance heat transfer during the charging and discharging rates of hydrogen in metal hydride tanks, Chung et al. (2013) found that the thermal resistance of their heat pipe ranged from 1.6 to 0.6 K/W at power inputs from 20 to 80 W.

Celata et al. (2010) investigated water filled stainless steel flat evaporators for space application. The authors plotted thermal resistance versus power input and found the overall thermal resistance decreased from 50.7 to 3.3 °C/W when power input increased from 20 to 80 W. Chang et al. (2012) investigated the thermal performance of two-phase water filled closed loop thermosyphons and plotted the thermal resistance versus power input. They found that the thermal resistance decreased from 0.30 to 0.05 K/W as the power input increased from 90 to 210 W. Solomon et al. (2013) determined the thermal performance of anodized two-phase closed thermosyphon by using a 16.5 mm ID x 350 mm long thermosyphon with a 150 mm long water cooled jacket and a 100 mm long resistance heater (1000W) at the evaporator section. The authors plotted the thermal resistance versus power input and found that the resistance reduced from 0.4 to 0.2 °C/W as power input increased from 50 to 250 W.

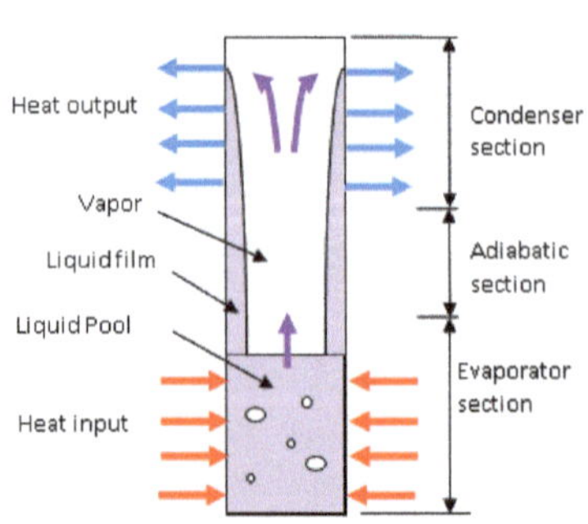

Fig. 1. Cross sectional view of a thermosyphon.

2. OBJECTIVE OF INVESTIGATION

All the experiments performed to determine the thermal performance of a thermosyphon involved measuring the rate of heat transfer from the evaporator to the condenser section via a water cooled jacket surrounding the latter. This entails high costs and requires extended testing times. A cheaper and quicker test method is required. This paper proposes a simple and quick method to obtain and to compare the performance of different types of thermosyphons.

3. METHODOLOGY

The proposed method involves, firstly, determining the transient temperature of an insulated container containing pre-heated hot water. Next the thermosyphon under investigation is inserted into the container and the temperature recorded. Heat is removed from the container by the thermosyphon throughout the cooling process and also through natural convection heat loss from the side walls of the container. The total heat removed from the container, $\dot{q}_b$, can be calculated from the transient bath temperature versus time results, viz.,

$$\dot{q}_b = W_b c_{pw} \left| \frac{dT_b}{d\theta} \right| \tag{1}$$

where, W_b is the weight of the water in the container, c_{pw} is the specific heat of water and $dT_b/d\theta$ is the rate of change of bath temperature with time.

In the case of a thermosyphon provided with a concentric pipe water cooling jacket, heat transfer from the condenser section, $\dot{q}_w$, of the thermosyphon to the cooling water is calculated from

$$\dot{q}_w = \dot{m}_w c_{pw} (T_{wo} - T_{wi}) \tag{2}$$

where, $\dot{m}_w$ is the mass flow rate of water, T_{wo} is the cooling water outlet temperature and T_{wi} is the inlet temperature.

The natural convection heat loss, $\dot{q}_{ins}$, from the container to the ambient can be estimated from

$$\dot{q}_{ins} = h_a A_{ins} (T_{ins} - T_a) \tag{3}$$

where, h_a is the natural convection heat transfer coefficient to the ambient, A_{ins} is the area of the insulation, T_{ins} is the insulation surface temperature and T_a is the surface ambient temperature.

The total heat removed from the container is the sum of the heat transfer to the coolant plus the natural convection heat loss, viz.,

$$\dot{q}_b = \dot{q}_w + \dot{q}_{ins} \tag{4}$$

A cooling index, γ, is introduced here, defined as the ratio of heat transfer to the coolant/total heat removed from the container, viz.,

$$\gamma = \frac{\dot{q}_w}{\dot{q}_b} \tag{5}$$

or

$$\gamma = \frac{\dot{q}_w}{\dot{q}_w + \dot{q}_{ins}} \tag{6}$$

In the absence of heat loss to the ambient, $\gamma = 1.00$.

The mean operating temperature difference between bath and coolant water is defined as

$$\Delta T_w = (T_b - T_w) \tag{7}$$

where, T_b is the mean bath temperature and T_w is the mean coolant water temperature defined by.

$$T_w = \frac{T_{wo} + T_{wi}}{2} \tag{8}$$

4. EXPERIMENTAL INVESTIGATION

The experimental set up is shown in Fig. 2. The equipment consisted of a 3.2 litre capacity hot water container (± 0.001 litre). All temperatures were measured using type T copper-constantan thermocouples (± 0.5 °C) connected to a data logger and logged every minute. Water temperature in the container was measured using two thermocouples immersed in the upper (280 mm from the base) and lower portions (90mm from the base) of the container. Temperature differences less than 3 °C were obtained. Mean container water temperature was taken as the arithmetic mean of the two positions. Two more thermocouples were attached to the outside of the insulated container to determine the external surface temperature of the container in order to calculate the natural convection heat loss from the external surface of the free-standing container. These thermocouples were insulated from the atmosphere by duct tape. Ambient temperature was measured with another thermocouple located nearby. Cross-pieces were attached to the top of each thermosyphon. It held a pressure gauge to measure the saturation pressure, a thermocouple to measure the saturation temperature and a filling valve. Inlet and outlet water temperatures were measured with thermocouples inserted inside the connecting plastic hoses of the cooling jacket. Experiments were conducted at two water flow rates, viz., at 0.9 and 50.0 ml/s (± 1 ml/s).

In order to demonstrate the effectiveness of the method, five copper thermosyphons were fabricated. Dimensions of the thermosyphon are shown in Table 1 and in Fig. 3 to Fig. 6. The first unit (HP # 1) is a water cooled type. The condenser section consisted of a 38 mm diameter x 127 mm long copper water cooled jacket. Thermocouples were mechanically attached to the adiabatic and condenser sections of the thermosyphons shown in Fig. 3 The second unit (HP # 2) is an air cooled type with a stainless steel spiral fin measuring 30 mm diameter x 0.5 mm thick stainless steel fins spaced 5 mm apart. The air cooled HPs # 3, 4, 5 were provided with parallel aluminum fins each 0.5 mm thick x 22.5 mm square x 2 mm pitch at the condenser section respectively. . The location of the thermocouples are shown in Figs. 4 to Fig. 6. Force air circulation rate was provided with an electric fan. Air flow was measured with a hot wire anemometer (± 0.1 m/s). For natural convection the fan was switched off.

Altogether eight experimental runs were conducted. The experimental procedure involved heating up the insulated container of water to a temperature of 100 °C and then allowing it to cool down. The water temperature was measured throughout the cooling process. Experiments were repeated by inserting and immersing the various thermosyphons to be tested into the container and measuring the temperature of the hot water as it cooled to 50 °C. Tests were conducted with various water flow rates and under natural or force air circulation. Each experimental run was repeated three times and the average results were obtained and plotted. Details of the test conditions under which the experiments were carried out are tabulated in Table 1.

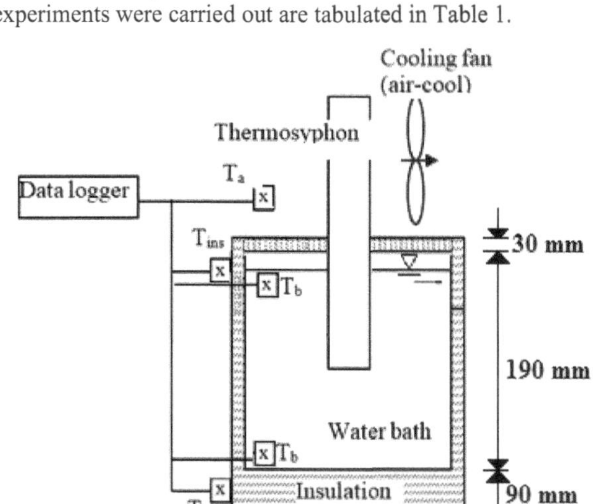

Fig. 2. Schematic of experimental set up.

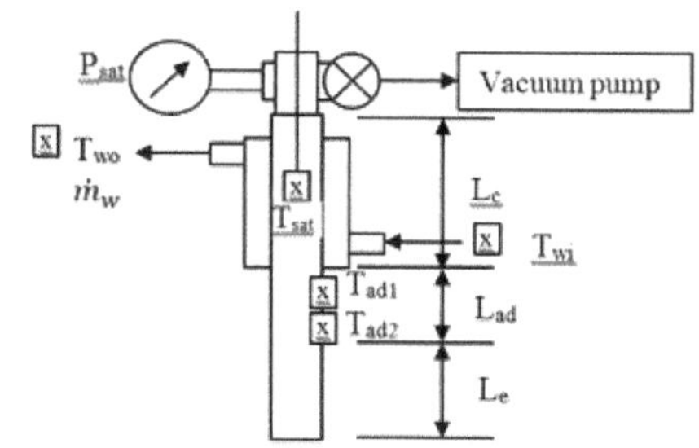

Fig. 3. Details of water cooled thermosyphon (HP1).

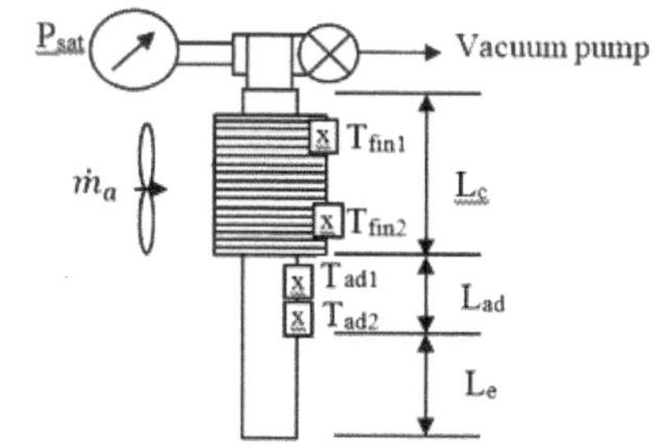

Fig. 4. Details of air cooled thermosyphon (HP2&3).

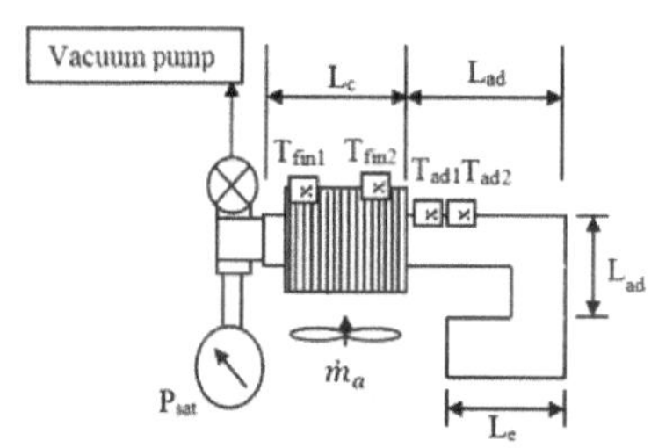

Fig. 5. Details of air cooled thermosyphon (HP4).

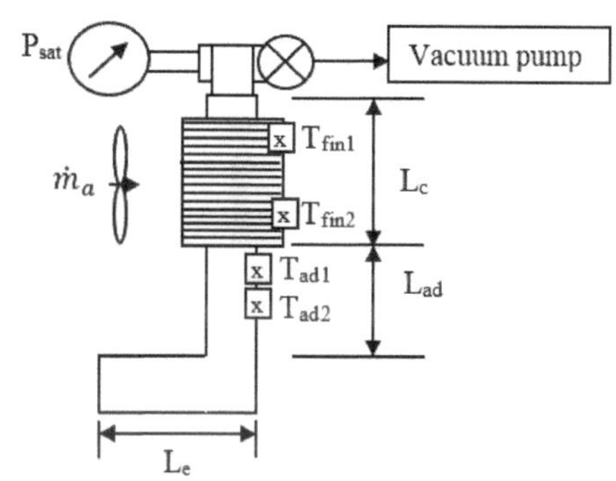

Fig. 6. Details of air cooled thermosyphon (HP5).

Table 1. Dimensions and experimental Runs conducted.

Run	HP #	Pipe shape	Coolant	Flow rate (ml/s)	D_e (cm)	L_e (cm)	L_c (cm)	L_{ad} (cm)	A_e (cm^2)	L_e/D_e	L_c/D_e	L_e/L_c
1	-	-	-	-	-	-	-	-	-	-	-	-
2	1	straight	water	0.9	2.54	12.7	12.7	5.0	101	5	5	1
3	1	straight	water	50.0	2.54	12.7	12.7	5.0	101	5	5	1
4	2	straight	air	0.0	1.2	12.7	12.7	5.0	48	11	11	1
5	2	straight	air	2.0	1.2	12.7	12.7	5.0	48	11	11	1
6	3	straight	air	2.0	0.9	12.7	25.4	5.0	36	14	28	0.5
7	4	J shape	air	2.0	0.9	12.7	25.4	75	36	14	28	0.5
8	5	L shape	air	2.0	0.9	12.7	25.4	75	36	14	28	0.5

5. EXPERIMENTAL RESULTS

5.1 Transient temperature

A typical set of transient temperature results obtained for Run 2 with minimum coolant flow rate of 0.9 ml/s is shown in Fig 7. Mean water container (T_b), coolant water inlet (T_{wi}), coolant water outlet (T_{wo}), mean insulation (T_{ins}) and ambient temperatures are plotted. The results show that it took more than 3 hours for the water to be cooled down from 100°C to 50°C with the type HP1 thermosyphon. The experiment was stopped after 3.33 hours. A quadratic temperature-time relationship ($T_b = 0.0013\ \theta^2 - 0.47\ \theta + 95$) was obtained for the mean container temperature with a regression coefficient of 0.9866. The coolant water outlet temperature dropped from 72°C initially to 30°C after 3.33 hours. The temperature difference between inlet and outlet of the coolant water was about 44°C initially and about 5°C after 2 hours of cooling indicating that not much cooling could be performed when the container temperature dropped to below about 50°C.

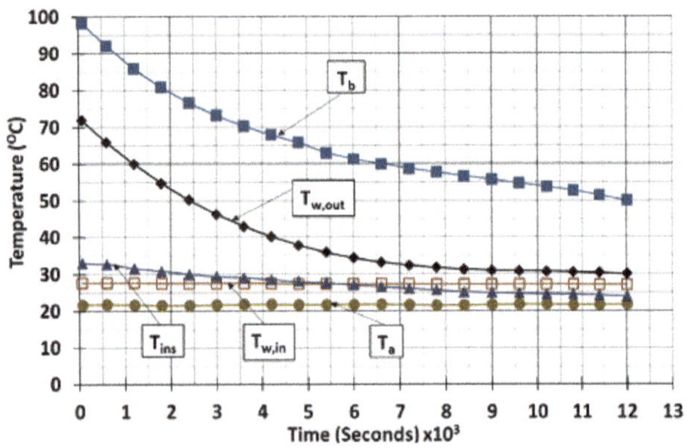

Fig. 7 Typical transient temperatures for Run 2

5.2 Heat transfer cooling rates

Typical heat transfer cooling rates for Run 2 are plotted in Fig. 8 for up to 2 hours of operation. The heat removed from the container ($\dot{q}_b$), heat transfer rate to the coolant water ($\dot{q}_w$) and the natural convection heat loss ($\dot{q}_{ins}$) are plotted together with the cooling index (γ). The results show that the cooling rate of the thermosyphon decreased with time or as the container cools which is to be expected. The heat loss from the container ($\dot{q}_{ins}$) is small. $\dot{q}_{ins} = 3$ W at the start of the experiment with $h_a = 1.5$ W/m²K. Towards the end of the experiment $\dot{q}_{ins} = 2$ W with $h_a = 1.5$ W/m²K. γ varied from 1.00 to 0.98. With $h_a = 2.0$ W/m²K γ varied from 0.98 to 0.93. The variation in the γ factor is attributed to the value of h_a assumed to calculate for heat loss to the ambient.

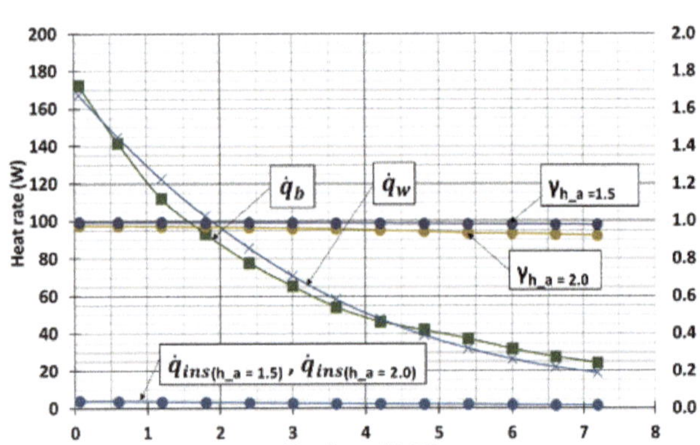

Fig. 8 Typical cooling heat rates for Run 2

5.3 Comparative performance of thermosyphons

The performances of the thermosyphons could be compared from the transient container temperature results shown plotted in Fig. 9. The rate at which container temperature decreases is a measure of the cooling efficiency of the thermosyphon. The result of natural cooling without any thermosyphon immersed in the container (Run 1) is shown as a base for comparison. It can be seen that the cooling rate is the slowest. All the results show that systems with the thermosyphons increased the cooling rate. For natural cooling (Run 1), it took nearly 2 hours to cool the water from 100 °C to 88 °C. With the type HP1 water cooled thermosyphon (Run 2) at the low coolant flow rate of 0.9 ml/s, 100 °C water was cooled to 60 °C in 2 hours. The same thermosyphon performed better at high coolant flow rate (Run 3) than with the low flow rate (Run 2) as expected. The results also show that force convection air cooling (Run 5) is better than natural convection air cooling (Run 4) for the type HP2 thermosyphon, as expected. From the results it would seem that the L-shaped type HP5 thermosyphon (Run 8) performed better than the straight pipe of type HP3 (Run 6) followed by the J-shape type HP4 (Run 7) unit. The difference in performance could be due to the shape of pipes. For the air cooled type HP2 & HP3 pipes, Run 6 with the longer condenser length performs better than the shorter one in Run 5. It should be pointed out here that the thermosyphons are of different sizes and the lengths of the evaporator sections that were immersed in the hot water varied from pipe to pipe as indicated in Table 1. Hence, in order to determine which geometry pipe performs better, it would be necessary to fabricate and compare pipes with the same equivalent sizes.

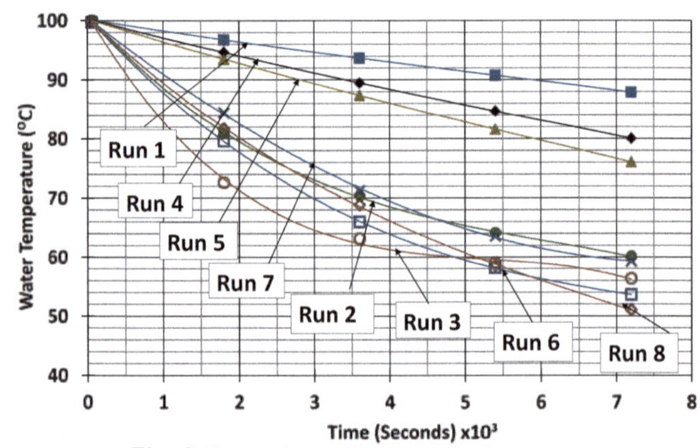

Fig. 9 Comparison of transient bath temperatures

5.4 Effect of coolant water flow rate and natural and force convection air cooling

The effect of coolant water flow rate and natural and force convection air cooling for Runs 2 to 5 are demonstrated in Fig. 10. Better performance results in lower container temperature as seen in Fig. 9 or greater heat removal rate from the container. The heat transfer rate results in Fig. 10 show that thermosyphons with higher coolant flow rate (Run 3) performed better than at low flow rate (Run 2). In the case of air cooling, thermosyphons with force convection air flow (Run 5) performed better than at natural convection (Run 4).

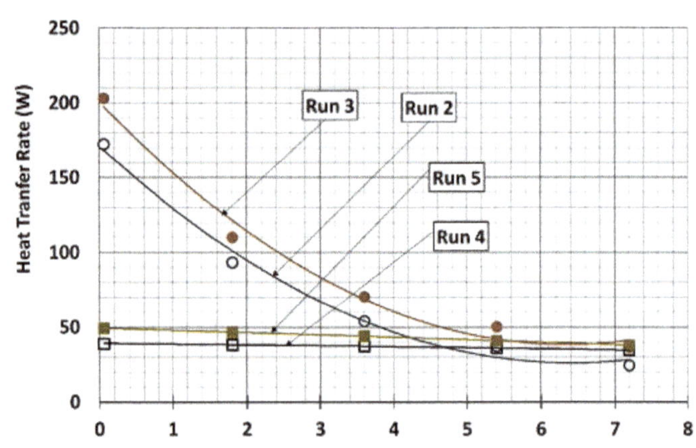

Fig. 10 Effect of coolant water flow rate and natural and force convection air cooling for Runs 2-5

6. Conclusions

A method to compare the performance of various thermosyphons rapidly and economically was proposed. The ease and simplicity of the procedure was demonstrated by cooling a container of hot water using various thermosyphons. From the results obtained it was found that force air convection cooling was better than natural convection and that high water flow rate was better than low flow rate.

NOMENCLATURE

A_{ins}	surface area of container (= 0.714 m^2)
A_e	surface area of evaporator section (m^2)
C_{pw}	specific heat of water (kJ/kg K)
D_e	external diameter of evaporator section (m)
h_a	natural convection heat loss (= 2 W/m^2 K)
L_{ad}	adiabatic length (m)
L_c	condenser length (m)
L_e	evaporator length (m)
T_a	ambient temperature (°C)
T_{ad1}, T_{ad2}	adiabatic temperatures (°C)
T_b	container water temperature (°C)
T_{fin1}, T_{fin2}	fin temperatures (°C)
T_{bu}, T_{bl}, T_w	water temperatures (°C)
Ti_{ns}	surface temperature of the insulation (°C)
T_{sat}	saturation temperature (°C)
T_{wi}, T_{wo}	coolant water inlet and outlet temperatures (°C)
T_w	mean coolant water temperature (°C)
W_b	mass of water in the container (kg)
$\dot{q}_b$	heat removed from container (kJ/s)
$\dot{q}_{ins}$	heat loss from container (kJ/s)
$\dot{q}_w$	heat transfer to coolant water (kJ/s)
ΔT_w, ΔT_m	mean difference between bath (evaporator) and coolant water temperature (condenser) (°C)
θ	inclination angle measured from the horizontal (degrees)
γ	ratio of heat transfer to the coolant/total heat removed from the containerl

REFERENCES

Anjakar, P.G. and Yarasu, D.R.B., 2012, "Experimental Analysis of Condenser Length Effect on the Performance of Thermosyphon," *International Journal of Emerging Technology and Advanced Engineering*, **2**, 494-499.

Buschmann, M. H. and Franzke, U., 2013, "Improvement of Thermosyphon Performance by Employing Nanofluid," *International Journal of Refrigeration*, **40**, 416-428.
http://dx.doi.org/10.1016/j.ijrefrig.2013.11.022

Celata, G.P., Cumo, M. and Furrer, M., 2010, "Experimental Tests of a Stainless Steel Loop Heat Pipe with Flat Evaporator," *Experimental Thermal and Fluid Science*, **34**, 866-878.
http://dx.doi.org/10.1016/j.expthermflusci.2010.02.001

Chang, S. W., Lo, D. C., Chiang, K. F. and Lin, C. Y., 2012, "Sub-atmospheric Boiling Heat Transfer and Thermal Performance of Two-Phase Loop Thermosyphon," *Experimental Thermal and Fluid Science*, **39**, 134-147.
http://dx.doi.org/10.1016/j.expthermflusci.2012.01.017

Huminic G., Huminic, A., Morjan, I. and Dunitrache, F., 2011, "Experimental Study of the Thermal Performance of Thermosyphon Heat Pipe Using Iron Oxide Nanoparticles," *International Journal of Heat and Mass Transfer*, **54**, 656-661.
http://dx.doi.org/10.1016/j.ijheatmasstransfer.2010.09.005

Huminic, G. and Huminic, A., 2011, "Heat Transfer Characteristics of a Two-phase Closed Thermosyphons Using Nanofluids," *Experimental Thermal and Fluid Science*, **35**, 550-557.
http://dx.doi.org/10.1016/j.expthermflusci.2010.12.009

Qu, J. and Wang, Q., 2013, "Experimental Study on the Thermal Performance of Vertical Closed-loop Oscillating Heat Pipes and Correlatiion Modeling," *Applied Energy*, **112**, 1154-1160.
http://dx.doi.org/10.1016/j.apenergy.2013.02.030

Shabgard, H., Xiao, B., Faghri, A., Gupta, R. and Weissman, W., 2014, "Thermal Characteristics of a Closed Thermosyphon under Various Filling Conditions," *International Journal of Heat and Mass Transfer*, **70**, 91-102.
http://dx.doi.org/10.1016/j.ijheatmasstransfer.2013.10.053

Solomon, A. B., Ramachandran, K. and Pillai B.C., 2012, "Thermal Performance of a Heat Pipe with Nanoparticles Coated Wick," *Applied Thermal Energy*, **36**, 106-112.
http://dx.doi.org/10.1016/j.applthermaleng.2011.12.004

Solomon, A. B., Mathew, A., Ramachandran, K., Pillai, B. C. and Karthikeyan, V. K., 2013, "Thermal Performance of Anodized Two Phase Closed Thermosyphon (TPCT)," *Experimental Thermal and Fluid Science*, **48**, 49-57.
http://dx.doi.org/10.1016/j.expthermflusci.2013.02.007

Wang J., 2009, "Experimental Investigation Of The Transient Thermal Performance of a Bent Heat Pipe with Grooved Surface," *Applied Energy*, **86**, 2030-2037.
http://dx.doi.org/10.1016/j.apenergy.2009.01.003

FLOW REGIME RECOGNITION IN TWO-PHASE THERMOSYPHON LOOPS USING PRESSURE PULSE ANALYSES

R.T. Dobson*, J.C. Ruppersberg

Department of Mechanical & Mechatronic Engineering, University of Stellenbosch, Stellenbosch, 7602, South Africa

ABSTRACT

The nuclear industry finds application for natural circulation in the form of passive heat removal systems used under accident conditions as well as for the removal of parasitic heat loss during normal operation. A high temperature gas-cooled reactor containment cooling system for instance can take the form of a series of closed-loop two-phase natural circulation thermosyphon type heat pipes. In this paper a means of remotely predicting the flow patterns in the two-phase regions of a closed loop heat pipe is proposed. The process involves the measurement of the working fluid pressure response in the single-phase flow region where safe access to the loop is possible from a nuclear radiation hazard point-of-view. The pressure signal is then decomposed and a frequency spectrum generated in order to find a characteristic frequency for each of the flow phenomenon observed. To establish the validity of the analytical method a 2.2 m high rectangular loop was built using 25 mm diameter copper pipe. Experiments were performed and a comparison of the frequency spectrums shows the ability of the proposed analytical method to correctly identify the presence of different flow patterns. It is thus provisionally concluded that the procedure as outlined in this paper can be used to remotely characterising the flow pattern associated with the different flow regimes in a closed loop two-phase thermosyphon-type heat pipe.

Keywords: *Two-phase flow, Natural circulation loops, Signal analysis*

1. INTRODUCTION

The concept of inherent safety is a fundamental part of the design philosophy of generation IV nuclear power reactors. In order to facilitate said safety philosophy, passive features are replacing active systems where possible. Passive features negate the need for machinery thereby reducing the risk of failure. This translates into an increase in system reliability whilst simultaneously reducing maintenance and operating costs.

It is the objective of this paper to show the application of this idea to the characterisation of a closed rectangular two-phase natural circulation loop. The pressure response of the system is measured in the single phase region of the loop rather than the two phase region, analysed and used to remotely predict the flow patterns in the two-phase regions of the evaporator and condenser sections. The single phase region being specifically selected since it allows for safe access to the loop from a nuclear radiation hazard point-of-view.

In the following sections the *operating modes* of thermosyphon-type heat pipes will be defined and described, the experimental setup will be discussed, the signal processing process described, results given and conclusions drawn.

2. BACKGROUND INFORMATION

Comprehensive reviews of natural circulation systems are given by Greif (1988) and Zvirin (1981). More recently the IAEA compiled a technical document with a number of authors contributing papers on a range of topics involving natural circulation systems in the nuclear industry (Reyes and Cleveland, 2005).

Tujikura et al. (2000) describes the development of a pressurised water reactor design using a hybrid safety system employing both active and passive systems. Each system is responsible for a given set of accident conditions. In addition the passive system serves as backup in case the active system fails. This concept can be taken one step further by designing an entirely passive system. Dobson (2006) for instance proposes the use of a closed natural circulation loop for use as a reactor cavity cooling system in a high temperature gas cooled nuclear reactor. Such systems are self-regulating in as much as an additional heat input from the reactor causes a corresponding increase in heat transferred to the heat sink.

Depending on the system configuration and heat input, the operating mode may be either in the single phase or two-phase flow regions. Given two-phase flow operation several flow patterns are likely to be present, all differently affecting the thermo-flow characteristics of the system. A method of determining the various flow conditions would therefore be useful.

Ding et al. (2007) proposed a pressure analysis method. The differential pressure is measured across two pressure tappings in horizontal pipes of three sizes while liquid and gas flows through the pipes. Gas velocities are varied to cause various identifiable flow patterns. Subsequent analysis of the signal shows that different energy spectrums exist for each flow pattern thus making them identifiable via pressure measurement.

3. OPERATING MODES AND THEORY

This section defines and describes a number of concepts peculiar to the operation of thermosyphon loops, such as the one depicted in figure 1. Consider a rectangular loop of internal diameter d, length (height) L,

* *Corresponding author: Email:rtd@sun.ac.za*

and width W, containing a working fluid and orientated in a vertical plane, relative to gravity, as shown in Fig.1. The loop may be operated at constant pressure using a relatively-large expansion tank maintained at a constant pressure P. The loop may be heated along the one vertical side and cooled along either the top horizontal side and/or the other vertical side, using tube-in-tube heat exchangers for instance. By introducing a temperature induced density gradient along the axial direction z around the loop the less dense side of the loop will tend to rise and the more-dense sink, due to the effect of gravity (buoyancy). [Heating along the horizontal bottom side and cooling along the top side has to be avoided as then, at least in theory, there will be either no flow, flow from left to right or from right to left. If the top portion is heated and the lower portion cooled there will be no flow.

If the left hand side of the loop is exposed to a heat source and the right hand side to a heat sink the fluid in the loop will begin to circulate and heat will be transferred to the heat sink. If there is no non-condensable gas in the system when the heat is applied the fluid will begin to flow as a liquid; this single phase flow regime may be called the *single phase operating mode*. As more heat is supplied bubbles begin to form towards the top of the heated section; in the top horizontal section the flow stratifies, and the vapour ultimately condenses at the top of the cooled section in what is reminiscent of so-called *annular* two-phase flow; this condition might thus be called the *two-phase operating mode*. As more heat is added to the heating section the bubbly flow ceases to occur and a *plug* flow regime is established, and as even more heat is added, so-called *churn* flow occurs. These flow patterns shown in Figure 4. In two-phase operating mode there is always stratified flow pattern in the top horizontal pipe with a relatively smooth-surfaced liquid-vapour interface, however as the heat load increases transverse waves emanating from just before the bend down into the condensing section move back towards the heating section.

In two-phase operating mode liquid is pushed into the expansion tank. If however the displaced liquid is prevented from flowing from the expansion tank back into the loop (by means of a suitably positioned shutoff valve) a *heat pipe operating mode* may be defined for the loop. By placing a valve at the highest portion of the loop and while operating in two-phase mode non-condensable gas may be expelled (together with a little working fluid). This non-condensable blow-off procedure is often called "burping" and eliminates the need for a vacuum pump when charging the loop for heat pipe operation.

The flow and heat transfer behaviour of a thermosyphon-type heat pipe loop may be theoretically simulated by dividing the loop into a series of control volumes and applying the equations of change, namely mass, motion and energy, to each control volume and making use of say by way of example the homogeneous two-phase flow models (Dobson and ruppersberg, 2007). Such techniques work well with single and two-phase operating modes. In heat pipe operating mode the liquid level in the loop would typically be between one-third and one-half the height of the loop. For the heat pipe to function properly in this operating mode the working fluid must first start boiling. When boiling starts liquid is blasted up into the liquid deficient higher regions of the heating section reminiscent of the behaviour of a geyser. In this case simple one-dimensional two-phase theory cannot capture the geysering behaviour and the boiling and condensation heat transfer coefficients for the heating and cooling sections, respectively, need to be experimentally determined.

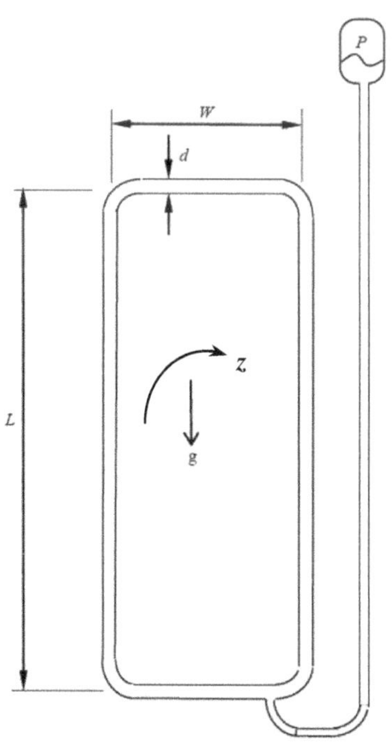

Fig. 1 Constant pressure closed loop natural circulation thermosyphon-type heat pipe

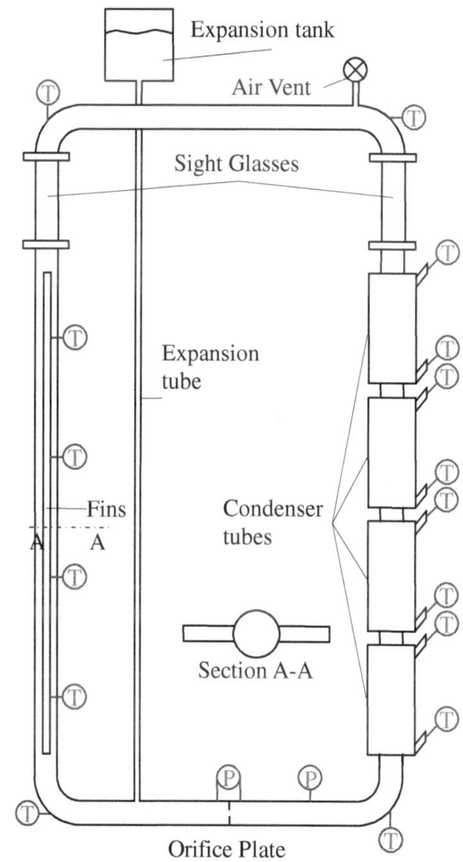

Fig. 2 Experimental closed loop thermosyphon

4. EXPERIMENTAL SETUP

A rectangular loop, 2.2 m high and 1 m wide, was built (see Fig. 2) using 25 mm inside diameter copper pipe with 0.22 m long sight glasses in the top of the evaporator and condenser sections of the loop. The working fluid of preference was water, chosen for availability, ease of use and low cost. An expansion tank was connected to the loop to allow the working fluid to contract and expand during the heating cycle. Heating elements were attached to the fins welded along the length of the evaporator tube. The input power was variable with an upper limit of 2 kW. Cooling was provided by tube-in-tube heat exchangers arranged along the length of the condenser tube. The cooling heat exchanger fluid was water delivered from a constant header tank at ambient temperatures.

Sensors were installed to measure both pressure and temperature at various points in the loop. The temperatures of the fins, working fluid and condenser fluid were measured with K-type thermocouples. The wall pressure in the single phase region was measured with a PCB 108A02 piezoelectric pressure transducer. Mass flow rate measurements across the orifice were performed with the aid of an Endress-Hauser PMD75 differential pressure transducer.

The signal from the PCB transducers were logged using a National Instruments NI cDAQ-9172 data logger while the thermocouple and differential pressure transducer signals were logged with a 34970A Agilent data logger. The power settings and condenser flow rates could be varied to generate data sets for comparison over a range of input conditions.

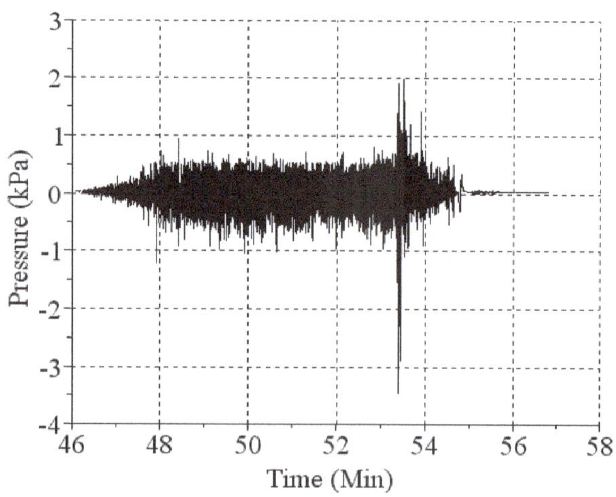

Fig. 3 Typical pressure signal as a result of two-phase (as recorded in single phase region)

4. SIGNAL PROCESSING

Due to the nature of two phase flow the measured pressure response signal is both nonlinear and non-stationary (frequency changes with time). A literature review showed the analysis method of Huang et al. (1998) to be applicable. This decomposition method is based on the local characteristic time scale of the signal, adapting the base functions rather than simply assuming sinusoidal functions thus allowing for the determination of a meaningful frequency spectrum. The method has been successfully applied by several authors for example Ding et al. (2007) in horizontal two-phase flow, Yang (2008) in the vibrations analysis of exhaust valves in an internal combustion engine and Ong et al. (2007) in structural vibration monitoring.

The Hilbert-Huang transformation method was applied to the experimental data sets and resulted in a range of frequencies being identified for each of the flow regimes. This was due to certain limitations associated with the decomposition method. The main limitation was the inability of the method to decompose the signal into mono-component functions under certain circumstances. These limitations have been addressed in literature by a number of authors, each proposing a method that adapts the original method to overcome the limitations. Among these the method of Ong et al. (2008) was chosen since it does not require prior knowledge of the frequencies under investigation, in contrast to many of the other proposed methods. The method involves the use of digital filtering prior to the use of the Hilbert-Huang transform resulting in a clearly defined frequency spectrum. For each experiment Ong's procedure was followed and a spectrum generated in order to find characteristic frequencies for each of the flow phenomenon observed.

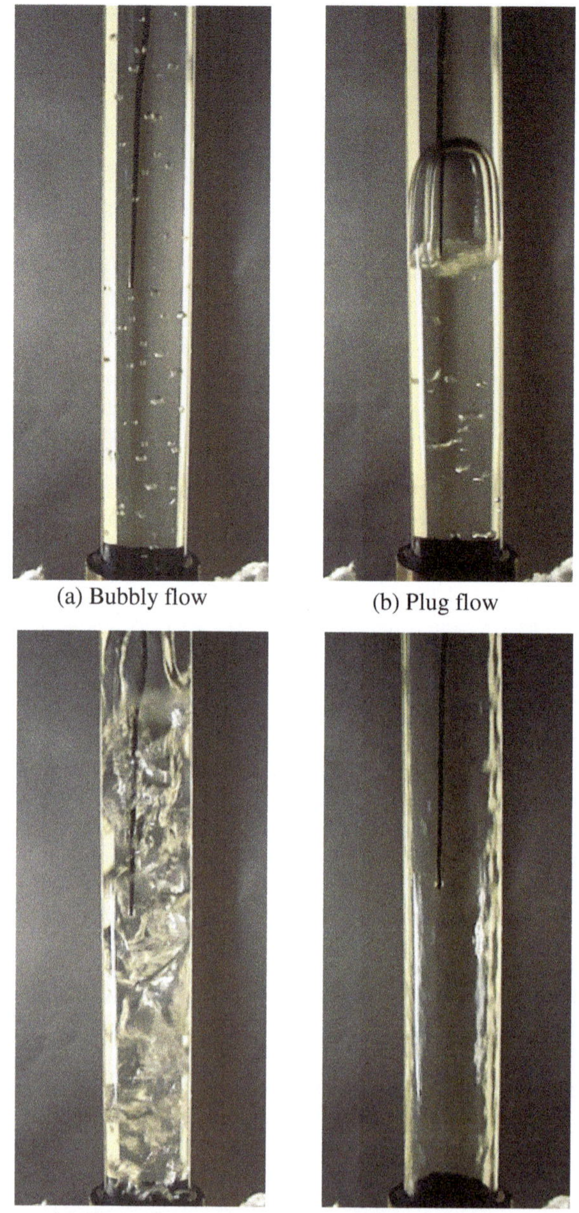

(a) Bubbly flow (b) Plug flow

(c) Churn flow (d) Annular flow

Fig. 4 Selected flow patterns (as observed)

5. RESULTS AND DISCUSSION

In a previous paper by Dobson and Ruppersberg (2010) the pressure signal was evaluated using alternative pressure transducers and produced the following results. The bubble flow regime showed a frequency of 0.5-0.6 Hz, the churn flow regime a frequency of 0.2 Hz and the plug flow regime a frequency of 0.14 Hz. The frequencies presented were very low. It was concluded that the transducer measured the oscillating static pressure in the lower leg of the loop. The oscillations being caused by the filling and draining of the relatively small sized expansion tank, which was dependent on the amount of vapour present in the loop, which in turn depended on the flow pattern present.

In order to find the frequencies directly linked to the flow patterns, rather than through the expansion tank, pressure transducers capable of a higher sampling frequency were installed. Fig. 3 shows the measured wall pressure signal for a single heating and cooling cycle in a single experiment using the PCB transducer.

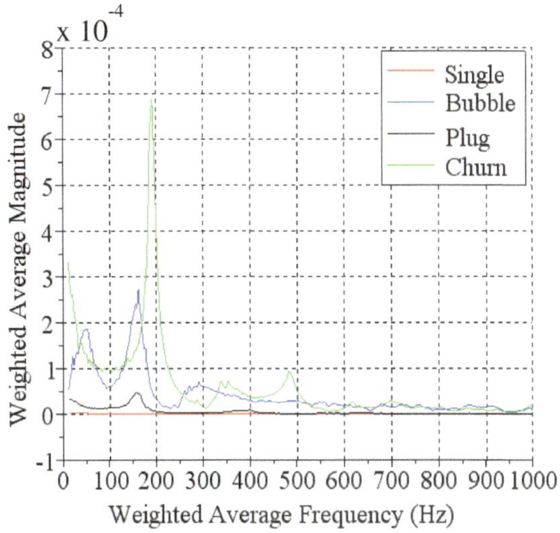

Fig. 5 Typical weighted average magnitude as a function of the weighted frequency for single phase, bubble, plug and churn flow patterns

During each thermal cycle the loop operated in the single phase region, transitioned to the two-phase region and back again. Figure 4 shows the examples of the flow patterns as typically observed through the evaporator sight glass. On the opposite side of the loop the general flow pattern in the condenser section was more-or-less characterised by annular flow independent of whether the flow was bubbly, plug, churn or turbulent in the evaporator section.

At the onset of the experiments it was unclear what frequency range would apply to the flow regimes. The logging was performed at the maximum rate of the transducer which was 25 kHz. In post processing, the data could be reduced to the most appropriate sampling rate. The signal analysis method requires an upper and lower boundary within which the characteristic frequency may be found. The boundaries were set to between 10 Hz and 3000 Hz. This range was broad enough to capture the characteristic frequencies without being overly computationally expensive.

Fig. 5 shows the frequency spectrum generated. Four lines are shown corresponding to the flow patterns present. Each line was produced by applying the analysis method to the pressure signal corresponding to the time period recorded during visual observation of a given flow pattern.

The line with no significant peaks (line just above zero amplitude) indicates the single phase flow region that had scant influence on the pressure signal. Bubble flow shows two distinct peaks at 50 Hz and 160 Hz and a minor peak at 300 Hz while plug flow has its main peak at 170 Hz with a minor peak at 400 Hz and churn flow has its main peak at 187 Hz with two more minor peaks at 350 Hz and 485 Hz. (The peak at 50 Hz was due to improper grounding of the 50 Hz electrical power frequency.

In addition to the characteristic frequencies that were found it can also be seen that each of the flow patterns produced significantly different weighted average magnitudes thus enhancing the method's ability to differentiate between the flow patterns.

These results are reproduced in Fig, 6 showing each of the flow patterns individually but with additional results from two more different experiments added for the sake of comparison, and thereby showing a reasonable degree of repeatability using this method. The results indicate the presence of distinct frequencies for each of the different flow patterns. This is however only the first step, several questions remain. What physical phenomena can explain the frequency values? Are these frequencies geometry dependant? What is the influence of flow obstructions? Does the working fluid have an influence on the results? Will signal attenuation play a role in larger loops?

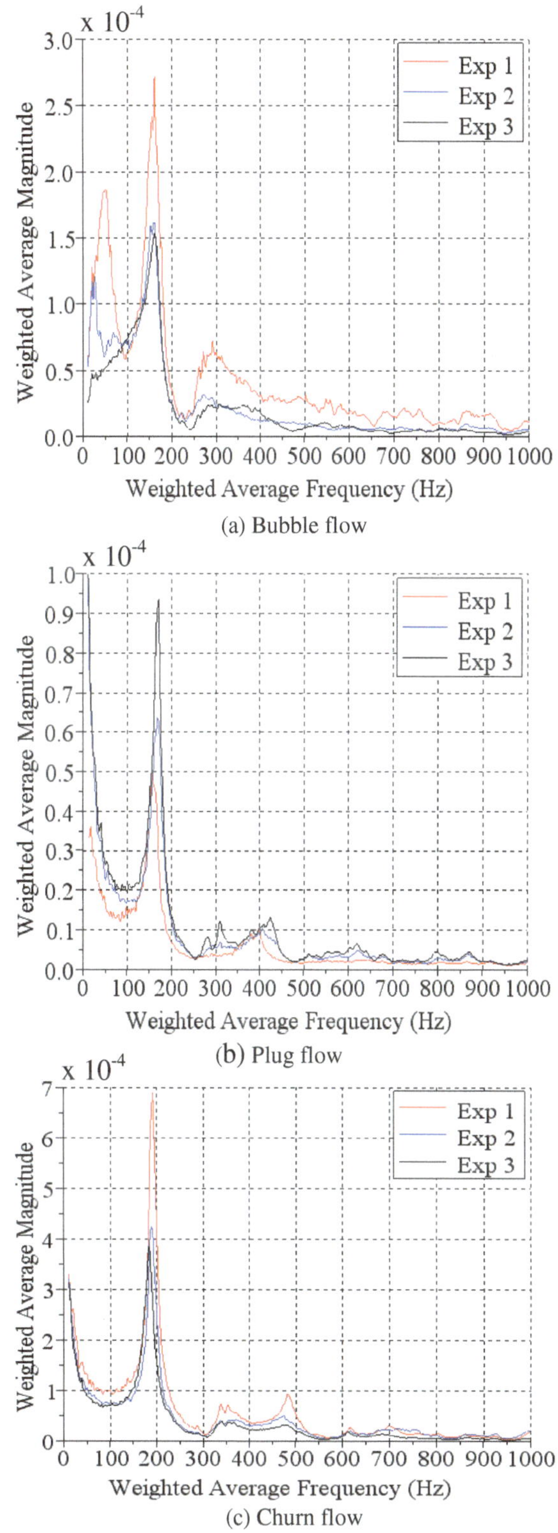

Fig. 6 Weighted average magnitude as a function of the weighted frequency for (a) bubble, (b) plug and (c) churn flow patterns

6. CONCLUSIONS

A means of remotely predicting the flow patterns in the two-phase regions of a closed loop heat pipe system applicable to the cooling of a HTGR citadel was proposed.

The pressure response in the single-phase working fluid was measured experimentally and a representative data set was presented. For signal analysis the local and adaptive method of Huang et al. (1998) with the proposed improvements of Ong et al. (2008) was chosen, described and successfully applied.

Characteristic frequencies for three different flow regimes were determined as present in a 2 m high rectangular closed copper loop. The main frequencies are 160 Hz for bubble flow, 170 Hz for plug flow and 187 Hz for churn flow.

The influence of various parameters on the specific frequencies is as yet unclear and will be investigated in future. To this end a second 8 m loop has been constructed and a third 26 m loop is under construction.

The viability of the adapted analytical procedure as outlined in this paper for the remote characterisation of the pressure pulse to identify the different flow regimes in a rectangular closed loop two-phase thermosyphon-type heat pipe has been provisionally proven.

ACKNOWLEDGEMENT

The authors would like to thank PBMR Pty LTD and the South African for their support.

NOMENCLATURE

d	diameter (m)
g	acceleration due to gravity (m^2/s)
L	length, height (m)
P	Pressure (Pa)
W	width (m)
z	axial direction

REFERENCES

Ding, H., Huang, Z., Song, Z., and Yan, Y., 2007, "Hilbert-Huang Transform Based Signal Analysis for the Characterization of Gas-liquid Two-phase Flow," Flow Measurement and Instrumentation, 18, 37-36.

Dobson, R.T., 2006, "A Novel Closed Loop Thermosyphon Heat Pipe Reactor Cavity Cooling System for a Pebble Bed Modular Reactor," Proceedings of the 8th International Heat Pipe Symposium, 24-27 Sept., 2006, Kumamoto, 384-397.

Dobson, R.T., and Ruppersberg, JC, 2010, "A Pressure Pulse Analysis Method to Characterise the Flow Patterns in a Two-Phase Closed-loop Natural Circulation Thermosyphon-type Heat Pipe," Proceedings of the 15th International Heat Pipe Conference, 25-30 Apr. 2010, Clemson..

Greif, R., 1988, "Natural Circulation Loops," Journal of Heat Transfer, 110, 1243-1257.

Huang, N.E., Shen, Z., Long, S.R., Wu, M.C., Shih, H.H., Zheng, Q., Yen, N.C., Tung, C.C., and Liu, H.H, 1998, "The Empirical Mode Decomposition and the Hilbert Spectrum for Nonlinear and Non-stationary Time Series Analysis," Proceedings of the Royal Society *London Series A*, 454, 903-995.

Ong, K.C.G., Wang, Z., and Maalej, M., 2008, "Adaptive magnitude spectrum algorithm for Hilbert-Huang transform based frequency identification", *Engineering Structures*, **30**, 33-41.

Reyes, J., and Cleveland, J., 2005, "Natural Circulation in Water Cooled Nuclear Power Plants, Phenomena, Models and Methodology for System Reliability Assessments," IAEA TECDOC, 1-51.

Ruppersberg, J.C. and Dobson, R.T., 2007, "Flow and Heat Transfer in a Closed Loop Thermosyphon Part I – Theoretical Simulation," Journal of Energy in Southern Africa, 18(4), 32-40.

Ruppersberg, J.C., and Dobson, R.T., 2007, "Flow and Heat Transfer in a Closed Loop Thermosyphon Part II – Experimental Simulation," Journal of Energy in Southern Africa, 18(4), 31-48.

Tujikura, Y., Oshibe, T., Kijima, K., and Tabuchi, K., 2000, "Development of Passive Safety Systems for Next-generation PWR in Japan," *Nuclear Engineering and Design*, 201, 61-70.

Yang, W.X., 2008, "Interpretation of Mechanical Signals using an Improved Hilbert-Huang Transform," *Mechanical Systems and Signal Processing*, 22, 1061-1071.

Zvirin, Y., 1981, "A review of Natural Circulation Loops in Pressurized Water Reactors and other Systems," *Nuclear Engineering and Design*, 67, 203-225.

THERMO-HYDRODYNAMIC BEHAVIOR OF METHANOL CHARGED CLOSED LOOP PULSATING HEAT PIPE

Rahul S. Borkar*, Pramod R. Pachghare

Department of Mechanical Engineering, Government College of Engineering, Amravati-444 604, India

ABSTRACT

Flow visualization experiment has conducted on the methanol charged vertical Closed Loop Pulsating Heat Pipe (PHP) to investigate the effect of different heat inputs on overall thermal resistance, average evaporator and condenser temperature and corresponding hydrodynamic behavior. A PHP has made from four parallel glass tubes forming adiabatic section, interconnected alternately by copper U tubes, having internal and external diameter 2 mm and 3.06 mm respectively. The length of evaporator, condenser and adiabatic section are 42 mm, 50 mm and 170 mm respectively. For experimentation, filling ratio (FR) 50% and vertical bottom heating mode position was maintained. Electrical heat input to the evaporator varies from 8W to 80W with interval of 8W after steady state was reached. The result shows that, thermal resistance of a PHP rapidly decreases with increasing heating power from 8W to 32W and thereafter reasonably steady for further heating powers. From visualization it is observed that, the thermal performance of a PHP is dependent on the flow patterns in PHP tubes. Continuous unidirectional flow circulation in the tubes of PHP gives best thermal performance than pulsation and dry-out conditions.

Keywords: *Pulsating, phase, passive, heat, flow pattern*

1. INTRODUCTION

The Pulsating Heat Pipe (PHP) is a two-phase passive heat transfer device proposed and patented by Akachi (1990). A PHP is simple in structure with a small diameter coil partially filled with certain working fluid in it and extended from the heat source to sink. PHP uses the technique of transporting the working fluid by means of differential pressure across liquid slugs and vapor plugs from evaporator to condenser and reverse. The fluid from the evaporator is pushed towards the condenser in the form of discrete liquid slugs and vapor bubbles. The vapor gets partially condensed at the condenser where it loses its latent heat and then returns to evaporator and completes the cycle. A PHP is being explored for thermal management of electronic devices to remove heat without any electrical power supply and it able to dissipate substantial amount of heat with small temperature drop, simple design and low cost.

Tong *et al*. (2001) observed that during the startup period the working fluid oscillate with large amplitude, after this period continuous circulation of the working fluid occurs. The direction of circulation for working fluid is consistent once circulation is obtained but the direction of circulation can be different for same experimental run. Smaller vapor bubbles have larger upward flow velocities and slower downward flow velocity.

Khandekar and Groll (2003) conducted experiments on a PHP made of copper capillary tube of 2 mm inner diameter for three different working fluids viz. water, ethanol and R-123. The PHP was tested in vertical (bottom heat mode) as well as horizontal orientation and indicated that, a 100% filled PHP (not working in the pulsating mode but instead as a single-phase buoyancy-induced thermosyphon) is thermally better performing than a partially filled pulsating mode device under certain operating conditions. Higher input heat fluxes result in a transition of slug flow to annular flow at the outlet of the evaporator U-bends. Since the evaporator U-sections experience convective boiling through the thin liquid film rather than nucleate type boiling in slug flow regime. Khandekar and Groll (2004) studied that, complete stop-over is in the loop occurs more frequently for filling ratio < 50% coupled with low heat input power. Stop-over phenomenon has also been observed for higher filling ratios. The 'self-sustained' oscillating character is then lost, such a behavior has never been reported for multi-turn PHPs because of alternating periods in which bubble plugs are moving rapidly (activity phase) and 'stopping' (static phase). Khandekar *et al.* (2009) observed that the quasi-steady state corresponding to best thermal performance consists of continuous unidirectional flow circulations, while the state corresponding to poor thermal performance is characterized by the intermittent bidirectional flow reversals. Oscillating Taylor bubble flows require immediate attention (Khandekar *et al*., 2010), unless the nuances of oscillatory confined bubbles in small capillaries is well discerned, the net pressure drop correlations for a PHP cannot be exhaustively constructed.

Dadong and Cui (2010) indicated that the thermal resistance decreases with the increase of the heating power at the same filling ratio. For the pure working fluid PHPs, the thermal resistance is decreases in the sequence of water, ethanol, methanol and acetone. No measurable difference has been recorded between a PHP running with the azeotropic mixture and a PHP running with pure ethanol, in terms of overall thermal resistance Mameli *et al.* (2011).

Pachghare and Mahalle (2012) conducted experiments on PHP to investigate the effect of binary mixture of working fluids on thermal performance of PHP, made from copper tube with 2 mm inner diameter and 10 numbers of turn, for different heat inputs with ethanol, methanol, acetone, water and their binary mixture with water as a working fluids at filling ratio 50%. Pure acetone is having lesser thermal resistance than working fluids methanol, ethanol, and water.

*Corresponding author Email: borkar44@gmail.com

No measurable difference has been recorded between the PHP running with pure and binary mixture working fluids, in terms of overall thermal resistance. Working fluid behavior is strongly depends on the thermo-physical properties, but latent heat of vaporization is the main property that strongly affects the thermal performance.

Narasimha *et al.* (2012) studied that, at atmospheric condition; the saturation temperature is higher compared to evacuated situation. Thus more liquid phase exists in the tube with a consequent increase in the heat transfer. Kammuang-lue *et al.* (2008) studied that, the higher latent heat of the working fluid the higher critical heat flux. Yang *et al.* (2009) pointed out that increasing heat load clearly improves the thermal performance. Meena *et al.* (2009) concluded that as working fluids change from R123 to Ethanol and water the critical heat flux decreased. The latent heat of vaporization affects the critical heat flux. The working fluid with the lower latent heat of vaporization exhibits a higher critical heat flux. Mass transfer characteristics are also affected by the local hydrodynamic properties of the flow (Bercic and Pintar, 1997).

To study the effect of flow pattern on thermal performance, an experimental set-up designed for pure convection model. Therefore, borosilicate glass tubes are used in the adiabatic section for visualization and heat transfer by pure convection. Various temperatures are recorded on the wall of PHP .Thermal performance is measured in terms of overall thermal resistance for various heat inputs also flow patterns are captured and analyzed with the thermal performance.

2. EXPERIMENTATION

2.1 Experimental set-up

The experimental set-up of PHP was developed as shown in Figure 1. The details of copper-glass tube assembly and evaporator tank as shown in Figure 2. The set-up consists of a closed loop PHP, temperature recorder, power supply unit, and water tank cooling system for condenser.

A PHP, consisting of 2 turns, the presence of two U-turns and four parallel channels increases the level of perturbations thereby diminishing the probability of phase recoiling and consequent complete stop-over of the fluid motion.

Evaporator sections (Length 42 mm) and condenser sections (Length 50 mm) are made of copper capillary tube having inner diameter is 2 mm and the outer diameter is 3.06 mm. The adiabatic section (Length 170 mm) made from borosilicate glass tube to understanding flow pattern for different heat inputs. Fluid motion and flow patterns have been captured with a camera (Nikon®, COOLPIXP60, 8.1 Megapixel, 5X Zoom) just above the evaporator section.

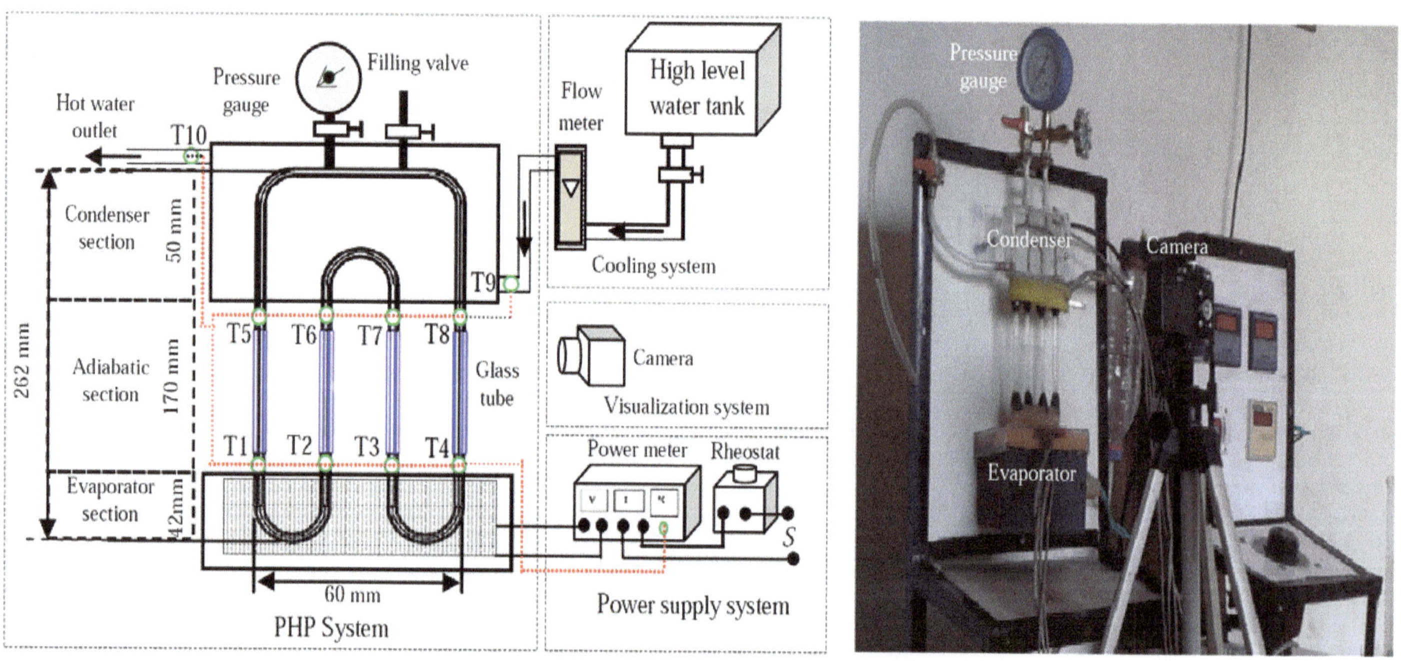

Fig. 1 Schematic and photograph of experimental set-up

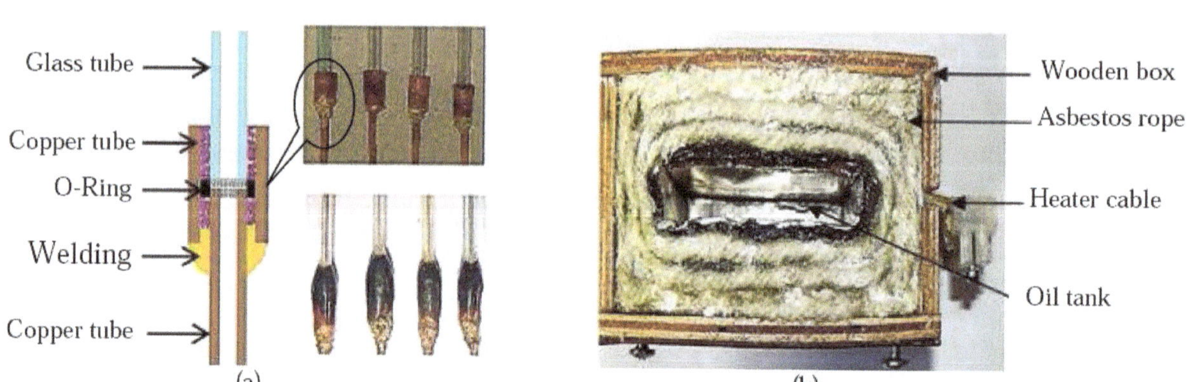

Fig. 2 Assembly details (a) copper and glass tubes (b) evaporator tank

The heating power is provided by a carefully designed power supply unit. Heating was done by oil bath and cooling by water tank with maintained 30 ml/min water flow rate through condenser. The power meter measures the AC voltage, the current and the corresponding temperature simultaneously. The filling ratio 50% was maintained. The heating configuration was bottom heat orientation. For temperature measurements, ungrounded sheathed thermocouples (K-Type, accuracy ±0.2°C and D = 0.5 mm) were used. All the thermocouples are mounted on the wall of the PHP tubes.

Filing of the PHP is only done when the internal vacuum level is at least lower or equal to 0.99 bar. Relation of surface tension and bouncy could be explained by the dimensionless formula in Eq. (1):

$$\text{E\"o} = \frac{gD_{cri}^2\left[\rho_{liq} - \rho_{vap}\right]}{\sigma} \tag{1}$$

When $\text{E\"o} = 4$, the bubble will get seized on both size of wall. At this condition, the terminal velocity becomes zero and the fluid slug flow is formed. The inner diameter of the tube must be small enough such that liquid vapor plugs and slugs exist. Such condition is ensured only if inner diameter of a tube is smaller than a critical diameter. The required diameter calculated by using following Eq. (2).

$$D_{cri} \le 2\times\sqrt{\frac{\sigma}{g\left[\rho_{liq} - \rho_{vap}\right]}} \tag{2}$$

All tubes have inner diameter (2 mm) less than critical diameter. Two smaller copper tubes has been brazed on the main tube of the condenser section in order to connect the vacuum/filling valve and pressure indicator as shown in Figure 1.

The glass tube attached to copper tubes in the condenser section and evaporator section by following procedure:

1. At each end of the U-turns a 5 mm inner diameter copper tubes are attached by welding (i.e. copper fitting).
2. The 4 mm inner diameter of glass tubes are inserted in the copper fitting and coupled to the copper tubes by mean of an O-ring.
3. Mixture of m-seal and araldite is fill in the remaining gap between the copper fitting and the glass tube above and below the O-ring, thus ensuring a good seal as shown in Figure 2-a.

The evaporator section was heated by oil bath for uniform heating. A flat flexible heater (450W) has been placed at one side of the evaporator oil tank. Insulation is provided on tank by two asbestos sheet (width: 3 mm) and the whole assembly is kept in wooden block, remaining gaff between oil tank and wooden block wall is filled with glass wool for better insulation and asbestos rope is kept on glass wool and fixed by araldite as shown in Figure 2-b.

2.2 Experimental procedure

The experimental set-up is used for experimentation and the following procedure is adopted during the present transient experiment:

1. Before filling the working fluid, it is ensured that there is no other fluid exists inside the tubes of PHP.
2. In order to obtain vacuum inside the PHP, a reciprocating vacuum pump is connected to the filling valve. Filing of the PHP is only done when the internal vacuum level is lower than or equals to 0.99 bar.
3. Once the evacuation of the PHP system is complete, the valves are closed. The required amount of working fluid is then filled by opening the one valve as shown in Figure 1. The PHP is filled with the working fluid with a filling ratio 50% at ambient temperature (25°C ± 2°C).
4. The cooling water is allowed to the condenser section of PHP from the constant water bath and the amount of cooling water 30 ml /min is maintained during entire experimentation.
5. The temperature indicator is then switched on to record the temperature.
6. The required input heat power supply from control panel.
7. Transient experiments are conducted with different heating input.
8. After a quasi-steady state was reached continuous movies were recorded and photographs were taken at specified times by digital camera. At steady state from inlet, outlet temperature and mass flow rate of coolant the heat transfer rate through PHP calculated.
9. Evaporator and condenser temperature recorded at each steady state and again increase heating power by 8W. Repeat this step until 80W.

3. RESULTS AND DICUSSIONS

Experimentation conducted on the methanol charged PHP at vertical bottom heat mode. Heat input to the evaporator section has been increased with steps of 8W, starting from 8W to 80W. When, the steady-state was reached for each heat input level then temperature for different points recorded and the overall thermal performance of the PHP has been estimated.

In the ideal measurement, the value assigned by the measurement would be the actual value of the physical variable intended to be measured. However, measurement errors bring on an uncertainty in the correctness of the value resulting from the measurement. To give some measure of confidence to the measured value, measurement errors must be identified, and their probable effect on the result estimated. Uncertainty is simply an estimate of a possible value for the error in the reported results of a measurement. There are four thermocouples mounted in the evaporator section and four in the condenser section and hence the uncertainty in the evaporator temperature and condenser temperature are evaluated about 5% (Kline *et al.*, 1953).

The overall thermal resistance of PHP defined as the ratio of the average temperature difference between the evaporator and condenser to the power input. It can be calculated as:

$$R_{th} = \frac{T_e - T_c}{Q} \tag{3}$$

Where, Q is heat input power to the PHP at evaporator by considering thermal loses.

3.1 Heat transfer capability

Figure 3 shows that the variation of thermal resistance at different heat inputs for methanol charged PHP. From Figure 3, it is clear that, the thermal resistance decreases with increase in heat input up to 40W. The thermal resistance is dependent on the temperature difference and applied power; as the input heat power increases the temperature difference between evaporator and condenser decreases.

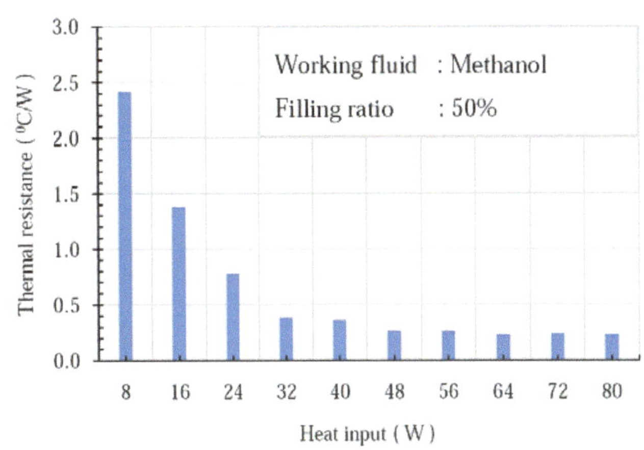

Fig. 3 Thermal resistance of methanol PHP for different heat inputs

With increasing heat input, the evaporator temperature rises resulting in a greater density gradient in the tubes. Simultaneously the liquid viscosity also drops diminishing the wall friction and evaporation of working fluid inside the tube of PHP. Methanol PHP most suitable for heat input 40 to 80W gives uniform performance.

3.2 Formation of slug-plug pattern

When working fluid filled in to the evacuated tube of PHP, the working fluid randomly distributed inside the tubes in the form of liquid slugs and vapor plugs train automatically. The coexistence of liquid slugs and vapor plugs of different lengths in their initial condition as shown in Figure 4-a. Due to the evacuation, the boiling temperature of working fluid fall down and partial evaporation occur. The tube is partially filled (50%) and has capillary dimension, surface tension force dominate over the gravitational force and train of liquid slugs and vapor plugs having menisci on their edges is formed , as shown in Figure 4-b. The amount of liquid phase in the PHP depends on filling ratio. Higher filling ratio implies that there is more liquid phase than vapor phase.

3.3 Startup behavior of PHP

Slug fluctuation in PHP (stage-1st): When heating start (8W) at evaporator section, the temperature of evaporator section increased but very small changes observed at temperature of condenser section. Heating power 8W is insufficient to start proper evaporation there for very less amount of heat transfer towards the condenser section from evaporator section by local axial fluctuation and phase changes in working fluid as shown in Figure 5. Further when heating power increased from 8W to 16W rate of fluctuation increased with rising the condenser temperature because of transmission of heat towards the condenser. The basic phenomenon behind fluctuation is nucleate boiling occur at liquid slugs, thus size of adjacent vapor plug increased consequently, at same time because of heating vapor plug expand cause pressure rise in evaporator section. This vapor plugs pushes the liquid column toward the low temperature condenser section. The temperature and pressure decrease in the condenser section because of to condensation. Therefore, constant unsteady internal pressure difference exists in the tube of PHP which is the driving force.

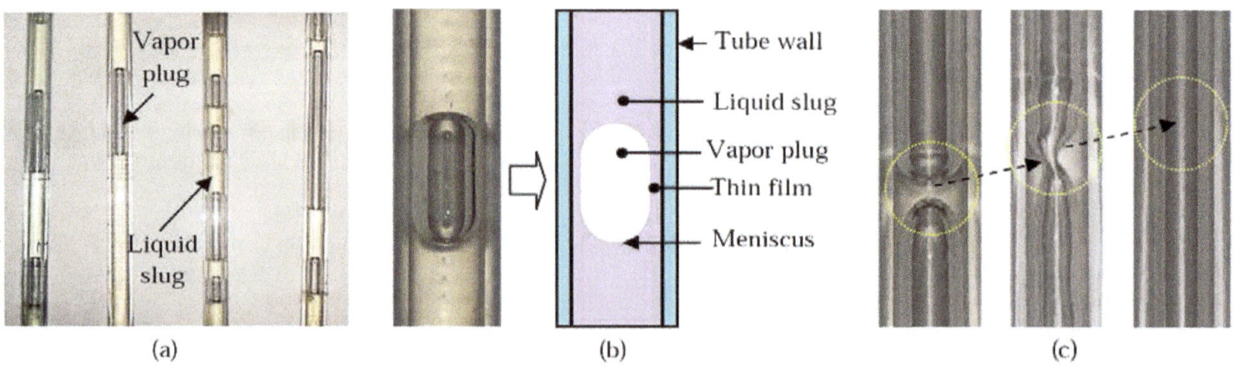

Fig. 4 Flow visualization (a) flow pattern after filling (b) plug and slug profile (c) transmission from slug to annular pattern

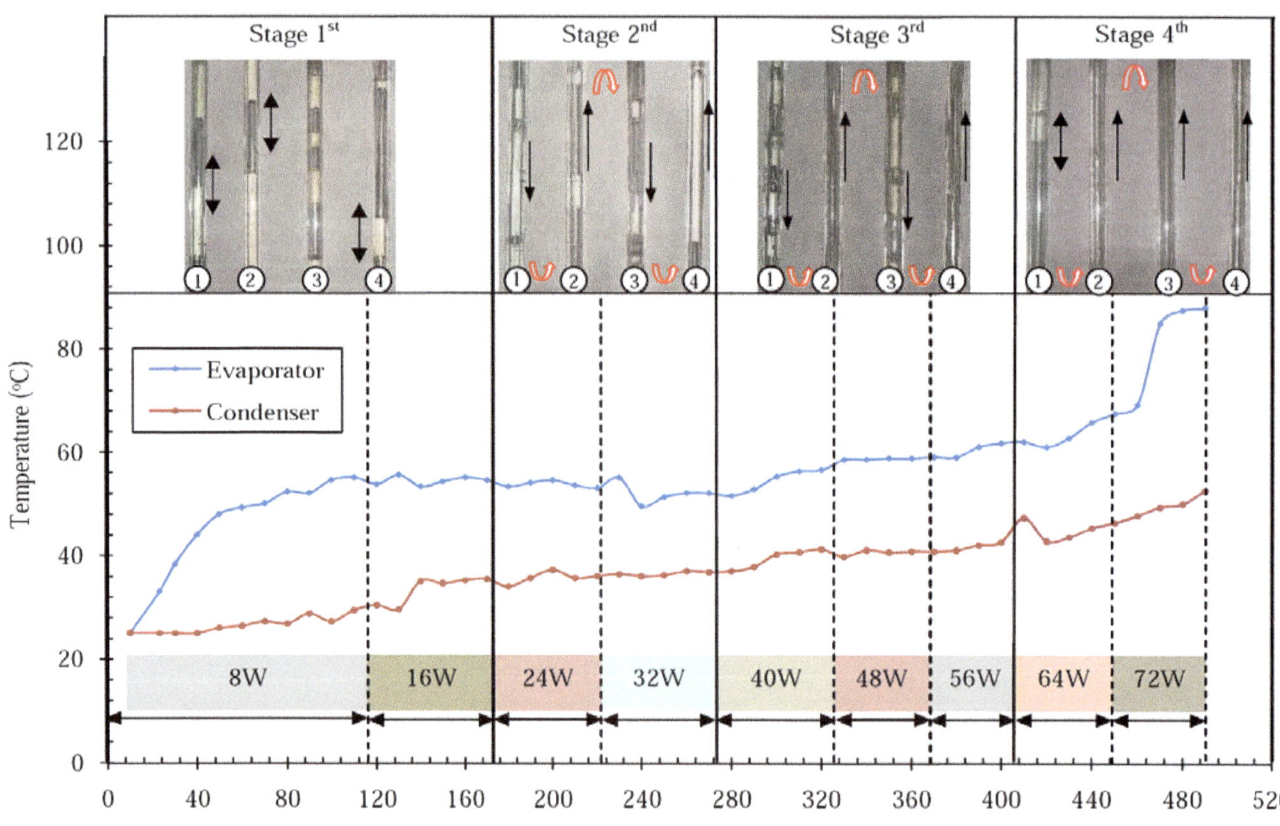

Fig. 5 Evaporator, condenser temperature and corresponding flow pattern at different heat inputs

The tubes are interconnected; therefore motion of liquid slugs and vapor plugs at one section of the tube toward the condenser also leads to the motion of slugs and plugs in the next section toward evaporator section this works as the restoring force. The inter-play between the driving force and the restoring force leads to pulsation of the vapor plugs and liquid slugs in the axial direction.

One directional circulation in PHP (Stage-2nd): Again when heating power increase (24W), within a short interval of time one directional oscillatory semi-annular circulation observed, this happened because pressure at a point "A" is higher than "B" then flow can be initiate as shown in Figure 6. One directional circulation starts automatically without any mechanical moving component in the system.

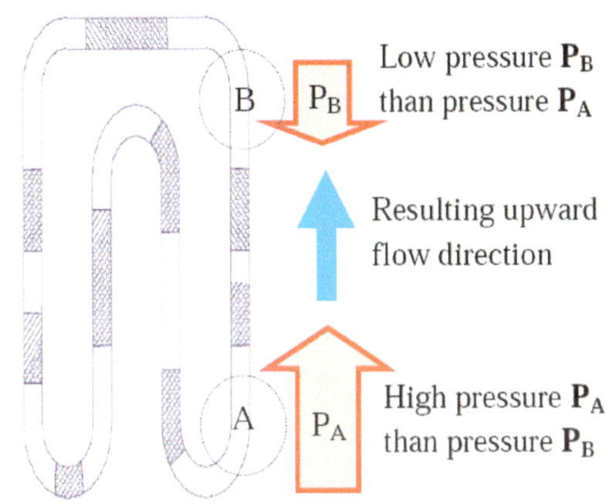

Fig. 6 Flow circulation startup

Flow pattern change from oscillation flow to oscillatory one directional circulation flow and thermal resistance decrease, it means, in case of oscillatory one directional circulation more heat transfer than only oscillation flow with evaporator and condenser temperature profile same as 16W. The term "oscillatory circulation" used because, during circulation liquid sludge oscillates axially. One directional circulation occurs because of following phenomenon.

If two-phase fluid come from condenser to evaporator in 1st tube convection and nucleate boiling occur in vapor plug and liquid slug respectively resulting size and pressure increase in vapor plug at evaporator section cause push the surrounding liquid slug in 2nd tube towards condenser section. Due to the symmetry the same phenomena are also happening in 3rd & 4th tube respectively as shown in Figure 5&7, during the circulation thin liquid film surrounding vapor plug is always available. The quantity of vapor in 2nd & 4th tube is higher than 1st and 3rd tube because of evaporation and condensation respectively. 1st & 3rd tube supply liquid slug with uncondensed vapor plug to evaporator for absorb heat and 2nd & 4th tube supply evaporated vapor to condenser for loss heat, cycle of the evaporation and condensation is continuous in entire operation.

At heating power 32W oscillatory circulation flow convert in to uniform circulation flow, evaporator temperature slightly increased but no effect on condenser temperature observed. Thermal performance is better than heating power 24W; hence heat transfer capability is more in case of uniform circulation than oscillatory one directional circulation. Similar results also recorded by Khandekar *et al.* (2009).

Flow pattern transition (Stage-3rd): When further heating load increased (40W) the evaporation rate increases, slug coming from condenser evaporate completely when reached to evaporator, because of high heat power. The vapor pressure in the evaporator section is now able to break most of the liquid slug menisci bridges (see Figure 4-c) and therefore up-comer flow pattern changing from semi-annular to pure annular known as "flow pattern transition". The liquid film is thick and wavy. The above flow pattern observed for input powers 40W to 56W with decreasing quantity of down comer liquid slug and slightly increasing evaporator and condenser temperature with respect to input heat power.

Figure 7 illustrates that the flow directions during the operation of methanol PHP at different input heat power. During one directional circulation the neighboring capillary tube always have an inverse direction. When PHP is trying to switch from one flow direction (anti clockwise) to another direction (clockwise) then flow direction in all capillary tubes was changed. The flow direction switch process observed at 40 and 56W input heat power.

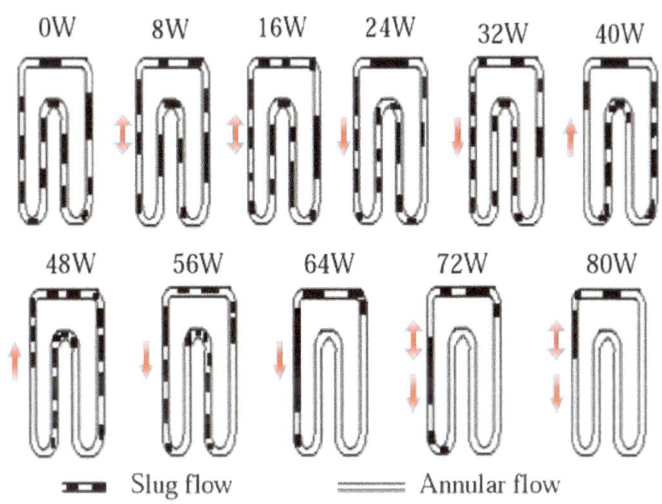

Fig. 7 Flow pattern, flow direction and switch phenomenon for different heat input at steady state

Before dry out condition (Stage-4th): Again when heat load increased, in these cases, the rate of evaporation is much higher than condensation there for the condenser unable to condense superheated vapor from evaporator. Slug fluctuate only in 1st tube and pure annular flow observed in 2nd, 3rd and 4th tubes (see Figure 5) with very thin liquid film on the inner surface of tube because of shortage of working fluid in liquid form at evaporator section, therefore evaporator temperature increase rapidly, resulting, evaporator-condenser temperature difference increases therefore increases in thermal resistance, this condition was observed at heat input 64W, 72W and 80W as shown in Figure 7 schematically. Shortage of liquid slug increase with increasing heat input because of higher rate of evaporation and insufficient condensation.

3.4 Flow pattern characterization

From the visualization of methanol PHP, it is recorded that different type of bubbles formed during the operation are characterized as fallows.

Dispersed bubbles are spherical shape and have less diameter than the inside diameter of the capillary tube. The dispersed bubbles are created by the continuous nucleated boiling, if the coalescence does not occur in the evaporator section. The dispersed bubbles are so small that they are quickly condensed and form slugs once they are out of the evaporator, that span almost the entire cross section of channel, resulting flow regime referred as slug flow as shown in Figure 8.

Vapor plugs are the bubbles in cylindrical form have length higher than the diameter of tube. A part of an expanding dispersed bubble breaks away from it and travels further to merge with another bubble encountered in the way form vapor plug as shown in Figure 9. Motion and shape of the vapor plugs depend on the exerted pressure on both sides of plugs, if pressure at upward direction (P_2) is higher than downward (P_1) then the plug moves upward direction and during motion the shape of vapor plug as shown in Figure 10.

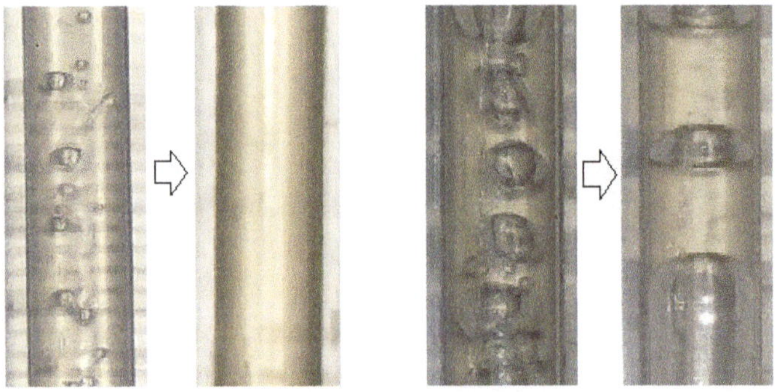

Fig. 8 Dispersed bubble

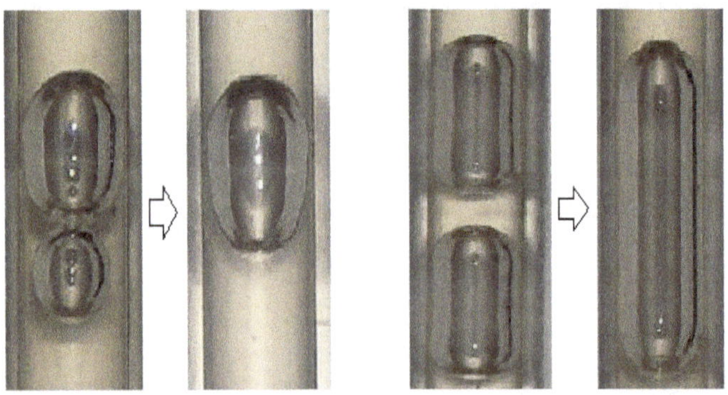

Fig. 9: Vapor plugs.

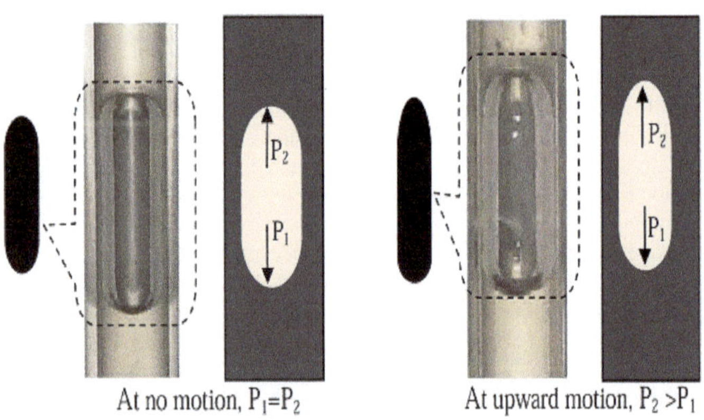

Fig. 10 Shape of bubble with pressure

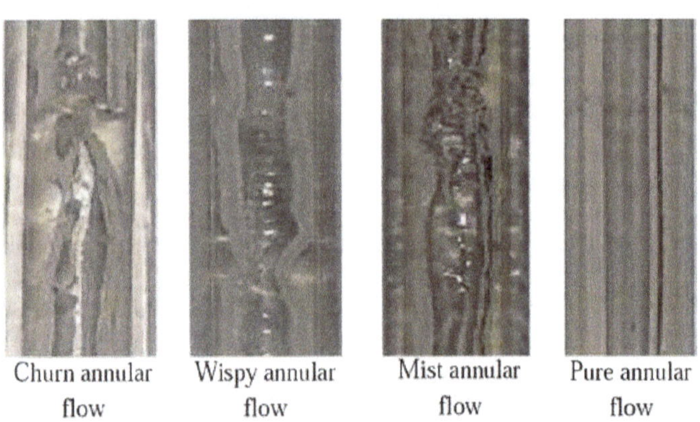

Fig. 11 Characterization of annular flow

Annular flow configuration is the two phase flow with most of the liquid flowing along the wall of the tube and vapor flow in the center core, this regime referred as annular flow. Annular flows are obtained at high evaporation level because of high flow rate of vapor and characterized as;

- **Churn annular flow.** At high velocity of the vapor flow, the structure of the flow becomes unstable with the fluid traveling up and down in an oscillatory fashion but with a net upward flow. The instability is the result of the relative parity of the gravity and shear forces acting in opposing directions on the thin film of liquid of Taylor bubbles. This flow pattern is in fact an intermediate regime between the slug flow and annular flow regimes. In small diameter tubes, churn flow may not develop at all and the flow passes directly from slug flow to annular flow.
- **Wispy annular flow.** When the flow rate of vapor is further increased, the entrained droplets may form transient coherent structures as clouds or wisps of liquid in the central vapor core.
- **Mist annular flow.** At very high vapor flow rates, the annular film is thinned by the shear of the vapor core on the interface until it becomes unstable and is destroyed, such that all the liquid in entrained as droplets in the continuous vapor phase, analogous to the inverse of the bubbly flow regime
- **Pure annular flow**. Once the interfacial shear of the high velocity vapor on the liquid film becomes dominant over gravity, the liquid is expelled from the center of the tube and flows as a thin film on the wall (forming an annular ring of liquid) while the vapor flows as a continuous phase up the center of the tube. The interface is disturbed by high frequency waves and ripples.

4. CONCLUSIONS

To understand the thermo-hydrodynamic behavior of PHP at different heat inputs, the experiment has conducted on methanol charged PHP with 50% filling ratio. The results of the present experimentation are summarized as:

1. Thermal performance of PHP is depends on behavior of working fluid with respect to heat input. The methanol PHP is more suitable for heat input between 40W to 80W.
2. From the visualization of PHP it conclude that thermal performance of device increase with heat input because of increasing rate of pulsation, maximum heat transfer during one dimensional circulation state than only pulsation mode.
3. During starting period, the rate of fluctuation of slug was slow and increase with respect to increasing heating power.
4. Overall thermo-hydrodynamic behavior of PHPs goes through four stages, stage-1st: slug fluctuate, stage-2nd: slug-annular up comer, stage-3rd: pure annular up comer, stage-4th: annular up comer at 2nd, 3rd and 4th tubes with intermittent slug fluctuation and downward flow observed in 1st tube.
5. The thermal performance of PHP is independent on the direction of flow circulation (clock wise or anti clock wise).

ACKNOWLEDGEMENTS

Special thanks to my brother Dhananjay Borkar for motivation and financial support.

NOMENCLATURE

D : Diameter (m)
Q : Heat input (W)
Eö : Eötvös number
R_{th} : Thermal resistance (°C/W)
T : Average temperature (°C)

Greek Symbols

σ : Surface tension (N/m)
ρ : Density (Kg/m^3)
g : Acceleration due to gravity (m/s^2)

Subscripts

e	: evaporator
c	: condenser
liq	: liquid
vap	: vapor
cri	: critical

REFERENCES

Akachi, H., 1990, "Structure of a Heat Pipe," U.S. Patent 4,921,041.

Berčič, G., and Pintar, A., 1997, "The Role of Gas Bubbles and Liquid Slug Lengths on Mass Transport in the Taylor Flow through Capillaries," *Chemical Engineering Science*, 52, 3709-3719.

Kammuang-lue, N., Charoensawan, P., Ritthidech, S., Booddachan, K. and Terdtoon, P., 2008, "Effects of Working Fluids on Heat Transfer Characteristics of a Closed-Loop Pulsating Heat Pipe at Critical State," *International Heat Pipe Conference.*

Khandekar, S. and Groll, M., 2003, "On the Definition of Pulsating Heat Pipe," *Proc. 5th Minsk International Seminar (Heat Pipes, Heat Pumps and Refrigerators),* Minsk, Belarus.

Khandekar, S., Dollinger N. and Groll M., 2003, "Understanding Operational Regimes of Closed Loop Pulsating Heat Pipes: An Experimental Study," *Applied Thermal Engineering*, 23, 707–719. http://dx.doi.org/10.1016/S1359-4311(02)00237-5

Khandekar, S., and Groll, M., 2004, "An Insight Into Thermo-Hydraulic Coupling in Pulsating Heat Pipes," *International Journal of Thermal Sciences*, 43(1), 13-20. http://dx.doi.org/10.1016/S1290-0729(03)00100-5

Khandekar, S., Gautam, A. and Sharma, P., 2009, "Multiple Quasi-Steady States in a Closed Loop Pulsating Heat Pipe," *International Journal of Thermal Sciences*, 48(3), 535-546. http://dx.doi.org/10.1016/j.ijthermalsci.2008.04.004

Kline, S., and Mcclintock, F., 1953, "The Descriptions of Uncertainties in Single Sample Experiments", *Mechanical Engineering, 75, 3-8.*

Mameli, M., Khandekar, S. and Marango, M., 2011, "An Exploratory Study of a Pulsating Heat Pipe Operated with a Two Component Fluid Mixture," *Proceedings of the 21st National and 10th ISHMT-ASME Heat and Mass Transfer conference*, IIT Madras, India.

Meena, P., Rittidech, S. and Tammasaeng, P., 2009, "Effect of Evaporator Section Lengths and Working Fluids on Operational Limit of Closed Loop Oscillating Heat Pipes with Check Valves (CLOHP/CV)," *American Journal of Applied Sciences*, 6 (1), 133-136. http://dx.doi.org/10.3844/ajassp.2009.133.136

Narasimha, K., Sridhara, S., Rajagopal, M. and Seetharamu, K., 2012, "Influenc e of Heat Input, Working Fluid and Evacuation Level on the Performance of Pulsating Heat Pipe", *Journal of Applied Fluid Mechanics*, Vol. 5/2, pp. 33-42.

Pachghare, P. and Mahalle M., 2012, "Thermal Performance of Closed Loop Pulsating Heat Pipe Using Pure and Binary Working Fluids," *Frontiers in Heat Pipes* (FHP), 3,033002. http://dx.doi.org/10.5098/fhp.v3.3.3002

Tong, B., Wong, T., and Ooi, K., 2001, "Closed Loop Pulsating Heat Pipe," *Applied Thermal Engineering*, ISSN 1359-4311, Vol. 21/18, pp. 1845-1862. http://dx.doi.org/10.1016/S1359-4311(01)00063-1

Wang, D. and Cui, X., 2010, "Experiment Research on Pulsating Heat Pipe With Different Mixtures Working Fluids," *The 21st International Symposium on Transport Phenomena,* Kaohsiung City, Taiwan.

Yang, H., Khandekar, S. and Groll, M., 2009, "Performance Characteristics of Pulsating Heat Pipes as Integral Thermal Spreaders," *International Journal of Thermal Sciences*, 48, 815–824. http://dx.doi.org/10.1016/j.ijthermalsci.2008.05.017

Permissions

All chapters in this book were first published in FHP, by Global Digital Central; hereby published with permission under the Creative Commons Attribution License or equivalent. Every chapter published in this book has been scrutinized by our experts. Their significance has been extensively debated. The topics covered herein carry significant findings which will fuel the growth of the discipline. They may even be implemented as practical applications or may be referred to as a beginning point for another development.

The contributors of this book come from diverse backgrounds, making this book a truly international effort. This book will bring forth new frontiers with its revolutionizing research information and detailed analysis of the nascent developments around the world.

We would like to thank all the contributing authors for lending their expertise to make the book truly unique. They have played a crucial role in the development of this book. Without their invaluable contributions this book wouldn't have been possible. They have made vital efforts to compile up to date information on the varied aspects of this subject to make this book a valuable addition to the collection of many professionals and students.

This book was conceptualized with the vision of imparting up-to-date information and advanced data in this field. To ensure the same, a matchless editorial board was set up. Every individual on the board went through rigorous rounds of assessment to prove their worth. After which they invested a large part of their time researching and compiling the most relevant data for our readers.

The editorial board has been involved in producing this book since its inception. They have spent rigorous hours researching and exploring the diverse topics which have resulted in the successful publishing of this book. They have passed on their knowledge of decades through this book. To expedite this challenging task, the publisher supported the team at every step. A small team of assistant editors was also appointed to further simplify the editing procedure and attain best results for the readers.

Apart from the editorial board, the designing team has also invested a significant amount of their time in understanding the subject and creating the most relevant covers. They scrutinized every image to scout for the most suitable representation of the subject and create an appropriate cover for the book.

The publishing team has been an ardent support to the editorial, designing and production team. Their endless efforts to recruit the best for this project, has resulted in the accomplishment of this book. They are a veteran in the field of academics and their pool of knowledge is as vast as their experience in printing. Their expertise and guidance has proved useful at every step. Their uncompromising quality standards have made this book an exceptional effort. Their encouragement from time to time has been an inspiration for everyone.

The publisher and the editorial board hope that this book will prove to be a valuable piece of knowledge for researchers, students, practitioners and scholars across the globe.

List of Contributors

Shung-Wen Kang
Mechanical and Electro-Mechanical Engineering, Tamkang University, Tamsui Dist., New Taipei City, 25137, Taiwan(R.O.C)

Yu-Tang Chen
Department of Mechanical Engineering, De Lin Institute of Technology, Tucheng Dist., New Taipei City, 236, Taiwan(R.O.C)

Chun-Hsiang Hsu
Mechanical and Electro-Mechanical Engineering, Tamkang University, Tamsui Dist., New Taipei City, 25137, Taiwan(R.O.C)

Jun-Ying Lin
Mechanical and Electro-Mechanical Engineering, Tamkang University, Tamsui Dist., New Taipei City, 25137, Taiwan(R.O.C)

Frédéric Lefèvre
Université de Lyon, CNRS
INSA-Lyon, CETHIL, UMR5008, F-69621, Villeurbanne, France
Université Lyon 1, F-69622, France

Stéphane Lips
Université de Lyon, CNRS
INSA-Lyon, CETHIL, UMR5008, F-69621, Villeurbanne, France
Université Lyon 1, F-69622, France

Romuald Rullière
Université de Lyon, CNRS
INSA-Lyon, CETHIL, UMR5008, F-69621, Villeurbanne, France
Université Lyon 1, F-69622, France

Jean-Baptiste Conrardy
Université de Lyon, CNRS
INSA-Lyon, CETHIL, UMR5008, F-69621, Villeurbanne, France
Université Lyon 1, F-69622, France

Martin Raynaud
Thales Communications, Thales, 92701 Colombes, France

Jocelyn Bonjour
Université de Lyon, CNRS
INSA-Lyon, CETHIL, UMR5008, F-69621, Villeurbanne, France
Université Lyon 1, F-69622, France

K.N. Shukla
Karunya University, Karunya Nagar, Coimbatore-641 114, India

A. Brusly Solomon
Karunya University, Karunya Nagar, Coimbatore-641 114, India

B.C. Pillai
Karunya University, Karunya Nagar, Coimbatore-641 114, India

Shanmuga Sundaram Anandan
Department of Mechanical Engineering, Rajalakshmi Institute of Technology, Chennai, Tamil Nadu, India-600124

Anirudh Bhaskaran
Department of Mechanical Engineering, Rajalakshmi Institute of Technology, Chennai, Tamil Nadu, India-600124

Sandesh S. Chougule
Indian Institute of Technology, Indore, M.P., 453446, India

Santosh Kumar Sahu
Indian Institute of Technology, Indore, M.P., 453446, India

Ashok T. Pise
Government College of Engineering, Karad, M.S., 415124, India

Brent S. Taft
Air Force Research Laboratory Space Vehicles Directorate, Kirtland AFB, NM, 87117, USA

R. Manoj
Avasarala Technologies Ltd, Bangalore, Karnataka, 560099, India
Centre for Emerging Technologies, Jain University, Bangalore, Karnataka, 562112, India

M. C. Kumar
Avasarala Technologies Ltd, Bangalore, Karnataka, 560099, India

R. NarasimhaRao
Avasarala Technologies Ltd, Bangalore, Karnataka, 560099, India

K. Rama Narasimha
Centre for Emerging Technologies, Jain University, Bangalore, Karnataka, 562112, India

P. V. S. Suresh
Avasarala Technologies Ltd, Bangalore, Karnataka, 560099, India

Pramod R. Pachghare
Assistant Professor, Department of Mechanical Engineering, Government College of Engineering, Amravati-444 604, India

Ashish M. Mahalle
Associate Professor, Department of General Engineering, Laxinarayan Institute of Technology, Nagpur-444 001, India

Shwin-Chung Won
Department of Power Mechanical Engineering, National Tsing Hua University, Taiwan, ROC

Ya-Chi Ho
Department of Power Mechanical Engineering, National Tsing Hua University, Taiwan, ROC

R. Senthilkumar
Department of Mechanical Engineering, Annamalai University, Annamalai Nagar, India, 608 002

S. Vaidyanathan,
Department of Mechanical Engineering, Annamalai University, Annamalai Nagar, India, 608 002

B. Sivaraman
Department of Mechanical Engineering, Annamalai University, Annamalai Nagar, India, 608 002

Taoufik Brahim
Laboratory of Thermal and Energy Systems Study (LESTE), National School of Engineers of Monastir (ENIM), University of Monastir (UM), Road Ibn Jazzar 5019 Monastir, Tunisia

Abdelmajid Jemni
Laboratory of Thermal and Energy Systems Study (LESTE), National School of Engineers of Monastir (ENIM), University of Monastir (UM), Road Ibn Jazzar 5019 Monastir, Tunisia

Gyana Ranjan Rana,
Department of Mechanical Engineering, Indian Institute of Technology Kanpur, Kanpur 208016 India

Basant Singh Sikarwar
Department of Mechanical Engineering, Indian Institute of Technology Kanpur, Kanpur 208016 India

Sameer Khandekar
Department of Mechanical Engineering, Indian Institute of Technology Kanpur, Kanpur 208016 India

P. K. Panigrahi
Department of Mechanical Engineering, Indian Institute of Technology Kanpur, Kanpur 208016 India

R Senthilkumar
Department of Mechanical Engineering, Annamalai University, Annamalai Nagar, India 608 002

J Prabakaran
Department of Mechanical Engineering, Annamalai University, Annamalai Nagar, India 608 002

M Karthikeyan
Department of Mechanical Engineering, Annamalai University, Annamalai Nagar, India 608 002

Amir Faghri
Department of Mechanical Engineering, University of Connecticut, Storrs, CT 06269, U.S.A

Yasushi Koito
Department of Advanced Mechanical Systems, Kumamoto University, Kumamoto-shi, Kumamoto, 860-8555, Japan

Masakazu Tsurumoto
Department of Advanced Mechanical Systems, Kumamoto University, Kumamoto-shi, Kumamoto, 860-8555, Japan

Fumiya Hatano
Department of Advanced Mechanical Systems, Kumamoto University, Kumamoto-shi, Kumamoto, 860-8555, Japan

Toshio Tomimura
Department of Advanced Mechanical Systems, Kumamoto University, Kumamoto-shi, Kumamoto, 860-8555, Japan

S.F. Wang
Key Laboratory of Enhanced Heat Transfer and Energy Conservation of the Ministry of Education, South China University of Technology, Guangzhou City, Guangdong Province, 510640, China

Z.R. Lin
Key Laboratory of Enhanced Heat Transfer and Energy Conservation of the Ministry of Education, South China University of Technology, Guangzhou City, Guangdong Province, 510640, China

Z.Y. Lee
Novark Technology Inc, Shenzhen City, Guangdong Province, 518172, China

L.W. Zhang
Novark Technology Inc, Shenzhen City, Guangdong Province, 518172, China

Guan-Wei Wu
Department of Mechanical Engineering, National Taiwan University, Taipei,10617, Taiwan

Sih-Li Chen
Department of Mechanical Engineering, National Taiwan University, Taipei,10617, Taiwan

Wen-Pin Shih
Department of Mechanical Engineering, National Taiwan University, Taipei,10617, Taiwan

R. Manimaran
Department of Mechanical Engineering, Pondicherry Engineering College, Puducherry, India

K. Palaniradja
Department of Mechanical Engineering, Pondicherry Engineering College, Puducherry, India

N. Alagumurthi
Department of Mechanical Engineering, Pondicherry Engineering College, Puducherry, India

Mohamed H. A. Elnaggar
Engineering Department, Palestine Technical College, Deir EL-Balah, Gaza Strip, Palestine

Rémi Bertossi
Institut Pprime, CNRS - ENSMA - Université de Poitiers, UPR 3346 Département Fluide, Thermique et Combustion 1, avenue Clément Ader 86961 uturoscope Chasseneuil Cedex, FRANCE

Cyril Romestant
Institut Pprime, CNRS - ENSMA - Université de Poitiers, UPR 3346 Département Fluide, Thermique et Combustion 1, avenue Clément Ader 86961 uturoscope Chasseneuil Cedex, FRANCE

Vincent Ayel
Institut Pprime, CNRS - ENSMA - Université de Poitiers, UPR 3346 Département Fluide, Thermique et Combustion 1, avenue Clément Ader 86961 uturoscope Chasseneuil Cedex, FRANCE

Yves Bertin
Institut Pprime, CNRS - ENSMA - Université de Poitiers, UPR 3346 Département Fluide, Thermique et Combustion 1, avenue Clément Ader 86961 uturoscope Chasseneuil Cedex, FRANCE

Markus Kley
Hochschule Aalen, Beethovenstr. 1, 73430 Aalen, Germany

Rüdiger Kretschmer
DHBC e.V., Schlossberg 12, 67366 Weingarten, Germany

Yorck Leschber
DHBC e.V., Schlossberg 12, 67366 Weingarten, Germany

Ralf Riegerb
DHBC e.V., Schlossberg 12, 67366 Weingarten, Germany

R.T Dobson
Department of Mechanical Engineering, University of Stellenbosch, Stellenbosch Private Bag X1, MATIELAND 7602, South Africa

R Laubscher
Department of Mechanical Engineering, University of Stellenbosch, Stellenbosch Private Bag X1, MATIELAND 7602, South Africa

Masataka Mochizuki
Fujkura Ltd. 1-5-1, Kiba, Koto-Ku, Tokyo 135-8512, Japan

Thang Nguyen
Fujkura Ltd. 1-5-1, Kiba, Koto-Ku, Tokyo 135-8512, Japan

Koichi Mashiko
Fujkura Ltd. 1-5-1, Kiba, Koto-Ku, Tokyo 135-8512, Japan

Yuji Saito
Fujkura Ltd. 1-5-1, Kiba, Koto-Ku, Tokyo 135-8512, Japan

Randeep Singh
Fujkura Ltd. 1-5-1, Kiba, Koto-Ku, Tokyo 135-8512, Japan

Tien Nguyen
Fujkura Ltd. 1-5-1, Kiba, Koto-Ku, Tokyo 135-8512, Japan

Vijit Wuttijumnong
Fujkura Ltd. 1-5-1, Kiba, Koto-Ku, Tokyo 135-8512, Japan

Jong Soo Kim
Department of Refrigeration and Air-Conditioning Engineering, Pukyong National University, Nam-gu, Busan, 608-739, Republic of Korea

Soo Jung Ha
Department of Refrigeration and Air-Conditioning Engineering, Pukyong National University, Nam-gu, Busan, 608-739, Republic of Korea

Sang Il Lee
Department of Refrigeration and Air-Conditioning Engineering, Pukyong National University, Nam-gu, Busan, 608-739, Republic of Korea

Bang Ho Jeong
Kyungunbeomyang Energy Co., LTD., Gangseo-gu, Busan, 618-280, Republic of Korea

K. S. Ong
Universiti Tunku Abdul Rahman, Jalan Universiti, 31900 Kampar, Malaysia

W. L. Tong
School of Engineering, Monash University Sunway Campus, Jalan Lagoon Selatan, 46150 Bandar Sunway, Malaysia

J. S. Ganb
School of Engineering, Monash University Sunway Campus, Jalan Lagoon Selatan, 46150 Bandar Sunway, Malaysia

N. Hishamb
School of Engineering, Monash University Sunway Campus, Jalan Lagoon Selatan, 46150 Bandar Sunway, Malaysia

R. Laubscher
Department of Mechanical and Mechatronic Engineering, University of Stellenbosch, Stellenbosch Private Bag X1, MATIELAND 7602, South Africa

R.T. Dobson
Department of Mechanical and Mechatronic Engineering, University of Stellenbosch, Stellenbosch Private Bag X1, MATIELAND 7602, South Africa

G. Kumaresan
Department of Mechanical Engineering, National Institute of Technology, Tiruchirappalli, Tamilnadu, India, 620 015

S. Venkatachalapathy
Department of Mechanical Engineering, National Institute of Technology, Tiruchirappalli, Tamilnadu, India, 620 015

Yiding Cao
Department of Mechanical and Materials Engineering, Florida International University, Miami, Florida 33174

Dmitry Khrustalev
ATK Space, Beltsville, MD, 20705, USA

Chuck Stouffer
ATK Space, Beltsville, MD, 20705, USA

Jentung Ku
NASA Goddard Space Flight Center, Greenbelt, MD, USA

Jon Hamilton
Lockheed Martin Space Systems, Palo Alto, CA, 94304, USA

Mark Anderson
Lockheed Martin Space Systems, Palo Alto, CA, 94304, USA

Fritz F. Laun
Air Force Research Labs, Albuquerque, New Mexico, 87108, USA

Brent S. Taft
Air Force Research Labs, Albuquerque, New Mexico, 87108, USA

K. S. Ong
Universiti Tunku Abdul Rahman, 31900 Kampar, Perak, Malaysia

Christopher Lim
Universiti Tunku Abdul Rahman, 31900 Kampar, Perak, Malaysia

R.T. Dobson
Department of Mechanical & Mechatronic Engineering, University of Stellenbosch, Stellenbosch, 7602, South Africa

J.C. Ruppersberg
Department of Mechanical & Mechatronic Engineering, University of Stellenbosch, Stellenbosch, 7602, South Africa

Rahul S. Borkar
Department of Mechanical Engineering, Government College of Engineering, Amravati-444 604, India

Pramod R. Pachghare
Department of Mechanical Engineering, Government College of Engineering, Amravati-444 604, India